Discrete Encounters

Chapman & Hall/CRC Cryptography and Network Security Series

Series Editors: Douglas R. Stinson and Jonathan Katz

https://www.crcpress.com/Chapman–HallCRC-Cryptography-and-Network-Security-Series/book-series/CHCRYNETSEC

Discrete Encounters

Craig P. Bauer

York College of Pennsylvania

CRC Press
Taylor & Francis Group
Boca Raton London New York

CRC Press is an imprint of the
Taylor & Francis Group, an **informa** business

A CHAPMAN & HALL BOOK

CRC Press
Taylor & Francis Group
6000 Broken Sound Parkway NW, Suite 300
Boca Raton, FL 33487-2742

First issued in paperback 2022

© 2020 by Taylor & Francis Group, LLC

CRC Press is an imprint of Taylor & Francis Group, an Informa business

No claim to original U.S. Government works

ISBN 13: 978-1-4987-3586-5 (hbk)
ISBN 13: 978-1-03-247448-9 (pbk)
ISBN 13: 978-0-429-40050-6 (ebk)

DOI: 10.1201/9780429400506

Visit the Taylor & Francis Web site at
www.taylorandfrancis.com

and the CRC Press Web site at
www.crcpress.com

To Brian J. Winkel, whose boundless energy and focus on

helping others is inspiring.

Contents

Acknowledgments

As will become evident to the reader, I made use of a tremendous number of print and online sources in the pages that follow in order to be able to offer the reader the most interesting examples and applications, as well as to flesh out the human side of the story and, in general, make this book as entertaining as possible. This would have been a much more time-consuming task if various people hadn't pointed me toward many of these sources, answered my questions when said sources were confusing, or helped me in other ways, such as granting reproduction permission. I thank Archives de l'Académie des sciences, Anthony Austin, June Barrow-Green, Morgen Bell, Megan Blount, Andrew Burbanks, California Institute of Technology Archives, Chris Christensen, George Csicsery, Jason Davies, John W. Dawson, Jr., A. W. F. Edwards, Kerry Emanuel, Enfield Publishing & Distribution Co., The Erwin Tomash Library on the History of Computing, José Ferreirós, Carl Feynman, Michelle Feynman, Ananth Ganesan, Gottfried Wilhelm Leibniz Bibliothek, Anastasia Goudy, Ronald Graham, Brian Gray, José Grimm, Josh Gross, Don Harris, Brian Hayes, David W. Henderson, Christian Hewicker, Andreas M. Hinz, Reed Hutchinson, János Bolyai Mathematical Society, Matthew Jones, Loma Karklins, Rory Kraft, Jr., Kathryn A. Lindsey, Linyuan Lu, Mathematical Association of America, MIT Museum, Joe Molinar, Princeton University Press, RAND Corporation, Joe Redish, David Rees, Karen Rice-Young, Mike Riley, The Rockefeller University, Royal Irish Academy, Frank Ruskey, Garrett Ruths, The Shelby White and Leon Levy Archives Center, Institute for Advanced Study, Princeton, NJ, Neil J. A. Sloane, Lance A. Snyder, Cody Spath, Springer/Birkhäuser, Ian Stewart, Emily Sullivan, Terence Tao, Alan D. Taylor, Robert Tubbs, University of Illinois Archives, Stan Wagon, Erik Whitelaw, Steve Wildish, Stephen Wolfram, Alyssa Yom, and Jim Yorke.

Introduction

There are four themes that I feel are important to include in any book dealing with mathematics. I introduce each of them below with a quote.

> I may add that in any treatise or higher text-book it is always desirable that references to the original memoirs should be given, and, if possible, short historic notices also. I am sure that no subject loses more than mathematics by any attempt to dissociate it from its history.
>
> James Whitbread Lee Glaisher[1]

Mathematics is discovered by people, so in addition to giving the results, I put some humanity into the story by providing details of the men and women behind the mathematics, their motivations, and a bit about their personalities.

> Unification, the establishment of a relationship between seemingly diverse objects, is at once one of the great motivating forces and one of the great sources of aesthetic satisfaction in mathematics.
>
> Philip J. Davis and Reuben Hersh[2]

The topics introduced in this book appear again and again in later chapters through connections with the new material. This should make it easier to retain the material, as one progresses toward the end of the book. It also gives a greater appreciation for the subject as a whole. Just as John Donne wrote "No man is an island." it seems that no mathematical result stands alone.

> We keep returning to the positive integers, with new methods, new insights, and a new appreciation for the elegant interplay that exists between the continuous and the discrete. In the end, it is perhaps not surprising for in the nineteenth century, when analysis was finally placed on a firm footing, it was arithmetic that served as the foundation.
>
> Robert M. Young[3]

While this book focuses on *discrete* mathematics, an important part of the story would be missing if continuous mathematics were to be ignored. Inclusion of the continuous mathematics that the context seems to demand provides the reader with a deeper appreciation for mathematics as a whole.

---

1 Quoted in Moritz, Robert Edouard, *Memorabilia Mathematica or The Philomath's Quotation-Book*, The MacMillan Company, New York, 1914, p. 96, which cites "Presidential Address British Association for the Advancement of Science, Section A," *Nature*, Vol. 42, 1890, p. 466.

2 Davis, Philip J. and Reuben Hersh, *The Mathematical Experience*, Houghton Mifflin Company, Boston, MA, 1981, p. 198.

3 Young, Robert M., *Excursions in Calculus: An Interplay of the Continuous and the Discrete*, The Mathematical Association of America, 1992, p. 368.

Look! It's moving. It's alive. It's alive... It's alive, it's moving, it's alive, it's alive, it's alive, it's alive, IT'S ALIVE!

Henry Frankenstein[4]

Nearly every topic in discrete mathematics, or just about any other area of mathematics for that matter, can be extended, generalized, put in another context, or in some other way lead to other problems. With many of these problems still unsolved, the subject is seen to still live and breathe. I include such problems in nearly every chapter. Just as a young Andrew Wiles saw the then unproven Fermat's last theorem in a book and went on, many years later, to solve it, so I hope one or more of these open problems will fall to someone who first encounters them here.

A note on sources: Part of my aim with this book is to deepen the readers' appreciation of mathematics and create, or increase, a desire to learn more about it. In expectation of this goal being met, I've included a large number of references, both in print and online. While the online sources offer the potential for quicker accessibility, they can't all be counted on to remain available. If you find that a link is broken, please also take a moment to check for an archived version of the page at https://archive.org/web/. For print sources, be sure to check https://archive.org/ for a free digital copy before making a purchase or going to a library.

---

4 Frankenstein, 1931, www.imdb.com/title/tt0021884/quotes.

Author

Craig P. Bauer is a professor of mathematics at York College of Pennsylvania. He's the editor-in-chief of *Cryptologia* and was the 2011–2012 Scholar-in-Residence at the National Security Agency's Center for Cryptologic History. He loves to carry out research, write, and lecture. His previous books are *Secret History: The Story of Cryptology* and *Unsolved! The History and Mystery of the World's Greatest Ciphers from Ancient Egypt to Online Secret Societies*. With the present book he stays true to his style, blending mathematics and history. Craig earned his Ph.D. in mathematics from North Carolina State University and did his undergraduate work at Franklin & Marshall College.

0

Continuous vs. Discrete

Definitions

If something can be broken into smaller and smaller pieces forever, that is, if it is infinitely divisible, then we say it is **continuous**. The real number line is an example. No matter how small of an interval you wish to take, it can be split into even smaller intervals. By contrast, anything that is not infinitely divisible is said to be **discrete**.

If something can eventually no longer be split, then that piece can be considered a fundamental (discrete) building block of the item being considered. Many things are obviously discrete. Examples include the balance in a checking account, where the indivisible unit is a penny and the enrollment at a college, which must be a whole number of students. Computers are discrete in the sense that everything they do boils down to 0s and 1s. In fact, discrete mathematics is so important in computer science that it's a required course for students majoring in the subject, even though it is often not required for mathematics majors! The latter encounter pieces of it here and there in various undergraduate classes.

A few lines from the film *Apocalypse Now* (1979) give what could be taken as a laymen's definition of discrete:

> Do you know what the man's saying? This is dialectics. It's very simple. Dialectics. One through nine. No Maybes, no supposeds, no fractions. You can't travel in space – you can't go out into space, without like you know uh, with fractions, what're ya gonna land on 1/4, 3/8? Whaddya gonna do when you go from here to Venus or something? That's dialectic physics okay. Dialectic logic is there's only love and hate. You either love somebody or you hate 'em.
> Photojournalist in *Apocalypse Now* played by Dennis Hopper[1]

Distinguishing between the discrete and the continuous may seem simple, but in some instances it has proven to be a very difficult question.

An Ancient Question

As simple as it may seem at first glance, trying to decide whether the universe we live in is discrete or continuous has resulted in thousands of years of debate, and a winner cannot yet be conclusively determined. The ancient Greek philosopher Democritus

---

1 *Apocalypse Now*, 1979, www.imdb.com/title/tt0078788/?ref_=fn_al_tt_1.

(460 BCE–370 BCE)[2] is often credited with the atomic theory, which is discrete. However, it is nothing like today's theory of the atom. Democritus simply suggested that matter can only be broken down so far until the pieces are no longer divisible. Others who held such ideas include Epicurus (341 BCE–270 BCE), Diodorus Cronus (unknown – c. 284 BCE), and even a Roman, Titus Lucretius Carus (99 BCE–c. 55 BCE).[3]

Epicurus noted how if a piece of matter was somehow infinitely divisible, then each component would have some weight, and so the weight of the original piece would have to be infinite. As this is not, in fact, the case, he felt he had provided a strong argument for the discrete nature of reality.

Michael J. White examined various ancient Greek arguments for both the continuous and the discrete, from the context of the times as well as from a modern viewpoint, in his book *The Continuous and the Discrete: Ancient Physical Theories from a Contemporary Perspective*. Some of his comments on Epicurus's argument follow:

> This passage can be interpreted as setting forth a version of what has, in recent times, come to be termed 'Zeno's paradox of measure'. The paradox, for which a fairly tenuous connection to Zeno of Elea can be established on the basis of several passages in Simplicius' commentary on Aristotle's Physics,[4] is set forth by Brian Skyrms as follows:
>
> > Suppose that the line-segment is composed of an infinite number of parts. Zeno claims that this leads to absurdity in the following way
> >
> > (I) Either the parts all have zero magnitude or they all have [the same] positive magnitude.
> >
> > (II) If they have zero magnitude, the line-segment will have zero magnitude, since the magnitude of the whole is the sum of the magnitudes of its parts.
> >
> > (III) If they have [the same] positive magnitude, then the line-segment will have infinite magnitude, for the same reason.[5,6]

The followers of Epicurus, known as Epicureans, believed not only in discrete magnitude, but also in discrete motion and discrete time.[7]

---

2 Some sources state his birth year as 490 BCE.

3 White, Michael J., *The Continuous and the Discrete: Ancient Physical Theories from a Contemporary Perspective*, Clarendon Press, Oxford, 1992, p. 194, states "According to most (but not all) scholars, such a quantum conception of spatial magnitude and time was developed by Epicurus; and it was certainly a central element of the thought of Diodorus Cronus." For information on Diodorus Cronus see White p. 263 and the references in the footnote on p. 265. For information on Lucretius, whose argument in favor of discreteness wasn't great, see White, p. 220.

4 Simplicius, *In phys.*, CAG 9, 139, 3–19 and 140. 27–141. 8.

5 "Zeno's Paradox of Measure," in Cohen, Robert S. and Larry Laudan, editors, *Physics, Philosophy and Psychoanalysis: Essays in Honor of A. Grünbaum*, Dordrecht, 1983, p. 226.

6 White, Michael J., *The Continuous and the Discrete: Ancient Physical Theories from a Contemporary Perspective*, Clarendon Press, Oxford, 1992, p. 199. Also see pp. 8–9 and 202. Page 200 gives a modern resolution to the paradox.

7 White, Michael J., *The Continuous and the Discrete: Ancient Physical Theories from a Contemporary Perspective*, Clarendon Press, Oxford, 1992, pp. 269–270.

These Greek atomists were opposed by Aristotle (384 BCE–322 BCE), who favored a continuous view of nature. He thought the universe (and time) was not only continuous, but also infinite.[8]

One argument from the continuous side considers the length of an indivisible piece of matter. If matter is not infinitely divisible, then the length will be greater than zero. Considered from the middle, this piece can be said to have a left side and a right side, but this offers a division![9]

Some things that are discrete, like the number of hairs on a man's head, have an interesting sort of interplay with the continuous. A paradox posed in ancient Greece, and attributed to Eubulides (4th century BCE), asks one to consider a man with a full head of hair, and to imagine removing hairs one at a time. At what point would we say he is bald? For the sake of this question, we take "bald" as we would in casual conversation and not to mean "completely bald." It is very difficult to decide which hair would finally make the difference![10]

This can be considered a variant of the sorites paradox that begins with a heap of sand. Grains of sand are then removed one at a time. At what point do you no longer have a heap? When only two grains remain, I wouldn't be comfortable calling it a heap, but at what exact point before that should I stop using the term heap?[11]

How Should Math Be Done?

If the nature of the universe could be proven to be discrete, then it would make sense to model the universe with discrete mathematics. Of course, continuous mathematics would seem to be the better choice if reality turned out to be continuous. Sadly, mathematics had to move on without a firm answer to this big question. The continuous mathematics of calculus and differential equations proved to be very successful and the undergraduate curriculum is now lopsided in that direction, but several great mathematicians have advocated for discrete foundations. One of these was Leopold Kronecker (Figure 0.1). His famous line is "Die ganzen Zahlen hat der liebe Gott gemacht, alles andere ist Menschenwerk" ("God made the integers, all else is the work of man").

Kronecker rejected mathematics that was not based on integers. This includes things as common as π, a number that cannot be expressed exactly as the ratio of two integers. Its value is approximately 3.14159265, but no block of digits repeats endlessly, without interruption, from some point on, as with ratios of integers like $22/7 = 3.142857142857142857 \ldots$ The digits of π look random.

---

8 See White, Michael J., *The Continuous and the Discrete: Ancient Physical Theories from a Contemporary Perspective*, Clarendon Press, Oxford, 1992, p. 169 for a quote to this effect. A summary of atomistic or quantum models and their supporters in the Hellenistic world is given on p. 191 and a terse summary of Aristotle's perspective is on p. 192.

9 Forrest, Peter, *The Necessary Structure of the All-pervading Aether: Discrete or Continuous? Simple or Symmetric?*, Philosophische Analyse/Philosophical Analysis, Volume 49, ontos verlag, Heusenstamm, Germany, 2012, p. 63: "The intuition that any extended thing should have left and right halves, or more generally be divisible in each direction was produced as an argument against the Greek atoms."

10 https://en.wikipedia.org/wiki/Eubulides#Paradoxes_of_Eubulides.

11 https://en.wikipedia.org/wiki/Sorites_paradox.

FIGURE 0.1
Leopold Kronecker (1823–1891).

Public domain.

Because π can't be expressed as a ratio of integers, it is called **irrational**. The first proof that this is the case was found by Johann Heinrich Lambert in 1761. Other irrational numbers go back much farther. Like Kronecker, the Pythagoreans tried to base everything on integers, but a figure as simple as a square with sides of length 1 caused trouble. The Pythagorean theorem reveals this square's diagonal to have length $\sqrt{2}$ and a proof was found that this number is irrational. According to legend, the man who found this result was taken out in a row boat and thrown overboard in the hope that his proof would drown with him.[12] This did not work.

So, irrational numbers were nothing new during Kronecker's time, but new facts were still being learned about π. For example, in 1882, Ferdinand von Lindemann proved that π is transcendental. This is a fancy name for a very simple property. If a number is a root of some polynomial with integer coefficients, we say it is **algebraic**. If it is not, we say it is **transcendental**, for it "transcends" polynomials in a way. Some irrational numbers, like π, are transcendental, while others are algebraic. For example, $\sqrt{2}$ is a root of $x^2 - 2 = 0$, so it is algebraic. Proving that a number is transcendental can be very difficult. Here's Kronecker's reaction to Lindemann's marvelous proof.

---

12 According to some accounts, he was not drowned for finding the result, but rather for sharing it with outsiders (i.e., non-Pythagoreans).

> What good your beautiful proof on [the transcendence of] π: Why investigate such problems, given that irrational numbers do not even exist?
>
> – Leopold Kronecker[13]

Oh well, at least Kronecker didn't try to drown him. As stated previously, Kronecker was a great mathematician, not a lunatic. And there have been others who took his seemingly extreme point of view. E. T. Bell, a great popularizer of mathematics (some of his work is detailed later in this book), wrote

> Here Kronecker incidentally denied the 'existence' of π, and he was less of a radical at that than some of his successors.[14]

David Hilbert, whom we will also meet again, favored a discrete interpretation of reality.

> [A] homogeneous continuum which admits of the sort of divisibility needed to realize the infinitely small is nowhere to be found in reality. The infinite divisibility of a continuum is an operation which exists only in thought. It is merely an idea which is in fact impugned by the results of our observations of nature and of our physical and chemical experiments.
>
> David Hilbert (1862–1943)[15]

Throwing out "big names" to support an argument isn't the best way to go. In fact, there's a logical fallacy called "argument from authority." While having Hilbert on your side of an argument can help, it is far from a proof. Great thinkers have been wrong often enough. And we can usually find another big name on the other side of a given debate. In the spirit of doing so, the next section offers Charles Fort's perspective.

An Argument for Continuity

So that the debate is not one-sided, I'm including excerpts from Charles Fort's 1919 work *The Book of the Damned*.[16] Fort (Figure 0.2) believed everything to be continuous. He was not a mathematician, but his ideas are interesting. We'll encounter him again when we look at the three-body problem.

> Or that red is not positively different from yellow: is only another degree of whatever vibrancy yellow is a degree of: that red and yellow are continuous, or that they merge in orange.

---

13 Quoted in Schroeder, Manfred, *Fractals, Chaos, Power Laws: Minutes from an Infinite Paradise*, W. H. Freeman and Company, New York, 1991, p. 237, reprinted by Dover, Mineola, New York, 2009.

14 Bell, Eric Temple, *Mathematics, Queen and Servant of Science*, Copyright, 1951, British Edition 1952, reprinted 1954, 1958, 1961. Printed in Great Britain by Butler and Tanner Ltd., Frome and London, p. 408.

15 Hilbert, David, "On the Infinite," in Benacerraf, Paul and Hilary Putnam, editors, *The Philosophy of Mathematics, Selected Readings*, second edition, Cambridge University Press, Cambridge, 1983, pp. 183–201, p. 186 quoted here. This is a reprinting of the 1925 paper.

16 Fort, Charles, *The Book of the Damned*, Boni and Liveright, New York, 1919. An annotated online edition is available at www.resologist.net/damn01.htm.

FIGURE 0.2
Charles Fort (1874–1932).

Public domain.

> So then that, if, upon the basis of yellowness and redness, Science should attempt to classify all phenomena, including all red things as veritable, and excluding all yellow things as false or illusory, the demarcation would have to be false and arbitrary, because things colored orange, constituting continuity, would belong on both sides of the attempted border-line.
>
> Charles Fort (1919)

Can you draw a line separating yellow from red that is not arbitrary? See color figure 1 **(See color insert)**.

Fort's argument is similar to those from ancient Greece asking how many grains of sand it takes to make a heap or how many hairs a man needs to have before you would no longer say he's bald. We need such definitions, aka classifications, before we can draw conclusions or make meaningful statements about things. Fort continued:

> "As we go along, we shall be impressed with this:
>   That no basis for classification, or inclusion and exclusion, more reasonable than that of redness and yellowness has ever been conceived of.
>   Science has, by appeal to various bases, included a multitude of data. Had it not done so, there would be nothing with which to seem to be. Science has, by appeal to various bases, excluded a multitude of data. Then, if redness is continuous with yellowness: if

every basis of admission is continuous with every basis of exclusion, Science must have excluded some things that are continuous with the accepted. In redness and yellowness, which merge in orangeness, we typify all tests, all standards, all means of forming an opinion –

Or that any positive opinion upon any subject is illusion built upon the fallacy that there are positive differences to judge by –"

Charles Fort (1919)

He posed the challenge of formulating a specific definition.

What is a house?

It is not possible to say what anything is, as positively distinguished from anything else, if there are no positive differences.

A barn is a house, if one lives in it. If residence constitutes houseness, because style of architecture does not, then a bird's nest is a house: and human occupancy is not the standard to judge by, because we speak of dogs' houses; nor material, because we speak of snow houses of Eskimos – or a shell is a house to a hermit crab – or was to the mollusk that made it – or things seemingly so positively different as the White House at Washington and a shell on the sea-shore are seen to be continuous.

So no one has ever been able to say what electricity is, for instance. It isn't anything, as positively distinguished from heat or magnetism or life. Metaphysicians and theologians and biologists have tried to define life. They have failed, because, in a positive sense, there is nothing to define: there is no phenomenon of life that is not, to some degree, manifest in chemism, magnetism, astronomic motions.

White coral islands in a dark blue sea.

Their seeming of distinctness: the seeming of individuality, or of positive difference one from another – but all are only projections from the same sea bottom. The difference between sea and land is not positive. In all water there is some earth: in all earth there is some water.

So then that all seeming things are not things at all, if all are inter-continuous, any more than is the leg of a table a thing in itself, if it is only a projection from something else: that not one of us is a real person, if, physically, we're continuous with environment; if, psychically, there is nothing to us but expression of relation to environment.

Charles Fort (1919)

In Chapter 18, you'll see how Benoit Mandelbrot investigated the lengths of coastlines and found that Fort was right when he wrote, "The difference between sea and land is not positive." But was Fort right about all of it? Is reality continuous? In any case, he had more to say on the matter.

When I say that there is nothing to prove, I mean that to those who accept Continuity, or the merging away of all phenomena into other phenomena, without positive demarcations one from another, there is, in a positive sense, no one thing. There is nothing to prove.

For instance nothing can be proved to be an animal – because animalness and vegetableness are not positively different. There are some expressions of life that are as much vegetable as animal, or that represent the merging of animalness and vegetableness. There is then no positive test, standard, criterion, means of forming an opinion. As distinct from vegetables, animals do not exist. There is nothing to prove. Nothing could be proved to be good, for instance. There is nothing in our 'existence' that is good, in a positive sense, or as really outlined from evil. If to forgive be good in times of peace, it is evil in wartime. There is nothing to prove: good in our experience is continuous with, or is only another aspect of evil.

Charles Fort (1919)

If you want to see how arbitrary the definition of "life" is, look it up in a current biology text. Wikipedia offers the following:

> The definition of life is controversial. The current definition is that organisms maintain homeostasis, are composed of cells, undergo metabolism, can grow, adapt to their environment, respond to stimuli, and reproduce. However, many other biological definitions have been proposed, and there are some borderline cases of life, such as viruses.[17]

Borderline cases – just as Fort argued there must always be! The evolution of viruses remains an open problem. Biologist David R. Wessner noted,

> Because of the great diversity among viruses, biologists have struggled with how to classify these entities and how to relate them to the conventional tree of life. They may represent genetic elements that gained the ability to move between cells. They may represent previously free-living organisms that became parasites. They may be the precursors of life as we know it.[18]

How strange that there may be "things" on Earth that can transition from living to non-living or non-living to living over the generations. Another biologist, Harald Brüssow, mentioned borderline cases on a larger scale:

> What about frogs that are completely frozen during hibernation? Are these metaboli-cally frozen frogs alive? What about plant seeds with little or no metabolic activity? Does it make sense to distinguish alive and dead by allowing an intermediate category of biological material that has the potential to return to life?[19]

Brüssow admitted, "Biologists have not yet elaborated a broadly accepted answer to the question 'What is life?'."[20] For those who reject viruses that require a living host, as being alive themselves, what do they make of the Sputnik virophage, which grows on another virus?[21]

In an update that Charles Fort probably did not anticipate, computer viruses could qualify as living, depending on your definition.

---

17 https://en.wikipedia.org/wiki/Life.

18 Wessner, David R., "The Origins of Viruses," *Nature Education*, Vol. 3, No. 9, 2010, pp. 37, www.nature.com/scitable/topicpage/the-origins-of-viruses-14398218#.

19 Brüssow, Harald, "The Not So Universal Tree of Life or the Place of Viruses in the Living World," *Philosophical Transactions of the Royal Society B*, Vol. 364, No. 1527, August 12, 2009, pp. 2263–2274, www.ncbi.nlm.nih.gov/pmc/articles/PMC2873004/pdf/rstb20090036.pdf.

20 Brüssow, Harald, "The Not So Universal Tree of Life or the Place of Viruses in the Living World," *Philosophical Transactions of the Royal Society B*, Vol. 364, No. 1527, August 12, 2009, pp. 2263–2274, www.ncbi.nlm.nih.gov/pmc/articles/PMC2873004/pdf/rstb20090036.pdf.

21 Brüssow, Harald, "The Not So Universal Tree of Life or the Place of Viruses in the Living World," *Philosophical Transactions of the Royal Society B*, Vol. 364, No. 1527, August 12, 2009, pp. 2263–2274, www.ncbi.nlm.nih.gov/pmc/articles/PMC2873004/pdf/rstb20090036.pdf.

Fort even challenged Isaac Newton, who was an atomist in the Greek style, by the way.

> Newton's three laws, and that they are attempts to achieve positiveness, or to defy and break Continuity, and are as unreal as are all other attempts to localize the universal:
>
> That, if every observable body is continuous, mediately or immediately, with all other bodies, it can not be influenced only by its own inertia, so that there is no way of knowing what the phenomena of inertia may be; that, if all things are reacting to an infinitude of forces, there is no way of knowing what the effects of only one impressed force would be; that if every reaction is continuous with its action, it can not be conceived of as a whole, and that there is no way of conceiving what it might be equal and opposite to –
>
> Charles Fort (1919)

Fort wrapped up his argument with the following:

> All sciences begin with attempts to define.
> Nothing ever has been defined.
> Because there is nothing to define.
> Darwin wrote "The Origin of Species."
> He was never able to tell what he meant by a "species."
>
> Charles Fort (1919)

This is not to be taken as a rejection of evolution, but rather yet another example of how elusive a definition can be, even when a branch of science badly needs the definition.

The State of Things Today

So, where do things stand now? Differential equations provide excellent models of nature, but do they really allow a glimpse into the underlying reality of nature or are they merely useful approximations? Many would argue that the infinitesimals, upon which both differentiation and integration rely, do not exist. Mathematically, there is no problem. Between any two real numbers one can always find another real number. Yet in real space there is a smallest distance to which we can probe without creating a black hole. This is called the Planck length and measures $\approx 10^{-35}$ cm.[22] It could be considered a fundamental unit of length. Similarly, the time it takes light to travel a distance of one Planck length, about 10^{-43} seconds, could be regarded as a fundamental unit of time.

Quantum mechanics forms the basis of reality, so let's review a definition Stephen Hawking gives.

> Quantum (plural: quanta) – The *indivisible* unit in which waves may be absorbed or emitted.[23]

---

22 Hawking, Stephen, *The Universe in a Nutshell*, Bantam Books, New York, 2001, p. 206. To get a sense of how small this is, go to http://htwins.net/scale2/.

23 Hawking, Stephen, *The Universe in a Nutshell*, Bantam Books, New York, 2001, p. 206.

So much for infinitesimals. It appears that reality is discrete.

Wait! String Theory assumes a continuous nature to the universe.[24] But the causal sets program assumes discreteness.[25] A case has been made that Loop Quantum Gravity, which initially seemed to support discreteness, actually demands continuity.[26] There are many other formulations for discrete space, and probably just as many arguments against them.[27]

For example, if space is discrete, we have something known as the "Weyl Tile Problem," posed in 1949 by Hermann Weyl.

> If a square is built up of miniature tiles, then there are as many tiles along the diagonal as there are along the sides; thus the diagonal should be equal in length to the side.[28]

Accepting a discrete universe might force us to make some consolations in what we think we know, but surely losing the Pythagorean theorem is too high a price to pay! We also lose the distance function we're used to from Euclidean geometry, because it's a consequence of the Pythagorean theorem. Fortunately, there are several ways around this problem. A discrete space doesn't have to boil down to squares or cubes. There are more exotic possibilities that avoid the Weyl Tile Problem. One of these, put forth By Charles Radin in the 1990s, based on work by John H. Conway, is the pinwheel tiling (Figure 0.3).[29]

These shapes yield a distance function that closely approximates what we're used to from Euclidean geometry. Notice that the small triangles, grouped by fives, make larger triangles with the same relative dimensions. The larger triangles are indicated with thicker lines. This can be done again to get even larger triangles. Such "self-similar at different scales" designs will be seen again in Chapter 18. Another intriguing tiling is presented in Chapter 13.

The pinwheel tiling can be generalized in various ways, including to three-dimensional space.[30]

---

24 Forrest, Peter, *The Necessary Structure of the All-pervading Aether: Discrete or Continuous? Simple or Symmetric?*, Philosophische Analyse/Philosophical Analysis, Volume 49, ontos verlag, Heusenstamm, Germany, 2012, pp. 2, 15–16, 137 and 146.

25 Forrest, Peter, *The Necessary Structure of the All-pervading Aether: Discrete or Continuous? Simple or Symmetric?*, Philosophische Analyse/Philosophical Analysis, Volume 49, ontos verlag, Heusenstamm, Germany, 2012, p. 137.

26 Forrest, Peter, *The Necessary Structure of the All-pervading Aether: Discrete or Continuous? Simple or Symmetric?*, Philosophische Analyse/Philosophical Analysis, Volume 49, ontos verlag, Heusenstamm, Germany, 2012, p. 16 and 151. For the argument for discretization, see Thiemann, Thomas, "Lectures on Loop Quantum Gravity," *Lecture Notes in Physics*, Vol. 631, 2003, pp. 41–135.

27 For a quick survey of discrete formulations, see Forrest, Peter, *The Necessary Structure of the All-pervading Aether: Discrete or Continuous? Simple or Symmetric?*, Philosophische Analyse/Philosophical Analysis, Volume 49, ontos verlag, Heusenstamm, Germany, 2012, pp. 44–53.

28 Weyl, Hermann, *Philosophy of Mathematics and Natural Sciences*, Princeton University Press, Princeton, NJ, 1949, p. 43.

29 Radin, Charles, "Symmetry of Tilings of the Plane," *Bulletin of the American Mathematical Society*, New Series, Vol. 29, No. 2, October 1993, pp. 213–217. Also see Radin, Charles, "The Pinwheel Tilings of the Plane," *Annals of Mathematics*, Second Series, Vol. 139, No. 3, May 1994, pp. 661–702 and Radin, Charles, "Symmetry and Tilings," *Notices of the American Mathematical Society*, Vol. 42, No. 1, January 1995, pp. 26–31.

30 Conway, John H. and Charles Radin, "Quaquaversal Tilings and Rotations," *Inventiones Mathematicae*, Vol. 132, 1988, pp. 179–188.

FIGURE 0.3
A pinwheel tiling.

Image created by Levochik. This file is licensed under the Creative Commons Attribution-Share Alike 3.0 Unported, 2.5 Generic, 2.0 Generic and 1.0 Generic license.

Another solution to the Weyl Tile Problem involves giving lines width, which makes perfect sense in a discrete geometry.[31]

So, those in favor of a discrete universe have several possible responses to Weyl. And they can pose a different problem to the continuous camp; if time is not discrete, there are supertask paradoxes with which the physicists must grapple. This challenge won't be detailed here, but you can follow the references.[32]

It seems that which theories the physicists favor depends, in part, on their preconceptions about how the universe should be. Those who prefer simplicity are more at home with the discrete view, while those favoring symmetry prefer continuity.[33]

---

31 Van Bendegem, Jean-Paul., "Zeno's Paradoxes and the Weyl Tile Argument," *Philosophy of Science*, Vol. 54, No. 2, 1987, pp. 295–302. Or see Van Bendegem, Jean-Paul, "Finitism in Geometry," *The Stanford Encyclopedia of Philosophy*, spring 2009 edition, Edward N. Zalta, editor, 2009. A substantive revision is available online at https://stanford.library.sydney.edu.au/archives/win2015/entries/geometry-finitism/.

32 Forrest, Peter, *The Necessary Structure of the All-pervading Aether: Discrete or Continuous? Simple or Symmetric?*, Philosophische Analyse/Philosophical Analysis, Volume 49, ontos verlag, Heusenstamm, Germany, 2012, pp. 171–172. See www.youtube.com/watch?v=ffUnNaQTfZE for a neat video on supertasks starring Michael Stevens.

33 Forrest, Peter, *The Necessary Structure of the All-pervading Aether: Discrete or Continuous? Simple or Symmetric?*, Philosophische Analyse/Philosophical Analysis, Volume 49, ontos verlag, Heusenstamm, Germany, 2012, p. 195.

If discreteness wins out, future generations may be modeling the universe with a completely different set of tools than those Isaac Newton invented/discovered for the purpose back in the 1600s.

Attempts to discretize the universe go back to the late 1920s. One interesting theory, put forth by Werner von Heisenberg in 1930, mostly flew under the radar of the journals, but the following did see print:

> [It would seem plausible] to introduce the radius r_0 [of the electron] in such a way that space is divided into cells of finite magnitude r_0^3, and the previous differential equations are replaced with difference equations. In such a lattice world the self-energy will, at any rate, be finite. However, although such a lattice world possesses remarkable properties, one must also observe that it leads to deviations from the present theory which do not seem plausible from the point of view of experiment. In particular, the assumption that a minimal length exists is not relativistically invariant, and one can see no way to bring the demand for relativistic invariance into conformity with the introduction of a fundamental length.[34]

Heisenberg had previously communicated his ideas on this topic to Niels Bohr in a letter dated February 26, 1930. Near the end of the letter, Heisenberg wrote, "I do not know if you find this radical attempt completely mad. But I have the feeling that nuclear physics is not to be had much more cheaply."[35] Bohr was skeptical. This may be part of the reason Heisenberg didn't put his ideas forth publicly at the time, although he did bring some of it up again later.

Albert Einstein (Figure 0.4) summed the matter up, addressing other work that Heisenberg did make public.

> To be sure, it has been pointed out that the introduction of a space-time continuum may be considered as contrary to nature in view of the molecular structure of everything which happens on a small scale. It is maintained that perhaps the success of the Heisenberg method points to a purely algebraical method of description of nature, that is to the elimination of continuous functions from physics. Then, however, we must also give up, by principle, the space-time continuum. It is not unimaginable that human ingenuity will some day find methods which will make it possible to proceed along such a path. At the present time, however, such a program looks like an attempt to breathe in empty space.[36]

Nevertheless, I think that it will happen. Amit Hagar ended his book, *Discrete or Continuous?: The Quest for Fundamental Length in Modern Physics*, with the following:

> We can finally formulate our closing statement: a thousand years from now, when the rigid adherence to continuum geometry will have been loosened by centuries of discrete math, algebra, information theory, and computer science, the resistance to the notion of fundamental length will probably look like just another case of human

---

34 Heisenberg, Werner, "Die Selbstenergie des Elektrons," *Zeitschrift für Physik*, Vol. 65, 1930, pp. 4–13. Taken here from Carazza, Bruno and Helge Kragh, "Heisenberg's Lattice World: The 1930 Theory Sketch," *American Journal of Physics*, Vol. 63, No. 7, 1995, pp. 595–605. The quote appears on p. 595.

35 Letter from Werner von Heisenberg to Niels Bohr, February 26, 1930. Taken here from Carazza, Bruno and Helge Kragh, "Heisenberg's Lattice World: The 1930 Theory Sketch," *American Journal of Physics*, Vol. 63, No. 7, 1995, pp. 595–605. The quote appears on p. 597.

36 Einstein, Albert, *Ideas and Opinions*, Crown, New York, 1919/1954, p. 319.

FIGURE 0.4

"Only two things are infinite, the universe and human stupidity, and I'm not sure about the former." – Albert Einstein (π day 1879–1955).

Public domain.

myopia. Until then, not only will it be difficult to persuade the skeptic (one hopefully without any personal stake in the matter) that spacetime is not continuous, it will also be difficult to persuade her that the question is actually an empirical one.

Difficult, but, as argued here to an extended length, not impossible[37]

What about Infinity?

If infinitesimals are simply a mathematical abstraction with no concrete manifestation, we still have another kind of infinity to deal with, the infinitely large. Are ∞ and –∞ real? I would argue that these are also mathematical concepts without any physical analog. For example, we cannot have an infinite velocity. The limit is $c \approx 3 \times 10^8$ m/s (the speed of light in a vacuum).[38] We cannot have anything colder than – 273° C (0° K),

---

37 Hagar, Amit, *Discrete or Continuous?: The Quest for Fundamental Length in Modern Physics*, Cambridge University Press, Cambridge, 2014, p. 238.

38 Okay, I have heard of tachyons and I am aware of experiments in which a beam of light was accelerated to go "faster than light" with the result that it traveled backwards in time. This is interesting stuff, but I am not qualified to discuss it! White, Michael J., *The Continuous and the Discrete: Ancient Physical Theories from*

known as absolute zero. So, there is a minimum temperature. It seems that there should also be a maximum temperature, as heat is caused by vibrations and the speed of light imposes a limit on the speed of these vibrations.

In *The Theory of Everything: The Origin and Fate of the Universe*, Stephen Hawking argues that the universe is finite with no edge or boundary. Of course, the big bang theory provides a Universe with a definite beginning, so we cannot speak of events infinitely far in the past.[39]

Interplay

For practical purposes it doesn't seem to matter if the universe is discrete or continuous, for we can model a discrete system with continuous mathematics and we can model a continuous system with discrete mathematics. We often use whichever model seems to be the easiest to handle, ignoring the underlying nature of the phenomena. Some examples follow.

In economics, one looks at things like marginal revenue. This is a function that takes the number of items being sold and calculates the increase in revenue attained by selling one more. There is, of course, a point at which the market becomes saturated. To ensure increased sales, price must be lowered. If we denote the revenue function by $R(x)$, then the marginal revenue at x is $R(x + 1) - R(x)$. Now compare the two expressions shown below:

$$\frac{R(x+1) - R(x)}{1}$$

$$\frac{R(x+h) - R(x)}{h}$$

The first expression is the marginal revenue at x. Dividing by 1 doesn't change anything. The second expression, where h takes the place of 1, should look familiar. It's a difference quotient. If we take the limit as h approaches to 0, we get the derivative $R'(x)$. The value 1 isn't especially close to 0, but in economics it is often close enough. $R'(x)$ is frequently used as an approximation of the marginal revenue function. A company manufacturing an item sold one-at-a-time is definitely in a discrete business; yet, continuous mathematics is useful. In addition to marginal revenue, economists and businessmen look at marginal cost and marginal profit.

Integrals also exhibit an interplay between the discrete and the continuous. A Riemann sum gives a discrete approximation of the area under the curve. In the abstract world of mathematics, a curve is truly continuous, but Riemann sums can give us values as precise as desired. Using a billion rectangles is still a discrete approximation, but it will likely be amazingly close to the true value. Taking the limit as the

_____

   a Contemporary Perspective, Clarendon Press, Oxford, 1992, pp. 39–40, notes that Aristotle thought there was no maximum speed.

39 White, Michael J., *The Continuous and the Discrete: Ancient Physical Theories from a Contemporary Perspective*, Clarendon Press, Oxford, 1992, p. 111 states that Aristotle "holds that the cosmos has no beginning in time."

number of rectangles approaches infinity (or equivalently, as their width approaches zero) is a simple enough step, but that is what moves us from the discrete to the continuous, giving the exact answer.

There's even a lot of interplay between the continuous and the discrete to be found outside of economics and mathematics. Let's consider a few examples that relate to our senses. When viewing a movie, if it is on film, we typically have 24 frames per second. This is more than the human eye can perceive individually, so we have the illusion of continuous motion on the screen, when it is actually discrete. Increasing the number of frames per second doesn't make it feel "more continuous," but does offer an improved experience if the viewer chooses to watch a scene in slow motion. Just as our eyes can be fooled into taking the discrete as continuous, so too can our ears. A note repeated rapidly enough is indistinguishable from a pure tone. Even though Aristotle favored a continuous interpretation of nature, he did admit that the number of gradations of various things that we can actually *perceive* is finite.[40]

The interplay between the continuous and the discrete is returned to again and again in this book. While the focus is on discrete mathematics, there are too many beautiful connections to the continuous to ignore!

Exercise Set 0

> There's nothing like the Eureka moment of discovering something that no one knew before. I won't compare it to sex, but it lasts longer.
>
> Stephen Hawking[41]

> We do not worry about being respected in towns through which we pass. But if we are going to remain in one for a certain time, we do worry. How long does this time have to be?[42]
>
> Blaise Pascal

1. How high does the resolution of an image have to be for the transition between pixels to be indistinguishable to the human eye?

2. Do a search and find the rate at which a single note (say on a guitar) would have to be played in order to sound continuous. That is, how quickly would the note have to be repeated in order to sound like a single tone that simply persists, as opposed to individual notes. Can anyone play that fast?

3. Provide an example not mentioned in this chapter of something that is discrete, but appears continuous.

4. What's *your* opinion as to the nature of reality? Do you think it is continuous or discrete? Support your view with the strongest argument you can find.

---

40 White, Michael J., *The Continuous and the Discrete: Ancient Physical Theories from a Contemporary Perspective*, Clarendon Press, Oxford, 1992, pp. 127–130. Also see pp. 134–135.

41 Roberts, Siobhan, *Genius at Play – The Curious Mind of John Horton Conway*, Bloomsbury USA, 2015, p. 57

42 Quoted in Auden, W. H. and Louis Kronenberger, editors, *The Viking Book of Aphorisms*, Barnes & Noble Books, New York, 1993, p. 135.

5. Look up the definition of "number" in a dictionary. According to this definition, is a complex number a "number"? Can you come up with a better definition?

6. Can you define mathematics? Give it your best shot! At the end of this book, you will encounter this question again. Your answer may change.

7. Find a definition of life online or in a book. State it and then try to find something that fits the definition that you do not think should be considered alive, or something that is alive that does not fit the definition.

8. On the microscopic scale there have existed things that have evolved back and forth from living to non-living, according to some definitions! Research this topic and provide some details.

9. Pluto used to be a planet. Find a definition for "planet" and share your thoughts as to its appropriateness.

10. Ask a friend to give a definition of some common object. Then see if you can nitpick his or her definition. Reproduce the conversation here, along with the response to your nitpicking!

11. What is the legal definition of "organic" in the United States? Is it what you expected?

12. What is the legal definition of "free range chicken" in the United States? Is it what you expected?

13. Give an example of a word whose definition has changed over time.

References/Further Reading

I love the smell of book ink in the morning.
– Umberto Eco

Amati, Daniele, Marcello Ciafaloni, and Gabriele Veneziano, "Can Spacetime Be Probed below the String Size?" *Physics Letters B*, Vol. 216, No. 1–2, January 1989, pp. 41–47.

Baez, John C., "The Quantum of Area?" *Nature*, Vol. 421, No. 6924, February 13, 2003, pp. 702–703.

Bailey, Cyril, *The Greek Atomists and Epicurus. A Study*, Russell & Russell, New York, 1964.

Bell, Eric Temple, *Mathematics, Queen and Servant of Science*, Copyright, 1951, British Edition 1952, reprinted 1954, 1958, 1961. Printed in Great Britain by Butler and Tanner Ltd., Frome and London.

Born, Max, "The Mysterious Number 137," *Proceedings of the Indian Academy of Science – Section A*, Vol. 2, No. 6, December 1935, pp. 533–561.

Brüssow, Harald, "The Not so Universal Tree of Life or the Place of Viruses in the Living World," *Philosophical Transactions of the Royal Society B*, Vol. 364, No. 1527, August 12, 2009, pp. 2263–2274. www.ncbi.nlm.nih.gov/pmc/articles/PMC2873004/pdf/rstb20090036.pdf.

Butterfield, Jeremy and Christopher Isham, "Spacetime and the Philosophical Challenge of Quantum Gravity," in Callender, Craig and Nick Huggett, editors, *Physics Meets Philosophy at the Planck Scale*, Cambridge University Press, Cambridge, 1999, pp. 33–39.

Callender, Craig and Nick Huggett, editors, *Physics Meets Philosophy at the Planck Scale: Contemporary Theories in Quantum Gravity*, Cambridge University Press, Cambridge, 1999.

Carazza, Bruno and Helge Kragh, "Heisenberg's Lattice World: The 1930 Theory Sketch," *American Journal of Physics*, Vol. 63, No. 7, July 1995, pp. 595–605.

Chew, Geoffrey F., "The Dubious Role of the Space-Time Continuum in Subatomic Physics," *Science Progress*, Vol. 51, 1963, pp. 529–539.

Coish, Harold R., "Elementary Particles in a Finite World Geometry," *Physical Review*, Vol. 114, April, 1959, pp. 383–388.

Conway, John H. and Charles Radin, "Quaquaversal Tilings and Rotations," *Inventiones Mathematicae*, Vol. 132, 1988, pp. 179–188.

Darrigal, Olivier, "The Origin of Quantized Matter Waves," *Historical Studies in the Physical Sciences*, Vol. 16, No. 2, 1986, pp. 198–253.

Denyer, Nicholas, "The Atomism of Diodorus Cronus," *Prudentia*, Vol. 13, 1981, pp. 33–45.

Einstein, Albert, *Ideas and Opinions*, Crown, New York, 1919/1954.

Fitzgerald, Janet, *Alfred North Whitehead's Early Philosophy of Space and Time*, University Press of America, Washington, DC, 1979.

Fogelin, Robert, "Hume and Berkeley on the Proofs of Infinite Divisibility," *Philosophical Review*, Vol. 97, No. 1, January 1988, pp. 47–69.

Forrest, Peter, "Is Spacetime Discrete or Continuous? – An Empirical Question," *Synthese*, Vol. 103, No. 3, June 1995, pp. 327–354.

Forrest, Peter, *The Necessary Structure of the All-pervading Aether: Discrete or Continuous? Simple or Symmetric?* Philosophische Analyse/Philosophical Analysis, Vol. 49, ontos verlag, Heusenstamm, Germany, 2012.

Fort, Charles, *The Book of the Damned*, Boni and Liveright, New York, 1919. An annotated edition is available online at www.resologist.net/damn01.htm.

Furley, David J., *Two Studies in the Greek Atomists*, Princeton University Press, Princeton, NJ, 1967.

Furley, David J., "Aristotle and the Atomists on Infinity," in Düring, Ingemar, editor, *Naturphilosophie bei Aristoteles und Theophrast*, 1969, pp. 85–96.

Furley, David J., *The Greek Cosmologists, Vol. I. The Formation of the Atomic Theory and Its Earliest Critics*, Cambridge University Press, Cambridge, 1987.

Graham, Ronald L., Donald E. Knuth, and Oren Patashnik, *Concrete Mathematics: A Foundation for Computer Science*, second edition, Addison-Wesley, Reading, MA, 1994. "Concrete" in the title is not used just in contrast to abstract, but also as a portmanteau for CONtinuous and disCRETE.

Hagar, Amit, "Length Matters: The Einstein–Swann Correspondence," *Studies in the History and Philosophy Science Part B, Studies in the History and Philosophy of Modern Physics*, Vol. 39, No. 3, September 2008, pp. 532–556.

Hagar, Amit, *Discrete or Continuous?: The Quest for Fundamental Length in Modern Physics*, Cambridge University Press, Cambridge, 2014. This book is written at a much higher level than the present text. In particular, it assumes a knowledge of elementary analysis, some topology, abstract algebra, and much physics.

Hawking, Stephen, *The Universe in a Nutshell*, Bantam Books, New York, 2001.

Hawking, Stephen, *The Theory of Everything: The Origin and Fate of the Universe*, New Millennium Press, Princeton, NJ, 2002.

Heisenberg, Werner, "Die Selbstenergie des Elektrons," *Zeitschrift für Physik*, Vol. 65, November 1930, pp. 4–13.

Heisenberg, Werner, "The Universal Length Appearing in the Theory of Elementary Particles," *Annalen Der Physik*, Vol. 32, 1938, pp. 20–33.

Heisenberg, Werner, "On the Mathematical Frame of the Theory of Elementary Particles," *Communications on Pure and Applied Mathematics*, Vol. 4, No. 1, June 1951, pp. 15–22.

Hilbert, David, "On the Infinite," in Benacerraf, Paul and Hilary Putnam, editors, *The Philosophy of Mathematics, Selected Readings*, second edition, Cambridge University Press, Cambridge, 1983, pp. 183–201. p. 186 quoted here. This is a reprinting of the 1925 paper.

Hilgevoord, J. and J. Uffink, "The Mathematical Expression of the Uncertainty Principle," in van der Merwe, Alwyn, Franco Selleri, and Gino Tarozzi, editors, *Microphysical Reality and Quantum Description*, Kluwer, Dordrecht, 1988, pp. 91–114.

Huang, Cary, *The Scale of the Universe 2*, 2012, available online at http://htwins.net/scale2/. Follow this link and have fun zooming in and out to get a sense of the relative size of things. In particular, make comparisons with the possible minimum length in our universe. Note: This website works with Internet Explorer, but not Google Chrome.

Kragh, Helge, "Arthur March, Werner Heisenberg, and the Search for a Smallest Length," *Revue d'Histoire Des Sciences (Paris)*, Vol. 48, No. 4, 1995, pp. 401–434.

Kragh, Helge and Bruno Carazza, "From Time Atoms to Spacetime Quantization: The Idea of Discrete Time, ca 1925–1936," *Studies in the History and Philosophy of Science Part A*, Vol. 25, No. 3, 1994, pp. 437–462.

Luria, Solomon Y., "Die Infinitesimaltheorie der Antiken Atomisten," *Quellen und Studien zur Geschichte der Mathematik, Astronomie und Physik, Series B*, Vol. 2, 1932, pp. 106–185.

Magueijo, João, "New Varying Speed of Light Theories," *Reports on Progress in Physics*, Vol. 66, No. 11, October 2003, pp. 2025–2068.

Makin, Stephen, "The Indivisibility of the Atom," *Archiv für Geschichte der Philosophie*, Vol. 71, No. 2, 1989, pp. 125–149.

Mau, Jürgen, *Zum Problem des Infinitesimalen bei den antiken Atomisten*, Akademie-Verlag, Berlin, 1954.

McDaniel, Kristopher, "Distance and Discrete Space," *Synthese*, Vol. 155, 2007, pp. 157–162.

McKinley, William A., "Search for a Fundamental Length in Microscopic Physics," *American Journal of Physics*, Vol. 28, No. 2, 1960, pp. 129–134.

Mead, C. Alden, "Observable Consequences of Fundamental-Length Hypotheses," *Physical Review*, Vol. 143, No. 4, March 1966, pp. 990–1005.

Mead, C. Alden, "Walking the Planck Length through History," *Physics Today*, Vol. 54, No. 11, November 2001, p. 15.

Mead, C. Alden, "Possible Connection between Gravitation and Fundamental Length," *Physical Review*, Vol. 135, No. 3, August 1964, pp. B849.

Miller, Jr., Fred D., "Aristotle against the Atomists," in Kretzmann, Norman, editor, *Infinity and Continuity in Ancient and Medieval Thought*, Cornell University Press, Ithaca, NY, 1982, pp. 87–111.

Moore, Cristopher, "Recursion Theory on the Reals and Continuous-Time Computation," *Theoretical Computer Science*, Vol. 162, No. 1, August 1996, pp. 23–44.

Moyer, Michael, "Is Space Digital?" *Scientific American*, Vol. 306, No. 2, February 2012, pp. 30–37.

Owen, Gwilym Ellis Lane, "Zeno and the Mathematicians," (1957), in Salmon, Wesley C., editor, *Zeno's Paradoxes*, Hacket Publishing Company, Indianapolis, IN, 2001, pp. 139–163.

Radin, Charles, "Symmetry of Tilings of the Plane," *Bulletin of the American Mathematical Society*, New Series, Vol. 29, No. 2, October, 1993, pp. 213–217.

Radin, Charles, "The Pinwheel Tilings of the Plane," *Annals of Mathematics*, Second Series, Vol. 139, No. 3, May, 1994, pp. 661–702.

Radin, Charles, "Symmetry and Tilings," *Notices of the American Mathematical Society*, Vol. 42, No. 1, January 1995, pp. 26–31.

Rogers, Ben, "On Discrete Spaces," *American Philosophical Quarterly*, Vol. 5, No. 2, April 1968, pp. 117–123.

Rovelli, Carlo and Lee Smolin, "Discreteness of Area and Volume in Quantum Gravity," *Nuclear Physics B*, Vol. 442, No. 3, May 1995, pp. 593–619.

Ruark, Arthur, "The Roles of Discrete and Continuous Theories in Physics," *Physical Review*, Vol. 37, No. 3, February 1931, pp. 315–326.

Schild, Alfred, "Discrete Spacetime and Integral Lorenz Transformations," *Physical Review*, Vol. 73, No. 4, February 1948, pp. 414–415.

Schild, Alfred, "Discrete Spacetime and Integral Lorenz Transformations," *Canadian Journal of Mathematics*, Vol. 1, No. 1, 1949, pp. 29–47.

Silberstein, L., *Discrete Spacetime. A Course of Five Lectures Delivered in the McLennan Laboratory*, University of Toronto Press, Toronto, 1936.

Skyrms, Brian, "Zeno's Paradox of Measure," in Cohen, Robert S. and Larry Laudan, editors, *Physics, Philosophy, and Psychoanalysis, Essays in Honor of Adolf Grünbaum*, D. Reidel Publishing Company, Dordrecht, Holland, 1983, pp. 223–254.

Snyder, Hartland S., "Quantized Space-time," *Physical Review*, Vol. 71, No. 1, January 1947, pp. 38–41.

Snyder, Hartland S., "The Electromagnetic Field in Quantized Space-Time," *Physical Review*, Vol. 72, No. 1, July 1947, pp. 68–71.

Solmsen, Friedrich, "Abdera's Arguments for the Atomic Theory," *Greek, Roman, and Byzantine Studies*, Vol. 29, No. 1, 1988, pp. 59–73.

Sorabji, Richard, "Atoms and Time Atoms," in Kretzmann, Norman, editor, *Infinity and Continuity in Ancient and Medieval Thought*, Cornell University Press, Ithaca, NY, 1982, pp. 37–86.

Sorkin, Rafael D., "Causal Sets: Discrete Gravity," in Gomberoff, Andrés and Don Marolf, editors, *Proceedings of the Valdivia Summer School*, held January 2002 in Valdivia, Chile, Plenum, 2005.

Steen, Lynn Arthur, "New Models for the Real Number Line," *Scientific American*, Vol. 224, No. 2, August 1971, pp. 92–99.

Steinhart, Eric, "Digital Metaphysics," in Bynum, Terrell and James Moor, editors, *The Digital Phoenix: How Computers are Changing Philosophy*, Basil Blackwell, New York, 1998, pp. 117–134.

Stevens, Michael, "Supertasks," *Vsauce*, available online at www.youtube.com/watch?v=ffUnNaQTfZE.

Szekeres, Peter, "Discrete Space-time," in Carey, Alan L., William J. Ellis, Paul A. Pearce, and Anthony W. Thomas, editors, *Confronting the Infinite: A Conference in Celebration of the 70th Years of H.S. Green and C.A. Hurst, University of Adelaide, Adelaide, Australia, February 14–17, 1994*, World Scientific, Singapore, 1995, pp. 293–303.

Thiemann, Thomas, "Lectures on Loop Quantum Gravity," *Lecture Notes in Physics*, Vol. 631, 2003, pp. 41–135.

University of Illinois at Urbana-Champaign, "Study Adds to Evidence that Viruses are Alive," *ScienceDaily*, September 25, 2015, available online at www.sciencedaily.com/releases/2015/09/150925142658.htm.

Van Bendegem, Jean-Paul, "Zeno's Paradoxes and the Weyl Tile Argument," *Philosophy of Science*, Vol. 54, No. 2, 1987, pp. 295–302.

Van Bendegem, Jean-Paul, "Ross' Paradox is an Impossible Super-Task," *The British Journal for the Philosophy of Science*, Vol. 45, No. 2, June 1994, pp. 743–748.

Van Bendegem, Jean-Paul, "Finitism in Geometry," in Zalta, Edward N., editor, *The Stanford Encyclopedia of Philosophy*, spring 2009 edition. A substantive revision is available online at https://stanford.library.sydney.edu.au/archives/win2015/entries/geometry-finitism/.

Vlastos, Gregory, "Minimal Parts in Epicurean Atomism," *Isis*, Vol. 56, No. 2, Whole No. 184 1965, pp. 121–147.

Wessner, David R., "The Origins of Viruses," *Nature Education*, Vol. 3, No. 9, 2010, p. 37, available online at www.nature.com/scitable/topicpage/the-origins-of-viruses-14398218#.

Weyl, Hermann, *Philosophy of Mathematics and Natural Sciences*, Princeton University Press, Princeton, NJ, 1949.

White, Michael J., *The Continuous and the Discrete: Ancient Physical Theories from a Contemporary Perspective*, Clarendon Press, Oxford, 1992.

Winnie, John A., "Deterministic Chaos and the Nature of Chance," in Earman, John and John D. Norton, editors, *The Cosmos of Science*, University of Pittsburgh Press, Pittsburgh, PA, 1997, pp. 299–324.

Young, Robert M., *Excursions in Calculus: An Interplay of the Continuous and the Discrete*, The Mathematical Association of America, 1992.

1

Logic

The fact that all Mathematics is Symbolic Logic is one of the greatest discoveries of our age; and when this fact has been established, the remainder of the principles of mathematics consists in the analysis of Symbolic Logic itself.

Bertrand Russell, *Principles of Mathematics*, 1903[1]

If you are engaged in an argument and forced to admit that your opponent's viewpoint is logical, you are in trouble. The only thing worse would be if he or she presents a proof. In that case, no response is possible. Mathematics is the only discipline that can claim proof. Therefore, it is important to develop a deeper understanding of what that means. The first step is to understand logic.

Credit for first establishing logic as a field of study goes to Aristotle (Figure 1.1). Socrates and Plato are sometimes cited, but they did not look at things as abstractly (generally) as Aristotle. Appeals may even be made going back farther to the Pythagoreans, for example, or others. This is not appropriate. To understand why we must make a subtle distinction. J. M. Bocheński explained, "Thinkers of these schools did indeed establish chains of inference, but logic consists in studying inference, not in inferring."[2]

Thus, Aristotle was the first to develop formal logic, intended as a set of rules by which one could reason correctly. He produced six works on logic. In collected form, they became known as the Organon.[3] Questions have been raised concerning how much of these works should really be attributed to Aristotle, as opposed to his followers.[4]

In any case, Aristotle's examples of logical reasoning are known as syllogisms. An example that is frequently given is

All men are mortal.
Socrates is a man.
Therefore, Socrates is mortal.

However, this is not the way Aristotle actually did things. His work is represented more accurately as

---

1 Taken here from http://www-history.mcs.st-and.ac.uk/Quotations/Russell.html, which has many other great quotes from Russell.

2 Bocheński, J. M., *A History of Formal Logic* (translated and edited by Ivo Thomas), University of Notre Dame Press, Notre Dame, IN, 1961, p. 11.

3 This collection can be downloaded from https://ia601407.us.archive.org/34/items/AristotleOrganon/Aristo tleOrganoncollectedWorks.pdf where it appears in a form translated under the editorship of W. D. Ross. Other works of Aristotle are included in this pdf as well.

4 Bocheński, J. M., *A History of Formal Logic* (translated and edited by Ivo Thomas), University of Notre Dame Press, Notre Dame, IN, 1961, pp. 40–41.

FIGURE 1.1
Aristotle (384 BCE–322 BCE).

Image created by jlorenz1. This file is licensed under the Creative Commons Attribution 2.5 Generic license.

If all men are mortal
and all Greeks are men,
then all Greeks are mortal.[5]

But he did much more than demonstrate such commonsensical (to us anyway) arguments. He looked at things abstractly. This process seems to have begun with abbreviations. Using single letters to stand for men, mortal, and Greek, we can make the leap to treating these letters as variables.[6] A pair of examples follow:[7]

1) Since if A (is predicated) of all B, and B of all C, A must be predicated of all C.
2) Similarly too if A (is predicated) of no B, and B of all C, it is necessary that A will belong to no C.

These are less clear (as often happens when things become more abstract), but much more general. Replacing the variables in 1) with specific choices gives[8]

---

5 Taken here from http://planetmath.org/aristotelianlogic.
6 Bocheński, J. M., *A History of Formal Logic* (translated and edited by Ivo Thomas), University of Notre Dame Press, Notre Dame, IN, 1961, p. 42.
7 Taken from Bocheński, J. M., *A History of Formal Logic* (translated and edited by Ivo Thomas), University of Notre Dame Press, Notre Dame, IN, 1961, p. 64.
8 Taken from Bocheński, J. M., *A History of Formal Logic* (translated and edited by Ivo Thomas), University of Notre Dame Press, Notre Dame, IN, 1961, p. 66.

If animal belongs to all man
and man belongs to all Greek,
then animal belongs to all Greek.

For 2), we have an example[9] given by

If stone belongs to no man
and man belongs to all Greek,
then stone belongs to no Greek.

The abstraction from specific examples to the general syllogisms stated by Aristotle was a very important step. He also tried to build his logic on an axiomatic foundation. That is, he found that some syllogisms followed logically from others.

This also happens in every area of mathematics. That is, there are some statements that are deductive consequences, by specified rules of inference, of others. Wouldn't it be nice if we could find rules of inference so that any statement that must be true whenever a given set of statements (taken as *axioms*) is true will be a deductive consequence of those axioms using those rules? This is the idea of an axiomatic system.

While axioms are not proven, in an axiomatic system everything follows from them. That is, if the axioms are regarded as true and the rules of inference preserve truth, then every deductive consequence of those axioms must also be true. For this reason the axioms need to be selected very carefully. We want them to be obvious, or as Thomas Jefferson phrased it, "We hold these truths to be self-evident." He put forth America's case in the Declaration of Independence in a manner not unlike a mathematical proof, beginning with axioms and logically building on them to get to his conclusion.[10]

You likely noticed that all of the syllogisms provided above (whether abstract or concrete) had certain similarities. Aristotle took three propositions at a time to construct his syllogisms. The **premise** consists of the first two propositions and the third is the **conclusion**. The propositions in the premise share exactly one term (in our examples above these are "men" and "man") and the conclusion does not contain this term, but does contain the other two terms.

The statements (after removing the ifs, ands, and thens) that make up the syllogisms are termed **propositions**. A proposition is either true or false. Looking closer at the propositions, each consists of a **subject** and a **predicate**. Aristotle allowed the subject to be modified by the words "all," "some," and "no" (**quantifiers**). The propositions then continue with "are/is" or "are/is not" (this is the **copula**) and conclude with the **predicate**.

Example:	All	men	are	mortal
	quantifier	subject	copula	predicate

9 Also from Bocheński, J. M., *A History of Formal Logic* (translated and edited by Ivo Thomas), University of Notre Dame Press, Notre Dame, IN, 1961, p. 66.

10 For more see "The Contribution of Scientific Men to American Independence," *The Scientific Monthly*, Vol. 25, August 1927, pp. 187 and 189. This is part of "The Progress of Science," a regular feature edited by Edwin E. Slosson and appearing on pp. 183–192 in this particular issue.

When propositions are replaced by letters such as A, B, and C, the letters are known as **propositional variables**. Some authors refer to **statements** and **statement variables** instead. Generally, when something has several different names, it's an indication of its importance. It often means that it has been found to be of interest to people working in different fields who feel no need to get together and standardize their terminology. You'll soon see that there are a wide range of notations in use for abstracting the laws of logic.

Logic itself may be divided into many different subfields, some of which overlap. While many of them are not of interest to us in the context of discrete mathematics, terms applied to the subject matter of this chapter include **propositional logic**, **formal logic**, **predicate logic**, and **symbolic logic**.

There's a lot more jargon that goes with the subject, and some of it is in Latin. We continue with a pair of simple examples. Aristotle's connection to these is disputed:

> ... "Aristotelian logic," as it was taught from late antiquity until the 20th century, commonly included a short presentation of the argument forms *modus (ponendo) ponens, modus (tollendo) tollens, modus ponendo tollens*, and *modus tollendo ponens*. In late antiquity, arguments of these forms were generally classified as "hypothetical syllogisms." However, Aristotle did not discuss such arguments, nor did he call any arguments "hypothetical syllogisms."[11]

On the other side:

> Aristotle expressly recognizes that neither *modus ponens* nor *modus tollens* is reducible to syllogisms in the narrow sense.[12]

Both quotes are from scholarly publications. One must always verify! Here, it is a "simple" matter to read all of the Organon and find the answer for oneself. I will do so when I have more time.

modus ponens (this translates from Latin to English as "mode that affirms")

Example:	If you have a library card, you can check out books.
	You have a library card.
	(Therefore) You can check out books.

Symbolically:	$p \rightarrow q$
	p
	$\therefore q$

---

11 Bobzien, Suzanne, "The Development of Modus Ponens in Antiquity: From Aristotle to the 2nd Century AD," *Phronesis*, Vol. 47, No. 4, 2002.

12 Woods, John and Andrew Irvine, "Aristotle's Early Logic," pp. 27–99, p. 53 quoted here, in Gabbay, Dov M., and John Woods, editors, *Handbook of the History of Logic, Volume 1: Greek, Indian and Arabic Logic*, Elsevier, Amsterdam, 2004.

modus tolens (this translates from Latin to English as "mode that denies")

Example: You can't check out books.
          If you have a library card, you can check out books.
          (Therefore) You don't have a library card.

Symbolically: ¬q
          p → q
          ∴ ¬p

Michael J. White mentioned a generalization in an interesting context:

> Another noteworthy example is found in Sextus Empiricus. Hypotyposeis 1.69, where the Stoic Chrysippus is presented as arguing that even a dog accepts a generalization of modus tollendo ponens: either A or B or C; but not A and not B; therefore, C. For, according to Chrysippus, a dog, coming to a triple fork in the road and not detecting the scent of its quarry in 2 of the 3 paths, will set off straightaway down the 1 remaining path without bothering to check for the scent.[13]

Some Criticism

Although Aristotle laid the basis for logic, he wasn't perfect. Petr Beckmann, in his lively book, *A History of π (Pi)*, shared his thoughts on the man.

> Aristotle, we are invariably told, was "antiquity's most brilliant intellect," and the explanation of this weird assertion, I believe, is best summarized in Anatole France's words: The books that everybody admires are the books that nobody reads. But on taking the trouble to delve into Aristotle's writings, a somewhat different picture emerges (see also Chapter 6). His ignorance of mathematics and physics, compared to the Greeks of his time, far surpasses the ignorance exhibited by this tireless and tiresome writer in the many subjects he felt himself called upon to discuss.[14]

In Chapter 6, Beckmann reveals:

---

13 White, Michael J., *The Continuous and the Discrete: Ancient Physical Theories from a Contemporary Perspective*, Clarendon Press, Oxford, 1992, p. 248 footnote.

14 Beckmann, Petr, *A History of π (Pi)*, Dorset Press, New York, 1989, p. 40. Aristotle was considered an authority on "everything from botany to metaphysics, zoology to astronomy, poetry to politics to how to have better sex" (quotation from Goldstone, Lawrence and Nancy Golstone, *The Friar and the Cipher: Roger Bacon and the Unsolved Mystery of the Most Unusual Manuscript in the World*, Doubleday, New York, 2005, p. 28).

[Plato and Aristotle] taught that experimentation was fit only for slaves, and that the laws of nature could be deduced merely by man's lofty intellect; and Aristotle used his lofty intellect to deduce that heavier bodies fall to the ground more rapidly; that men have more teeth than women; that the earth is the center of the universe; that heavenly bodies never change; and much more of such wisdom, for he was a very prolific writer. As a matter of fact, Aristotle was defeated on his own grounds, by sheer intellectual deduction unaided by experimental observation. Long before Galileo Galilei dropped the wooden and leaden balls from the leaning tower of Pisa, he asked the following question: "If a 10 lb stone falls ten times as fast as a 1 lb stone, what happens if I tie the two stones together? Will the combination fall faster than the 10 lb stone because it weighs 11 lbs, or will it fall more slowly because the 1 lb stone will retard the 10 lb stone?"[15]

I include these criticisms for a very important reason: we make progress in science and mathematics by questioning authority. If we put those who have made contributions to these fields up on pedestals, we may fail to challenge them in areas where they were wrong.

[Aristotle's] teachings held up the progress of science for close to 2,000 years.[16]

How could this be? Is it an exaggeration? I'll present a few facts and let you decide for yourself, hopefully only after reading some other sources. Over a millennium after his death, Aristotle's teachings were adopted by the Catholic Church, in a modified way, and the man himself was said to be infalliable.[17] Saint Thomas Aquinas (1225–1274), another favorite philosopher of the church, rejected experimental science:[18]

Because Aquinas made no distinction between method and conclusion, many of Aristotle's scientific conclusions that were just plain wrong, such as the geocentric universe and the division of matter into earth, air, fire, and water, became cornerstones of natural science within the Christian religion. Since it was often hard to define where permissible scientific inquiry ended and the truth of faith began, all of these dubious truths were now off-limits and thus inviolable.[19]

This sorry situation lasted a long time:

A statute of Oxford University in 1583 provided that any master or bachelor who deviated from the doctrine of Aristotle on any point should pay a fine of five shillings

---

15 Beckmann, Petr, *A History of π (Pi)*, Dorset Press, New York, 1989, p. 70. Galileo is often credited with showing that heavier objects fall at the same speed as lighter objects by dropping two balls, of different weights, from the leaning tower of Pisa. However, we don't know for sure whether the experiment was carried out or is just legendary. See Cooper, Lane, "Galileo and Scientific History," *The Scientific Monthly*, Vol. 43, August 1936, pp. 163–167. Cooper is also the author of *Aristotle, Galileo, and the Tower of Pisa*. In these works he tried to get at the historical truth of what happened.

16 Beckmann, Petr, *A History of π (Pi)*, Dorset Press, New York, 1989, p. 40.

17 Goldstone, Lawrence and Nancy Golstone, *The Friar and the Cipher: Roger Bacon and the Unsolved Mystery of the Most Unusual Manuscript in the World*, Doubleday, New York, 2005, p. 42.

18 Goldstone, Lawrence and Nancy Golstone, *The Friar and the Cipher: Roger Bacon and the Unsolved Mystery of the Most Unusual Manuscript in the World*, Doubleday, New York, 2005, p. 89.

19 Goldstone, Lawrence and Nancy Golstone, *The Friar and the Cipher: Roger Bacon and the Unsolved Mystery of the Most Unusual Manuscript in the World*, Doubleday, New York, 2005, p. 126.

for each such offense. I do not know whether the rule has been repealed yet or not. Probably not. Oxford rarely repeals. But while it was enforced a teacher could not often afford the luxury of mentioning that the earth moves.[20]

Finally, men like Nicolaus Copernicus, Galileo Galilei, Johannes Kepler, and Isaac Newton got the ball rolling once again by daring to challenge Aristotle's entrenched ideas. In a similar manner, Einstein's rebellious attitude emboldened him to improve dramatically upon Newton's work (which only approximates reality well when velocities are small compared to that of light). Today's researchers should not shy away from challenging Einstein, if their research suggests his conception of the universe is flawed.

Raphael depicted Aristotle speaking with Plato (his teacher) in this detail from *The School of Athens* (Figure 1.2). The depictions do not mesh with what we know of the appearance of these men.

FIGURE 1.2
Plato (left) and Aristotle (right) as imagined by Raphael.

Public domain.

---

20 "Legislative Orthodoxy," *The Scientific Monthly*, Vol. 21, August 1925, p. 223. This is part of "The Progress of Science," a regular feature edited by Edwin E. Slosson appearing on pp. 217–224 in this particular issue.

Plato, the former wrestler, was a manly man who dressed simply. He probably growled. Aristotle, on the other hand, was skinny, dressed in the most fashionable togas, wore lots of rings, and spoke with a lisp. Plato is known to have referred to him as "a mind on legs," and scrawny legs at that.[21]

Next Steps

For centuries after Aristotle, there was little progress in logic. The next great leap forward was made by Gottfried Wilhelm Leibniz (Figure 1.3). His best-known contribution to mathematics is his discovery of calculus. Isaac Newton had hit upon it first, but Leibniz was the first to publish. Leibniz contributed to many other fields and also left much important work unpublished. While he pushed logic ahead, he didn't take it as far as he thought possible. He wrote:

> As I was keenly occupied with this study, I happened unexpectedly upon this remarkable idea, that an alphabet of human thought could be devised, and that everything could be discovered (*inveniri*) and distinguished (*dijudicari*) by the combination of the letters of this alphabet and by the analysis of the resulting words.[22]

Leibniz believed his approach could be used to resolve *all* disputes:

> Then, in case of a difference of opinion, no discussion between two philosophers will be any longer necessary, as (it is not) between two calculators. It will rather be enough for them to take pen in hand, set themselves to the abacus, and (if it so pleases, at the invitation of a friend) say to one another: *Calculemus!*[23]

Leibniz was not afraid to tackle big questions. On the title page of his *Dissertatio de Arte Combinatoria* (Dissertation on the Combinatorial Art), written when he was only 20, he claimed that he had proved, with complete mathematical certainty, the existence of God.[24]

Leibniz even thought calculations could resolve the basic disagreements between Catholics and Protestants.[25] He didn't publish his findings. In fact, his work in this

---

21 Goldstone, Lawrence and Nancy Golstone, *The Friar and the Cipher: Roger Bacon and the Unsolved Mystery of the Most Unusual Manuscript in the World*, Doubleday, New York, 2005, p. 32.

22 Bocheński, J. M., *A History of Formal Logic* (translated and edited by Ivo Thomas), University of Notre Dame Press, Notre Dame, IN, 1961, p. 274.

23 Bocheński, J. M., *A History of Formal Logic* (translated and edited by Ivo Thomas), University of Notre Dame Press, Notre Dame, IN, 1961, p. 275.

24 Leibniz, G. W., *Dissertatio de Arte Combinatoria, in qua ex Arithmeticae Fundamentis Complicationum ac Transpositionum Doctrina Novis Praeceptis Exstruitur, et Usus Ambarum per Universum Scientiarum Orbem Ostenditur, Nova etiam Artis Meditandi seu Logicae Inventionis Semina Sparguntur*, Fick & Seubold, Leipzig, 1666. It was reprinted in *G. W. Leibniz, Samtliche Schriften und Briefe*, Vol. 1, Akademie der Wissenschaften der DDR, Series VI, second edition, Berlin, 1990, pp. 163–230. For an English translation of the claimed proof, see Loemker, Leroy E., editor and translator, *Gottfried Wilhelm Leibniz, Philosophical Papers and Letters*, Vol. I, The University of Chicago Press, Chicago, Illinois, 1956, pp. 117–119.

25 This was in an era when mathematicians were still deeply religious. One of the accomplishments Isaac Newton was proudest of was his translation of portions of the Bible. Another thing Leibniz and Newton had in common was a strong interest in alchemy. Leibniz spent the last hours of his life discussing alchemy with his doctor – see Berlinski, David, *The Advent of the Algorithm*, Harcourt Inc., New York, 2000, p. 2.

FIGURE 1.3
Gottfried Wilhelm Leibniz (1646–1716).

Public domain.

area didn't see print until 1901.[26] In the meanwhile, it had been rediscovered by others in the 1800s, most notably George Boole. The rediscoverers could likely have done far more, if they could simply have built upon what Leibniz did instead of having to exert the effort to obtain it from scratch again.

On another note, Leibniz is often credited with being the first to represent numbers in binary. Interestingly, an ordering of chapters used in some versions of the *I Ching* led him to mistakenly conclude that Chinese mathematicians preceded him.[27]

> For most of its history, logic had been studied, and taught, quite separately from mathematics. But in the 1800s, there began to be connections.
>
> Stephen Wolfram[28]

---

26 Bocheński, J. M., *A History of Formal Logic* (translated and edited by Ivo Thomas), University of Notre Dame Press, Notre Dame, IN, 1961, p. 267.

27 Wilson, Robin and John J. Watkins, editors, *Combinatorics: Ancient & Modern*, Oxford University Press, Oxford, 2013, p. 5. For appearances of binary between the *I Ching* and Leibniz, see Bruderer, Herbert, "The Binary System Was Created Long Before Leibniz," Communications of the ACM, BLOG@ACM, https://cacm.acm.org/blogs/blog-cacm/221749-the-binary-system-was-created-long-before-leibniz/full text, October 10, 2017.

28 Wolfram, Stephen, 100 Years since *Principia Mathematica*, November 25, 2010, http://blog.stephenwolfram.com/2010/11/100-years-since-principia-mathematica/.

The growth of logic in the 1800s represented greater abstraction, and this trend continued into the next century, leading to Bertrand Russell and Alfred North Whitehead's publication of volume 1 of *Principia Mathematica* (1910). A portion of the often reproduced page 362 of this work follows below (Figure 1.4).

In *Logicomix*, a wonderful graphic novel relating the history of logic, a child asks Bertrand Russell, "But why *all* these pages … to prove 1+1=2?" Russell responds, "Hm … How shall I put it? It's the price you pay for being *truly certain*."[29] One can establish this result with even greater rigor by following methods proposed by Bourbaki, but defining 1 takes a bit longer via this method. Adrian R. D. Mathias took a close look at this problem in a paper titled "A Term of Length 4,523,659,424,929."[30] He showed that the number in the title of the paper is the number of symbols needed to rigorously define 1, following Bourbaki's approach. This does not include the 1,179,618,517,981 links between symbols needed to disambiguate the expression.

---

362   PROLEGOMENA TO CARDINAL ARITHMETIC   [PART II

$*54{\cdot}42. \quad \vdash :: \alpha \,\epsilon\, 2 . \supset :. \beta \subset \alpha . \exists ! \beta . \beta \neq \alpha . \equiv . \beta \,\epsilon\, \iota``\alpha$

     Dem.

$\vdash . *54{\cdot}4 . \quad \supset \vdash :: \alpha = \iota`x \cup \iota`y . \supset :.$

$\qquad\qquad \beta \subset \alpha . \exists ! \beta . \equiv : \beta = \Lambda . \mathbf{v} . \beta = \iota`x . \mathbf{v} . \beta = \iota`y . \mathbf{v} . \beta = \alpha : \exists ! \beta :$

$[*24{\cdot}53{\cdot}56.*51{\cdot}161] \qquad \equiv : \beta = \iota`x . \mathbf{v} . \beta = \iota`y . \mathbf{v} . \beta = \alpha \qquad (1)$

$\vdash . *54{\cdot}25 . \text{Transp} . *52{\cdot}22 . \supset \vdash : x \neq y . \supset . \iota`x \cup \iota`y \neq \iota`x . \iota`x \cup \iota`y \neq \iota`y :$

$[*13{\cdot}12] \qquad \supset \vdash : \alpha = \iota`x \cup \iota`y . x \neq y . \supset . \alpha \neq \iota`x . \alpha \neq \iota`y \qquad (2)$

$\vdash . (1) . (2) . \supset \vdash :: \alpha = \iota`x \cup \iota`y . x \neq y . \supset :.$

$\qquad\qquad \beta \subset \alpha . \exists ! \beta . \beta \neq \alpha . \equiv : \beta = \iota`x . \mathbf{v} . \beta = \iota`y :$

$[*51{\cdot}235] \qquad\qquad\qquad\qquad \equiv : (\exists z) . z \,\epsilon\, \alpha . \beta = \iota`z :$

$[*37{\cdot}6] \qquad\qquad\qquad\qquad \equiv : \beta \,\epsilon\, \iota``\alpha \qquad (3)$

$\vdash . (3) . *11{\cdot}11{\cdot}35 . *54{\cdot}101 . \supset \vdash . \text{Prop}$

$*54{\cdot}43. \quad \vdash :. \alpha, \beta \,\epsilon\, 1 . \supset : \alpha \cap \beta = \Lambda . \equiv . \alpha \cup \beta \,\epsilon\, 2$

     Dem.

$\qquad \vdash . *54{\cdot}26 . \supset \vdash :. \alpha = \iota`x . \beta = \iota`y . \supset : \alpha \cup \beta \,\epsilon\, 2 . \equiv . x \neq y .$

$[*51{\cdot}231] \qquad\qquad\qquad\qquad \equiv . \iota`x \cap \iota`y = \Lambda .$

$[*13{\cdot}12] \qquad\qquad\qquad\qquad \equiv . \alpha \cap \beta = \Lambda \qquad (1)$

$\qquad \vdash . (1) . *11{\cdot}11{\cdot}35 . \supset$

$\qquad\qquad \vdash :. (\exists x, y) . \alpha = \iota`x . \beta = \iota`y . \supset : \alpha \cup \beta \,\epsilon\, 2 . \equiv . \alpha \cap \beta = \Lambda \qquad (2)$

$\qquad \vdash . (2) . *11{\cdot}54 . *52{\cdot}1 . \supset \vdash . \text{Prop}$

From this proposition it will follow, when arithmetical addition has been defined, that $1 + 1 = 2$.

FIGURE 1.4

Part of page 362 of *Principia Mathematica*.

Public domain.

---

29 Doxiadis, Apostolos and Christos H. Papadimitriou, *Logicomix: An Epic Search for Truth*, Bloomsbury, New York, 2009, p. 185.

30 Mathias, Adrian R. D., "A Term of Length 4 523 659 424 929," *Synthese*, Vol. 133, 2002, pp. 75–86.

José Grimm had a paper published in 2010 showing that it's really worse than this.[31] See Figure 1.5!

Other contexts in which slight changes lead to huge consequences will be seen in Chapter 18.

In 2017, Grimm redid the computations and found the value to be the slightly larger, 5733067044018381256614727294093073067849676375305221319180433.[32]

1. INTRODUCTION

1.1 Normal Form of One in Bourbaki

When I was young, I was intrigued by the following quote of Bourbaki [Bou68, p. 158]: The mathematical *term denoted* (Chapter 1, § 1, no. 1) by the symbol "1" is of course not to be confused with the *word* "one" in ordinary language. The term denoted by "1" is equal, by virtue of the definition above, to the term denoted by the symbol

$$\tau_Z((\exists u)(\exists U)(u = (U, \{\emptyset\}, Z) \text{ and } U \subset \{\emptyset\} \times Z \text{ and } \cdots)) \tag{*}$$

As a rough estimate, the term so *denoted* is an assembly of several tens of thousands of signs (each of which is one of τ, $\Box$, $\lor$, $\neg$, $=$, $\in$, $\supset$).

This assembly is too big to be written on a sheet of paper, but computing its size presents no difficulty; one can check that the size of $\emptyset$ is 12, the size of $\{\emptyset\}$ is 513, the size of $U \subset \{\emptyset\} \times Z$ is $3\,000\,000$, etc. Thus, Bourbaki underestimated the size which can be shown to be of the order of 10^{15}. Some years later, I used a computer algebra system to get the exact number (see [Gri09a] for details); later on I discovered that the English edition of Bourbaki (1968) differs from the French one (1970, [Bou70]) by the removal of the sign $\supset$, that constructs the ordered pair (x, y), and the axiom A3 that governs the use of this symbol; a new definition has been added, namely $(x, y) = \{\{x\}, \{x, y\}\}$. This naturally increases somehow the size of "1", which becomes

5733067044017980337582376403672241161543539419681476659296689.

In [Mat02], the author says "$\{x\}$ is the term $\tau_y \forall z(z \in y \Leftrightarrow z = x)$ slightly simplified from the actual definition as $\{x, x\}$". The actual term is $\tau_y \forall z(z \in y \Leftrightarrow z = x \text{ or } z = x)$. With this definition $\{\emptyset\}$ has size 217, and "1" is of size 4.10^{12} (a thousand times smaller than our computations). He has also computed the size of "1" for the 1970 edition: it is smaller by a factor 10^9 to our estimation. This shows that a slight change in one definition may induce a huge change in subsequent constructions.

FIGURE 1.5
The new cost of certainty.

Courtesy of José Grimm.

---

31 Grimm, José, "Implementation of Bourbaki's Elements of Mathematics in Coq: Part One, Theory of Sets," *Journal of Formalized Reasoning*, Vol. 3, No. 1, 2010, pp. 79–102, portions of pp. 79–80 shown here.

32 Grimm, José, Implementation of Bourbaki's Elements of Mathematics in Coq: Part Two; Ordered Sets, Cardinals, Integers. [Research Report] RR-7150, Inria Sophia Antipolis; INRIA. 2018, pp. 826. https://hal.inria.fr/inria-00440786/en/.

We won't go to such lengths here, however. What you really need to know follows.

> No matter how correct a mathematical theorem may appear to be, one ought never to be satisfied that there was not something imperfect about it until it also gives the impression of being beautiful.[33]
>
> George Boole

> Pure mathematics was discovered by George Boole in his work published in 1854.
>
> Bertrand Russell

Russell's claim is clearly an exaggeration, but it's impressive that the man who earned this praise (Figure 1.6) was largely self-taught. And this is far from a unique situation in the world of mathematics. Ramanujan, some of whose work you'll encounter in Chapters 16 and 22; Isaac Newton, the discoverer of calculus and so much more; and many other great mathematicians were also autodidacts. Newton didn't take any

FIGURE 1.6
George Boole (1815–1864).

The Illustrated London News, **January 21, 1865. Public domain.**

---

33 http://www-gap.dcs.st-and.ac.uk/~history/Quotations/Boole.html.

mathematics classes until his junior year in college! Despite lacking a university degree, Boole obtained a position as a mathematics professor at Queen's College, Cork. Only Boole's work on logic is detailed here, but he also wrote texts on differential and difference equations. In addition to his books, he published approximately 50 papers. His most famous work is *An investigation into the Laws of Thought, on Which are founded the Mathematical Theories of Logic and Probabilities* (1854).

The three main models to which the ideas in this chapter may be applied are the algebra of logic, the algebra of sets (set theory is covered in Chapter 4), and the algebra of electrical circuits. The last is the most pleasant of surprises. Boole had no idea that his work would find great use in computer science. Applications of mathematics tend to be unpredictable.

In the logic developed in this section, only two values are possible, TRUE and FALSE. Hence, we have a two-valued logic. It follows that any test on Boolean algebra must, of necessity, be a TRUE/FALSE test. Multi-valued logics, in which more than two values are possible, also exist.

Letting p be a proposition that is either TRUE or FALSE, we can form NOT p, which would be the opposite. NOT p may be denoted by $\sim p$, $\neg p$, $!p$, or $\bar{p}$. We get the following **truth table** (**truth matrix** in some older books):

p	$\sim p$
TRUE	FALSE
FALSE	TRUE

The first description of truth values was given by Gottlob Frege (1848–1925), although the basic idea goes back to ancient Greece.[34] Credit for the table form shown here belongs to C. S. (S.) Peirce (1839–1914), but it didn't catch on until it was independently rediscovered in the 1920s, after Peirce's death.[35]

You should be able to negate statements expressed symbolically as well as in normal English. The symbolic case is actually easier, as you'll see! A few examples in English are given below.

Example 1
Statement: Coca-Cola originally contained cocaine.
Negation: Coca-Cola did not originally contain cocaine.

Example 2
Statement: Suicide accounted for at least 39,000 deaths in the US in 2010.[36]
Negation: Suicide did not account for at least 39,000 deaths in the US in 2010.

---

34 See Bocheński, J. M., *A History of Formal Logic* (translated and edited by Ivo Thomas), University of Notre Dame Press, Notre Dame, IN, 1961, p. 327 and p. 330.

35 Bell, Eric Temple, *Mathematics, Queen and Servant of Science*, Copyright, 1951, British Edition 1952, reprinted 1954, 1958, 1961, Printed in Great Britain by Butler and Tanner Ltd., Frome and London, pp. 61–62. Also see Bocheński, J. M., *A History of Formal Logic* (translated and edited by Ivo Thomas), University of Notre Dame Press, Notre Dame, IN, 1961, p. 330.

36 This proposition is true. The number is given as 39,518 at http://www.cdc.gov/nchs/fastats/leading-causes-of-death.htm.

This may also be phrased as: Suicide accounted for less than 39,000 deaths in the US in 2010.

Example 3
Statement: Tyrion did not kill Joffrey.
Negation: Tyrion did kill Joffrey.

In classical logic, negating a negative has the effect of cancelling both nots out.[37] If we take the negation of any of the negated statements above, the result is the original statement again.

Because NOT is a function of p alone, it is called a **unary operator**, in contrast to a **binary operator** such as AND, which is examined next. AND may be written as ∧.

p	q	p ∧ q
TRUE	TRUE	TRUE
TRUE	FALSE	FALSE
FALSE	TRUE	FALSE
FALSE	FALSE	FALSE

p ∧ q is also called the **conjunction** of **p** and **q**.
Another binary operator is OR, which may be written as ∨.

p	q	p ∨ q
TRUE	TRUE	TRUE
TRUE	FALSE	TRUE
FALSE	TRUE	TRUE
FALSE	FALSE	FALSE

p ∨ q is also called the **disjunction** of **p** and **q**.
This is often referred to as the inclusive OR. It was realized early on that "or" has two possible interpretations. To illustrate this, consider some everyday examples of or.

1) Your meal comes with a choice of soup or salad.
2) You may take MATH 201, if you have had MATH 200 or its equivalent.

In 1), experience tells you that or means one or the other, but not both. But in 2) it is clear that you will be allowed to take MATH 201, if you have had MATH 200, its

---

37 In *intuitionistic* logic the double negation of a statement need not always have the same meaning as the statement itself (although a triply-negated statement is equivalent to a singly-negated one). The reason is that for an intuitionist to assert, for example, that $f(x) = 0$ has a solution means that he or she can exhibit such a solution, whereas the double negation of $f(x) = 0$ would mean that the assumption that $f(x)$ is never zero leads to a contradiction, which does not necessarily yield an actual solution.

equivalent, or both. So, in some situations or excludes the possibility of both, and in other instances it allows both to hold. Above, we included the possibility of both, so TRUE ∨ TRUE has the value TRUE.

In the 14th century, there was still disagreement about how OR should be defined. Walter Burleigh (c. 1275–1344/5) wrote:

> Some say that for the truth of a disjunctive it is always required that one part be false, because if both parts were false it would not be a true disjunctive; for disjunction does not allow those things which it disjoins to be together, as Boethius says. But I do not like that. Indeed I say that if both parts of a disjunctive are true, the whole disjunctive is true. And I prove it thus. If both parts of a disjunctive are true, one part is true; and if one part is true, the disjunctive is true. Therefore (arguing) from the first to the last: if both parts of a disjunctive are true, the disjunctive is true.
>
> Further, a disjunctive follows from each of its parts, but it is an infallible rule that if the antecedent is true, the consequent is true; therefore if each part is true the disjunctive is true.
>
> I say therefore, that for the truth of a disjunctive it is not required that one part be false.[38]

What Burleigh didn't realize was that there's no reason to restrict ourselves to a single OR. That is, we don't have to choose. Later on, a separate or, referred to as exclusive or (aka XOR), was defined. Which is appropriate depends on context. They each have their uses.

The exclusive or, XOR, is denoted by ⊕. It only differs from OR in the case of both **p** and **q** being TRUE.

p	**q**	**p ⊕ q**
TRUE	TRUE	FALSE
TRUE	FALSE	TRUE
FALSE	TRUE	TRUE
FALSE	FALSE	FALSE

Using 1 in place of TRUE and 0 in place of FALSE, ⊕ is seen to be identical to addition modulo 2. That is, add the values of **p** and **q**, then divide by 2 and write the remainder.

Just as we can change one of the results in OR to get a new binary operator, XOR, any set of results can be considered and named.

For any logical statement there are two choices, T or F, for the value of p and two choices for the value of q. Thus, there are four distinct values possible for the pair p and q. It

---

38 Taken here from Bocheński, J. M., *A History of Formal Logic* (translated and edited by Ivo Thomas), University of Notre Dame Press, Notre Dame, IN, 1961, p. 197. The original source appears to be *The Shorter Treatise on the Purity of the Art of Logic*. Note: the name also appears as Burley.

follows that there are $2^4 = 16$ possible distinct truth tables with two variables. They can be realized as follows:

p	q	T	p ∨ q	q → p	p → q	p | q	p	q	p ↔ q
T	T	T	T	T	T	F	T	T	T
T	F	T	T	T	F	T	T	F	F
F	T	T	T	F	T	T	F	T	F
F	F	T	F	T	T	T	F	F	T

p	q	p ⊕ q	~q	~p	p ∧ q	p ∧ ~q	~p ∧ q	p ↓ q	F
T	T	F	F	F	T	F	F	F	F
T	F	T	T	F	F	T	F	F	F
F	T	T	F	T	F	F	T	F	F
F	F	F	T	T	F	F	F	T	F

The most common of these have already been detailed, but there is much left to say. The operator **T** that sends everything to T is sometimes called **Tautology**. And the operator **F** that sends everything to F is sometimes called **Contradiction**.

There's also some terminology associated with p → q that should be mentioned. The → means **implication** and is read "implies." The proposition p that appears before it is called the **antecedent** (aka **hypothesis** or **premise**) and the statement q that follows is called the **consequent** or **consequence** (aka **conclusion**). The whole relation is called the **conditional**. Looking closely at the truth table for the conditional, we notice something strange.

p	q	p → q
T	T	T
T	F	F
F	T	T
F	F	T

A False statement can imply something True as well as False. That is, F → T is true and F → F is also True. This sounds even stranger in an example.

Example: If I fought The Rock (Figure 1.7), then I kicked his ass.
This is True. It doesn't matter whether the proposition "I kicked his ass." is True or not, because the proposition "I fought The Rock." is False. At least, so far. This sort of statement is known as **vacuously true**, because the condition has never been met.

Example – False Can Imply Anything
Starting out with a False statement like 1 = −1, we can add 1 to each side to get another False statement (2 = 0) or square both sides to get a True statement (1 = 1). So, False really can imply either True or False, depending on the steps that are made.[39]

If you're still fretting about False → True, know that you're not the first. E. T. Bell discussed the resistance in one of his popular books:

---

39 This example was taken from Stewart, Ian, *Significant Figures*, Basic Books, New York, 2017, p. 153.

FIGURE 1.7
The Rock.

Image created by Mandy Coombes. This file is licensed under the Creative Commons Attribution-Share Alike 2.0 Generic license.

The only one of these that seems when interpreted to diverge from traditional common sense is p ⇒ q, which can be read "if p, then q." The column for p ⇒ q when interpreted says that p implies q if p is false or q is true, and incidentally that a false proposition implies any proposition. For example, "the moon is made of green cheese" implies that "twice two is four." But "twice two is four" does not imply that "the moon is made of green cheese." Some of the classical logicians objected to this definition of implication on metaphysical grounds, overlooking the fact that it is futile to quarrel with a definition. Implication was defined precisely and formally as above because the definition was found expedient in actual work in mathematical logic. It does not contradict tradition but supplements it.[40]

Bell went on to explain how. Consider the **biconditional** p ↔ q. Clearly, we only want this to be true when p and q are both True or both False. But p ↔ q is the same as (p → q) ∧ (q → p). We need both implications to hold in order to have the biconditional.

---

40 Bell, Eric Temple, *Mathematics, Queen and Servant of Science*, Copyright, 1951, British Edition 1952, reprinted 1954, 1958, 1961, Printed in Great Britain by Butler and Tanner Ltd., Frome and London, pp. 62–63.

p	q	p → q	q → p	p → q ^ q → p	p ↔ q
T	T	T	T	T	T
T	F	F	T	F	F
F	T	T	F	F	F
F	F	T	T	T	T

In order for p ↔ q to be False when p and q fail to match, the value of False → True must differ from that of True → False.

The symbol ↔ may be read as "if and only if." Definitions are always to be taken as if and only if. For example, we define an integer to be prime if it has exactly two distinct positive divisors. The unspoken portion of the definition is that if it does not have exactly two distinct positive divisors, then we do not say it is prime.

When stating theorems, we do not assume the biconditional. For example, we have a theorem in calculus that states differentiability → continuity, but the other direction is differentiability ← continuity. Switching sides, it becomes continuity → differentiability, which does not hold in general. If we wish it to be known that a theorem is in fact a biconditional, then we must explicitly state it as if and only if. Many times this is the case. In fact, it happens often enough that the abbreviation iff has been introduced to mean if and only if. So, if you see iff, it is probably not just a typo of if.

When we have p ↔ q, we can say that p and q are **logically equivalent**. This may also be written as p⇔q or p ≡ q.

NOR and NAND

p ↓ q is called **NOT OR** and can be written **p NOR q**. The ↓ is referred to as the **Peirce arrow**.

p | q (also written **p ↑ q**) is called **NOT AND**. This can be written **p NAND q**. When the vertical line notation | is used, it may be referred to as the **stroke**, or less tersely the **Sheffer stroke**, after Henry M. H. Sheffer, who introduced it in 1928.[41] This isn't a great notation today, as it may be confused with ||, which is used for OR in some programming languages.

Tautologies

We have various tautologies that can be verified. A **tautology** is simply a formula that is true for all possible values. Outside of logic, they are usually called identities.

---

41 Bocheński, J. M., *A History of Formal Logic* (translated and edited by Ivo Thomas), University of Notre Dame Press, Notre Dame, IN, 1961, p. 319 and p. 344. Bocheński incorrectly gives the name as M. H. Sheffer here. http://en.wikipedia.org/wiki/Henry_M._Sheffer gives the date as 1913.

Commutative laws	$p \wedge q \Leftrightarrow q \wedge p$
	$p \vee q \Leftrightarrow q \vee p$
Associative laws	$p \wedge (q \wedge r) = (p \wedge q) \wedge r$
	$p \vee (q \vee r) = (p \vee q) \vee r$
Distributive laws	$p \wedge (q \vee r) = (p \wedge q) \vee (p \wedge r)$
	$p \vee (q \wedge r) = (p \vee q) \wedge (p \vee r)$
De Morgan's laws[42]	$\sim (p \wedge q) = (\sim p) \vee (\sim q)$
	$\sim (p \vee q) = (\sim p) \wedge (\sim q)$

E. T. Bell described British mathematician Augustus De Morgan (1806–1871) as "a born nonconformist always on the lookout for something unusual."[43] In addition to the work he is famous for, he carried out paranormal investigations, despite being an atheist.[44]

Excluded middle	$p \vee \sim p \Leftrightarrow T$
Contradiction	$p \wedge \sim p \Leftrightarrow F$
Double negation law	$\sim (\sim p) \Leftrightarrow p$

In October 1978, *OMNI* magazine announced a contest in which readers were asked to design non-verbal signs.[45] The first-place winner, which may have been inspired by the double negation law, is shown in color figure 2 **(See color insert)**.

---

42 Actually discovered far earlier by Leibniz and others.

43 Bell, Eric Temple, *Mathematics, Queen and Servant of Science*, Copyright, 1951, British Edition 1952, reprinted 1954, 1958, 1961, Printed in Great Britain by Butler and Tanner Ltd., Frome and London, p. 151.

44 See Brandon, Ruth, *The spiritualists: The Passion for the Occult in the Nineteenth and Twentieth Centuries*, Weidenfeld and Nicholson, 1983, p. 56. Augustus's wife, Sophia, wrote a book titled *Ten Years' Experience in Spirit Manifestations*, to which Augustus penned a lengthy preface. An excerpt from it follows:

> I am perfectly convinced that I have both seen, and heard in a manner which should make unbelief impossible, things called spiritual which cannot be taken by a rational being to be capable of explanation by imposture, coincidence, or mistake. So far I feel the ground firm under me. But when it comes to what is the cause of these phenomena, I find I cannot adopt any explanation which has yet been suggested.

Augustus feared that making such statements openly could harm his career, so neither his name nor that of his wife appeared on the book. In their places were the initials A. B. and C. D.

45 Morris, Scot, "Games," *OMNI*, Vol. 1, No. 1, October 1978, pp. 174–175.

Contrapositive law[46] $p \rightarrow q \Leftrightarrow \sim q \rightarrow \sim p$

Conditional as disjunction $p \rightarrow q \Leftrightarrow \sim p \vee q$

Negation of conditional $\sim (p \rightarrow q) \Leftrightarrow p \wedge \sim q$

Biconditional as implication $(p \leftrightarrow q) \Leftrightarrow (p \rightarrow q) \wedge (q \rightarrow p)$

Idempotent laws $p \wedge p \Leftrightarrow p$
                                      $p \vee p \Leftrightarrow p$

Absorption laws $p \wedge (p \vee q) \Leftrightarrow p$
                                      $p \vee (p \wedge q) \Leftrightarrow p$

Dominance laws $p \wedge F = F$
                                      $p \vee T = T$

Exportation laws $p \rightarrow (q \rightarrow r) \Leftrightarrow (p \rightarrow q) \rightarrow r$

Identity laws $p \vee F \Leftrightarrow p$
                                      $p \wedge T \Leftrightarrow p$

Truth tables can be used to verify these identities.

Example: Distributive law: $p \wedge (q \vee r) = (p \wedge q) \vee (p \wedge r)$

p	q	r	q ∨ r	p ∧ (q ∨ r)	(p ∧ q)	(p ∧ r)	(p ∧ q) ∨ (p ∧ r)
T	T	T	T	T	T	T	T
T	T	F	T	T	T	F	T
T	F	T	T	T	F	T	T
T	F	F	F	F	F	F	F
F	T	T	T	F	F	F	F
F	T	F	T	F	F	F	F
F	F	T	T	F	F	F	F
F	F	F	F	F	F	F	F

Because columns five and eight are seen to be identical, $p \wedge (q \vee r)$ is logically equivalent to $(p \wedge q) \vee (p \wedge r)$ and the identity holds. You should try to verify other identities to make sure you have the hang of it. The identities can sometimes be used to simplify expressions.

Example: Suppose you want a block of computer code to execute if and only if exactly two of the three propositions, p, q, and r, are true. You could write it as

---

46 This one gives us a proof technique known as "proof by contrapositive," which is examined in Chapter 2.

$$\text{If } (p \wedge q \wedge \sim r) \vee (p \wedge \sim q \wedge r) \vee (\sim p \wedge q \wedge r),$$

But there is a shorter representation. Find it.

Solution:
First note that the Distributive law, $p \wedge (q \vee r) = (p \wedge q) \vee (p \wedge r)$, can also be written (by switching sides) as $(p \wedge q) \vee (p \wedge r) = p \wedge (q \vee r)$.

Now rewrite the given $(p \wedge q \wedge \sim r) \vee (p \wedge \sim q \wedge r) \vee (\sim p \wedge q \wedge r)$, using more parentheses, as

$$= \mathbf{(p \wedge (q \wedge \sim r)) \vee (p \wedge (\sim q \wedge r))} \vee (\sim p \wedge q \wedge r)$$

The portion that the Distributive law will be applied to is in boldface to make matters clearer. We get

$$= \mathbf{(p \wedge ((q \wedge \sim r) \vee (\sim q \wedge r)))} \vee (\sim p \wedge q \wedge r)$$

Now observe that we only want $((q \wedge \sim r) \vee (\sim q \wedge r))$ to be true if exactly one of q or r is true. But this is $q \oplus r$. Thus, our expression is

$$= (p \wedge (q \oplus r)) \vee (\sim p \wedge q \wedge r)$$

The original formulation used eight binary operators and three unary operators. The shortened form uses five binary operators and one unary operator.

Many tech companies don't see college transcripts as the best mean to determine whom the best people to hire are. So, they ask applicants questions of varying levels of difficulty and see how they respond. Google is famous for this, but the approach is fairly common. At stackoverflow.com a poster shared a question he encountered, "An interviewer recently asked me this question: given three Boolean variables, a, b, and c, return true if at least two out of the three are true."[47] He came up with a solution, but it wasn't the most efficient one. The problem is left to you as an exercise.

Boole found "The Laws of Thought," but it was Claude Shannon (Figure 1.8) who applied them to show that *machines* can think.

In 1937 Claude Shannon produced "A Symbolic Analysis of Relay and Switching Circuits," which has been described as "possibly the most important, and also the most noted, master's thesis of the century."[48] In it, he showed how Boolean algebra can be used to model digital circuits. Shannon's biographers detailed the connection.

Every single concept from Boole's algebra had its physical counterpart in an electric circuit. An on switch could stand for "true" and an off switch for "false," and the

---

47 http://stackoverflow.com/questions/3076078/check-if-at-least-two-out-of-three-booleans-are-true.
48 Gardner, Howard, *The Mind's New Science: A History of the Cognitive Revolution*, Basic Books, 1987, p. 144.

FIGURE 1.8
Claude Shannon (1916–2001).

Courtesy of MIT Museum.

whole thing could be represented in 1's and 0's. More important, as Shannon pointed out, the logical operators of Boole's system – AND, OR, NOT – could be replicated exactly as circuits. A connection in series becomes AND, because the current must flow through two switches successively and fails to reach its destination unless both allow it passage. A connection in parallel becomes OR, because the current can flow through either switch, or both. The current flows through two closed switches in parallel and lights a light; $1 + 1 = 1$.[49]

Trial and error in circuit design were no longer necessary – all of the work could be done with algebra, following simple rules to create a design that would be as elegant and efficient as possible.

Over the decades following Shannon's great insight, the hardware evolved. Relay circuits became vacuum tube circuits, tubes were replaced by transistors and solid-state circuitry, and circuits placed on silicon chips have shrunk to amazingly small dimensions. Yet, the mathematical connection Shannon found has endured unscathed.

To be clear, Shannon didn't invent the mathematics of circuits. He simply noticed that there was a product already on the shelves (put there by George Boole) that

---

49 Soni, Jimmy and Rob Goodman, *A Mind at Play: How Claude Shannon Invented the Information Age*, Simon & Schuster, New York, 2017, pp. 38–39.

could serve this purpose. Shannon himself referred to this portion of his work as "trivial."[50] Still, nobody else had seen the connection. Why was Shannon different? I think the answer is because he really was different. He had an extremely creative and playful mind and seemed to take little notice of whether others approved or not. I believe worrying too much about people's opinions and being accepted can hamper creativity.

When Shannon worked at Bell Labs, he was often seen riding a unicycle in a long hallway and juggling at the same time! He had designed a special unicycle for just this purpose. The off-set wheel keeps the rider steady while he juggles. He also built machines that juggle.[51] Other inventions of his include a rocket-powered Frisbee, machines to play chess and solve Rubik's cube, a flame-throwing trumpet, and a mysterious box with a switch on it. When someone saw this box sitting on his desk and flipped the switch, the box would open and a mechanical hand would reach out and flip the switch back again. After this, the hand would pull back into the box and the lid would close, returning the system to its original state.[52]

Claude Shannon and Alan Turing had much in common. Shannon worked on SIGSALY, a voice encryption system developed in America that allowed President Roosevelt to speak securely with Prime Minister Winston Churchill during World War II.[53] He later served on the National Security Agency's Scientific Advisory Board. Alan Turing beta tested SIGSALY and is well-known for his role in breaking Nazi Enigma ciphers. In addition to their shared expertise in cryptology, both men were atheists and both thought machines would soon be made that could "simulate entirely the human brain."[54] Shannon even expected the computers to eventually become superior to humans.

> I visualize a time when we will be to robots what dogs are to humans, and I'm rooting for the machines.
>
> Claude Shannon[55]

Quantifiers

Although Aristotle's logic provided a useful basis, there are many statements that cannot be expressed in such a **categorical logic**, as it is termed. The only quantifiers it has are all, some, and no, which is a drawback. We are stuck with what are known as

---

50 Soni, Jimmy and Rob Goodman, *A Mind at Play: How Claude Shannon Invented the Information Age*, Simon & Schuster, New York, 2017, p. 42.

51 Soni, Jimmy and Rob Goodman, *A Mind at Play: How Claude Shannon Invented the Information Age*, Simon & Schuster, New York, 2017, pp. 61–62.

52 Bauer, Craig P., *Secret History: The Story of Cryptology*, Chapman & Hall/CRC, Boca Raton, Florida, 2013, p. 338. I devoted Chapter 10 of this book to some of the work Shannon did that plays an important role in today's cryptology.

53 Soni, Jimmy and Rob Goodman, *A Mind at Play: How Claude Shannon Invented the Information Age*, Simon & Schuster, New York, 2017, p. 101.

54 Soni, Jimmy and Rob Goodman, *A Mind at Play: How Claude Shannon Invented the Information Age*, Simon & Schuster, New York, 2017, p. 46 and p. 107.

55 Liversidge, Anthony, "Claude Shannon Interview: Father of the Electronic Information Age," *OMNI*, August 1987, Vol. 9, No. 11, pp. 60-62, 64–66, 110, p. 61 quoted here.

the AEIO forms of propositional claims. Letting S designate the subject and P the predicate. These forms are:

All S are P.	(A form)
All S are not P.	(E form)
Some S are P.	(I form)
Some S are not P.	(O form)

First-order logic introduces the **existential quantifier**, which is read "there exists" or "there is" and noted as $\exists$. This is needed, for Aristotelian logic says nothing about existence. For example, statements like "All unicorns are white." would be vacuously true, and without being able to comment on unicorns' existence, or lack thereof, we could have problems.

As an example of the existential quantifier, consider the statement $\exists\ x\ x^2 = 2$. This means that there exists an x such that $x^2 = 2$. It doesn't indicate for how many values of x the equality holds. If we want to claim a unique solution, we write $\exists!\ x\ x^2 = 2$. That is, $\exists!$ means "there exists a unique..." We call $\exists!\ x\ P(x)$ the uniqueness quantification of $P(x)$. Another notation used to designate a unique value is $\exists_1$. It may be read as "there exists exactly one," but it means the exact same thing as $\exists!$. Of course, there are two values (one positive and one negative) whose squares are equal to 2, so the claim used to illustrate this idea is false. A true claim is given by $\exists_2\ x\ x^2 = 2$. Just as $\exists_1$ specified exactly one solution, $\exists_2$ specifies exactly two solutions. However, this notation is not seen often. We're typically just interested in specifying if the solution exists, or if it exists and is unique.

In some cases, a proposition holds for all values. When this is true, we use the symbol $\forall$, known as the **universal quantifier**. It means "for all" or "for every."

Thus, we can write $\forall\ x\ x + 0 = x$ to express the fact that 0 is the additive identity.

More generally, $\forall\ x\ P(x)$ is the universal quantification of the propositional function[56] $P(x)$. It means that $P(x)$ is true for all x in the domain of discourse. Sometimes this domain will be clear from context, but at other times it should be explicitly stated. A propositional function can have more than one variable. The notation $P(x_1,...,x_n)$ may be used in this case, where the n variables are $x_1, ..., x_n$.

Note that the general form $\exists x\ P(x)$, which is the existential quantification of $P(x)$, only requires $P(x)$ to be true for a single value x, although it may be true for several or even all values.

The expansion of our quantifiers, provided by first-order logic, allows more sophisticated arguments. That is, we can move beyond the syllogisms.

It was mentioned earlier in this chapter that negating a statement expressed symbolically is even easier than negating the corresponding statement expressed in plain English. The reason is that we do not have to think about the meaning of the quantifiers. We simply replace each $\exists$ with a $\forall$ and replace each $\forall$ with $\exists$. The rest of the statement is handled as before.

---

56 The term "propositional function" was brought to the worlds of math and science by Bertrand Russell, but the idea goes back farther.

Example: Negate the proposition $\forall \, x \; x + 0 = x$.

Solution: $\exists \, x \; x + 0 \neq x$.

Example: Negate the proposition $\exists \, x \; x = 1/x$.

Solution: $\forall \, x \; x \neq 1/x$.

Breaking the Law

Aristotle put forth the **principle of the excluded middle**. It is also known as the **law of the excluded middle**, for it sounds rock-solid. If you want it to sound more pretentious, you can refer to it as *principium tertii exclusi* or *tertium non datur*. In any case, it simply means that given a proposition p and its negation ~p, we must have that one is True and the other is False. In other words, we have a two-valued logic (True and False), in which every proposition must take one of these values. There's no "maybe." It's not claimed that we can always determine the value, but it cannot turn out that the value is neither True nor False, for example, or both True and False.

This seems obvious, but some quantum physicists reject the law of the excluded middle in certain situations. This is not a purely theoretical matter. An application exists where Aristotle's law should apparently be rejected! Quantum computers operate in a manner that is fundamentally different from traditional (Turing) machines. Instead of being based on bits that are each either 0 or 1 (in accordance with Aristotle's law), they make use of quantum bits (aka qbits, pronounced "cue bits") that can take individual values of 0, 1, or both.

Mathematics tends to race ahead of applications, calling itself "pure mathematics." So it shouldn't be surprising that prior to anyone attempting to build quantum computers, multi-valued logics were studied. In particular, Jan Łukasiewicz proposed a three-valued logic in 1917:[57]

> H. Reichenbach has shown that the theory of Quantum Mechanics can be axiomatized on the basis of Łukasiewicz's three-valued logic, which cannot be done on the basis of two-valued logic.
>
> J. M. Bocheński[58]

Once the door to logics with more than one value was opened, it wasn't long until logics with four, five, and more values appeared. There are even logics with an infinity of values, both denumerable and nondenumerable.[59]

While three-valued logic will be useful in modeling quantum computers, computers, in turn, have contributions to make to logic. This is not a new breakthrough. An impressive result was actually obtained back in the 1950s. A trio of researchers – Allen Newell, J. C. Shaw, and Herbert A. Simon – wrote a program called Logic Theorist that attempted to prove theorems using the same mathematical tools Russell and Whitehead

---

57 http://en.wikipedia.org/wiki/Jan_%C5%81ukasiewicz.

58 Bocheński, J. M., *A History of Formal Logic* (translated and edited by Ivo Thomas), University of Notre Dame Press, Notre Dame, IN, 1961, p. 407.

59 Bell, Eric Temple, *Mathematics, Queen and Servant of Science*, Copyright, 1951, British Edition 1952, reprinted 1954, 1958, 1961, Printed in Great Britain by Butler and Tanner Ltd., Frome and London, p. 64.

used in their *Principia Mathematica*. The tools were modus ponens, substituting for propositional variables, and replacing defined symbols with their definitions.[60] The researchers set their program loose on a JOHNNIAC[61] with the same five axioms that Russell and Whitehead used. JOHNNIAC managed to prove 38 of the first 52 results.[62] But the proofs weren't all identical to those of the two humans; some were better![63]

The Dream

The mathematician's dream, going back to Leibniz, of being able to resolve everything through symbolic logic was not met by Boole, although his work found great applications through Shannon, nor was it met by Russell and Whitehead, or any of the other researchers mentioned in this chapter. In 1900, at the International Congress of Mathematicians, held in Paris, France, David Hilbert gave an important speech. He presented 10 unsolved problems that he felt it was important for mathematics to address in the new century.[64] His prefatory comments, prior to detailing the problems, included the following:

> This conviction of the solvability of every mathematical problem is a powerful incentive to the worker. We hear within us the perpetual call: There is the problem. Seek its solution. You can find it by pure reason, for in mathematics there is no *ignorabimus*.[65]
>
> David Hilbert

It turns out that logic isn't enough to attain the solvability of every mathematical problem. Could logic plus set theory suffice? Set theory is an extremely important area of modern mathematics and it's covered in Chapter 4, so if you don't know how this story ends, you'll have to get through a few more chapters to find out! Before getting to set theory, Chapters 2 and 3 investigate various methods of mathematical proof. This chapter closes with a glimpse at the illogical.

Logical Fallacies

Aristotle didn't just demonstrate valid lines of reasoning. He also looked at informal fallacies such as *argumentum ad hominem*.[66] This is often referred to as an *ad hominem*

---

60 Davis, Martin, "The Early History of Automated Deduction," in Robinson, Alan and Andrei Voronkov, *Handbook of Automated Reasoning* 1, Elsevier, Amsterdam, 2001, pp. 3–15, p. 6 cited here.

61 This is an abbreviation for John v. Neumann Numerical Integrator and Automatic Computer. The man the machine is named after appears in Chapter 19.

62 Newell, A., J. Shaw, and Herbert A. Simon, "Empirical explorations with the logic theory machine: A case study in heuristics," in Feigenbaum, E. A. and J. Feldman, editors *Computers and Thought*, McGraw-Hill, New York, 1963, pp. 109–133, p. 122 cited here.

63 McCorduck, Pamela, *Machines who Think – A Personal Inquiry into the History and Prospects of Artificial Intelligence*, A K Peters, Ltd., Natick, MA, 2004, p. 168. Also see https://en.wikipedia.org/wiki/Logic_Theorist.

64 He published a list of 23 problems, but only had time to present 10 at the conference.

65 A transcript can be found at https://mathcs.clarku.edu/~djoyce/hilbert/problems.html.

66 http://www.newworldencyclopedia.org/entry/History_of_logic.

attack. Its use consists of attacking the person making the argument, rather than the argument itself.

An example of this is provided by a billboard I saw in rural Pennsylvania. It showed a headshot of Robert Redford, along with the text "Demands green living. Flies on private jets." Instead of addressing his arguments about going green, his opponents, in this instance, chose to make an *ad hominem* attack, implying that he's a hypocrite. Whether he's a hypocrite or not has absolutely nothing to do with the correctness of his arguments.

Political billboards are a rich source of logical fallacies. Another showed Ted Kaczynski, aka the Unabomber, and the words "I still believe in Global Warming. Do you?" Well, 98% of scientists believe in global warming – is the man pictured really representative of the group?[67]

Sadly, it's a common tactic in politics to single out the most extreme people on the other side of the debate and portray them as representative. Rush Limbaugh spent a lot of time talking about the most extreme feminists. He referred to them as "feminazis," and though he did sometimes say that there were only about six of these in the world, he sure spent a lot of time talking about them.

Descriptions and examples of 42 kinds of logical fallacies can be found at www.nizkor.org/features/fallacies/.

Exercise Set 1

> As soon as I realized that I was intelligent, I determined to achieve something of intellectual importance if it should be at all possible, and throughout my youth I let nothing whatever stand in the way of this ambition.
>
> Bertrand Russell[68]

1. Form the negation of "Every superhero wears a cape."
2. Form the negation of "There's a new car with a sticker price under \$5,000."
3. Form the negation of "Paul Erdös wrote over 1,500 math papers."
4. Form the negation of "Fred Butler does not have a tattoo of Leibniz on his lower back."
5. Form the negation of "It's over 9,000 miles from Los Angeles to Annandale, Virginia."
6. Form the negation of $\forall x \forall y \ x + y = y + x$.
7. Form the negation of $\exists x \ x^2 + 1 = 0$.
8. Form the negation of $\exists x > 0 \ x - x = x$.
9. Form the negation of $\forall x > 0 \ 1/x > 0$.
10. Form the negation of $\forall \varepsilon > 0 \ \exists \delta > 0 : |x - c| < \delta \Rightarrow |f(x) - L| < \varepsilon$. This is the definition of the limit, taken from continuous mathematics (laying the foundation for calculus).

---

67 You can see a picture of this billboard at http://mediamatters.org/blog/2012/05/04/billboards-illustrate-heartlands-approach-to-sc/185773.

68 Russell, Bertrand, *The Autobiography of Bertrand Russell, Vol. I: 1872–1914*, Little, Brown and Company, Boston, Massachusetts, 1967, p. 39.

11. Which OR does the expression "take it or leave it" use, inclusive or exclusive?

12. Which OR is being used when a waiter asks if you'd like french fries or a potato with your steak, inclusive or exclusive?

13. Which OR is used in "You have an increased risk for diabetes, if your mother or father is diabetic," inclusive or exclusive?

14. Which OR is used in "The use of alcohol or drugs in the workplace is prohibited," inclusive or exclusive?

15. Simplify the following expression: $(p \wedge \sim p) \vee (p \wedge q)$.

16. Find an expression that will evaluate to TRUE if and only if exactly one of p, q, and r is true.

17. Use a truth table to verify: DeMorgan's law $\sim (p \wedge q) = (\sim p) \vee (\sim q)$.

18. Use a truth table to verify: the Contrapositive law $p \to q \Leftrightarrow \sim q \to \sim p$.

19. Use a truth table to verify: Negation of conditional $\sim (p \to q) \Leftrightarrow p \wedge \sim q$.

20. Use a truth table to verify: Biconditional as implication $(p \leftrightarrow q)$ $\Leftrightarrow (p \to q) \wedge (q \to p)$.

21. Give an example, not included in this chapter, of something that Aristotle got wrong.

22. So you can see that it isn't hard to do, make it an *argumentum ad hominem* against Isaac Newton or Albert Einstein. This may involve looking over biographies of these great men until you find some dirt. Once you have something to criticize, go for it! Does your argument invalidate $F = ma$ or $E = mc^2$?

23. Make an *argumentum ad hominem* against Bertrand Russell.

24. Go to www.nizkor.org/features/fallacies/and read about logical fallacies. Then select one particular fallacy and provide an example of it that you encountered somewhere in your personal life.

25. Many of today's magazines contain ads as laughable as the billboards described in this chapter. Find a current example and explain the logical fallacy connected with it.

26. Give an example of a logical fallacy used in a political ad.

References/Further Reading

> I cannot live without books.
> – Thomas Jefferson.

"Aristotelian Logic," *PlanetMath*, available online at http://planetmath.org/aristotelianlogic, has clear explanations of some of the material covered in this chapter, as well as other material not included here.

Bauer, Craig P., *Secret History: The Story of Cryptology*, Chapman & Hall/CRC, Boca Raton, FL, 2013. I devoted Chapter 10 of this book to some of the work Shannon did that plays an important role in today's cryptology.

Beckmann, Petr, *A History of π (Pi)*, Dorset Press, New York, 1989.

Bell, Eric Temple, *Mathematics, Queen and Servant of Science*, Copyright, 1951, British Edition 1952, reprinted 1954, 1958, 1961, Printed in Great Britain by Butler and Tanner Ltd., Frome and London.

Berlinski, David, *The Advent of the Algorithm*, Harcourt Inc., New York, 2000.

Bobzien, Suzanne, "The Development of Modus Ponens in Antiquity: From Aristotle to the 2nd Century AD," *Phronesis*, Vol. 47, No. 4, 2002, pp. 359–394.

Bocheński, Joseph. M., *A History of Formal Logic* (translated and edited by Ivo Thomas), University of Notre Dame Press, Notre Dame, IN, 1961. There's also a 2012 reprint.

Brandon, Ruth, *The Spiritualists: The Passion for the Occult in the Nineteenth and Twentieth Centuries*, Weidenfeld and Nicholson, London, 1983.

Bruderer, Herbert, "The Binary System was Created Long before Leibniz," *Communications of the ACM, BLOG@ACM*, available online at https://cacm.acm.org/blogs/blog-cacm/221749-the-binary-system-was-created-long-before-leibniz/fulltext, October 10, 2017.

Carroll, Lewis (pseudonym of Charles Lutwidge Dodgson), *The Game of Logic*, Macmillan & Co., London, 1886.

Carroll, Lewis (pseudonym of Charles Lutwidge Dodgson), *Symbolic Logic*, Macmillan and Co, London, 1896, available online at www.gutenberg.org/files/28696/28696-h/28696-h.htm.

The two books referenced above are also available in an omnibus edition from Dover.

Davis, Martin, "The Early History of Automated Deduction," in Robinson, Alan and Andrei Voronkov, editors, *Handbook of Automated Reasoning*, Vol. 1, Elsevier, London, 2001, pp. 3–15.

De Morgan, Sophia, preface by De Morgan, Augustus, *Ten Years' Experience in Spirit Manifestations*, Spotisswood and Co., London, p. v, available online at https://archive.org/details/frommattertospir00demorich.

Doxiadis, Apostolos and Christos H. Papadimitriou, *Logicomix: An Epic Search for Truth*, Bloomsbury, New York, 2009. This is the story of logic presented in graphic novel (comic book) form. It's great.

Gabbay, Dov M., Francis Jeffry Pelletier, and John Woods, editors, *Handbook of the History of Logic*, North Holland, an imprint of Elsevier, Amsterdam, 2004. This is a series of books (11 volumes so far) with contributions by various authors. For a listing of individual volumes and contents, go to www.sciencedirect.com/science/journal/18745857.

Gardner, Howard, *The Mind's New Science: A History of the Cognitive Revolution*, Basic Books, New York, 1987.

Goldstone, Lawrence and Nancy Golstone, *The Friar and the Cipher: Roger Bacon and the Unsolved Mystery of the Most Unusual Manuscript in the World*, Doubleday, New York, 2005.

Grimm, José, "Implementation of Bourbaki's Elements of Mathematics in Coq: Part One, Theory of Sets," *Journal of Formalized Reasoning*, Vol. 3, No. 1, 2010, pp. 79–126.

Grimm, José, Implementation of Bourbaki's Elements of Mathematics in Coq: Part Two; Ordered Sets, Cardinals, Integers. [Research Report] RR-7150, Inria Sophia Antipolis; INRIA. 2018, https://hal.inria.fr/inria-00440786/en/.

History and Philosophy of Logic, This journal has been published since 1980. See www.tandfonline.com/toc/thpl20/current#.VKYD8ivF9ps for contents.

King, Peter and Stewart Shapiro, "The History of Logic," in *The Oxford Companion to Philosophy*, Oxford University Press, 1995, pp. 496–500, available online at http://individual.utoronto.ca/pking/miscellaneous/history-of-logic.pdf.

Kneale, William and Martha Kneale, *The Development of Logic*, Clarendon Press, Oxford, 1962.

Liversidge, Anthony, "Claude Shannon Interview: Father of the Electronic Information Age," *OMNI*, Vol. 9, No. 11, August 1987, pp. 60–62, 64–66, 110.

Logical Analysis and History of Philosophy/Philosophiegeschichte und logische Analyse, available online at https://dbs-lin.ruhr-uni-bochum.de/philosophy/pla/?q=home-call. This series began in 1998 and runs papers in English and German.

Mathias, Adrian R. D., "A Term of Length 4 523 659 424 929," *Synthese*, Vol. 133, 2002, pp. 75–86.

McCorduck, Pamela, *Machines Who Think – A Personal Inquiry into the History and Prospects of Artificial Intelligence*, 2nd ed., 25th anniversary update, A K Peters, Ltd., Natick, MA, 2004.

Newell, Allen, John Shaw, and Herbert A. Simon, "Empirical Explorations with the Logic Theory Machine," in *Proceedings of the Western Joint Computer Conference*, 1957, pp. 218–239.

Newell, Allen, John Shaw, and Herbert A. Simon, "Empirical Explorations with the Logic Theory Machine: A Case Study in Heuristics," in Feigenbaum, E. A. and J. Feldman, editors, *Computers and Thought*, McGraw-Hill, New York, 1963, pp. 109–133.

Newell, Allen, John Shaw, and Herbert A. Simon, "Empirical Explorations with the Logic Theory Machine," in Siekmann, Jörg and Graham Wrightson, editors, *Automation of Reasoning, Classical Papers on Computational Logic*, Vol. 1, Springer, Berlin, 1983, pp. 49–73.

Perkins, Jr., Ray, *Logic and Mr. Limbaugh: A Dittohead's Guide to Fallacious Reading*, Open Court, Chicago, IL, 1998.

Pitowsky, Itamar, *Quantum Probability – Quantum Logic*, Springer Verlag, Berlin, 1989.

Pitowsky, Itamar, "George Boole's 'Conditions of Possible Experience' and the Quantum Puzzle," *British Journal for the Philosophy of Science*, Vol. 45, No. 1, March 1994, pp. 95–125.

Pitowsky, Itamar, "From Logic to Physics: How the Meaning of Computation Changed Over Time," in Cooper, S. Barry, Thomas F. Kent, Benedikt Löwe, and Andrea Sorbi, editors, *Computation and Logic in the Real World Third Conference on Computability in Europe, CiE 2007, Siena, Italy, June 18–23, 2007. Proceedings*, Lecture Notes in Computer Science, Vol. 4497, 2007, pp. 621–631.

Rosen, Kenneth H., editor, *Handbook of Discrete and Combinatorial Mathematics*, CRC Press, Boca Raton, FL, 2000.

Russell, Bertrand, *Why I Am Not a Christian*, 1927, available online at https://users.drew.edu/~jlenz/whynot.html. Although this is really just a pamphlet, arising from a lecture Russell gave, it's shown up on *book* lists, both "best books" and "worst books." You can guess why.

Russell, Bertrand, *The Autobiography of Bertrand Russell, Vol. I: 1872–1914*, Little, Brown and Company, Boston, MA, 1967. There are two more volumes, but this one covers his mathematically productive years. It seems that the effort required to produce *Principia Mathematica* broke him, mathematically.

Scripta Mathematica: A Quarterly Journal Devoted to the Philosophy, History, and Expository Treatment of Mathematics, Yeshiva University, New York, 1932–1973. This journal was aimed at a general audience.

Shenefelt, Michael and Heidi White, *If A, Then B: How the World Discovered Logic*, Columbia University Press, New York, 2013. This book places the historical development of logic in the full context of the times.

Soni, Jimmy and Rob Goodman, *A Mind at Play: How Claude Shannon Invented the Information Age*, Simon & Schuster, New York, 2017.

Stewart, Ian, *Significant Figures*, Basic Books, New York, 2017.

The Nizkor Project, Fallacies, available online at www.nizkor.org/features/fallacies/. This website provides an extensive listing of logical fallacies.

White, Michael J., *The Continuous and the Discrete: Ancient Physical Theories from a Contemporary Perspective*, Clarendon Press, Oxford, 1992.

Whyte, James, *Crimes against Logic: Exposing the Bogus Arguments of Politicians, Priests, Journalists, and Other Serial Offenders*, McGraw-Hill, New York, 2004.

Wikipedia contributors, "Automated Theorem Proving," *Wikipedia, The Free Encyclopedia*, available online at https://en.wikipedia.org/w/index.php?title=Automated_theorem_proving&oldid=847874642.

Wikipedia contributors, "Logic Theorist," *Wikipedia, The Free Encyclopedia*, available online at https://en.wikipedia.org/w/index.php?title=Logic_Theorist&oldid=849490280.

Wilson, Robin and John J. Watkins, editors, *Combinatorics: Ancient & Modern*, Oxford University Press, Oxford, 2013.

Wolfram, Stephen, "Dropping in on Gottfried Leibniz," *Stephen Wolfram Blog*, May 14, 2013, available online at http://blog.stephenwolfram.com/2013/05/dropping-in-on-gottfried-leibniz/.

Wolfram, Stephen, "100 Years since *Principia Mathematica*," *Stephen Wolfram Blog*, November 25, 2010, available online at http://blog.stephenwolfram.com/2010/11/100-years-since-principia-mathematica/

Woods, John and Andrew Irvine, "Aristotle's Early Logic," in Gabbay, Dov M. and John Woods, editors, *Handbook of the History of Logic, Volume 1: Greek, Indian and Arabic Logic*, Elsevier, Amsterdam, 2004, pp. 27–99.

Logic in India

Readers aware of the development of logic in India may notice that I did not include it here. My reason for this was that it reads quite differently (although it led to some of the same problems and solutions) and was not an influence on those (Boole, et al.) who brought us what needs to be included in today's texts. A classic example of Indian logic follows below, for those who are curious.[69]

> Proposition: There is fire on the mountain;
> Reason: Because there is smoke on the mountain;
> Example: As in a kitchen – not as in a lake;
> Application: It is so;
> Conclusion: Therefore it is so.

Some discussion/explanation of this can be found at https://itunes.apple.com/us/book/role-example-drstanta-in-classical/id517684044?mt=11.

---

69 Bocheński, J. M., *A History of Formal Logic* (translated and edited by Ivo Thomas), University of Notre Dame Press, Notre Dame, IN, 1961, p. 428.

2

Proof Techniques

The moving power of mathematical invention is not reasoning but imagination.

Augustus De Morgan[1]

Imagination is more important than knowledge.

Albert Einstein[2]

This chapter details some basic proof techniques and provides guidelines as to which techniques you might try to have the greatest chance of success. However, there is generally no algorithm to follow to construct a proof. The best mathematicians find their proofs by being as creative as the great artists. The imagination that Einstein so highly rated is the key.

Direct Proof

We have a direct proof when the claim follows from axioms and/or already-established theorems.

Theorem: The angle in a semicircle is a right angle. This result is from Thales (624–547 BCE).[3]

Examples:

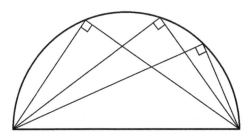

FIGURE 2.1
Some angles inscribed in a semicircle.

Image created by Garrett Ruths.

---

1 Taken here from Burger, Edward B. and Michael Starbird, *The Heart of Mathematics*, third edition, Wiley, Hoboken, New Jersey, 2010, p. 360.
2 This quote is from "What Life Means to Einstein: An Interview by George Sylvester Viereck," *The Saturday Evening Post*, October 26, 1929, pp. 17, 110, 113–114, 117, p. 117 cited here.
3 Early dates are only approximate.

Three such angles are shown in Figure 2.1, but these examples do not constitute a proof. There are infinitely many ways we could draw the angle, so we could never sketch them all, and even if we could, the fact that they all appear to be right angles does not mean that they are. Perhaps they are 89.995° or 90.0002°. A proof follows.

Proof:

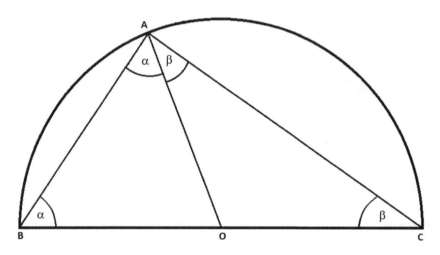

FIGURE 2.2
A sketch to help understand the proof.

Image created by Garrett Ruths.

In Figure 2.2, letting O denote the center of the circle, OA and OB must be of the same length (both are equal to the radius of the circle). Because we have an isosceles triangle OBA, we can label the two angles as α (they are equal). The same idea works for the isosceles triangle OCA. We also have the large triangle ABC, whose angles must sum to 180°. That is 2α + 2β = 180°. So, α + β = 90°. **QED**

An ox was sacrificed in celebration of this discovery.[4]

As was done above, the abbreviation "QED" is often placed at the end of a proof. It stands for "*quod erat demonstrandum,*" which is variously translated as "this was to be demonstrated," "as was to be demonstrated," or "which was to be demonstrated." If you want to be pompous you can say this conclusively at the end of an argument that you've won and it will basically mean "Not only am I right, while you are wrong, but I also speak Latin and you do not."

Euclid used the (original) Greek version of the Latin words abbreviated as QED at the ends of proofs of theorems, but after justifying a construction he used words that translate to the Latin "*quod erat faciendum,*" which is abbreviated QEF. In English, this becomes "that which was to be done."[5]

---

4 See *Lives of Eminent Philosophers* by Diogenes Laertius.
5 http://mathworld.wolfram.com/QEF.html, which, in turn, cites Heath, T. L., *The Thirteen Books of the Elements,* 2nd ed., Vol. 1: Books I and II. Dover, New York, 1956, pp. 124–129.

Proof by Construction

We are sometimes able to prove that something exists without actually finding it. Examples you may encounter include demonstrations that certain differential equations have solutions. In a proof by construction, we explicitly build (or construct) the desired solution.

Theorem: There are arbitrarily long sequences of consecutive integers (i.e., without skipping any) that do not contain any primes.

Proof: This proof relies on the factorial function. It is denoted by an exclamation point, !, and simply means that the given (integer) value is to be multiplied by all smaller positive integers. For example, $5! = 5 \cdot 4 \cdot 3 \cdot 2 \cdot 1 = 120$. This function will be examined in more detail in Chapter 8, but right now we know all we need to about it to prove the present theorem. If you want n integers in a row, all of which are composite (nonprime), here you are!

$$(n+1)! + 2, \ (n+1)! + 3, \ (n+1)! + 4, \ \cdots (n+1)! + n, \ (n+1)! + n + 1$$

The first is divisible by 2, the second by 3, and so on.   □

   This time, instead of placing QED at the end of the proof, a "□" was used. In this context, the symbol is called the tombstone or the Halmos, after Paul Halmos, who was the first to bring it to mathematics. It's just another way to indicate that the proof is done.

Proof by Contrapositive

As we saw in the previous chapter, the contrapositive of an implication is logically equivalent to the original implication. Sometimes, the contrapositive is easier to prove. An example follows below.

Theorem: n^2 even implies n is even.

Proof: The claim is: n^2 even $\Rightarrow n$ is even, so the contrapositive is: n not even $\Rightarrow n^2$ is not even. This can be expressed more conveniently as n odd $\Rightarrow n^2$ is odd. This is what we will prove. If n is odd, then $n = 2k + 1$ for some integer k. Thus, $n^2 = (2k+1)^2 = 4k^2 + 4k + 1 = 2(2k^2 + 2k) + 1$. As this is one more than a multiple of 2, this is also an odd number. That is, n^2 is odd. This establishes the contrapositive as true, and therefore the original claim must also be true.   □

Hempel's Paradox

An interesting philosophical debate has arisen connected with this method of proof. Suppose we are trying to prove that all crows are black. If we search for crows and find thousands of them, and they are all black, we can't say we have a proof, but we could claim to have *evidence* in support of our claim. Now, let's consider the contrapositive of our claim. This, logically equivalent, statement is "everything that is not black is not a crow." If we search for objects that are not black and find thousands of them and none of them are crows, should this count as evidence that our claim is true? In other words, is a white cat evidence that all crows are black? A large number of papers have appeared in academic journals debating this.[6]

Proof by Exhaustion

This is also known as proof by cases or brute force. In such a proof, two or more separate possibilities (aka cases) are looked at. Collectively, they must cover all of the possibilities allowed in the statement of the claim.

Theorem: $|a + b| \leq |a| + |b|$. This is known as the triangle inequality.

Proof: As indicated, we will split the proof into several different cases, each of which will then be examined individually.

Case 1: If $a = 0$, the claim reduces to $|b| \leq |b|$, which is true (we actually have equality in this case).

Case 2: Similarly, if $b = 0$, the claim reduces to $|a| \leq |a|$, which we know is true.

The remaining cases are when a and b are both positive, both negative, and a mix of positive and negative.

Case 3: If $a > 0$ and $b > 0$, then $a + b = |a + b|$ and $a + b = |a| + |b|$. (The absolute values make no difference in this case.) So we have $|a + b| \leq |a| + |b|$.

Case 4: If $a < 0$ and $b < 0$, then $|a + b| = -(a + b) = (-a) + (-b) = |a| + |b|$.

Case 5: Suppose $a > 0$ and $b < 0$.

Clearly, in this case $a > b$. But if we compare the absolute values of a and b, either one could be larger. We split this case into two subcases, 5a and 5b.

Subcase 5a: Suppose $|a| \geq |b|$. This is true when, for example, a = 5 and b = –2. It then follows that $|a + b| = |a| - |b|$. But $|a| - |b| < |a| < |a| + |b|$, so by transitivity, we have $|a + b| < |a| + |b|$. Note that this case gave us a strict inequality, while we only need to verify $\leq$ (a weaker claim) in general.

Subcase 5b: Suppose $|b| > |a|$. This is true when, for example, $a = 5$ and $b = -7$. It then follows that $|a + b| = |b| - |a| < |b| < |a| + |b|$, so again we get $|a + b| < |a| + |b|$.

Using subcases makes it a proof by exhaustion within a proof by exhaustion!

Case 6: If $a < 0$ and $b > 0$, we argue the same way as in Case 5 (and its subcases). Simply take Case 5 and change every a to b and every b to a.

Because the six cases above address every possibility, we may now conclude that the triangle inequality holds for all real numbers a and b. □

As with many theorems, we have other ways to prove this one.[7] I'm not suggesting that the proof given above is the simplest of these. It is included here simply as a way to illustrate proof by exhaustion with an important result.

---

6 Hempel, Carl G., "Le problème de la vérité," *Theoria* (a Swedish journal), Vol. 3, 1937, pp. 206–246. Hempel used ravens, where I use crows. Martin Gardner brought the paradox to a wider audience through his column in *Scientific American*. The piece was reprinted in Gardner, Martin, *Hexaflexagons and Other Mathematical Diversions, The First Scientific American Book of Puzzles & Games*, The University of Chicago Press, Chicago, Illinois, 1988. Gardner often updated his books with new material, so it is not the first edition that is referenced here. See pp. 52–54 for the discussion, pp. 175–176 for appearances of this paradox before Hempel's paper, and pp. 191–194 for a long list of references.

7 See http://math.stackexchange.com/questions/307348/proof-of-triangle-inequality for more.

Proof by Induction

> In the Theory of Numbers it happens rather frequently that, by some unexpected luck, the most elegant new truths spring up by induction.
>
> Carl Friedrich Gauss[8]

There are many patterns that hold true for some of the natural numbers (1, 2, 3, etc.), but we can never prove they *always* hold by simply giving a bunch of examples where they work. A fairly long-lived pattern is given by the function $f(x) = x^2 + x + 41$, found by the Swiss mathematician Leonhard Euler (Figure 2.3).[9] Let's look at a table of its values.

x	$x^2 + x + 41$	x	$x^2 + x + 41$
1	43	11	173
2	47	12	197
3	53	13	223
4	61	14	251
5	71	15	281
6	83	16	313
7	97	17	347
8	113	18	383
9	131	19	421
10	151	20	461

What are some properties that the numbers in the second and fourth columns all have in common? Will all of the outputs of this function have these properties? This is left as an exercise, but, in the meanwhile, let's consider some lines Euler placed in a paper concerning another result, after he verified it for the integer values 1 through 20:

> I think these examples are sufficient to discourage anyone from imagining that it is mere chance that my rule is in agreement with the truth.[10]

After further demonstrating the claim for the values 101 and 301, Euler wrote:

---

8 Polya, George, *Mathematics and Plausible Reasoning, Vol. I, Induction and Analogy in Mathematics*, Princeton University Press, Princeton, New Jersey, 1954, p. 59. I like this quote even if Gauss was referring to inductive reasoning rather than proof by induction.

9 For a list of similar polynomials see http://mathworld.wolfram.com/Prime-GeneratingPolynomial.html.

10 Taken here from Young, Robert M., *Excursions in Calculus: An Interplay of the Continuous and the Discrete*, The Mathematical Association of America, 1992, p. 362. It previously appeared in Polya, George, *Mathematics and Plausible Reasoning, Vol. I, Induction and Analogy in Mathematics*, Princeton University Press, Princeton, New Jersey, 1954, pp. 91–98.

FIGURE 2.3
Leonard Euler (1707–1783), as seen on a banknote from Switzerland.

Image from http://www2.physics.umd.edu/~redish/Money/, courtesy of Joe Redish.

> The examples that I have just developed will undoubtedly dispel any qualms which we might have had about the truth of my formula.[11]

Later in the paper, in explaining how he found the formula in the first place, he wrote:

> It seems impossible that the law which has been discovered to hold for 20 terms, for example, would not be observed in the terms that follow.[12]

Euler did make it clear, however, that he knew these examples did not constitute a proof. The result for which Euler made these remarks is examined in Chapter 17.

Euler was one of the best mathematicians ever. Still, he remained humble. He actually claimed that others could have made the same discoveries he had, if they had only devoted as much time to their investigations as he had. While nearly all of today's mathematicians would disagree with this assessment, they can agree that Euler did devote a tremendous amount of time to his mathematical work. Upon losing the use of his right eye Euler remarked, "Now I will have less

---

11 Taken here from Young, Robert M., *Excursions in Calculus: An Interplay of the Continuous and the Discrete*, The Mathematical Association of America, 1992, p. 364. It previously appeared in Polya, George, *Mathematics and Plausible Reasoning, Vol. I, Induction and Analogy in Mathematics*, Princeton University Press, Princeton, New Jersey, 1954, pp. 91–98.

12 Taken here from Young, Robert M., *Excursions in Calculus: An Interplay of the Continuous and the Discrete*, The Mathematical Association of America, 1992, p. 365. It previously appeared in Polya, George, *Mathematics and Plausible Reasoning, Vol. I, Induction and Analogy in Mathematics*, Princeton University Press, Princeton, New Jersey, 1954, pp. 91–98.

distraction."[13] He eventually went blind in his left eye as well; yet, his mathematical output then increased! Some rank Paul Erdös as the most prolific mathematician in history, but others claim Euler holds this title. While we can regard Euler with awe, generations of mathematicians have managed to improve upon much of his work. Many of his proofs would not be considered rigorous (a fancy word mathematicians use to mean sufficiently detailed to convince readers of their correctness) by today's standards.

If we can't verify a claim that something holds true for all natural numbers by simply testing various values, then how can we verify it? The answer is a technique known as "proof by induction." It's a rather unfortunate name for this method to go by, for it has nothing to do with induction. In fact, it's a deductive method. To be clear, let's review these definitions.

Deductive reasoning – a process in which necessary conclusions are drawn from premises (think axioms) and/or other conclusions that, in turn, followed from the premises.

Inductive reasoning – a process in which generalizations are drawn from specific instances in a way that extrapolates beyond what was initially known.

> Theorems proved by mathematical induction are deductive and certain, while conclusions depending upon scientific induction are only probable at best.
>
> John Allen Paulos[14]

So, while induction is a common (and fantastic!) way to form conjectures in mathematics,[15] all mathematical proofs are deductive. Other, more appropriate, names exist for "proof by induction." Henri Poincaré called it "proof by recurrence," for example.[16] Sadly, none of the alternate names have really caught on.

Whatever we choose to call it, the idea behind proof by induction is analogous to verifying that a chain of dominos will all fall.

Step 1: Verify that the first domino will fall.

Step 2: Make sure that any particular domino, if knocked over, will cause the one after it to fall.

That's all! If the first domino falls and each one knocks down the one that follows, then domino number 2 must fall and it must knock down domino number 3, and so on. They all fall down. We usually refer to checking that the first one falls as the **base case**. Step 2 is referred to as the **induction step**.

Example 1

Theorem: $1 + 2 + 3 + 4 + 5 + \cdots + n = n(n + 1)/2$

---

13 http://www-gap.dcs.st-and.ac.uk/~history/Quotations/Euler.html, which took it from Eves, Howard Whitley, *In Mathematical Circles*, Prindle Weber & Schmidt, Boston, Massachusetts, 1969.

14 Paulos, John Allen, *Beyond Numeracy*, Vintage Books, New York, 1992, p. 122.

15 This is probably the kind of induction Gauss had in mind when offering the quote that begins this section of the chapter.

16 Poincaré, Henri, *Science and Hypothesis*, Dover, New York, 1952, pp. 9 and 13. This is a republication of the first English translation, published in 1905 by the Walter Scott Publishing Company, Ltd.

Proof: We first establish the base case by verifying the claim for $n = 1$. If we sum only the first of the positive integers, we get 1. Compare this to the right-hand side, when $n = 1$. We get $1(1+1)/2 = 2/2 = 1$. Hence, the claim is valid for $n = 1$.

Now suppose the claim is true for $n = k$. That is,

$$1 + 2 + 3 + 4 + 5 + \cdots + k = k(k+1)/2$$

We can add $(k + 1)$ to both sides of the above to get

$$
\begin{aligned}
1 + 2 + 3 + 4 + 5 + \cdots + k + (k+1) &= k(k+1)/2 + (k+1) \\
&= (k^2 + k)/2 + (2k+2)/2 \\
&= (k^2 + 3k + 2)/2 \\
&= (k+1)(k+2)/2 \\
&= (k+1)((k+1)+1)/2,
\end{aligned}
$$

which shows the claim holds for $n = k +1$. This completes the induction step and shows that the claim must hold for all natural numbers. $\square$

A great many things may be proven by induction. Often the proofs are very easy (mechanical – just follow the steps), so authors will omit them with the phrase "it follows by induction that…" Don't try this on a test!

There's a second form of proof by induction. For this we first establish a base case, as before, but for the induction step, we assume the claim holds for *all* natural numbers less than or equal to k and show this implies the claim holds for $k + 1$. This second form is sometimes called "strong induction," while the first form may be referred to as "weak induction."

Example 2

Theorem: Every natural number is interesting.[17]

Proof: To establish the base case, we need to show that "1" is interesting. Pick a reason! There are many. The number 1 is the smallest natural number and the only natural number that is neither a prime nor a product of primes. So, the base case is easy to establish.

Now for the induction step. Assume that all natural numbers less than or equal to k are interesting. If $k +1$ happens to be interesting, we're done. Suppose $k + 1$ is not interesting. Then $k + 1$ is the smallest natural number that is not interesting. Hey! That's interesting. Therefore, $k + 1$ must be interesting and it follows by induction that all natural numbers are interesting. $\square$

In David Wells's book *The Penguin Dictionary of Curious and Interesting Numbers*, the entry for 51 is as follows:

---

17 For slightly different versions of this argument see Burger, Edward B. and Michael Starbird, *The Heart of Mathematics*, third edition, Wiley, Hoboken New Jersey, 2010, pp. 56–57 and Gardner, Martin, *Hexaflexagons and Other Mathematical Diversions, The First Scientific American Book of Puzzles & Games*, The University of Chicago Press, Chicago, Illinois, 1988, p. 148.

> This appears to be the first uninteresting number, which of course makes it an especially interesting number, because it is the smallest number to have the property of being uninteresting. It is therefore also the first number to be simultaneously interesting and uninteresting.[18]
>
> David Wells

Okay, that example was a bit of a joke, but it does illustrate how the second form of induction may be used.

Proof by Induction in Geometry

Dominos are typically used to explain proof by induction, but we can go one better and use trominoes (aka triominoes) as an example.[19]

To avoid confusion between the dominoes in the analogy I used for proof by induction and the physical trominoes we are about to examine, let's consider another analogy. We can think of the positive integers as rungs on a ladder, with the bottom rung being 1, or the base case. To show, by mathematical induction, that we can climb the ladder to as great a height as we please, we first show that we can get on the bottom rung (this is the base case). Then, all that is left is to show that from any given rung, we can get to the next rung. This is the induction step. If we can verify both of these steps, then it is established that we can climb arbitrarily high.

Now, returning to the world of trominoes, consider a grid that measures 2^n by 2^n for some value of n. In Figure 2.4, $n = 3$ gives an 8 by 8 grid, but n can take any natural number value. Once you have the grid, pick one square to leave out and then attempt to cover the rest with trominoes.

See color figure 3 **(See color insert)** for a tiling where the square in row 6, column 2 was left out. Color figure 4 **(See color insert)** gives another example, where the missing square is in row 5, column 5.

It turns out that you can remove any one square and it will always be possible to cover the rest of the squares with trominoes. This holds not just for the 8 by 8 grid, but also for 2^n by 2^n grids for every natural number n. The proof of this fact is easily achieved with induction. Can you find it?

The proof was first found by Solomon Golomb and published in 1954.[20] An applet by Alexander Bogomolny demonstrating the tilings can be found at www.cut-the-knot.org/Curriculum/Games/TrominoPuzzle.shtml.

A much more detailed history of this problem, along with variations and generalizations, is given by Norton Starr at www3.amherst.edu/~nstarr/trom/intro.html. A paragraph from this website follows below. Follow the link to see the citations he provides:

---

18 Wells, David, *The Penguin Dictionary of Curious and interesting Numbers*, revised edition, Penguin Books, London, 1997, p. 107.

19 See http://www.cut-the-knot.org/Curriculum/Geometry/Tromino.shtml and http://www.mathdemos.org/mathdemos/tromino/tromino.html for more on this topic.

20 Golomb, Solomon W., "Checker Boards and Polyominoes," *American Mathematical Monthly*, Vol. 61, No. 10, December 1954, pp. 675–682.

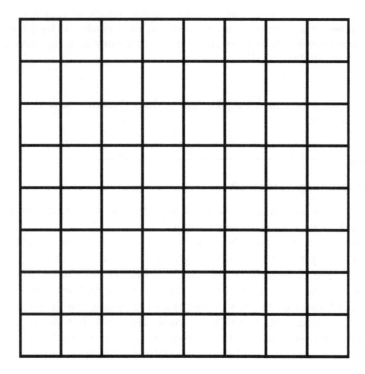

FIGURE 2.4
An 8 by 8 grid.

Image created by Josh Gross.

Another direction is three dimensional. Consider a cube of side length 2^n, containing 2^{3n} unit cells, one of which is occupied (single deficiency.) Can the remaining cells be tiled with three dimensional trominoes (three cubes in an L-shape, with two of them meeting the third on two adjacent faces of the latter)? The necessary condition that $2^n = 3k + 1$ turns out to be sufficient as well. [23, Chapter 6: Norton Starr's 3-Dimensional Tromino Tiling], [24, pp. 72–87], and [25] The case of a 4×4×4 cube presents some modest challenges that may amuse young puzzlers.

The two-dimensional problem has a puzzle-like feel to it, and indeed it has been made into a game. See www.gamepuzzles.com/polycub2.htm#V21 for the commercial version, which puts some twists on the original problem.

A Proof by Induction Candidate?

Consider the following improper integrals.[21]

---

21 Taken from Chamberland, Marc, *Single Digits: In Praise of Small Numbers*, Princeton University Press, Princeton, New Jersey, 2015, pp. 188–189.

$$\int_0^\infty \frac{\sin(x)}{x}\,dx = \frac{\pi}{2}$$

$$\int_0^\infty \frac{\sin(x)}{x}\frac{\sin(x/3)}{x/3}\,dx = \frac{\pi}{2}$$

$$\int_0^\infty \frac{\sin(x)}{x}\frac{\sin(x/3)}{x/3}\frac{\sin(x/5)}{x/5}\,dx = \frac{\pi}{2}$$

$$\int_0^\infty \frac{\sin(x)}{x}\frac{\sin(x/3)}{x/3}\frac{\sin(x/5)}{x/5}\frac{\sin(x/7)}{x/7}\,dx = \frac{\pi}{2}$$

$$\int_0^\infty \frac{\sin(x)}{x}\frac{\sin(x/3)}{x/3}\frac{\sin(x/5)}{x/5}\frac{\sin(x/7)}{x/7}\frac{\sin(x/9)}{x/9}\,dx = \frac{\pi}{2}$$

$$\int_0^\infty \frac{\sin(x)}{x}\frac{\sin(x/3)}{x/3}\frac{\sin(x/5)}{x/5}\frac{\sin(x/7)}{x/7}\frac{\sin(x/9)}{x/9}\frac{\sin(x/11)}{x/11}\,dx = \frac{\pi}{2}$$

$$\int_0^\infty \frac{\sin(x)}{x}\frac{\sin(x/3)}{x/3}\frac{\sin(x/5)}{x/5}\frac{\sin(x/7)}{x/7}\frac{\sin(x/9)}{x/9}\frac{\sin(x/11)}{x/11}\frac{\sin(x/13)}{x/13}\,dx = \frac{\pi}{2}$$

Based on these examples, there's an obvious conjecture that can be made. Can it be proved by induction? Absolutely not. How can I be so sure? Maybe there's a really clever way to apply the technique, even if another method might be easier. Well, I'm sure that a proof by induction will never be found, because the conjecture isn't true. The very next case gives a counterexample.

$$\int_0^\infty \frac{\sin(x)}{x}\frac{\sin(x/3)}{x/3}\frac{\sin(x/5)}{x/5}\frac{\sin(x/7)}{x/7}\frac{\sin(x/9)}{x/9}\frac{\sin(x/11)}{x/11}\frac{\sin(x/13)}{x/13}\frac{\sin(x/15)}{x/15}\,dx$$

$$\approx 0.499999999992647\pi.$$

This example is given, not as an example of proof by induction, but as another example for why proof by induction is sometimes needed. We can't assume patterns continue. If we don't have a proof, we don't really know!

Who Deserves the Credit? Who Deserves the Blame?[22]

Who deserves credit (or blame, depending on your perspective) for coming up with the idea of proof by induction? Many writers attribute it to Blaise Pascal (1623–1662), who certainly used it, but it can be found in work done earlier.

> Although discovery of the method is usually attributed to Pascal, it appears that the first person to apply mathematical induction to rigorous proofs was the Italian scientist Francesco Maurolico in his *Arithmetica* of 1575. (See, for example, Giovanni Vacca, "Maurolycus, the first discoverer of the principle of mathematical induction," *Bulletin of the Amer. Math. Society* 16 (1909–1910), 70–73, and also W. H. Bussey, "The origin of mathematical induction," *American Math. Monthly* 24 (1917), 199–207.) Further improvements were made in the early 17th century by Pierre de Fermat.[23] Pascal in one of his letters acknowledged Maurolico's introduction of the method and used it himself in his Traité du triangle arithmétique (1665), in which he presents what is now called "Pascal's triangle"..."The phrase 'mathematical induction' was apparently coined by De Morgan in the early 19th century. Of course the method is implicit even in Euclid's proof of the infinitude of the primes. For he shows"..."that if there are n primes, there must be $n + 1$ primes; and since there is a first prime, the number of primes must be infinite." For further discussion of mathematical induction, see the great works of George Pólya, *Mathematical Discovery* and *Mathematics and Plausible Reasoning*.
>
> Robert M. Young[24]

Or should credit go to Levi ben Gerson (1288–1344)? This Jewish mathematician and astronomer lived in France and is known for his work in combinatorics as well as in philosophy and Biblical commentary.[25] Gerson carried out what are essentially induction proofs by doing the induction step first, and then showing that there's some small number for which the claim holds true. Thus, the claim holds for all values greater than or equal to that small value. Counting against giving priority to Gerson is the fact that, for the induction step, he had no notation to represent the arbitrary value that is typically expressed with a k in today's induction proofs. So, what he actually did at this stage is use a specific value and try to make it clear that there was nothing special about this particular value.[26] A critic could say that despite his good intentions, he really just had two base cases.

Some claim induction was used *centuries* before Levi ben Gerson:

22　This is a line from the song *Lobachevsky* by Tom Lehrer, performing artist, NSA mathematician, and inventor of the Jell-O shot.

23　Note: Fermat's method of infinite descent resembles induction in a way. The technique was used by Euclid, long before Fermat – see Bell, Eric Temple, *Mathematics, Queen and Servant of Science*, Copyright, 1951, British Edition 1952, reprinted 1954, 1958, 1961, Printed in Great Britain by Butler and Tanner Ltd., Frome and London, p. 232 and footnote.

24　Young, Robert M., *Excursions in Calculus: An Interplay of the Continuous and the Discrete*, The Mathematical Association of America, 1992, p. 15, footnote

25　Wilson, Robin and John J. Watkins, editors, *Combinatorics: Ancient & Modern*, Oxford University Press, Oxford, 2013, p. 115.

26　Wilson, Robin and John J. Watkins, editors, *Combinatorics: Ancient & Modern*, Oxford University Press, Oxford, 2013, p. 116.

Al-Karaji [953–1029] was the first to use the method of proof by mathematical induction to prove his results, by proving that the first statement in an infinite sequence of statements is true, and then proving that, if any one statement in the sequence is true, then so is the next one.[27]

One priority dispute that is much easier to resolve concerns the first appearance of an "Inducktion Proof." It's in the lighthearted textbook *Discrete Mathematics with Ducks*.[28]

Proof by Contradiction

> ...*reductio ad absurdum*, which Euclid loved so much, is one of a mathematician's finest weapons. It is a far finer gambit than any chess gambit: a chess player may offer the sacrifice of a pawn or even a piece, but a mathematician offers *the game*.
>
> G. H. Hardy[29]

While medical school is a better place for those wishing to use Latin terms, mathematicians have some options besides the previously introduced *"quod erat demonstrandum."* Both *"reductio ad absurdum"* (reduction to absurdity) and *"argumentum ad absurdum"* (argument to absurdity) may be used in place of the English "proof by contradiction." It's illustrated below with a slight variant of a proof included by Euclid in the *Elements*. Euclid's original did not actually use contradiction, but rather was constructive.

Theorem: There are infinitely many primes.

Proof (by contradiction): Suppose there are only finitely many primes. Let S be the set of all primes, p_1 through p_n. Now consider $n = p_1 p_2 \cdots p_n + 1$ (one more than the product of all the primes). This number is certainly larger than any in S, so if it is prime we have a contradiction. We also have a contradiction if it is not prime. Because all of the numbers in S leave a remainder of 1 when divided into n, the prime factors of n cannot be in S. Hence, the set S cannot contain every prime and we see that there must be infinitely many primes. □

For the Pythagoreans, math wasn't just math. It was deeply entwined with their philosophy. One could say it was their religion. And this mathematical religion was built entirely on the integers. Ratios of integers were okay, but irrationals weren't even believed to exist. It was thought that absolutely everything could be expressed using just integers and ratios of integers. So, the following theorem concerning a far from exotic number (the length of the diagonal for a square with sides of length one) was devastating. The proof is by contradiction.

Theorem: $\sqrt{2}$ is irrational.

Proof: Assume $\sqrt{2}$ is rational. Then for some integers a and b, with $b \neq 0$, we have $\sqrt{2} = a/b$. Furthermore, we can assume that a/b is in reduced form, so a and b are not both even. Squaring both sides gives us $2 = a^2/b^2$. We can then multiply through by b^2 to get $2b^2 = a^2$. The left-hand side, $2b^2$, is even (because it is a multiple of 2), so the right-hand side, a^2, must also be even. But, we saw previously (our example for proof by contrapositive) that a^2 even

---

27 http://www.storyofmathematics.com/islamic.html.

28 belcastro, sarah-marie, *Discrete Mathematics with Ducks*, CRC Press, Boca, Raton, Florida, 2012, pp. 103–104.

29 Hardy, G. H., *A Mathematician's Apology*, Cambridge University Press, Cambridge, 1940, p. 94.

implies a is even. Because any even number can be expressed as an integer multiple of 2, we have $a = 2k$ for some integer k. So, $2b^2 = a^2$ can be rewritten as $2b^2 = (2k)^2$, which can also be written as $2b^2 = 4k^2$. Dividing through by 2 then gives $b^2 = 2k^2$. Now we repeat a previous argument in this proof, but using new letters/constants. That is, $2k^2$ is even, so b^2 must be even. But this implies that b is even. So we now have both a and b are even, which contradicts our assumption that they are not both even (i.e., that the fraction could be expressed in reduced form, as is the case for all rational numbers). Thus, $\sqrt{2}$ is not rational.   □

There are several such indirect proofs that $\sqrt{2}$ is irrational.[30] We don't have a "first edition" of the proof and it's been speculated that the ancient Greek proof was likely a geometric argument.[31]

Eli Maor did a good job of giving the context that shows how upsetting this result could be.

> Thus, Pythagoras reasoned that if music is based on rational numbers, surely the entire universe must be too. Rational numbers thus dominated the Greek view of the world, just as rational thinking dominated their philosophy (indeed, the Greek word for rational is *logos*, from which the modern word *logic* derives).
>
> Eli Maor[32]

Indeed, the Pythagoreans did not take it well. This was mentioned briefly in Chapter 0, but we now take a deeper look.

> The discovery that $\sqrt{2}$ is irrational left the Pythagoreans in a state of shock, for here was a quantity that could clearly be measured and even constructed with a straightedge and compass, yet it was not a rational number. So great was their bewilderment that they refused to think of $\sqrt{2}$ as a number at all, in effect regarding the diagonal of a square as a numberless magnitude! (This distinction between arithmetic number and geometric magnitude, which in effect contradicted the Pythagorean doctrine that number rules the universe, would henceforth become an essential element of Greek mathematics.) True to their pledge of secrecy, the Pythagoreans vowed to keep the discovery to themselves. But legend has it that one of them, a man named Hippasus, resolved to go his own way and reveal to the world the existence of irrational numbers. Alarmed by this breach of loyalty, his fellows conspired to throw him overboard the ship they were sailing on.
>
> Eli Maor[33]

In some accounts, it was the discoverer of irrationals who was killed to keep the secret, instead of the man who intended to divulge it. Of course, the two could have been the same person. If they were distinct individuals, it's possible that it was Pythagoras himself who made the discovery. It's hard to tell at this late date, for several reasons,

---

30 See Chamberland, Marc, *Single Digits: In Praise of Small Numbers*, Princeton University Press, Princeton, New Jersey, 2015, pp. 55–56 for two alternate proofs that √2 is irrational. One uses contradiction and the other does not.

31 Maor, Eli, *e: The Story of a Number*, ninth printing with new material on pp. 183–186, Princeton University Press, Princeton, New Jersey, 1998, p. 188.

32 Maor, Eli, *e: The Story of a Number*, ninth printing with new material on pp. 183–186, Princeton University Press, Princeton, New Jersey, 1998, p. 188.

33 Maor, Eli, *e: The Story of a Number*, ninth printing with new material on pp. 183–186, Princeton University Press, Princeton, New Jersey, 1998, p. 190.

including the fact that *all* discoveries made by Pythagoreans were attributed to Pythagoras. In any case, no punishment, however severe, can change mathematical truth, and the truth was out.

> Even Aristotle, though shockingly ignorant of the science of his time, was dimly aware of the proof that $\sqrt{2}$ is irrational.
>
> Petr Beckmann[34]

Ancient Greek mathematicians went on to prove that the values of several other square roots of integers are also irrational. We now know that the square root of a positive integer is always one extreme or another, an integer (as with $\sqrt{1}$, $\sqrt{4}$, $\sqrt{9}$, etc.) or an irrational (as with $\sqrt{2}$, $\sqrt{3}$, $\sqrt{5}$, etc.). We never have a root that is a rational value with a denominator other than 1, giving an integer.

While other numbers, including π and e, were shown to be irrational in the 1700s, the question of rationality has yet to be resolved for many simple values, including $\pi + e$, π/e, and $\log \pi$.[35] It's kind of funny that while we can't find a single proof for the rationality or irrationality of many numbers, we have multiple proofs for the irrationality of $\sqrt{2}$. This illustrates how important imagination is in mathematics. We can't always clone familiar proofs when attempting to prove conjectures, however similar the conjectures might be to the established theorems.

Grumblings

I began the section on proof by contradiction with a quote from Hardy, but not all mathematicians felt as he did about it. There were even objections from one of the first computer programmers, long before Hardy praised the approach.

> You should never select an <u>indirect</u> proof, when a <u>direct</u> one can be given.
>
> Ada Lovelace[36]

Ada Lovelace (Figure 2.5) was a friend of Charles Babbage, the 19th-century computer designer. His difference engine was a special purpose machine, but his later analytic engine was actually programmable. Lovelace's knowledge of this machine's potential may have exceeded Babbage's for she's the one who wrote programs for it. For example, she wrote a program to calculate Bernoulli numbers, using nested loops.[37] She realized how loops could be used to reduce the number of lines of code in a program.[38] At the time, the lines took the form of punched cards. The only flaw in this team-up was Babbage's inability to bring any of his massive projects to

---

34 Beckmann, Petr, *A History of π (Pi)*, Dorset Press, New York, 1989, p. 166.

35 Taken from Young, Robert M., *Excursions in Calculus: An Interplay of the Continuous and the Discrete*, The Mathematical Association of America, 1992, p. 238, which quotes Borwein, J. M., P. B. Borwein, and D. H. Bailey, "Ramanujan, Modular Equations, and Approximations to Pi or How to Compute One Billion Digits of Pi," *American Mathematical Monthly*, Vol. 96, No. 3, March 1989, pp. 201–219, particularly p. 203.

36 Quote taken from Wolfram, Stephen, *Idea Makers*, Wolfram Media, Inc., Printed by Friesens, Manitoba, Canada, 2016, p. 53.

37 Wolfram, Stephen, *Idea Makers*, Wolfram Media, Inc., Printed by Friesens, Manitoba, Canada, 2016, pp. 79–82.

38 Wolfram, Stephen, *Idea Makers*, Wolfram Media, Inc., Printed by Friesens, Manitoba, Canada, 2016, p. 78.

FIGURE 2.5
Ada Lovelace (1815–1852).

Public domain.

a conclusion. Although he began the work, he never completed either machine. Hence, Lovelace's programs could not be put to the test during her lifetime. For what it's worth, Babbage also began writing a book on cryptology, but it too was never completed.

During and after her time working with Babbage, Lovelace was tutored by Augustus De Morgan.[39] Babbage described her as "Enchantress of Number."[40] Stephen Wolfram speculated, "In person, beyond the polished Victorian upper-class exterior, I suspect she was something of a nerd, complete with math jokes and everything."[41] Despite all of

---

39 Wolfram, Stephen, *Idea Makers*, Wolfram Media, Inc., Printed by Friesens, Manitoba, Canada, 2016, pp. 54–55.
40 Wolfram, Stephen, *Idea Makers*, Wolfram Media, Inc., Printed by Friesens, Manitoba, Canada, 2016, p. 67.
41 Wolfram, Stephen, *Idea Makers*, Wolfram Media, Inc., Printed by Friesens, Manitoba, Canada, 2016, p. 95.

these appealing qualities, as a female she was still discriminated against and denied access to Royal Society's library.[42]

By contrast, Babbage received many degrees and honors during his lifetime. So many that they filled six lines when he listed them after his name; yet, he still complained of his lack of recognition.[43] Sadly, the work of Babbage and Lovelace was far ahead of its time, and Lovelace died of cancer well before her time at age 36. Their work was largely forgotten and only rediscovered after computers of a different sort began to make an impact in the 20th century.

Lovelace's preference for direct proofs is shared by most mathematicians, but for the Dutch mathematician Luitzen E. J. Brouwer (Figure 2.6) it was no mere preference. He claimed that indirect proofs were not even valid. His take on proof by contradiction differed from that of classical logic as follows. If an assumption not-A led to a contradiction, he regarded that as a proof of not-not-A, but not as a proof of

FIGURE 2.6
Luitzen Egbertus Jan Brouwer (1881–1966).

Public domain.

42 Wolfram, Stephen, *Idea Makers*, Wolfram Media, Inc., Printed by Friesens, Manitoba, Canada, 2016, p. 69.
43 Wolfram, Stephen, *Idea Makers*, Wolfram Media, Inc., Printed by Friesens, Manitoba, Canada, 2016, p. 59, a picture of the six lines is included here.

A. That is, he did not accept the double negation of a statement as being logically equivalent to the original statement. This is a rejection of the law of excluded middle. Brouwer's approach to mathematics is called **intuitionism**.

> Most works in mathematics, including Brouwer's own famous earlier work, do not live up to the intuitionistic standards of mathematical proofs. Brouwer's ideas were regarded as revolutionary and, while on his lecture tours, he was received with an enthusiasm not usually associated with mathematics.
>
> Gil Kalai[44]

Brouwer explained his viewpoint as follows:

> The use of the Principle of the Excluded Middle is *not permissible* as part of a mathematical proof.... [It] has only scholastic and heuristic value, so that theorems which in their proof cannot avoid the use of this principle lack all mathematical content.[45]

David Hilbert responded with,

> taking the Principle of the Excluded Middle from the mathematician is the same as ... prohibiting the boxer the use of his fists.[46]

In an excellent chapter in *The Search for Truth*, E. T. Bell explained how questioning axioms, what other's take for self-evident truths, can lead to mathematical revolutions far more valuable than what the mathematician might have added to human knowledge from simply playing along with everyone else.[47] Non-Euclidean geometries offer one example. In *The Scientific Monthly*, James Byrnie Shaw commented on Brower and the situation in logic:

> Brouwer said, "Let there be numbers which are neither rational nor yet irrational," and logic has been deprived of the law of excluded middle. We have seen logics created in which the law of identity has vanished. There are logics in which the law of contradiction has also gone.[48]

Bell also commented on Schopenhauer's orneriness when it came to the concept of proof:

> Indeed the philosopher A. Schopenhauer (1788–1860) objected most peevishly to the Pythagorean theorem. The proof, he declared, was as deceptive as a mousetrap into which he had been enticed by too easy assumptions. In his mathematics Schopenhauer was what is today called an intuitionist.[49]

---

44 https://gilkalai.wordpress.com/2008/12/02/godel-hilbert-and-brouwer/.
45 From a paper presented to the Royal Academy of Sciences, December 18, 1920. See Mancosu, Paolo, *From Brouwer to Hilbert*, Oxford University Press, New York, 1998, p. 23.
46 Reid, Constance, *Hilbert*, Springer, New York, 1970, p. 149.
47 Bell, Eric Temple, *The Search for Truth*, Reynal & Hitchcock, New York, 1934, Chapter XIV, pp. 204–217.
48 Shaw, James Byrnie, "The Unity of Mathematics," *The Scientific Monthly*, Vol. 45, November 1937, pp. 402–411, p. 410 quoted here.
49 Bell, Eric Temple, *Mathematics, Queen and Servant of Science*, Copyright, 1951, British Edition 1952, reprinted 1954, 1958, 1961. Printed in Great Britain by Butler and Tanner Ltd., From and London, p. 264

Terence Tao

If all of the above proof techniques fail, there's another approach that can be taken to get a proof for a conjecture: get Terence Tao (Figure 2.7) interested in the problem.

Tao got an early start in mathematics and was taking college-level math classes by the age of 9. He not only rose to the top in the International Mathematical Olympiad, he set records as the youngest person to win at each level along the way: bronze (10 years, 363 days), silver (11 years, 364 days), and gold (13 years, 4 days).[50] He earned his BS at age 16, his MS at 17, and his Ph.D. from Princeton University at age 20.[51] He went on to become the youngest-ever full professor at UCLA. In 2006, he won a Field's Medal. He's written over 350 papers and 17 books. A bit of his work will be detailed later in this chapter. For now, suffice it to say if you can get Tao interested in your problem, you might soon find that you have a proof.

FIGURE 2.7
Paul Erdös with a young Terence Tao (1975–).

Courtesy of Terence Tao.

50 https://en.wikipedia.org/wiki/List_of_International_Mathematical_Olympiad_participants.
51 https://www.math.ucla.edu/~tao/preprints/cv.html.

Take Your Pick

For many theorems, mathematicians have found a variety of proof techniques that can be used to verify them. Some theorems already have hundreds of distinct proofs. John W. Dawson has written a wide-ranging book on alternative proofs and what they have to tell us.[52] Far more focused is *The Pythagorean Proposition* by Elisha Scott Loomis. This volume claims to present 367 different proofs of the Pythagorean theorem, although some are not really different and several are fallacious.[53]

With diverse approaches leading to success in so many instances, how can we say that there's a "best" way to attempt to prove a conjecture? Simply put, we cannot. When I was a Scholar-in-Residence with the National Security Agency's Center for Cryptologic History (despite only having degrees in mathematics!), I had a very interesting conversation with a gentleman who had headed mathematical research for many years. In this conversation, I expressed my surprise concerning a paper from the 1940s that described using Fourier series to attack cipher machines like Enigma. These machines are all discrete. The type of math applied in this paper seemed completely inappropriate to the task! The response I got was essentially "It all works." He elaborated that the kind of math you apply to model or solve a problem depends on what kind of mathematician you are. If you're an algebraist, you use algebra, if you're a graph theorist, you use graph theory, etc. He went on to give me an example of how topology can be applied in the world of modern cryptology. This is good news, but if you want a simple algorithm to approach theorem proving, it doesn't exist. Nevertheless, I can provide some general techniques.

General Techniques

While these techniques do not provide a proof in and of themselves, they will often be useful in helping you find a proof of the types covered above.

1. **Draw a picture.** No matter how complicated or abstract a problem is, you can likely find a way to represent it with an image of some sort. The image might provide some insight into the problem that you'd be unlikely to achieve without it. You'll see many examples of this in the chapters to come. In some cases, the picture can only be efficiently created with the help of a computer.

2. **Look at simpler cases.** In physics classes, simplified models are encountered first. They ignore things like air resistance and other forms of friction. Ropes are often assumed to be massless. It's only later that these considerations are brought into play. So, why should we attempt to solve math problems in their full complexity right from the start? It's often easier to start with simplified problems and see how the answer changes as the problem grows in complexity. You may notice a pattern

---

52 Dawson, Jr., John W., *Why Prove it Again?: Alternative Proofs in Mathematical Practice*, Springer, Heidelberg, Germany, 2015.

53 Loomis, Elisha Scott, *The Pythagorean Proposition*, 1927, second edition 1940. The NCTM issued the second edition again in 1968.

that suggests the solution for the case you're actually interested in. That is, the easier problems you are able to answer may indicate a generalization that you couldn't see without them.

3. **Look at a more general case.** Weird as it sounds, it's sometimes easier, when trying to prove something specific to instead generalize it to a much bigger claim and then prove that. For example, if you're trying to prove a claim holds for $n = 35$, you might instead try to prove it for any positive integer n. This suggestion may seem to contradict technique 2 above, but while you won't find both techniques useful for the same problem, each will come in handy at times.

4. **Imitate another proof.** This is about as far from "mathematical magic" as you can get. Proving something by this technique won't make you feel like a creative genius, but it does often work. Your skillset should include the ability to find similar results that have already been established and alter the details of the steps so that they confirm the result you're interested in.

5. **Work backwards.** You cannot assume what you hope to prove. That is not a valid proof technique. However, your scratch work won't be seen, so you can do as you please at that stage of the attack. If you work backwards from what you hope to prove, you may find that some of the steps can be reversed and that you have the core of a proof. If you're very lucky, your work will lead from the conclusion to what you were given, and all of the steps will be reversible. Many readers will have experienced this when trying to verify trigonometric identities.

6. **Rewrite the problem.** Much of integral calculus seen in a typical calculus II class consists of rewriting the integrand. The techniques include partial fraction decomposition, completing the square, long division of polynomials, etc. An example from discrete math is given here:[54]
   Calculate the sum

   $$S = 100^2 - 99^2 + 98^2 - 97^2 + \cdots + 2^2 - 1^2$$

   We rewrite it as $S = \left(100^2 - 99^2\right) + \left(98^2 - 97^2\right) + \cdots + \left(2^2 - 1^2\right)$ and then rewrite it again by factoring to get $S = (100 + 99)(100 - 99) + (98 + 97)(98 - 97) + \cdots + (2 + 1)(2 - 1)$
   Now performing the subtractions inside the parenthesis turns it into

   $$S = (100 + 99)(1) + (98 + 97)(1) + \cdots + (2 + 1)(1)$$
   $$= (100 + 99) + (98 + 97) + \cdots + (2 + 1)$$
   $$= 100 + 99 + 98 + 97 + \cdots + 2 + 1.$$

   This is a problem we already know how to handle from the section on proof by induction. The answer is 5,050.

7. **Go to Extremes.** When attempting to disprove a claim, look at the boundaries or at the extremes. Just as real-world objects usually break at these locations, false claims

---

54 Hathout, Dean, *Wearing Gauss's Jersey*, CRC Press, Boca Raton, Florida, 2013, p. 10. This is far from the first appearance of this proof, but I'm referencing it because it's a great book for people who love mathematics.

do as well. For potential counterexamples, try things like a very large number, a very small number, a number with a large number of factors, a prime, 0, a big negative number, 1, etc.

8. **Find a new perspective.** If what you're trying doesn't seem to be working, try something different! Don't get stuck in a rut, but rather try to find as many completely different ways of looking at the problem as possible. Think of the students in the film *Dead Poets Society* standing on their desks!

9. **"Don't wait for the whole solution. Do what you can."**[55] It's a common approach and it simply asks you to chip away at the problem. You (or someone else) might be able to build on this work later to find a full solution. Professional mathematicians do this all the time. You'll see many examples of big names in mathematics satisfying themselves (temporarily!) with some little amount of progress on a problem.

The Structure of Proofs

Larger proofs are often laid out in the following format: Lemma, Theorem, Corollary. Each component will have a proof that immediately follows. The theorem portion is, of course, necessary, but lemmas and corollaries are not always present. The following paragraphs explain when each may appear.

If a theorem has a special case that is of great interest, it may be singled out and stated after the theorem as a **corollary**. But corollaries are not always special cases. Sometimes they're new theorems, but their proofs follow as straightforward consequences of the main theorem.

Sometimes in a proof there's a technical detail that's needed to make the proof work, but it seems like a bit of an aside. That is, it may distract from the main flow of the proof. In such cases, it's normal to isolate this detail and prove it first to get it out of the way. When this is done, the detail is labeled as a **lemma**. A proof can have more than one lemma.

Israeli mathematician Doron Zeilberger (Figure 2.8) proved the alternating sign matrix conjecture in an 84-page paper that made use of lemmas, sublemmas, subsublemmas, and more![56] For example, on page 29 he wrote:

> We now need the following $(sub)^6$ lemma:
>
> Subsubsubsubsubsublemma 1·2·1·2·1·1·1: Let U_j, $j = 1, \ldots, l$, be quantities in an associative algebra, then:

$$1 - \prod_{j=1}^{l} U_j = \sum_{j=1}^{l} \left\{ \prod_{h=1}^{j-1} U_h \right\} (1 - U_j).$$

---

55 This approach was suggested in Burger, Edward B. and Michael Starbird, *The Heart of Mathematics*, third edition, Wiley, Hoboken, New Jersey, 2010, p. 398.

56 Zeilberger, Doron, "Proof of the Alternating Sign Matrix Conjecture." *The Electronic Journal of Combinatorics*, Vol. 3, No. 2, R13, pp. 1–84, 1996, available online at http://www.combinatorics.org/ojs/index.php/eljc/article/view/v3i2r13.

FIGURE 2.8
Doron Zeilberger (1950–).

Public domain.

Zeilberger's proof saw print in 1996, but prior to this he sent portions of the paper to 88 experts, asking each to vouch for the correctness of the portion received. His paper thanks all of them and gives very brief (but interesting) biosketches of each.

A shorter proof (the paper was only 10 pages long this time) was soon found.[57]

While Zeilberger is the sole author on his proof of the alternating sign matrix conjecture, he does collaborate with others. One frequent coauthor is Shalosh B. Ekhad. On Ekhad's website, he refers to himself as "Doron Zeilberger's servant" and calls Zeilberger "my beloved master."[58] The two formed a journal, *The Personal Journal of*

---

57 Kuperberg, Greg, "Another Proof of the Alternating-Sign Matrix Conjecture," *International Mathematics Research Notices*, No. 3, pp. 139–150, 1996, available online at http://arxiv.org/pdf/math/9712207.pdf.

58 "Welcome to the Home Page of Doron Zeilberger's servant," available online at http://sites.math.rutgers.edu/~zeilberg/ekhad.html.

Shalosh B. Ekhad and Doron Zeilberger, which consists solely of papers with at least one of Ekhad and Zeilberger among the authors.[59]

Zeilberger has made impressive solo efforts, as well as collaborations with humans (and others), but some extremely long proofs have resulted from *massive* collaborations. A record-setter was the (successful) effort to classify all finite simple groups, a problem in abstract algebra. The individual papers on this topic sum to about 15,000 pages.[60] Daniel Gorenstein, who led the effort, referred to it as "The Thirty Years' War."[61]

Proof by Computer

This is the last general proof technique to examine. Of course, the computer is not really offering a new method of proof; it simply carries out steps that we could do at a much faster rate and with less chance of error. But the speed differential between man and machine is so large, many proofs obtained by computer could never be obtained without them, at least not by the same method.

The 1976 proof of the four-color map theorem was, for a time, the most extreme example. A computer played a vital role in the proof, which was such that no human could check it all. The story behind it is told more fully in Chapter 20. Despite the angst expressed by some mathematicians over such an inelegant proof, even more extreme computer proofs were in the future. The record for the biggest computer proof has now been broken repeatedly.

A recent record-breaker came on May 3, 2016, when a paper was posted online by Marijn Heule (The University of Texas at Austin), Oliver Kullmann (Swansea University), and Victor Marek (University of Kentucky).[62] The paper solved the Boolean Pythagorean Triples problem.

Pythagorean triples are positive integers, a, b, c, that satisfy $a^2 + b^2 = c^2$. An example is 3, 4, 5. There are infinitely many such triplets. The problem the trio of researchers solved is whether or not it's possible to assign each positive integer one of two colors (either blue or red, for example), such that no Pythagorean triple consists of numbers all having the same color. For example, if 3 and 4 are assigned the color blue, then 5 would have to be red. Another Pythagorean triple is 5, 12, 13. So, having made 5 red, either 12 or 13 would have to be blue. This is an example of a problem in an area called Ramsey theory. The proof of a much simpler Ramsey theory problem is given in Chapter 20.

The proof took up 200 terabytes. Anyone wanting to verify it could download a compressed form, occupying only 68 gigabytes. The catch is that the verification

59 "The Personal Journal of Shalosh B. Ekhad and Doron Zeilberger," available online at http://sites.math. rutgers.edu/~zeilberg/pj.html.

60 Wilson, Robin, and John J. Watkins, editors, *Combinatorics: Ancient & Modern*, Oxford University Press, Oxford, 2013, p. 361.

61 Roberts, Siobhan, *Genius at Play – The Curious Mind of John Horton Conway*, Bloomsbury USA, 2015, p. 251.

62 Heule, Marijn J. H., Oliver Kullmann, and Victor W. Marek, "Solving and Verifying the Boolean Pythagorean Triples problem via Cube-and-Conquer," May 3, 2016, available online at https://arxiv.org/abs/1605.00723.

takes about 30,000 hours of processor time.[63] This huge proof allowed the team to collect a prize from Ronald Graham, dating back to the 1980s, of $100.[64]

The previous record-holder for largest computer-assisted proof was from 2014 and took up 13 gigabytes to prove a special case of the Erdös discrepancy problem.[65] The more impressive general case was solved a year later, by hand, by Terence Tao (Figure 2.9).[66]

FIGURE 2.9
Terence Tao – better than the machines.

Reed Hutchinson/UCLA.

63 Lamb, Evelyn, "Two-Hundred-Terabyte Maths Proof Is Largest Ever," *Nature*, Vol. 534, No. 7605, May 26, 2016, available online at https://www.nature.com/news/two-hundred-terabyte-maths-proof-is-largest-ever-1.19990.

64 Lamb, Evelyn, "Two-Hundred-Terabyte Maths Proof Is Largest Ever," *Nature*, Vol. 534, No. 7605, May 26, 2016, available online at https://www.nature.com/news/two-hundred-terabyte-maths-proof-is-largest-ever-1.19990.

65 Lamb, Evelyn, "Two-Hundred-Terabyte Maths Proof Is Largest Ever," *Nature*, Vol. 534, No. 7605, May 26, 2016, available online at https://www.nature.com/news/two-hundred-terabyte-maths-proof-is-largest-ever-1.19990.

66 Tao, Terence, "The Erdős Discrepancy Problem," September 17, 2015, available online at https://arxiv.org/pdf/1509.05363.pdf.

Perhaps surprisingly, Tao is down to Earth and well-rounded with a wife and two children. A student of his remarked, "They will never make a movie about him. He doesn't have a troubled life. He has a family, and they seem happy, and he's usually smiling."[67] There's nothing bizarre other than his extreme mathematical talent to draw in viewers.

While Tao provides a ray of hope for those who fear the growth of computer proofs that no person can wrap his or her head around and check every detail, there are other places to turn for those with a preference for the simple.

Short Proofs

Some prime numbers are of the form $4n + 1$. That is, they are one more than a multiple of 4. Examples include $13 = 4 \cdot 3 + 1$ and $29 = 4 \cdot 7 + 1$. It turns out that all such primes can be written as the sum of two squares. For our examples, we have $13 = 2^2 + 3^2$ and $29 = 2^2 + 5^2$.

It requires some mathematics beyond the scope of this book, but there's a proof that this works for all primes of the form $4n + 1$ and this proof can be written in a single sentence of reasonable length.[68]

Israeli mathematician Saharon Shelah (Figure 2.10) went a bit too far, in my opinion, when it came to brevity, in some his proofs. The paper in question appeared in the second volume of a two-volume tribute to Paul Erdös. The claims and their proofs are reproduced, in full, below so you can judge for yourself.[69]

CLAIM 1.4:

(1) $J_{<\lambda}[\bar{\lambda}]$ is an ideal (of $\mathcal{P}(\kappa)$ i.e. on κ, but the ideal may not be proper).

(2) if $\lambda \leq \mu$, then $J_{<\lambda}[\bar{\lambda}] \subseteq J_{<\mu}[\bar{\lambda}]$

(3) if λ is singular, $J_{<\lambda}[\bar{\lambda}] = J_{<\lambda^+}[\bar{\lambda}] = J_{\leq\lambda}[\bar{\lambda}]$

(4) if $\lambda \notin \mathrm{pcf}(\bar{\lambda})$, then $J_{<\lambda}[\bar{\lambda}] = J_{\leq\lambda}[\bar{\lambda}]$.

(5) If $A \subseteq \kappa$, $A \notin J_{<\lambda}[\bar{\lambda}]$, and $f_\alpha \in \prod \bar{\lambda} \restriction A$, $\langle f_\alpha: \alpha < \lambda \rangle$ is $<_{J_{<\lambda}[\bar{\lambda}]}$-increasing cofinal in $(\prod \bar{\lambda} \restriction A)/J_{<\lambda}[\bar{\lambda}]$ then $A \in J_{\leq\lambda}[\bar{\lambda}]$. Also this holds when we replace $J_{<\lambda}[\bar{\lambda}]$ by any ideal J on κ, $I^* \subseteq J \subseteq J_{\leq\lambda}[\bar{\lambda}]$.

(6) The earlier parts hold for $J_{<\lambda}[\bar{\lambda}, \Gamma]$ too.

Proof: Straight.

---

67 Cook, Gareth, "The Singular Mind of Terry Tao," *The New York Times Magazine*, July 24, 2015, available online at https://www.nytimes.com/2015/07/26/magazine/the-singular-mind-of-terry-tao.html.

68 Young, Robert M., *Excursions in Calculus: An Interplay of the Continuous and the Discrete*, The Mathematical Association of America, 1992, p. 119 reproduces a one-sentence proof that every prime number of the form $4n + 1$ is the sum of two squares. The original publication was Zagier, Don, "A One-Sentence Proof That Every Prime $p \equiv 1 \pmod 4$ is a Sum of Two Squares," *American Mathematical Monthly*, Vol. 97, No. 2, February 1990, p. 144.

69 Graham, Ronald L., Jaroslav Nešetřil, and Steve Butler, editors, *The Mathematics of Paul Erdös*, Vol. II, second edition, Springer, New York, 2013, portions of p. 448 and p. 472 reproduced here. The paper is also available online at https://arxiv.org/pdf/math/9502233.pdf.

CLAIM 3.8: *If $D \subseteq E$ are filters on κ then*

$$|\prod_{i<\kappa}\lambda_i/D| \leq |\prod_{i<\kappa}\lambda_i/E| + \sup_{A\in E\smallsetminus D}|\prod_{i<\kappa}\lambda_i/(D+(\kappa\smallsetminus A))| + (2^\kappa/D) + \aleph_0.$$

We can replace $2^\kappa/D$ by $|\mathcal{P}|$ if $\mathcal{P}$ is a maximal subset of E such that $A \neq B \in \mathcal{P} \Rightarrow (A\smallsetminus B)\cup(B\smallsetminus A) \neq \emptyset \bmod D$.

Proof: Think.

FIGURE 2.10
Saharon Shelah.

For some theorems, we can do even better than a one-word proof (without sacrificing understanding!):

Proofs without Words

Theorem:

$$\frac{1}{4} + \frac{1}{16} + \frac{1}{64} + \cdots = \frac{1}{3}$$

Proof:[70]

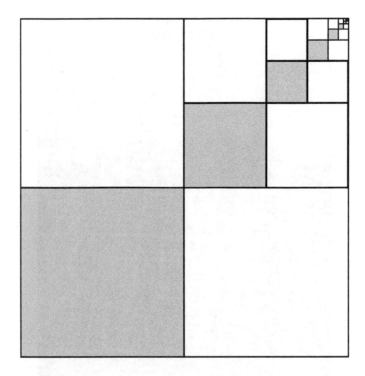

FIGURE 2.11
A proof without words.

Image created by Garrett Ruths.

Proofs like that of Figure 2.11 have been a regular feature in *Mathematics Magazine* since 1975 and three collections of them have seen print. See the References/Further Reading list for details.

---

70 This example was taken from https://jeremykun.com/tag/proofs-without-words/page/2/.

Exercise Set 2

> I think and think for months and years. Ninety-nine times, and the conclusion is false. The hundredth time I am right.
>
> Albert Einstein[71]

> Take it as axiomatic that you are stupid. If you think you have proved something, think again. Find the holes in your own proof.
>
> John Conway[72]

1. Prove that given any rational numbers, a and b, with $a < b$, there is another rational number, n, such that $a < n < b$.

2. Use the contrapositive to prove the following claim: If the product of two positive numbers a and b is greater than 100, then a is greater than 10 or b is greater than 10.

3. Referring back to $f(x) = x^2 + x + 41$. Will this function yield a prime for every natural number? Prove it does or provide a counterexample.

4. Prove that $4^n - 1$ is divisible by 3 for all $n \in \mathbb{N}$.

5. Use induction to prove

$$1^2 + 2^2 + 3^2 + 4^2 + 5^2 + \cdots + n^2 = n(n+1)(2n+1)/6$$

6. Use induction to prove $2^n \geq n + 10$ for $n \geq 4$.

7. Use induction to show that the sum of the first n even integers is $n^2 + n$.

8. Use induction to prove Bernoulli's inequality: $(1 + x)^n \geq 1 + nx$, if $x > -1$ and n is a nonnegative integer.

9. What is a terse way of representing $1^2 + 3^2 + 5^2 + \cdots + (2n - 1)^2$? No fair just compressing it using summation notation!

10. Use contradiction to prove that the sum of a rational and an irrational must be irrational.

11. Prove $\sqrt{3}$ is irrational.

12. Prove that $\sqrt{2} + \sqrt{3}$ is irrational.

13. Prove or disprove: The sum of two irrational numbers is always an irrational number.

14. Let a and b be irrational numbers. Show that either $a + b$ or $a - b$ is irrational.

15. Generalize the following examples and then use induction to prove your generalization holds.

$$1 = 1^2$$
$$1 + 3 = 2^2$$
$$1 + 3 + 5 = 3^2$$
$$1 + 3 + 5 + 7 = 4^2$$

---

71 Quote taken from *OMNI*, September 1981, p. 50.

72 Roberts, Siobhan, *Genius at Play – The Curious Mind of John Horton Conway*, Bloomsbury USA, 2015, p. 348.

16. Use the fact that $\sum_{p \text{ prime}} \frac{1}{p}$ diverges (i.e., eventually exceeds any given value)[73,74] to prove that there are infinitely many primes.

17. Suppose you have dominoes that are big enough for each to cover two squares on a chess board. Can you use 31 of these to completely cover the board, with the exception of two squares at opposite corners? It may not be obvious how to approach this problem, but that is the beauty of mathematical thinking – you can apply it to solve problems unlike any you've previously encountered.

18. Give a full, but completely different, proof of one of the results demonstrated in this chapter.

19. What is the flaw in the following "proof" by induction of the Fundamental Theorem of Arithmetic?

Theorem: Every integer $n \geq 2$ can be written uniquely as a product of prime numbers, up to order and sign.

The "up to order and sign part" means that, for example, while 35 can be written as $5 \cdot 7$, or $7 \cdot 5$, or even $(-5) \cdot (-7)$, there is no way to get 35 using other primes. The only fidgeting we can do is reordering the factors or changing the signs in pairs.

Base Case: The base case that needs to be established is for $n = 2$. Because 2 is a prime, we have an easy representation as the "product" of primes, with just one factor. Also, 2 is the smallest prime, so it cannot be expressed as the product of any other primes. Hence, its representation as a product of primes is unique. This establishes the base case.

Induction step: Suppose the statement is true for $n = 2, 3, 4, \ldots, k$, for some $k \geq 2$.

If $k + 1$ is prime, we have the unique representation of $k + 1$ as a product of primes given by itself.

If $k + 1$ is not prime, then it has a smallest prime divisor, call it p. We then have $k + 1 = p \cdot N$, for some $N < k$. But our induction assumption is that every integer between 2 and k, inclusive, can be expressed uniquely as a product of primes. Now, knowing that N can be written uniquely as a product of primes, we can include the factor p to have a unique representation for $k + 1$ as a product of primes. $\square$

References/Further Reading

> The more I study, the more insatiable do I feel my genius for it to be.
> – Ada Lovelace.

Beckmann, Petr, *A History of π (Pi)*, Dorset Press, New York, 1989.
belcastro, sarah-marie, *Discrete Mathematics with Ducks*, CRC Press, Boca, Raton, FL, 2012.
Bell, Eric Temple, *The Search for Truth*, Reynal & Hitchcock, New York, 1934.

---

73 For a proof that this sum does, in fact, diverge see Euler, Leonhard, "Variae observationes circa series infinitas" [Various observations concerning infinite series], *Commentarii Academiae Scientiarum Petropolitanae*, 1737, Vol. 9, pp. 160–188.

74 Taken here from Young, Robert M., *Excursions in Calculus: An Interplay of the Continuous and the Discrete*, The Mathematical Association of America, 1992, pp. 289–290. Pages 290–291 explain colorfully how slowly this summation diverges.

Bell, Eric Temple, *Mathematics, Queen and Servant of Science,* Copyright, 1951, British Edition 1952, reprinted 1954, 1958, 1961, Printed in Great Britain by Butler and Tanner Ltd., Frome and London.

Borwein, Jonathan (Jon) M., Peter B. Borwein, and David H. Bailey "Ramanujan, Modular Equations, and Approximations to Pi or How to Compute One Billion Digits of Pi," *American Mathematical Monthly,* Vol. 96, No. 3, March 1989, pp. 201–219.

Burger, Edward B. and Michael Starbird, *The Heart of Mathematics,* third edition, Wiley, Hoboken, NJ, 2010.

Chaitin, G., "Randomness and Mathematical Proof," *Scientific American,* Vol. 232, No. 5, May 1975, pp. 47–52.

Chamberland, Marc, *Single Digits: In Praise of Small Numbers,* Princeton University Press, Princeton, NJ, 2015.

Chang, Kenneth, "In Math. Computers Don't Lie. Or Do They?" *New York Times,* April 6, 2004, available online at www.nytimes.com/2004/04/06/science/in-math-computers-don-t-lie-or-do-they.html.

Cook, Gareth, "The Singular Mind of Terry Tao," *The New York Times Magazine,* July 24, 2015, available online at www.nytimes.com/2015/07/26/magazine/the-singular-mind-of-terry-tao.html.

Dawson, Jr., John W., *Why Prove It Again?: Alternative Proofs in Mathematical Practice,* Springer, Heidelberg, Germany, 2015, pp. 45–46 are on inductive proofs of the Fundamental Theorem of Arithmetic.

Doyle, Tim, Lauren Kutler, Robin Miller, and Albert Schueller, "Proofs without Words and Beyond," *Convergence,* August, 2014, available online at www.maa.org/press/periodicals/convergence/proofs-without-words-and-beyond.

Gardner, Martin, *Hexaflexagons and Other Mathematical Diversions, the First Scientific American Book of Puzzles & Games,* The University of Chicago Press, Chicago, IL, 1988.

Golomb, Solomon W., "Checker Boards and Polyominoes," *American Mathematical Monthly,* Vol. 61, 1954, pp. 675–682.

Graham, Ronald L., Jaroslav Nešetřil, and Steve Butler, editors, *The Mathematics of Paul Erdös,* Vol II, second edition, Springer, New York, 2013.

Grünbaum, A., "Whitehead's Method of Extensive Abstraction," *British Journal of the Philosophy of Science,* Vol. 4, No. 15, November 1953, pp. 215–226.

Halmos, Paul R., "How to Write Mathematics," *L'Enseignement Mathématique,* Vol. 16, 1970, pp. 123–152.

Hardy, G. H., *A Mathematician's Apology,* Cambridge University Press, Cambridge, 1940.

Hathout, Dean, *Wearing Gauss's Jersey,* CRC Press, Boca Raton, FL, 2013.

Hellman, Hal, *Great Feuds in Mathematics: Ten of the Liveliest Disputes Ever,* John Wiley & Sons, Inc., Hoboken, NJ, 2006.

Hempel, Carl G., "Le problème de la vérité," *Theoria (A Swedish Journal),* Vol. 3, 1937, pp. 206–246.

Krantz, Steven G., *The History and Concept of Mathematical Proof,* 2007, pp. 239–268, available online at www.math.wustl.edu/~sk/eolss.pdf.

Krantz, Steven G., *The Proof is in the Pudding: The Changing Nature of Mathematical Proof,* Springer, New York, 2011.

Lamb, Evelyn, "Two-hundred-terabyte Maths Proof Is Largest Ever," *Nature,* Vol. 534, No. 7605, May 26, 2016, www.nature.com/news/two-hundred-terabyte-maths-proof-is-largest-ever-1.19990.

Loomis, Elisha Scott, *The Pythagorean Proposition,* 1927, second edition 1940. The NCTM issued the second edition again in 1968.

Mancosu, Paolo, *From Brouwer to Hilbert,* Oxford University Press, New York, 1998.

Maor, Eli, *e: The Story of a Number,* ninth printing with new material on pp. 183–186, Princeton University Press, Princeton, NJ, 1998.

McKubre-Jordens, Maarten, "Constructive Mathematics," *Internet Encyclopedia of Philosophy,* available online at www.iep.utm.edu/con-math/.

Nelsen, Roger B., *Proofs without Words: Exercises in Visual Thinking*, MAA, Washington, DC., 1993.

Nelsen, Roger B., *Proofs without Words II: More Exercises in Visual Thinking*, MAA, Washington, DC, 2001.

Nelsen, Roger B., *Proofs without Words III: Further Exercises in Visual Thinking*, MAA, Washington, DC, 2016.

Paulos, John Allen, *Beyond Numeracy*, Vintage Books, New York, 1992.

Poincaré, Henri, *Science and Hypothesis*, Dover, New York, 1952. This is a republication of the first English translation, published in 1905 by the Walter Scott Publishing Company, Ltd.

Polya, George, *Mathematics and Plausible Reasoning, Volume I: Induction and Analogy in Mathematics*, Princeton University Press, Princeton, NJ, 1954.

Polya, George, *Mathematics and Plausible Reasoning: Volume II Patterns of Plausible Inference*, Princeton University Press, Princeton, NJ, 1954.

Polya, George, and John H. Conway, *How to Solve It: A New Aspect of Mathematical Method*, Princeton University Press, Princeton, NJ, reprint edition, 2014.

Reid, Constance, *Hilbert*, Springer, New York, 1970.

Roberts, Siobhan, *Genius at Play – The Curious Mind of John Horton Conway*, Bloomsbury, USA, 2015. This book includes Conway's involvement in classifying the finite simple groups.

Shaw, James Byrnie, "The Unity of Mathematics," *The Scientific Monthly*, Vol. 45, November 1937, pp. 402-411.

Tao, Terence, www.math.ucla.edu/~tao/.

Vellemen, Daniel J., *How to Prove it: A Structured Approach*, third edition, Cambridge University Press, Cambridge, 2019.

Wells, David, *The Penguin Dictionary of Curious and Interesting Numbers*, revised edition, Penguin Books, London, 1997.

Wilson, Robin and John J. Watkins, editors, *Combinatorics: Ancient & Modern*, Oxford University Press, Oxford, 2013.

Wolfram, Stephen, *Idea Makers*, Wolfram Media, Inc., Printed by Friesens, Manitoba, Canada, 2016.

Young, Robert M., *Excursions in Calculus: An Interplay of the Continuous and the Discrete*, The Mathematical Association of America, 1992.

Zeilberger, Doron, "Welcome to the Home Page of Doron Zeilberger's Servant," Rutgers University, http://sites.math.rutgers.edu/~zeilberg/ekhad.html.

Zeilberger, Doron, "The Personal Journal of Shalosh B. Ekhad and Doron Zeilberger," Rutgers University, available online at http://sites.math.rutgers.edu/~zeilberg/pj.html.

3

Practice with Proofs

In the last chapter, induction was used to prove $1 + 2 + 3 + 4 + 5 + \cdots + n = n(n + 1)/2$. At a very young age,[1] Carl Friedrich Gauss (1777–1855) and his classmates were challenged by their teacher with a more specific task, evaluating the sum $1 + 2 + 3 + 4 + 5 + \cdots + 99 + 100$. Of course, none of the children knew the formula, although it was old news even back then. They were supposed to do their calculations on the individual slates they had, turning them in at the teacher's desk when done. Gauss finished very quickly and placed his slate face down on the desk. The teacher didn't believe any kid could add the 100 numbers so rapidly and figured that Gauss was blowing off the assignment, but when he actually looked at the slate, he saw the correct answer, 5,050.

Gauss had approached the problem by realizing that there was a method much quicker than adding the integers one at a time, starting with 1. Instead he looked at $1 + 100$, $2 + 99$, $3 + 98$, etc. Each of these pairs summed to 101 and there were a total of 50 such pairs, so the answer had to be $(101)(50) = 5,050$.

This approach generalizes nicely. To find the sum of $1 + 2 + 3 + 4 + 5 + \cdots + n$, simply take the pairs $1 + n$, $2 + (n - 1)$, $3 + (n - 2)$, etc. Each pair sums to $n + 1$ and there are $n/2$ pairs, so the total must be $(n + 1)(n/2)$, which is $n(n + 1)/2$. Approaching the problem in this manner is not only shorter than the proof by induction, but is also better in another way – we don't need to know the result ahead of time. Proofs like this are known as **derivations**. They have nothing to do with the derivatives of calculus, but refer to the fact that the result is derived or "figured out" as the proof progresses. A big drawback of proof by induction is that you need to know the result before you can even begin the proof.

Some readers might object to the derivation above, pointing out that it doesn't work if n is odd. They are right! When n is odd, we can't pair all of the numbers being summed; one will be left over. Fortunately, there's a similar approach that works whether n is even or odd. Start by writing the sum out twice, first in ascending order, then in descending order:

$$1 + \quad 2 \quad + \quad 3 \quad + \cdots + (n - 2) + (n - 1) \quad + n$$
$$n + (n - 1) + (n - 2) + \cdots + \quad 3 \quad + \quad 2 \quad + 1$$

Then add each number in the top row to the number directly below. The sum is always $(n + 1)$. There are n such pairs, so the total is $(n + 1)(n)$. But we wrote the original list of numbers twice, so our result is twice the value we seek. Halving it, we get the correct answer of $(n + 1)(n)/2$.

---

[1] The age varies, depending on the source. Gauss may have been 7 or 8.

There are other ways to derive this result that are completely different. A geometric proof is presented next, followed by a method that can be used to establish many more results.

A Geometric Proof[2]

Although the original problem is not about circles or rectangles, we can attack it with geometry. The numbers 1, 2, 3, 4, 5 can be arranged like so:

```
        *
      * *
     * * *
    * * * *
   * * * * *
```

The first (vertical) column has one star, the second two, and so on, ending with five stars in the fifth column. This makes a triangle. Now placing an identical (but flipped upside-down) triangle on top of it, as shown below, we get a rectangle:

```
     * * * * *
     * * * * *
     * * * * *
     * * * * *
     * * * * *
     * * * * *
```

Because it's a 5 by 6 rectangle, the total number of stars is $5 \cdot 6 = 30$. We are only interested in the original stars, which are half of the total, so we divide 30 by 2 to get 15. We've used geometry to establish that $1 + 2 + 3 + 4 + 5 = 15$. This generalized nicely. If we wanted to sum the first n positive integers, our rectangle would have n columns and $n + 1$ rows, so the total number of stars would be $(n)(n + 1)$ and we would once again divide by 2 to count the stars we are interested in. This establishes that $1 + 2 + 3 + 4 + 5 + \cdots + n = n(n + 1)/2$.

Because the sum of the first n integers can always be drawn as a triangle, these numbers are called **triangular**. Similarly, if a given number of stars can be arranged to make a rectangle, pentagon, hexagon, etc., then we say that the number is **rectangular, pentagonal, hexagonal**, and so on. The Pythagoreans cared about such things and found many theorems concerning connections between such **figurate numbers** (a general term). This may seem silly and antiquated.[3] Surely, today's mathematicians don't concern themselves with the "shapes" of numbers. Or do they? Have you ever heard someone refer to a number like 4, 9, or 25 as "a perfect square"? Some problems concerning figurate numbers are given in the exercises and we'll encounter them several more times in later chapters.

---

2 This proof has been described in many other books. See, for example, Chamberland, Marc, *Single Digits*, Princeton University Press, Princeton, NJ, 2015, pp. 71–72.

3 It wasn't just the Pythagoreans. Other ancients took up such problems. For example, Diophantus of Alexandria wrote a book titled *On Polygonal Numbers* around 250 CE.

A Proof with Potential

There's a proof that's lengthier, but has an advantage the others do not. It begins with an expression that doesn't look useful at first glance. It sums *squares* of consecutive integers:

$$\sum_{k=1}^{n} (k-1)^2.$$

We can expand the binomial to get

$$\sum_{k=1}^{n} (k-1)^2 = \sum_{k=1}^{n} (k^2 - 2k + 1) = \sum_{k=1}^{n} k^2 - 2\sum_{k=1}^{n} k + \sum_{k=1}^{n} 1.$$

The left-hand side is $0^2 + 1^2 + 2^2 + 3^2 + 4^2 + 5^2 + \cdots + (n-1)^2$ and the first summation on the right-hand side is $1^2 + 2^2 + 3^2 + 4^2 + 5^2 + \cdots + n^2$, so nearly all of these terms cancel out. Equating these two sides after performing the cancellations gives

$$0 = n^2 - 2\sum_{k=1}^{n} k + \sum_{k=1}^{n} 1.$$

We can then perform a few steps to isolate the summation that we are interested in:

$$2\sum_{k=1}^{n} k = n^2 + \sum_{k=1}^{n} 1$$

$$\sum_{k=1}^{n} k = \frac{n^2}{2} + \frac{1}{2}\sum_{k=1}^{n} 1.$$

The rightmost summation is just n ones added together, which gives n. We have

$$\sum_{k=1}^{n} k = \frac{n^2}{2} + \frac{1}{2}n.$$

A bit more arithmetic gives us the familiar result:

$$\sum_{k=1}^{n} k = \frac{n^2}{2} + \frac{n}{2} = \frac{n^2 + n}{2} = \frac{(n+1)(n)}{2}.$$

As mentioned previously, this result was far from new in Gauss's day. It was known, for example, to Abraham ben Meir ibn Ezra (1092–1167), a rabbi and a mathematician who lived in Muslim Spain, although he certainly wasn't aware of all of the proofs shown above.[4]

Next Steps

Which of the proofs above did you like best? Gauss's proof and generalizations thereof are very simple, as is the geometric proof, but there isn't an obvious way to adapt them to derive a formula for $1^2 + 2^2 + 3^2 + \cdots + n^2$. Induction offers a straightforward proof, but we would need to know the formula in order to apply that proof technique. I like the last proof presented above the best because it's easy to modify it to tackle this new problem. That is, the same approach can be used to derive the higher power formulas. To get the formula for the sum of the consecutive squares, we start off with a summation of *cubes*:

$$\sum_{k=1}^{n} (k-1)^3.$$

We can expand the binomial to get

$$\sum_{k=1}^{n} (k-1)^3 = \sum_{k=1}^{n} (k^3 - 3k^2 + 3k - 1) = \sum_{k=1}^{n} k^3 - 3\sum_{k=1}^{n} k^2 + 3\sum_{k=1}^{n} k - \sum_{k=1}^{n} 1.$$

The left-hand side is $0^3 + 1^3 + 2^3 + 3^3 + \cdots + (n-1)^3$ and the first summation on the right-hand side is $1^3 + 2^3 + 3^3 + \cdots + n^3$, so nearly all of these terms cancel out. Equating these two sides after performing the cancellations gives

$$0 = n^3 - 3\sum_{k=1}^{n} k^2 + 3\sum_{k=1}^{n} k - \sum_{k=1}^{n} 1.$$

We can then perform a few steps to isolate the summation that we are interested in:

$$3\sum_{k=1}^{n} k^2 = n^3 + 3\sum_{k=1}^{n} k - \sum_{k=1}^{n} 1$$

$$\sum_{k=1}^{n} k^2 = \frac{n^3}{3} + \sum_{k=1}^{n} k - \frac{1}{3}\sum_{k=1}^{n} 1.$$

The rightmost summation is just n ones added together, which gives n. We have

---

4 Katz, Victor J., "Jewish combinatorics," in Wilson, Robin and John J. Watkins, editors, *Combinatorics: Ancient & Modern*, Oxford University Press, Oxford, 2013, pp. 109–121, p. 114 cited here. For more on ibn Ezra see http://www-history.mcs.st-andrews.ac.uk/Biographies/Ezra.html.

$$\sum_{k=1}^{n} k^2 = \frac{n^3}{3} + \sum_{k=1}^{n} k - \frac{n}{3}.$$

We still have one summation left on the right-hand side, but this is just a sum of consecutive integers, a sum that we already evaluated by the method we are now applying, so plugging in that result we get

$$\sum_{k=1}^{n} k^2 = \frac{n^3}{3} + \frac{(n+1)(n)}{2} - \frac{n}{3}.$$

A bit more arithmetic gives us the simplified result:

$$\sum_{k=1}^{n} k^2 = \frac{2n^3}{6} + \frac{3(n+1)(n)}{6} - \frac{2n}{6} = \frac{2n^3 + 3(n+1)(n) - 2n}{6} = \frac{2n^3 + 3n^2 - n}{6}.$$

Factoring an n out of the numerator on the right-hand side, and then factoring the remaining quadratic, gives

$$\sum_{k=1}^{n} k^2 = \frac{(n)\left(2n^2 + 3n + 1\right)}{6} = \frac{(n)(2n+1)(n+1)}{6}.$$

This formula is very old and was known, for example, to Ch'on Huo (1011–1075) and Yang Hui (about 1260), although they did not prove it in the manner shown above.[5] Notice that determining the sums of the squares by this method required knowing the sums of the first powers. Similarly, if we try this approach for the sums of the cubes, we will need to make use of the previously obtained formulas for the sum of the squares and the sum of the first powers. This isn't a problem, if we wish to generate the various formulas in order. We can generate as many of them as they wish, but each will require having correctly obtained all previous formulas. While this is tedious, it doesn't require us to know any of the formulas we are trying to find in advance, like induction does. We are *deriving* them, one a time.

When we determine new results for a problem by using previous results, as above, we are making use of **recursion**. This is a very powerful technique and it will be seen throughout this book, with a close look taken in the chapters on difference equations, also known as recurrence relations. The recursive approach to the sums of powers of consecutive integers was first taken by Fermat in 1636. The method used above, however, is much closer to the manner in which Pascal derived the formulas in 1654.[6]

---

5 Hofmann, Joseph Ehrenfried, *The History of Mathematics*, translated from the German *Geschichte der Mathematik*, The Philosophical Library, New York, 1957, p. 46.

6 See Young, Robert M., *Excursions in Calculus: An Interplay of the Continuous and the Discrete*, The Mathematical Association of America, 1992, pp. 82–83 for the methods of both Fermat and Pascal. Also see https://nrich.maths.org/267 for Pascal's approach.

The results are as follows.[7] Our previous results are repeated for completeness. All may be derived by the method shown above. Using induction to prove these formulas has been called "simple but sterile:"[8]

$$1 + 2 + 3 + \cdots + n = n(n+1)/2$$

$$1^2 + 2^2 + 3^2 + \cdots + n^2 = n(n+1)(2n+1)/6$$

$$1^3 + 2^3 + 3^3 + \cdots + n^3 = n^2(n+1)^2/4$$

$$1^4 + 2^4 + 3^4 + \cdots + n^4 = n(n+1)(2n+1)(3n^2 + 3n - 1)/30$$

$$1^5 + 2^5 + 3^5 + \cdots + n^5 = n^2(n+1)^2(2n^2 + 2n - 1)/12$$

$$1^6 + 2^6 + 3^6 + \cdots + n^6 = n(n+1)(2n+1)(3n^4 + 6n^3 - n^2 - 3n + 1)/42$$

$$1^7 + 2^7 + 3^7 + \cdots + n^7 = n^2(n+1)^2(3n^4 + 6n^3 - n^2 - 4n + 2)/24$$

$$1^8 + 2^8 + 3^8 + \cdots + n^8 = n(n+1)(2n+1)(5n^6 + 15n^5 + 5n^4 - 15n^3 - n^2 + 9n - 3)/90$$

$$1^9 + 2^9 + 3^9 + \cdots + n^9 = n^2(n+1)^2(2n^6 + 6n^5 + n^4 - 8n^3 + n^2 + 6n - 3)/20$$

$$1^{10} + 2^{10} + 3^{10} + \cdots + n^{10} = n(n+1)(2n+1)(n^2 + n - 1)(3n^6 + 9n^5 + 2n^4 - 11n^3 + 3n^2 + 10n - 5)/66.$$

While it's difficult to determine when some of these formulas were first found (by whatever means), we do know that the sum of consecutive cubes formula, like the simpler cases, goes back to antiquity.[9] It was used by Ibn al-Haytham (965–1040) in the special case $n = 4$, when he tried to find the volume of a paraboloid, and more than 800 years earlier it was already known to Nicomachus (60–120).[10] The formula for the sum of fourth powers was known to al-Kachi (1394–1437), who may have been the first to find it.[11]

Of course, if we wish to jump ahead to find the formula for the sum of the 100th powers, we cannot do so with recursion, we must find the previous 99 formulas first. But, if you are clever, examining the first 10 formulas could be sufficient for you to find a shortcut. We will return to this idea soon, but for now we take a look at some places where sums of powers of integers arise, answering the question of "When would I ever need formulas like these?"

7 Rosen, Kenneth H., editor, *Handbook of Discrete and Combinatorial Mathematics*, CRC Press, Boca Raton, FL, 2000, p.200.

8 Young, Robert M., *Excursions in Calculus: An Interplay of the Continuous and the Discrete*, The Mathematical Association of America, 1992, p. 78.

9 Young, Robert M., *Excursions in Calculus: An Interplay of the Continuous and the Discrete*, The Mathematical Association of America, 1992, p. 78.

10 https://oeis.org/A000537.

11 https://oeis.org/A000538.

Applications

Example 1 – Counting Squares on a Chess Board[12]

How many squares are there on a chessboard? A simple answer could be 64, but suppose we wish to count not just the individual squares, but larger squares that can be formed from them, such as 2 × 2s (consisting of four small squares), 3 × 3s (consisting of nine small squares), all the way up to the 8 × 8 square (the entire board). Then how many are there?

The answer is $1^2 + 2^2 + 3^2 + \cdots + 8^2 = 204$. This result generalizes nicely. If we're using a board that is not 8 × 8, but rather $n \times n$, then the total number of squares is $1^2 + 2^2 + 3^2 + \cdots + n^2 = \frac{1}{3}n^3 + \frac{1}{2}n^2 + \frac{1}{6}n$.

Example 2 – Stacking Cannonballs

There are various ways to stack cannonballs, one of which is to form a square for the base and then place smaller squares of cannonballs on top of them, repeatedly, to finally end with a single cannonball on top. An example of this kind of stacking is shown in Figure 3.1.

It could be difficult to try to count cannonballs in such an arrangement one at a time, as many of them are not visible, but the total at each level is an integer squared. So, our

FIGURE 3.1
One way to stack cannonballs.

---

12 I found this problem in the magazine *OMNI*. The question was posed in the May 1982, issue, p. 153 and answered in June 1982 on p. 147, both appearances being part of Scot Morris's "Games" column. This is not the first place the problem appeared.

grand total is $1^2 + 2^2 + 3^2 + 4^2 + 5^2 = 55$. Like the previous example, this result generalizes to any size pile.

Example 3 – Predicting a Noble Gas

While the formula I'm about to present is not identical to any we've examined thus far, it is similar. I found it in a 1928 issue of *The Scientific Monthly*.[13] Back then, only 92 elements were known. Scientists didn't yet have the ability to create any of the elements beyond uranium, number 92 on the periodic table. With this in mind, read the reproduced page from *The Scientific Monthly* that follows (Figure 3.2).

HYPON—A HYPOTHETICAL ELEMENT AND A POSSIBLE SOURCE OF STELLAR ENERGY

By W. S. ANDREWS
THE GENERAL ELECTRIC COMPANY

THE source of the energy that is continuously radiated into space chiefly in the form of heat and light by our sun and other stars is an unsettled problem although many ingenious theories have been devised to account for it. The evolution of temporary stars or "novas" is also still enshrouded in mystery. It was formerly believed that a nova was a cosmic conflagration caused by the collision of two heavenly bodies, but modern study has ascribed its appearance to the actual explosion of a star from within itself, like a gigantic bombshell. A few facts—and fancies—relating to these matters are here presented to the reader.

It is well known that the atomic numbers of six gases—helium 2, neon 10, argon 18, krypton 36, xenon 54 and radon 86—agree with the first six terms of the following simple number series:
$2(1^2 + 2^2 + 2^2 + 3^2 + 3^2 + 4^2 + 4^2 + \ldots)$.

That is to say:

$2(1^2) = 2$, helium's number
$2(1^2 + 2^2) = 10$, neon's number
$2(1^2 + 2^2 + 2^2) = 18$, argon's number
$2(1^2 + 2^2 + 2^2 + 3^2) = 36$, krypton's number
$2(1^2 + 2^2 + 2^2 + 3^2 + 3^2) = 54$, xenon's number
$2(1^2 + 2^2 + 2^2 + 3^2 + 3^2 + 4^2) = 86$, radon's number

This remarkable agreement in numbers, up to and including the sixth term of the number series, invites attention to the seventh term, which is 118 $[2(1^2 + 2^2 + 2^2 + 3^2 + 3^2 + 4^2 + 4^2) = 118]$. There are, however, only ninety-two known chemical elements, and uranium being the heaviest is atomic number 92. It is, therefore, plain that 118 can not be counted as an atomic number unless, indeed, twenty-six new and hypothetical elements, all heavier than uranium, are arbitrarily added to the regular stand-

grumbled that they did not see how there could be such an influence at a distance; but it was generally agreed that the influence was communicated through the ocean and might be better understood when more was known about the nature of water."

FIGURE 3.2
Discrete math applied to chemistry.

Public domain.

13 Andrews, W. S., "Hypon – A Hypothetical Element and a Possible Source of Stellar Energy," *The Scientific Monthly*, Vol. 27, December 1928, pp. 535–537. Note: Niels Bohr made the same prediction in 1922. See https://en.wikipedia.org/wiki/Oganesson#cite_note-leach-18, which, in turn, cites Leach, Mark R. "The INTERNET Database of Periodic Tables," available online at http://www.meta-synthesis.com/webbook/ 35_pt/pt_database.php?PT_id=285.

What hutzpah! Extrapolating all the way out from 92 to 118 solely on the basis of a mathematical pattern! *If 26 more elements are discovered/created (!!!)*, then here's what the last of these will be like … Well, the mathematical prediction may have turned out to be correct. Element 118 was synthesized in 2002 and named Oganesson, in 2016, after the Russian nuclear physicist Yuri Oganessian. Its properties are still being investigated.

Example 4 – Counting Rectangles on a Chess Board[14]

Example 1 considered the number of squares that can be found on a chess board. Loosening the requirement to merely finding rectangles, there are clearly many more possibilities. It turns out that the sum of consecutive cubes formula answers this geometric question.

On a chess board we have $1^3 + 2^3 + 3^3 + \cdots + 8^3 = 1,296$ rectangles in total. The result generalizes. For an $n \times n$ board, there are $1^3 + 2^3 + 3^3 + \cdots + n^3 = \frac{1}{4}n^4 + \frac{1}{2}n^3 + \frac{1}{4}n^2$ rectangles.

A Connection with Calculus

Sequences of sums of powers of consecutive integers arise when using Riemann sums to approximate the area under curves of the form $f(x) = x^n$. This is a discrete problem, but taking the limit as the number of rectangles approaches infinity (equivalently, as their width approaches zero) turns it into a problem in continuous mathematics.

As a quick example, consider the area under the curve $f(x) = x^2$ between $x = 0$ and $x = 1$. We can approximate this area with 10 rectangles, as shown in Figure 3.3.

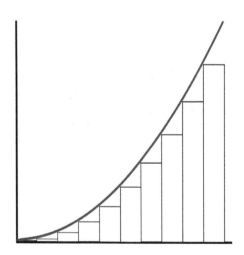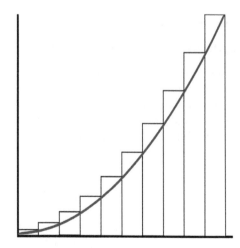

FIGURE 3.3
Left (left) and right (right) Riemann sums for $f(x) = x^2$ with ten rectangles.

Image created by Josh Gross.

---

14 https://oeis.org/A000537.

The sketch on the left uses a left Riemann sum and the sketch on the right uses a right Riemann sum. Because the function $f(x) = x^2$ is strictly increasing, the left sum is an underestimate and the right sum is an overestimate. The calculations are as follows:

$$\left(\frac{1}{10}\right)\left(\left(\frac{0}{10}\right)^2 + \left(\frac{1}{10}\right)^2 + \left(\frac{2}{10}\right)^2 + \cdots + \left(\frac{9}{10}\right)^2\right) < \text{area}$$

$$\text{area} < \left(\frac{1}{10}\right)\left(\left(\frac{1}{10}\right)^2 + \left(\frac{2}{10}\right)^2 + \left(\frac{3}{10}\right)^2 + \cdots + \left(\frac{10}{10}\right)^2\right).$$

This simplifies to

$$\frac{1}{10^3}\left(0^2 + 1^2 + 2^2 + \cdots + 9^2\right) < \text{area} < \frac{1}{10^3}\left(1^2 + 2^2 + 3^2 + \cdots + 10^2\right)$$

A calculator then reveals $57/200 < \text{area} < 77/200$.
If we used 100 rectangles (not shown due to resolution difficulties!), we get

$$\frac{1}{100^3}\left(0^2 + 1^2 + 2^2 + \cdots + 99^2\right) < \text{area} < \frac{1}{100^3}\left(1^2 + 2^2 + 3^2 + \cdots + 100^2\right).$$

I wouldn't trust myself to correctly enter such sums of squares into a calculator! But the formula we found provides a nice shortcut that is easily managed. Calculus provides a different sort of shortcut, via limits and continuous mathematics.[15]

Patterns in Summation Formulas

What patterns can you find behind these summation formulas? When looking for patterns, it is often a good idea to represent the sample data you have in as many different ways as possible. One representation might suggest a pattern much more readily than another. For our summation formulas, an alternate representation would be looking at the answers as polynomials, as in the applications, instead of in factored form:

$$1 + 2 + 3 + \cdots + n = \frac{1}{2}n^2 + \frac{1}{2}n$$

$$1^2 + 2^2 + 3^2 + \cdots + n^2 = \frac{1}{3}n^3 + \frac{1}{2}n^2 + \frac{1}{6}n$$

---

15 See Young, Robert M., *Excursions in Calculus: An Interplay of the Continuous and the Discrete*, The Mathematical Association of America, 1992, pp. 84–85, for how Fermat and Pascal reasoned about this, before Newton and Leibniz.

$$1^3 + 2^3 + 3^3 + \cdots + n^3 = \frac{1}{4}n^4 + \frac{1}{2}n^3 + \frac{1}{4}n^2$$

$$1^4 + 2^4 + 3^4 + \cdots + n^4 = \frac{1}{5}n^5 + \frac{1}{2}n^4 + \frac{1}{3}n^3 - \frac{1}{30}n$$

$$1^5 + 2^5 + 3^5 + \cdots + n^5 = \frac{1}{6}n^6 + \frac{1}{2}n^5 + \frac{5}{12}n^4 - \frac{1}{12}n^2$$

$$1^6 + 2^6 + 3^6 + \cdots + n^6 = \frac{1}{7}n^7 + \frac{1}{2}n^6 + \frac{1}{2}n^5 - \frac{1}{6}n^3 + \frac{1}{42}n$$

$$1^7 + 2^7 + 3^7 + \cdots + n^7 = \frac{1}{8}n^8 + \frac{1}{2}n^7 + \frac{7}{12}n^6 - \frac{7}{24}n^4 + \frac{1}{12}n^2$$

$$1^8 + 2^8 + 3^8 + \cdots + n^8 = \frac{1}{9}n^9 + \frac{1}{2}n^8 + \frac{2}{3}n^7 - \frac{7}{15}n^5 + \frac{2}{9}n^3 - \frac{1}{30}n$$

$$1^9 + 2^9 + 3^9 + \cdots + n^9 = \frac{1}{10}n^{10} + \frac{1}{2}n^9 + \frac{3}{4}n^8 - \frac{7}{10}n^6 + \frac{1}{2}n^4 - \frac{3}{20}n^2$$

$$1^{10} + 2^{10} + 3^{10} + \cdots + n^{10} = \frac{1}{11}n^{11} + \frac{1}{2}n^{10} + \frac{5}{6}n^9 - 1n^7 + 1n^5 - \frac{1}{2}n^3 - \frac{5}{66}n.$$

The German mathematician Johann Faulhaber (Figure 3.4) studied these sums and published formulas for them all the way up to the 17th power, a record at the time, without clearly explaining the general pattern. Can you do more with less?[16]

It was a Swiss mathematician, Jacob Bernoulli (Figure 3.5), who finally found and published a nice solution.[17]

Bernoulli presented the equations above, using his own notation and, unfortunately, including a mistake (corrected here), in his book *Ars conjectandi*. It saw print in 1713, after his death, so if anyone complained about the errors, Bernoulli wasn't around to hear it. Instead of explicitly giving the rule he found to continue with equations for higher powers, he wrote

> Whoever will examine the series as to their regularity may be able to continue the table.[18]

Can you do it? If not, try to at least chip away at the problem. If you cannot completely determine the rule for 11th powers, can you at least say something about it? Many published math papers merely offer partial solutions, so do *something*!

---

16 See https://www.maa.org/press/periodicals/convergence/sums-of-powers-of-positive-integers-introduc tion and Wilson, Robin and John J. Watkins, editors, *Combinatorics: Ancient & Modern*, Oxford University Press, Oxford, 2013, p. 157.

17 Jacob is sometimes spelled Jakob or Jacobi.

18 Taken here from Young, Robert M., *Excursions in Calculus: An Interplay of the Continuous and the Discrete*, The Mathematical Association of America, 1992, p. 86.

FIGURE 3.4
Johann Faulhaber (1580–1635).

Image created by Clemens Ammon. This file is licensed under the Creative Commons Attribution-Share Alike 3.0 Unported license.

Sums of Reciprocal Powers

It turns out that summing powers of integers is, in general, way easier than summing powers of *reciprocals* of integers. For example, finding a formula for

$$\frac{1}{1^2} + \frac{1}{2^2} + \frac{1}{3^2} + \cdots$$

proved to be a tremendous challenge. It was Euler who finally found the surprising answer $\frac{\pi^2}{6}$.

FIGURE 3.5
Jacob Bernoulli (1654–1705).

Public domain.

Open Problems

Triangular and Factorial

The triangular numbers 1, 6, and 120 have a special property. They can all be expressed as factorials. We have $1 = 1!$, $6 = 3!$, and $120 = 5!$. Are there any other triangular numbers that have this property? No one knows![19]

Always a Prime?

The sequence of triangular numbers begins 1, 3, 6, 10, 15, 21, 28, 36, ... Between 1 and 3 we have the prime number 2. Between 3 and 6, we have the prime 5. Between 6 and 10, we have the prime 7. The next pair of triangular numbers, 10 and 15, have two primes

---

19 See the comments section of https://oeis.org/A000217 for a mention of the conjecture.

between them, namely 11 and 13. It seems that we can find at least one prime between any pair of consecutive triangular numbers. Or can we? It's still only a conjecture. Maybe there's an exception far out in the sequence.[20]

Just as we don't know if there's always a prime between consecutive triangular numbers, we don't know if there is always a prime between consecutive squares.[21]

Exercise Set 3

> I have yet to see a problem, however complicated, which when you looked at it in the right way, did not become still more complicated.
>
> Poul Anderson

1. Letting T_n represent the nth triangular number, prove $T_n + T_{n+1} = (n + 1)^2$.
2. Prove $\left(T_n - T_{n-1}\right)^2 = n^2$.
3. Prove $T_n^2 - T_{n-1}^2 = n^3$.
4. Prove $T_{a+b} = T_a + T_b + ab$
5. Prove $T_{ab} = T_a T_b + T_{a-1} T_{b-1}$.
6. Prove $1^3 + 2^3 + 3^3 + \cdots + n^3 = (1 + 2 + 3 + \cdots + n)^2$ algebraically.
7. Explain how color figure 5 **(See color insert)** can be used to prove that

$$(1 + 2 + 3 + \cdots + n)^2 = 1^3 + 2^3 + 3^3 + \cdots + n^3.$$

This is known as Nicomachus's theorem, after Nicomachus of Gerasa (c. 60–120). The proof method suggested by the figure was known to Arab mathematicians almost 1,000 years ago.[22]

8. One can go from $\sum_{i=1}^{n} i$ to $\sum_{i=1}^{n} i^2$ and $\sum_{i=1}^{n} i^3$ and other powers, but there are other directions in which the original summation can be taken. For example, we have $\sum_{i=1}^{n} \frac{(i)(i+1)}{2!} = \frac{n(n+1)(n+2)}{3!}$ and $\sum_{i=1}^{n} \frac{(i)(i+1)(i+2)}{3!} = \frac{n(n+1)(n+2)(n+3)}{4!}$. Prove these using induction.[24,25]

9. On p. 312 of *La Verite des Sciences contre les Sceptiques ou Pyrrhoniens* (The Truth of the Sciences against the Sceptics or Pyrrhonians), Mersenne gave a table of figurate numbers (Figure 3.6).[23]

    Fill in another row and column.

---

20 See the comments section of https://oeis.org/A000217 for a mention of the conjecture, as well as https://oeis.org/A065383 for a related sequence (the sequence of smallest primes between triangular numbers).

21 Hoffman, Paul, *Archimedes' Revenge: The Joys and Perils of Mathematics*, W. W. Norton & Company, New York, 1988.

22 Young, Robert M., *Excursions in Calculus: An Interplay of the Continuous and the Discrete*, The Mathematical Association of America, 1992, p. 81.

23 https://books.google.com/books?id=FF44pHy9x_sC&printsec=frontcover&hl=nl&source=gbs_ge_summary_r&cad=0#v=onepage&q&f=false.

24 Although they are expressed in modern notation, these patterns were first found by Fermat. He did not publish a proof of them.

25 Taken here from Young, Robert M., *Excursions in Calculus: An Interplay of the Continuous and the Discrete*, The Mathematical Association of America, 1992, pp. 76–77.

Ordre na. 1. 2. 3. 4. 5. 6. 7. 8. 9. 10. 11.
Triangu. 1. 3. 6. 10. 15. 21. 28. 36. 45. 55. 66.
Quarrez 1. 4. 9. 16. 25. 36. 49. 64. 81. 100. 121.
Pentago. 1. 5. 12. 24. 35. 51. 70. 92. 117. 145. 176.
Exagones 1. 6. 15. 30 45. 60. 91. 120. 153. 190. 241.
Heptago. 1. 7. 18. 36. 55. 81. 112. 148. 189. 235. 296.
Octogon. 1. 8. 21. 42 65. 96. 133. 176. 225. 280. 251.

FIGURE 3.6
Mersenne's figurate numbers table.

Public domain.

References/Further Reading

I find television very educating. Every time somebody turns on the set, I go into the other room and read a book.

– Groucho Marx.

Andrews, W. S., "Hypon – A Hypothetical Element and a possible Source of Stellar Energy," *The Scientific Monthly*, Vol. 27, December 1928, pp. 535–537.

Beery, Janet, "Sums of Powers of Positive Integers – Introduction," *Convergence*, July 2010, available online at www.maa.org/press/periodicals/convergence/sums-of-powers-of-positive-inte gers-introduction.

Bernoulli, Jacobi, *Ars Conjectandi*, Thurnisii fratres, Basel, 1713.

Chamberland, Marc, *Single Digits*, Princeton University Press, Princeton, NJ, 2015.

Dawson, Jr., John W., *Why Prove It Again?: Alternative Proofs in Mathematical Practice*, Springer, Heidelberg, Germany, 2015, pp. 13–18 have a geometric proof of the sum of the squares of the first *n* positive integers.

Hayes, Brian, "Gauss's Day of Reckoning," *American Scientist*, Vol. 94, No. 3, May–June 2006, pp. 200–205.

Hoffman, Paul, *Archimedes' Revenge: The Joys and Perils of Mathematics*, W. W. Norton & Company, New York, 1988.

Hofmann, Joseph Ehrenfried, *The History of Mathematics*, translated from the German *Geschichte der Mathematik*, The Philosophical Library, New York, 1957.

Katz, Victor J., "Jewish Combinatorics," in Wilson, Robin and John J. Watkins, editors, *Combinatorics: Ancient & Modern*, Oxford University Press, Oxford, 2013, pp. 109–121.

Knuth, Donald E., "Johann Faulhaber and Sums of Powers," *Mathematics of Computation*, Vol. 61, No. 203, July 1993, pp. 277–294.

Nelsen, Roger B., *Proofs without Words: Exercises in Visual Thinking*, MAA, Washington, DC, 1993, p. 77 has a geometric proof of the sum of the squares of the first *n* positive integers.

Pólya, George, *Mathematical Discovery*, 2 volumes, Wiley, New York, 1962. See volume 1, p. 60, for Gauss anecdote.

Rosen, Kenneth H., editor, *Handbook of Discrete and Combinatorial Mathematics*, CRC Press, Boca Raton, FL, 2000.

Sloane, N. J. A., "Square Pyramidal Numbers," *The On-Line Encyclopedia of Integer Sequences*, available online at https://oeis.org/A000330. This is the sequence for sums of consecutive squares.

Sloane, N. J. A., "Sum of First n Cubes; or n-th Triangular Number Squared," *The On-Line Encyclopedia of Integer Sequences*, available online at https://oeis.org/A000537.

Sloane, N. J. A., "Triangular Numbers," *The On-Line Encyclopedia of Integer Sequences*, available online at https://oeis.org/A000217.

Wikipedia contributors, "Faulhaber's Formula," *Wikipedia, The Free Encyclopedia*, available online at https://en.wikipedia.org/w/index.php?title=Faulhaber%27s_formula&oldid=818365118.

Young, Robert M., *Excursions in Calculus: An Interplay of the Continuous and the Discrete*, The Mathematical Association of America, 1992.

4

Set Theory

A **set** is a collection of distinct objects.

Simple as this definition is, it seems to have been a long time in coming. Indeed, our very language shows that this was an alien concept, a level of abstraction not achieved!

We speak of a *school* of fish, a *pod* of dolphins, a *pack* of wolves, a *flock* of geese, a *pride* of lions, a *rookery* of penguins, etc. They are all sets. Why have a different word for each kind of animal that can be grouped? Without the abstract concept of a set, each group of animals may have been thought of as a distinct concept. The use of such collective nouns extends beyond animals. We also have a *flight* of stairs, a *deck* of cards, a *cluster* of computers, etc.[1]

Typically, credit for developing set theory is given to Georg Cantor (Figure 4.1), even though others had hit upon a few of his ideas years, even centuries earlier. Those who anticipated the subject will be mentioned later in this chapter. First on this chapter's

FIGURE 4.1
Georg Cantor (1845–1918).

Public domain.

---

1 The origins of collective nouns vary from attempts to sound sophisticated to jokes (e.g., A rash of dermatologists). See www.colleenpatrickgoudreau.com/a-gaggle-of-geese-a-pride-of-lions-a-school-of-fish-and-more-collective-animal-nouns-2/ for a long list of collective nouns for animals. Some of their origins are explained in a podcast at this webpage, which is from the vegan perspective. Also see https://en.wikipedia.org/wiki/List_of_English_terms_of_venery,_by_animal, https://en.wiktionary.org/wiki/Appendix:Glossary_of_collective_nouns_by_collective_term, https://en.wiktionary.org/wiki/Appendix:English_collective_nouns, www.newyorker.com/humor/daily-shouts/a-compiled-list-of-collective-nouns (includes some humorous examples, like "An angry group of pedestrians is called New York."), and www.writeawriting.com/grammar/collective-nouns/.

agenda[2] of items is Cantor's work in set theory and his development of transfinite numbers. As we will see, these ideas led to some important paradoxes. Today, set theory is the foundation of mathematics.

> Later generations will regard Mengenlehre (set theory) as a disease from which one has recovered.
>
> Henri Poincaré[3]

> No one shall expel us from the Paradise that Cantor has created.
>
> David Hilbert

Set theory has gone by several names, including *Mannigfaltigkeitslehre* (German), *theorie des ensembles* (French), and theory of aggregates (English). But which is it a disease or paradise? While the Poincaré quote given above is disputed by some,[4] the extreme viewpoint it represents certainly existed. Before it can be understood how such an extreme reaction could have been generated, some basics (i.e., definitions and notations) need to be examined.

Sets are usually denoted by uppercase letters and elements of the sets by lowercase letters. For example, we may write $a \in S$ to indicate that a belongs to (is a member of) the set S. The expression $b \notin S$ means that b does not belong to (is not a member of) the set S. If a set S consists of a small number of elements, we may define S by explicitly listing all of its members.

Example 1. If S is the set of single digit prime numbers, then S = {2, 3, 5, 7}. Note the use of curly braces around the elements of S. This is a standard notation. One should not use () or [] when listing the elements that belong to a set.

For sets that contain a large number of elements, we hope to be able to avoid writing them all out. Ideally, we're able to find a condition $P(x)$ that holds for exactly the elements of our set. We then use this condition to tersely express the set.

Example 2. M = $\{x \mid x^2 > 1\}$. Here we have the condition $x^2 > 1$ as our $P(x)$. Because $P(x)$ is true only when x is less than -1 or greater than 1, our set M consists of exactly these real numbers. In interval notation, M is expressed as $(-\infty, -1) \cup (1, \infty)$. When we write $\{x \mid P(x)\}$, we are using what is known as **set builder notation**. The horizontal line, $\mid$, can be read as "such that." You may also see a colon, :, used to mean the same thing.

When using set builder notation it's important to make sure that it is clear what elements should be considered. In the example above, M is $(-\infty, -1) \cup (1, \infty)$ if the domain of discourse is the real numbers, but it would be reduced to {2, 3, 4,...} if the domain of discourse was restricted to the natural numbers. We may explicitly state the elements that should be considered like so: M = $\{x \mid x^2 > 1, x \in \text{real numbers}\}$.

When mathematicians find themselves writing something frequently, they usually come up with a shorthand notation for it. We have the following convenient notations for the various classes of numbers:

---

2 See how I cleverly used a collective noun here?

3 Poincare also said, "Cantorism [promises] the joy of a doctor called to follow a fine pathological case."

4 See Gray, Jeremy, "Did Poincare Say 'Set Theory Is a Disease'?" *Mathematical Intelligencer*, Vol. 13, No. 1, December 1991, pp. 19–22. Also, see https://en.wikiquote.org/wiki/Henri_Poincar%C3%A9.

$\mathbb{N}$ = natural numbers

$\mathbb{Z}$ = integers

$\mathbb{Q}$ = rational numbers

$\mathbb{R}$ = real numbers

$\mathbb{C}$ = complex numbers.

The set we were looking at can be represented more tersely as $M = \{x \mid x^2 > 1, x \in \mathbb{R}\}$

Sometimes we want to specify only the positive numbers from one of the sets above. In that case, we append a small plus sign, like an exponent. That is,

$\mathbb{Z}^+$ = positive integers

$\mathbb{Q}^+$ = positive rational numbers

$\mathbb{R}^+$ = positive real numbers.

In most books, the natural numbers are taken to be $\{1, 2, 3, \ldots\}$, so it would be redundant to place a + in the upper right to indicate the positives. However, some authors include 0 as a natural number, so you may see the notation $\mathbb{N}^+$ in a book by one of them to indicate what I simply label as $\mathbb{N}$.

The integers, in any book you care to look at, are the natural numbers, their additive inverses (the negatives integers) and zero.

The rationals are fractions or ratios of integers, hence the name. The only restriction is that the denominator can never be zero. In set builder notation, we can write

$$\mathbb{Q} = \{a/b : a \in \mathbb{Z} \wedge b \in \mathbb{Z} \wedge b \neq 0\}.$$

Some of the modified letters shown above make sense, like $\mathbb{R}$ for reals. The symbol $\mathbb{Q}$ for rationals arises from the rationals being *quotients* of integers. The Germans are to blame for the use of $\mathbb{Z}$. The German word for integers is *Zahlen*.

Often in mathematics certain notations have become standard and we no longer think about them. We'll see that x, y, and z tend to be variables, while a, b, and c tend to be constants, following the example of Descartes. This standardization helps make things less confusing. Sometimes letters are taken from the middle of the alphabet. Many books on set theory will have sentences like "Let M be the set of..." Why M? S seems more natural. The answer is that the German word for set is *Menge* and the ideas of set theory (*Mengenlehre*) were first published in German. So, in that language, it was natural to denote a set by M and later translators did not bother to change this. A similar explanation reveals why a "field" (from abstract algebra) would be denoted by K instead of F. The German word is *Körper*.

Why do so many notations trace back to the Germans? Well, prior to Big Mistake Number 2, Germany was home to some of the best mathematicians (and physicists) in the world and had been for decades. When German texts were translated into English and other languages, the words were translated, but the numbers and the notations were kept the same. That's how we got so many standard notations that seem mismatched with what they represent. For some older notations Latin is to blame, but for more modern mathematics it is often German. Fortunately, the next bit makes sense in English (by coincidence).

In set theory, the domain of discourse is often called the universal set and designated by an uppercase U. It is important not to confuse U with ∪, the symbol used to designate the union of sets.

Just as we can write equations with numbers, like $4 = 2^2$, $6 \neq 9$, and $5 < 7$, we can write equations for sets. We have S = M if the sets S and M contain the exact same elements and S ≠ M if one set has one or more elements not contained in the other. The symbols ⊂ and ⊆ play the role of < and ≤ for sets.

Example 3. Let A = {1, 2, 3, 4}, B = {1, 3}, and C = {2, 5}.

Because every element of B is contained in A, we say that B is a **subset**[5] of A. We can express this symbolically as B ⊂ A.

Because C contains an element that A does not, we can write C ⊄ A.

Although ⊂ indicates proper inclusion (i.e., that for A ⊂ B there is some element x ∈B such that x ∉A) some authors wish to emphasize this feature by using the symbol ⊊.

Just as we can write 5 < 7 or 7 > 5, for sets A and B such that A ⊂ B, we may also write B ⊃ A. If the two sets might be the same, we can write A ⊆ B; we may also write B ⊇ A.

We can show that A is not a subset of B by writing A ⊄ B.

The complement of a set A refers to all of the elements contained in U, but not contained in A. In the following example, the complement of A is represented by A^C. It can also be represented as A′.

Example 4. Let U = {1, 2, 3, 4, 5, 6, 7, 8, 9}. If A = {3, 4, 7, 8} then A^C = {1, 2, 5, 6, 9}.

There are many notations for the complement of a set. They include A^C, ∼ A, ¬ A, !A, and $\bar{A}$. As the latter notations indicate, the complement is the set theoretic equivalent of NOT in logic.

Note: If A = U, then A^C = ∅, where ∅ denotes the set with no elements, called **the empty set**.

We also have $\emptyset^C$ = U.

Transfinite Sets

In some areas of mathematics, like calculus, infinity is something that can only be approached. For example, we never plug ∞ into an equation. If it appears as a limit in an (improper) integral, we sneak up on it like so: $\int_1^\infty \frac{1}{x}dx = \lim_{b\to\infty} \int_1^b \frac{1}{x}dx$

Our new upper limit, b, is a real number and can be plugged in after integrating. We then let b approach infinity and see where the result heads. We do not actually plug in ∞. That is, we are dealing with a *potential* infinity, not an *actual* or *completed* infinity. This feels like notational trickery to avoid treating infinity like an actual number. When Cantor dealt with infinite sets, he handled infinity in a way that offended many mathematicians. How can infinity be bound in a set? Cantor treated infinity like a number. In his mathematics there were finite numbers, which mathematicians were very comfortable with, and transfinite numbers. For the latter, some amazing properties arise when considering one-to-one correspondences. Before looking at these, consider the example below, using finite numbers.

---

5 Back in Cantor's day, a subset was sometimes called a "partial aggregate."

Example 5. In Figure 4.2, without counting, you can see by the one-to-one correspondence indicated with the arrows that there are the same number of (public domain) icons[6] on the left as there are (public domain) emojis[7] on the right.

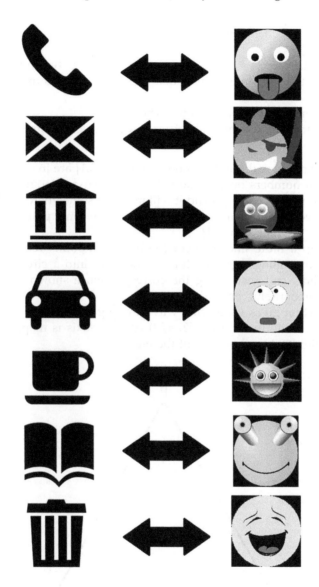

FIGURE 4.2
A one-to-one correspondence between finite sets.

Image created by Craig P. Bauer.

But consider what can be done if we carry the idea over to the transfinites:

---

6 http://publicicons.org/envelope-icon/.
7 https://publicdomainvectors.org/en/tag/emoji.

Example 6.

$$\cdots -3, -2, -1, 0, 1, 2, 3, \cdots$$
$$\cdots \; | \; \; | \; \; \; | \; | \; | \; | \; | \; \cdots$$
$$\cdots -6, -4, -2, 0, 2, 4, 6, \cdots$$

The above shows that there's a one-to-one correspondence between the integers and the even integers, although one might think that there are twice as many integers as even integers.

How can a proper subset be equal to the whole?

We have a similar result for intervals of real numbers, as Example 7 shows.

Example 7.

Figure 4.3 shows that the real numbers on the interval [0,1] are in one-to-one correspondence with the real numbers on the interval [0,4]. Given any point q on [0,1] we may draw the straight line defined by pq and it will intersect somewhere on the interval [0,4]. Furthermore, no two distinct points on [0,1] will hit the same point on [0,4] and every point on [0,4] has a preimage on [0,1].

We can also map a finite line segment (with the endpoints removed) to the entire real number line, by rolling the finite line segment up into a circle and placing it on the real number line, and then using the correspondence shown in Figure 4.4.[8] This map is called a stereographic projection.

The tangent function gives us another way of doing this.

In Figure 4.5, the finite interval $(-\pi/2, \pi/2)$ on the x-axis is mapped by the tangent function to the infinite interval $(-\infty, \infty)$ of the y-axis.

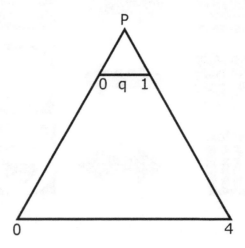

FIGURE 4.3
A strange one-to-one correspondence.

Image created by Josh Gross.

---

8 Adapted from Burger, Edward B. and Michael Starbird, *The Heart of Mathematics*, third edition, Wiley, Hoboken, NJ, 2010, pp. 217–219.

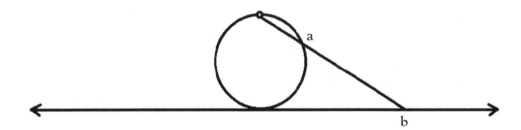

FIGURE 4.4
A stereographic projection.

Image created by Craig P. Bauer.

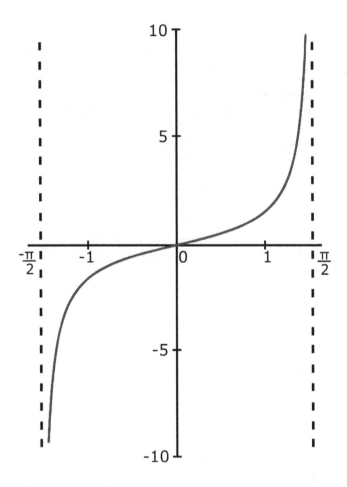

FIGURE 4.5
y = tan(x).

Image created by Josh Gross.

Are there different kinds of infinity? A first reaction might be "no" and the examples above showed that what seemed like different infinite sets could, in fact, be put into one-to-one correspondence. Let's investigate further!

> The infinite. No other great question has ever moved the spirit of man so powerfully; no other has stimulated his intellect so fruitfully.
>
> David Hilbert[9]

Cantor's Theorem of the Non-denumerablity of the Continuum (1874)

All finite sets are said to be **countable**, because we could, given enough time, list all of their elements. The natural numbers form an infinite set, but they can also be listed, like so:

1, 2, 3, 4, 5, 6, 7, 8, 9, 10, …

We know exactly where in the list any given natural number will occur. So, we also say that this set is countable. Because a countable set can be finite or infinite, it is nice to have a way to distinguish between the two possibilities. The terms **denumerable** and **enumerable** refer to an infinite countable set. If a set is infinite, but not denumerable, then we call it **non-denumerable**. Such sets are also **uncountable**.

In 1874, when Cantor published his first paper on set theory, he proved that the real numbers are non-denumerable. The proof presented here is a nicer approach that Cantor used in a later paper, from 1891.[10]

If we could write forever it would be a simple matter to list all of the integers in an orderly manner. 0, −1, 1, −2, 2, −3, 3, −4, 4, … This set is countably infinite, because it can be enumerated in this manner. In other words, a set is said to be countably infinite if and only if it can be put into one-to-one correspondence with the natural numbers. Suppose we tried to do this for the real numbers in the interval (0,1). Our list would have the following form:

1. $.n_{1,1}, n_{1,2}, n_{1,3}, n_{1,4}...$

2. $.n_{2,1}, n_{2,2}, n_{2,3}, n_{2,4}...$

3. $.n_{3,1}, n_{3,2}, n_{3,3}, n_{3,4}...$

4. $.n_{4,1}, n_{4,2}, n_{4,3}, n_{4,4}...$

9 Quoted in Keyser, Cassius Jackson, *Mathematics as a Culture Clue and Other Essays*, Volume I of The Collected Works of Cassius Jackson Keyser, Published by Scripta Mathematica, Yeshiva University, New York, 1947, p. 42.

10 Cantor, Georg, "Ueber eine elementare Frage der Mannigfaltigkeitslehre," *Jahresbericht der Deutschen Mathematiker-Vereinigung*, Vol. 1, 1890–1891, pp. 75–78. English translation: Ewald, William B., editor, *From Immanuel Kant to David Hilbert: A Source Book in the Foundations of Mathematics, Volume 2*, Oxford University Press, 1996, pp. 920–922.

where $n_{i,j}$ is the jth digit after the decimal place of the ith number. Now consider the number $.m_1, m_2, m_3, m_4, \ldots$ where m_i is chosen so that it is unequal to $n_{i,i}$. Because this number is not in our list (it differs from every element in the list in at least one place), our list cannot be complete. Thus, the set of real numbers in the interval (0, 1) is said to be uncountably infinite. We can shorten this and simply say the set is uncountable, since uncountable implies infinite.

Now that we know there are different kinds of infinity, we have to be careful how we talk about them. Instead of speaking of the "size" of a set, we refer to its **cardinality**. If the set is finite, this refers to the *number* of elements in the set. If the set is infinite, we need to be more precise. The integers and the set of real numbers in the interval (0, 1) cannot be put into one-to-one correspondence. Hence, they cannot have the same cardinality.

It follows that we should have a notation to indicate these different kinds of infinities. Cantor provided this. He introduced the Hebrew letter $\aleph$ (aleph) with various subscripts for this purpose. This has led some to conclude that Cantor was a Jew, but he was not.[11]

$\aleph_0$ = number of integers

$\aleph_1$ = number of real numbers

$\aleph_2$ = number of all geometrical curves

Cantor called $\aleph_0$ "Aleph-zero," but today it's also referred to as aleph-naught or aleph-null.[12] While the use of Greek letters is common in mathematics, Hebrew letters are rare.

What about the rational numbers? One would expect their cardinality to lie between that of the integers and the reals, perhaps $\aleph_{0.5}$. This is not the case. In fact, the rationals can be put into one-to-one correspondence with the natural numbers, so their cardinality is $\aleph_0$. The diagram shown in Figure 4.6 provides a simple proof. For what it's worth, Cantor thought mathematics should be carried out algebraically, so this diagram does not appear in any of his papers. Yet, it is by far the most common way for the proof to be presented today.

The zigzag lines, sometimes called "the snake," eventually pass through every point of the form (x, y), where x and y are positive integers. But every positive rational number can be written in the form x/y, so the snake hits positions representing every positive rational number. It will actually encounter each positive real infinitely many times. For example, the points (1, 2), (2, 4), (3, 6), etc., all represent 1/2. If we want a one-to-one correspondence, we must skip over these repeats. Doing so, the correspondence begins

---

11 Cantor's father was Lutheran and his mother was Roman Catholic. See Letter from Georg Cantor to Cardinal Franzelin, January 22, 1886, available at www.schillerinstitute.org/fid_91-96/943_transfinite.html.

12 Cantor, Georg, *Contributions to the Founding of the Theory of Transfinite Numbers*, Dover, New York, 1955, pp. 103–104.

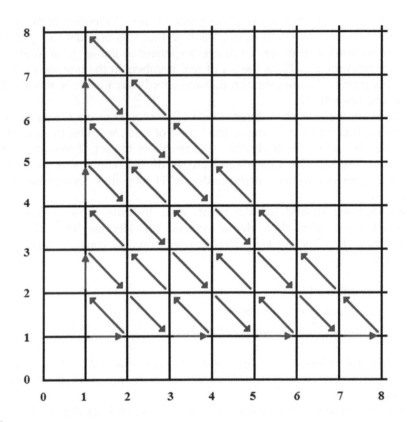

FIGURE 4.6
The snake.

Image created by Craig P. Bauer.

1	2	3	4	5	6	7	8	9	10	11	12	13...
↕	↕	↕	↕	↕	↕	↕	↕	↕	↕	↕	↕	↕
1/1	2/1	1/2	1/3	3/1	4/1	3/2	2/3	1/4	1/5	5/1	6/1	5/2...

A more elegant solution that doesn't require skipping numbers is given in Chapter 17. You may have noticed that I left out the negative rationals. They can be accommodated like so:

1	3	5	7	9	11	13	15	17	19	21	23	25...
↕	↕	↕	↕	↕	↕	↕	↕	↕	↕	↕	↕	↕
1/1	2/1	1/2	1/3	3/1	4/1	3/2	2/3	1/4	1/5	5/1	6/1	5/2...

2	4	6	8	10	12	14	16	18	20	22	24	26...
↕	↕	↕	↕	↕	↕	↕	↕	↕	↕	↕	↕	↕
−1/1	−2/1	−1/2	−1/3	−3/1	−4/1	−3/2	−2/3	−1/4	−1/5	−5/1	−6/1	−5/2...

Because the rationals are the same kind of infinity as the natural numbers and the reals are a larger kind of infinity, a randomly chosen real number is almost certainly (with probability 1) irrational.

Now, recall from Chapter 0 that a number is algebraic if it's the root of a polynomial with integer coefficients. The set of algebraic numbers includes all of the rationals and many irrationals such as $\sqrt{2}$. The question is, are there enough irrational algebraic numbers to give us a larger infinity than that of the natural numbers and the rationals? The answer is no, for the algebraic numbers can be paired with the naturals, as Cantor showed in his 1874 paper.[13] The proof begins with a definition. The **height** of the polynomial

$p(x) = a_n x^n + a_{n-1} x^{n-1} + \cdots + a_2 x^2 + a_1 x + a_0$ is given by
$N = n - 1 + |a_n| + |a_{n-1}| + \cdots + |a_2| + |a_1| + |a_0|.$

Each algebraic number arises from a polynomial of some positive height. And there are a finite number of polynomials of any specific height, each of which has n roots. So, there are a finite number of algebraic numbers associated with any given height. To list all algebraic numbers, we can begin with those associated with polynomials having height 1, listing them from smallest to largest, then continue with those associated with polynomials of height 2, and so on. Duplicates can be skipped over to make the correspondence with the natural numbers one-to-one.

Another way to make the correspondence one-to-one is to only consider polynomials that are irreducible (cannot be factored into polynomials of lesser degree) and for which the greatest common divisor of the coefficients is 1. The result will be that there are no duplicates.

So, a randomly chosen real number is not just nearly certain to be irrational, it is also nearly certain to be transcendental (non-algebraic).

With a small upgrade, we can make this proof show that the complex algebraic numbers are also denumerable. The upgrade concerns how we list the complex algebraic numbers associated with a given height, once all of these numbers are known. A solution is to use the lexicographic order. That is, numbers of the form $a + bi$ can be ordered as in a dictionary, if a and b are thought of as the first and second "letters" in the "word." Numbers with a smaller value for the real component a would come before the others. If two numbers have the same real component, then the number with the smaller value for the complex portion comes first.

In comparing infinities, there's no need to restrict ourselves to lines (e.g. numbers that appear on the real number line). Complex numbers are graphed in the plane, so the above result encourages us to look at surfaces.

In 1874, Dedekind received a letter from Cantor asking:

> Can a surface (say a square that includes the boundary) be uniquely referred to a line (say a straight line segment that includes the end points) so that for every point on the surface there is a corresponding point of the line and, conversely, for every point of the line there is a corresponding point of the surface? I think that answering this question would be no easy job, despite the fact that the answer seems so clearly to be "no" that proof appears almost unnecessary.[14]

---

13 Cantor, Georg, *Contributions to the Founding of the Theory of Transfinite Numbers*, Dover, New York, 1955, pp. 38–39 of Philip E. B. Jordain's introduction.

14 Havil, Julian, *Impossible?: Surprising Solutions to Counterintuitive Conundrums*, Princeton University Press, Princeton, NJ, 2008, p. 80.

The answer was "yes." In a later letter to Dedekind, after he found the answer, Cantor wrote, "I see it, but I don't believe it!"

The basic idea behind the one-to-one correspondence between the square and the line segment is very simple, but it requires a little patching to make it work perfectly. If we have a point on the square, it can be represented as the ordered pair (x, y). The point that it corresponds to on the line segment is simply the interweaving of the digits of x and y. For example,

$(x, y) = (0.5772156649015328606065120900824024310421593359392...,$
$\qquad\qquad 0.3183098861837906715377675267450287240689192914809...)$

corresponds to

0.5371782310596868469108135739208667016503675716270592060784254002284732140046281951993239519438909921...

What beautiful simplicity! It seems that different points on the square must go to different points on the line segment. And it should work the other way too. Separating the digits in the even and odd position of a point on the plane, to get a pair of coordinates, ought to lead to a different point on the square for each point on the line segment.

However, there are a couple of snags. Trouble is caused by the fact that 0.5 is the same as 0.4999999...., if we let the 9s continue forever. Having two different representations for the same number is not desirable here! Patching this is easy. We simply choose to represent every number whose decimal representation eventually ends by decreasing the digit in the last position by 1 and following it by an infinite string of 9s.

The second snag is that some points on the line segment are actually left out of the correspondence. An example is 0.41909090909090 ... The point on the square that ought to be paired with it is (0.4999999...., 0.1000000...), but 0.1000000 is written as 0.0999999 ... to avoid the previous snag. As a consequence, this point on the square is paired with 0.40999999999999 ... and 0.41909090909090 ... is left alone.

To address this second problem, the idea of strictly alternating the digits to go from the square to the line segment must be modified. We need to think of strings like 01, 06, 008 as single digits. Anytime you encounter a 0 or a string of consecutive 0s do not alternate them. Pretend that these 0s are attached to the first non-zero digit that appears after them. With this new approach,

$(x, y) = (0.5772156649015328606065120900824024310421593359392...,$
$\qquad\qquad 0.3183098861837906715377675267450287240689192914809...)$

corresponds to

0.5371782310958686641980135739206876106506357172609700852246027443510204827125490638395199329991124...

Because the square's x-value contained more 0s than its y-value, the digits I provided for x were all used, while the last four digits given for y, 8091, have not yet been applied.

There's an important lesson to learn here. You can present the original correspondence I showed above to many people, including some with degrees in mathematics, and convince them that it's correct. But it isn't! It has subtle flaws. When you construct a proof, look hard for such flaws. Look at extreme cases and see if any exceptions can be found. And if they are found, try to patch. Subjecting your proofs to intense scrutiny will help you catch some of your errors.[15]

The points on a line segment can be put into one-to-one correspondence with those of a cube just as easily, and so on into higher dimensions.

Is there another infinity beyond $\aleph_2$? A simple concept from set theory will help answer this obvious question.

Given a set S we define the **power set**, P(S), to be the set of all subsets of S.

Example 8. If S = $\{a, b, c\}$, then P(S)=$\{\{a\},\{b\},\{c\},\{a,b\},\{a,c\},\{b,c\},\{a,b,c\},\oslash\}$.
Recall that $\oslash$ denotes the empty set, the set containing no elements. There's actually a neat way to generate power sets. It was published by Peter Nicholson in 1818 and attributed by him to English mathematician and astronomer Thomas Harriot (1560–1621).[16] This is all long before set theory was a thing! Naturally the idea was expressed in different terms, that of combinations. In any case, to carry out the method, simply add x to each element and then multiply them all together.

$$(x + a)(x + b)(x + c) = x^3 + (a + b + c)x^2 + (bc + ac + ab)x + abc.$$

The constant term is abc, corresponding to the set $\{a, b, c\}$. The linear term has coefficient $bc + ac + ab$, corresponding to the sets $\{b, c\}$, $\{a, c\}$, and $\{a, b\}$. The quadratic term has coefficient $a + b + c$, corresponding to the sets $\{a\}$, $\{b\}$, and $\{c\}$. Finally, the cubic term has coefficient 1, which does not involve any of the element of our set S, so it corresponds to the empty set, $\oslash$.

An alternative multiplication that gives the elements of the power set is

$$(1 + ax)(1 + bx)(1 + cx) = 1 + (a + b + c)x + (bc + ac + ab)x^2 + (abc)x^3$$

There's nothing special about a three-element set here. The results generalize to n-element sets.

If S is a finite set with n elements, then P(S) will contain exactly 2^n elements. When forming a subset we have two choices for each element, whether to include it or not. Thus, by the multiplication principle (discussed in detail in Chapter 7), there are 2^n choices when considering a set of n elements. But what happens when we take P($\aleph_0$)? What cardinality is $2^{\aleph_0}$? This is yet another problem that Cantor tackled.

---

15 I saw the alternating digit version of the one-to-one correspondence between the square and a line segment in several books, but first encountered the second snag and its resolution in Burger, Edward B. and Michael Starbird, *The Heart of Mathematics*, third edition, Wiley, Hoboken, NJ, 2010, p. 223. The other authors were either unaware of, or chose to ignore, the difficulty.

16 Nicholson, Peter, *Essays on the Combinatorial Analysis*, Longman, Hurst, Rees, Orme, and Brown, 1818, pp. v–vi. I first saw the result in Wilson, Robin and John J. Watkins, editors, *Combinatorics: Ancient & Modern*, Oxford University Press, Oxford, 2013, pp. 286–287.

Cantor's Theorem

There are many neat results due to Cantor to describe in this chapter, but this is one of the bigger ones, so when someone refers to "Cantor's theorem," it is likely the following.

If S is any nonempty set, then the cardinality of S is less than the cardinality of P(S).

Letting O(S) denote the order (aka cardinality or size) of a set S, Cantor's theorem says O(S) < O(P(S)). Taking what we know concerning the order of infinite sets, and then applying Cantor's theorem, we have

$$O(\aleph_0) < O(\aleph_1) < O(P(\aleph_1)) < O(P(P(\aleph_1))) < O(P(P(P(\aleph_1)))).$$

The sequence continues forever. There is no greatest infinity. We have infinitely many different kinds of infinity.

And there's more. Cantor explained,

> But even the unlimited sequence of cardinal numbers
>
> $$\aleph_0, \aleph_1, \aleph_2, \ \ldots \aleph_\nu, \ldots$$
>
> does not exhaust the conception of transfinite cardinal number. We will prove the existence of a cardinal number which we denote by $\aleph_\omega$ and which shows itself to be the next greater to all the numbers $\aleph_\nu$, out of it proceeds in the same way as $\aleph_1$ out of $\aleph_0$ a next greater $\aleph_{\omega+1}$, and so on, without end.[17]

Another common notation for the order of a set should be pointed out before moving on. It simply uses vertical lines around the set, like an absolute value function. So, you may encounter the result above expressed as $|\aleph_n| < |P(\aleph_n)|$.

The Cost

Cantor died of a heart attack while hospitalized in a sanatorium for mental illness. What were the contributing factors to his poor mental health? Which did more harm, his mathematical results or the criticism heaped on them by various other mathematicians?

Much of the criticism against Cantor arose from the subtle distinction between the "potential" infinite and the "actual" or "completed" infinite. Aristotle favored the potential infinite and, as we saw, his influence carried a long way. The potential infinite is something that can never be reached.

Cantor dared describe sets containing an infinite number of elements. How can the infinite be embraced in a set? This actual infinity that could be handled like finite values

---

17 Cantor, Georg, *Contributions to the Founding of the Theory of Transfinite Numbers*, Dover, New York, 1955, p. 109. Cantor gives references on pp. 109–110.

was very controversial. Bucking Aristotle's legacy was still a big deal and it was not something Cantor did lightly. He explained:

> I was logically forced, almost against my will, because in opposition to traditions which had become valued by me in the course of scientific researches extending over many years, to the thought of considering the infinitely great, not merely in the form of the unlimitedly increasing, and in the form, closely connected with this, of convergent infinite series, but also to fix it mathematically by numbers in the definite form of a "completed infinite." I do not believe, then, that any reasons can be urged against it which I am unable to combat.[18]

It wasn't just the shade of Aristotle with whom Cantor had to contend. He mentioned other authority figures in a letter.

> It is now exactly two years ago, that Mr. Rudolf Lipschitz in Bonn called my attention to a certain passage in the correspondence between Gauss and Schumacher, where the former declares himself against *any* bringing into play of the Actual-Infinite in mathematics (letter of July 12, 1831); I have answered in detail, and have *in this point* dismissed the authority of Gauss, of which I think so highly in all other respects, as I reject today the testimony of Cauchy and, in my short paper "Grundlagen einer allgemeinen Mannigfaltigkeitslehre, Leipzig, 1883," among others also the authority of Leibniz, who in this question has committed a peculiar inconsistency.[19]

Cantor reflected on his transfinite numbers as the new irrationals. There were parallels both mathematically and sociologically (in terms of people's negative reactions).[20]

Despite there being so many big names who objected, there were others, before Cantor, who anticipated some of his work. The next three sections take a look at what they did.

Duns Scotus (c. 1266–1308)

This philosopher/theologian from the Middle Ages recognized that one-to-one correspondences can be made between regions that do not represent the same quantities. He demonstrated the fact with concentric circles, as shown in Figure 4.7.

The next contributor to the drama of the infinite is much better known than Duns Scotus. He is famous enough to be recognized by a single name, Galileo (Figure 4.8).

---

18 Cantor, Georg, *Contributions to the Founding of the Theory of Transfinite Numbers,* Dover, New York, 1955, p. 53.
19 From a letter from Georg Cantor to Mr. G. Eneström, available online at www.schillerinstitute.org/fid_91-96/943_transfinite.html.
20 Cantor, Georg, *Contributions to the Founding of the Theory of Transfinite Numbers,* Dover, New York, 1955, p. 77.

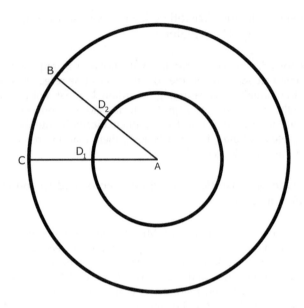

FIGURE 4.7
A pair of circles with different diameters, but the same number of points.

Image created by Josh Gross.

FIGURE 4.8
Galileo Galilei (1564–1642).

Public domain.

Galileo Galilei

An extract from Galileo's *Discourses and Mathematical Demonstrations Relating to Two New Sciences*, written while he was under house arrest and published in 1638, demonstrates his understanding of how proper subsets of the integers can be put into one-to-one correspondence with the integers.[21] In this work Salviatus is the wise man, while Simplicius is his intellectual punching bag. The inquisition suspected that Simplicius was Galileo's caricature of the Pope.[22]

SALV: ... an Indivisible, added to another Indivisible, produceth not a thing divisible; for if that were so, it would follow, that even the Indivisibles were divisible ...

SIMP: Here already riseth a doubt, which I think unresolvable... . Now this assigning an Infinite bigger than an Infinite is, in my opinion, a conceit that can never by any means be apprehended.

[To make the infinite plain even to Simplicius, Salviatus patiently explains what a square integer is before proceeding as follows.]

SALT: Farther questioning, if I ask how many are the Numbers Square, you can answer me truly, that they be as many, as are their proper roots; since every Square hath its Root, and every Root its Square, nor hath any Square more than one sole Root, or any Root more than one sole Square.

[This is the kernel of the matter: the one-one correspondence between a part of an infinite class (here, that of all the natural numbers) and one of its subclasses (here, that of all the integer squares). Continuing the argument, Salviatus compels Simplicius to surrender.]

SIMP: What is to be resolved on this occasion?

SALV: I see no other decision that it may admit, but to say, that all Numbers are infinite; Squares are infinite; and that neither is the multitude of Squares less than all Numbers, nor this greater than that: and in conclusion, that the Attributes of Equality, Majority, and Minority have no place in Infinities, but only in terminate quantities ...

[In modern terminology, two classes which can be placed in one-one correspondence are said, as we have noted, to be equivalent or similar. In Galileo's example, the class of all square integers is equivalent to the class of all positive integers.]

[The extracts are from the first English translation, London, 1685, of Galileo Galilei, *Ditcorti e dimonstrasiane matematiche intorno a due nuove scienze*, Leida, 1638. The 1665 translation, Galileus Galileus, *Mathematical Discourse and Demonstrations, and so forth*, is much sharper than subsequent translations. Unfortunately, this edition is an excessive rarity, owing to the Great Fire of London, among other causes. It should be reproduced. Even the Library of Congress does not have a copy.][23]

---

21 See https://en.wikipedia.org/wiki/Two_New_Sciences for more information about this interesting book.

22 Bell, Eric Temple, *Mathematics, Queen and Servant of Science*, Copyright, 1951, British Edition 1952, reprinted 1954, 1958, 1961, Printed in Great Britain by Butler and Tanner Ltd., Frome and London, p. 399. The quotation from Galileo's work that follows is taken via Bell here. Asides indicated in [] are from Bell.

23 Bell, Eric Temple, *Mathematics, Queen and Servant of Science*, Copyright, 1951, British Edition 1952, reprinted 1954, 1958, 1961, Printed in Great Britain by Butler and Tanner Ltd., Frome and London, pp. 399–400.

So, the wild one-to-one correspondences Cantor found had antecedents, but nothing more was done with these ideas until much later.

Bernard Bolzano

Although he was of Italian descent, Bernard Bolzano[24] (Figure 4.9) was born in Prague, now part of the Czech Republic, and grew up speaking German. Hence, he used this language for the 1851 book in which he put forth his ideas, *Paradoxien des Unendlichen* (The Paradoxes of the Infinite).[25] Sadly, he did not live to see its publication. It is in this important book that the word *Menge* (translated in English as "set") was used for the first time.

In addition to creating terminology that is still used today, Bolzano gave more examples like Galileo's in which an infinite set is put into a one-to-one correspondence with one of its proper subsets.[26] This cannot be done with any finite set and so this unusual property came to be used as one way to define infinite sets.

FIGURE 4.9
Bernard Bolzano (1781–1848).

Public domain.

---

24 His full name is Bernard Placidus Johann Nepomuk Bolzano, but mathematicians always refer to him as Bolzano.

25 Bolzano, Bernard, *Paradoxien des Unendlichen*, C. H. Reclam sen, Leipzig, 1851, available online at https:// archive.org/details/paradoxiendesune00bolz and https://dml.cz/bitstream/handle/10338.dmlcz/400242/ Bolzano_14-1851-1_5.pdf. For an English translation, see Bolzano, Bernard, *Paradoxes of the Infinite*, Yale University Press, New Haven, CT, 1950.

26 See http://en.wikipedia.org/wiki/The_Paradoxes_of_the_Infinite, www-history.mcs.st-and.ac.uk/Biogra-phies/Bolzano.html, and www-history.mcs.st-and.ac.uk/HistTopics/Beginnings_of_set_theory.html for more.

Definitional Difficulties

Set theory faces some challenges right from the start. Recall that the definition was given like so:

A **set** is a collection of distinct objects.

But "collection" is a synonym for set. Therefore, the definition is circular! This is not good. Even though mathematicians have a clear sense of what is meant by "set," it is very hard to define in a noncircular manner.

A definition that Cantor offered in 1895, with the word "class" meaning set in this translation, is no better. In fact, I'd say it's worse, for it uses more words, but conveys no more information.

> By a class we understand any collection into a single whole of definite well-distinguished objects of our intuition or of our thought.[27]

Again, we have a circular definition. Bad as this sounds, we have worse problems.

Paradoxes in Set Theory

Having failed to achieve the mathematicians' dream of being able to prove or disprove any statement with logic alone, it was realized that set theory was needed as well. Among those working on the problem was Gottlob Frege (1848–1925), a German mathematician and philosopher. He was well-suited for the task, having previously done important work in logic. His new project was titled *Grundgesetze der Arithmetik* (The Basic Laws of Arithmetic). Volume I appeared in 1893 and, a decade later, volume II was about to appear, collectively representing Frege's life's work, when he received a letter from Bertrand Russell.

The letter detailed a paradox Russell had found in 1901.[28] It arose from Frege's Basic Law V which appeared in volume I. To understand it we must first recognize that some sets can contain themselves as elements. As an example, consider S = the set of all sets that can be defined in 100 words or less. The definition of S used less than 100 words, so S is a member of S. Another example is the set of all sets containing more than 10 elements. So, some sets contain themselves and some do not. No problem. But, Russell asked about another set, call it R, defined to be the set of all sets that do not contain themselves. This causes a paradox for when we ask if R is contained in R we get a contradiction no matter which answer is given.

Following receipt of Russell's letter, Frege quickly added an appendix to volume II of *Grundgesetze der Arithmetik* that began:

> Hardly anything more unfortunate can befall a scientific writer than to have one of the foundations of his edifice shaken after the work is finished. This was the position

---

27 Taken here from Bell, Eric Temple, *Mathematics, Queen and Servant of Science*, Copyright, 1951, British Edition 1952, reprinted 1954, 1958, 1961, Printed in Great Britain by Butler and Tanner Ltd., Frome and London, p. 402.
28 https://plato.stanford.edu/entries/russell-paradox/.

I was placed in by a letter of Mr. Bertrand Russell, just when the printing of this volume was nearing its completion.[29]

In what followed, Frege described the paradox and made an attempt to eliminate it by altering part of what he had done in volume I.

On the other side, the paradoxes seemed to be breeding. Some are simply reformulations, but all are interesting, and so, a few are now detailed.

The Set of All Sets

We can attempt to define the largest possible set by calling it "the set of all sets." It would contain every possible set, including itself. What could possibly be bigger? Well, Cantor showed that P(S) is always bigger than S, so we immediately have an even larger set, contradicting that S was the biggest!

The Barber Paradox

Russell also gave a less abstract version of the paradox that rocked Frege's world.[30]

You can define the barber as "one who shaves all those, and those only, who do not shave themselves." The question is, does the barber shave himself?[31]

To convert this back to the language of set theory, we can ask, if S is the set of all people whom the barber shaves, is the barber in S?

The Displaced Mayor Paradox

Imagine a country where some of the mayors live in the cities they serve and some do not. One day, a new nation-wide law is passed that requires all mayors who do not live in the cities they serve to live in a new city, created just for them. Only these mayors are allowed to reside there. This new city turns out to have a population large enough to require a mayor. When one is elected, where should he or she live?[32]

---

29 Van Heijenoort, Jean, *From Frege to Gödel: A Source Book in Mathematical Logic, 1879–1931*, Harvard University Press, 1967; reprinted with corrections, 1977.

30 Although Russell presented it, he claimed that it was created by someone else, whom he did not name.

31 Russell, Bertrand, "The Philosophy of Logical Atomism," *The Monist*, Vol. 29, No. 3, July 1919, pp. 345–380, p. 355 quoted here, reprinted in *The Collected Papers of Bertrand Russell, 1914–19*, Vol. 8, quote on p. 228.

32 Adapted from Paulos, John Allen, *Beyond Numeracy*, Vintage Books, New York, 1992, p. 215. I don't know who first posed this particular paradox.

Berry's Paradox

First put into print by Bertrand Russell, it was created by another. Russell noted, "This paradox has been suggested to me by Mr. G. G. Berry of Oxford's Bodleian Library. It has the merit of not exceeding the finite numbers."[33] It can be expressed as "the smallest integer that cannot be expressed in less than 22 syllables."

In case this isn't clear, consider some integers. In one syllable we can express the integer "three." Saying, "four hundred twenty" requires five syllables, so we're still under the limit of 22. Of course, there are terser ways to express some numbers. "The smallest hundred-digit prime" uses only eight syllables, far less than used in actually naming the number in question, but the shorter form still specifies the integer in question. The paradox here is that "the smallest integer that cannot be expressed in less than twenty-two syllables" *is expressed* by that very phrase, which has only 21 syllables.

More to Come

It's not hard to make up paradoxes. One of mine is presented in the exercise set at the end of this chapter. Another exercise asks you to create a paradox. Many more are likely to appear in the future. An example from David Rees's comic strip *Get Your War On* is given in Figure 4.10

Mathematicians are not all in agreement concerning the proper resolutions for various paradoxes. Some of the suggested resolutions are complex. Another point of view completely ignores the paradoxes and allows any collection of objects to be considered a valid set. If you'd like, you can simply conclude that some words that seem to define sets do not and get on with your life.

FIGURE 4.10
A comic by a man who greatly enjoyed his discrete mathematics class.

Courtesy of David Rees.

---

33 Russell, Bertrand, "Les Paradoxes de la Logique," *Revue de Métaphysique et de Morale*, T. 14, No. 5, September 1906, pp. 627–650, Published by Presses Universitaires de France. Although Russell's first language was English, he published this paper in French. My translation of the paradox required changing the number of syllables. There are other variants of this paradox where the restriction is on the number of words or letters.

Although his creation of set theory led to some thorny paradoxes, as we've seen, Cantor also left mathematicians with an open question that he likely expected a definite resolution to.

Cantor's Continuum Hypothesis: There is no transfinite cardinal falling strictly between $\aleph_0$ and $\aleph_1$.

We saw that the rationals and the algebraic numbers have the same cardinality as the natural numbers, while the irrationals and the transcendentals have the same cardinality as the reals. Cantor's continuum hypothesis reasonably posits that any set we try to create with a cardinality between the naturals and the reals will actually have a one-to-one correspondence with one or the other. Although no exceptions were found, neither was anyone able to prove the conjecture. Before describing how this problem was resolved, we need to consider a firming up of set theory. That is, we will examine the axiomatization of the field.

The Zermelo–Fraenkel (ZF) Axioms for Set Theory[34]

The axioms shown below resulted from modifications made by Abraham Fraenkel (Figure 4.12) to the first axiomatic system for set theory, proposed by Ernst Zermelo (Figure 4.11) in 1908.[35]

Almost all of mathematics is built upon the axioms of set theory. Without going into great details, these axioms are now given:

1. **Extensionality (equality):** Two sets with the same elements are equal.
2. **Pairing:** For every a and b, the set $\{a, b\}$ exists.
3. **Separation (subset):** If A is a set and P(x) is a predicate with free variable x, then the subset of A exists that consists of those elements $c \in A$ such that P(c) is true. (The separation axiom guarantees that the intersection of two sets exists.)

   The intersection of two sets A and B is represented by A ∩ B.

   Example 9. If A = {1, 2, 3, 4} and B = {2, 4, 6, 8}, then A ∩ B = {2, 4}.

   For intersections of three or more sets, we take only the elements that are in all of the given sets.

   Example 10. If A = {2, 3, 4}, B = {2, 4, 6, 8}, and C = {2, 3, 5}, then A ∩ B ∩ C = {2}.

   For the intersection of a large number of sets, we can use ellipsis to indicate missing sets, as in $A_1 \cap A_2 \cap A_3 \cap \ldots A_{99} \cap A_{100}$. But there's a terser way to show this, namely

$$\bigcap_{i=1}^{100} A_i.$$

   In this expression, i stands for index. The $i = 1$ tells us where to start the index and the 100 tells us where to stop. It is understood to be incremented by ones in its

---

34 Taken here from Rosen, Kenneth H., editor, *Handbook of Discrete and Combinatorial Mathematics*, CRC Press, Boca Raton, FL, 2000, p. 30.

35 https://en.wikipedia.org/wiki/Zermelo%E2%80%93Fraenkel_set_theory.

FIGURE 4.11
Ernst Zermelo (1871–1953).

Public domain.

journey from 1 to 100. The intersection of an infinite number of sets can also be represented in this manner:

$$\bigcap_{i=1}^{\infty} A_i$$

or as

$$\bigcap_{i=\mathbb{N}} A_i,$$

where $\mathbb{N}$ is the set of natural numbers. An advantage of this latter notation is that it is easily adapted for when we want the intersection of a larger infinity of sets. For example, we can write

$$\bigcap_{i \in S} A_i,$$

FIGURE 4.12
Abraham Fraenkel (1891–1965).

Public domain.

which tells us to take the intersection of all sets A_i where i is an element of S. The set S can be anything, even an uncountably infinite set.

4. **Union:** The union of a set (i.e., the set of all the elements of its elements) exists. (The union axiom together with the pairing axiom implies the existence of the union of two sets.) The union of two sets A and B is represented by A ∪ B.

Example 11. If A = {1, 2, 3} and B = {7, 8, 9}, then $A \cup B = \{1, 2, 3, 7, 8, 9\}$.

Example 12. If A = {1, 2, 3} and B = {3, 4, 5}, then $A \cup B = \{1, 2, 3, 4, 5\}$.

Even though 3 is an element of both A and B, we only write it once when listing the elements of A ∪ B.

We can write the union of three sets, A, B, and C, as A ∪ B ∪ C, but as with intersections, we have alternatives. We can use ellipsis to indicate missing sets, if the pattern is clear, as in $A_1 \cup A_2 \cup A_3 \cup \ldots A_{99} \cup A_{100}$. And we can also write

$$\bigcup_{i=1}^{100} A_i, \ \bigcup_{i=1}^{\infty} A_i, \ \bigcup_{i \in \mathbb{N}} A_i, \text{ and } \bigcup_{i \in S} A_i$$

where the meanings are analogous to the similar appearing intersections above.

5. **Power set:** The power set (the set of all subsets) of a set exists.

6. **Empty set:** The empty set exists.

    This set plays a role similar to that of zero in set theory and is denoted by the similar symbol ∅. Some students, especially those in computer science, like to distinguish between zero and the letter "o" by placing a diagonal line through zero. This can cause trouble if set theory is the subject, because it may cause zero to be confused with the empty set. While it plays a similar role, it's different and the distinction is important.

7. **Regularity (foundation):** Every nonempty set contains a "foundational" element; that is, every nonempty set contains an element that is not an element of any other element in the set. (The regularity axiom prevents anomalies such as a set being an element of itself.)

8. **Replacement:** If f is a function defined on a set A, then the collection of images $\{f(a) \mid a \in A\}$ is a set. The replacement axiom (together with the union axiom) allows the formation of large sets by expanding each element of a set into a set.

9. **Infinity:** An infinite set, such as ω exists.

There is also the **axiom of choice**, which may be appended or left off this list, depending on the preference of the mathematician. The axiom list is often denoted by ZF or ZFC depending on the exclusion or inclusion of the axiom of choice.

Axiom of Choice:[36] If A is any set whose elements are pairwise disjoint nonempty sets, then there exists a set X that has as its elements exactly one element from each set in A.

Disjoint sets are sets that have no elements in common. Their intersection is the empty set. Basically, this axiom says that if we have a whole bunch of sets, then we can choose one element from each of them and then group them all together in a new set. An objection to accepting this axiom is that it doesn't tell us how we can select an element from each set. If the sets consist of positive integers, then we can simply take the smallest element from each, but if the sets are open intervals of the real number line, this rule for selecting elements won't work. If we can't give a general rule for how the elements are selected, are we justified in saying that it can be done?

The axiom of choice is equivalent to the following, to which similar objections can be made.

The well-ordering principle: Every set can be well-ordered; that is, for every set A there exists a total ordering on A such that every subset of A contains a smallest element under this ordering.

Generalized axiom of choice (functional version): If A is any collection of nonempty sets, then there is a function f whose domain is A, such that $f(X) \in X$ for all $X \in A$.

Zorn's lemma: Every nonempty partially ordered set in which every chain (totally ordered subset) contains an upper bound (an element greater than all the other elements in the chain) has a maximal element (an element that is less than no other element).

---

36 It's called the Axiom of Selection in some older works. It was put forth by Ernst Zermelo, who credited Erhard Schmidt. See Zermelo, Ernst, "Beweis, daß jede Menge wohlgeordnet werden kann," *Mathematische Annalen.* Vol. 59, No. 4, 1904, pp. 514–516, reprinted in English translation as "Proof that every set can be well-ordered," van Heijenoort, 1976, pp. 139–141. Both Cantor and G. H. Hardy had previously made unconscious use of it. See Cantor, Georg, *Contributions to the Founding of the Theory of Transfinite Numbers*, Dover, New York, 1955, pp. 204–205.

The Hausdorff maximal principle: Every chain in a partially ordered set is contained in a maximal chain (a chain that is not strictly contained in another chain).

Trichotomy: Given any two sets A and B, either there is a one-to-one function from A to B, or there is a one-to-one function from B to A; that is, either $|A| \le |B|$ or $|B| \le |A|$.

In addition to the previously stated objection, the axiom of choice is controversial because it can be used to prove ridiculous things like the Banach-Tarski theorem (1924).[37] A special case of this theorem says that in three-dimensional space, a ball can be divided up into finitely many pieces and those pieces reassembled using only rigid motions to form a pair of balls identical to the original. This theorem is so contrary to common sense that it's usually referred to as the Banach-Tarski *Paradox*.

The axiom of choice truly is an axiom, as it is consistent with, but independent of, the other axioms of set theory. This was established in pieces, with the first contribution made by Kurt Gödel and the second by Paul Cohen. So, we don't have to worry about contradictions arising from the inclusion of this axiom. It's just that we get weird results. A simple way to eliminate these weird results is by dropping the axiom of choice from our foundation for set theory.

A Resolution

I've kept you in suspense regarding Cantor's continuum hypothesis long enough. Is there an infinity strictly between $\aleph_0$ and $\aleph_1$? The answer lies in the following paragraph (sort of).

Cantor was unable to prove or disprove this hypothesis. In 1900, Hilbert listed this as the first of 23 important unsolved problems in mathematics. In 1940, Kurt Gödel (Figure 4.13) proved that the continuum hypothesis could not be disproven using the set-theoretic axioms. In 1963, Paul Cohen showed that it couldn't be proven using the set-theoretic axioms. Thus, the hypothesis is independent of the principles of set theory. John H. Conway described the proof as "the work of an alien being" and Gödel was moved by the proof to write to Cohen.[38] Gödel's letter included the following:

> I think that in all essential respects you have given the best possible proof and this does not happen frequently. Reading your proof had a similarly pleasant effect on me as seeing a really good play.[39]

The continuum hypothesis turned out to be unprovable, but it should not have surprised mathematicians that such a conclusion could be drawn, because a result addressing such matters had been published by Gödel in 1931.

---

37 Banach, Stefan and Alfred Tarski, "Sur la décomposition des ensembles de points en parties respectivement congruentes," *Fundamenta Mathematicae*, Vol. 6, 1924, pp. 244–277.

38 Roberts, Siobhan, *Genius at Play – The Curious Mind of John Horton Conway*, Bloomsbury, New York, 2015, p. 209.

39 Roberts, Siobhan, *Genius at Play – The Curious Mind of John Horton Conway*, Bloomsbury, New York, 2015, p. 209.

FIGURE 4.13
Kurt Gödel (1906–1978).

Courtesy of John W. Dawson, Jr.

Gödel's Incompleteness Theorem

> The mathematical logician Kurt Gödel was one of the preeminent intellectual giants of the twentieth century, and, assuming the survival of the species, will probably be one of the few contemporary figures remembered in 1,000 years.
>
> John Allen Paulos[40]

As was explained in Chapter 1, in the attempt to develop a mathematics in which every conjecture could potentially be proven or disproven using tools built from a small set of axioms, it was realized that logic was not enough. The attempted patch was logic plus set theory. Today, the view of the majority of mathematics is that of functions acting on sets. So, the tools had changed, but Hilbert's demand still stood. In 1900 it was simply expressed:

> This conviction of the solvability of every mathematical problem is a powerful incentive to the worker. We hear within us the perpetual call: There is the problem.

---

40 Paulos, John Allen, *Beyond Numeracy*, Vintage Books, New York, 1992, p. 95.

Seek its solution. You can find it by pure reason, for in mathematics there is no *ignorabimus*.[41]

David Hilbert

In 1930, the demand was repeated.

We must not believe those, who today, with philosophical bearing and deliberative tone, prophesy the fall of culture and accept the *ignorabimus*. For us there is no *ignorabimus*, and in my opinion none whatever in natural science. In opposition to the foolish *ignorabimus* our slogan shall be: Wir müssen wissen – wir werden wissen (We must know – we will know.)

David Hilbert[42]

This remained the dream of the mathematicians. That every well-stated mathematical statement could be proven or disproven, if only the mathematicians could be clever enough to figure out how. This is the dream that Gödel destroyed only a year later.

In 1931, Gödel proved that in any axiomatic system sufficient to do arithmetic, there will always be statements that can be neither proven nor disproven. That is, such statements are independent of the axioms. This is known as Gödel's incompleteness theorem. A key step in the proof is a diagonalization argument just like Cantor used in his later proof of the non-denumerability of the reals. This is a superb illustration of the potential value of reproving already-established theorems. If Cantor hadn't offered this "extra" proof, which was seen by Gödel, then Gödel might not have come up with the diagonalization approach on his own and hence never have proven his big result. Another key step in the proof was the use of a paradox. Basically, it was the statement "This statement is not provable." This statement leads to a contradiction, whether it is considered provable or not. A later proof of Gödel's incompleteness theorem found by the American computer scientist Gregory Chaitin made use of Berry's paradox and complexity theory.[43] As Leibniz remarked, "...there is hardly a paradox without utility."[44]

Now, what's to be done with undecidable statements like the continuum hypothesis? One could simply assume that the statement is true. That is, make it an axiom. But just as reasonably, one could assume that its negation is true and make that statement the axiom. Having one extra axiom is not such a big deal. The problem is that Gödel's incompleteness theorem guarantees there will be more undecidable statements. Even if we let our set of axioms number in the millions, there will remain statements that are undecidable within the greatly expanded axiomatic system. This is discouraging. However, it did lead to something that has improved all of our lives. But before getting to that, it's appropriate to take a quick look at the personal life of Kurt Gödel, considered by some, including myself, to be the greatest logician of all time.

Born in Austria in 1906, Gödel originally trained as a physicist. However, he earned his PhD in mathematics from the University of Vienna in 1930. It was Moritz Schlick who had sparked Gödel's interest in logic. This scholar was murdered in 1936. And the death sparked another obsession in Gödel's life. He began to exhibit paranoia, in

---

41 A transcript can be found at https://mathcs.clarku.edu/~djoyce/hilbert/problems.html.
42 https://en.wikipedia.org/wiki/Ignoramus_et_ignorabimus.
43 Paulos, John Allen, *Beyond Numeracy*, Vintage Books, New York, 1992, p. 96.
44 Quoted in Mandelbrot, Benoit, *The Fractal Geometry of Nature*, Updated and Augmented, W. H. Freeman and Company, New York, 1983, p. 405.

particular a fear of being poisoned, and had to spend months in a sanitarium. In 1938, he married Adele Nimbursky, an exotic dancer.[45] Gödel had spent time working in America in previous years and shortly after World War II began, he and Adele made what turned out to be a permanent move to the United States. Employed by Princeton's Institute for advanced studies, he was able to pursue mathematics and have frequent conversations with his friend Albert Einstein. Historians of mathematics would love to know what the two great men talked about, but there are few records. If they had lived far apart there would be a chance of a large correspondence being preserved, but their close proximity meant that the exchanges were face to face and left no paper trail. Gödel became an American citizen in 1947. His work shifted from mathematics to physics (Einstein probably inspired this) and philosophy. In the 1970s, he penned what he claimed was a proof of the existence of God. His paranoia was also present in this decade. To avoid being poisoned, he had Adele prepare all of his meals. When Adele was hospitalized for six months in 1977, Gödel faced a dilemma. He resolved it by not eating. The result was that he starved to death.

Mental illness, like suicide and drug addiction, is a problem that many people don't want to talk about. But when we do discuss such things, it becomes clear that they affect a great many people. Historically, those affected have included some of the best minds modern civilization has seen. More openness concerning such problems could lead to greater understanding and a better chance of people who need help actually receiving it.

The mathematician John H. Conway, whom you'll see again in Chapter 19, is an atheist; yet, in one of his interviews with his biographer Siobhan Robert, he talked about meeting God. For once we have an apparent paradox with an easy resolution! God was a nickname for Gödel. Other religious nicknames of mathematicians were Christ (Georg Kreisel), the Church (Alonzo Church), and the Bishop (Errett Bishop).[46]

The Silver Lining

With Gödel having shown that there will always be mathematical statements that are undecidable, and an example being known, there was a natural desire to find a way, ideally an efficient way, to identify which statements fell into this unfortunate category. That is, can we decide whether or not a given statement is decidable? This was known as the **decision problem**, although more commonly referred to by its German name, the **Entsheidungsproblem**. In the current form, it represented a generalization of Hilbert's 10th problem, from his famous talk in 1900.

In the wake of Gödel's proof, Alan Turing took on the decision problem. His landmark paper "On Computable Numbers, with an Application to the Entsheidungsproblem" was submitted in 1936 and published in 1937. Once again, the answer was disappointing. Turing proved that there can be no general process for determining if a given statement is provable or not.

---

45 Roberts, Siobhan, *Genius at Play – The Curious Mind of John Horton Conway*, Bloomsbury, New York, 2015, p. 209.
46 Roberts, Siobhan, *Genius at Play – The Curious Mind of John Horton Conway*, Bloomsbury, New York, 2015, p. 211.

Unknown to Turing, Alonzo Church had established this result shortly before him. The Church's paper saw print in 1936.[47]

So, how can there be a silver lining? It seems like the news keeps getting worse! Well, Turing's proof differed from Church's. It included a description of what is now known as a Turing machine. And this machine provides a theoretical foundation for modern computers. Thus, it's one of the key steps leading to the information age. As I indicated before, it's very hard to predict future applications of pure mathematics. In this case, a problem in pure mathematics led to the development of the computer! Later on, an application was found for Gödel's friend's theory of relativity – GPS. The clock in a GPS unit on the surface of the Earth runs slower than the clocks in the satellites, because of the Earth's gravitational field. If this relativistic effect is not taken into consideration, the accuracy of the unit would be reduced. It might indicate you are on a road parallel to the one on which you are actually driving.

All of this should be remembered when determining how much government funding should be allocated to pure research, and to which researchers it should go. Fields that seem to be of no practical value might have large payoffs!

The Big Picture

Set theory can be seen as the basis of everything in mathematics. The two definitions of mathematics that I'm most comfortable with are (briefly)

1. functions acting on sets
2. pattern detection and verification

A closer look at what is meant by the first definition is given in the next chapter. My second definition will make more sense to you as you continue to explore mathematics and form conjectures.

Exercise Set 4

> There is an infinite set A that is not too big.[48]
> John von Neumann

1. Let S be the set of even natural numbers. Let M be the set of prime numbers. What is S ∩ M?

2. Let S be the set of triangular numbers. Let M be the set of factorial numbers. What elements are known to be contained in S ∩ M?

---

47 Church, Alonzo, "An Unsolvable Problem of Elementary Number tTeory," *American Journal of Mathematics*, Vol. 58, 1936. Also see Church, Alonzo, "A Note on the Entsheidungsproblem," *The Journal of Symbolic Logic*, Vol. 1, No. 1, March 1936.

48 Von Neumann, John, "An Axiomatization of Set Theory," 1925. Reproduced in Van Heijenoort, Jean, *From Frege to Gödel: A Source Book in Mathematical Logic, 1879–1932*, pp. 393–413, p. 401 quoted here. Also see www-groups.dcs.st-and.ac.uk/~history/Quotations/Von_Neumann.html.

3. Let U = {1, 2, 3, 4, 5, 6, 7, 8, 9, 10}, A = {1, 4, 7, 9}, B = {2, 3, 5, 7}, and C = {4, 6, 8, 9}. Find the following:

   a. $A \cap B$

   b. $A \cup B$

   c. $B \cap C$

   d. B^C

   e. $(A \cap C)^C$

   f. $A \cap (B \cup C)$

4. Give a real-world example of two sets that are disjoint.

5. Is there a set that is a subset of every set?

6. Let S = {a, b, c, d}. How many elements does P(S) have? List them.

7. If S = {1, 2, 3}, how many elements are in each of the following sets?

   a. P(S)

   b. P(P(S))

   c. P(P(P(S)))

   d. P(P(P(P(S)))) Can your calculator handle this number?

8. How big does S have to be for P(P(P(S))) to contain more than a googol (1 followed by 100 zeros) elements?

9. Give a one-to-one correspondence between the set of even integers and the set of odd integers.

10. How many infinitely long binary strings are there?

11. What is the number of possible sequences of heads and tails that could arise if a coin were flipped repeatedly, forever?

12. Consider the top row of an infinitely wide piece of graph paper (Figure 4.14). How many ways are there to color some of the square black and leave the others white?

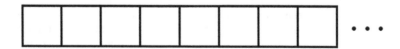

FIGURE 4.14
Too long!

Image created by Craig P. Bauer.

13. What's the flaw in the following "proof" that the natural numbers are in 1:1 correspondence with the reals in the interval (0, 1)?

    First list all of the reals with a single digit after the decimal point, 0.1, 0.2, 0.3, 0.4, 0.5, 0.6, 0.7, 0.8, and 0.9. Next, list all of the numbers with a pair of digits after the decimal point, skipping over any that are equal to previously listed numbers, 0.01, 0.02, 0.03, ..., 0.99. Then list all real numbers with three digits after the decimal point, again skipping over any previously encountered. Continuing in this manner, listing the real numbers in each finite set, in order, every real number will eventually appear on the list. Hence, the natural numbers are in 1:1 correspondence with the reals.

14. Find a function, expressed algebraically, without any trigonometric functions, that gives a one-to-one correspondence between the open interval (0, 1) and the reals.

15. **Bauer's Paradox:** Is it possible to make a New Year's resolution to break all of your New Year's resolutions and to keep it?

16. Make up a paradox of your own.

References/Further Reading

> I did not, however, commit suicide, because I wished to know more of mathematics.
> – Bertrand Russell.[49]

Banach, Stefan and Alfred Tarski, "Sur la décomposition des ensembles de points en parties respectivement congruentes," *Fundamenta Mathematicae*, Vol. 6, 1924, pp. 244–277.

Bell, Eric Temple, *Mathematics, Queen and Servant of Science*, Copyright, 1951, British Edition 1952, reprinted 1954, 1958, 1961, Printed in Great Britain by Butler and Tanner Ltd., Frome and London.

Bolzano, Bernard, *Paradoxes of the Infinite*, Translated from the German edition by Dr. Fr. Prihonský with a historical introduction by Donald A. Steele, Routledge and Kegan Paul, London, 1950.

Bolzano, Bernard, *Paradoxien des unendlichen*, (*The Paradoxes of the Infinite*), C. H. Reclam sen, Leipzig, 1851. The German (original) version is available online at https://archive.org/details/para doxiendesune00bolz and http://dml.cz/bitstream/handle/10338.dmlcz/400242/Bol zano_14-1851-1_5.pdf.

Burger, Edward B. and Michael Starbird, *The Heart of Mathematics*, third edition, Wiley, Hoboken, NJ, 2010.

Cantor, Georg, "Ueber eine elementare Frage der Mannigfaltigkeitslehre," *Jahresbericht der Deutschen Mathematiker-Vereinigung*, Vol. 1, 1890–1891, pp. 75–78. English translation: Ewald, William B. (ed.), *From Immanuel Kant to David Hilbert: A Source Book in the Foundations of Mathematics, Volume 2*, Oxford University Press, Oxford, 1996, pp. 920–922.

Cantor, Georg, *Contributions to the Founding of the Theory of Transfinite Numbers*, Dover, 1955. This is a reprint edition, translated by Philip E. B. Jordain, that contains two memoirs from *Mathematische Annalen* that first appeared in 1895 (vol. xlvi, pp. 481–512) and 1897 (vol. xlix, pp. 207–246) under the title "Beiträge zur Begründung der transfiniten Mengenlehre." Note: this book is 211 pages long, counting the index, but the introduction by Jordain accounts for 82 of these pages. There are also seven pages of notes at the end, after Cantor's two memoirs.

Church, Alonzo, "An Unsolvable Problem of Elementary Number Theory," *American Journal of Mathematics*, Vol. 58, No. 2, 1936, pp. 345–363.

Church, Alonzo, "A Note on the Entscheidungsproblem," *The Journal of Symbolic Logic*, Vol. 1, No. 1, March 1936, pp. 40–41.

Cohen, Paul J. and Reuben Hersh, "Non-Cantorian Set Theory," *Scientific American*, Vol. 217, No. 6, December 1967, pp. 104–116.

---

49 Russell, Bertrand, *The Autobiography of Bertrand Russell, Vol. I: 1872–1914*, Little, Brown and Company, Boston, MA, 1967, p. 50.

Dauben, Joseph W., *Georg Cantor: The Battle for Transfinite Set Theory*. A joint AMS–MAA lecture presented in Atlanta, GA, USA, January 1988. Videotape (NTSC; 60 min. VHS). Originally released on VHS, this talk is now available on DVD. See www.ams.org/bookstore-getitem/item=DVD-1.

Dauben, Joseph Warren, *Georg Cantor: His Mathematics and Philosophy of the Infinite*, Princeton University Press, Princeton, NJ, 1990.

Davis, Martin, *The Undecidable*, Dover, New York, 1958.

Devlin, Keith, *The Joy of Sets: Fundamentals of Contemporary Set Theory*, second edition, Springer, New York, 1993.

Dougherty, Randall and Matthew Foreman, "Banach-Tarski Paradox Using Pieces with the Property of Baire," *Proceedings of the National Academy of Sciences*, Vol. 89, No. 22, November 15, 1992, pp. 10726–10728.

Ewald, William B., editor, *From Immanuel Kant to David Hilbert: A Source Book in the Foundations of Mathematics, Volume 2*, Oxford University Press, Oxford, 1996, pp. 920–922.

Ferreirós, José, *Labyrinth of Thought: A History of Set Theory and its Role in Modern Mathematics*, second revised edition, Birkhäuser, Basel, Switzerland, 2007.

Forrest, Peter, "Grit or Gunk: Implications of the Banach-Tarski Paradox," *The Monist*, Vol. 87, No. 3, July 2004, pp. 351–370.

Frege, Gottlob, *Grundgesetze der Arithmetik*, Band I, 1893, Band II, 1903, Verlag Hermann Pohle, Jena. A revision in modern formal notation is *Grundgesetze der Arithmetik – Begriffsschriftlich abgeleitet. Band I und II: In moderne Formelnotation transkribiert und mit einem ausführlichen Sachregister versehen*, edited by T. Müller, B. Schröder, and R. Stuhlmann-Laeisz, Paderborn: mentis, 2009. And an English version is *Basic Laws of Arithmetic*, translated and edited with an introduction by Philip A. Ebert and Marcus Rossberg, Oxford University Press, Oxford, 2013.

Gray, Jeremy, "Did Poincare Say 'Set Theory is a Disease'?," *Mathematical Intelligencer*, Vol. 13, No. 1, 1991, pp. 19–22.

Halmos, Paul R., *Naïve Set Theory*, D. Van Nostrand, Princeton, NJ, 1960.

Hartnett, Kevin, "Mathematicians Measure Infinities and Find They're Equal," *Quantamagazine*, September 12, 2017. This article was reprinted on ScientificAmerican.com and Spektrum.de. It's available online at https://www.quantamagazine.org/mathematicians-measure-infinities-find-theyre-equal-20170912/ and https://tinyurl.com/ydftfuvz. If you have a long URL that you would like to direct people to using print, as opposed to a link, you can go to https://tinyurl.com/ to create a much shorter address that will work just as well.

Havil, Julian, *Impossible?: Surprising Solutions to Counterintuitive Conundrums*, Princeton University Press, Princeton, NJ, 2008.

Irvine, Andrew David, "Russell's Paradox," *Stanford Encyclopedia of Philosophy*, First published December 8, 1995; substantive revision October 9, 2016, https://plato.stanford.edu/entries/russell-paradox/.

Johnson, Phillip E., *A History of Set Theory*, Prindle, Weber & Schmidt, Boston, MA, 1972.

Kanamori, "The Mathematical Development of Set Theory from Cantor to Cohen," *Bulletin of Symbolic Logic*, Vol. 2, No. 1, March 1996, pp. 1–71.

Meschkowski, Herbert, *Georg Cantor: Leben, Werk und Wirkung*, Bibliographisches Institut, Mannheim, 1983.

Nicholson, Peter, *Essays on the Combinatorial Analysis*, Longman, Hurst, Rees, Orme, and Brown, London, 1818.

Paulos, John Allen, *Beyond Numeracy*, Vintage Books, New York, 1992.

Roberts, Siobhan, *Genius at Play – The Curious Mind of John Horton Conway*, Bloomsbury, New York, 2015.

Rosen, Kenneth H., editor, *Handbook of Discrete and Combinatorial Mathematics*, CRC Press, Boca Raton, FL, 2000.

Russell, Bertrand, "Les Paradoxes de la Logique," *Revue de Métaphysique et de Morale*, Vol. 14, No. 5, September 1906, pp. 627–650, Published by Presses Universitaires de France.

Russell, Bertrand, "The Philosophy of Logical Atomism," *The Monist*, Vol. 29, No. 3, July 1919, pp. 345–380.

Stevens, Michael, "The Banach-Tarski Paradox," *Vsauce*, available online at www.youtube.com/ watch?v=s86-Z-CbaHA. This video is accessible to a general audience.

Taylor, Alan D. and Stan Wagon, "A Paradox Arising from the Elimination of A Paradox," *American Mathematical Monthly*, Vol. 126, No. 4, April 2019. This paper explains how replacing the axiom of choice with an alternative to eliminate the Banach–Tarski paradox yields a system in which another paradox arises.

Tiles, Mary, *The Philosophy of Set Theory: An Historical Introduction to Cantor's Paradise*, Dover, Mineola, NY, 2004.

Van Heijenoort, Jean, *From Frege to Gödel: A Source Book in Mathematical Logic, 1879–1931*, Harvard University Press, Cambridge MA, 1967, reprinted with corrections, 1977.

Vilenkin, N. Ya., *Stories about Sets*, Academic Press, New York, 1968.

Von Neumann, John, "An Axiomatization of Set Theory," 1925. Reproduced in Van Heijenoort, Jean, *From Frege to Gödel: A Source Book in Mathematical Logic, 1879–1931*, pp. 393–413.

Wagon, Stan, *The Banach-Tarski Paradox*, Cambridge University Press, Cambridge, 1985.

Wallace, David Foster, *Georg Cantor: Der Jahrhundertmathematiker und die Entdeckung des Unendlichen*, Piper, Munich, 2007.

Wilson, Robin and John J. Watkins, editors, *Combinatorics: Ancient & Modern*, Oxford University Press, Oxford, 2013.

Zermelo, Ernst, "Beweis, daß jede Menge wohlgeordnet werden kann," *Mathematische Annalen*, Vol. 59, No. 4, 1904, pp. 514–516, reprinted in English translation as "Proof that every set can be well-ordered," van Heijenoort 1976, pp. 139–141.

5

Venn Diagrams

Venn diagrams are usually drawn as intersecting circles, as in Figure 5.1.

The circles labeled A, B, and C represent sets. The intersections show the possibility of elements being held in common by two or more of the sets. If we want to emphasize that a pair of sets have nothing in common, we can depict them as in Figure 5.2.

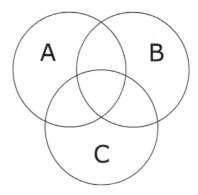

FIGURE 5.1
A three-set Venn diagram.

Image created by Josh Gross.

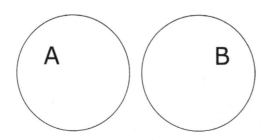

FIGURE 5.2
Disjoint sets.

Image created by Josh Gross.

Such sets are said to be disjoint. However, when an intersection is depicted, as in Figure 5.1, it doesn't necessarily mean that the sets have elements in common; it just allows for the possibility.

Venn diagrams are named after John Venn (Figure 5.3), but he's definitely not their creator. I'll detail their history later in this chapter. For now, let's examine a pair of problems they can be used to solve.

Verifying Identities

In Chapter 1, many tautologies were presented and I showed how they could be verified using truth tables. These identities have analogs in set theory that can be proven with Venn diagrams.

Example 1. Verifying a Distributive Law

Claim: $A \cap (B \cup C) = (A \cap B) \cup (A \cap C)$.

Proof: We first construct the left-hand side. Figure 5.4 shows this being done in small steps that should be easy to follow.

FIGURE 5.3
John Venn (1834–1923).

Public domain.

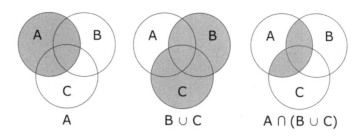

FIGURE 5.4
Constructing $\mathbf{A} \cap (\mathbf{B} \cup \mathbf{C})$.

Image created by Josh Gross.

Constructing the right-hand side of our proposed identity in the same step by step manner (Figure 5.5), it is seen to consist of the same areas.

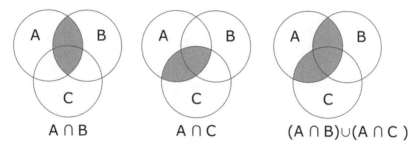

FIGURE 5.5
Constructing $(\mathbf{A} \cap \mathbf{B}) \cup (\mathbf{A} \cap \mathbf{C})$.

Image created by Josh Gross.

Hence, the two sides of the proposed identity are equal and the claim is proven. □

Example 2. Verifying One of De Morgan's Laws

Claim: $(\mathbf{A} \cap \mathbf{B})^{\mathbf{C}} = \mathbf{A}^{\mathbf{C}} \cup \mathbf{B}^{\mathbf{C}}$.

Proof: Because this identity involves complements, it's convenient to place the sets in the Venn diagram inside a square designating the universal set U. This offers a domain of discourse with which to take our complements in respect to.

The left-hand side can be constructed like so (Figure 5.6).

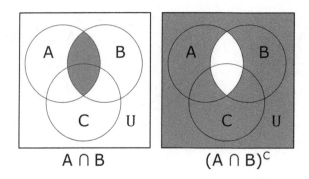

FIGURE 5.6
Constructing $(A \cap B)^C$.

Image created by Josh Gross.

The right-hand side can be built up as shown in Figure 5.7.

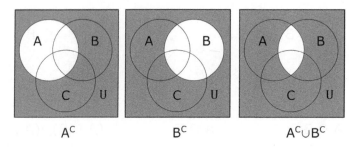

FIGURE 5.7
Constructing $A^C \cup B^C$.

Image created by Josh Gross.

Again, the two final diagrams match, so the identity is established.   □

A list of many of the laws of set theory is provided on page 139 for handy reference. Can you verify all of them using Venn diagrams?

Another application of Venn diagrams is solving problems like those in Examples 3 and 4 below.

Example 3.
If 34 students in a music class like Beethoven, 27 like Wagner, and 20 like both, how many like Beethoven or Wagner?

Commutative laws	$A \cap B = B \cap A$
	$A \cup B = B \cup A$
Associative laws	$A \cap (B \cap C) = (A \cap B) \cap C$
	$A \cup (B \cup C) = (A \cup B) \cup C$
Distributive laws	$A \cap (B \cup C) = (A \cap B) \cup (A \cap C)$
	$A \cup (B \cap C) = (A \cup B) \cap (A \cup C)$
De Morgan's laws	$(A \cap B)^C = A^C \cup B^C$
	$(A \cup B)^C = A^C \cap B^C$
Complement laws	$A \cap A^C = \varnothing$
	$A \cup A^C = U$
Double complement law	$\left(A^C\right)^C = A$
Idempotent laws	$A \cap A = A$
	$A \cup A = A$
Absorption laws	$A \cap (A \cup B) = A$
	$A \cup (A \cap B) = A$
Dominance laws	$A \cap \varnothing = \varnothing$
	$A \cup U = U$
Identity laws	$A \cup \varnothing = A$
	$A \cap U = A$

Solution: We start by filling in the intersection (Figure 5.8).

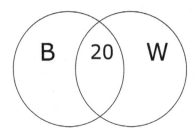

FIGURE 5.8
The number of students who like both is 20.

Image created by Josh Gross.

Because a total of 34 students like Beethoven, there must be another 14 people in the "No Wagner" section of Beethoven's circle of fans (Figure 5.9).

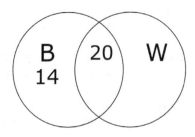

FIGURE 5.9
The number of students who only like Beethoven is 14.

Image created by Josh Gross.

Finally, the rest of Wagner's 27 fans must be accounted for in Figure 5.10. We get:

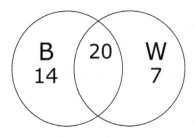

FIGURE 5.10
The number of students who only like Wagner is 7.

Image created by Josh Gross.

The solution to the original question is the sum $14 + 20 + 7 = 41$.

Suppose the question is changed slightly to "How many like Beethoven or Wagner, but not both?" This is changing the or in "Beethoven or Wagner" from the inclusive or to the exclusive or, aka XOR. With the above Venn diagram, it's easy to answer this new question. We simply ignore the area of intersection and take $14 + 7$ to get 21.

Using our sets B and W, this new problem is asking the number of elements in $B \oplus W$, which is called the **symmetric difference** of B and W. In general, the symmetric difference of two sets is the set consisting of the elements in exactly one of the sets – in other words, the elements in either of the sets, but not both.

Example 4.
In a class of 100 students, 30 have read *Starship Troopers*, 20 have read *Red Mars*, 25 have read *Valis*, 8 have read *Valis* and *Red Mars*, 10 have read *Red Mars* and *Starship Troopers*, 6 have read *Valis* and *Starship Troopers*, and 2 have read all three books.

a. Draw a Venn diagram and label all of the sections.
b. How many students only read *Valis*?

c. How many students read none of the three?

d. How many students read *Valis*, but not *Red Mars*? (That is, *Mars* is unred.)

Solution

Again, we start by filling in the intersection of all of the sets (Figure 5.11) and then gradually work our way out (Figures 5.12 through 5.17).

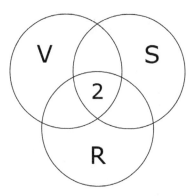

FIGURE 5.11
Two have read all three books.

Image created by Josh Gross.

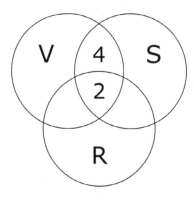

FIGURE 5.12
Six have read *Valis* and *Starship Troopers*.

Image created by Josh Gross.

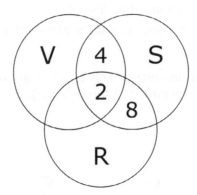

FIGURE 5.13
Ten have read *Red Mars* and *Starship Troopers*.

Image created by Josh Gross.

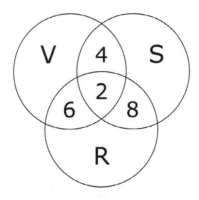

FIGURE 5.14
Eight have read *Valis* and *Red Mars*.

Image created by Josh Gross.

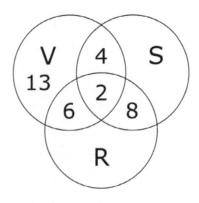

FIGURE 5.15
Twenty-five have read *Valis*.

Image created by Josh Gross.

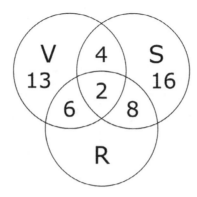

FIGURE 5.16
Thirty have read *Starship Troopers*.

Image created by Josh Gross.

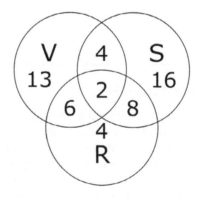

FIGURE 5.17
Twenty have read *Red Mars*.

Image created by Josh Gross.

Now, we started with 100 students and the numbers entered in the Venn diagram only add up to 53. Therefore, there must be 47 students who have not read any of these novels. We may box the diagram and place 47 within the box, but outside the circles, to indicate these students (Figure 5.18).

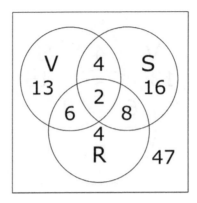

FIGURE 5.18
The completed Venn diagram.

Image created by Josh Gross.

It's now easy to answer questions b through d.
b. 13
c. 47
d. 17

The (Dirichlet) Pigeon–Hole Principle

When placing numbers in sections of a Venn diagram, it should surprise no one that if there are more numbers than sections, one number must be left over or one section would have to have two numbers placed in it, which would not be appropriate. What may be surprising is that this is considered a mathematical principle and is important enough to have a name attached to it. If we wish to credit its "discoverer," we call it the **Dirichlet Pigeon-hole principle**. Dirichlet formalized this result in 1834, and it can be stated as follows:

If we distribute m items among n boxes, and $m > n$, then at least one box must contain more than one item.

Dirichlet called this the *Schubfachprinzip* and it has been referred to by many other names since then, including **Dirichlet's box principle** and **Dirichlet's drawer principle**.[1] As the following example shows, the principle saw use long before it had a name.

Example 1 – Big Hairy Deal!

DIRICHLET'S "BOXES" IN THE 17TH CENTURY. [Pierre] Nicole, one of the authors of the famous Logic of Port-Royal, relates the following: "One day I told Madame de Longueville that I could prove that there are at least two people living in Paris with the same number of hairs on their heads. She asserted that I could never prove this without counting them first. My premisses are these. No head has more than 200,000 hairs, and the worst provided has one.

1 http://jeff560.tripod.com/p.html and https://en.wikipedia.org/wiki/Pigeonhole_principle.

Consider 200,000 heads, none having the same number of hairs. Then each must have a number of hairs equal to some number from 1 to 200,000 both included. Of course if any have the same number of hairs my bet is won. Now take one more person, who has not more than 200,000 hairs on his head. His number must be one of the numbers 1 to 200,000 included. As the inhabitants of Paris are nearer 800,000 than 200,000, there are many heads with an equal number of hairs."[2]

Like nearly all of the other ideas presented in this book, we can further develop the pigeon-hole principle. An easy generalization is:

If we distribute $mn + 1$ items among n boxes, and $m > n$, then at least one box must contain more than m items.

However, much more has been done!

This simple idea was developed almost beyond recognition in the 20th century, by the logician F. P. Ramsey and others, to prove the existence of substructures of a required type in a larger structure, provided that the original structure is sufficiently large.

Ian Anderson[3]

Ramsey theory is covered in Chapter 20.

Some History of Venn Diagrams

As mentioned earlier, John Venn was not the first to make use of diagrams like those above. They were used, for example, by Gottfried Wilhelm Leibniz (1646–1716).[4] Figures 5.19 and 5.20 are unfortunately sloppy pages in Leibniz's hand showing his use of the diagrams. Notice how he made sketches with rectangles, as well as the circles that are more common today. Like much of his other work, these diagrams were not published during Leibniz's lifetime. In fact, his work on the subject didn't see print until 1903.[5]

Euler diagrams also came before Venn diagrams. However, they were different in an important way. Areas of intersection are only shown in Euler diagrams if the intersection is nonempty, while for Venn diagrams intersections and other sections can be empty (Figure 5.21).

---

2 "Gleanings from Far and Near," *The Mathematical Gazette*, Vol. 32, No. 300, July 1948, p. 159. These "gleanings" consist of short amusing passages used to fill the blank space that would otherwise follow the last lines of the individual journal articles.

3 Anderson, Ian, "Combinatorial Set Theory," in Wilson, Robin and John J. Watkins, editors, *Combinatorics: Ancient and Modern*, Oxford University Press, Oxford, pp. 309–328, p. 310 quoted here.

4 Bocheński, J. M., *A History of Formal Logic* (translated and edited by Ivo Thomas), University of Notre Dame Press, Notre Dame, IN, 1961, pp. 260 and 262. Some "Venn" diagrams drawn by Leibniz are reproduced on the page facing page 260.

5 Bocheński, J. M., *A History of Formal Logic* (translated and edited by Ivo Thomas), University of Notre Dame Press, Notre Dame, IN, 1961, pp. 260 and 262. See Leibniz, Gottfried Wilhelm [ca. 1690]. "De Formae Logicae per linearum ductus," in Couturat, Louis. *Opuscules et fragmentes inedits de Leibniz* (in Latin), 1903, pp. 292–321 for the first publication.

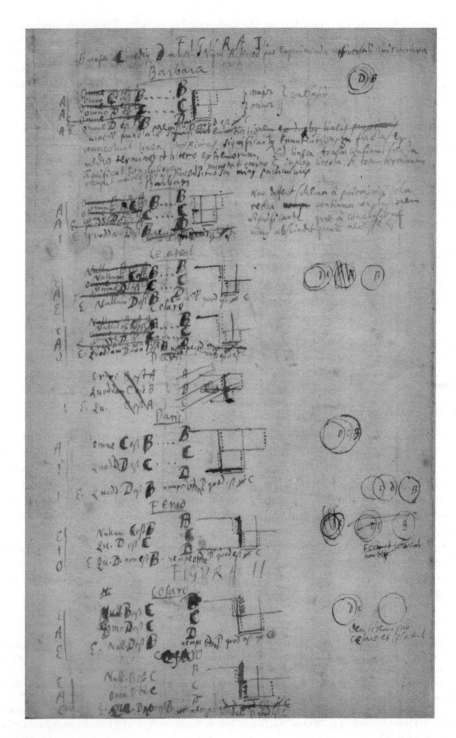

FIGURE 5.19
Leibniz's "Venn" diagrams.

Gottfried Wilhelm Leibniz Bibliothek – Niedersächsische Landesbibliothek, Hannover.

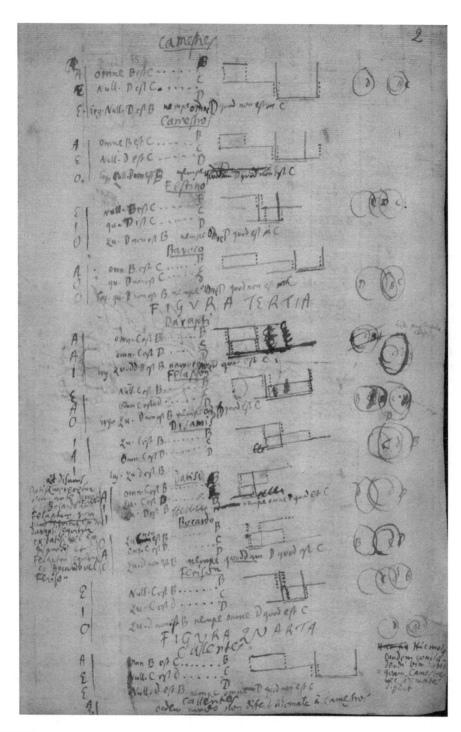

FIGURE 5.20

More of Leibniz's "Venn" diagrams.

Gottfried Wilhelm Leibniz Bibliothek – Niedersächsische Landesbibliothek, Hannover.

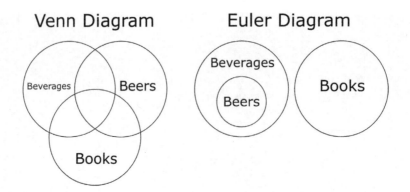

FIGURE 5.21
A Venn diagram contrasted with an Euler diagram.

Image created by Josh Gross.

Insisting that the diagrams consist only of circles would be a mistake, because if we wish to consider all possible intersections of four sets, a diagram is not even possible with circles alone. It can, however, be done with ellipses, as shown by John Venn and reproduced in Figure 5.22.

There are other solutions using a mix of shapes (Figures 5.23 and 5.24).

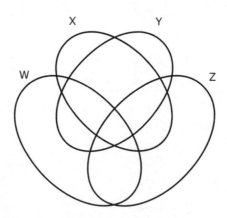

FIGURE 5.22
John Venn's use of ellipses for a four-set diagram.

Image created by Josh Gross, based on Bocheński, J. M., *A History of Formal Logic* (translated and edited by Ivo Thomas), University of Notre Dame Press, Notre Dame, IN, 1961, p. 261.

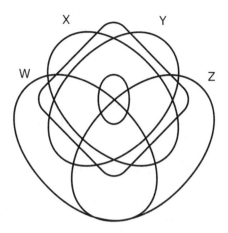

FIGURE 5.23
Another diagram from Venn.

Image created by Josh Gross, based on Bocheński, J. M., *A History of Formal Logic* (translated and edited by Ivo Thomas), University of Notre Dame Press, Notre Dame, IN, 1961, p. 261.

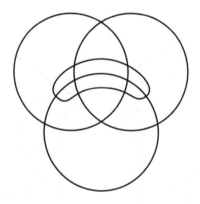

FIGURE 5.24
A sad-looking diagram from Venn for four sets.

Image created by Josh Gross, based on Bocheński, J. M., *A History of Formal Logic* (translated and edited by Ivo Thomas), University of Notre Dame Press, Notre Dame, IN, 1961, p. 261.

Some of the possible intersections are missing in this last diagram. An exercise asks you to identify them.

From an aesthetic point of view, the loss of rotational symmetry that these four-set diagrams entail is not especially pleasing. But progress has been made since Venn's day. And some of it has been made in a way that is typical in mathematics, by various mathematicians chipping away at a problem.

The quest for rotationally symmetric Venn diagrams got off to a slow, and disappointing, start. In 1963, David W. Henderson showed that such a four-set Venn diagram cannot exist. His result was more general though. He found that if a rotationally symmetric diagram exists for n sets, then n must be prime.[6] Because 4 is not prime, we don't have a chance. But his theorem was not an "if and only if." That is, it was left as an open question whether every prime number of sets does indeed have a rotationally symmetric Venn diagram. He showed they exist for the prime 5 with the two examples shown in Figure 5.25.

Henderson also noted, "A symmetric Venn diagram for seven classes has been found, the regions being irregular hexagons, but the existence of symmetric Venn diagrams for primes greater than seven has not been demonstrated."[7] He did not include an image of the seven-set diagram in his 1963 paper.

Decades later a beautiful seven-set rotationally symmetric diagram was discovered by A. W. F. Edwards (see color figure 6 in color insert). Edwards later wrote:

> I cannot see a sense in which, to take an example from the book, the Venn diagrams with sevenfold rotational symmetry did not exist before they were found. We did not know they existed – Professor Branko Grünbaum, the authority on Venn diagrams, had even come to doubt the possibility – but the moment they were uncovered they seemed ageless and eternal. My discovery in December 1992 of the one I christened "Adelaide" (for that is where I found it) was a source of infinite pleasure to me that nothing can take away and is in no way diminished by the later knowledge that Grünbaum had found it earlier in the same year (but not published it).[8]

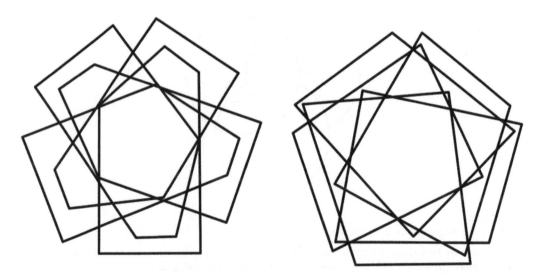

FIGURE 5.25
Henderson's five-set rotationally symmetric Venn diagrams.

Image created by Josh Gross, based on Henderson, David W., "Venn Diagrams for More than Four Classes, *American Mathematical Monthly,* **Vol. 70, No. 4, April 1963, pp. 424–426.**

6 Henderson, David W., "Venn diagrams for more than four classes," *American Mathematical Monthly*, 70, No. 4, April 1963, pp. 424–426.
7 Henderson, David W., "Venn diagrams for more than four classes," *American Mathematical Monthly*, 70, No. 4, April 1963, pp. 424–426, p. 425 cited here.
8 Edwards, A. W. F., *Cogwheels of the Mind: The Story of Venn Diagrams*, Foreword by Ian Stewart, The Johns Hopkins Press, Baltimore and London, 2004, p. xvi.

It is not uncommon for results to appear nearly simultaneously for decades-old problems!

Such examples may increase confidence of rotationally symmetric solutions existing for all primes, but examples do not constitute a proof that it can always be done. About a decade later, the next person to chip away at the problem was Peter Hamburger. In 2002, he found a rotationally symmetric Venn diagram for the next prime number of sets, 11.[9]

Progress might have continued in this manner with new constructions being found for various prime numbers of sets every few years, but in 2004 a paper appeared that brought the problem to a close. Jerrold Griggs, Charles E. Killian, and Carla D. Savage showed that the rotationally symmetric Venn diagrams do indeed exist for all prime number of sets.[10] See color figure 7 **(See color insert)** for examples of symmetric seven-set diagrams found by Frank Ruskey.

While the work described above took care of one conjecture concerning Venn diagrams, there are many other directions in which research can be carried out. For example, a six-set diagram cannot be constructed with rotational symmetry, but Edwards showed that it can still be attractively drawn (Figure 5.26).

Diagrams such as these can be used to solve problems like Examples 3 and 4 above, when more sets are involved. However, it becomes easier to make mistakes when there are so many areas to choose from! Fortunately, there's another approach to such problems that doesn't require a diagram.

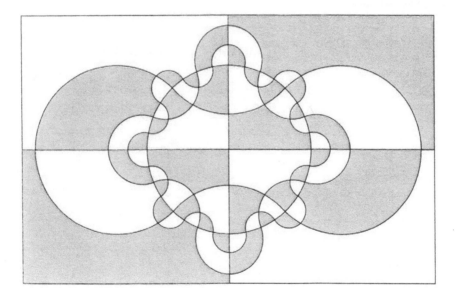

FIGURE 5.26
A six-set Edwards–Venn diagram.

Courtesy of A. W. F. Edwards.

---

9 Hamburger, Peter, "Doodles and Doilies, Non-simple Symmetric Venn Diagrams," *Discrete Mathematics*, Vol. 257, No. 2–3, November 28, 2002, pp. 423–439.

10 Griggs, Jerrold, Charles E. Killian, and Carla D. Savage, "Venn Diagrams and Symmetric Chain Decompositions in the Boolean Lattice," *The Electronic Journal of Combinatorics*, 11, No. 1, R2, 2004.

The Inclusion–Exclusion Principle

If n(S) denotes the number of elements in the set S, we have

$$n(A \cup B) = n(A) + n(B) - n(A \cap B)$$

We can use a Venn diagram to see that this formula is correct.
We first sketch the two sets, A and B (Figure 5.27).

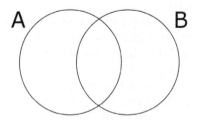

FIGURE 5.27
Our two sets.

Image created by Josh Gross.

We then check off the sections that are part of n(A) (Figure 5.28).

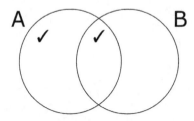

FIGURE 5.28
Counting elements in A.

Image created by Josh Gross.

Next we check off the sections that are part of n(B) (Figure 5.29).

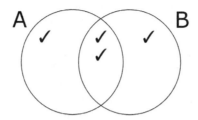

FIGURE 5.29
Counting elements in B.

Image created by Josh Gross.

Figure 5.29 represents n(A) + n(B), but it clearly has too many check marks in the section for A ∩ B. So, to get the correct total, we must subtract n(A ∩ B). The representation for n(A) + n(B) − n(A ∩ B) is shown in Figure 5.30.

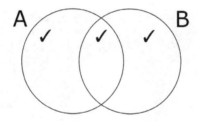

FIGURE 5.30
Discarding elements counted twice.

Image created by Josh Gross.

This is exactly what we want. Each section is counted exactly once. Thus, the formula $n(A \cup B) = n(A) + n(B) - n(A \cap B)$ is correct.

In the same manner, we can verify formulas for three and four sets. For three sets, we have

$$n(A \cup B \cup C) = n(A) + n(B) + n(C) - n(A \cap B) - n(A \cap C) - n(B \cap C) + n(A \cap B \cap C)$$

For four sets, we have

$$\begin{aligned} n(A \cup B \cup C \cup D) = {} & n(A) + n(B) + n(C) + n(D) \\ & - n(A \cap B) - n(A \cap C) - n(A \cap D) - n(B \cap C) - n(B \cap D) - n(C \cap D) \\ & + n(A \cap B \cap C) + n(A \cap B \cap D) + n(A \cap C \cap D) + n(B \cap C \cap D) \\ & - n(A \cap B \cap C \cap D) \end{aligned}$$

Do you notice the pattern? If not, take some time to recognize it for yourself before continuing.

For four sets, the pattern amounts to first summing (or including) the orders of the sets taken one at a time, then we subtract (or exclude) the orders of intersections of sets taken two at a time. It's then time for inclusion again, summing the intersections of the sets taken three at a time. We end by excluding the intersection of all four sets.

What would the formula be for five sets? It would alternate in the same way between inclusion (adding in values) and exclusion (subtracting values out again). The difference is that the calculation would run for one extra step and end with including the intersection of all five sets. Also, each stage, except the very last, would have at least one more term in it. It would be a little messy to sketch and confirm, as we did for the two-set version, but a proof could be given in this manner.

The pattern continues for all $n \in \mathbb{N}$.[11]

It's obvious from the examples above why the rule is known as the inclusion–exclusion principle, but it hasn't always gone by this name. It was sometimes called the **cross-classification principle**.[12]

A religious refugee, Abraham de Moivre (Figure 5.31), is to thank for this result. He was born in France, but, as a protestant, decided to flee to England in 1685 and

FIGURE 5.31
Abraham de Moivre (1667–1754).

Public domain.

---

11 See Young, Robert M., *Excursions in Calculus: An Interplay of the Continuous and the Discrete*, The Mathematical Association of America, 1992, p. 32, for a sketch of the proof for the general case.

12 Wilson, Robin, and John J. Watkins, editors, *Combinatorics: Ancient & Modern*, Oxford University Press, Oxford, 2013, p. 313.

remained there for the rest of his life.[13] de Moivre applied the inclusion–exclusion principle in his 1718 book *The Doctrine of Chances*, when investigating derangements, as is shown in Chapter 9.

Venn diagrams are the source of much humor. An example is reproduced in Figure 5.32 and two more appear as color figures 8 and 9 **(See color insert)**.

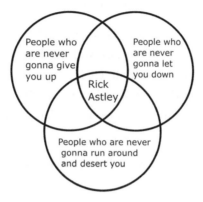

FIGURE 5.32
Rickrolled!

Image created by Josh Gross, based on an internet meme.

Exercise Set 5

> Everything is theoretically impossible, until it is done.
> Robert A. Heinlein

1. Use Venn diagrams to prove

$$A \cup (B \cap C) = (A \cup B) \cap (A \cup C)$$

2. Use Venn diagrams to prove

$$(A \cup B)^C = A^C \cap B^C$$

3. If 28 people like Guinness, 19 people like Yuengling, and 10 people like both, how many like Guinness or Yuengling?

---

13 Wilson, Robin, and John J. Watkins, editors, *Combinatorics: Ancient & Modern*, Oxford University Press, Oxford, 2013, p. 311

4. In a group of 60 people, 35 support marijuana legalization, 42 support gay marriage, and 29 support both. How many support neither?

5. In a group of 79 college graduates, 31 failed a class, 34 were athletes, 20 liked the cafeteria food, 16 failed a class and were athletes, 10 were athletes and liked the cafeteria food, 11 failed a class and liked the cafeteria food, and 7 failed a class, were athletes, and liked the cafeteria food. Draw a Venn diagram to represent this information and then answer the following questions.

   a. How many athletes liked the cafeteria food?
   b. How many students failed a class and liked the cafeteria food?
   c. How many students weren't athletes?
   d. How many athletes didn't fail a class or like the cafeteria food?
   e. How many of the students were not athletes, didn't fail a class, and didn't like the cafeteria food?

6. What's wrong with the attempt to draw a Venn diagram for four sets shown in Figure 5.33?

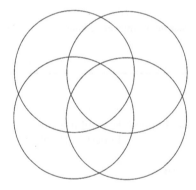

FIGURE 5.33
Four-set FAIL.

Image created by Craig P. Bauer.

7. What's wrong with the attempt to draw a Venn diagram for four sets shown in Figure 5.34?

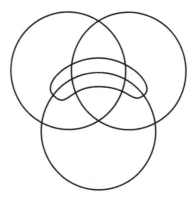

FIGURE 5.34
Another FAIL.

Image created by Josh Gross, based on Bocheński, J. M., *A History of Formal Logic* **(translated and edited by Ivo Thomas), University of Notre Dame Press, Notre Dame, IN, 1961, p. 261.**

8. Fill in the missing area of the Venn diagram in Figure 5.35.

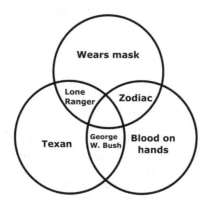

FIGURE 5.35
Who is the mystery man in the center?

Image created by Josh Gross.

9. Make your own humorous Venn diagram. It must have three (or more) sets.
10. Suppose the heights of 120 people are determined to the closest inch. Must there have been two people in the group to whom the same height is assigned?
11. Suppose the weights of 1,500 people are determined to the closest pound. Must there be two people in the group to whom the same weight is assigned?

12. Now, suppose that for 1,500 people both height and weight are determined, again to the closest inch and pound. Must there be two people assigned both the same height and weight?

13. In the fall 2017 semester, University of California, Berkeley, had 29,311 students.[14] Must there have been two with matching initials (first middle and last)?

References/Further Reading

Bauer's law: There's always room for another bookshelf.

Anderson, Ian, "Combinatorial Set Theory," in Wilson, Robin and John J. Watkins, editors, *Combinatorics: Ancient and Modern*, Oxford University Press, Oxford, 2013, pp. 309–328.

Baron, Margaret E., "A Note on the Historical Development of Logic Diagrams," *The Mathematical Gazette*, Vol. 53, No. 384, May 1969, pp. 113–125.

Bocheński, J. M., *A History of Formal Logic* (translated and edited by Ivo Thomas), University of Notre Dame Press, Notre Dame, IN, 1961. There's also a 2012 reprint.

Cipra, Barry, "Diagram Masters Cry 'Venn-I, Vidi, Vici'," *Science*, Vol. 299, No. 5607, January 31, 2003, p. 651.

Edwards, A. W. F., *Cogwheels of the Mind: The Story of Venn Diagrams*, Foreword by Ian Stewart, The Johns Hopkins Press, Baltimore and London, 2004.

Euler, Leonhard, *Lettres à une princesse d'Allemagne sur divers sujets de physique et de philosophie* (Letters to a German Princess on Various Physical and Philosophical Subjects), l'Academie Impériale des Sciences, Saint Petersburg, Russia, 1768, Vol. 2, pp. 95–126.

Griggs, Jerrold, Charles E. Killian, and Carla D. Savage, "Venn Diagrams and Symmetric Chain Decompositions in the Boolean Lattice," *The Electronic Journal of Combinatorics*, Vol. 11, No. 1, 2004, #R2.

Hamburger, Peter, "Doodles and Doilies, Non-simple Symmetric Venn Diagrams," *Discrete Mathematics*, Vol. 257, No. 2–3, November 28, 2002, pp. 423–439.

Henderson, David W., "Venn Diagrams for More than Four Classes," *American Mathematical Monthly*, Vol. 70, No. 4, April 1963, pp. 424–426.

Mac Queen, Gailand, *The Logic Diagram*, Master of Arts Thesis, McMaster University, October 1967, available online at https://macsphere.mcmaster.ca/bitstream/11375/10794/1/fulltext.pdf.

Moran, Lee, "Scott Walker Gets Mercilessly Mocked by Twitter Users Over Venn Diagram Fail," *HUFFPOST*, December 15, 2018, available online at www.huffingtonpost.com/entry/scott-walker-venn-diagram-meme_us_5c14e6d5e4b05d7e5d8258cc.

Ruskey, Frank, Carla D. Savage, and Stan Wagon, "The Search for Simple Symmetric Venn Diagrams," *Notices of the AMS*, Vol. 53, No. 11, December 2006, pp. 1304–1311.

Venn, John, "On the Diagrammatic and Mechanical Representation of Propositions and Reasonings," *The London, Edinburgh, and Dublin Philosophical Magazine and Journal of Science*, Vol. 9, 1880, pp. 1–18.

Wilson, Robin, and John J. Watkins, editors, *Combinatorics: Ancient & Modern*, Oxford University Press, Oxford, 2013.

Young, Robert M., *Excursions in Calculus: An Interplay of the Continuous and the Discrete*, The Mathematical Association of America, 1992.

---

14 https://blog.prepscholar.com/the-biggest-colleges-in-the-united-states

6

The Functional View of Mathematics

> It was in 1694 that Leibniz introduced that most useful word, function (or its Latin equivalent), into mathematics.
>
> E. T. Bell[1]

Today the concept of a function is taken for granted and many readers will likely be surprised that the idea doesn't go back to ancient Greece. In fact, the Greeks based their mathematics largely on ratios (proportions) instead, and the functional view really is a modern perspective.

We can think of nearly all of mathematics as functions acting on sets. What exactly those functions and sets are varies depending on the kind of mathematics being examined, but the approach holds. The previous chapter introduced set theory and this chapter covers the concept of functions and how they can act on sets.

> Since the time of Leibniz the idea has so grown and developed in importance and power that the function concept may now be said, without exaggeration, to have attained supremacy alike in the enterprise of mathematics and in that of science.
>
> Cassius Jackson Keyser[2]

Such an important concept should be defined carefully. Some books offer a definition for function like this:

> A **function** is a rule that takes certain numbers as inputs and assigns to each a definite output number. The set of all input numbers is called the **domain** of the function and the set of all resulting output numbers is called the **range** of the function.[3]

The range is sometimes referred to as the **codomain** of the function. What we plug into the function is called the "argument." This brings us to a problem with the definition given above. It is not the most general definition, but rather the definition for a *real-valued* function. It suffices for some mathematics classes, although the concept of function is actually far broader. A few examples illustrating this follow.

---

1 Bell, Eric Temple, *Mathematics, Queen and Servant of Science*, Copyright, 1951, British Edition 1952, reprinted 1954, 1958, 1961, Printed in Great Britain by Butler and Tanner Ltd., Frome and London, p. 321.

2 Keyser, Cassius Jackson, *Mathematics as a Culture Clue and Other Essays, Volume 1 in The Collected Works of Cassius Jackson Keyser*, Scripta Mathematic, Yeshiva University, New York, 1947, p. 95.

3 Hughes-Hallett, Deborah, Andrew M. Gleason, William G. McCallum, et al., *Calculus*, third edition, John Wiley & Sons, Inc., New York, 2002, p. 2.

Example 1 – The Caesar Shift Cipher

This cipher, which was used by Julius Caesar back in the day, is a function that assigns words to enciphered words like so, by shifting each letter forward three positions in the alphabet:

MATHEMATICS → PDWKHPDWLFV

So, a function can take words as input and give meaningless strings of letters as output. We need to make sure our rule is well-defined for all elements of the domain. So, we must have a way of enciphering words like ZERO. How do we shift Z forward by three positions? Simple – loop back to the start. Z is shifted through A, B, and finally lands at C. We have ZERO → CHUR. Whatever rule we wish to consider, it must unambiguously associate an element of the range with each element of the domain, if it's to be called a function. However, there's no requirement that a function has a simple mathematical interpretation.

Example 2 – Garage Sale Pricing

A function can be represented as an explicit list of what is sent to what. We don't have to express it algebraically.

Book → 50 cents

Hammer → 1 dollar

Shirt → 75 cents

DVD → 1 dollar

Shot Glass → 50 cents

This function represents the prices a person placed on items to sell at a garage sale. The domain of the function consists of the items and the range consists of the prices.

Example 3 – Proposal Probabilities

Example: $f(A, B)$ = probability of person A proposing to person B.

A list of some of the points of the domain and where they're sent by f might look like this:

Alice, Bob → 0.0001

Bob, Alice → 0.92

Bob, Connie → 0.05

Bob, David → 0.00

David, Alice → 0.10

This models a situation where Alice and Bob are in a serious relationship, but Alice can't imagine popping the question herself. That's okay, because Bob is likely to propose, if things continue to go well. On the other hand, if it doesn't work out, Bob also likes Connie and he sees a possibility of proposing to her in that case. Bob doesn't see any chance of proposing to his best friend David, but unbeknownst to Bob, David is strongly attracted to Alice and will try to date her if it doesn't work out for Bob. So, David believes there is a probability of 0.10 that he'll end up proposing to Alice one day.

As the above example shows, the domain can consist of ordered pairs of names and the range numbers. Order matters in the domain because, as seen above, A may be far more likely to propose to B than B is to propose to A. That is, $f(A, B) > f(A, B)$.

If we want to emphasize the set theory underlying this example, we can define S to be the set of people we are considering and say that our function maps the Cartesian product $S \times S$ to the interval $[0, 1]$.

Example 4 – A Placement Test
In the above examples the range consisted of a string of letters or numerical values, but that needn't be the case. A placement test could be used to assign a student to a particular level of a foreign language.

90–100 → German 202

80–89 → German 201

60–79 → German 102

0–59 → German 101

Here the range consists of particular German language classes.

Example 5 – Anniversary Gifts
If one of the potential proposals of Example 3 was actually made and accepted, we also have a function suggesting what sort of material each anniversary gift should consist of based on which anniversary is being celebrated:[4]

1ST → Paper	9TH → Pottery	25TH → Silver
2ND → Cotton	10TH → Tin, Aluminum	30TH → Pearl
3RD → Leather	11TH → Steel	35TH → Coral
4TH → Linen, Silk	12TH → Silk	40TH → Ruby
5TH → Wood	13TH → Lace	45TH → Sapphire
6TH → Iron	14TH → Ivory	50TH → Gold
7TH → Wool, Copper	15TH → Crystal	55TH → Emerald
8TH → Bronze	20TH → China	60TH → Diamond

The list really is arbitrary and many would now consider ivory to be in bad taste.

Example 6 – Differential Calculus
As an illustration of the power of this broad view of function, consider differential calculus. Certainly many functions are seen in this class, but the class itself consists of a detailed study of just one function, namely $\frac{d}{dx}(f(x))$. That is, the course examines the differentiation function, a function that maps a function to another function. We have

$$x^2 \to 2x \qquad \sin(x) \to \cos(x) \qquad \ln(x) \to 1/x$$

The domain of the function $\frac{d}{dx}(f(x))$ consists of all differentiable functions and the range consists of all integrable functions.

---

4 https://en.wikipedia.org/wiki/Wedding_anniversary is the source of the list used here.

Example 7 – Integral Calculus
The definite integral is a function that takes an ordered triplet (an integrable function, a lower limit of integration, and an upper limit of integration) as input and yields a real number as output. For example, we have

$$(x^2,\ 0,\ 1) \to 1/3.$$

There are many ways to show how a function maps the elements of one set to another. The explicit listing of the pairings of domain and range elements, as seen in many of the examples above, is clear but impractical except when the domain consists of a very small number of elements.

Representing a Function by a Rule (i.e., Algebraically)

This is not new to you. We're used to representing functions by rules such as $f(x) = \sqrt{x}$ and $g(x) = \cos(x)$. Often, when a function is defined algebraically, we also wish to explicitly state its domain and range. A typical format is shown below.

$$f(x) = e^x$$

$$f : \mathbb{R} \to \mathbb{R}^+.$$

The first line tells us the mathematical rule for associating an element of the domain with an element of the range. The second line tells us the domain is $\mathbb{R}$, the set of real numbers, and the range is $\mathbb{R}^+$, the set of positive real numbers.

We often indicate the domain more subtly by using a particular letter for the variable. For example, $f(n)$ designates the natural numbers as the domain, $f(x)$ designates the reals, and $f(z)$ designates the complex numbers. This is tradition and it is unfortunate that z is similar in appearance to $\mathbb{Z}$, which represents the integers.

Onto Functions, One-to-one Functions

In general, we may write $f : A \to B$ to indicate that the domain of the function is the set A and the range is the set B. However, in practice, the range is often a *subset* of the set B. For example, we can write $f(x) = \sin(x)$, $f : \mathbb{R} \to \mathbb{R}$. This shows that we're interested in $\sin(x)$ as a real-valued function (as opposed to complex) and is not meant to indicate that the range of $\sin(x)$ is all of $\mathbb{R}$. The range is just $[-1, 1]$ and we could write $f(x) = \sin(x)$, $f : \mathbb{R} \to [-1, 1]$, but this is often not done. If a function is such that each value in the indicated range is actually achieved for some value in the domain, as in the function $f(x) = x^3$, $f : \mathbb{R} \to \mathbb{R}$, we say that the function is **onto** or that the mapping is **surjective** or a **surjection**.

Independent of whether this happens or not, we may also ask if there are ever two distinct values in the domain, say x_1 and x_2 such that $f(x_1) = f(x_2)$. That is, do two different points ever get sent to the same value by the function? If the answer is yes, we say that the function is **one-to-one** and we may call the mapping **injective** or an **injection**. We sometimes write 1:1 as a terser expression of one-to-one.

A function that is both one-to-one and onto is called **bijective** or a **bijection**.

Because of the importance of these concepts there are yet more names for them, but they're typically reserved for use in particular fields of mathematics. So, unless you are in an abstract algebra or topology course, it's best to stick to the terminology used here.

Representing a Function Graphically

Example 8 – The Bell Curve

The name of this curve arises from the fact that it's shaped like a bell. It has nothing to do with E. T. Bell. One indication that something is important is that it can be referred to by several different names. This curve is also known as the normal curve, the normal distribution, and the Gaussian distribution. A sketch appears in Figure 6.1.

It's easiest to understand (or explain) this function by looking at its graphical representation. By contrast, the algebraic formula, $f(x) = (1/\sqrt{2\pi\sigma^2})e^{-(x-\mu)^2/2\sigma^2}$, is much less clear to people who don't already have a lot of experience with mathematics.

For now, note that when a curve is represented graphically, it's a function if and only if every vertical line that can be drawn on the graph hits the curve at no more than one point. This is just another way of saying that there's never more than one value of y associated with each value of x. This is clearly the case for the bell curve.

Example 9 – The Limaçon of Pascal

The graph in Figure 6.2 arises from the equation $(x^2 + y^2 - 2ax)^2 = b^2(x^2 + y^2)$.

It's a fine graph, but it fails the vertical line test and is therefore not a function. It is an example of a relation, a topic covered in Chapter 22.

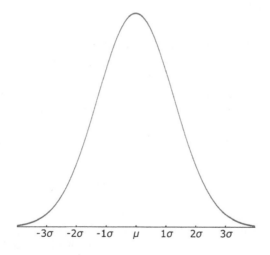

FIGURE 6.1

The bell curve peaks at $x = \mu$ and is symmetrical.

Image created by Josh Gross.

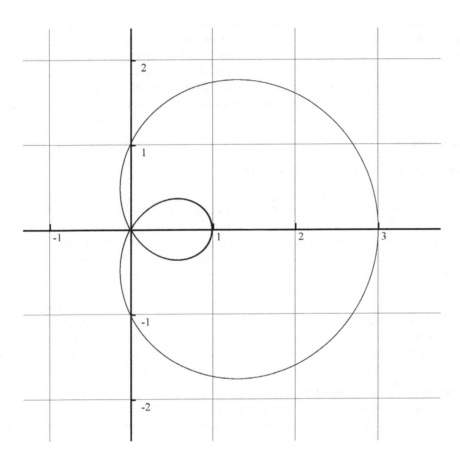

FIGURE 6.2
The Limaçon of Pascal.

Image created by Garrett Ruths.

Example 10 – The Number of Divisors of an Integer
A mathematician searching for a pattern in the number of distinct positive divisors of the natural numbers could graph early values, as shown in Figure 6.3.

Because the domain consists of natural numbers, the graph is simply a bunch of dots, known as a **scatterplot**. Sometimes in such a discrete plot the points are connected by straight lines, but this is not meant to indicate that there are in-between values. Because every value on the horizontal axis has no more than one corresponding value on the vertical axis, this graph represents a function.

Example 11 – Graphing a Function of Two Variables
As Figure 6.4 shows, we can also represent functions of two variables graphically.

We may express this function algebraically as either $z = x^2 + y^2$ or $f(x,y) = x^2 + y^2$.

There is much science that consists of collecting data (experimentally) and then searching for a simple expression of a function that produces data very close to what was collected. Hope is placed in the predictive value of the function, that is, that output values it yields for input values not collected would be close to what such collected outputs would actually be. An example shows this process.

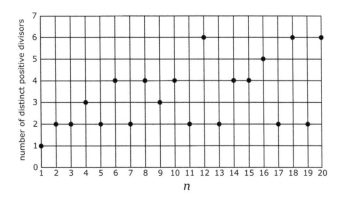

FIGURE 6.3
An example of a scatterplot.

Image created by Josh Gross.

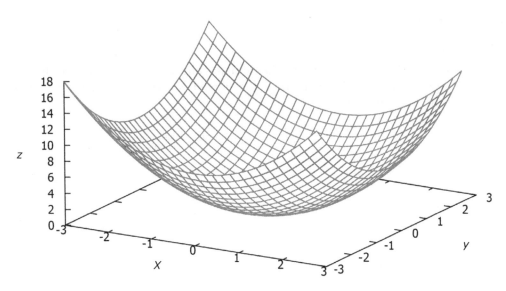

FIGURE 6.4
A function of two variables drawn in the plane to create the illusion of depth.

Image created by Josh Gross.

Example 12 – Bode's Law[5]

A mathematical model that goes back to the 18th century attempted to express the distances of the planets from the sun in a simple way. Known as Bode's law, after the German astronomer Johann Elert Bode, it can be expressed as[6]

---

5 Titius (and others) had versions of this law before Bode. It is now often called the Titius–Bode law.
6 There are other, equivalent, formulations. The one used here can be found at https://en.wikipedia.org/wiki/ Titius%E2%80%93Bode_law and elsewhere.

$$a = 0.4 + (0.3)(2^m),$$

where a is the distance from the sun in astronomical units[7] and m takes the values $-\infty$, 0, 1, 2, ...

This formula worked well for the planets known at the time, but the test of its value is in its predictive ability. Plugging in $-\infty$, 0, 1, and 2 gave good values for the distances of Mercury, Venus, Earth, and Mars, but there was nothing corresponding to $m = 3$. Jupiter was represented by $m = 4$ and Saturn by $m = 5$. Well, there was no undiscovered planet to be found between Mars and Jupiter. In fact, nothing was believed to be there. But, in what the formula might claim as a victory, the asteroid belt, when discovered, was found to be at the distance given when $m = 3$. The discovery of Uranus handed Bode's law another victory; it corresponded to $m = 6$. However, Neptune was found to be at a distance that was off from the prediction by over 20%. So, while the "law" got off to a decent start, it is no longer valued. This is often the course that science takes. While we have proof in mathematics, a scientific theory is only as good as its most recent test. Figure 6.5 shows the failure of Bode's law.

In Figure 6.5, the asteroid belt, represented by Ceres, is included, as is Pluto, which is also not presently considered to be a planet. That's why parentheses were used for each of these.

Moving Backwards

We are sometimes interested in finding the domain values that are sent to specific range values by a given function. The notation $f^{-1}(x)$ is read "f inverse of x" and refers to the values that the function f sends to x.

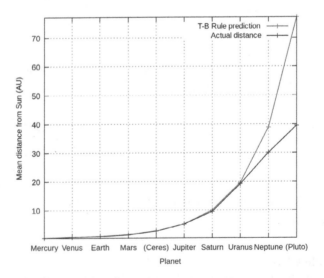

FIGURE 6.5
Bode's "law."

---

7 1 astronomical unit (AU) = the average distance from the Earth to the Sun.

Returning to Example 1, if someone asks for f^{-1}(SLCCD), the answer is PIZZA. If the function is not one-to-one, then it is not invertible. The problem is caused by $f^{-1}(x)$ not having a unique value for every x. Example 2 illustrates this. f^{-1}(50 cents) could be either "Book" or "Shot Glass."

We sometimes use the notation $f^{-1}(x)$ even when f is not invertible. In this instance it refers to the set of all elements that are sent to x by f. We can even consider $f^{-1}(M)$, where M is a set. This consists of all elements of the domain of f that are sent to an element of M by f. In set builder notation, we have $f^{-1}(M) = \{x \mid f(x) \in M\}$.

In a similar way, we can evaluate a function on an entire set, S. We have $f(S) = \{x \mid f(s) = x$ for some $s \in S\}$. Of course, S must be a subset of the domain of f for this to make sense. The set $f(S)$ is referred to as the image of S under f.

New Functions from Old

Given two functions f and g, we can define new functions as follows:

Sum of functions: $(f + g)(x) = f(x) + g(x)$
Difference of functions: $(f - g)(x) = f(x) - g(x)$
Product of functions: $(fg)(x) = f(x) \cdot g(x)$
Quotient of functions: $(f/g)(x) = f(x)/g(x)$, provided that $g(x) \neq 0$
Composition of functions: $(f \circ g)(x) = f(g(x))$.

Such new functions arise in many diverse situations. In the business world, we have a sum of functions when considering costs: Total Costs = Fixed Costs + Variable Costs. A difference of functions arises as Profit = Revenue – Costs. New functions arising from a product, quotient, and composition are examined in Examples 13–15.

Example 13 – Damped Oscillations
Something that oscillates may be modeled by a function like $\sin(x)$ or $\cos(x)$ (Figure 6.6). Examples include a child being pushed on a swing or a buoy in a pool moving under the action of a wave machine.

However, if the pushing stops (and the person on the swing is not contributing to the motion), or if the wave machine is turned off, then the amplitude will diminish. This can be modeled mathematically by taking the product of two functions, $\sin(x)$ or $\cos(x)$ and the exponential decay function, $A_0 e^{-kx}$. That is, a formula for damped oscillation is $y = A_0 e^{-kx} \sin(x)$ (Figure 6.7).

Example 14 – Surface-area-to-volume Ratio
This ratio is important in several fields. In chemistry, it's a factor in how quickly a chemical reaction will occur. The change can be dramatic when the ratio reaches an extreme. For example, grain is stable, but grain dust is explosive. In biology, marine life is affected by this factor in terms of water drag. A higher ratio reduces drag, making it easier to stay near the water's surface.[8]

For a sphere, this quotient of functions is formed from $SA = 4\pi r^2$ and $V = (4/3)\pi r^3$. So, we have $SA : V = 4\pi r^2/(4/3)\pi r^3 = 3/r$. The ratio is different for other shapes.

---

8 https://en.wikipedia.org/wiki/Surface-area-to-volume_ratio.

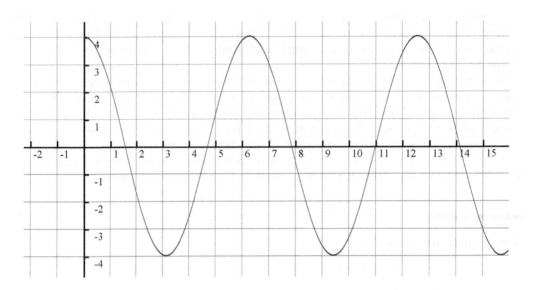

FIGURE 6.6
Oscillatory motion.

Image created by Garrett Ruths.

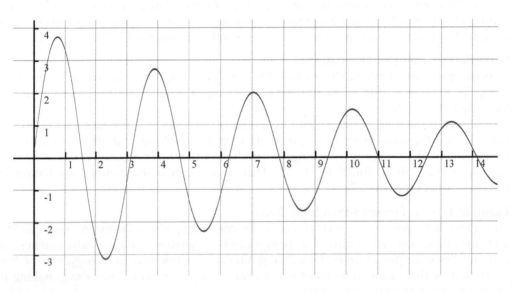

FIGURE 6.7
Damped oscillation.

Image created by Garrett Ruths.

Example 15 – Updating Newton

Isaac Newton's second law, $F = ma$, is an excellent approximation for force, if the velocity, v, of the object being considered is small compared to the speed of light, c. However, if it is not, then treating the mass as a constant can cause significant error, depending on how close the velocity is to that of light. The mass of an object traveling at velocity v is given by $m = \dfrac{m_0}{\sqrt{1-\frac{v^2}{c^2}}}$, where m_0 is the rest mass.[9] So, to correctly evaluate the force, we should compose the functions F and m.

Special Functions

Just as we have 0 as the additive identity and 1 as the multiplicative identity, so we also have a function that "does nothing." It's known as the identity function and it simply sends every element of its domain to itself. It is sometimes denoted by the Greek letter ι (iota, the Greek version of i). We have $\iota(x) = x$ for all x.

Sometimes we wish to modify a function so that the output is always an integer. One approach is to round off the result to the closest integer. In such cases we often follow the arbitrary rule of rounding up numbers halfway between integer values. So, 3.5 would become 4, for example. Other useful ways to convert from reals to integers are

The **ceiling function**, $\lceil x \rceil$ = the least integer greater than or equal to x.
The **floor function**, $\lfloor x \rfloor$ = the greatest integer less than or equal to x.

Because of the great importance of these functions, they're built into many programming languages and computer algebra systems, where they have names like ceil(x) and floor(x). The modern names and notation are due to Kenneth E. Iverson, who first used the notations in his 1962 book *A Programming Language*.[10] Prior to this, less intuitive names and notations had been used by Legendre and Gauss, as the concepts were needed for proofs they carried out and a name or notation of some kind was desired.[11] Graphs of these functions are shown in Figures 6.8 and 6.9. Note that they are discontinuous functions.

When evaluating a negative number in the floor function it's easy to make a mistake. Please note, for example, that $\lfloor -3.2 \rfloor \neq -3$. It may feel like this is rounding down (from 3.2 to 3), but rounding down means moving to the left on the real number line, so $\lfloor -3.2 \rfloor = -4$.

The floor and ceiling functions often come in handy for expressing other mathematical functions and concepts. For example, the Dirichlet Pigeon-hole principle, from Chapter 5, can be stated more efficiently using the ceiling function, like so:

---

9 Some physicists, including Einstein, were opposed to the idea of relativistic mass and expressed a preference for working through such ideas using momentum instead.

10 http://mathworld.wolfram.com/FloorFunction.html, which cites Graham, Ronald L., Donald E. Knuth, and Oren Patashnik. "Integer Functions." Ch. 3 in *Concrete Mathematics: A Foundation for Computer Science*, second edition. Addison-Wesley Publishing Company, Reading, MA, 1994, pp. 67–101. Also see https://en.wikipedia.org/wiki/Floor_and_ceiling_functions.

11 https://en.wikipedia.org/wiki/Floor_and_ceiling_functions.

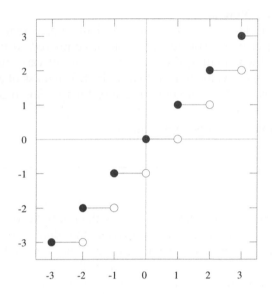

FIGURE 6.8

A graph of the floor function.

Image created by Omegatron. This file is licensed under the Creative Commons Attribution-Share Alike 3.0 Unported, 2.5 Generic, 2.0 Generic and 1.0 Generic license.

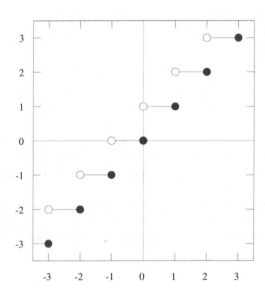

FIGURE 6.9

A graph of the ceiling function.

Image created by Omegatron. This file is licensed under the Creative Commons Attribution-Share Alike 3.0 Unported, 2.5 Generic, 2.0 Generic and 1.0 Generic license.

If n objects are put into m boxes, some box must contain $\geq \lceil n/m \rceil$ objects, and some box must contain $\leq \lfloor n/m \rfloor$.[12]

Some simple functions, like the absolute value function (Figure 6.10), have a sharp corner.

Could there be a continuous curve that has infinitely many sharp corners, or must the number remain finite in order to preserve continuity? You are asked to consider this problem in the exercises below, but a tougher question is whether or not there can be a continuous function that has a sharp corner *everywhere*. That question will be answered in the chapter on fractals.

The absolute value function can be expressed in several ways:

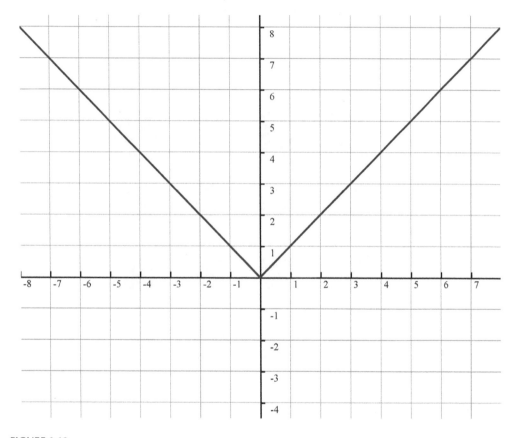

FIGURE 6.10
A graph of the absolute value function.

Image created by Garrett Ruths.

12 Graham, Ronald L., Donald E. Knuth, and Oren Patashnik, *Concrete Mathematics*, second edition, Addison-Wesley Publishing Company, Reading, MA, 1994, p. 95.

1. $f(x) = |x|$ This notation was first used by Karl Weierstrass in 1841.[13]
2. $f(x) = \begin{cases} x & \text{if } x \geq 0 \\ -x & \text{if } x < 0 \end{cases}$ This is known as a **piece-wise defined** function. The domain is split into two pieces and the function is defined separately on each piece. It offers the advantage of not requiring the introduction of any new notation.
3. $f(x) = \sqrt{x^2}$ This representation avoids both new notation and the piece-wise format.

The triangle inequality, $|a + b| \leq |a| + |b|$, from Chapter 2 is an example of a rule that is most conveniently expressed using the absolute value function with Weierstass's notation.

While attending the PA-SSHE-MA 2005 Annual Conference at Millersville University, I saw the presentation "Absolutely!" by Slippery Rock University's Robert E. Buck, who dazzled attendees with some unexpected graphs, the most spectacular of which is shown in Figure 6.11.[14]

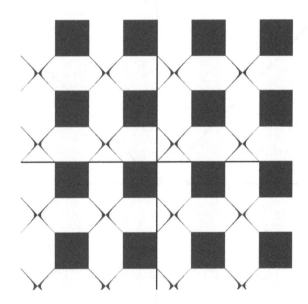

FIGURE 6.11
A graph of $|\sin(y)| + \sin(y) = |\sin(x)| + \sin(x)$.

Image redrawn by Josh Gross.

13 https://en.wikipedia.org/wiki/Absolute_value, which cites Higham, Nicholas J., *Handbook of Writing for the Mathematical Sciences*, SIAM, p. 25.

14 This stands for Pennsylvania State System of Higher Education – Mathematics. I was not working at a college that is part of the state system, but I attended the conference anyway.

I presented this result to some of my students and asked them to explore the wide world of functions involving absolute value signs and see what they could find.

The quickest way to investigate such equations is to simply create a pair of nested loops that test each pixel in a rectangular area of a computer screen to see if, at that (x, y) pair, the absolute value of the difference between the right and left sides of the equations is less than some small positive number ε. For larger values of ε, thicker lines result and there's a greater potential for "false hits" (plotted points that do not belong on the graph). Two neat graphs that students found are reproduced in Figures 6.12 and 6.13

While the absolute value function does not appear in the equation for Figure 6.12, it was used in the code to generate the image. Kelly Bursey used $\varepsilon = 0.01$, so what you see is really a plot of $|\sec(x) - \sec(y)| < 0.01$.

Brad Dembo investigated a variant of the graph I saw at the conference by squaring the variables. The result is my new favorite.

Feel free to explore this graph yourself for various values of ε. You can also try to make it even prettier by creating a rule that allows several colors to be used. For example, let the color be a function of how close the difference between the two sides of the equation is to zero.

An exercise at the end of this chapter challenges you to come up with an equation involving one or more absolute value functions whose graph is prettier or more surprising than those above.

FIGURE 6.12
$\sec(x) = \sec(y)$.

Investigated by Kelly Bursey, redrawn by Josh Gross.

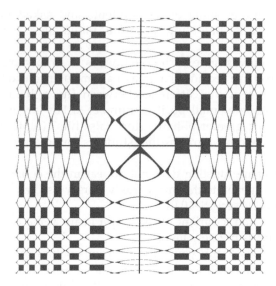

FIGURE 6.13

$|\sin(y^2)| + \sin(y^2) = |\sin(x^2)| + \sin(x^2)$.

Investigated by Brad Dembo, redrawn by Josh Gross.

More Art Appreciation

A positive integer n is prime if and only if[15]

$$\sum_{m=1}^{\lfloor \sqrt{n} \rfloor + 1} \left(\left\lfloor \frac{n}{m} \right\rfloor - \left\lfloor \frac{n-1}{m} \right\rfloor \right) = 1.$$

An Open Problem

It is not known whether or not there is a positive integer $k \geq 6$ such that

$$3^k - 2^k \left\lfloor \left(\frac{3}{2} \right)^k \right\rfloor > 2^k - \left\lfloor \left(\frac{3}{2} \right)^k \right\rfloor - 2.$$

Kurt Mahler established, in a 1957 paper, that the number of solutions must be finite, and to this day none have been found.[16]

---

15 Found at https://en.wikipedia.org/wiki/Floor_and_ceiling_functions, which cites Crandall, Richard, and Carl Pomerance, *Prime Numbers: A Computational Perspective*, Springer, New York, 2001, Ex. 1.3, p. 46.

16 https://en.wikipedia.org/wiki/Floor_and_ceiling_functions, which, in turn, cites Mahler, Kurt, "On the Fractional Parts of the Powers of a Rational Number (II)," *Mathematika*, Vol. 4, No. 2, 1957, pp. 122–124.

Exercise Set 6

> I came, I saw, I conquered.
> Julius Caesar

1. Find or create a function that sends words to numbers.

2. Find or create a function that sends numbers to words.

3. Find or create a function that sends words to words.

4. Find or create a function that sends pairs of words to numbers.

5. In 1897 Amos E. Dolbear had a paper published in which he pointed out that the chirping of crickets (the species was not specified) becomes synchronized at night and that there's a relationship between this rate and the temperature.[17] Answer the question that concludes the paper and is reproduced below. Dolbear's answer is left out here.

   One may express this relation between temperature and chirp rate thus.
   Let T stand for temperature and N, the rate per minute.

$$T = 50 + \frac{N - 40}{4}$$

   For example, what is the temperature when the concert of crickets is 100 per minute?

   This formula was referenced in an episode (Season 3, Episode 2, "The Jiminy Conjecture") of the American TV sitcom *The Big Bang Theory* (although Sheldon referred to Amos Dolbear as Emile Dolbear and gave the year of publication as 1890). It is also referenced in two episodes ("Highs and Lows", "Jungles") of the British comedy show QI.[18]

6. Find a graphical representation of a function online or in print that takes an ordered pair of real numbers as input and yields a real number as output.

7. Evaluate the following:

   a. $\lfloor 4.3 \rfloor$

   b. $\lfloor 4.9 \rfloor$

   c. $\lfloor \sqrt{2} \rfloor$

   d. $\lfloor \pi \rfloor$

   e. $\lfloor e \rfloor$

8. Evaluate the following:

   a. $\lfloor -5.7 \rfloor$

   b. $\lfloor -8.2 \rfloor$

   c. $\lfloor -\sqrt{2} \rfloor$

---

17 Dolbear, Amos E., "The Cricket as a Thermometer," *The American Naturalist*, Vol. 31, 1897, pp. 970–971, p. 971 quoted here, available online at www.journals.uchicago.edu/doi/pdfplus/10.1086/276739.
18 https://en.wikipedia.org/wiki/Dolbear%27s_law#cite_note-dolbear-3.

    d. $\lfloor -\pi \rfloor$

    e. $\lfloor -e \rfloor$

9. Evaluate the following:

    a. $\lceil 7.2 \rceil$

    b. $\lceil 7.8 \rceil$

    c. $\lceil \sqrt{2} \rceil$

    d. $\lceil \pi \rceil$

    e. $\lceil e \rceil$

10. Evaluate the following:

    a. $\lceil -5.1 \rceil$

    b. $\lceil -8.7 \rceil$

    c. $\lceil -\sqrt{2} \rceil$

    d. $\lceil -\pi \rceil$

    e. $\lceil -e \rceil$

11. Show that if $n \in \mathbb{Z}$, then $\lfloor x + n \rfloor = \lfloor x \rfloor + n$.

12. Does $\lfloor nx \rfloor = n \lfloor x \rfloor$ for all $n \in \mathbb{N}$? Prove that it does or provide a counterexample.

13. Prove that $\max\{a, b\} = \frac{1}{2} (a + b + |a - b|)$.

14. Could there be a continuous curve that has infinitely many sharp corners, or must the number remain finite in order to preserve continuity?

15. Find an equation involving one or more absolute value functions whose graph is prettier or more surprising than those above.

References/Further Reading

> I guess there are never enough books.
>   – John Steinbeck

Bell, Eric Temple, *Mathematics, Queen and Servant of Science*, Copyright, 1951, British Edition 1952, reprinted 1954, 1958, 1961, Printed in Great Britain by Butler and Tanner Ltd., Frome and London.

Crandall, Richard and Carl Pomerance, *Prime Numbers: A Computational Perspective*, Springer, New York, 2001.

Dolbear, Amos E., "The Cricket as a Thermometer," *The American Naturalist*, Vol. 31, November 1897, pp. 970–971, www.journals.uchicago.edu/doi/pdfplus/10.1086/276739.

Graham, Ronald L., Donald E. Knuth, and Oren Patashnik, *Concrete Mathematics: A Foundation for Computer Science*, second edition, Addison-Wesley Publishing Company, Reading, MA, 1994.

Iverson, Kenneth E., *A Programming Language*, John Wiley and Sons, Inc., New York, 1962.

Keyser, Cassius Jackson, *Mathematics as a Culture Clue and Other Essays, Volume 1 in the Collected Works of Cassius Jackson Keyser*, Scripta Mathematic, Yeshiva University, New York, 1947.

Mahler, Kurt, "On the Fractional Parts of the Powers of a Rational Number (II)," *Mathematika*, Vol. 4, No. 2, 1957, pp. 122–124.

7

The Multiplication Principle

No other idea in combinatorial mathematics is as simple, yet as far-reaching as this bland sounding principle.

John Allen Paulos[1]

An Ancient Problem

This chapter is the first of several to address combinatorics. The most straightforward problems in this area of mathematics simply ask us to find the number of ways something can happen. The multiplication principle is the natural starting point for learning how to grapple with such problems, but before looking at it let's go back thousands of years to the time of Archimedes (c. 287 BCE–c. 212 BCE), for problems in combinatorics, where the principle is used, appear to be at least that old.

I say "appear" because much of what we have from the ancient world is fragmentary and leaves us a bit shy of making absolute statements. In this case, the ambiguity arises from a problem Archimedes related concerning Figure 7.1.

The problem is known as the stomachion (pronounced sto-MOCK-yon) or ostomachion. As the image shows, it involves 14 pieces. They're scaled to fit a 12 by 12 square. Modern scholars thought that Archimedes was investigating how the pieces could be reassembled to make various shapes. This would make the puzzle very much like Tangram, which consists of seven pieces. However, Reviel Netz, a professor of classics (and also a professor of philosophy) at Stanford University, eventually came to a different conclusion, which he published in 2003. Netz's story begins when a retired businessman kindly mailed him a model of the Stomachian made out of blue cut-glass.[2] Like normal people, mathematicians sometimes receive gifts, but in this case it wasn't as random as it sounds. Netz had established himself as an expert on Archimedes, so it was a very appropriate gift. Examining the model, Netz noticed that the pieces were not arranged in the same manner as on a page of Archimedes's work that he was transcribing.[3] This and another look at the fragmentary writings of Archimedes on the puzzle gave Netz the insight that the real puzzle was to find all possible arrangements of the pieces that yielded the 12 by 12 square.

---

1 Paulos, John Allen, *Beyond Numeracy: Ruminations of a Numbers Man*, Vintage Books, A Division of Random House, Inc., New York, 1992, p. 150.

2 Kolata, Gina, "In Archimedes' Puzzle, a New Eureka Moment," *The New York Times*, December 14, 2003, available online at www.nytimes.com/2003/12/14/us/in-archimedes-puzzle-a-new-eureka-moment.html.

3 Kolata, Gina, "In Archimedes' Puzzle, a New Eureka Moment," *The New York Times*, December 14, 2003, available online at www.nytimes.com/2003/12/14/us/in-archimedes-puzzle-a-new-eureka-moment.html.

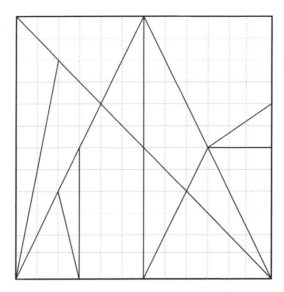

FIGURE 7.1
The Stomachion of Archimedes.

Public domain.

Netz believed that Archimedes would not have posed the question, if he hadn't already found the answer.[4] Despite this belief, the solution, if it was ever put to paper, no longer existed, and calculating it anew was challenging for the modern mathematicians who were led by Netz's claim to investigate it for themselves.

Netz was able to interest an all-star team in the problem. It consisted of two husband and wife pairs: Ronald Graham (Figure 7.2) and Fan Chung (University of California, San Diego, mathematicians)[5] and Persi Diaconis and Susan Holmes (Stanford statisticians).[6] It took this foursome, working together, six weeks to find the answer.[7] As Diaconis related, "It was hard."[8]

As mentioned in Chapter 2, how you approach a problem depends on what kind of mathematician you are. So, when William H Cutler, a computer scientist working in industry, approached the problem independently, it was by writing a program.[9] Happily, Cutler and the team of four both arrived at the same answer, 17,152. This number counts

---

4 Kolata, Gina, "In Archimedes' Puzzle, a New Eureka Moment," *The New York Times*, December 14, 2003, available online at www.nytimes.com/2003/12/14/us/in-archimedes-puzzle-a-new-eureka-moment.html.

5 Because of their prior work in combinatorics, these two could be labeled "combinatorists," but they're really much broader mathematically than that.

6 Kolata, Gina, "In Archimedes' Puzzle, a New Eureka Moment," *The New York Times*, December 14, 2003, available online at www.nytimes.com/2003/12/14/us/in-archimedes-puzzle-a-new-eureka-moment.html.

7 Kolata, Gina, "In Archimedes' Puzzle, a New Eureka Moment," *The New York Times*, December 14, 2003, available online at www.nytimes.com/2003/12/14/us/in-archimedes-puzzle-a-new-eureka-moment.html.

8 Kolata, Gina, "In Archimedes' Puzzle, a New Eureka Moment," *The New York Times*, December 14, 2003, available online at www.nytimes.com/2003/12/14/us/in-archimedes-puzzle-a-new-eureka-moment.html.

9 Kolata, Gina, "In Archimedes' Puzzle, a New Eureka Moment," *The New York Times*, December 14, 2003, available online at www.nytimes.com/2003/12/14/us/in-archimedes-puzzle-a-new-eureka-moment.html.

FIGURE 7.2
Ronald Graham (1935-).

Courtesy of Ronald Graham.

as distinct solutions that are mirror images or rotations of each other. Excluding such possibilities, the answer is 536.

After his battle with the problem, Diaconis made a guess as to why the puzzle was named Stomachion:

> It comes from "stomach turner." If you get involved with it, that's what happens.[10]

If you wish to search the literature and learn more about this problem, it can also be found under the names "the Loculus of Archimedes" and "Archimedes' Box."[11]

The present text is not intended to prepare you for solving such a difficult problem. The Stomachion is only presented to show that combinatorics is an ancient field and problems of this type can be both easy to state and very difficult to solve.

The Multiplication Principle

Although most of the history of combinatorics deals with "counting problems," the field is now much broader. If we want to specifically refer to problems of this type, we can use the phrase **enumerative combinatorics**. But whatever label we use, we don't want to approach these problems by counting directly, especially when the final answer is likely

---

10 Kolata, Gina, "In Archimedes' Puzzle, a New Eureka Moment," *The New York Times*, December 14, 2003, available online at www.nytimes.com/2003/12/14/us/in-archimedes-puzzle-a-new-eureka-moment.html.
11 https://en.wikipedia.org/wiki/Archimedes.

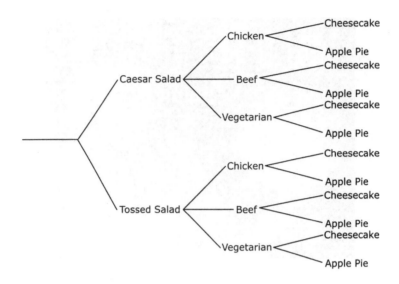

FIGURE 7.3
A tree diagram.

Image created by Josh Gross.

to be large. That is, we try to find shortcuts whenever possible. This is why combinatorics has been informally described as "the art of counting without counting." The first and simplest rule in this area is the multiplication principle. It will be motivated through an example.

Example 1

Suppose a banquet is being held that offers a three-course meal. It begins with a choice of a Caesar salad or a tossed salad. The main course may be chosen from chicken, beef, or an unspecified vegetarian dish.[12] Finally, dessert is either cheesecake or apple pie. How many distinct meals may be ordered?

We may construct what is known as a **tree diagram** (Figure 7.3) to help us solve this particular problem, as well as gain insight into more general problems.

If you turn your head to the right, the splits sort of look like the branches of a tree, hence the name "tree diagram." Each path along the tree represents a different meal. If you like the Many-Worlds Interpretation of quantum mechanics, you can imagine each endpoint as a distinct universe. Because there's only one path that leads to any particular dessert listed on the far right-hand side (the leaves, if you will), counting the entries in the dessert column gives the total number of possible meals. The answer is 12.

Notice that the salad choice splits the diagram to give two branches. The main course then splits each of those branches into three more. Finally each of those branches is split into two by the choice of dessert.

---

12 I think whoever prepares the options, knowing a vegetarian will not choose the chicken or beef, typically figures there's no need to be more specific.

The net result is that to solve such a problem, we simply multiply the numbers:

$$2 \quad \times \quad 3 \quad \times \quad 2 \quad = \quad 12$$
(*salad choices*) × (*main course choices*) × (*dessert choices*)

This is known as the multiplication principle.

Athanasius Kircher (1602–1680), although not the first to use the principle, stated it very clearly:

> If there are k successive choices to be made, and if the ith choice can be made in n_i ways, for $1 \le i \le k$, then the total number of ways of making these choices is the product $n_1 \times n_2 \times \cdots \times n_k$.[13]

Example 2 – Numbers to Letters

On some old telephones, the numbers 2 through 9 (inclusive) each have three letters associated with them. The numbers 0 and 1 have no associated letters. If your phone number does not contain any 0s or 1s, how many distinct combinations of letters represent your number on such phones? Ignore the area code.

Solution:

Each of the seven digits has three possible letters that can represent it. Hence, the solution is $3 \cdot 3 \cdot 3 \cdot 3 \cdot 3 \cdot 3 \cdot 3 = 3^7 = 2,187$. To see a list of the possibilities that form words for a particular number, go to http://phonespell.org/.

Example 3 – Math Fashion

Suppose a math professor has 2 pairs of shoes, 4 pairs of pants, and 12 shirts. Assuming that he isn't concerned about his clothes matching, how many possible outfits does he have?

Solution: $2 \cdot 4 \cdot 12 = 96$

Example 4 – Secure or Insecure?

Consider a lock with three tabs, like the one shown in Figure 7.4. Each tab may be set at a number between 0 and 9, inclusive. How many possibilities are there for the correct position to open the lock? If two possibilities can be tested every second, how long will it take *on average* to open the lock without knowing the proper setting in advance?

Solution: There are 10 choices for each of the three tabs, so the total number of possibilities is $10 \cdot 10 \cdot 10 = 1,000$. If we can test 120 positions every minute (2 per second), it will take $1,000/120 \approx 8.33$ minutes to check all of them. On average, only half need to be checked before the lock opens, so the final answer is $8.33/2 \approx 4.16$ minutes. This lock is "TSA approved," by the way.

Applying the multiplication principle to some other problems leads to strange results. For example, it appears that for a 16-digit credit card number there are 10^{16} possibilities. This is way larger than is necessary. In fact, this number is large enough for every person on the planet (there are about 7.7 billion of us) to have 1,298,701 different credit cards!

---

13 Taken here from Wilson, Robin and John J. Watkins, editors, *Combinatorics: Ancient & Modern*, Oxford University Press, Oxford, 2013, p. 137.

FIGURE 7.4
A good lock?

Image created by Craig P. Bauer.

Another example is provided by the International Standard Book Number (ISBN). These now consist of 13 digits, giving a whopping 10^{13} possibilities. So far only an estimated 1.3×10^8 books have been published.[14] This estimate covers all languages and the entire history of writing. Once again, we seem to have far more room to grow than is necessary. Why don't credit cards, books, and other items bear shorter, more convenient, identification numbers?

When I ask students to guess what the reason is, they often suggest that it's to make it harder to guess a valid number. They think it would be bad if credit card numbers were so short that a random number of the appropriate length was likely to appear on someone's card. While this sounds reasonable, shorter numbers really wouldn't help criminals. It's not enough to come up with a credit card number

---

14 https://mashable.com/2010/08/05/number-of-books-in-the-world/#GLoCFjL.imqZ.

that's in use; the name that goes with that number is also needed to make purchases with it.

The real reason credit card numbers are so long is that they contain **check digits**. These extra digits help detect errors that can be made when the number is conveyed to make a purchase, whether it be typed into a web form or read over a phone. Some simpler check digit schemes are considered below. Exploring credit card check digits is left as an exercise.

UPC Identification Numbers

UPC stands for Universal Product Code and it does indeed appear on almost everything we purchase. Commonly referred to as "bar codes," the error-checking scheme they use is based on a calculation that is seen in many areas of mathematics. In calculus III classes it's called the dot product or inner product. In linear algebra classes it's often called the scalar product. It takes a pair of ordered n-tuples, multiplies each pair of corresponding positions together, and then adds them all up. Algebraically, it looks like this:

$$(a_1, a_2, a_3, a_4, \ldots, a_k).(b_1, b_2, b_3, b_4, \ldots, b_k) = a_1b_1 + a_2b_2 + a_3b_3 + a_4b_4 + \cdots + a_{k-1}b_{k-1} + a_kb_k$$

In terms of matrices, it can be written as

$$\begin{pmatrix} a_1 & a_2 & \ldots & a_{k-1} & a_k \end{pmatrix} \begin{pmatrix} b_1 \\ b_2 \\ \vdots \\ b_{k-1} \\ b_k \end{pmatrix} = a_1b_1 + a_2b_2 + \cdots + a_{k-1}b_{k-1} + a_kb_k$$

The first digit indicates the kind of UPC,[15] the next five identify the manufacturer, and the five after that identify the product. The 12th digit is the check digit. It's chosen to make the inner product of $(3, 1, 3, 1, 3, 1, 3, 1, 3, 1, 3, 1)$, called the **weighting vector**, and the UPC come out to a multiple of 10. It's always possible to do this, because the last digit in the UPC is multiplied by 1 when taking the inner product. That is, a 12-digit UPC, $a_1a_2a_3a_4 \cdots a_{12}$, will satisfy

$$(3, 1, 3, 1, \ldots, 3, 1).(a_1, a_2, a_3, a_4, \ldots, a_{12}) = 0 \text{ mod } 10.$$

Example 5
To check if the UPC in Figure 7.5 is legitimate, we calculate

---

15 See https://en.wikipedia.org/wiki/Universal_Product_Code#Number_system_digit for an explanation of the different types.

FIGURE 7.5
A sample UPC.

Public domain.

$$3 \cdot 0 + 1 \cdot 2 + 3 \cdot 0 + 1 \cdot 6 + 3 \cdot 2 + 1 \cdot 6 + 3 \cdot 7 + 1 \cdot 1 + 3 \cdot 2 + 1 \cdot 3 + 3 \cdot 5 + 1 \cdot 4 = 70 = 0 \bmod 10.$$

It works!

Now, let's see what happens if an error is made. Suppose that the UPC was incorrectly read as 02068671235. The error here is in position 5. Calculating the dot product gives

$$3 \cdot 0 + 1 \cdot 2 + 3 \cdot 0 + 1 \cdot 6 + 3 \cdot 8 + 1 \cdot 6 + 3 \cdot 7 + 1 \cdot 1 + 3 \cdot 2 + 1 \cdot 3 + 3 \cdot 5 + 1 \cdot 4 = 88 \neq 0 \bmod 10.$$

Because we did not get a multiple of 10, we know a mistake was made.

If an incorrect digit appears in an even position, where it's multiplied by a 1, it will change the value of the dot product so that it's no longer a multiple of 10. This will also happen if an error is made in one of the odd positions, because the digits 0 through 9, when multiplied by 3, all result in numbers having different values in the ones position. So, this scheme will catch every possible single digit error.

A check digit scheme using other numbers might not work as well. For example, consider

$$(5, 1, 5, 1, \ldots, 5, 1) \cdot (a_1, a_2, a_3, a_4, \ldots, a_{12}) = 0 \bmod 10.$$

The 12-digit number 274651947238 checks with this modified scheme, because

$$5 \cdot 2 + 1 \cdot 7 + 5 \cdot 4 + 1 \cdot 6 + 5 \cdot 5 + 1 \cdot 1 + 5 \cdot 9 + 1 \cdot 4 + 5 \cdot 7 + 1 \cdot 4 + 5 \cdot 3 + 1 \cdot 8 = 180 = 0 \bmod 10.$$

But if an error was made and the first digit was taken to be an 8, then the check scheme gives

$$5 \cdot 8 + 1 \cdot 7 + 5 \cdot 4 + 1 \cdot 6 + 5 \cdot 5 + 1 \cdot 1 + 5 \cdot 9 + 1 \cdot 4 + 5 \cdot 7 + 1 \cdot 4 + 5 \cdot 3 + 1 \cdot 8 = 210 = 0 \bmod 10,$$

which does not indicate an error.

Exercise 14 asks which numbers, if any, can take the place of the 3s in this check scheme and successfully detect any instance of a single error.

The UPC check scheme will also detect another common type of error, transposing two adjacent numbers, 90% of the time.[16] Exercise 15 asks you to find one of the exceptions.

It would be nice to be able to catch all transposition errors, but that would require more than one check digit being appended to the number.

---

16 https://en.wikipedia.org/wiki/Universal_Product_Code#Number_system_digit.

A highly publicized error of this type recently occurred when Lt. Col. Oliver North gave United States Secretary of State Elliott Abrams an incorrect Swiss bank account number for depositing $10 million for the contras. The correct account number began with 386; the number North gave Abrams began with 368.

Joseph Gallian[17]

ISBN – International Standard Book Number

This number, used to identify books, comes in two versions with two different check schemes.

ISBN-13 works like the UPC check scheme, taking the dot product of the 13-digit number with (3, 1, 3, 1, 3, 1, 3, 1, 3, 1, 3, 1, 3). The last digit of the ISBN-13 is chosen to make the result a multiple of 10.

ISBN-10 takes the dot product with (10, 9, 8, 7, 6, 5, 4, 3, 2, 1) and selects the last digit to make the result a multiple of 11. In some cases this can only be achieved if the last "digit" is 10. When this happens, an X is used to denote 10.

Example 6
Is 0691167672 a valid ISBN-10?

To check, we compute the dot product $(10, 9, 8, 7, 6, 5, 4, 3, 2, 1) \cdot (0, 6, 9, 1, 1, 6, 7, 6, 7, 2) = 10 \cdot 0 + 9 \cdot 6 + 8 \cdot 9 + 7 \cdot 1 + 6 \cdot 1 + 5 \cdot 6 + 4 \cdot 7 + 3 \cdot 6 + 2 \cdot 7 + 1 \cdot 2 = 231 = 0 \bmod 11$.

Thus, this ISBN-10 checks.

Note: Saying "ISBN number" is redundant, because number is what the N in ISBN stands for. Other common redundant phrases include "ATM machine," "PIN number," and "DC Comics."

Florida Drivers' Licenses[18]

In Florida, the last three digits of the driver's license number of a female with birth month m and birth date b are represented by $40(m - 1) + b + 500$. For both males and females, the fourth and fifth digits from the end give the year of birth.

Missouri Drivers' Licenses[19]

In Missouri, the last three digits of the driver's license number of a female with birth month m and birth date b are represented by $63m + 2b + 1$.

17 Gallian, Joseph, *Contemporary Abstract Algebra*, second edition, D. C. Heath and Company, Lexington, MA, 1990, p.43.

18 Gallian, Joseph, *Contemporary Abstract Algebra*, second edition, D. C. Heath and Company, Lexington, MA, 1990, p. 18.

19 Gallian, Joseph, *Contemporary Abstract Algebra*, second edition, D. C. Heath and Company, Lexington, MA, 1990, p.18.

CAS Registry Number[20]

The Chemical Abstracts Service (CAS) assigns numbers to chemical substances. Including a check digit helps prevent mistakes. The scheme is not quite as easy to explain as the previous example, because the numbers are made up of three groups of digits with the first group being anywhere from two to seven digits long. At least the second group is fixed at two digits. The third group consists of a single digit, the check digit. If we represent the number as $N_i \cdots N_4 N_3 - N_2 N_1 - R$, where $N_i \cdots N_4 N_3$ is the first group of digits, $N_2 N_1$ is the second group of digits, and R is the check digit, then the check digit must satisfy $R = 1 \cdot N_1 + 2 \cdot N_2 + 3 \cdot N_3 + 4 \cdot N_4 + \cdots + i \cdot N_i$ mod 10.

Example 7

Does 1972-08-3 pass the CAS Registry Number check scheme?

To see, calculate $1 \cdot 8 + 2 \cdot 0 + 3 \cdot 2 + 4 \cdot 7 + 5 \cdot 9 + 6 \cdot 1 = 8 + 0 + 6 + 28 + 45 + 6 = 93 = 3$ mod 10. Thus, the check digit should be 3. Indeed, that's what we find in the check digit position of the given number, so it passes the test.

Check Check Digit

This is a check digit scheme for actual checks.

The bank identification number (the first nine digits at the bottom left) has a check digit in position 9 to make $(7, 3, 9, 7, 3, 9, 7, 3, 9) \cdot (a_1, a_2, a_3, a_4, \ldots, a_9) = 0$ mod 10. Applying this to the check shown in Figure 7.6, we get $7 \cdot 1 + 3 \cdot 2 + 9 \cdot 2 + 7 \cdot 0 + 3 \cdot 0 + 9 \cdot 0 + 7 \cdot 6 + 3 \cdot 6 + 9 \cdot 1 = 100 = 0$ (mod 10). It checks!

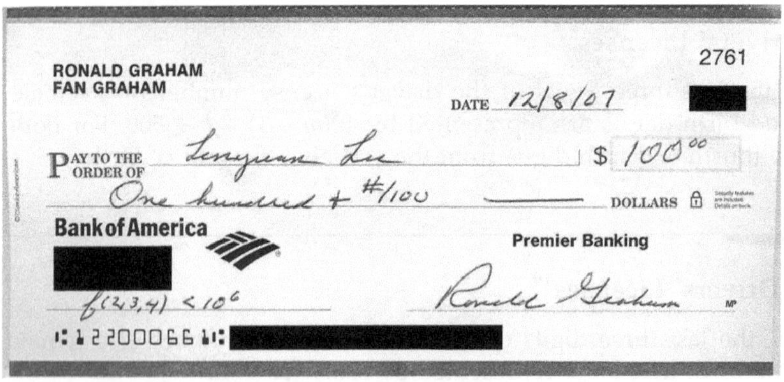

FIGURE 7.6
A small mathematical prize – part of Paul Erdös's legacy, maintained by Ronald Graham.

Courtesy of Linyuan Lu. The account number, address, and phone number have been blacked out in this reproduction. See Chapter 20 for a full explanation of such prize problems.

---

20 https://en.wikipedia.org/wiki/CAS_Registry_Number

Social Security Numbers

Now, consider social security numbers. These are nine digits long, so there appear to be $10^9 = 1$ billion possibilities. This may seem excessive as the population of America is much smaller, but one digit less would only give 100 million possibilities, which is too small. So, social security numbers can't contain a check digit. Numbers issued prior to June 25, 2011 do contain some information, though. The first three digits indicate the location at which the card was issued. The next two, although not issued sequentially, did follow a pattern and reveal roughly when the card was issued.[21]

Beginning on June 25, 2011, the scheme was changed and became closer to being random. This is good, because it makes the digits much harder to guess for someone with bad intentions. Still, the system is not completely random. For example, the number groups 000, 666, and 900–999 are not used to start SSNs.[22]

Venn Diagrams and Error-Correcting Codes

The codes discussed above can detect when an error has occurred, but they can't fix the error. All we know is that something went wrong. In 1950, Richard Hamming made the next step and devised error-*correcting* codes. His ground-breaking idea was not illustrated with Venn diagrams when it first appeared, but we can use them to more easily present the idea.

Given a string of bits, $b_1 \, b_2 \, b_3 \, b_4 \, \ldots$ that we wish to be able to correct, if errors appear, we take four at a time and place them in a three-set Venn diagram (Figure 7.7).

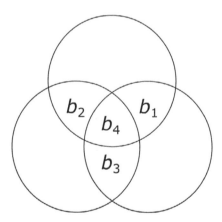

FIGURE 7.7
A Venn diagram with the data bits.

Image created by Josh Gross.

---

21 https://web.archive.org/web/20060204142439/ www.academic.marist.edu/mwa/sccard.htm.
22 https://www.ssa.gov/employer/randomization.html.

We then place check bits, c_1 c_2 c_3, in the still vacant sections such that each set contains an even number of 1s (Figure 7.8).

As an example, suppose the b bits are 1011. Then the Venn diagram begins as shown in Figure 7.9.

Next, to make each set contain an even number of 1s, we must assign $c_1 = 0$, $c_2 = 1$, and $c_3 = 0$, as shown in Figure 7.10.

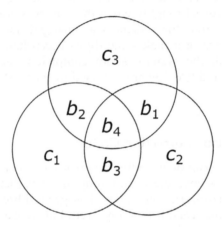

FIGURE 7.8
A Venn diagram with the data and check bits.

Image created by Josh Gross.

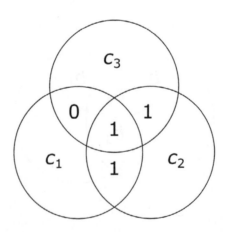

FIGURE 7.9
The data bits 1011 and check bits c_1, c_2, and c_3.

Image created by Josh Gross.

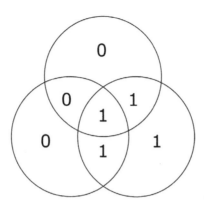

FIGURE 7.10
The data bits 1011 and check bits 010.

Image created by Josh Gross.

With the additional three check bits, the original four-bit string becomes the seven-bit string 1011010.

As an illustration of how this method can not only detect, but also correct errors, suppose we receive the following string representing four bits of data: 0010011. We start by filling these values into a Venn diagram (Figure 7.11).

The set shown on the bottom right of Figure 7.11 has an even number of 1s, but the other two sets do not. This shows that if there's just one error, then the error must be where these two sets intersect, but not the area where all three sets intersect. We therefore correct a single bit to get the Venn diagram shown in Figure 7.12.

The corrected string is 0110011. Stripping away the error-correcting bits, the relevant portion is 0110.

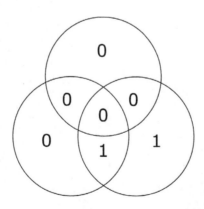

FIGURE 7.11
Data and check bits with an error somewhere.

Image created by Josh Gross.

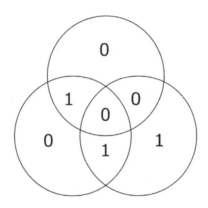

FIGURE 7.12
The corrected bits.

Image created by Josh Gross.

Richard Hamming (Figure 7.13) was an American mathematician who, prior to developing the error-correcting code shown above, was part of the Manhattan project to develop the atomic bomb. His contributions included programming the computers for the physicists there. As early as 1961, Hamming expressed

FIGURE 7.13
Richard Hamming (1915–1998).

Courtesy of the University of Illinois Archives, Richard Hamming, 1938, from RS 2/5/15, Box 120.

concern for the potential threat that computers posed to people in terms of loss of privacy.[23]

Coding Theory

Error-detecting and error-correcting codes are part of an area of mathematics known as **coding theory** that seeks to minimize errors in data transmission. More sophisticated check schemes, requiring more than one extra digit, can catch far more errors. For example, compact discs (CDs) use a scheme such that 192 bits of music, once coded to correct errors, become 588 bits on the disc. That's a lot of redundancy! The point of it is …

> In theory, the combination of parity and interleaving in a CD player can detect and correct a burst error of up to 4000 bad bits – or a physical defect 2.47 mm long. Interpolation can conceal errors up to 13,700 or physical defects up to 8.5 mm long.[24]

When I first saw this claim, I was skeptical. After all, I've encountered problems with CDs having much smaller defects. I finally realized that there's a difference between what is theoretically recoverable and what my cheap CD player can actually recover. Theoretically recoverable simply means that if the bits, with whatever errors may be present, are analyzed mathematically, then the incorrect bits can be identified and fixed. A special program could do this, but my CD player cannot do it as the disc plays. In some cases a damaged CD can be recovered by simply making a copy, if the copying program applies an error-correcting algorithm as it copies. The original remains damaged, but the copy may be error free.

Coding theory is a book-length subject and must now be left behind to consider other topics.

The Potential of Combinatorics

This chapter introduced the topic of combinatorics and the next few chapters delve into it in greater detail, but there's much more that could be said. To sum up the importance of this material tersely, for now:

> [This subject] has a relation to almost every species of useful knowledge that the mind can be employed upon.
>
> Jacob Bernoulli, *Ars Conjectandi*, 1713.[25]

---

23 http://stuffnobodycaresabout.com/2014/12/13/harmful-effects-computer-technology-predicted-1961/.
24 http://www.mayfieldmastering.com/cd-r-masters-vs-ddp-masters/.
25 Taken here from Wilson, Robin and John J. Watkins, editors, *Combinatorics: Ancient & Modern*, Oxford University Press, Oxford, 2013, p. 3.

Exercise Set 7

> Try again. Fail again. Fail better.
> Samuel Beckett[26]

1. How many license plates are possible, if each consists of three letters followed by four digits?

2. My Pennsylvania driver's license has an eight-digit identification number. How many possibilities are there for this number?

3. The following is problem 79 from the Rhind papyrus (ca. 1650 BCE).[27] Find the solution.

   > In each of 7 houses are 7 cats, each cat kills 7 mice, each mouse would have eaten 7 ears of spelt, each ear of spelt will produce 7 hekat of grain; how much grain is thereby saved?

4. How many three-letter code words are possible? Note: code words need not be real words.

5. How many four-letter code words are possible if the letters U and V are not allowed?

6. How many five-digit numbers are there that contain at least one 8?

7. "Haiku is, today, a 17-syllable verse form consisting of three metrical units of 5, 7, and 5 syllables."[28] Using an estimate (or an exact value if you can find it!) of the number of distinct syllables in English, establish an upper bound for the number of possible haiku in this language.

8. In describing a multiple-choice questionnaire that had four options for each of 25 questions, Anne Klein, writing for *OMNI*, concluded "a little multiplication on the back of an envelope ($25 \times 25 \times 25 \times 25$) showed that there were 9,765,625 possible ways to respond."[29] Was she correct?

9. If a laptop screen has a resolution of 1,366 by 768 pixels and the number of possible colors for each pixel is $2^{24} = 16{,}777{,}216$ (24-bit color), how many distinct images can be displayed?

10. Does the UPC shown in Figure 7.14 pass the check digit scheme?

11. Does the ISBN of the randomly chosen book shown in Figure 7.15 pass the check digit scheme?

12. Does the ISBN of the randomly chosen book shown in Figure 7.16 pass the check digit scheme?

---

26 Roberts, Siobhan, *Genius at Play – The Curious Mind of John Horton Conway*, Bloomsbury, New York, USA, 2015, p. 119.

27 Taken here from *The Scientific Monthly*, Vol. 42, 1936, p. 359. See Paulos, John Allen, *Beyond Numeracy: Ruminations of a Numbers Man*, Vintage Books, A Division of Random House, Inc., New York, 1992, p. 150 for the history of this problem.

28 http://www.toyomasu.com/haiku/#whatishaiku.

29 Klein, Anne, "Mind: Therapy by Mail," *OMNI*, April 1982, p. 24. Also see Greg Lowery's letter "Numbers Runner" to *OMNI* on this calculation, appearing in the August 1982 *OMNI*, p. 10.

FIGURE 7.14
A correct UPC?

Public domain.

FIGURE 7.15
A correct ISBN?

Public domain.

FIGURE 7.16
A correct ISBN?

From *UNSOLVED! The History and Mystery of the World's Greatest Ciphers from Ancient Egypt to Online Secret Societies* by Craig P. Bauer. Princeton University Press, 2017.

13. I explained how the ISBN check scheme takes the dot product with (10, 9, 8, 7, 6, 5, 4, 3, 2, 1), but other sources use (1, 2, 3, 4, 5, 6, 7, 8, 9, 10) instead.[30] Was a mistake made or is this simply another way to do it?

---

30 For example, Burger, Edward B. and Michael Starbird, *The Heart of Mathematics*, third edition, Wiley, Hoboken New Jersey, 2010, p. 109.

14. In the UPC check digit scheme which, if any, numbers can take the place of the 3s and still allow any single error to be detected?

15. The UPC check scheme will also detect transposition of two adjacent numbers, 90% of the time.[31] Find one of the exceptions.

16. An easier check digit scheme for UPCs would be to simply sum the first eleven digits and choose the 12th such that the sum is equal to zero modulo ten. This would serve to reveal any time a single digit is incorrect; yet, the slightly more complex scheme that is in use is better. Explain why.

17. Find an explanation of a credit card check digit scheme online and explain it in your own words.

References/Further Reading

> The odd thing about people who had many books was how they always wanted more.
> – Patricia A. McKillip, *The Bell at Sealey Head.*

Burger, Edward B. and Michael Starbird, *The Heart of Mathematics*, third edition, Wiley, Hoboken, NJ, 2010.

"Check Digit Verification of CAS Registry Numbers," *CAS, A Division of the American Chemical Society*, available online at http://support.cas.org/content/chemical-substances/checkdig.

De Smet, Alan, "Unique ID, the Numbers that Control Your Life," available online at www.high programmer.com/alan/numbers/.

Gallian, Joseph, *Contemporary Abstract Algebra*, second edition, D. C. Heath and Company, Lexington, MA, 1990, pp. 42–45.

Graham, Fan Chung and Ron Graham, "A Tour of Archimedes' Stomachion," available online at www.math.ucsd.edu/~fan/stomach/.

"Identification Numbers & Check Digit Schemes, Credit Card Numbers," *Marist College, Mathematics & Writing in ACTION*, available online at web.archive.org/web/20060204131740/www.academic.marist.edu:80/mwa/cccard.htm.

"Identification Numbers & Check Digit Schemes, Social Security Numbers" *Marist College, Mathematics & Writing in ACTION*, available online at web.archive.org/web/20060204142439/www.academic.marist.edu/mwa/sccard.htm.

"Identification Numbers & Check Digit Schemes, Washington State Drivers License Numbers," *Marist College, Mathematics & Writing in ACTION*, available online at web.archive.org/web/20060204093323/www.academic.marist.edu/mwa/wsdln.htm.

Kolata, Gina, "In Archimedes' Puzzle, a New Eureka Moment," *The New York Times*, December 14, 2003, available online at www.nytimes.com/2003/12/14/us/in-archimedes-puzzle-a-new-eureka-moment.html.

Kuhn, Kelin J., "Audio Compact Disk – Writing and Reading the Data," *EE 498*, available online at web.archive.org/web/20060314080153/www.ee.washington.edu/conselec/CE/kuhn/cdaudio2/95x7.htm.

Lahanas, Michael, "Archimedes and Combinatorics (The Loculus of archimedes)," available online at www.hellenicaworld.com/Greece/Science/en/ArchimedesComb.html.

---

31 https://en.wikipedia.org/wiki/Universal_Product_Code#Number_system_digit

Netz, Reviel, Homepage at Stanford University, available online at https://web.stanford.edu/dept/HPS/netz.html.

Netz, Reviel, Fabio Acerbi, and Nigel Wilson, "Towards a Reconstruction of Archimedes' *Stomachion*," *SCIAMVS*, Vol. 5, 2004, pp. 67–99.

Paulos, John Allen, *Beyond Numeracy: Ruminations of a Numbers Man*, Vintage Books, A Division of Random House, Inc., New York, 1992.

Pegg, Jr., Ed, "The Loculus of Archimedes, Solved," November 17, 2003. This article is available online at www.mathpuzzle.com/MAA/04-Loculus%20of%20Archimedes/mathgames_11_17_03.html. It includes images of all 536 basic solutions.

Roberts, Siobhan, *Genius at Play – The Curious Mind of John Horton Conway*, Bloomsbury, New York, USA, 2015.

"Social Security Numbers," *Rob's SearchEngineZ*, available online at http://searchenginez.com/socsec_usa.html. This website provides the mapping between the first three digits of old social security numbers and the states where the cards were issued.

"UPC Search," available online at www.upc-search.org/.

Wilson, Robin and John J. Watkins, editors, *Combinatorics: Ancient & Modern*, Oxford University Press, Oxford, 2013.

8

Permutations

In combinatorics, the order of what you are trying to count sometimes matters and sometimes does not. When order matters we have a permutation, the subject of this chapter. Combinations, in which order does not matter, are examined in Chapter 9.

Example 1 – Queuing Up

If we want to know how many ways five people can line up at the checkout of a grocery store, order matters, because a given person will care where he or she is in line. We have five choices for who is first. There are then four people remaining, so there are four choices for who is second, and so on. The entire line is depicted below with the checkout at the left-hand side.

Checkout ____   ____   ____   ____   ____

       5 choices   4 choices   3 choices   2 choices 1 possibility

By the multiplication principle, the total number of ways for five people to wait in line is $5 \cdot 4 \cdot 3 \cdot 2 \cdot 1 = 120$. Thus, there are 120 permutations on a set of five objects.

Multiplying a chain of integers that decrease from n to 1, one unit at a time, is a common event in combinatorics. So common, in fact, that a special notation has been created to represent it. We typically abbreviate $5 \cdot 4 \cdot 3 \cdot 2 \cdot 1$ as 5!, which is then read as five factorial. This notation was also presented in Chapter 2.

Yet another way of expressing products is to use the uppercase Greek letter pi:

$$\prod_{i=1}^{5} i$$

This expression indicates that we should multiply i by itself repeatedly while it takes all integer values between 1 and 5 inclusive. So, this expression is equivalent to 5! It's overkill to use the product notation for such a simple expression, but it is the nicest way to express more complicated products. For example,

$$\frac{2}{1} \cdot \frac{2}{3} \cdot \frac{4}{3} \cdot \frac{4}{5} \cdot \frac{6}{5} \cdot \frac{6}{7} \cdot \frac{8}{7} \cdot \frac{8}{9} \cdots = \prod_{n=1}^{\infty} \left(\frac{2n}{2n-1} \cdot \frac{2n}{2n+1} \right) = \frac{\pi}{2}$$

This result was found by John Wallis in 1655.

The table below gives $n!$ for the integers 0 through 10. The one value that might seem unclear is 0!. For now, consider the following question. In how many ways can zero people line up? There's only one way – by not doing it! Hence, 0! = 1. If you don't like this line of reasoning, that's okay! We'll soon see other reasons why it makes sense to define 0! as 1.

n	$n!$
0	1
1	1
2	2
3	6
4	24
5	120
6	720
7	5,040
8	40,320
9	362,880
10	3,628,800

Hence, 10 people can line up at the grocery store in 3,628,800 ways. This number may be larger than you expected for such a simple problem. Indeed, the French mathematician Christian Kramp introduced the "!" notation in 1808 because the function $n!$ grows surprisingly fast.[1] It was not, as Steven Wright claimed, an attempt to make mathematics look exciting.

The concept of factorial existed long before Kramp's notation. It can be traced back to, for example, the music theorist Śārṅgadeva (1175–1247), as evidenced by his Sanskrit text *Saṅgīta Ratnākara* (translated as Jewel-Mine of Music or Ocean of Music and Dance).[2]

Although the ! notation has been around since 1808 and is the most commonly used notation today, another notation is to be found in the literature of the intervening years. I've encountered the notation $\lfloor n$ for $n!$ in the *American Mathematical Monthly* (1894),[3] and *The Scientific Monthly* (1946).[4] It was first suggested in 1827, and in the old days, the ∟ shape would have to be written by hand on the text, for there was no key for it on typewriters. Alan Turing used this notation, but on at least one occasion forgot to pen it in when he completed his work. This and other errors made portions of

1 Ghahramani, Saeed, *Fundamentals of Probability*, Prentice-Hall, Upper Saddle River, NJ, 1996, p. 45.
2 See Sections 1.4.60–71. Sources: Wilson, Robin and John J. Watkins, editors, *Combinatorics: Ancient & Modern*, Oxford University Press, Oxford, 2013, p. 21, https://en.wikipedia.org/wiki/Sarangadeva, and https://en.wikipedia.org/wiki/Sangita_Ratnakara. For the original, see Christian Kramp's preface to *Elements d'arithmétique universelle*, De l'imprimerie de Th. F. Thiriart, et se vend chez Hansen, libraire, Cologne, 1808.
3 Lilley, George, "Pascal's Arithmetic Triangle," *American Mathematical Monthly*, Vol. 1, No. 12, December 1894, p. 426.
4 Wylie, Jr., Clarence R., "Mathematical Allusions in Poe," Vol. 63, September 1946, pp. 227–235, p. 231 cited here.

the paper "hard to decipher," as Cliff B. Jones put it.[5] Other notations were used, as well.[6]

With computing power being so prevalent today, there's no need to expand the table of factorials for higher values. However, in the past such a feature was considered useful in books dealing with this material. In 1625, Marin Mersenne (1588–1648) gave a table for the values 1! through 50! in his *La Verite des Sciences contre les Sceptiques ou Pyrrhoniens* (The Truth of the Sciences against the Sceptics or Pyrrhonians).[7] Later, in his 1648 book, *Harmonicorum libri XII*, Mersenne expanded his table "up to 64!, a ninety-digit number and the largest factorial ever calculated up to then, although his table contains a number of errors."[8]

The factorials wouldn't be worthy of inclusion here if the only problem they were suited for was to calculate the number of ways a group of people can arrange themselves in a grocery store checkout lane. The concept is valuable because it arises in a wide range of contexts. Mersenne, for example, used 5! = 120 to answer the question of how many five note songs can be made from the set of five distinct notes ut, re, mi, fa, sol. He went on to list them all.[9] He did the same for six notes, explicitly listing 6! = 720 songs.[10]

Although the work detailed here may seem merely tedious, Mersenne investigated other combinatorial problems and has been described as "by far the most important Renaissance author in the history of combinatorics before Gottfried Wilhelm Leibniz"[11] Also, Mersenne's explicit listing of hundreds of items raises an important issue. It's one thing to calculate how many of something there can be, but it is quite another matter to develop an algorithm for generating all of the possibilities. The latter is a task that often confronts computer programmers.

There's a great deal of fascinating material connected with the concept of factorials. The pages that follow alternate between examples of this and some simple problems in combinatorics.

Elementary Functions

There are some basic functions that should be covered in any precalculus class. They are polynomials, roots, exponentials, logarithms, and trigonometric functions and their

---

5 Jones, Cliff B., in Cooper, Barry S. and Jan van Leeuwen, editors, *Alan Turing: His Work and Impact*, Elsevier, 2013, p. 457.

6 For a survey of them, see Cajori, Florian, *A History of Mathematical Notations*, Two Volumes Bound as One, Dover Publications, Mineola, NY, 1993, Vol. II, pp. 71–77, available online at https://monoskop.org/images/2/21/Cajori_Florian_A_History_of_Mathematical_Notations_2_Vols.pdf.

7 The table appeared on pages 549–551. It is available online at https://books.google.com/books?id=FF44pHy9x_sC&printsec=frontcover&hl=nl&source=gbs_ge_summary_r&cad=0#v=onepage&q&f=false.

8 Wilson, Robin and John J. Watkins, editors, *Combinatorics: Ancient & Modern*, Oxford University Press, Oxford, 2013, p. 133. Mersenne, Marin, *Harmonicorum libri XII*, Baudry, 1648, pp. 116–117.

9 Wilson, Robin and John J. Watkins, editors, *Combinatorics: Ancient & Modern*, Oxford University Press, Oxford, 2013, p. 133. See pp. 546–547 of *La Verite des Sciences contre les Sceptiques ou Pyrrhoniens* (The Truth of the Sciences against the Sceptics or Pyrrhonians) by Mersenne

10 Wilson, Robin and John J. Watkins, editors, *Combinatorics: Ancient & Modern*, Oxford University Press, Oxford, 2013, p. 133.

11 Knobloch, Eberhard, "Renaissance combinatorics," in Wilson, Robin and John J. Watkins, editors, *Combinatorics: Ancient & Modern*, Oxford University Press, Oxford, 2013, pp. 122–145, p. 132 quoted here.

inverses. Depending on the course, the hyperbolic trig functions might also be encountered. These all fall under the heading of **elementary functions**. Combinations of these functions, whether by addition, subtraction, multiplication, division, or composition, are also elementary functions. This is not a complete list, but it is close enough for the purpose here.

The absolute value function, $f(x) = |x|$, doesn't look like any of the functions in the list above, but it can be rewritten as $f(x) = \sqrt{x^2}$, so the absolute value function *is* an elementary function.

There are other more complicated functions that can be rewritten and seen to be elementary, but $f(n)=n!$ is not one of them.[12] It's not that mathematicians just haven't been clever enough to find it yet; it simply doesn't exist. Does it seem strange to you that the sum of the first n integers has the lovely representation $n(n+1)/2$, but there's no elementary function for the product of these same integers?

Stirling's formula

Although it will never be exact, we can get close, in a sense, to $n!$ with an elementary function. In 1730, Abraham de Moivre (1667–1754) discovered $n! \sim (c\sqrt{n})\left(\frac{n}{e}\right)^n$ for some constant c. That same year, James Stirling (1692–1770) found the formula independently and determined the value of c as $\sqrt{2\pi n}$. de Moivre doesn't typically get any credit.[13] We have what is known as Stirling's formula.

$$n! \sim \left(\sqrt{2\pi n}\right)\left(\frac{n}{e}\right)^n$$

This provides a handy approximation of $n!$ for large n.[14] It is unfortunate that the symbol "$\sim$" has several different meanings. In physics, for example, it means "proportional to." In the context of the formula above, however, it is read as "asymptotically approaches" or "is asymptotically equal to" and means

$$\lim_{n \to \infty} \frac{n!}{\left(\sqrt{2\pi n}\right)\left(\frac{n}{e}\right)^n} = 1.$$

This does not mean that $\left(\sqrt{2\pi n}\right)\left(\frac{n}{e}\right)^n$ gets "closer and closer" to $n!$ as n grows. It only means that the ratio gets closer to 1. In other words, it addresses the relative or percentage error (which approaches zero), not the absolute error. A table of values helps illustrate this.[15]

12 Young, Robert M., *Excursions in Calculus: An Interplay of the Continuous and the Discrete*, The Mathematical Association of America, 1992.

13 There's a discrepancy in the secondary literature as to whether Stirling was familiar with de Moivre's work or not, as well as the years the results were obtained. I'm following Young, Robert M., *Excursions in Calculus: An Interplay of the Continuous and the Discrete*, The Mathematical Association of America, 1992, p. 266 here. Young wrote, "If, however, we moderate our demands and ask only for an approximation to $n!$, then the problem is entirely reasonable and a solution was found independently in 1730 by James Stirling and Abraham de Moivre."

14 For a proof see Young, Robert M., *Excursions in Calculus: An Interplay of the Continuous and the Discrete*, The Mathematical Association of America, 1992, p. 267.

15 To generate the values for Stirling's approximation in this table π and e were used with 15 digits of accuracy.

n	n!	Stirling's approx.	Stirling's approx./n!	n! – Stirling's approx.
1	1	0.922137009	0.922137009	0.077862991
10	3,628,800	3,598,695.619	0.99170404	30,104.38126
20	2.4329E+18	2.42279E+18	0.995842347	1.01152E+16
30	2.65253E+32	2.64517E+32	0.997226179	7.35764E+29
40	8.15915E+47	8.14217E+47	0.997918879	1.69802E+45
50	3.04141E+64	3.03634E+64	0.998334744	5.06473E+61
60	8.32099E+81	8.30944E+81	0.998612088	1.15488E+79
70	1.1979E+100	1.1964E+100	0.99881024	1.42516E+97
80	7.1569E+118	7.1495E+118	0.998958881	7.4512E+115
90	1.4857E+138	1.4843E+138	0.999074506	1.375E+135
100	9.3326E+157	9.3248E+157	0.999167017	7.7739E+154
150	5.7134E+262	5.7102E+262	0.9994446	3.1732E+259

As the last column shows, the absolute error quickly grows very large. On the other hand, the second to last column shows the percent accuracy growing from about 92.2% to about 99.94%. Not all asymptotic formulas have an absolute error that grows so rapidly. In some cases, there are several different asymptotic formulas for the same function, with some better than others. However, all have relative errors that converge to zero.

A better approximation is given in this instance by what is known as the Stirling series.[16] If we only use the first term in the parenthesis, we get the formula investigated above, but if we take more terms, the absolute error is less:

$$n! \sim \sqrt{2\pi n}\left(\frac{n}{e}\right)^n\left(1 + \frac{1}{12n} + \frac{1}{288n^2} - \frac{139}{51840n^3} - \frac{571}{2488320n^4} + \cdots\right)$$

A 2010 paper by Gergő Nemes gave a formula for the coefficients in the Stirling series.[17]

Every topic in this book runs deep and the research continues.

Example 2 – Knights of the Round Table
Suppose six people wish to sit at a round table. How many different ways can this be done, if each person only cares who is to his or her right and left and rotations of all of the people about the table are not counted? What if there are seven people? What if there are n people?

Solution
There are several ways to approach this problem. What we want to avoid is trying to list all of the possibilities for the case of six people. It would be too easy to miss

---

16 https://en.wikipedia.org/wiki/Stirling%27s_approximation.
17 Nemes, Gergő, "On the Coefficients of the Asymptotic Expansion of n!" *Journal of Integer Sequences*, Vol. 13, No. 6, 2010. See http://www.maths.ed.ac.uk/~gnemes/and https://scholar.google.com/citations?user=Nw0zmiwAAAAJ&hl=en&oi=sra for more about the author.

one or include one twice. One approach consists of looking at smaller, simpler cases. For one person, there's obviously just 1 way to seat him. For two people, the answer is again 1, for if we switch them, the result is the same as rotating the table (and those seated with it!) by 180°. Moving on to three and four people, we can sketch out the possibilities without much difficulty and see that the answers are 2 and 6, respectively. Placing the answers we have obtained thus far in a table, we have

Number of people	Number of seatings
1	1
2	1
3	2
4	6

The pattern may already be clear to you, but if it isn't, adding one more row might help. The values for the number of seatings are simply the factorials of one less than the number of people being seated. That is, there are 5! distinct ways for six people to sit at a round table and $(n - 1)!$ distinct ways for n people to be seated. Now, this does not constitute a proof. As we have seen in previous chapters, there are plenty of patterns that look good at first but turn out to not continue forever. Still, having a suspected answer often leads to an insight as to how a proof can be constructed. In this instance, we see that one of the people doesn't seem to count. How can this be? Well, the first person can sit anywhere without eliminating any of the possible seatings. It's only when we place a second person next to him or her (moving clock-wise around the table, say) that we are making a real choice and selecting a particular subset of seatings from the class of all that are possible. Similarly, selecting the 3rd, 4th, 5th, etc., persons to place at the table continues to make a difference. Figure 8.1 makes this clearer.

So, the total number of possibilities for n people is (the number of choices for the 2nd person to be seated)·(the number of choices for the 3rd person to be seated)·(the number of choices for the 4th person to be seated)· ⋯ ·(the number of choices for the nth person to be seated) = $(n - 1)(n - 2)(n - 3)\cdots(1) = (n - 1)!$

Another way to think about this problem is by treating the table like a line. Pick some position to be the "head" of the table and then line up everyone, starting at the head and continuing clockwise. There will be $n!$ possibilities. But, of course, a round table doesn't have a head, so we must divide by the number of equivalent positons for this starting point, which is the number of seats, n. Equivalently, you can take the $n!$ we originally obtained and divide by n to eliminate rotations of the seatings, which are considered the same. Either way, we end up with $n!/n = (n - 1)!$.

If you followed one of the arguments above, but don't think you could have figured it out on your own, don't feel bad. Some big names worked in combinato-rics, but it took Gottfried Wilhelm Leibniz, one of the co-discoverers of calculus to

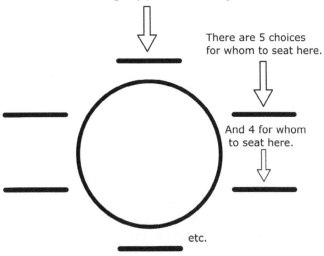

We can sit anyone here without eliminating any potential seatings.

There are 5 choices for whom to seat here.

And 4 for whom to seat here.

etc.

FIGURE 8.1
Round table seating possibilities.

Image created by Josh Gross.

knock this problem out. He encountered these permutations again when he studied determinants.[18]

Generalize!

The factorial function, as defined above, only makes sense for positive integer values, but we can broaden our definition. With a little calculus, we can expand our definition to cover a much larger class of numbers. What follows is known as Euler's[19] Gamma function:

$$\Gamma(t) = \int_0^\infty x^{t-1} e^{-x} \, dx.$$

Using integration by parts to evaluate $\Gamma(t+1)$ with $u = x^t$ and $dv = e^{-x} \, dx$ we get

---

18 Wilson, Robin and John J. Watkins, editors, *Combinatorics: Ancient & Modern*, Oxford University Press, Oxford, 2013, p. 151.

19 So, we meet Euler again! If you ever get all the way through a course without hearing about Euler, it wasn't a mathematics course.

$$\Gamma(t+1) = \int_0^\infty x^t e^{-x} dx = -x^t e^{-x} \Big|_0^\infty + \int_0^\infty t x^{t-1} e^{-x} dx = 0 + t \int_0^\infty x^{t-1} e^{-x} dx = t\Gamma(t).$$

If t is an integer $n = 1, 2, 3, \ldots$, then

$$\begin{aligned}
\Gamma(n) &= (n-1)\Gamma(n-1) \\
&= (n-1)(n-2)\Gamma(n-2) \\
&\cdots \\
&= (n-1)(n-2)\cdots(1)\Gamma(1) \\
&= (n-1)!, \quad \text{because } \Gamma(1) = 1.
\end{aligned}$$

So the Gamma function reduces to the factorial for a positive integer argument. Once again (as with Stirling's formula), we have connections between the continuous and the discrete. A graph of the Gamma function is given in Figure 8.2. Notice that it's left undefined for negative integers.

The Bohr–Mollerup Theorem (1922)

There are many functions $f(x)$ defined for all positive real numbers that agree with the values of the factorials on the positive integers and satisfy $f(1) = 1$ and $f(x+1)=x \cdot f(x)$. What makes Euler's solution unique is that $\Gamma(x)$ is log convex. This means that if we

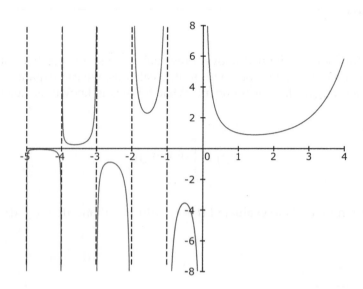

FIGURE 8.2
A graph of Euler's Gamma function.

Image created by Josh Gross.

graph the function $\log(\Gamma(x))$, we can then select any two points on the curve and the straight line that connects them will lie entirely above the curve. The Bohr–Mollerup theorem simply asserts the uniqueness of the solution offered by $\Gamma(x)$, once this extra condition is imposed.[20]

Harald Bohr and Johannes Mollerup published this theorem in a complex analysis text they wrote. Usually the first appearance of new results is in papers published in journals. The most important of these find their way into textbooks, but this comes later. The reason for the break with tradition in this case was that Bohr and Mollerup didn't know that it was a new result; they thought it had already been proven![21]

Applications!

> There is no branch of mathematics, however abstract, which may not some day be applied to phenomena of the real world.
>
> Nikolai Lobatchevsky (1792–1856)

Euler's Gamma function led to many surprises. It turns out that an n-dimensional hypersphere with radius R has volume $V_n = \dfrac{\pi^{n/2}}{\Gamma((n/2)+1)}R^n$.[22] It's reasonable for π to make an appearance in a formula for a sphere, but what does a generalization of the factorial concept have to do with it?

A much bigger application arose from another function described by Euler, the Euler Beta function. This can be expressed using his Gamma function, like so

$$B(x,y) = \frac{\Gamma(x)\Gamma(y)}{\Gamma(x+y)}.$$

In 1968, Gabriele Veneziano, an Italian physicist, used this function to explain the strong nuclear force.[23] Leonard Susskind, an American physicist, examined it and this led him to create string theory. He later detailed his experience in submitting a paper describing his discovery:[24]

> I was completely convinced that when it came back it was going to say, "Susskind is the next Einstein," or maybe even, "the next Newton." And it came back saying, "this paper's not very good, probably shouldn't be published." I was truly knocked off my

---

20 See Chamberland, Marc, *Single Digits*, Princeton University Press, Princeton, NJ, 2015, pp. 20–21 or https://en.wikipedia.org/wiki/Bohr%E2%80%93Mollerup_theorem for more.

21 https://en.wikipedia.org/wiki/Bohr%E2%80%93Mollerup_theorem The citation for the text is Mollerup, Johannes and Harald Bohr, *Lærebog i Kompleks Analyse, Vol. III*, Copenhagen, 1922.

22 https://en.wikipedia.org/wiki/Factorial and https://en.wikipedia.org/wiki/Volume_of_an_n-ball.

23 Veneziano, Gabriele, "Construction of a crossing-symmetric, Regge-behaved amplitude for linearly rising trajectories," *Nuovo Cimento A*, Vol. 57, 1968, pp. 190–197.

24 https://sites.google.com/site/thephilosoraptorsays/spooky-physics-and-einstein-v-bohr-bohr-v-einstein/science-that-makes-you-go-wtf-explored which seems to quote from "The Elegant Universe." This PBS program is available online at http://www.pbs.org/wgbh/nova/physics/elegant-universe.html in three parts. The relevant portion begin at 20:18 in part 2.

chair. I was depressed, I was unhappy. I was saddened by it. It made me a nervous wreck, and the result was I went home and got drunk.

Leonard Susskind

Susskind was simply too far ahead of the reviewers for them to recognize his genius. He did, however, get the recognition he deserved later on.

Another example of pure mathematics finding unexpected applications, many years later, was pointed out by Marc Chamberland in his fun book *Single Digits*:

> The octonians [an eight dimensional generalization or extension of the complex numbers] languished for many years because they apparently lacked connection to mathematical physics. Glimpses appeared in the 1930s, but only in the 1980s was the tie made to the area of string theory.[25]

So, some forgotten math (Euler's Gamma function) made string theory possible, and then string theory made some other forgotten math (octonians) relevant.

Art Appreciation

In addition to the serious physics, the Gamma function yields some beautiful formulas, such as

$$\left(-\frac{1}{2}\right)! = \sqrt{\pi}.$$

Our favorite numbers frequently arise in unexpected places. This is actually why numbers like π and e are so well-liked. They're ubiquitous. We also have[26]

$$\Gamma(x)\Gamma(1-x) = \frac{\pi}{\sin(\pi x)}.$$

A Power Series

The factorial notation also comes in handy for power series. For example,

$$e^x = \frac{1}{0!} + \frac{x}{1!} + \frac{x^2}{2!} + \frac{x^3}{3!} + \frac{x^4}{4!} + \cdots = \sum_{n=0}^{\infty} \frac{x^n}{n!}.$$

---

25 Chamberland, Marc, *Single Digits*, Princeton University Press, Princeton, NJ, 2015, p. 202; a reference is given on p. 219: Baez, John, "The Octonians," *Bulletin of the American Mathematical Society*, Vol. 39, 2002, pp. 145–205. Errata in Bulletin of the American Mathematical Society, Vol. 42, 2005, p. 213, available online at http://math.ucr.edu/home/baez/octonions/.
26 Havil, Julian, *Gamma: Exploring Euler's Constant*, Princeton University Press, Princeton, NJ, 2010, p. 59.

The first term may be written as 1 (Recall: 0! = 1). I expressed it using 0! to emphasize the pattern of the terms. Similarly, the first two numerators could have been written as x^0 and x^1, respectively. For the special case $x = 1$, this gives us $e = 1/0! + 1/1! + 1/2! + 1/3! + \cdots$

Transcendental Numbers

Factorials came in handy in yet other, seemingly unrelated, areas of mathematics. Recall from Chapter 0 that a number is said to be algebraic if it is the root of a polynomial with integer coefficients. All rational numbers are algebraic, but at least some irrationals are as well. The number $\sqrt{2}$ is irrational, but it is a root of $x^2 - 2 = 0$, so it is algebraic. Even complex numbers can be algebraic. We have $i = \sqrt{-1}$ as a root of $x^2 + 1 = 0$, so i is algebraic. It was Joseph Liouville (1809–1882), a French mathematician, who in 1844 first proved the existence of nonalgebraic numbers.[27] Euler referred to such numbers as **transcendental**, but he did not prove that any actually existed.[28] Prior to Euler, it was Leibniz who first brought the word transcendental to mathematics, although he used it to describe functions, like $\sin(x)$, which cannot be expressed algebraically in a finite number of operations.[29]

In 1851, Liouville presented the following transcendental number, now known as Liouville's number or Liouville's constant.[30]

$$\frac{1}{10^{1!}} + \frac{1}{10^{2!}} + \frac{1}{10^{3!}} + \frac{1}{10^{4!}} + \cdots$$

Note the use of the factorial symbol in each term.

Investigating $n!$ has led to some fascinating results, but not all permutation problems are as straightforward as taking $n!$, where n is the number of objects involved.

Example 3 – Back to the Grocery Store

Let's now return to some simpler mathematics and consider a modification of the grocery store problem (Example 1). There are now 10 people at a grocery store heading for the checkout. As before only one lane is open. However, this time, after the first six people get in line, the remaining four, unwilling to wait that long, set their items down and walk out. In how many ways can 6 of the 10 people be lined up?

Solution: We solve this problem with the same diagram used in Example 1.

27 Maor, Eli, *e: The Story of a Number*, ninth printing with new material on pp. 183–186, Princeton University Press, Princeton, NJ, 1998, p. 191.

28 https://en.wikipedia.org/wiki/Transcendental_number.

29 https://en.wikipedia.org/wiki/Transcendental_number.

30 Liouville, Joseph, "Sur des classes très-étendues de quantités dont la valeur n'est ni algébrique, ni même réductible à des irrationelles algébriques," *Journal de Mathématiques Pures et Appliquées*, Vol. 16, 1851, pp. 133–142.

Checkout ____   ____   ____   ____   ____   ____
                10 choices   9 choices   8 choices   7 choices   6 choices   5 choices

So, the answer is $10 \cdot 9 \cdot 8 \cdot 7 \cdot 6 \cdot 5 = 151{,}200$. We can't express the answer as 10!, because 4, 3, 2, and 1 are not present in the multiplication. This doesn't mean that the factorial notation needs to be abandoned. The answer may be written as $10!/4!$.

In general if we are interested in the number of permutations on r objects chosen from a group of n, we write nPr and the answer is $\frac{n!}{(n-r)!}$.

Example 4 – Olympic Medals
In how many ways can the gold, silver, and bronze medals be awarded to three of ten athletes?

Solution: Order matters, as getting the silver (the agony of defeat) is quite different from getting the gold (the thrill of victory). Hence, we have a permutation. The answer is $_{10}P_3 = \dfrac{n!}{(n-r)!} = \dfrac{10!}{(10-3)!} = \dfrac{10!}{7!} = 10 \cdot 9 \cdot 8 = 720$.

Example 5 – A Die Game
Suppose 10 people are given an opportunity to bet on the result of a ten-sided die roll. The first person may pick one number, more than one number, or decide to place no bet. The second person may then do the same, but cannot pick a number already selected by the first player. Players 3 through 10 place their bets, in order, in the same manner. It is possible that all numbers have been chosen before some of the latter players get a chance to bet. At the other extreme, players 1 through 9 could decline to place bets, in which case player 10 could select every number. In fact, to make sure all of the numbers are covered, player 10 is required to take all numbers, if any are left, that have not yet been bet on when it comes to his turn. Several questions can be posed about this game:

  a) In how many distinct ways can the bets be placed?
  b) In how many ways can each player bet on exactly one number?
  c) In how many ways can the bets be placed so that at least one player bets on more than one number?

Solution:

  a) Instead of focusing on each player and what he or she does, it's easier to focus on the numbers. For 1, there are 10 possibilities as to which player placed a bet on it. For 2, there are also 10 choices as to who bet on it. The same holds for 3 through 10. Thus, the total number of possibilities, by the multiplication principle, is 10^{10}.

  b) If each player bets on just one number, then we can imagine them lined up in order by the numbers they bet on. Each line-up will correspond to a different answer. Because 10 people can line up in 10! ways, the answer is $10! = 3{,}628{,}800$. This can also be calculated as $_{10}P_{10}$.

  c) We could try to directly determine the number of ways that one or more player could bet on more than one number. This approach could lead to the analysis of many different cases, which, if disjoint and collectively covering all of the possibilities, could be summed to get the correct answer. But there's a far simpler approach. Our answer to part a is the "anything goes" value and our answer to part b is exactly the results we are *not* interested in now, the ones where no better selects more than one number. So, the easiest approach is to just subtract the answer to b from the answer to a. We have $10^{10} - 3,628,800 = 9,996,371,200$ ways in which at least one player can bet on more than one number. The approach used here is the popular trick of looking at the complement. When there are a large number of cases to consider, you should always look at the complement to see if it would be easier to calculate that instead, and then subtract that value from the total to get your desired result.

The approach used in part c generalizes and was described by Mersenne. He found that if p objects are to be selected from n choices, with repeats allowed, then the number of possibilities with at least one repeat is $n^p - {_nP_p}$.[31]

Example 6 – Polya's Problem (counting necklaces)
Suppose you want to create a necklace by placing n beads on a string. Also, suppose that you have k different kinds of beads, in large enough quantities that the entire necklace can consist solely of any particular kind of bead. There are no restrictions on how beads can be used. The only rule is that the completed necklace must have n beads on it. Ignoring the clasp, rotations of a given necklace are considered the same, as is the result obtained by flipping the necklace upside down. So, this problem is similar to Example 2 – Knights of the Round Table, with the differences that the table can be flipped upside down, along with the knights, who can be repeated!
Solution: Despite the similarity, the solution to this new problem bears no resemblance to our previous answer. It is given by

$$\frac{1}{n} \sum_{d|n} \phi(d) k^{n/d}$$

where $d \mid n$ indicates that the summation is to be taken over values of d that divide evenly into n and the $\phi(d)$, put forth by Euler and known as Euler's totient function, is the number of positive integers less than d and relatively prime to d. By definition, $\phi(1) = 1$. As an example of using this solution, if a six-bead necklace is desired and there are four kinds of beads to choose from, then $n = 6$, $k = 4$, and we have

$$\frac{1}{n} \sum_{d|n} \phi(d) k^{n/d} = \frac{1}{6} \sum_{d|6} \phi(d) 4^{6/d} = \frac{1}{6} \left(\phi(1) 4^{6/1} + \phi(2) 4^{6/2} + \phi(3) 4^{6/3} + \phi(6) 4^{6/6} \right)$$

$$= \frac{1}{6} \left(1 \cdot 4^6 + 1 \cdot 4^3 + 2 \cdot 4^2 + 2 \cdot 4 \right)$$

$$= \frac{1}{6} (4096 + 64 + 32 + 8)$$

$$= 700.$$

31 Wilson, Robin and John J. Watkins, editors, *Combinatorics: Ancient & Modern*, Oxford University Press, Oxford, 2013, p. 134.

It's not clear who was the first to solve this problem, but George Polya gave a solution in 1936.[32] My aim in presenting this is not for you to understand where the formula came from; I'm happy if you are just able to apply it. The point is to show how two similar-looking problems can have wildly different solutions and how difficult some easy-sounding problems turn out to be!

Also, this problem turns out to have many applications:

> [T]his formula can be used to solve a difficult problem in the theory of Lie algebras, which in turn has a deep effect on contemporary physics.... [and it has] been applied to the most disparate problems of enumeration in mathematics, physics, and chemistry (where, for example, the formula gives the number of isomers of a given molecule).[33]

Example 7 – Permuting Letters
In how many ways can the letters in the word COMPUTER be arranged?

Solution: We are taking all of the objects (letters), so we could write this as
$$_8P_8 = \frac{n!}{(n-r)!} = \frac{8!}{(8-8)!} = \frac{8!}{0!} = \frac{8!}{1} = 8! = 40,320.$$

Alternatively, we could have solved this one by observing that all eight letters of COMPUTER are distinct, so the answer is simply 8! = 40,320.

In *The Sepher Yetsirah* (The Book of Creation), written by an anonymous Jewish author sometime in the wide range from the 2nd to the 8th century, there's a passage in which smaller cases of the problem above are solved.

> Two stones [letters] build two houses [words], three build six houses, four build twenty-four houses, five build one hundred and twenty houses, six build seven hundred and twenty houses, seven build five thousand and forty houses. From thence further go and reckon what the mouth cannot express and the ear cannot hear.[34]

So, I guess I've followed his instructions and reckoned "what the mouth cannot express and the ear cannot hear"!

Example 8 – Say What?
In how many ways can the letters in the word COHOMOLOGY be arranged?

Solution: This strange word from the world of mathematics has 10 letters and 4 of them are Os. If we try to use the alternate approach from the previous example, we have 10!, but this is not correct as it treats the Os as all distinct, when they are actually interchangeable. Having "overcounted" we need to divide out by the number of ways of rearranging the Os to eliminate the duplicates. Because there are 4! ways to arrange the Os, the correct final answer is 10!/4! = 3,628,800/24 = 151,200.

---

32 Kac, Mark, Gian-Carlo Rota, and Jacob T. Schwartz, *Discrete Thoughts: Essays on Mathematics, Science, and Philosophy*, Birkhäuser, Boston, MA, 1986, p. 54.

33 Kac, Mark, Gian-Carlo Rota, and Jacob T. Schwartz, *Discrete Thoughts: Essays on Mathematics, Science, and Philosophy*, Birkhäuser, Boston, MA, 1986, p. 54.

34 Karlisch, Isidor, editor and translator, *The Sepher Yezirah* (The Book of Formation), Heptangle Books, Gillette, NJ, 1987, p. 23. This is discussed in Katz, Victor J., "Jewish Combinatorics," in Wilson, Robin and John J. Watkins, editors, *Combinatorics: Ancient & Modern*, Oxford University Press, Oxford, 2013, pp. 108–121, p. 111 has the quote given here.

In general, if a word has n letters and m of them appear repeatedly, $k_1, k_2,..., k_m$ times, respectively, then the number of distinct permutations of the letters in this word is

$$\frac{n!}{k_1!k_2!\ldots k_m!}.$$

This expression appears often enough, in different kinds of problems, that there's a special notation for it:

$$\binom{n}{k_1, k_2, \ldots, k_m}$$

It is known as a **multinomial coefficient**, because these numbers arise as the coefficients in the expansion of expressions of the form $(x_1 + x_2 + \cdots + x_m)^n$.

The formula $\frac{n!}{k_1!k_2!\ldots k_m!}$ appeared in *Līlāvatī*, written by Bhāskara II in 1150 CE, although it was not notated in this manner.[35] Mersenne was the first Westerner to publish it.[36] He did so in 1636.

Neat Facts

Here are three more cool results that connect with factorials.

1. **Wilson's theorem:** A natural number $n > 1$ is prime if and only if $(n-1)! \equiv -1 \pmod{n}$.

    This result goes back to about 1000 CE, when it was stated by Ibn al-Haytham, a mathematician who was born in Iraq, but later moved to Egypt where he carried out much work in mathematics, astronomy and physics.[37] The English mathematician John Wilson rediscovered the theorem by 1770,[38] but it was not until 1771 that a proof was put forth by Lagrange.[39] It might seem fairer to name the theorem after Ibn al-Haytham or Lagrange, but it is known as Wilson's theorem.[40] It could

---

35 Wilson, Robin and John J. Watkins, editors, *Combinatorics: Ancient & Modern*, Oxford University Press, Oxford, 2013, p. 54 and p. 171.

36 Wilson, Robin, and John J. Watkins, editors, *Combinatorics: Ancient & Modern*, Oxford University Press, Oxford, 2013, p. 179.

37 https://en.wikipedia.org/wiki/Ibn_al-Haytham and O'Connor, John J. and Edmund F. Robertson, "Abu Ali al-Hasan ibn al-Haytham," MacTutor History of Mathematics archive, University of St Andrews, available online at http://www-history.mcs.st-andrews.ac.uk/Biographies/Al-Haytham.html.

38 https://en.wikipedia.org/wiki/Wilson%27s_theorem, which adds in a note, "Edward Waring, *Meditationes Algebraicae* (Cambridge, England: 1770), page 218 (in Latin). In the third (1782) edition of Waring's *Mediationes Algebraicae*, Wilson's theorem appears as problem 5 on page 380. On that page, Waring states: "Hanc maxime elegantem primorum numerorum proprietatem invenit vir clarissimus, rerumque mathematicarum peritissimus Joannes Wilson Armiger." (A man most illustrious and most skilled in mathematics, Squire John Wilson, found this most elegant property of prime numbers.)

39 https://en.wikipedia.org/wiki/Wilson%27s_theorem, which adds in a note, Joseph Louis Lagrange, "Demonstration d'un théorème nouveau concernant les nombres premiers" (Proof of a new theorem concerning prime numbers), *Nouveaux Mémoires de l'Académie Royale des Sciences et Belles-Lettres*, Berlin, Vol. 2, 1771, pp. 125–137.

40 https://en.wikipedia.org/wiki/Wilson%27s_theorem.

even be named after Leibniz, for it appears that he knew of the theorem, as well, but as happened too often with this great researcher, he did not publish it.[41]

2. **A result from Leibniz:** A natural number $n > 1$ is prime if and only if $(n - 2)! \equiv 1 \pmod{n}$.[42] If you multiply both sides of this equality by $(n - 1)$ and reduce the right-hand side modulo n, by subtracting n, this result is seen to be identical to Wilson's theorem.

3. **The "36 Officers Puzzle" – a puzzle with no solution:**[43] In 1782, Leonhard Euler put forth a problem that he conjectured could not be solved.[44] It involved six regiments of officers, each of which was made up of six officers of distinct ranks. The challenge was to arrange them in a square such that no row or column contains officers having the same rank or belonging to the same regiment. What makes the puzzle interesting is its resistance to proof despite its seeming simplicity. It might seem like an easy matter to just test all of the possibilities. But there are $36! \approx 3.7 \times 10^{41}$ ways to fill in the grid, so brute-force is not a practical approach to this problem. It wasn't until 1901 that a definitive answer was given.[45] It turned out that Euler was right. Amateur French mathematician Gaston Terry showed that it couldn't be done.

Open Problems

1. **Triangular factorials**
   These are numbers that are both triangular and factorial. Some examples are 1, 6, and 120. Are there others? No one knows!

---

41 https://en.wikipedia.org/wiki/Wilson%27s_theorem, which adds in a note, "Giovanni Vacca (1899) "Sui manoscritti inediti di Leibniz" (On unpublished manuscripts of Leibniz), *Bollettino di bibliografia e storia delle scienze matematiche* ... (Bulletin of the bibliography and history of mathematics), Vol. 2, pp. 113–116; see page 114 (in Italian). Vacca quotes from Leibniz's mathematical manuscripts kept at the Royal Public Library in Hanover (Germany), vol. 3 B, bundle 11, page 10:
   Original: Inoltre egli intravide anche il teorema di Wilson, come risulta dall'enunciato seguente:
   "Productus continuorum usque ad numerum qui antepraecedit datum divisus per datum relinquit 1 (vel complementum ad unum?) si datus sit primitivus. Si datus sit derivativus relinquet numerum qui cum dato habeat communem mensuram unitate majorem."
   Egli non giunse pero a dimostrarlo.
   Translation: In addition, he [Leibniz] also glimpsed Wilson's theorem, as shown in the following statement:
   "The product of all integers preceding the given integer, when divided by the given integer, leaves 1 (or the complement of 1?) if the given integer be prime. If the given integer be composite, it leaves a number which has a common factor with the given integer [which is] greater than one."
   However, he didn't succeed in proving it.
   See also: Giuseppe Peano, editor, *Formulaire de mathématiques*, Vol. 2, No. 3, 1897, p. 85.
42 Young, Robert M., *Excursions in Calculus: An Interplay of the Continuous and the Discrete*, The Mathematical Association of America, 1992, p. 118.
43 Wilson, Robin and John J. Watkins, editors, *Combinatorics: Ancient & Modern*, Oxford University Press, Oxford, 2013. See pp. 257–269 for a broader discussion of the 36 Officers Puzzle in the context of Latin squares.
44 Euler, Leonhard, *Recherches sur une nouvelle espece de quarres magiques*, 1782.
45 See Tarry, Gaston, "Le Probléme de 36 Officiers," *Compte Rendu de l'Association Française pour l'Avancement des Sciences*. Secrétariat de l'Association, Vol. 1, 1900, pp. 122–123 and Tarry, Gaston, "Le Probléme de 36 Officiers," *Compte Rendu de l'Association Française pour l'Avancement des Sciences, Secrétariat de l'Association*, Vol. 2, 1901, pp. 170–203.

2. **Primes with special forms**
Are there infinitely many prime numbers of the form $n! + 1$? Are there infinitely many prime numbers of the form $n! - 1$? No one knows the answer to either of these questions![46] In the exercise set for this chapter you are asked to prove that there are infinitely many composite numbers of the forms $n! + 1$ and $n! - 1$. This might sound like the same question, but it is not. An infinite subset of the natural numbers can have a complement that is either finite or infinite.

3. **Evaluating $\Gamma(x)$ exactly**
The exact values of $\Gamma(1/3)$, $\Gamma(1/4)$, and infinitely many others remain unknown.[47]

4. **Self-avoiding Walks (SAWs)**
Imagine standing at the origin and only being able to move exactly one unit, that is, one step, at a time in one of the four directions – north, south, east, or west. How many n step walks are possible such that you never return to a point you stepped to previously? Values can be calculated for small n, but no exact formula is known and the problem is thought to be NP-hard.[48] As is usually the case, mathematicians have chipped away at the problem by finding ways to approximate solutions, as well as finding exact formulas for special cases. For example, the number of self-avoiding walks from the origin to (m, n) that only take steps in the directions north and east is given by $\binom{m+n}{m,\,n} = \frac{(m+n)!}{m!n!}$.

A related set of open questions deals with results for random walks, in which the decision as to which direction to step each time is made randomly. There's a lot of experimental data, but few mathematical results connected with this problem.[49]

Exercise Set 8

Each of us assumes everyone else knows what he is doing. They all assume we know what we are doing. We don't.

Philip K. Dick

Note: Some problems may only require techniques from earlier chapters:[50]

1. In how many ways may four professors from a group of five be selected to teach the following classes (one professor for each class): Calculus III, Discrete Mathematics, Abstract Algebra, and Cryptology?

2. In C++ a variable name may only consist of the following characters:

   letters (upper and lower case are distinct)

---

46 Young, Robert M., *Excursions in Calculus: An Interplay of the Continuous and the Discrete*, The Mathematical Association of America, 1992, p. 118.
47 Havil, Julian, *Gamma: Exploring Euler's Constant*, Princeton University Press, Princeton, NJ, 2010, p. 55.
48 https://en.wikipedia.org/wiki/Self-avoiding_walk.
49 Kac, Mark, Gian-Carlo Rota, and Jacob T. Schwartz, *Discrete Thoughts: Essays on Mathematics, Science, and Philosophy*, Birkhäuser, Boston, MA, 1986, p. 55.
50 The reasoning behind this is: 1. It is good to review and 2. Real life problems are not found in a context that indicates what technique to use – you must decide yourself.

digits (0 through 9 inclusive)

the underscore character (_)

Additionally, a variable name may not begin with an underscore (in order to avoid conflicts with operating system variables). You cannot begin with a number either. Given these rules, how many variables of length four are possible? (Don't worry about the few "reserved words" or "keywords" that are not allowed.)

3. How many distinct arrangements of the letters in the word GOLDEN are there?

4. Suppose the word we are anagramming has one or more letters repeated? How many distinct anagrams (counting the original ordering) of the word INFINITY are there?

5. Mersenne, whom you will hear more about, took the time to list all of the distinct arrangements of the letters in IESUS (Jesus). How many are there? Please *do not* list them.[51]

6. John Wallis, the greatest British mathematician before Isaac Newton, listed all of the arrangements of the letters in MESSES.[52] How many are there?

7. What is 10! seconds in terms of weeks?

8. In 1494, Luca Pacioli calculated 11! = 39,916,800.[53] Was Pacioli correct?

9. While it is not known if there are infinitely many prime numbers of the form $n! + 1$, it has been established that there are infinitely many composite numbers of the form $n! + 1$. Prove it.

10. While it is not known if there are infinitely many prime numbers of the form $n! - 1$, it has been established that there are infinitely many composite numbers of the form $n! - 1$. Prove it.

11. A committee made up of 10 people decides to form a subcommittee. If the subcommittee must consist of at least one person, but not more than five, how many possibilities are there?

12. Jay Fox Band had a total of 25 songs on their first two albums. If they wish to perform 12 of them in concert, how many possible orderings are there?

13. Consider the following formulas:

$$\sum_{n=1}^{\infty} \frac{1}{2^n} = 1! \qquad \sum_{n=1}^{\infty} \frac{n}{2^n} = 2! \qquad \sum_{n=1}^{\infty} \frac{n^2}{2^n} = 3!$$

Prove that the pattern continues or give a counterexample.

---

51 Inspired by Wilson, Robin and John J. Watkins, editors, *Combinatorics: Ancient & Modern*, Oxford University Press, Oxford, 2013, p. 10.

52 Inspired by Wilson, Robin and John J. Watkins, editors, *Combinatorics: Ancient & Modern*, Oxford University Press, Oxford, 2013, pp. 10–11. Original source: Wallis, John, *Discourse of Combinations*, 1685, published as a supplement to his *Treatise of Algebra*.

53 Wilson, Robin and John J. Watkins, editors, *Combinatorics: Ancient & Modern*, Oxford University Press, Oxford, 2013, p. 126, citing Pacioli, Luca, *Summa de Arithmetica Geometria Proportioni e Proportionalita* (Comprehensive Treatise on Arithmetic, Geometry, Proportions, and Proportionality), Venice, 1494, 2nd distinction, 5th treatise.

14. In John Sandford's novel *The Devil's Code*, a simple way to get past a door with a numerical combination is detailed. The first step involves spraying the keypad with deodorant. Later, after someone has entered the correct number (and hopefully left again!), the would-be intruders may return. We pick up the novel at this point below.

> LuEllen had programmed Bloch's phone number into her cell phone earlier in the evening and dialed the number as she got out of the car. The hallway leading past Bloch Tech was empty. I walked to the door. LuEllen close behind, and mimed a knock: we could hear the phone ringing inside. No answer.
>
> As I mimed another knock, LuEllen turned off her phone and pointed a little battery-operated black light, the kind teenagers used to buy in head shops, at the keypad. The powdery crystals in the deodorant fluoresced in the light – except for the three that had no powdery crystals.
>
> "Four-six-seven," she said. "But there are four digits in a Vermond lock. In this model. So they repeated one of them."
>
> Nobody in the hall: I took a dime notebook out of my pocket and began scrawling number combinations as quickly as I could write, calling them out as I jotted them down. The thing about number pads is, with ten digits, there are 10,000 possible combinations. Getting inside with a brute-force attack is tough. And a few locks, but not this one, were alarmed, or would lock up, after a certain number of incorrect combinations. Then they could only be opened with a key.
>
> But if you know the four digits involved in the combination… ah, then there were only twenty-four possible combinations. If one of the digits is repeated, like it was here, and you don't know which, the number goes up to thirty-six. But most people start their combination with the lowest number, in this case a four. We started with four-four-six-seven, and went to four-four-seven-six, and to four-six-four-seven, and so on. We were lucky, hit it on the eighth combination, and pushed into the darkened office.[54]

How is Sandford's mathematics?

References/Further Reading

The only thing that you absolutely have to know, is the location of the library.
– Albert Einstein.

Artin, Emil, *The Gamma Function*, Holt, Rinehart, Winston, New York, 1964.

Baez, John, "The Octonians," *Bulletin of the American Mathematical Society*, Vol. 39, 2002, pp. 145–205. Errata in *Bulletin of the American Mathematical Society*, Vol. 42, 2005, p. 213, available online at http://math.ucr.edu/home/baez/octonions/.

Bohr, Harald and Johannes Mollerup, *Lærebog I Kompleks Analyse, Vol. III*, Gjellerup, Copenhagen, 1922. This is a textbook in Complex Analysis.

Cajori, Florian, *A History of Mathematical Notations*, Two Volumes Bound as One, Dover Publications, Mineola, NY, 1993, available online at https://monoskop.org/images/2/21/Cajori_Florian_A_History_of_Mathematical_Notations_2_Vols.pdf. This is a reprint of the volumes from 1928 to 1929.

Chamberland, Marc, *Single Digits*, Princeton University Press, Princeton, NJ, 2015.

---

54 Sandford, John, *The Devil's Code*, G. P. Putnam's Sons, New York, 2000, p. 125.

Courant, Richard and Herbert Robbins, "Liouville's Theorem and the Construction of Transcendental Numbers," §2.6.2 in *What Is Mathematics?: An Elementary Approach to Ideas and Methods*,
second edition, Oxford University Press, Oxford, 1996, pp. 104–107.

Davis, Philip J., "Leonhard Euler's Integral: A Historical Profile of the Gamma Function," *American
Mathematical Monthly*, Vol. 66, No. 10, December 1959, pp. 849–869. This paper was reprinted in
Abbott, J. C., editor, *The Chauvenet Papers: A Collections of Prize-winning Papers in Mathematics*,
Volume II, Mathematical Association of America, Washington, DC, 1978, pp. 332–351.

Euler, Leonhard, *Recherches sur une nouvelle espece de quarres magiques*, Zeeuwsch Genootschao,
Middelburg, Netherlands, 1782.

Ghahramani, Saeed, *Fundamentals of Probability*, Prentice-Hall, Upper Saddle River, NJ, 1996.

Ginsburg, J., "Rabbi Ben Ezra on Permutations and Combinations," *The Mathematics Teacher*, Vol. 15,
No. 6, October 1922, pp. 347–356.

Havil, Julian, *Gamma: Exploring Euler's Constant*, Princeton University Press, Princeton, NJ, 2010.
Don't confuse the constant gamma with the Gamma function. We distinguish the two gammas
by always using upper case for the function, Γ, and lower case, γ, for the constant. Although the
title refers to the constant gamma, pages 53–60 discuss Euler's Gamma function.

Hayes, Brian, "How to Avoid Yourself," *American Scientist*, Vol. 86, No. 4, July–August 1998,
pp. 314–319.

Huber, Greg, "Gamma Function Derivation of n-Sphere Volumes," *American Mathematical Monthly*,
Vol. 89, No. 5, May 1982, pp. 301–302.

Kac, Mark, Gian-Carlo Rota, and Jacob T. Schwartz, *Discrete Thoughts: Essays on Mathematics, Science, and
Philosophy*, Birkhäuser, Boston, MA, 1986, pp. 54–55 have more information on Polya's problem.

Kempner, Aubrey J., "On Transcendental Numbers," *Transactions of the American Mathematical
Society*, Vol. 17, No. 4, October 1916, pp. 476–482.

Knobloch, Eberhard, "Renaissance Combinatorics," in Wilson, Robin and John J. Watkins, editors,
Combinatorics: Ancient & Modern, Oxford University Press, Oxford, 2013, pp. 123–145.

Lilley, George, "Pascal's Arithmetic Triangle," *American Mathematical Monthly*, Vol. 1, No. 12,
December 1894, p. 426.

Liouville, Joseph, "Mémoires et communications des membres et des correspondants de l'Académie," *Comptes Rendus de l'Académie des Sciences*, Paris, Vol. 18, 1844, pp. 883–885.

Liouville, Joseph, "Sur des classes très-étendues de quantités dont la valeur n'est ni algébrique,
ni même réductible à des irrationelles algébriques," *Journal de Mathématiques Pures et Appliquées*, Vol. 16, 1851, pp. 133–142.

Liouville, Joseph, "Nouvelle démonstration d'un théorème sur les irrationalles algébriques,
inséré dans le *Compte Rendu* de la dernière séance," *Comptes Rendus de l'Académie des
Sciences*, Paris, Vol. 18, 1844, pp. 910–911.

Liśkiewicz, Maciej, Mitsunori Ogihara, and Seinosuke Toda, "The Complexity of Counting
Self-avoiding Walks in Subgraphs of Two-dimensional Grids and Hypercubes," *Theoretical
Computer Science*, Vol. 304, No. 1–3, July 2003, pp. 129–156.

Maor, Eli, *e: The Story of a Number*, ninth printing with new material on pp. 183–186, Princeton
University Press, Princeton, NJ, 1998.

Nemes, Gergő, "On the Coefficients of the Asymptotic Expansion of $n!$," *Journal of Integer Sequences*,
Vol. 13, Article 10.6.6, 2010, pp. 1–5.

Nielsen, Niels, *Handbuch der Theorie der Gammafunktion*, Teubner, Leipzig, 1906.

Sandford, John, *The Devil's Code*, G. P. Putnam's Sons, New York, 2000.

Tarry, Gaston, "Le Probléme de 36 Officiers," *Compte Rendu de l'Association Française pour l'Avancement
des Sciences*, Secrétariat de l'Association, Vol. 1, 1900, pp. 122–123.

Tarry, Gaston, "Le Probléme de 36 Officiers," *Compte Rendu de l'Association Française pour l'Avancement
des Sciences*, Secrétariat de l'Association, Vol 2, 1901, pp. 170–203.

"The Elegant Universe" This PBS program from 2012 is available online at www.pbs.org/wgbh/
nova/physics/elegant-universe.html. It mentions the unexpected application of Euler's
Gamma function to string theory. If you're in a hurry to watch this bit, start with the second
part of the video at 16: 25.

Turetzky, Morris, "Permutations in the 16th Century Cabala," *The Mathematics Teacher*, Vol. 16, No. 1, January 1923, pp. 29–34.

Veneziano, Gabriele, "Construction of a Crossing-symmetric, Regge-behaved Amplitude for Linearly Rising Trajectories," *Nuovo Cimento A*, Vol. 57, 1968, pp. 190–197.

Wilson, Robin and John J. Watkins, editors, *Combinatorics: Ancient & Modern*, Oxford University Press, Oxford, 2013.

Wylie, Jr., Clarence R., "Mathematical Allusions in Poe," *The Scientific Monthly*, Vol. 63, September 1946, pp. 227–235.

Young, Robert M., *Excursions in Calculus: An Interplay of the Continuous and the Discrete*, The Mathematical Association of America, 1992.

9

Combinations

Chapter 8 examined permutations, problems in combinatorics where order does not matter. This chapter considers problems where order does matter. These are known as **combinations**.

Example 1 – Overtraining
Suppose three students from a group of eight are to be assigned extra homework. In how many ways can this be done? It doesn't matter who is the first or the last selected. The net result is the same. Thus, we have a combination problem.

Solution: It is convenient to continue using the lines from the checkout problem of Chapter 8.

_____	_____	_____
8 choices	7 choices	6 choices

This suggests an answer of $8 \cdot 7 \cdot 6$, but I could have picked these three people in any order and the result would be the same. Order doesn't matter here. Because those selected can be ordered in 3! ways, dividing the suggested answer by 3! yields the correct value. That is, $8 \cdot 7 \cdot 6/(3!) = 8!/(5! \cdot 3!) = 56$.

In general, the number of combinations of n items taken r at a time is given by $\frac{n!}{(n-r)!r!}$. The $(n-r)!$ is in the denominator to cancel the tail end of the $n!$ from the numerator, if we are not selecting all n items, and the $r!$ is to eliminate reorderings of the items we are selecting. A common notation for this is $\binom{n}{r}$. Be sure to not put a line between the two values. This does NOT represent a fraction. Richard K. Guy suggested what has become the standard way to pronounce it: "n choose r."[1] It can also be denoted as nCr or $C(n, r)$.

This is very old mathematics. Gerolamo Cardano knew this formula by 1550,[2] but it was also known to Rabbi Levi ben Gerson (see his _Maasei Hoshev_, 1321)[3], Bhāskara II (see sections 274 and 275 of his _Līlāvatī_, ca. 1150 CE), and Mahāvīra in the 9th century (see his _Gaṇitasārasaṅgraha_, ca. 850), although these mathematicians did not use the factorial notation.[4] Possibly earlier, Śrīdhara covered the special case of $n = 6$ for the various

---

1 He took credit for this in his biography at the end of the paper Guy, Richard K., "John Horton Conway: Mathematical Magus," _The Two-Year College Mathematics Journal_, Vol. 13, No. 5, November 1982, pp. 290–299. He claimed that he wrote enough "so-called research papers" to ensure a raise each year, but that the pronunciation "n choose r" was his only lasting contribution to mathematics.

2 Wilson, Robin and John J. Watkins, editors, _Combinatorics: Ancient & Modern_, Oxford University Press, Oxford, 2013, pp. 127–128.

3 Wilson, Robin and John J. Watkins, editors, _Combinatorics: Ancient & Modern_, Oxford University Press, Oxford, 2013, p. 120.

4 Wilson, Robin and John J. Watkins, editors, _Combinatorics: Ancient & Modern_, Oxford University Press, Oxford, 2013, p. 12 and pp. 53–54.

values of r in his *Pāṭīgaṇita*, but much is in dispute, or simply not known, concerning this mathematician's life, including when he lived.[5]

The Arab world was also aware of the formula before Cardano published it. Ibn al-Bannā al-Marrākushī (1256–1321) gave it, again without a factorial notation, in *Tanbīh al-albāb* (Warning to Intelligent [People]) and the *Raf^c al-Ḥijāb* (Lifting of the Veil).[6]

After Cardano's time, the formula was independently discovered in China. Chen Houyao (1648–1722) presented it in *Cuozong fayi* (The Meaning of Methods for Alternation and Combination).[7]

Example 2 – Ordering Pizza

It's Friday night and you're planning to watch a movie with a couple of friends, but first you need to order pizza. Panucci's Pizza offers ten toppings to tempt your taste buds. After much discussion you decide on the toppings: Pepperoni, Ham, and Olives. You phone in the order. Returning to your living room, a friend asks, "You told him Pepperoni, Ham, and Olives, right?" Oh shit! You ordered Ham, Pepperoni, and Olives! Now what? Should you call back? Fortunately, the order of the toppings doesn't matter. You're dealing with a combination, not a permutation. However, you could have accidentally replaced one of the toppings with another. How many different three-topping pizzas are possible? Assume that all three toppings must be distinct. That is, you cannot order something like triple anchovy.

Solution: Because we are dealing with a combination, it is simply

$$_{10}C_3 = \frac{10!}{(10-3)!3!} = \frac{10!}{7!3!} = \frac{10 \cdot 9 \cdot 8}{3 \cdot 2 \cdot 1} = 120.$$

Example 3 – Combination Locks?

Now let's consider a combination lock (Figure 9.1). Hey, wait a minute! If the combination is 17-29-32, the lock won't open if I try 32-29-17. So, order matters. But if order matters, then we have a permutation, not a combination. Therefore, a "combination lock" should really be referred to as a permutation lock[8] and our solution only requires techniques from previous chapters. It's good to review, so let's continue. Nomenclature aside, how many possibilities must be tried (on average) to open such a lock, if you do not know the numbers involved?

Solution: The digits 0 through 39 provide 40 choices for each of the three numbers. The multiplication principle then tells us that the number of combinations is 40·40·40 = 64,000. But there's a shortcut – have you ever used one of these locks and accidentally gone one past the correct number? It opened anyway! If the combination is 12-27-09, then 12-28-09 will also work (at least with the locks I've used). Using this fact, the numbers that need to be tried for each position are reduced to 1 (takes care of 0, 1, and 2), 4 (takes care of 3, 4, and 5), 7, 10, 13,

---

5 Wilson, Robin and John J. Watkins, editors, *Combinatorics: Ancient & Modern*, Oxford University Press, Oxford, 2013, p. 53, www.britannica.com/biography/Shridhara, and www-history.mcs.st-andrews.ac.uk/Biographies/Sridhara.html.

6 Wilson, Robin and John J. Watkins, editors, *Combinatorics: Ancient & Modern*, Oxford University Press, Oxford, 2013, pp. 99–100. For biographical information see https://en.wikipedia.org/wiki/Ibn_al-Banna'_al-Marrakushi.

7 Wilson, Robin and John J. Watkins, editors, *Combinatorics: Ancient & Modern*, Oxford University Press, Oxford, 2013, pp. 70–71.

8 For similar reasons, the World Series should be called the North American Finite Sequence.

FIGURE 9.1
A "Combination" lock.

Image created by Craig P. Bauer.

16, 19, 22, 25, 28, 31, 34, 37, 39. Since there are only 14 of these, the number of possibilities to check reduces to 14·14·14 = 2,744. On average, we need only try half of these before the lock opens, so the amount of time required would average 1,372t, where t is the length of time required to try a particular possibility. Taking t to be 4 seconds gives 5,488 seconds ≈ 91.47 minutes, about an hour and a half.

One could force the lock open much more rapidly by using tools, but depending on the circumstances, this may arouse suspicion.

Nobel Prize winning physicist Richard Feynman (Figure 9.2) made something of a hobby of lock picking. He noted that lock combinations are often left at their factory settings, which were 25-0-25 or 50-25-50 (at that time, for certain manufacturers).[9] In a similar way, passwords are often left at their default settings by computer users.

If a user has set his or her own combination (or password), psychology can be applied. Feynman used this approach in Los Alamos, New Mexico, to open Frederic de Hoffman's locked filing cabinets, which "protected" the secrets of the Manhattan project (the development of the atomic bomb). After a few failed attempts, he opened

---

9 Feynman, Richard, as told to Ralph Leighton, *"Surely You're Joking, Mr. Feynman!" Adventures of a Curious Character*, W. W. Norton & Company, New York, 1985, p. 155.

FIGURE 9.2
Richard Feynman, safecracker.

Courtesy of Michelle Feynman and Carl Feynman.

them with 27-18-18. He reasoned that Freddy was *"just the kind of guy to use a mathematical constant for a safe combination."* He was right and the correct constant was e.[10]

The number e will pop up frequently in this book and it's not long at all before its next appearance, but for now we take a look at a pair of questions connected with gambling. Many problems in combinatorics arise naturally in poker.

Example 4 – Poker Hands
How many possible hands are there in five-card stud? Assume that a standard 52-card deck without jokers is used.

Solution: The order in which you get the cards doesn't matter, so we have a combination, namely $_{52}C_5 = \frac{52!}{(52-5)!5!} = \frac{52!}{47!5!} = \frac{52 \cdot 51 \cdot 50 \cdot 49 \cdot 48}{5 \cdot 4 \cdot 3 \cdot 2 \cdot 1} = 2,598,960$.

Example 5 – Full House
How many of the above hands are full houses?

---

10 Feynman, Richard, as told to Ralph Leighton, *"Surely You're Joking, Mr. Feynman!"* Adventures of a Curious Character, W. W. Norton & Company, New York, 1985, pp. 147-151.

Solution: We must choose the value of the three of a kind. The possibilities are ace, 2, 3, 4, 5, 6, 7, 8, 9, 10, Jack, Queen, King. Because there are 13 options, we start off with $_{13}C_1$. These three cards must be chosen from among the four suits, so we multiply by $_4C_3$. Similarly, for the pair, there are now 12 values to choose from (we cannot have the three of a kind and the pair both of the same value, as there are only four cards of each value). We have $_{12}C_1$. Choosing the two cards from the four suits may be done in $_4C_2$ ways. Thus, the total is $_{13}C_1 \cdot _4C_3 \cdot _{12}C_1 \cdot _4C_2 = 13 \cdot 4 \cdot 12 \cdot 6 = 3,744$. A full house is quite rare, in comparison to the total number of possible hands.

Combinations with Repetition Allowed

Consider the pizza problem of Example 2, but suppose that repeated toppings are allowed.

One way to approach this new problem is to break it into three cases:

1) No toppings are repeated (as in Example 2);
2) One topping is chosen twice, and the third topping is different (e.g., Ham, Ham, Olives);
3) All three toppings are the same.

We saw the answer for case 1 is 120.

For case 2, we have 10 choices for which topping is repeated, and then only 9 choices for the third topping, which must be different. By the multiplication principle, this gives 10·9 = 90 possibilities.

For case 3, we pick the same one of the 10 toppings each time, so there are only 10 possibilities.

Because the three cases cannot happen at the same time, but collectively cover all possibilities, we add these numbers together to get our final answer. It is 120 + 90 + 10 = 220 possible pizzas.

But what if the numbers were bigger? Suppose there are 30 toppings and you're selecting 10 (maybe this is the kind of menu item where you can get your picture on the wall, and your meal on the house, if you can eat it all). We could try to solve it by cases again, but the number of cases is daunting. A less tedious approach is desired.

Fortunately, there's another way to look at the initial problem that does not become more tedious as the numbers grow. A picture will help greatly. Returning to selecting three toppings from a total of ten, allowing repetition, nine vertical lines (bars) can be drawn such that the spaces between them (and on the very ends) represent the ten toppings:

$$| \quad | \quad | \quad | \quad | \quad | \quad | \quad | \quad |$$

Stars can then be placed between the bars to indicate which toppings are chosen.
For example,

$$* \ | \quad | \quad | \ * \ | \quad | \quad | \quad | \quad | \ * \ |$$

indicates toppings 1, 4, and 9 are selected and

$$| \quad | ** | \quad | \quad | \quad | \quad | \quad | \quad | *$$

indicates topping 3 was selected twice and topping 10 was selected once.

Because there's a one-to-one correspondence between the number of permutations of the three stars and the nine bars and the possible pizzas, we have reduced the problem to a much simpler one. We have a total of 12 objects (stars and bars), so imagine making 12 generic marks and then selecting 3 of them to be stars. In this manner, every possibility can be obtained. So, the answer is just $C(12,3) = \frac{12!}{(12-3)!3!} = \frac{12!}{9!3!} = \frac{12 \cdot 11 \cdot 10}{3 \cdot 2 \cdot 1} = \frac{1320}{6} = 220$. This matches the answer obtained by adding the three separate cases together, filling us with confidence that it's correct, and it offered the advantage of only requiring us to look at one case.

Notice that if there are n items to choose from and we wish to select k of them, we will have $n - 1$ bars and k stars, for a total of $n + k - 1$ objects. Because we choose k of these to be stars, the general solution is $C(n+k-1, k)$. While the revised problem, selecting 10 toppings from 30, would be a pain by cases, it's handled as easily as the first question when using our new formula. We have $C(30 + 10 - 1, 10) = C(39, 10) = 635,745,396$.

This wonderful formula was first found by Sebastián Izquierdo (1601–1681), who published it in *Pharus Scientiarum* (Lighthouse of Sciences) in 1659.[11]

There are many different sounding problems that can be handled by this same formula. The formula goes by the name "stars and bars" (if thought of as presented here), or "stars and stripes" or "balls and urns," if thought of slightly differently.

Example 6 – Distributing Extra Credit Points

Suppose a professor has 15 students in his class and every time they meet he offers a one-point bonus on the final average for the first student who can correctly answer some tough question. If the class meets 28 times, how many possible outcomes are there for distributing the 28 points among the 15 students?

Solution: For this problem, we can imagine 14 bars separating the 15 students. We then place 28 points (the stars) among the bars somehow. Because there are a total of 14 + 28 objects and we need 28 of them to be stars, our answer is $C(14 + 28, 28) = C(42, 28) = 52,860,229,080$.

If you found this section of the text challenging, don't feel bad! Remember, it wasn't until 1659 that a solution was published, and even after that mathematicians were still struggling with it. Also, some big names had tried to solve the problem before 1659, but failed.

> Combinations with repetition were not well understood until Jacob Bernoulli's *Ars Conjectandi* (The Art of Conjecturing) came out in 1713. In Part 2, Chapter 5, Bernoulli simply listed the possibilities in lexicographic order, and showed that the formula $C(m + n - 1, n)$ follows by induction as an easy consequence. Niccolò

---

11 Izquierdo, Sebastián, *Pharus Scientiarum*, 2 volumes, Lyon, 1659. See sections 48–51. Pages 319–358 cover combinatorics. My source was Wilson, Robin and John J. Watkins, editors, *Combinatorics: Ancient & Modern*, Oxford University Press, Oxford, 2013, p. 19.

Tartaglia had, incidentally, come close to discovering this formula in 1556 in his *General Trattato di Numeri, et Misure* (General Treatise of Numbers and Measures); so had the Maghreb mathematician Ibn Mun^{c}im in his 13th-century *Fiqh al-Ḥisāb*.[12]

Partitions

Partitions are covered in far more detail in Chapter 22, but one special case can be nicely addressed now. Suppose we have n distinct objects that we wish to split into groups of distinct size $s_1, s_2, s_3, ..., s_m$. In how many ways can this be done?

This category of problem was solved by Mersenne, and later by Leibniz, in a new way, which is detailed here.[13] We first select the elements to be in the group of size s_1. Because the order within this group doesn't matter, the number of ways to do this is $C(n, s_1)$. We now have $n - s_1$ objects remaining, from which we select s_2 for the next group. This can be done in $C(n - s_1, s_2)$ ways. There are now $n - s_1 - s_2$ objects remaining, from which we select s_3. This can be done in $C(n - s_1 - s_2, s_3)$ ways. Continuing in this manner, our answer is seen to be

$$C(n, s_1) \cdot C(n - s_1, s_2) \cdot C(n - s_1 - s_2, s_3) \cdots C(n - s_1 - s_2 - \cdots - s_{m-1}, s_m).$$

This may be rewritten as

$$\frac{n(n - 1)(n - 2) \cdots (n - s_1 - s_2 - s_3 \cdots - s_m + 1)}{s_1! s_2! s_3! \cdots s_m!}.$$

Example 7 – Forming Teams

Suppose we have 20 students and we wish to form a baseball team (9 players), a hockey team (6 players), and a basketball team (5 students).[14] Assume that we are not concerned with the positions the students will be assigned. We only care which sport each student is playing. In how many distinct ways can the teams be formed?

Solution: We have $C(20, 9)$ ways to form the baseball team. Once that is done, there are 11 students remaining, so there are $C(11, 6)$ ways to form the hockey team. The five remaining students will constitute the basketball team. We can write this as $C(5, 5)$, for completeness, but it is equal to 1. Our answer is $C(20, 9) \cdot C(11, 6) \cdot C(5, 5) = 167{,}960 \cdot 462 \cdot 1 = 77{,}597{,}520$.

---

12 Wilson, Robin and John J. Watkins, editors, *Combinatorics: Ancient & Modern*, Oxford University Press, Oxford, 2013, p. 20. Also see p. 157: "[Jacob] Bernoulli [(1654–1705)] seems to be the first to have investigated more difficult questions involving combinations, obtaining the independent expression $C(n + k - 1, k)$ for combinations with repetition taken k at a time." This contradicts p. 19, which says that Izquierdo was first.

13 Wilson, Robin and John J. Watkins, editors, *Combinatorics: Ancient & Modern*, Oxford University Press, Oxford, 2013. See Chapter 5 (pp. 134–135) for Mersenne's method and p. 150 for Leibniz's method.

14 The number of players represents those on the field. There is no one left on the benches in these minimal formations.

An extra step *may* be needed if two or more of the teams are of the same size. If these teams can be interchanged, we need to divide the factorials of the numbers of teams (not their sizes) that can be interchanged. To make this clearer, suppose there are six teams being formed and two of them require five players, while the other four require eight players. If the sports are all different, then it certainly makes a difference if the members of two teams are switched (they care what sport they are playing!), so we do not do any extra divisions. However, if each size team is for the same sport, we would divide by both 2! (for the two teams of the same size) and 4! (for the four teams of the same size).

In combinatorics you often have to be very careful in reading the problem and interpreting the context correctly.

John Wallis, $_nC_0$, and 0!

John Wallis, who was mentioned in Chapter 8, considered a seemingly silly combinatorics problem in his 1685 book *Discourse of Combinations*,[15] namely the number of combinations of n things taken 0 at a time. He concluded that $_nC_0 = 1$ for all $n \geq 0$.[16] Using our formula for $_nC_0$, this gives $\frac{n!}{(n-0)!0!} = 1$, which quickly becomes $\frac{n!}{n!0!} = 1$, and then, after canceling the $n!$, $\frac{1}{0!} = 1$. From this we conclude that $0! = 1$, as we decided made sense in Chapter 8. While this conclusion is easy to draw from what Wallis wrote, he did not explicitly state that $0! = 1$.

Pretty Patterns

In addition to solving problems in combinatorics, the formula $C(n, k) = \frac{n!}{(n-k)!k!}$ leads to some pretty results. A very simple example, recognized by Rabbi Levi ben Gerson (way back in 1321),[17] Gerolamo Cardano (1570),[18] and others is $C(n,k) = C(n, n-k)$.

Mystical Connections

For the Jews, calculating permutations and combinations was no mere mathematical amusement. They believed that God had created the world and everything in it through the power of the spoken word (e.g., "And God said, Let there be light: and there was light."). So, by knowing the number of possible words, in Hebrew, they would know the number of things that could exist. This problem appears in the *Sefer Yetsirah* (Book of

---

15 Wallis, John, *Discourse of Combinations*, London, 1685. See pages 110 and 113.

16 Wallis, John, *Discourse of Combinations*, London, 1685. See pages 110 and 113.

17 See his *Maasei Hoshev* (The Art of the Calculator), 1321, proposition 68. Note: The title is sometimes transcribed as *Maaseh Hoshev*.

18 In *Opus Novum de Proportionibus*, as pointed out in Wilson, Robin and John J. Watkins, editors, *Combinatorics: Ancient & Modern*, Oxford University Press, Oxford, 2013, pp. 127–129, 140.

Creation), the oldest known Jewish work on combinatorics, produced sometime between the 2nd and 8th centuries, and investigations continued into the 16th century.[19]

It was not just a single number that was sought. The questions piled up. How many words of each length are possible? How many with letters repeated? How many with no letters repeated? With numerical values assigned to the letters, many more questions can be asked. Although this seems to be getting away from mathematics into numerology (or Kabbalah[20], as it was called by the Jews), eventually the investigations were bound to generate some interesting mathematics. An example is given by derangements.

Derangements (aka Subfactorials)

Derangements are rearrangements of the letters of a word such that no letter occupies its original position.

The number of arrangements of n items is simply $n!$ Calculating derangements, notated by $!n$ and sometimes called **subfactorials**,[21] is a bit harder. One way to approach the problem is by explicitly finding all possibilities for small numbers and looking for a pattern in these values.

For one item, a derangement is not possible, and for two items, say **ab**, the only possible derangement is **ba**.

Now consider the case with three items, **abc**. The possible derangements are **bca** and **cab**. Check for yourself that any other arrangement fixes at least one of the three letters.

With four items, **abcd**, we have the following possibilities:

badc	cadb	dabc
bcda	cdab	dcab
bdac	cdba	dcba

Here we notice a symmetry. Three possibilities begin with **b**, three with **c**, and three with **d**. Because it doesn't matter which letter we start with (as long as it isn't **a**), we could count the number of possibilities starting with **b** and multiply by 3 to account for the other possibilities and get the answer.

We use this approach with the case $n = 5$. **abcde** may be deranged as

badec	bcaed	bdaec	beacd
baecd	bcdea	bdeac	bedac
	bcead	bdeca	bedca

19 Wilson, Robin and John J. Watkins, editors, *Combinatorics: Ancient & Modern*, Oxford University Press, Oxford, 2013, p. 110.

20 aka Kabbala, Cabala, and Qabala(h).

21 $!n$ is sometimes denoted by n followed by an upside-down exclamation mark. W. Allen Whitworth coined the term subfactorial in 1878. See Cajori, Florian, *A History of Mathematical Notations*, Two Volumes Bound as One, Dover Publications, Mineola, NY, 1993, Vol II, p. 77, available online at https://monoskop.org/images/2/21/Cajori_Florian_A_History_of_Mathematical_Notations_2_Vols.pdf. Cajori cites Whitworth, W. Allen, *Messenger of Mathematics*, Vol. VII, 1878, p. 145 and Whitworth, W. Allen, *Choice and Chance*, Cambridge, 1886, Preface, p. xxxiii.

We have 11 possibilities. However, the first letter could have been anything other than **a**, so we must multiply by 4. The final answer is 44.

The values calculated thus far have been placed in the table below. Because it will prove interesting, a column showing the ratio of arrangements to derangements has also been included.

n	Arrangements, $n!$	Derangements, $!n$	Arr/Der
1	1	0	NA
2	2	1	2
3	6	2	3
4	24	9	2.66666…
5	120	44	2.72727…

How many more rows can you add to this table? Examining your expanded table, what value would you expect to get for Arr/Der in the limit as n approaches infinity? Can you prove the value of the limit is equal to your guess? If you have time, this would be a good point to stop reading and take the challenge of seeing how much you can figure out on your own. If you can't do anything else, at least try to accurately find !6. Remember, much mathematical research consists of just chipping away at a problem, so at least do *something*.

A good first step in expanding the table is to find a nice formula for derangements, so their values can be found more efficiently than by the increasingly tedious approach used above. There are several, quite different, formulas to be found. One is developed below and another is derived in Chapter 17.

Abraham de Moivre demonstrated how derangements can be calculated via the case $n = 6$ in his book *The Doctrine of Chances* (1718).[22] He approached the problem with the inclusion–exclusion principle of Chapter 5.

Taking a, b, c, d, e, and f as the six objects to be deranged, de Moivre defined A to be the set of arrangements in which a is fixed, B to be the set of arrangements in which b is fixed, and so on for C, D, E, and F. It should be noted that while the arrangements in A fix a, some of them will fix other letters as well. The same can be said for the arrangements in B, C, D, E, and F. In other words, the sets A, B, C, D, E, and F are not disjoint; they have a lot of overlap.

If we take all 6! possible arrangements and subtract out the ones for which at least one letter stays fixed, that is, $6! - n(A \cup B \cup C \cup D \cup E \cup F)$, our result will be the number of arrangements in which no letter stays put. This is another application of counting by looking at the complement. Applying the inclusion–exclusion principle to evaluate $n(A \cup B \cup C \cup D \cup E \cup F)$, gives

22 de Moivre, Abraham, *The Doctrine of Chances*, Pearson, 1718, p. 96. Taken here from Wilson, Robin and John J. Watkins, editors, *Combinatorics: Ancient & Modern*, Oxford University Press, Oxford, 2013, p. 312. Note: this result had already appeared elsewhere. Pierre Rémond de Montmort published it in the second edition of his *Essay d'Analyse sur les Jeux de Hazard*: de Montmort, Pierre Rémond, *Essay d'Analyse sur les Jeux de Hazard*, second edition, J. Quillau, Paris, 1713 (reprinted by Chelsea, 1980). But de Montmort was still not the first. He got it from Nicolaus Bernoulli (see Wilson et al., p. 160).

$$n(A \cup B \cup C \cup D \cup E \cup F) = n(A) + n(B) + n(C) + n(D) + n(E) + n(F)$$
$$- n(A \cap B) - n(A \cap C) - \cdots - n(E \cap F)$$
$$+ n(A \cap B \cap C) + n(A \cap B \cap D) + \cdots + n(D \cap E \cap F)$$
$$- n(A \cap B \cap C \cap D) - n(A \cap B \cap C \cap E) - \cdots - n(C \cap D \cap E \cap F)$$
$$+ n(A \cap B \cap C \cap D \cap E) + \cdots + n(B \cap C \cap D \cap E \cap F)$$
$$- n(A \cap B \cap C \cap D \cap E \cap F)$$

Sets created from the same number of intersected sets will have the same order here because of the symmetry in how A, B, C, D, E, and F are defined. Their sizes are given by $k!$, where k is the number of sets not in the intersection (i.e., the number of letters not forced to go to themselves). To simplify the above expression, we also need to know how many sets we have for each number of intersections. Because the k sets that are intersected are chosen from n, the answer to this is just $C(n, k)$. Putting all of this together, the above is

$$= C(6,1) \cdot 5! - C(6,2) \cdot 4! + C(6,3) \cdot 3! - C(6,4) \cdot 2! + C(6,5) \cdot 1! - C(6,6)$$
$$= 6 \cdot 120 - 15 \cdot 24 + 20 \cdot 6 - 15 \cdot 2 + 6 \cdot 1 - 1$$
$$= 720 - 360 + 120 - 30 + 6 - 1$$
$$= 455$$

Subtracting these from the total of $6!$ possible arrangements gives 265 derangements. That is, $D_6 = 265$.

de Moivre's result generalizes to

$$!n = n! - \binom{n}{1}(n-1)! + \binom{n}{2}(n-2)! - \cdots \pm \binom{n}{n}0!.$$

The $n!$ term can be rewritten as $\binom{n}{0}(n-0)!$, which allows us to rewrite the above result as

$$!n = \binom{n}{0}(n-0) - \binom{n}{1}(n-1)! + \binom{n}{2}(n-2)! - \cdots \pm \binom{n}{n}0!.$$

This can be expressed more tersely using summation notation:

$$!n = \sum_{i=0}^{n} (-1)^i \binom{n}{i}(n-i)!.$$

But this isn't the last word! We can simplify by making the substitution $\binom{n}{i} = \frac{n!}{(n-i)!i!}$ and then performing some algebra. First the substitution

$$!n = \sum_{i=0}^{n} (-1)^i \frac{n!}{(n-i)!i!} (n-i)!$$

The $(n-i)!$ can be canceled to give

$$!n = \sum_{i=0}^{n} (-1)^i \frac{n!}{i!}.$$

Because the $n!$ appears in every term of the summation, it can be pulled out to give

$$!n = n! \sum_{i=0}^{n} \frac{(-1)^i}{i!}.$$

This is a lot prettier than the original result from the inclusion–exclusion principle. Expanding out the summation, it takes the form

$$!n = n! \left(1 - \frac{1}{1!} + \frac{1}{2!} - \frac{1}{3!} + \cdots + \frac{(-1)^n}{n!} \right).$$

With a formula in hand for derangements, it's an easy to expand the table.

n	Arrangements, $n!$	Derangements, $!n$	Arr/Der
1	1	0	NA
2	2	1	2
3	6	2	3
4	24	9	2.666666667
5	120	44	2.727272727
6	720	265	2.716981132
7	5,040	1,854	2.718446602
8	40,320	14,833	2.718263332
9	362,880	133,496	2.718283694
10	3,628,800	1,334,961	2.718281658
11	39,916,800	14,684,570	2.718281843
12	479,001,600	176,214,841	2.718281827

From this table, it appears that the ratio of arrangements to derangements is converging to e, the combinations to the filing cabinets at Los Alamos! But such numerical evidence does not constitute a proof.

In attempting a proof that

$$\lim_{n \to \infty} \frac{n!}{!n} = e,$$

it's tempting use the fact that

$$!n = n! \sum_{i=0}^{n} \frac{(-1)^i}{i!}$$

to write

$$\lim_{n \to \infty} \frac{n!}{!n} = \lim_{n \to \infty} \frac{n!}{n! \sum_{i=0}^{n} \frac{(-1)^i}{i!}}$$

and then see if it can be simplified to e. But after canceling the $n!$ that appears in the numerator and denominator, we don't have a nice next step. Using the identity

$$!n = n! \sum_{i=0}^{n} \frac{(-1)^i}{i!}$$

is actually a good idea, but instead of forming the ratio of arrangements to derangements, which is what we're interested in, it's better to look at its reciprocal:

$$\frac{!n}{n!} = \frac{n! \sum_{i=0}^{n} \frac{(-1)^i}{i!}}{n!}.$$

We can then cancel $n!$ to get

$$\frac{!n}{n!} = \sum_{i=0}^{n} \frac{(-1)^i}{i!}.$$

Recalling from Chapter 8 that

$$e^x = \sum_{i=0}^{\infty} \frac{x^i}{i!}.$$

and letting $x = -1$ gives

$$e^{-1} = \sum_{n=0}^{\infty} \frac{(-1)^i}{i!}.$$

So, returning to our expression for derangements over arrangements and taking the limit as n approaches infinity, we get

$$\lim_{n \to \infty} \frac{!n}{n!} = \lim_{n \to \infty} \sum_{i=0}^{n} \frac{(-1)^i}{i!} = e^{-1}.$$

That is,

$$\lim_{n \to \infty} \frac{!n}{n!} = \frac{1}{e}.$$

Taking the reciprocals of both sides turns this into

$$\lim_{n \to \infty} \frac{n!}{!n} = e$$

and the proof is complete.

An amazing property of $\frac{n!}{!n}$ is how close it is to e for small values of n. We can use this to determine the value of D_n for any natural number n. D_n is actually the nearest integer to $n!/e$.

Another formula for D_n makes use of the floor function from Chapter 6. We have $D_n = \lfloor \frac{n!}{e} + \frac{1}{2} \rfloor$. This is an example of how the floor function can be used to simplify the expression of a function. With this formula we don't have to worry about which integer is the nearest. We can always just round down.

The discussion above connected derangements to the inclusion–exclusion principle, e (via the ratio with arrangements), and the floor function. There is more to come in Chapters 12 and 17, but for now consider one more connection with something you've already seen in this text:

$$!n = \frac{\Gamma(n+1, \ -1)}{e},$$

where $\Gamma(n+1, \ -1)$ is the incomplete Gamma function.[23] Recall from last chapter that

$$\Gamma(t) = \int_0^\infty x^{t-1} e^{-x} \, dx.$$

The incomplete Gamma function takes an extra argument reflecting a change in the limits of integration. In the case of $\Gamma(n+1, \ -1)$, the bottom limit is –1.

When I first saw this, I thought it was cool that the Gamma function connected with $!n$, but then I quickly realized that the Gamma function is just a generalization of the concept of factorial, and I already knew that $!n \approx \frac{n!}{e}$, so for a second I was less impressed. It seemed like a rewording of a result I was already familiar with. But then after another second I realized that this new equation had an "=" in place of the "≈" present in the other formula. All of a sudden it was exciting again!

The results obtained here concerning derangements will be applied to a game of chance known as *jeu de recontre* in Chapter 12.

Sudoku

Do you like Sudoku? Are you afraid you do them so frequently that you might run out one day? Don't worry!

---

23 http://oeis.org/wiki/Number_of_derangements.

It has been calculated that there are 6 670 903 752 021 072 936 960 distinct Sudoku squares. This means that only about one in a million 9×9 Latin squares is a Sudoku square. Many of these can be obtained by performing simple operations on others, and the total number of *essentially different* Sudoku squares is 5 472 730 538.[24]

It's fun to count things. Once you master combinatorics, it can be applied to a wide range of problems.

A Connection to Set Theory

There's a nice problem that connects set theory, combinatorics, and the floor function. A pair of definitions need to be given first, though.

Given a set S, a collection of subsets $A_1, A_2, A_3, \ldots, A_k$ is called a **chain** if $A_1 \subset A_2 \subset A_3 \subset \ldots \subset A_k$.

Example
Let $S = \{1, 2, 3, 4\}$. Then $A_1 = \{2\}$, $A_2 = \{2, 4\}$, $A_3 = \{1, 2, 4\}$, $A_4 = \{1, 2, 3, 4\}$ is a chain.

Going to the other extreme, an **antichain** is a collection of set such that no set is contained in another.

Example Letting $S = \{1, 2, 3, 4\}$ again, an antichain is given by $A_1 = \{1, 2\}$, $A_2 = \{1, 3\}$, $A_3 = \{1, 4\}$, $A_4 = \{2, 3\}$, $A_5 = \{2, 4\}$, $A_6 = \{3, 4\}$.

A natural question to ask is, given a set containing n elements, what is the maximum number of sets that can form an antichain?

The answer, published by Emanuel Sperner in 1928, turned out to be $C\left(n, \left\lfloor \frac{1}{2}n \right\rfloor\right)$.[25] For the example $S = \{1, 2, 3, 4\}$, $n = 4$, so we have $C\left(4, \left\lfloor \frac{1}{2}4 \right\rfloor\right) = C(4, \lfloor 2 \rfloor) = C(4, 2) = 6$. Hence, there's no antichain for S that's bigger than the one given above. Looking at the next larger set, with $n = 5$, we have $C\left(5, \left\lfloor \frac{1}{2}5 \right\rfloor\right) = C(5, \lfloor 2.5 \rfloor) = C(5, 2) = 10$.

This problem marked the birth of an area of mathematics known as combinatorial set theory.[26]

Having already presented a bit of the mathematics behind combination locks, this chapter concludes with a glimpse of the physics of safecracking.

Safecracking and Thermodynamics

The law that entropy always increases – the second law of thermodynamics – holds, I think, the supreme position among the laws of Nature. If someone points

---

24 Wilson, Robin and John J. Watkins, editors, *Combinatorics: Ancient & Modern*, Oxford University Press, Oxford, 2013, pp. 279–280. Page 283 has "See www.afjarvis.staff.shef.ac.uk/sudoku/sudgroup.html for the calculations and an explanation of what is meant by 'essentially different'."

25 Sperner, Emanuel, "Ein Satz uber Untermengen einer endlichen Menge," *Mathematische Zeitschrift*, Vol. 27, 1928, pp. 544–548. For a (historical) discussion in English, see Wilson, Robin and John J. Watkins, editors, *Combinatorics: Ancient & Modern*, Oxford University Press, Oxford, 2013, pp. 309–310. Page 323 gives some applications.

26 Wilson, Robin and John J. Watkins, editors, *Combinatorics: Ancient & Modern*, Oxford University Press, Oxford, 2013, p. 310.

out to you that your pet theory of the universe is in disagreement with Maxwell's equations – then so much the worse for Maxwell's equations. If it is found to be contradicted by observation – well, these experimentalists do bungle things sometimes. But if your theory is found to be against the second law of thermodynamics I can give you no hope; there is nothing for it but to collapse in deepest humiliation.

Sir Arthur S. Eddington[27]

A bit about Feynman's safecracking exploits was shared earlier in this chapter. While we're on the subject of safecracking, it's worth mentioning a neat application of the second law of thermodynamics in this direction.

The second law basically says that entropy (the energy in a system that can't be used) must increase over time in a closed system. In other words, the usable energy decreases. That is, things wind down. A pair of examples will make this clearer.

If you're in a room that has a desk with a hot cup of coffee on it, you may use the heat energy in the cup to do some work, such as warm your hands, or scald an enemy. However, as time goes on, the coffee will slowly cool to room temperature. Actually, it warms the room slightly as it cools, and the room and the coffee settle on some temperature between their original values. If the room is large, compared to the volume of coffee, the final temperature of the system will be only a tiny bit higher than the original room temperature. The heat energy of the coffee is now uniformly distributed throughout the room and cannot be used.

The room is considered to be a closed system. If the coffee cup had a heating coil in it that was plugged into a wall outlet, it would be another matter. The energy coming into the system would continue to heat the coffee and usable energy would not decrease. But, the second law is only claimed to apply to closed systems.

Now, for another example, that brings us closer to safecracking, consider the jar shown in Figure 9.3.

If you keep assorted hardware in a jar for later use, and continue to add to it from time to time, eventually a new item won't fit. The solution is not to find a larger jar, but rather to put the lid on the original jar, shake it, remove the lid, and then try to fit the new item in again. This will likely work. Why? Consider the energy of the system. Every item in the jar has a center of mass that is some distance from the bottom of the jar. If we think of the bottom of the jar as ground level, then each item has a potential energy equal to the product of the item's mass, its center of mass's distance from "ground level," and the acceleration caused by gravity. Summing these potential energies for all of the items in the jar provides the usable energy for the system. If we shake the jar, the items in it will shift, settling to positions that decrease the total potential energy of the system. This process tends to create more space at the top.

We also see the consequences of this phenomenon when we open a bag of cheese curls or potato chips. Or perhaps something else is at work. The manufacturer would like you to believe a bag that was full when it left the factory has, upon being jostled in route to the store, and a bit more on the way to your home, settled (as a consequence of the laws of physics, which the manufacturer is not yet powerful enough to alter) down to about twenty percent of the bag's volume. To see how this phenomenon may be applied to safecracking, we need to look inside a typical combination lock.

---

27 Eddington, Sir Arthur S., *The Nature of the Physical World*, Cambridge at the University Press, 1918, p. 74.

FIGURE 9.3
We all have these.

Image created by Craig P. Bauer.

In Figure 9.4 you can see that the lock cannot be opened. However, it can be opened as shown in Figure 9.5. The difference is that the wheel at the center has turned so that the piece holding the lock closed can pivot to fall into the notch that's cut in the wheel. A typical three-number combination lock has three wheels, all of which must simultaneously be properly aligned to allow movement of the piece that will free the lock to be opened.

The lowest energy state for a wheel with a piece missing is when the missing piece is at the top. So, shaking the lock should settle all three wheels to this position (in accordance with the second law of thermodynamics), allowing it to be opened. However, we can't do this as effectively by hand, as we can for a jar of hardware. Instead, a hand-held vibrator should be used. For larger safes, operating on the same principle, industrial vibrators can be used by those who do not possess the combination. However, an expert summed up the disadvantages of this technique:

> I suppose the reason that this hasn't become really popular among safecrackers is because industrial vibrators are terribly loud, and they numb your hands after holding them a few minutes.[28]

Also, safe designers have found ways to block such attacks.

---

28 Yeager, Wayne B., *Techniques of Safecracking*, Loompanics Unlimited, Port Townsend, Washington, 1990, pp. 31–32.

FIGURE 9.4
The innards of a closed lock.

Image created by Craig P. Bauer.

FIGURE 9.5
The innards of an open lock.

Image created by Craig P. Bauer.

Exercise Set 9

> I told him, of course, that I didn't know – which is my answer to almost every question.
>
> Richard Feynman[29]

1. A meal at a restaurant comes with three side dishes. If there are 11 options, how many different combinations of sides can be selected, if repeats are not allowed.

2. A meal at a restaurant comes with three side dishes. If there are 11 options, how many different combinations of sides can be selected, if repeats are allowed.

3. If you have six quarters and walk into an arcade with nine video games, each costing one quarter to play, in how many ways can you select the games to play, if repeats are not allowed and order does not matter.

4. If you have six quarters and walk into an arcade with nine video games, each costing one quarter to play, in how many ways can you select the games to play, if repeats are allowed and order does not matter.

5. In how many ways can 10 out of 15 people be selected to form a chorus?

6. To determine $C(n,p)$ Mersenne used the rule $C(n,p) = \frac{P(n,p)}{P(p,p)}$.[30] Is this correct?

7. Suppose 12 pizza toppings are available. How many three-topping pizzas can be ordered, if toppings can be repeated?

8. How many distinct outcomes are possible if three dice are thrown at once, if we are not interested in the sum, but rather the numbers that appear, no matter the order of the dice. That is, if die #1 and die #2 both show 5 and die #3 shows 2, it is considered the same as die #1 showing 2 and die #2 and die #3 showing 5.[31]

9. How many ways are there to get two pairs in five-card stud? Careful: the card that isn't part of either pair must be of a distinct value!

10. Solve the following problem from Chen Houyao's *Cuozong fayi*.[32]

   > Let us suppose one has thirty playing cards. All cards have a [different] form. Each hand consists of nine cards. How many combinations can one obtain by drawing one hand?

11. The French mathematician Pierre Hérigone (?–*c*.1643) posed the question:

   > How many combinations without repetitions of n things taken k at a time contain a certain element?[33]

29 Feynman, Richard, "What Do You Care What Other People Think?," W.W. Norton, New York, 1988, p. 61.

30 Wilson, Robin and John J. Watkins, editors, *Combinatorics: Ancient & Modern*, Oxford University Press, Oxford, 2013, p. 134.

31 Inspired by Wilson, Robin and John J. Watkins, editors, *Combinatorics: Ancient & Modern*, Oxford University Press, Oxford, 2013, pp. 13–14. Also see Wilson et al., p. 125, bottom two paragraphs, for more historical context and related problems.

32 Taken here from Wilson, Robin and John J. Watkins, editors, *Combinatorics: Ancient & Modern*, Oxford University Press, Oxford, 2013, p. 72.

33 Taken here from Wilson, Robin and John J. Watkins, editors, *Combinatorics: Ancient & Modern*, Oxford University Press, Oxford, 2013, p. 141.

and gave the answer $C(n-1, k-1)$. Was he correct? After answering this, there's another mystery to consider. Very little is known about Hérigone. It's been conjectured that this name was merely a pseudonym for Clément Cyriaque de Mangin.[34] Will anyone be able to resolve this mystery?

12. André Tacquet (1612–1660), a Jesuit priest and mathematician, claimed that $P(24, 24)$ is so large that even if a billion people attempted to write out all of these permutations, filling 40 pages, with 40 permutations each, every day, for a billion years, they would still not produce all of them.[35] Was he correct?

13. A group of 12 students is going on a road trip. The three cars they are taking seat 6, 4, and 2. In how many distinct ways can they divide themselves up among these vehicles? Note: we don't care who is in which position, that is, driver, shotgun, etc. All we're concerned about is which car a particular person ends up in.

14. How many ways are there to form a single tennis doubles team, if there are 10 players? Don't worry about who plays close to the net. This problem only asks in how many ways the players can be chosen.

15. How many five card stud hands are there that contain at least one heart?

16. If there are ten students in a class, how many different lists are possible at the end of the semester of the students who made a B or higher? Note: It's possible that there are zero students on the list, ten students on the list, or any value between zero and ten.

17. In 1321, Rabbi Levi ben Gerson (1288–1344), in his *Maasei Hoshev*, proved each of the following.[36] How many can you prove?

    a) $P(n+1) = (n+1)P(n)$.

    b) $P(n, k) = n \times (n-1) \times (n-2) \times \cdots \times (n-k+1)$. Prove this one as the Rabbi did, by first establishing directly that $P(n, 2) = n \times (n-1)$, and then using induction on k.

18. Prove that if S is a set containing n elements, then P(S), the power set of S, contains 2^n elements.

19. Prove that $C(n,k) + C(n,k+1) = C(n+1,k+1)$. Cardano explained this in his *Opus Novum de Proportionibus* (1570).[37]

20. In *Opus Novum de Proportionibus* (1570), Gerolamo Cardano claimed $C(n,k) = \frac{n-(k-1)}{k} \times C(n,k-1)$.[38] Is this correct? Give a proof or a counterexample.

21. Prove that

$$C(n, s_1) \cdot C(n-s_1, s_2) \cdot C(n-s_1-s_2, s_3) \cdot \cdots \cdot C(n-s_1-s_2- \quad -s_{m-1}, s_m)$$

34 Taken here from Wilson, Robin and John J. Watkins, editors, *Combinatorics: Ancient & Modern*, Oxford University Press, Oxford, 2013, p. 141.

35 Taken here from Wilson, Robin and John J. Watkins, editors, *Combinatorics: Ancient & Modern*, Oxford University Press, Oxford, 2013, p. 141.

36 Wilson, Robin and John J. Watkins, editors, *Combinatorics: Ancient & Modern*, Oxford University Press, Oxford, 2013, pp. 110, 119–120, and 171.

37 Wilson, Robin and John J. Watkins, editors, *Combinatorics: Ancient & Modern*, Oxford University Press, Oxford, 2013, p. 129.

38 Wilson, Robin and John J. Watkins, editors, *Combinatorics: Ancient & Modern*, Oxford University Press, Oxford, 2013, pp. 128–129.

$$= \frac{n(n-1)(n-2)\cdots(n-s_1-s_2-s_3-\cdots-s_m+1)}{s_1!s_2!s_3!\cdots s_m!}.$$

22. Suppose there's a sale on beer: 50 bottles for $50. If there are only 12 different types of beer being offered, how many different ways are there to make the purchase? Repeats are, of course, allowed.

23. In the above problem, what is the minimum number of bottles that must be purchased of the most purchased beer in the set of 50? In other words, you must have at least this many of some particular beer.

24. How many 12-topping pizzas are there if 25 distinct toppings are offered and repeats are allowed?

25. How many derangements are there on a set of 13 objects?

26. What is the length of the longest antichain that can be formed from a set with 10 elements?

27. Consider the game shown in Figure 9.6. It was put out by Milton Bradley in 1970 under the name Drive Ya Nuts. Each of the seven hexagonal pieces can be rotated or swapped with any of the others. The object of the game is to position all seven such that touching sides are marked with the same numbers. The hexagon in the center prevents the arrangement depicted below from being

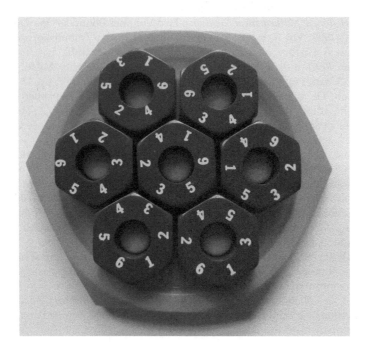

FIGURE 9.6
Drive Ya Nuts.

Image created by Craig P. Bauer.

a solution. Could we simply solve this puzzle by trying all of the possibilities? How many are there, if we don't consider rotations of the entire board as distinct?

28. In Chapter 4, work attributed to Thomas Harriott showed how to find the elements of the power set. It can also be used to count the number of elements in the power set having each given order.[39] For example, letting $a = b = c = 1$ in $(x+a)(x+b)(x+c)$ gives $(x+1)^3 = 1 + 3x + 3x^2 + 1x^3$. The coefficients of this expansion give the number of subsets of size 0, 1, 2, and 3 for a set of size 3. Carry out the analogous work for sets of size 4 and 5 and confirm that the correct answers are given.

References/Further Reading

It was amazing how many books one could fit into a room, assuming one didn't want to move around very much.

– Brandon Sanderson, *The Well of Ascension*.

ben Gerson, Rabbi Levi, *Maasei Hoshev* (The Art of the Calculator), 1321.

Biggs, Norman L., "The Roots of Combinatorics," *Historia Mathematica*, Vol. 6, No. 2, May 1979, pp. 109–136.

Borrel (Buteo), Jean, *Logistica, Quae et Arithmetica Dicitur in Libros Quinque Digesta*, Lyon, 1559. This includes an analysis of a combination-type lock, where the combination is formed by six rotating rings, each of which bears six letters.

Cajori, Florian, *A History of Mathematical Notations*, Two Volumes Bound as One, Dover Publications, Mineola, NY, 1993, Vol II, p. 77, available online at https://monoskop.org/images/2/21/Cajori_Florian_A_History_of_Mathematical_Notations_2_Vols.pdf.

Cardano, Gerolamo, *Practica Arithmeticae Generalis* (Practice of Arithmetic), Milan, 1539, reprinted in *Opera Omnia*, Charles Sponi, editor, 10 volumes, Lyons, 1663, Vol. IV, pp. 13–216.

Cardano, Gerolamo, *De Subtilitate* (On Subtlety), Nuremberg, 1550, reprinted in *Opera Omnia*, Charles Sponi, editor, 10 volumes, Lyons, 1663, Vol. III, pp. 352–672. The construction of locks made of rotating numbered rings is detailed in Chapter 17. Also see Cardano's *Liber de Ludo Aleae*.

Cardano, Gerolamo, *Opus Novum de Proportionibus* (New Work on Proportions), Basel, 1570, reprinted in *Opera Omnia*, Charles Sponi, editor, 10 volumes, Lyons, 1663, Vol. IV, pp. 463–601.

Cardano, Gerolamo, *Liber de Ludo Aleae* (Book on the Game of Dice), written circa 1550, first printed in *Opera Omnia*, Charles Sponi, editor, 10 volumes, Lyons, 1663, Vol. I, pp. 262–276. This book includes a combinatorial analysis of the locks described in Cardano's *de Subtilitate*.

Colbourn, Charles J. and Jeffrey H. Dinitz, editors, *The CRC Handbook of Combinatorial Designs*, CRC Press, Boca Raton, FL, 1996.

"Combination with Repetitions," *Mathematics StackExchange*, available online at https://math.stackexchange.com/questions/208377/combination-with-repetitions. This is a discussion of the stars and bars formula.

Courtney, Charles, in collaboration with Thomas M. Johnson, *Unlocking Adventure*, Whittlesey House, a division of McGraw-Hill, New York, 1942. This book is aimed at a general audience and contains many historical tales.

---

[39] Wilson, Robin and John J. Watkins, editors, *Combinatorics: Ancient & Modern*, Oxford University Press, Oxford, 2013, p. 287.

de Moivre, Abraham, *The Doctrine of Chances*, Pearson, London, 1718.

de Montmort, P. R., *Essay d'Analyse sur les Jeux de Hazard*, second edition, J. Quillau, Paris, 1713, reprinted by Chelsea, 1980.

Eddington, Sir Arthur S., *The Nature of the Physical World*, Cambridge at the University Press, Cambridge, 1918.

Electronic Journal of Combinatorics, www.combinatorics.org/.

Feynman, Richard, as told to Ralph Leighton, *"Surely You're Joking, Mr. Feynman!" Adventures of a Curious Character*, W. W. Norton & Company, New York, 1985.

Graham, Ronald L., Martin Grötschel, and László Lovász, editors, *Handbook of Combinatorics*, two volumes, North Holland, an imprint of Elsevier, Amsterdam, 1995.

Guy, Richard K., "John Horton Conway: Mathematical Magus," *The Two-Year College Mathematics Journal*, Vol. 13, No. 5, November 1982, pp. 290–299.

Hassani, Mehdi, "Derangements and Applications," *Journal of Integer Sequences*, Vol. 6, No. 1, Article 03.1.2, 2003, pp. 1–8.

Izquierdo, Sebastián, *Pharus Scientiarum*, 2 volumes, Lyon, 1659. See sections 48–51. Pages 319–358 cover combinatorics.

lock.jpg, *Education Database Online*, available online at https://web.archive.org/web/20160407132244/www.onlineeducation.net/lock/lock.jpg. This graphic presents a shortcut for determining the combination of a combination lock.

OEIS Wiki contributors, "Number of Derangments," *The On-Line Encyclopedia of Integer Sequences® (OEIS®) Wiki*, available online at http://oeis.org/wiki/Number_of_derangements.

Ore, Oystein, *Cardano, the Gambling Scholar*, with a translation from the Latin of Cardano's Book on Games of Chance, by S. H. Gould, Princeton University Press, Princeton, NJ, 1953, reprinted by Dover, 1965.

Regardie, Israel, editor, *777 and Other Qabalistic Writings of Aleister Crowley*, Samuel Weiser, Inc., York Beach, ME, 1997. This is where I first came upon derangements. It's interesting how mathematical results were once considered to have a mystical nature. Such beliefs date back to the Pythagoreans.

Sperner, Emanuel, "Ein Satz uber Untermengen einer endlichen Menge," *Mathematische Zeitschrift*, Vol. 27, 1928, pp. 544–548.

Tobias, Marc Weber, *Locks, Safes and Security: An International Police Reference*, second edition, Charles C. Thomas, Springfield, IL, 2000, two volumes. These are the definitive volumes. If you want the complete story, read these. If you happen to be in law enforcement, the author can share even more with you.

Wallis, John, *Discourse of Combinations*, printed by John Playford, for Richard Davis, bookseller, in the University of Oxford, London, 1685.

Whitworth, W. Allen, *Choice and Chance*, Deighton, Bell and Co., Cambridge, 1886.

Wilson, Robin and John J. Watkins, editors, *Combinatorics: Ancient & Modern*, Oxford University Press, Oxford, 2013.

Yeager, Wayne B., *Techniques of Safecracking*, Loompanics Unlimited, Port Townsend, WA, 1990. At only 88 pages, this is a quick introduction. As the title indicates, it consists of techniques, not anecdotes.

10

Pascal and the Arithmetic Triangle

"The more I see of men, the better I like my dog."

Blaise Pascal

Great mathematicians are often as interesting as their best results. So, this chapter begins with some background on the man for whom the famous arithmetic triangle is named.

Forbidden Fruit and Early Calculators

Blaise Pascal (Figure 10.1) was home-schooled by his father, Étienne Pascal, a lawyer who had some strange ideas. He thought that Blaise shouldn't study any mathematics until he was 15. As the forbidden fruit is always the most inviting, Blaise began studying geometry in secret when he was 12. Of course, his father found out. Fortunately, when Étienne realized how much Blaise wanted to study mathematics, he allowed him to continue, and even bought him a copy of Euclid's Elements. Upon hearing this story, I had to wonder if Étienne really wanted his son to take an interest in mathematics and only forbid it in the hopes of the ban making the subject more appealing – a sort of reverse psychology. An alternate explanation would be to take the story at face value. Perhaps Étienne hated mathematics, forbid it for that reason, and simply relented when he saw how much his son was enjoying it.

FIGURE 10.1
Blaise Pascal (1623–1662).

Public domain.

Well, it turns out that the real explanation is "none of the above." Remember the Limaçon of Pascal from Chapter 6? It was not discovered by Blaise Pascal, but rather Étienne, who in addition to being a lawyer was also an amateur mathematician. He feared that his son would love math, as he did, and devote too much time to it; hence the ban.

Free of restrictions, Blaise Pascal progressed quickly, inventing the world's first digital calculator when he was 18. It was called the Pascaline and resembled a mechanical calculator of the 1940s. "Digital" in this instance simply means that it used the digits 0 through 9. The Pascaline was the second mechanical calculator; Wilhelm Schickard had made one in 1624.[1] Leibniz, in 1671, made a machine that could add, subtract, multiply, and divide. Sadly, none of these devices had much impact. The means to mass produce them did not yet exist, so they're mere historical curiosities. The abacus and the slide rule, for example, were both far more important. Leibniz's use of binary could have been a more important contribution, but again it was too far ahead of its time and didn't catch on.[2]

As mentioned in Chapter 2, even in the 1800s work on computing machinery carried out by Charles Babbage was still too early to catch on (or even be completed).

Pascal vs. Descartes

Pascal worked in areas besides calculating machinery and mathematics. By 1647, he had convinced himself, through experiments, that vacuums existed. René Descartes, the famous philosopher who had written "I think, therefore I am" and invented analytic geometry (the *x–y* axis), both in his 1637 classic *Discourse on the Method*, visited Pascal in September 1647. He stayed for two days and there was much arguing about vacuums, which Descartes did not believe in. Descartes wrote in a letter to Huygens that Pascal "… has too much vacuum in his head." Had not the "great" Aristotle said, "Nature abhors a vacuum," implying they could not exist?

When it became clear that Pascal was correct in claiming that atmospheric pressure decreases with height and that a vacuum existed above the atmosphere, Descartes wrote to Pierre de Carcavi about Pascal's experiments saying:

> It was I who two years ago advised him to do it, for although I have not performed it myself, I did not doubt of its success…

Descartes disagreed with Pascal on other matters. For example, he encouraged Pascal to follow his example and stay in bed every day till 11. It was on such lazy mornings that Descartes did his best thinking. Ignoring this advice saved Pascal's life. A housemaid came with a knife early one morning to kill Pascal, but was unable, as he had already

---

1 www-groups.dcs.st-and.ac.uk/~history/Mathematicians/Pascal.html.

2 Bell, Eric Temple, *Mathematics, Queen and Servant of Science*, Copyright, 1951, British Edition 1952, reprinted 1954, 1958, 1961. Printed in Great Britain by Butler and Tanner Ltd., Frome and London, p. 248. On page 249 Bell details what he called Leibniz's "bizarre theology": "The binary scale may also be the ultimate secret of the creation of the universe. For the fact that zeros and ones suffice in binary arithmetic for the expression of any integer convinced Leibniz that God had created the universe (1) out of nothing (0). Leibniz was not only a great mathematician but a great philosopher as well. However, he confused 'nothing' with 'zero' – a remarkable feat for a mathematician and logician."

left.[3] Later in his life, Pascal had another close brush with death. More on that in Chapter 12! For now, consider Pascal's triangle.

Pascal's Triangle

Blaise Pascal was not the first to discover this wonderful triangle of numbers, nor was it his father, Étienne. It predates both of these men by centuries. By now, this is unlikely to surprise you. It is, in fact, such a common phenomenon, both within and outside of mathematics, that we have what is known as Stigler's law of eponymy.

Stigler's Law of Eponymy: No scientific discovery is named after its original discoverer.[4]
This law was, of course, stated by others before Stigler.

The triangle came to be named after Pascal, following his discussion of it in *Traite du Triangle Arithmetique* (*Treatise on the Arithmetical Triangle*), written in 1654, but not published until 1665, after Pascal's death.

The first eight rows of the triangle are given below:

$$
\begin{array}{ccccccccccccccc}
 & & & & & & & 1 & & & & & & & \\
 & & & & & & 1 & & 1 & & & & & & \\
 & & & & & 1 & & 2 & & 1 & & & & & \\
 & & & & 1 & & 3 & & 3 & & 1 & & & & \\
 & & & 1 & & 4 & & 6 & & 4 & & 1 & & & \\
 & & 1 & & 5 & & 10 & & 10 & & 5 & & 1 & & \\
 & 1 & & 6 & & 15 & & 20 & & 15 & & 6 & & 1 & \\
1 & & 7 & & 21 & & 35 & & 35 & & 21 & & 7 & & 1
\end{array}
$$

It begins with a triangle of 1s. As it grows, each new row begins and ends with a 1 and all other entries are the sum of the two numbers above it a bit to the left and right. Actually, the 1s can be said to follow this rule, if we imagine 0s lying beyond the 1s on each side of the triangle. So, it's very easy to generate new rows for this triangle. All you have to be able to do is add. The rows continue on forever. The following sections detail why we would want to generate more rows.

Combinations

The triangle can be used to solve problems in combinatorics.

Example 1 – Pizza Time!
How many distinct three-topping pizzas can be ordered if six toppings are available and the three chosen must all be different?

Solution: We could solve this by calculating $C(6, 3)$, as detailed in Chapter 9, but the triangle provides another means of solution. To find the answer we go to the row that

---

3 For the story of *why* she was trying to kill Pascal, see Muir, Jane, *Of Men and Numbers: The Story of the Great Mathematicians*, Dover, New York, 1965, p. 98.
4 https://en.wikipedia.org/wiki/Stigler%27s_law_of_eponymy.

begins 1, 6, … and read across counting 0 toppings, 1 topping, 2, toppings 3 toppings, which takes us to 20 (the answer).

In general, to find the value of $C(n, r)$ using Pascal's triangle, simply go to row $n + 1$ and moving from left to right go over to the entry in position $r + 1$. Some authors label the initial row of Pascal's triangle as "row 0." So, you may encounter instructions that look different from mine, for where the appropriate value can be found. In any case, this approach always works. The triangle can be expressed entirely in terms of $C(n, r)$ values, as indicated below:

$$
\begin{array}{ccccccccccccc}
& & & & & & C(0,0) & & & & & & \\
& & & & & C(1,0) & & C(1,1) & & & & & \\
& & & & C(2,0) & & C(2,1) & & C(2,2) & & & & \\
& & & C(3,0) & & C(3,1) & & C(3,2) & & C(3,3) & & & \\
& & C(4,0) & & C(4,1) & & C(4,2) & & C(4,3) & & C(4,4) & & \\
& C(5,0) & & C(5,1) & & C(5,2) & & C(5,3) & & C(5,4) & & C(5,5) & \\
C(6,0) & & C(6,1) & & C(6,2) & & C(6,3) & & C(6,4) & & C(6,5) & & C(6,6) \\
C(7,0) \; C(7,1) & & C(7,2) & & C(7,3) & & C(7,4) & & C(7,5) & & C(7,6) & & C(7,7)
\end{array}
$$

The symmetry in the triangle means that $C(n, r) = C(n, n - r)$ for $0 \leq r \leq n$.

The rule used to generate new entries can be expressed as $C(n, r) = C(n{-}1, r{-}1) + C(n{-}1, r)$, or, in another notation, as $\binom{n}{r} = \binom{n-1}{r-1} + \binom{n-1}{r}$.

Binomial Coefficients

The triangle provides the coefficients for the binomial expansions, $(x + y)^n$, $n \in \mathbb{N}$, in its rows.

Example 2 – Calculate $(x + y)^4$

Solution: We could write $(x + y)^4 = (x + y)(x + y)(x + y)(x + y)$ and then carefully multiply it all out. However, the row of the triangle beginning $1, 4, …$ gives the coefficients of the final result. The rest of each term is just descending powers of x from x^4 down to $x^0 = 1$ multiplied by ascending powers of y from $y^0 = 1$ up to y^4. We have

$$
(x + y)^4 = 1x^4 + 4x^3 y + 6x^2 y^2 + 4xy^3 + 1y^4.
$$

The first row is no exception: $(x + y)^0 = 1$, the top row of the triangle.

This is easy to demonstrate for any $n \in \mathbb{N}$, as is demonstrated below.

The Binomial Theorem

$$
(x + y)^n = \sum_{k=0}^{n} \binom{n}{k} x^{n-k} y^k
$$

Proof

Consider the term $x^a y^b$ in the expansion of $(x+y)^n = (x+y)(x+y) \cdots (x+y)$. Because the power of x is a, we must select x exactly a times when choosing an x or a y from each $(x+y)$. The rest of the time the selection must be y (there's no other choice). There are n sets of parentheses, so we must have $a + b = n$. If we denote b by k, then the exponent on x must be $n - k$. Now, how many times will we get this particular combination? Well, we have n sets of parentheses, so if we want to pick $n - k$ parentheses to take an x from, there are $C(n, n-k)$ ways to do it. Thus, the coefficient of $x^{n-k}y^k$ must be $C(n, n-k)$. This is equal to $C(n, k)$, so we have $C(n, k)x^{n-k}y^k$. There was nothing special about this particular term, so the complete expansion is just the sum of terms of this form. We have

$$(x+y)^n = \sum_{k=0}^{n} \binom{n}{k} x^{n-k} y^k$$

as was to be shown. □

The binomial theorem is very old. An understanding of special cases (i.e., low values for the exponent n) came first. Various authors published more and more of these, and the full-blown binomial theorem seems to arise gradually. How many cases need to be expressed to give the sense that the author understood the big picture? He may not have been the first, but Al-Karaji (953–1029) definitely understood it, for he had a proof.[5]

One of a mathematicians' favorite activities is generalizing and when Isaac Newton tackled the binomial theorem, this is exactly what he did. In a 1676 letter, he gave a formula that works for all rational exponents. Changing his notation to make it closer to the form shown above, we have

$$(x+y)^r = \sum_{k=0}^{\infty} \binom{r}{k} x^{r-k} y^k.$$

The fact that the summation has infinitely many terms is a big change, but what requires an explanation is the meaning of $\binom{r}{k}$ when r can be any rational number. It is defined by

$$\binom{r}{k} = \frac{r(r-1)(r-2) \cdots (r-(k-1))}{k!}.$$

Newton found his generalized binomial theorem useful for evaluating square roots. To do so, he took the exponent to be ½ and let the number he wished to find the square root of be represented by the number in parenthesis, expressed here as $x + y$. The expansion has infinitely many terms, but they diminish rapidly, so Newton could simply calculate a few terms, the exact number depending on how precise he wanted his estimate to be.

---

5 https://en.wikipedia.org/wiki/Al-Karaji, which cites O'Connor, John J. and Edmund F. Robertson, "Abu Bekr ibn Muhammad ibn al-Husayn Al-Karaji," MacTutor History of Mathematics archive, University of St Andrews, http://mathshistory.st-andrews.ac.uk/Biographies/Al-Karaji.html.

Powers of 11

Isaac Newton also noticed a pattern that produces rows of the triangle.[6]

$$11^0 = 1$$
$$11^1 = 11$$
$$11^2 = 121$$
$$11^3 = 1,331$$
$$11^4 = 14,641$$

Recall that not all patterns continue forever. If we wish to prove that running the numbers together in rows of Pascal's triangle yields consecutive powers of 11, no number of examples will suffice. Instead, we should try to apply proof by induction or some other approach that can handle an infinite number of possibilities. However, any such approach is doomed to failure, for $11^5 = 161,051$, while the corresponding row of Pascal's triangle is 1 5 10 10 5 1.

But there is a way to modify our conjecture to make it work! Reading the row 1 5 10 10 5 1 from right to left, when we get to the 10, we can carry the 1 in the 10s place over to the next 10 to get 1 5 11 0 5 1. If we then look at the 11 and carry the 1 in the 10s position once more, we get 1 6 1 0 5 1. Running all of these digits together we get $161,051 = 11^5$.

So, while simply running the digits together will not always yield a power of 11, if we make the extra step of carrying digits, then it *will* always work. Again, the example doesn't prove this, but it can be proven.

Row Sums

Consider the sums of the entries in the rows of Pascal's triangle.

						1						1
					1		1					2
				1		2		1				4
			1		3		3		1			8
		1		4		6		4		1		16
	1		5	10		10		5		1		32
1		6	15	20	15		6		1		64	
1	7	21	35	35	21	7	1				128	

They are all powers of 2. Algebraically, this result may be expressed as $\sum\limits_{i=0}^{n} \binom{n}{i} = 2^n$.

Having previously established $\sum_{i=0}^{n} \binom{n}{i} x^{n-i} y^i = (x+y)^n$, all we need to do is let $x = y = 1$ and we have a proof that $\sum_{i=0}^{n} \binom{n}{i} = 2^n$.

6 Young, Robert M., *Excursions in Calculus: An Interplay of the Continuous and the Discrete*, The Mathematical Association of America, 1992, pp. 257–258.

Fibonacci Numbers

Instead of summing the entries along the rows, we can instead sum the "oblique" or "shallow" diagonals, as indicated in Figure 10.2. When we do this, the Fibonacci numbers appear. These numbers are examined in detail in Chapter 13.

It's easier to see exactly which numbers lie on each diagonal if we left justify the rows of the triangle, as in Figure 10.3.

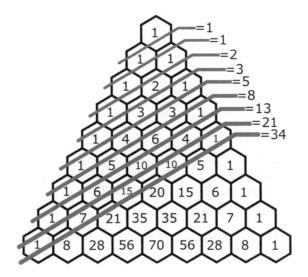

FIGURE 10.2
Pascal meets Fibonacci.

Image created by Josh Gross.

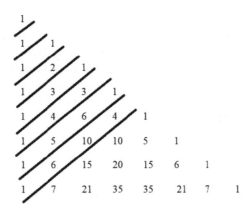

FIGURE 10.3
Another perspective on the Fibonacci numbers.

Image created by Craig P. Bauer.

This pattern can be expressed algebraically as

$$F_{n+1} = \binom{n}{0} + \binom{n-1}{1} + \binom{n-2}{2} + \cdots$$

The three dots at the end are not intended to indicate that the sum continues forever. It stops when we get to a value of the form $\binom{k}{k}$, where k is even, which would be a 1 on the right-hand side of the triangle, or a value of the form $\binom{k}{k-1}$, where k is odd, which would be k. This explanation can be avoided by using the handy floor function of Chapter 6:[7]

$$F_{n+1} = \sum_{k=0}^{\lfloor n/2 \rfloor} \binom{n-k}{k}$$

The pattern was found by Édouard Lucas, and published in 1876, although he does not give the name Fibonacci in this paper.[8] It was previously stated by the Indian scholar Hemachandra around 1150, about 50 years before Fibonacci posed the problem from which the so-called Fibonacci numbers arose in 1202![9]

Proof

The proof presented here is by strong induction. Before beginning, it's useful to split the claim into two cases, when the index of the Fibonacci number is even and when it is odd:

$$F_{2n} = \binom{2n-1}{0} + \binom{2n-2}{1} + \binom{2n-3}{2} + \cdots + \binom{2n-(n-1)}{n-2} + \binom{2n-n}{n-1}$$
$$F_{2n+1} = \binom{2n}{0} + \binom{2n-1}{1} + \binom{2n-2}{2} + \cdots + \binom{2n-(n-1)}{n-1} + \binom{2n-n}{n}$$

It is trivial to prove the base case, as $F_1 = F_2 = 1$ and the sum of the first two shallow diagonals d_1 and d_2 are easily seen to be 1. Using strong induction, we now prove that the claim must hold for a given shallow diagonal, if it holds for all previous shallow diagonals. This is split into two cases, depending on whether the diagonal we're attempting to prove the claim for is in an even or odd position. The odd case is dealt with first. If it's an odd diagonal, it can be labeled d_{2N+1} for some $N \in \mathbb{N}$. We have

7 From http://mathworld.wolfram.com/PascalsTriangle.html.
8 Lucas, Édouard, "Note sur la triangle arithmétique de Pascal et sur la serie de Lamé," *Nouvelle correspondance mathématique*, tome 2, 1876, pp. 70–75. More of Lucas's work is detailed in Chapter 14.
9 Kak, Subhash, "The Golden Mean and the Physics of Aesthetics," in Yadav, Bhuri Singh and Man Mohan, editors, *Ancient Indian Leaps into Mathematics*, Birkhäuser, New York, 2011, pp. 111–119. Also available at https://arxiv.org/abs/physics/0411195.

$$d_{2N+1} = \binom{2N}{0} + \binom{2N-1}{1} + \binom{2N-2}{2} + \cdots + \binom{2N-(N-1)}{N-1} + \binom{2N-N}{N}$$

$$= \binom{2N-1}{0} + \left[\binom{2N-2}{0} + \binom{2N-2}{1}\right] + \left[\binom{2N-3}{1}\binom{2N-3}{2}\right] + \cdots$$

$$+ \left[\binom{2N-(N-1)-1}{N-2}\binom{2N-(N-1)-1}{N-1}\right] + \binom{2N-(N+1)}{N-1}$$

In the above equality, the first and last terms were replaced with terms of equal value. Note that $2N - (N + 1) = N - 1$. All inner terms were replaced with pairs of terms according to the identity that defines Pascal's triangle, $\binom{n}{k} = \binom{n-1}{k-1} + \binom{n-1}{k}$. The terms can be reordered as

$$= \binom{2N-2}{0} + \binom{2N-3}{1} + \cdots + \binom{2N-(N-1)-1}{N-2} + \binom{2N-(N+1)}{N-1}$$

$$+ \binom{2N-1}{0} + \binom{2N-2}{1} + \binom{2N-3}{2} + \cdots + \binom{2N-(N-1)-1}{N-1}$$

$$= F_{2N} + F_{2N-1} \text{ (by the induction assumption).}$$

This shows that $d_{2N+1} = F_{2N} + F_{2N-1}$ and therefore $d_{2N+1} = F_{2N+1}$, as was to be shown. The even case is left as an exercise for the reader (see #7).[10]

Hockey Stick Patterns

Another lovely pattern can be found by picking any 1 on the sides of the triangle and then moving down in a diagonal direction through as many numbers as desired. Sum all of the numbers you pass through, including the original 1 and the number you stop at. You'll find that number directly below the number you stopped at, in the opposite diagonal direction. A few examples follow. Figure 10.4 makes it clear how these became known as hockey stick patterns.

The patterns inside the shaded hockey sticks are $1 + 5 + 15 + 35 + 70 + 126 = 152$, $1 + 8 + 36 + 120 + 330 = 495$, and $1 + 12 = 13$.

Some special cases of hockey stick patterns were recognized by Rabbi Abraham ibn Ezra (1090–1167), having arisen from his work in astrology.[11] However, there's no record of him giving the pattern in the general form. For the case of a hockey stick that begins with a 1 on the right-hand side of the triangle, it is:

$$C(n, k) = C(n - 1, k - 1) + C(n - 2, k - 1) + \cdots + C(k - 1, k - 1).$$

---

10 For another proof see Vilenkin, N. Ya, *Combinatorics*, Academic Press, New York and London, 1971, pp. 119–121.

11 Wilson, Robin, and John J. Watkins, editors, *Combinatorics: Ancient & Modern*, Oxford University Press, Oxford, 2013, pp. 113–114.

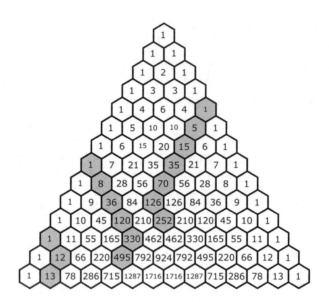

FIGURE 10.4
Hockey stick patterns.

Image created by Josh Gross.

Proof (By Induction on the Length of the Hockey Stick's Shaft)

The smallest possible hockey stick is one that has a shaft of length one, and this is achieved when $n = k$. So, the base case that must be established is $C(k, k) = C(k - 1, k - 1)$. This is clearly true, as both sides evaluate to 1.

Now, for the induction hypothesis, suppose that the claim holds for some particular $n \geq k$ (giving a shaft of length $n - k + 1$). We have

$$C(n, k) = C(n - 1, k - 1) + C(n - 2, k - 1) + \cdots + C(k - 1, k - 1).$$

Adding $C(n, k - 1)$ to both sides gives

$$C(n, k-1) + C(n, k) = C(n, k - 1) + C(n - 1, k - 1) + C(n - 2, k - 1) + \cdots + C(k - 1, k - 1).$$

The right-hand side now represents a hockey stick that has a shaft of length $n - k + 2$, which is one unit longer.

By Pascal's identity, $C(n, k - 1) + C(n, k) = C(n +1, k)$, so the left-hand side can be rewritten as $C(n + 1, k)$.

We then have $C(n + 1, k) = C(n, k - 1) + C(n - 1, k - 1) + C(n - 2, k - 1) + \cdots + C(k - 1, k - 1)$. This establishes the claim for all n. □

Ibn Ezra gave his results in *Sefer ha-Olam* (Book of the World), 1147–1148, which was originally written in Hebrew.[12] In 1281, a Latin edition appeared giving signs of having been translated from an Arabic edition that is no longer known to exist.[13]

Niccolò Tartaglia made use of the formula for a hockey stick that begins with a 1 on the left-hand side of the triangle, $\sum_k C(n + k - 1, k) = C(n + k, k)$, when investigating the number of distinct results of throwing up to eight dice.[14] He published his result in *General Trattato di Numeri et Misure* (General Treatise on Numbers and Measures) in 1556.[15]

Leonhard Euler used the pattern in his investigation of a gambling game known as *jeu de recontre*.[16] He later found a simpler way to obtain the same result. This is detailed in Chapter 12.

It's exciting when the same mathematics becomes relevant to problems in fields as diverse as astrology and gambling!

Triangular Numbers and More

If you start a hockey stick pattern with the 1 at the very top of the triangle and move to the right, you'll simply get a sum of ones. So, by taking sticks of different lengths, the sums produce all of the natural numbers, seen on the diagonal directly below the rightmost diagonal of ones.

If you start at the 1 that begins the sequence of natural numbers, the hockey stick patterns sum the first n natural numbers, which yields the triangular numbers, seen in Chapter 3. The sequence of triangular numbers appears as the shaded entries in Figure 10.5.

If we use the diagonal of triangular numbers to form hockey stick patterns, we get tetrahedral numbers (a sort of three-dimensional triangle) as the sums. Pascal himself noted the appearance of the triangular and tetrahedral numbers in the triangle.[17] This was not a new result, though. Indeed, almost everything he presented had been done before; some of it was even well known. It was his style of presentation, with proofs, that made his treatise important.[18]

Polygons are two dimensional. When we move to three dimensions, we have polyhedra. The name in higher dimensions (or just more generally) is **polytope**. In all cases, the associated numbers are **figurate**. As the last example from Pascal's triangle showed, we get tetrahedral (or pyramidal) numbers in three dimensions by summing triangular numbers.

---

12 https://en.wikipedia.org/wiki/Abraham_ibn_Ezra.

13 Wilson, Robin, and John J. Watkins, editors, *Combinatorics: Ancient & Modern*, Oxford University Press, Oxford, 2013, p. 115. For the Latin translation see Clagett, Marshall, *Nicole Oresme and the Medieval Geometry of Qualities and Motions*, University of Wisconsin Press, Madison, WI, 1968, pp. 445–446. For an English translation see Ginsburg, Jekuthiel and David Eugene Smith, "Rabbi ben Ezra on permutations and combinations," *The Mathematics Teacher*, Vol. 15, No. 6, October 1922, pp. 347–356.

14 Wilson, Robin and John J. Watkins, editors, *Combinatorics: Ancient & Modern*, Oxford University Press, Oxford, 2013, p. 129.

15 Tartaglia, Niccolò Fontana, *General Trattato di Numeri, et Misure*, per Curtio Troiano de i Nauò, Venice, 1556.

16 Hinz, Andreas M., Sandi Klavžar, Uroš Milutinović, and Ciril Petr, *The Tower of Hanoi – Myths and Maths*, Birkhäuser, 2013, p. 16.

17 Wilson, Robin and John J. Watkins, editors, *Combinatorics: Ancient & Modern*, Oxford University Press, Oxford, 2013, p. 168.

18 Wilson, Robin and John J. Watkins, editors, *Combinatorics: Ancient & Modern*, Oxford University Press, Oxford, 2013, p. 148.

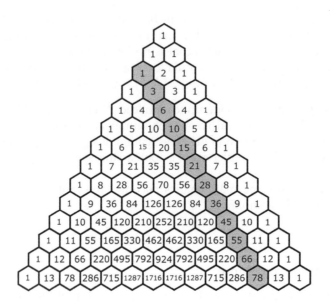

FIGURE 10.5
Triangular numbers in Pascal's triangle.

Image created by Josh Gross.

We can sum tetrahedral numbers to get four-dimensional "triangles." These also appear as a diagonal in Pascal's triangle, directly under the tetrahedral numbers. The following diagonals give five-dimensional, six-dimensional, etc., triangles. The procedure can be continued like this forever, using hockey stick patterns, and all of the numbers appear in consecutive diagonals of Pascal's triangle.

Hexagons and Perfect Squares

In Figure 10.6, pick any number that is neither a 1 nor in the bottom row.

Now, take the product of all of the numbers in the cells that border it. For example, if you picked 28, your product would be $7 \cdot 21 \cdot 56 \cdot 84 \cdot 36 \cdot 8 = 199{,}148{,}544 = 14{,}112^2$. The fact that we got a perfect square is not a coincidence. It will always be a perfect square. I excluded the bottom row, because the values directly below those entries are needed for the product, but are not shown. If you take the extra step of calculating them, you will see the pattern holds for the bottom row as well. Indeed, it works everywhere.[19] A proof of this fact follows.

Represent the chosen entry as $\binom{n}{r}$. The six numbers that surround it can be separated into two subsets, indicated by shading in Figure 10.7.

---

19 I first saw this theorem and its proof in Chamberland, Marc, *Single Digits*, Princeton University Press, Princeton, NJ, 2015, pp. 164–165.

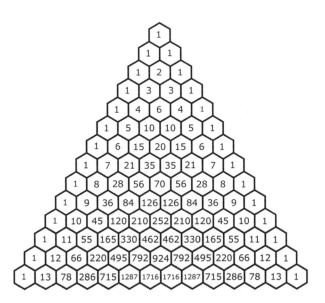

FIGURE 10.6
Pascal's triangle, again.

Image created by Josh Gross.

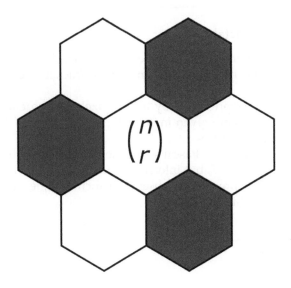

FIGURE 10.7
An entry in the triangle and the numbers around it.

Image created by Josh Gross.

The product of the shaded positions is $\binom{n-1}{r}\binom{n}{r-1}\binom{n+1}{r+1}$ and the product of

the white positions is $\binom{n-1}{r-1}\binom{n}{r+1}\binom{n+1}{r}$.

Evaluating the product of the shaded positions, we have

$$\binom{n-1}{r}\binom{n}{r-1}\binom{n+1}{r+1} = \frac{(n-1)!}{(n-1-r)!r!}\frac{n!}{(n-(r-1))!(r-1)!}\frac{(n+1)!}{(n+1-(r+1))!(r+1)!}$$

$$= \frac{(n-1)!}{(n-r-1)!r!}\frac{n!}{(n-r+1)!(r-1)!}\frac{(n+1)!}{(n-r)!(r+1)!}$$

And evaluating the product of the white positions, we have

$$\binom{n-1}{r-1}\binom{n}{r+1}\binom{n+1}{r} = \frac{(n-1)!}{(n-1-(r-1))!(r-1)!}\frac{n!}{(n-(r+1))!(r+1!)}\frac{(n+1)!}{(n+1-r)!(r)!}$$

$$= \frac{(n-1)!}{(n-r)!(r-1)!}\frac{n!}{(n-r-1)!(r+1)!}\frac{(n+1)!}{(n-r+1)!r!}$$

It is easy to see that both products have the same factors in the numerator and denominator, albeit in different orders. Hence, their values are the same, and, because they are both integers, their product is a perfect square.   □

Pascal's Triangle as a Fractal

If we shade the odd numbers in the triangle, the pattern shown in Figure 10.8 results. It's known as Sierpiński's triangle. This and other fractals are examined closely in Chapter 18.

Letting each number be represented by a tiny pixel, we can cram many more rows into a picture. The examples are given in color figures 10 through 15 **(See color insert)** with each shade having 1,000 rows.

If you compare the shadings of Pascal's triangle, you'll notice the black triangles are solid for the moduli 2, 3, 5, and 7, but interrupted for moduli 4 and 6. Can you use these few examples to predict the presence or absence of such triangles for other moduli? This is Exercise 6 at the end of this chapter.

Perfect Numbers

Examining Figure 10.8 closer, a neat set of numbers arises.

The first even value, 2, is by itself in a white block. The next white triangle, directly below it, contains 6 numbers. The larger white triangle below that contains 28 values). Do the numbers 6 and 28 look familiar? They're known as **perfect numbers**, because they're equal to the sums of their proper positive divisors. That is, the proper positive divisors of 6 are 1, 2, and 3 and we have 1 + 2 + 3 = 6. Similarly for 28. But the sequence of the number of elements in each white triangle began with 1 and this

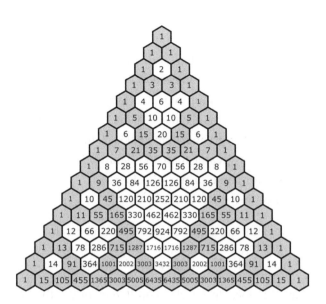

FIGURE 10.8
An intriguing shading of Pascal's triangle.

Image created by Josh Gross.

isn't a perfect number! That's true, Pascal's triangle doesn't always yield perfect numbers in this manner, but every even perfect number does appear somewhere in this sequence. This is because the number of elements in each white triangle is given by $2^{n-1}(2^n - 1)$. For $n = 1$, we get 1. Making $n = 2$ or 3 gives 6 and 28, respectively. Every even perfect number is of this form, but not every number of this form is perfect. What about odd perfect numbers? It's an open question as to whether or not there are any odd perfect numbers. Nobody knows! We don't even know if the number of perfect numbers is finite or infinite.

Some progress has been made though. As has been mentioned repeatedly, mathematicians often just chip away at the hardest problems. Benjamin Peirce (1809-1880) showed that if there is an odd perfect number, then it must have at least four distinct prime factors. In 1888, J. J. Sylvester, apparently unaware of Peirce's work, presented the same result, but later that year he raised the minimum to five.[20] This number continued to grow as other mathematicians chipped away at the problem. The minimum size of a potential odd perfect number continued to grow as well. It presently stands at 10^{1500}.[21] In addition to these constraints, several more were added to the list by various mathematicians. The following list shows requirements that any odd perfect number must satisfy:

20 Gimbel, Steven and John H. Jaroma, "Sylvester: Ushering in the Modern era of Research on Odd Perfect Numbers," *Integers: Electronic Journal of Combinatorial Number Theory*, Vol. 3, 2003, p. A16.
21 Ochem, Pascal and Michaël Rao, "Odd Perfect Numbers Are Greater than 10^{1500}," *Mathematics of Computation*, Vol. 81, No. 279, July 2012, pp. 1869–1877.

1. Of the form $p^{4b+1}Q^2$, where p is a prime of the form $4m+1$ (Euler)
2. Greater than 10^{1500} (Ochem and Rao[22])
3. Less than 2^{4^k}, where k is the number of distinct prime factors (Nielsen[23])
4. At least 9 distinct prime factors (Nielsen[24])
5. At least 75 prime factors, counting repeated factors (Hare[25])
6. Largest prime factor greater than 100,000,000 (Goto and Ohno[26])
7. Second largest prime factor greater than 10,000 (Iannucci[27])
8. Third largest prime factor greater than 100 (Iannucci[28])
9. Divisible by $q^b > 10^{20}$ for some prime q (Cohen[29])
10. Of the form $12k + 1$ or $36k + 9$ (Touchard[30]).

Although we're done with the topic of perfect numbers for now, I'm introducing some notation before moving on. We write $\sigma(n)$ to indicate the sum of the positive divisors of n, including n itself. Thus, for perfect numbers we have $\sigma(n) = 2n$. When $\sigma(n) < 2n$, we say n is deficient (it falls short of being perfect), and when $\sigma(n) > 2n$, we say n is abundant (it goes too far!). You'll encounter the σ notation again in Chapters 17 and 22, in which a formula for $\sigma(n)$ is given in terms of values of $\sigma(k)$, where $k < n$. Sadly, this formula doesn't answer the big questions about perfect numbers discussed here.

Back to Pascal

Blaise Pascal wrote, "It is extraordinary how fertile in properties this [triangle] is. Everyone can try his hand."[31] Indeed, many more examples could be given of interesting

22 Ochem, Pascal and Michaël Rao, "Odd Perfect Numbers Are Greater than 10^{15000}," *Mathematics of Computation*, Vol. 81, No. 279, July 2012, pp. 1869–1877.
23 Nielsen, Pace P., "An Upper Bound for Odd Perfect Numbers," *Integers: Electronic Journal of Combinatorial Number Theory*, Vol. 3, 2003, #A14, available online at www.kurims.kyoto-u.ac.jp/EMIS/journals/INTE GERS/papers/d14/d14.pdf.
24 Nielsen, Pace P., "Odd Perfect Numbers Have at least Nine Distinct Prime Factors," *Mathematics of Computation*, Vol. 76, No. 260, October 2007, pp. 2109–2126.
25 Hare, Kevin G., "New Techniques for Bounds on the Total Number of Prime Factors of an Odd Perfect Number," *Mathematics of Computation*, Vol. 76, No. 260, October 2007, pp. 2241–2248.
26 Goto, Takeshi and Yasuo Ohno, "Odd Perfect Numbers Have a Prime Factor Exceeding 10^8," *Mathematics of Computation*, Vol. 77, No. 263, July 2008, pp. 1859–1868.
27 Iannucci, Douglas E., "The Second Largest Prime Divisor of an Odd Perfect Number Exceeds Ten Thousand," *Mathematics of Computation*, Vol. 68, No. 228, May 1999, pp. 1749–1760.
28 Iannucci, Douglas E., "The Third Largest Prime Divisor of an Odd Perfect Number Exceeds One Hundred," *Mathematics of Computation*, Vol. 69, No. 230, May 2000, pp. 867–879.
29 Cohen, Graeme L., "On the Largest Component of an Odd Perfect Number," *Journal of the Australian Mathematical Society*, Series A, Vol. 42, 1987, pp. 280–286.
30 Touchard, Jacques, "On Prime Numbers and Perfect Numbers," *Scripta Mathematica*, Vol. 19, 1953, pp. 35–39.
31 Young, Robert M., *Excursions in Calculus: An Interplay of the Continuous and the Discrete*, The Mathematical Association of America, 1992, p. 78, which, in turn, cites Pascal's "*Traité du Triangle Arithmétique*." See *Oeuvres de Blaise Pascal*, edited by Léon Brunschvieg and Pierre Boutroux, Hachette, Paris, 1904–1914, reprinted by Kraus, Nendeln/Lichtenstein, 1976–1978, Vol. 3, p. 465.

properties that are present in Pascal's triangle, but there are too many other topics to cover. Why not try your hand at discovering some of them yourself?

Pascal did important work in other areas of mathematics, and his contributions to the founding of probability are examined in Chapter 12. While he wrote about the triangle that bears his name, he was far from the first to discover it. Earlier appearances of the triangle serve to illustrate some important historical lessons.

Early Appearances of "Pascal's" Triangle

The earliest European appearance was in 1527 when the German mathematician Petrus Apianus (1495–1552) used it on the frontispiece of *Ein newe und wolgegründete underweisung aller Kauffmanns Rechnung in dreyen Büchern, mit schönen Regeln und fragstücken begriffen*, [A handbook of commercial arithmetic] (Figure 10.9).[32]

Another German, Michael Stifel, gave his version of the triangle in 1544 (Figure 10.10).[33] Stifel viewed the entries as figurate numbers. He was motivated to generate the triangle by the problem of approximating roots.[34]

A mere six rows of the triangle appeared in 1556 in Niccolò Tartaglia's *General Trattato di Numeri et Misure* (General Treatise on Numbers and Measures).[35] Well aware of their man's priority over Pascal, some Italians refer to "Tartaglia's triangle."[36]

In 1570, another Italian, the prolific author Gerolamo Cardano, included the triangle in his book *Opus Novum de Proportionibus Numerorum* (New Work on the Proportions of Numbers).[37]

Marin Mersenne must have enjoyed calculating. You may recall from Chapter 8 how he made a table of factorials that went all the way to 50! and later extended it to 64!. Well, in 1636, between the appearance of the two books containing these tables, he came out with the biggest arithmetic triangle to see print (Figure 10.11). It appeared in his *Harmonicorum Libri XII* (Twelve Books of Harmonic Principles). He arranged the entries in it in his own style, and it would more accurately be called an arithmetic rectangle.

---

32 Apianus, Petrus, *Ein newe und wolgegründete underweisung aller Kauffmanns Rechnung in dreyen Büchern, mit schönen Regeln und fragstücken begriffen* (A Handbook of Commercial Arithmetic), Ingolstadt, 1527.

33 Wilson, Robin and John J. Watkins, editors, *Combinatorics: Ancient & Modern*, Oxford University Press, Oxford, 2013, p. 172.

34 Wilson, Robin and John J. Watkins, editors, *Combinatorics: Ancient & Modern*, Oxford University Press, Oxford, 2013, p. 172.

35 Wilson, Robin and John J. Watkins, editors, *Combinatorics: Ancient & Modern*, Oxford University Press, Oxford, 2013, p. 171. This page also gives the context of his discovery – the problem he was investigating that led to his generating the triangle. Page 172 has a picture of Tartaglia's triangle.

36 Wilson, Robin and John J. Watkins, editors, *Combinatorics: Ancient & Modern*, Oxford University Press, Oxford, 2013, p. 171.

37 See Wilson, Robin and John J. Watkins, editors, *Combinatorics: Ancient & Modern*, Oxford University Press, Oxford, 2013, p. 173 for info and a picture of Cardano's triangle.

FIGURE 10.9

The first European appearance of the famous triangle.

Public domain.

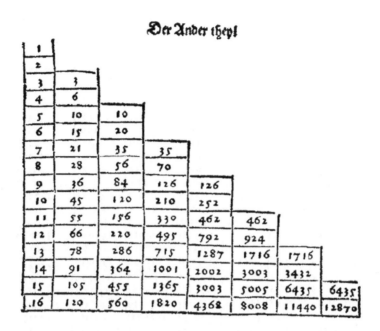

FIGURE 10.10
Michael Stifel's arithmetic triangle.

Public domain.

Blaise Pascal and his father had visited Mersenne and likely learned of the triangle from him.[38] As was mentioned before, Pascal put what he knew about the triangle on paper in 1654, but it was not published until 1665, after his death.

In 1666, a year after Pascal's treatise appeared, Leibniz included the triangle in his *Dissertatio de Arte Combinatoria*, independently duplicating some of Pascal's work.[39]

Although the European history of the triangle may seem extensive, Jacob Bernoulli (1655–1705) apparently did not encounter it in any book, but instead generated it independently.[40] As was the case with Pascal's treatise, Bernoulli's triangle was published posthumously, finally appearing in his *Ars Conjectandi* (Art of Conjecturing), 1713.[41] It's fun to read Bernoulli's enthusiastic description of its importance:

---

38 Wilson, Robin and John J. Watkins, editors, *Combinatorics: Ancient & Modern*, Oxford University Press, Oxford, 2013, p. 174.

39 Wilson, Robin and John J. Watkins, editors, *Combinatorics: Ancient & Modern*, Oxford University Press, Oxford, 2013, p. 179.

40 Wilson, Robin and John J. Watkins, editors, *Combinatorics: Ancient & Modern*, Oxford University Press, Oxford, 2013, p. 178 and Young, Robert M., *Excursions in Calculus: An Interplay of the Continuous and the Discrete*, The Mathematical Association of America, 1992, p. 78.

41 Wilson, Robin and John J. Watkins, editors, *Combinatorics: Ancient & Modern*, Oxford University Press, Oxford, 2013, p. 178.

Tabella pulcherrima & vtilißima Combinationis duodecim Cantilenarum.

	I.	II.	III.	IV.	V.	VI.	VII.	VIII.	IX.	X.	XI.	XII.
1	1	1	1	1	1	1	1	1	1	1	1	1
2	3	4	5	6	7	8	9	10	11	12	13	14
3	6	10	15	21	28	36	45	55	66	78	91	105
4	10	20	35	56	84	120	165	220	286	364	455	560
5	15	35	70	126	210	330	495	715	1001	1365	1820	2380
6	21	56	126	252	462	792	1287	2002	3003	4368	6188	8568
7	28	84	210	462	924	1716	3003	5005	8008	12376	18564	27132
8	36	120	330	792	1716	3432	6435	11440	19448	31824	50388	77520
9	45	165	495	1287	3003	6435	12870	24310	43758	75582	125970	203490
10	55	220	715	2002	5005	11440	24310	48620	92378	167960	293930	497420
11	66	286	1001	3003	8008	19448	43758	92378	184756	352716	646646	1144066
12	78	364	1365	4368	12376	31824	75582	167960	352716	705432	1352078	2496144
13	91	455	1820	6188	18564	50388	125970	293930	646646	1352078	2704156	5200300
14	105	560	2380	8568	27132	77520	203490	497420	1144066	2496144	5200300	
15	120	680	3060	11628	38760	116280	319770	817190	1961256	4457400	9657700	
16	136	816	3876	15504	54264	170544	490314	1307504	3268760	7726160	17383860	
17	153	969	4845	20349	74613	245157	735471	2042975	5311735	13037895	30421755	
18	171	1140	5985	26334	100947	346104	1081575	3124550	8436285	21474180	51895935	
19	190	1330	7315	33649	134596	480700	1562275	4686825	13123110	34597290	86493225	
20	210	1540	8855	42504	177100	657800	2220075	6906900	20030010	54627300	141120525	
21	231	1771	10626	53130	230230	888030	3108105	10015005	30045015	84672315	225792840	
22	253	2024	12650	65780	296010	1184040	4292145	14307150	44352165	129024480	354817320	
23	276	2300	14950	80730	376740	1560780	5852925	20160075	64512240	193536720	548354040	
24	300	2600	17550	98280	475020	2035800	7888725	28048800	92561040	286097760	834451800	
25	325	2925	20475	118755	593775	2629575	10518300	38567100	131128140	417225900	1251677700	

FIGURE 10.11

Mersenne's massive "triangle."

Public domain.

This Table has truly exceptional and admirable properties; for besides concealing within itself the mysteries of combinations, as we have seen, it is known by those expert in the higher parts of mathematics also to hold the foremost secrets of the whole of the rest of the subject.[42]

So, given all of the famous European mathematicians who presented the triangle, with Pascal somewhere in the middle, as well as non-European sources to be discussed shortly, how did it get named after Pascal? The attribution is probably due to Pierre de Montmort's *Essay d' Analyse sur les Jeux de Hazard* (Essay on the Analysis of Games of Chance). It appeared late in this story, 1713, but devoted 72 pages to combinations and included the following in its introduction:[43]

---

42 Wilson, Robin and John J. Watkins, editors, *Combinatorics: Ancient & Modern*, Oxford University Press, Oxford, 2013, p. 178. Bernoulli, Jacob, *Ars conjectandi*, Basel, 1713, reprinted in *Die Werke von Jakob Bernoulli*, Vol. 3, pp. 107–286, Birkhäuser, Basel, 1975, p. 159 quoted here.

43 Wilson, Robin and John J. Watkins, editors, *Combinatorics: Ancient & Modern*, Oxford University Press, Oxford, 2013, pp. 176–177.

> Pascal has proceeded furthest, as is clear from his treatise *The Arithmetical Triangle*, which is full of observations and discoveries on the figurate numbers of which I believe him to be the originator, since he does not cite any other person.[44]

Long before its European debut in 1527, the triangle was known in China.

China

It was in 1261 that Yang Hui, in his book *Xiangjie jiu zhang suanfa* (Detailed Explanations of The Nine Chapters on Mathematical Methods), first put forth the triangle in China.[45] But he credited it to Jia Xian from around the year 1100.[46] Sadly, this earlier work is now lost. The purpose the triangle served for these early Chinese mathematicians was not combinatorial, but rather for root extraction.

The Chinese triangle in Figure 10.12 is from the 1303 work *Siyuan yujian* (Jade Mirror of Four Elements) by Zhu Shijie.[47] At this time it was used "in the context of interpolation techniques, solutions of polynomial equations, and the construction of finite arithmetical series," but not yet for combinatorics![48] Compare it with the triangle used as the frontispiece of Pascal's treatise (Figure 10.13). Notice how different they look!

In 1867, Chinese mathematician Li Shanlan (1811–1882) published the following equality, which he had deduced from the famous triangle.

$$\sum_{k=1}^{n} k^3 = \sum_{k=1}^{n} C(k+2,3) + 4\sum_{k=1}^{n} C(k+1,3) + \sum_{k=1}^{n} C(k,3).^{49}$$

This would've fit nicely into Chaper 3's discussion of sums of powers of consecutive integers, but neither Pascal's triangle nor combinations had been introduced by that point. Discrete mathematics is so rich in connections between the various topics that we could start almost anywhere and reach just about everything else.

David Eugene Smith, noting that the triangle was known in Baghdad at the end of the 11th century, just prior to its appearance in China, suggested that the Arab world might well have been their source for this knowledge.[50]

---

44 Wilson, Robin and John J. Watkins, editors, *Combinatorics: Ancient & Modern*, Oxford University Press, Oxford, 2013, pp. 176–177.

45 Wilson, Robin and John J. Watkins, editors, *Combinatorics: Ancient & Modern*, Oxford University Press, Oxford, 2013, p. 73.

46 See https://en.wikipedia.org/wiki/Yang_Hui and https://en.wikipedia.org/wiki/Jia_Xian.

47 Wilson, Robin and John J. Watkins, editors, *Combinatorics: Ancient & Modern*, Oxford University Press, Oxford, 2013, p. 73.

48 Wilson, Robin and John J. Watkins, editors, *Combinatorics: Ancient & Modern*, Oxford University Press, Oxford, 2013, p. 73.

49 Wilson, Robin and John J. Watkins, editors, *Combinatorics: Ancient & Modern*, Oxford University Press, Oxford, 2013, p. 78.

50 Eugene Smith, David, "Unsettled Questions Concerning the Mathematics of China," *The Scientific Monthly*, Vol. 33, September 1931, pp. 244–250.

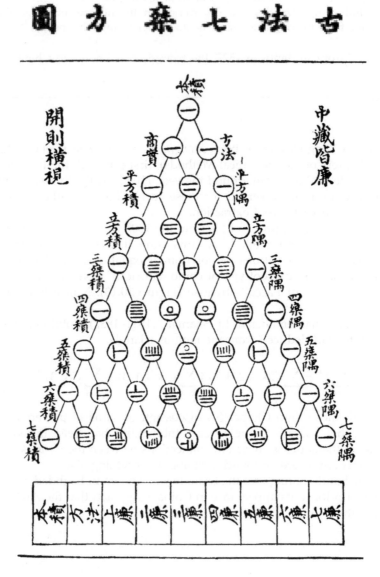

FIGURE 10.12

A Chinese version of "Pascal's" triangle.

Public domain.

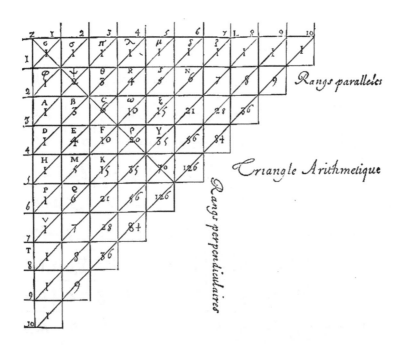

FIGURE 10.13
The frontispiece to Pascal's treatise.

Public domain.

The Arab World

Indeed, the triangle was known to al-Karajī (c. 953–c. 1029), a mathematician who had been active in Baghdad[51] and, as in Europe, there were later rediscoveries of the triangle and its applications. Among these was Ibn Munᶜim (?–1228), a mathematician who was active in Marrakech, Morocco. In the reproduction of Ibn Munᶜim's triangle given in Figure 10.14, note his use of Hindu-Arabic numerals that later made their way to the West, replacing the less convenient Roman numerals.

> "It is to be regretted that, fearing 'excesses and length', Ibn Munᶜim decided against expounding the 'extraordinary properties' (as he put it) that he knew how to obtain from a simple comparison of elements in the table."[52]

Perhaps most interesting, an investigation into factoring integers led al-Fārisī (d. 1320) to create a truncated version of Pascal's triangle.[53]

---

51 Wilson, Robin and John J. Watkins, editors, *Combinatorics: Ancient & Modern*, Oxford University Press, Oxford, 2013, p. 82 (picture) and pp. 89–91. Also see https://en.wikipedia.org/wiki/Al-Karaji.

52 Wilson, Robin and John J. Watkins, editors, *Combinatorics: Ancient & Modern*, Oxford University Press, Oxford, 2013, p. 96.

53 Wilson, Robin and John J. Watkins, editors, *Combinatorics: Ancient & Modern*, Oxford University Press, Oxford, 2013, p. 93.

FIGURE 10.14
Ibn Mun^c^im's triangle.

Public domain.

India

As we travel around the world, the triangle keeps aging and the process continues in India. There's actually a text from about 200 BCE titled *Chandaḥśāstra*, or sometimes *Piṅgala-sutras*, after its author, Piṅgala, that contains the seeds of the triangle. Portions of the text have been described as "very cryptic"[54] and "not fully comprehensible without a commentary"[55] – in other words, a lot like many of today's mathematics textbooks! In any case, a 10th-century commentary on this work by Halāyudha gives a description of the triangle.[56]

---

54 Wilson, Robin and John J. Watkins, editors, *Combinatorics: Ancient & Modern*, Oxford University Press, Oxford, 2013, p. 168.

55 https://en.wikipedia.org/wiki/Pingala.

56 Kak, Subhash, "The Golden Mean and the Physics of Aesthetics," in Yadav, Bhuri Singh and Man Mohan, editors, *Ancient Indian Leaps into Mathematics*, Birkhäuser, New York, NY, 2011, pp. 111–119, available online at https://arxiv.org/abs/physics/0411195.

What distinguishes it is its great age and that it contains, in addition to the oldest known description of the famous triangle, the first known appearances of the Fibonacci numbers and the binary number system![57] The presentation was made clearer by commentaries that appeared centuries later, specifically one by Varāhamihira in 505 CE and the previously mentioned commentary by Halayudha.[58]

At this point, you might expect a short section on how a recently discovered example of Paleolithic cave art dating to 38,000 BCE seems to depict a portion of the triangle. Well, maybe there'll be an update in the second edition. For now we turn to some much more recent history.

Pascal's Triangle Comes to America

Consider the page from an 1894 issue of the *American Mathematical Monthly* given in Figure 10.15.[59]

I reproduced this page to show how poor American mathematics was in 1894. Today, America is a world leader in mathematics and science. Indeed, English has become the *language* of science. Yet, in the 19th century she was centuries behind the Europeans. Pascal's triangle only came to America 200 plus years after Pascal's death!

America's rise from zero to hero in the world of mathematics can be attributed to the immigrants of the 20th century. Many of these were refugees from Europe during World War II. In addition to the persecution of Jews, homosexuals, and other groups, the Nazis also had a hatred of intellectuals. In a relatively short time Germany went from being among the very best nations for math and physics to a place the brightest minds wanted to escape. America also benefited from the talent that left various eastern European countries and the Soviet Union, as communism collapsed. In addition to all of these men and women, others came from various developing nations that, while not necessarily under the yoke of fascism or communism, failed to offer the opportunities that were available in America.

An easy way to see what a high percentage of America's best were born elsewhere is by looking at winner's lists for various mathematical prizes. Check the biographies for these individuals and you'll see that it really is immigrants who made the lion's share of the contribution toward America becoming a mathematical powerhouse. You'll see several examples over the course of this book, so I will only point out one recent example here.

Maryam Mirzakhani, a Stanford University professor, won mathematics' highest honor in 2014, the Fields Medal. Often referred to as the equivalent of a Nobel Prize, it's actually a greater honor, as it's only awarded every four years. She had earned her PhD at Harvard a decade earlier, but her undergraduate degree was from Sharif University of Technology, in her birth country – Iran. While Maryam's contributions to mathematics give her a sort of immortality, her life ended at the age of 40, when she died of cancer.

---

57 https://en.wikipedia.org/wiki/Pingala.
58 Wilson, Robin and John J. Watkins, editors, *Combinatorics: Ancient & Modern*, Oxford University Press, Oxford, 2013, p. 169.
59 Lilley, George, Pascal's Arithmetic Triangle, *American Mathematical Monthly*, Vol. 1, No. 12, December 1894, p. 426. Public domain.

PASCAL'S ARITHMETICAL TRIANGLE.

By GEORGE LILLEY, Ph. D., LL.D., Ex-President of Washington State Agricultural College and School of Science, Portland, Oregon.

I do not remember of having ever seen an account of this interesting device in any of our American text books, and, so far as I am able to ascertain, it has not been published in this country. The accompaning diagram explaines itself.

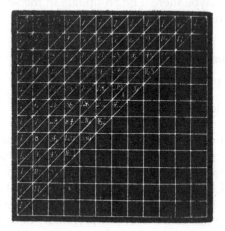

To find any number, in a triangle, take the sum of the number immediately above and the number immediately to the left of the required number, or take the sum of the numbers immediately above and to the left of the required number. Thus, the 7th number in the 4th row $= 28 + 56 = 84$, or

$$28 + 21 + 15 + 10 + 6 + 3 + 1 = 84.$$

The numbers on the diagonals are the coefficients of the expansion of a binomial.

The mth number in the nth row is given by the formula

$$\frac{|m+n-2}{|(m-1)|(n-1)} \ .$$

Thus, the 7th number in the 5th row

$$= \frac{|7+5-2}{|(7-1)|(5-1)} = \frac{|\,10}{|6\ |4} = \frac{1.2.3.4.5.6.7.8.9.10}{1.2.3.4.5.6.1.2.3.4} = 210., \text{ etc.}$$

FIGURE 10.15
The triangle arrives in America.

It's impossible for anyone carrying out research in mathematics in America to fail to notice the diversity among his or her colleagues. Unfortunately, not all politicians welcome the essential role such people play in keeping America competitive. In 2017, when Donald J. Trump attempted to ban travel to the United States by people from specific countries, he had some support from certain demographics, but was generally opposed by mathematicians and scientists. Apple, Google, Microsoft, Facebook, and 93 other companies filed a brief against the travel ban explaining that the ban "inflicts significant harm on American business."[60]

The mathematical capital of the world has occupied many different places over the millennia. Prior to arriving in America, it was located in Germany and France. In ancient

---

60 http://fortune.com/2017/02/06/donald-trump-travel-ban-apple-google-companies/.

times it existed in Greece and Egypt. Where will it be in 100 years? The answer depends on actions that are taken (or not taken) now.

And Now for Something Completely Different...

Let's consider a new problem. Suppose we place points on the perimeter of a circle and connect them all to each other in such a way as to maximize the number of divisions created within the circle. How many divisions can we get for each number of points? To answer this, let's look at a few examples (Figure 10.16) and hunt for a pattern.

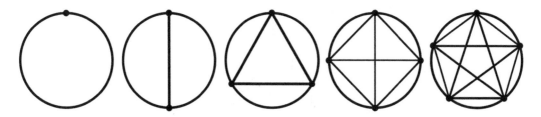

FIGURE 10.16
Dividing circles with straight lines.

Image created by Josh Gross.

As the figure shows, for just one point, we cannot draw any lines, so have 1 region, for two points, we may draw a line to get 2 regions, for three points, we get 4 regions, for four points, we get 8 regions, and for five points, we get 16 regions.

So we have the sequence 1, 2, 4, 8, 16, ... What will the next term be?

The sums of the rows of Pascal's triangles give us powers of 2, so it seems like we have a connection. Wouldn't it be lovely if the next value was 32? Well, it turns out that it's not! The next value is 31, as is demonstrated by Figure 10.17.

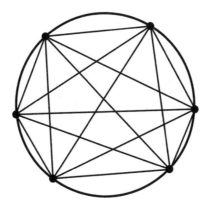

FIGURE 10.17
A Division into 31 regions.

Image created by Josh Gross.

This is the best we can do. It is not possible to get 32 regions, no matter how we position the six points about the circles perimeter.

What a strange sequence 1, 2, 4, 8 16, 31, ... It seems to have nothing to do with Pascal's triangle!

Wait a minute! If we sum only the elements appearing to the right of the diagonal line in Figure 10.18, we get this sequence! So, Pascal's triangle allows us to solve this problem after all. Six points give us 31 regions, seven points give us 57 regions, eight points give us 99 regions, and so on. It's possible to position the points to get fewer regions, in some cases, but we can never have more than the numbers given by the triangle in the manner described above.

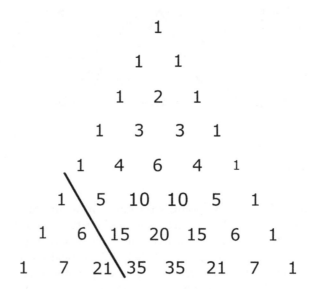

FIGURE 10.18
Pascal's triangle solves another problem!.

Image created by Josh Gross.

There are many more patterns, but we'll stop here, for

> Binomial coefficients satisfy literally thousands of identities, and for centuries their amazing properties have been continually explored. In fact, there are so many relations present that when someone finds a new identity, not many people get excited about it anymore, except the discoverer.[61]
>
> Donald E. Knuth

61 Knuth, Donald E., *The Art of Computer Programming*, Vol. 1, Fundamental Algorithms, third edition, Addison-Wesley, New York, 1997, p. 54.

Open Problems

1. **Haven't I Seen You Somewhere Before?**
   Some numbers appear repeatedly in Pascal's triangle. For example, 120 appears six times, although, as it turns out, this isn't anything special. David Singmaster used Fibonacci numbers to prove that there are infinitely many numbers that show up at least six times in the triangle. Letting F_n be the nth Fibonacci number and i any positive integer, defining $N = \begin{pmatrix} F_{2i+2}F_{2i+3} \\ F_{2i}F_{2i+3} \end{pmatrix}$, we have[62]

$$\begin{pmatrix} N \\ 1 \end{pmatrix} = \begin{pmatrix} N \\ N-1 \end{pmatrix} = \begin{pmatrix} F_{2i+2}F_{2i+3} \\ F_{2i}F_{2i+3} \end{pmatrix} = \begin{pmatrix} F_{2i+2}F_{2i+3}-1 \\ F_{2i}F_{2i+3}+1 \end{pmatrix} = \begin{pmatrix} F_{2i+2}F_{2i+3} \\ F_{2i+1}F_{2i+3} \end{pmatrix} = \begin{pmatrix} F_{2i+2}F_{2i+3}-1 \\ F_{2i}F_{2i+3}-2 \end{pmatrix}$$

   Singmaster also made the following conjecture.

   CONJECTURE. No binomial coefficient is repeated more than 10 times. (Perhaps the right number is 8 or 12?)[63]

   So far, the record for the number of appearances is 8, achieved by 3,003.[64] Singmaster checked all numbers up to 2^{48}.[65]

2. **Odd Perfect Numbers**
   As was indicated above, it remains open whether there are any odd perfect numbers.

3. **Catalan's Conjecture (1888)**
   Recall, if we let $\sigma(n)$ denote the sum of all of the positive divisors of n, then a number is perfect if and only if $\sigma(n) = 2n$. Note: we include n as a divisor of n here.
   Now, consider a new function $f(n) = \sigma(n) - n$. This simply sums the proper divisors of n.
   An unsolved problem posed by Eugène Charles Catalan asks whether the infinite sequence n, $f(n)$, $f(f(n))$, $f(f(f(n)))$, ... is eventually periodic for all starting values $n \in \mathbb{N}$.[66] Such sequences, whether they prove to all be periodic or not, are called aliquot sequences.[67] A few examples follow.
   Starting with $n = 10$, we have

   $f(10) = \sigma(10) - 10 = 18 - 10 = 8$
   $f\,f(10) = f(8) = \sigma(8) - 8 = 15 - 8 = 7$
   $f\,f\,f(10) = f(7) = \sigma(7) - 7 = 8 - 7 = 1$
   $f\,f\,f\,f(10) = f(1) = \sigma(1) - 1 = 0.$

---

62 Chamberland, Marc, *Single Digits*, Princeton University Press, Princeton, NJ, 2015, pp. 196–197.
63 Singmaster, David, "Repeated Binomial Coefficients and Fibonacci Numbers." *Fibonacci Quarterly*, Vol. 13, No. 4, 1975, pp. 295–298, p. 298 quoted from here.
64 Chamberland, Marc, *Single Digits*, Princeton University Press, Princeton, NJ, 2015, pp. 196–197.
65 Singmaster, David, "Repeated Binomial Coefficients and Fibonacci Numbers." *Fibonacci Quarterly*, Vol. 13, No. 4, 1975, pp. 295–298.
66 Catalan, Eugène Charles, "Propositions et questions diverses," *Bulletin de la Sociéte Mathematique de France*, Vol. 16, 1888, pp. 128–129. See item 3 – "Quelques théorèmes empiriques."
67 https://en.wikipedia.org/wiki/Aliquot_sequence.

The sum of the positive divisors of 0 is 0, because there are none. Hence, all further iterations yield 0. This is a period 1 solution.

Starting with $n = 220$ we have

$$f(220) = \sigma(220) - 220 = 504 - 220 = 284$$
$$f\,f(220) = f(284) = \sigma(284) - 284 = 504 - 284 = 220$$

Iterating again takes us back to 284, so we are stuck in a cycle of period 2.

For many numbers, we don't know what the long-term behavior is like. https://oeis.org/A131884 gives the following (incomplete) list of "numbers conjectured to have an infinite, aperiodic, aliquot sequence:"

276, 306, 396, 552, 564, 660, 696, 780, 828, 888, 966, 996, 1074, 1086, 1098, 1104, 1134, 1218, 1302, 1314, 1320, 1338, 1350, 1356, 1392, 1398, 1410, 1464, 1476, 1488, 1512, 1560, 1572, 1578, 1590, 1632, 1650, 1662, 1674, 1722, 1734, 1758, 1770, 1806, 1836.

Although Catalan's conjecture had not been proven, Leonard Eugene Dickson made an even stronger conjecture that every aliquot sequence either becomes periodic (with period greater than 1) or goes to a prime or a perfect number.[68] Of course, every prime is followed by 1 in the next iteration, and then 0 forever, and perfect numbers are followed by themselves. This stronger statement is called the Catalan–Dickson conjecture.

Exercise Set 10

A favorite theme used by Pascal to justify steps in his proofs that his peers objected to was that the heart intervenes to make this work clear.[69]

Robert M. Young

1. Use Pascal's triangle to quickly expand $(x + y)^5$.
2. Use Pascal's triangle to quickly expand $(x + y)^6$.
3. Use Pascal's triangle to find the 10th triangular number.
4. Prove $C(n, k) = C(n, n - k)$.
5. Use induction to prove $\sum_{i=0}^{n} \binom{n}{i} = 2^n$.
6. This chapter included the following:

   If you compare the shadings of Pascal's triangle, you'll notice the black triangles are solid for the moduli 2, 3, 5, and 7, but interrupted for moduli 4 and 6. Can you use these few examples to predict the presence or absence of this large triangle for other moduli?

   If you didn't make a conjecture at the time, do so now.

---

68 Dickson, Leonard Eugene, "Theorems and Tables on the Sums of Divisors of a Number," *Quarterly Journal of Pure and Applied Mathematics*, Vol. 44, 1913, pp. 264–296.
69 Young, Robert M., *Excursions in Calculus: An Interplay of the Continuous and the Discrete*, The Mathematical Association of America, 1992, p. 85, which cites Boyer, Carl, *The History of the Calculus and its Conceptual Development*, Dover, New York, 1959, p. 150.

7. Verify the claim for the even case of the sum of the shallow diagonals of Pascal's triangle being Fibonacci numbers.

8. Prove $\binom{n}{0}^2 + \binom{n}{1}^2 + \binom{n}{2}^2 + \cdots + \binom{n}{n}^2 = \binom{2n}{n}$.

9. Investigate the following sum for various values of n and k, satisfying $0 \le k \le n$ and form a conjecture as to a simple way to represent the sum in any case.

$$\binom{0}{k} + \binom{1}{k} + \binom{2}{k} + \cdots + \binom{n}{k}$$

10. Prove that the conjecture you formed for Exercise 9 is correct.

11. Investigate the following sum for various values of n and form a conjecture as to a simple way to represent it:

$$\binom{n}{1} + 2\binom{n}{2} + 3\binom{n}{3} + \cdots + n\binom{n}{n}.$$

12. Find a simple expression for

$$\binom{n}{0} - \binom{n}{1} + \binom{n}{2} - \binom{n}{3} + \cdots + (-1)^k\binom{n}{k}$$

For $n, k \in \mathbb{N}$, $k \le n$.

13. Prove that $\sum_{k=0}^{n} \binom{2k}{k}\binom{2n-2k}{n-k} = 4^n$.

An example of this, for $n = 3$, is $\binom{0}{0}\binom{6}{3} + \binom{2}{1}\binom{4}{2} + \binom{4}{2}\binom{2}{1} + \binom{6}{3}\binom{0}{0} = 4^3$.

14. Prove that for all $n \in \mathbb{N}$, $\binom{n+2}{3} - \binom{n}{3} = n^2$.

15. Use the formula given in Exercise 14 to evaluate $1^2 + 2^2 + 3^2 + \cdots + n^2$. This gives an alternate proof for a result you saw in Chapter 3.

16. Define $S_k = 1^k + 2^k + 3^k + \cdots + n^k$. Confirm that

$n = S_0$
$n^2 = 2S_1 - S_0$
$n^3 = 3S_2 - 3S_1 + S_0$
$n^4 = 4S_3 - 6S_2 + 4S_1 - S_0$.

17. How would n^5 and n^6 be expressed in the manner of Exercise 16?

18. Prove that the pattern you found in Exercise 17 continues forever.

19. Confirm that

$$\binom{1}{0} = \binom{2}{0}$$

$$\binom{1}{0} + \binom{2}{1} = \binom{3}{1}$$

$$\binom{1}{0} + \binom{2}{1} + \binom{3}{2} = \binom{4}{2}$$

$$\binom{1}{0} + \binom{2}{1} + \binom{3}{2} + \binom{4}{3} = \binom{5}{3}.$$

20. Prove that the pattern exhibited in Exercise 19 continues forever.

21. Can a perfect square be a perfect number? Give an example or prove that it is not possible.

22. As you'll see in Chapter 11, there are many triangles of numbers in mathematics. For now, a non-Pascal example is provided by Leibniz. It's known as his **Harmonic Triangle**.

```
1/1
1/2    1/2
1/3    1/6    1/3
1/4    1/12   1/12   1/4
1/5    1/20   1/30   1/20   1/5
1/6    1/30   1/60   1/60   1/30   1/6
```

Find some patterns in this triangle and give the next row.

23. Find some pattern in Pascal's triangle not mentioned in the text or exercise portions of this chapter. Explain it as clearly as you can, using diagrams, if appropriate.

References/Further Reading

> The reading of all good books is like conversation with the finest men of past centuries
> – René Descartes.

Apianus, Petrus, *Ein newe und wolgegründete underweisung aller Kauffmanns Rechnung in dreyen Büchern, mit schönen Regeln und fragstücken begriffen* (A Handbook of Commercial Arithmetic), Ingolstadt, 1527.

"The Arguments against," *OddPerfect.org*, available online at https://web.archive.org/web/20190727184025/http://oddperfect.org/against.html.

Bell, Eric Temple, *Mathematics, Queen and Servant of Science*, Copyright, 1951, British Edition 1952, reprinted 1954, 1958, 1961, Printed in Great Britain by Butler and Tanner Ltd., Frome and London.

Bernoulli, Jacob, *Ars Conjectandi*, Basel, 1713, reprinted, *Die Werke von Jakob Bernoulli*, Vol. 3, pp. 107–286, Birkhäuser, Basel, 1975.

Bernoulli, Jacob, *The Art of Conjecturing*, Together with *Letter to a Friend on Sets in Court Tennis*, translated with an introduction and notes by Edith Dudley Sylla, Johns Hopkins University Press, Baltimore, MD, 2006. This is an English translation of the work listed above.

Boyer, Carl, *The History of the Calculus and Its Conceptual Development*, Dover, New York, 1959.

Carr, Avery, "Odd Perfect Numbers: Do They Exist?" *AMS Blogs*, July 25, 2013, available online at https://blogs.ams.org/mathgradblog/2013/07/25/odd-perfect-numbers-exist/.

Catalan, Eugène Charles, "Propositions et questions diverses," *Bulletin de la Société Mathematique de France*, Vol. 16, 1888, pp. 128–129. See item 3 – "Quelques théorèmes empiriques".

Chamberland, Marc, *Single Digits*, Princeton University Press, Princeton, NJ, 2015.

Clagett, Marshall, *Nicole Oresme and the Medieval Geometry of Qualities and Motions*, University of Wisconsin Press, Madison, WI, 1968.

Cohen, Graeme L., "On the Largest Component of an Odd Perfect Number," *Journal of the Australian Mathematical Society*, Series A, Vol. 42, 1987, pp. 280–286.

Creyaufmueller, Wolfgang, "Aliquot Sequences," updated May 29, 2016, available online at www.aliquot.de/aliquote.htm#276.

Dickson, Leonard Eugene, "Theorems and Tables on the Sums of Divisors of a Number," *Quarterly Journal of Pure and Applied Mathematics*, Vol. 44, 1913, pp. 264–296.

Edwards, Anthony William Fairbank *Pascal's Arithmetical Triangle*, Charles Griffin & Company Limited, London, 1987.

Edwards, Anthony William Fairbank, *Pascal's Arithmetical Triangle: The Story of a Mathematical Idea*, The Johns Hopkins University Press, Baltimore, MD, 2002. This is a paperback, second edition, of the 1987 hardcover. The epilogue gives sources that were missed in the first edition, as well as new sources that only saw print after the first edition appeared. A new appendix, "Commentary on *Ars conjectandi*," is also included.

Gimbel, Steven and John H. Jaroma, "Sylvester: Ushering in the Modern Era of Research on Odd Perfect Numbers," *Integers: Electronic Journal of Combinatorial Number Theory*, Vol. 3, 2003, p. A16.

Ginsburg, Jekuthiel and David Eugene Smith, "Rabbi Ben Ezra on Permutations and Combinations," *The Mathematics Teacher*, Vol. 15, No. 6, October 1922, pp. 347–356.

Goto, Takeshi and Yasuo Ohno, "Odd Perfect Numbers Have a Prime Factor Exceeding 10^8," *Mathematics of Computation*, Vol. 77, No. 263, July 2008, pp. 1859–1868.

Hare, Kevin G., "New Techniques for Bounds on the Total Number of Prime Factors of an Odd Perfect Number," *Mathematics of Computation*, Vol. 76, No. 260, October 2007, pp. 2241–2248.

Heroor, Venugopal D., *The History of Mathematics and Mathematicians of India*, Vidya Bharathi, Karnataka, Bangalore, 2006.

Hilton, Peter and Jean Pedersen, "Looking into Pascal's Triangle: Combinatorics, Arithmetic, and Geometry," *Mathematics Magazine*, Vol. 60, No. 5, December 1987, pp. 305–316.

Hinz, Andreas M., Sandi Klavžar, Uroš Milutinović, and Ciril Petr, *The Tower of Hanoi – Myths and Maths*, Birkhäuser, Basel, Switzerland, 2013.

Iannucci, Douglas E., "The Second Largest Prime Divisor of an Odd Perfect Number Exceeds Ten Thousand," *Mathematics of Computation*, Vol. 68, No. 228, May 1999, pp. 1749–1760.

Iannucci, Douglas E., "The Third Largest Prime Divisor of an Odd Perfect Number Exceeds One Hundred," *Mathematics of Computation*, Vol. 69, No. 230, May 2000, pp. 867–879.

Ibn Ezra, Abraham, *Sefer Ha-Olam* (Book of the World), pp. 1147–1148.

Kak, Subhash, "The Golden Mean and the Physics of Aesthetics," in Yadav, Bhuri Singh and Man Mohan, editors, *Ancient Indian Leaps into Mathematics*, Birkhäuser, New York, NY, 2011, pp. 111–119, available online at https://arxiv.org/abs/physics/0411195.

Knuth, Donald E., *The Art of Computer Programming*, Vol. 1, Fundamental Algorithms, third edition, Addison-Wesley, New York, 1997.

Koshy, Thomas, *Triangular Arrays with Applications*, Oxford University Press, New York, 2011.

Lilley, George, "Pascal's Arithmetic Triangle," *American Mathematical Monthly*, Vol. 1, No. 12, December 1894, p. 426.

Lucas, Édouard, "Note sur la triangle arithmétique de Pascal et sur la serie de Lamé," *Nouvelle correspondance mathématique*, tome 2, 1876, pp. 70–75.

Muir, Jane, *Of Men and Numbers: The Story of the Great Mathematicians*, Dover, New York, 1965.

Myers, Andrew and Bjorn Carey, "Maryam Mirzakhani, Stanford Mathematician and Fields Medal Winner, Dies," *Stanford News*, July 15, 2017, available online at https://news.stanford.edu/2017/07/15/maryam-mirzakhani-stanford-mathematician-and-fields-medal-winner-dies/.

Nielsen, Pace P., "An Upper Bound for Odd Perfect Numbers," *Integers: Electronic Journal of Combinatorial Number Theory*, Vol. 3, 2003, #A14, available online at www.kurims.kyoto-u.ac.jp/EMIS/journals/INTEGERS/papers/d14/d14.pdf.

Nielsen, Pace P., "Odd Perfect Numbers Have at Least Nine Distinct Prime Factors," *Mathematics of Computation*, Vol. 76, No. 260, October 2007, pp. 2109–2126.

Ochem, Pascal and Michaël Rao, "Odd Perfect Numbers Are Greater than 10^{15000}," *Mathematics of Computation*, Vol. 81, No. 279, July 2012, pp. 1869–1877.

Pappas, Theoni, "Pascal's Triangle, the Fibonacci Sequence & Binomial formula," "Chinese Triangle," and "Probability and Pascal's Triangle," in *The Joy of Mathematics*, Wide World Publ./Tetra, San Carlos, CA, 1989, pp. 40–41, 88, and 184–186.

Pascal, Blaise, *Traité du Triangle Arithmétique* (Treatise on the Arithmetical Triangle), 1653, published by Guillaume Desprez, Paris, France, in 1665, available online at www.lib.cam.ac.uk/Rare Books/PascalTraite/.

Pascal, Blaise, *Œuvres de Blaise Pascal*, edited by Léon Brunschvieg and Pierre Boutroux, Hatchette, Paris, France, 1904–1914. Reprinted by Kraus, Nendeln/Lichtenstein, 1976–1978. This, of course, contains *Traité du Triangle Arithmétique*. See Vol. 3, pp. 433–598.

Pascal, Blaise, translated by Thomas M'Crie, W. F. Trotter and Richard Scofield, *Books of the Western World 33: The Provincial Letters, Pensées, Scientific Treatises by Blaise Pascal*, Encyclopaedia Britannica, 1955, 1963. This set of English translations includes Pascal's *Traité du Triangle Arithmétique*, as well as his correspondence with Fermat.

Peitgen, Heinz-Otto, Hartmut Jürgens, and Dietmar Saupe, *Chaos and Fractals: New Frontiers of Science*, pp. 80–86 are on Pascal's triangle. There's also a lot of other great material.

Posamentier, Alfred S. and Ingmar Lehmann, *The (Fabulous) Fibonacci Numbers*, Prometheus Books, Amherst, New York, 2007. This is where I first saw the sequence 1, 2, 4, 8 16, 31, ...

Rhodan, Maya, "These 97 Companies Filed a Brief against President Trump's Travel Ban," *Fortune*, February 6, 2017, available online at http://fortune.com/2017/02/06/donald-trump-travel-ban-apple-google-companies/.

Singmaster, David, "Repeated Binomial Coefficients and Fibonacci Numbers," *Fibonacci Quarterly*, Vol. 13, No. 4, 1975, pp. 295–298.

Smith, David Eugene, "Unsettled Questions Concerning the Mathematics of China," *The Scientific Monthly*, Vol. 33, September 1931, pp. 244–250.

Sylvester, James Joseph, "Sur l'impossibilitié de l'existence d'un nombre parfait qui ne contient pas au 5 diviseurs premiers distincts," *Comptes Rendus*, Vol. CVI, 1888, pp. 522–526.

Tartaglia, Niccolò Fontana, *General Trattato di Numeri, et Misure*, Venice, 1556.

Touchard, Jacques, "On Prime Numbers and Perfect Numbers," *Scripta Mathematica*, Vol. 19, 1953, pp. 35–39.

Uspenskii, Vladimir Andreevich, *Pascal's Triangle*, University of Chicago Press, 1975. This is a 35 page "book."

Vilenkin, N. Ya, *Combinatorics*, Academic Press, New York and London, 1971.

White, Christopher and Christopher Schwanner, *Some Sine and Cosine Identities Obtained from Pascal's Triangle*, Dorrance Publishing Co. Inc., 2012. Hundreds of years later, new results related to the triangle continue to appear! This "book" is 42 pages long.

Wilson, Robin and John J. Watkins, editors, *Combinatorics: Ancient & Modern*, Oxford University Press, Oxford, 2013.

Wolfram, Stephen, "Geometry of Binomial Coefficients," *The American Mathematical Monthly*, Vol. 91, No. 9, November 1984, pp. 566–571.

Young, Robert M., *Excursions in Calculus: An Interplay of the Continuous and the Discrete*, The Mathematical Association of America, 1992.

11

Stirling and Bell Numbers

James Stirling[1] (1692–1770) was mentioned in Chapter 8 in connection with Stirling's formula, an asymptotic approximation for $n!$, but he made many other contributions to mathematics. This chapter explores an interesting and widely applicable set of numbers known as **Stirling numbers of the second kind** that he introduced in 1730.[2] It was Niels Nielsen (1865–1931) who attached Stirling's name to the numbers, though.[3] I think only a narcissist would name something after himself.

Stirling numbers of the second kind give the number of ways a set of n objects can be partitioned into k disjoint nonempty subsets. They are notated $S_n^{(k)}$ or $S(n,k)$ or $S_2(n,k)$. Yet another popular notation is or $\left\{ {n \atop k} \right\}$. This one was put forth by Jovan Karamata, a Serbian mathematician, in 1935.[4] In this format, the numbers are sometimes read as "n heap k."[5] Be careful! The n and k are in different positions in the first two notations provided.

Example 1

Given the set $\{1, 2, 3\}$, we can split it into three disjoint nonempty subsets in only one way:

$$\{\{1\}, \{2\}, \{3\}\}$$

However, we can split it into two subsets in three ways:

$$\{\{1, 2\}, \{3\}\}$$
$$\{\{1, 3\}, \{2\}\}$$
$$\{\{1\}, \{2, 3\}\}$$

We can split it into one subset in, of course, only one way:

$$\{\{1, 2, 3\}\}$$

So we have

---

1 Ideally there'd be a picture of him here, but I couldn't find one. It might not exist. Some online images purporting to be Stirling are simply mislabeled.

2 Stirling, James, *Methodus differentialis: sive tractatus de summatione et interpolatione serierum infinitarum*, William Bowyer, London, 1730. Yes, there are also Stirling numbers of the first kind. They're addressed later in this chapter.

3 Nielsen, Niels, *Handbuch der Theorie der Gammafunktion*, Teubner, Leipzig, 1906. This is a 326-page book. See pp. 66–69.

4 Graham, Ronald L., Donald E. Knuth, and Oren Patashnik, *Concrete Mathematics*, Addison-Wesley, New York, 1994, p. 257.

5 Sloane, N. J. A., and Simon Plouffe, *The Encyclopedia of Integer Sequences*, Academic Press, San Diego, CA, 1995.

$$\left\{ \begin{array}{c} 3 \\ 3 \end{array} \right\} = 1, \qquad \left\{ \begin{array}{c} 3 \\ 2 \end{array} \right\} = 3, \qquad \left\{ \begin{array}{c} 3 \\ 1 \end{array} \right\} = 1.$$

Stirling numbers of the second kind may be calculated for larger values of n by using the formula

$$\left\{ \begin{array}{c} n \\ k \end{array} \right\} = \left\{ \begin{array}{c} n-1 \\ k-1 \end{array} \right\} + k \left\{ \begin{array}{c} n-1 \\ k \end{array} \right\}.$$

Stirling never gave this formula. In fact, he didn't even apply his numbers to combinatorics.[6] It turns out that they have many different applications and he found them when attacking a completely different problem, which will be addressed later in this chapter.

In any case, once finding this formula, it's natural to use it to form a triangle of values for the Stirling numbers of the second kind. The result is unlike Pascal's triangle in that it's asymmetrical. Stirling had to generate it in another manner. It begins with

```
                         1
                      1     1
                   1     3     1
                1     7     6     1
             1    15    25    10     1
          1    31    90    65    15     1
       1    63   301   350   140    21     1
    1   127   966  1701  1050   266    28     1
 1   255  3025  7770  6951   2646   462   36     1
```

Today, this is known as Stirling's triangle.

It shows increasing values for n as we head down and increasing values for k as we read across the rows. Thus, $S(6, 2) = 31$.

We now take a closer look at the first row in which the asymmetry is manifested. Given the set $\{1, 2, 3, 4\}$, we can split it into two disjoint nonempty subsets in seven ways:

$$\{\{1\}, \{2, 3, 4\}\}$$
$$\{\{2\}, \{1, 3, 4\}\}$$
$$\{\{3\}, \{1, 2, 4\}\}$$
$$\{\{4\}, \{1, 2, 3\}\}$$
$$\{\{1, 2\}, \{3, 4\}\}$$
$$\{\{1, 3\}, \{2, 4\}\}$$
$$\{\{1, 4\}, \{2, 3\}\}$$

---

6 Wilson, Robin, and John J. Watkins, editors, *Combinatorics: Ancient & Modern*, Oxford University Press, Oxford, 2013, p. 162.

Splitting the set into three disjoint nonempty subsets can be done in only six ways:

$$\{\{1\}, \{2\}, \{3, 4\}\}$$
$$\{\{1\}, \{3\}, \{2, 4\}\}$$
$$\{\{1\}, \{4\}, \{2, 3\}\}$$
$$\{\{2\}, \{3\}, \{1, 4\}\}$$
$$\{\{2\}, \{4\}, \{1, 3\}\}$$
$$\{\{3\}, \{4\}, \{1, 2\}\}$$

We define $S(0, 0) = 1$ and $S(n, 0) = 0$, for all $n > 0$.
   A non-recursive formula is given by

$$\left\{ \begin{matrix} n \\ k \end{matrix} \right\} = \frac{1}{k!} \sum_{i=0}^{k} (-1)^i \binom{k}{i} (k-i)^n$$

This shows a nice connection between Stirling numbers of the second kind and binomial coefficients ("Pascal" numbers).

Example 2
Calculate $\left\{ \begin{matrix} 6 \\ 3 \end{matrix} \right\}$ using the non-recursive formula.

Solution:

$$\left\{ \begin{matrix} 6 \\ 3 \end{matrix} \right\} = \frac{1}{3!} \sum_{i=0}^{3} (-1)^i \binom{3}{i} (3-i)^6$$

$$= \frac{1}{3!} \left[(-1)^0 \binom{3}{0} (3-0)^6 + (-1)^1 \binom{3}{1} (3-1)^6 + (-1)^2 \binom{3}{2} (3-2)^6 \right.$$

$$\left. + (-1)^3 \binom{3}{3} (3-3)^6 \right]$$

$$= \frac{1}{3!} [1 \cdot 1 \cdot 729 - 1 \cdot 3 \cdot 64 + 1 \cdot 3 \cdot 1 - 1 \cdot 1 \cdot 0]$$

$$= \frac{1}{3!} [729 - 192 + 3 - 0]$$

$$= \frac{1}{3!} [540]$$

$$= 90$$

Math by Imitation

Given that Stirling numbers of the second kind form a triangle and have a connection with Pascal's triangle, as the formula above shows, it makes sense to investigate what other properties these sets of numbers share. To do this, we simply try to imitate the results that were obtained in the chapter on Pascal's triangle.

Stirling's Triangle as a Fractal

One approach to Pascal's triangle that is easy to imitate here is shading the entries according to their values modulo various numbers. This is done in color figure 16 **(See color insert)**, for the modulus 2, for 1,000 rows. We get a result similar to the shadings of Pascal's triangle, but with a twist! color figures 17 through 21 **(See color insert)** show the results for other moduli. Note that this illustration starts with $\left\{ {n \atop 0} \right\} = 0$ for each row.

Factorial Powers

Stirling numbers of the second kind arise in many ways. Back in 1928, Jekuthiel Ginsburg noted:

> They have been discovered and rediscovered, cropping out again and again under various disguises, in almost every branch of mathematics.[7]

And much more was learned about these numbers and their applications in the 90 years that followed this remark.

It's now time to look at their first appearance. Back in 1730, James Stirling posed the question: How can powers of x be expressed as sums of what we might call "polynomial factorials"?[8] As is often the case, this is best illustrated with an example.

Example 3
Expressing x^4 as a sum of factorials.

$$\begin{aligned} x^4 = {} & 1x \\ & + 7x(x-1) \\ & + 6x(x-1)(x-2) \\ & + 1x(x-1)(x-2)(x-3) \end{aligned}$$

Expressions like $x(x-1)(x-2)(x-3)$ "count down" in the same way that $4! = 4(4-1)(4-2)(4-3)$ counts down, so it's a natural definition of factorial for this new context. It was not until 1818, however, that Peter Nicholson applied the name "factorial" to this idea.[9] They are now often called **falling factorials** or **factorial powers**, in contrast to **rising factorials** of the form $x(x+1)(x+2)(x+3)$. Notations used for

---

7 Ginsburg, Jekuthiel, "Note on Stirling's Numbers," *The American Mathematical Monthly*, Vol. 35, No. 2, February 1928, pp. 77–80, p. 77 quoted here.

8 Stirling, James, *Methodus differentialis: sive tractatus de summatione et interpolatione serierum infinitarum*, William Bowyer, London, 1730, p. 8. Marks the appearance of the Stirling numbers of the second kind in answer to this question. This is discussed in Wilson, Robin, and John J. Watkins, editors, *Combinatorics: Ancient & Modern*, Oxford University Press, Oxford, 2013, pp. 161–162.

9 Nicholson, Peter, *Essays on the Combinatorial Analysis*, Longman, Hurst, Rees, Orme, and Brown, 1818, Essay IV, p. 49.

falling factorials include $(x)_n$ and $x^{\underline{n}}$ and notations for rising factorials include $x^{(n)}$ and $x^{\overline{n}}$.

The coefficients of the new factorials that summed to x^4 were 1, 7, 6, and 1. This is a row of Stirling's triangle, and it is not a coincidence – the pattern continues forever.

The first few factorial powers are

$$1$$

$$x$$

$$x(x - 1)$$

$$x(x - 1)(x - 2)$$

$$x(x - 1)(x - 2)(x - 3)$$

$$\vdots$$

Using the notation $x^{\underline{0}}, x^{\underline{1}}, x^{\underline{2}}, x^{\underline{3}}, x^{\underline{4}}, \ldots$, the connection with Stirling numbers of the second kind is expressed as:

$$x^n = \sum_{k=0}^{n} \left\{ {n \atop k} \right\} x^{\underline{k}}.$$

We also have the following analogue of the binomial theorem:[10] $(x + y)^{\underline{n}} = \sum_{k=0}^{n} \binom{n}{k} x^{\underline{k}} y^{\underline{n-k}}$

Factorial powers are useful in the calculus of finite differences, a discrete form of the calculus you're already familiar with. In it, the difference operator Δ is defined like so:

$$\Delta f(x) = f(x + 1) - f(x).$$

We then get many results that mirror those of (continuous) calculus, such as the finite calculus power rule:

$$\Delta x^{\underline{n}} = n x^{\underline{n-1}} \text{ for } n = 1, 2, \ldots$$

In addition to giving the convenient name "factorial" to the expressions Stirling was making use of, Nicholson generalized Stirling's original problem. He considered any expression of the form $(x + n_1)(x + n_2)(x + n_3) \cdots (x + n_k)$, where the n_i are in arithmetic progression, to be a factorial. Nicholson gave many examples, in Essay III of his 1818 book, of expressing products of linear factors as sums of constant multiples of such factorials, but not the special case that Stirling examined.[11] To give you a better sense of Nicholson's work, his example III showed

10 Young, Robert M., *Excursions in Calculus: An Interplay of the Continuous and the Discrete*, The Mathematical Association of America, 1992, p. 72.

11 Wilson, Robin, and John J. Watkins, editors, *Combinatorics: Ancient & Modern*, Oxford University Press, Oxford, 2013, pp. 291–292.

$$(2x + 3)(x - 2)(x + 3)(3x + 7) = 6(x)(x + 1)(x + 2)(x + 3) - 7(x)(x + 1)(x + 2)$$
$$-37(x)(x + 1) - 102(x) - 126(1).$$

Sums of Products of Consecutive Integers

We looked at sums of consecutive integers (triangular numbers) in Chapter 3 and products of consecutive integers (factorials) in Chapter 8. We now consider sums of products of consecutive integers.

Example 4 – Sums of products of 1, 2, 3
Sum one at a time: $1 + 2 + 3 = 6$
Sum two at a time: $(1)(1) + (1)(2) + (1)(3) + (2)(2) + (2)(3) + (3)(3) = 25$
Sum three at a time: $(1)(1)(1) + (2)(2)(2) + (3)(3)(3) + (1)(1)(2) + (1)(1)(3) + (2)(2)(1)$
$+ (2)(2)(3) + (3)(3)(1) + (3)(3)(2) + (1)(2)(3) = 90$

So we get 6, 25, 90, which are all (surprise!) Stirling numbers of the second kind. Notice the pattern they follow, indicated by boldfacing, in Stirling's triangle:

$$
\begin{array}{ccccccccc}
 & & & & 1 & & & & \\
 & & & 1 & & 1 & & & \\
 & & 1 & & 3 & & 1 & & \\
 & 1 & & 7 & & \mathbf{6} & & 1 & \\
1 & & 15 & & \mathbf{25} & & 10 & & 1 \\
\end{array}
$$

1   1
1   1
1   3   1
1   7   **6**   1
1   15   **25**   10   1
1   31   **90**   65   15   1
1   63   301   350   140   21   1
1   127   966   1701   1050   266   28   1
1   255   3025   7770   6951   2646   462   36   1

We can also look at sums of *powers* of consecutive integers.

$$\sum_{i=1}^{n} i = 1 \cdot 0! \binom{n}{1} + 1 \cdot 1! \binom{n}{2}$$

$$\sum_{i=1}^{n} i^2 = 1 \cdot 0! \binom{n}{1} + 3 \cdot 1! \binom{n}{2} + 1 \cdot 2! \binom{n}{3}$$

$$\sum_{i=1}^{n} i^3 = 1 \cdot 0! \binom{n}{1} + 7 \cdot 1! \binom{n}{2} + 6 \cdot 2! \binom{n}{3} + 1 \cdot 3! \binom{n}{4}$$

$$\sum_{i=1}^{n} i^4 = 1 \cdot 0! \binom{n}{1} + 15 \cdot 1! \binom{n}{2} + 25 \cdot 2! \binom{n}{3} + 10 \cdot 3! \binom{n}{4} + 1 \cdot 4! \binom{n}{5}$$

I haven't been able to determine who was first to obtain this result, despite the fact that the formula is often stated in books and papers and has even been referred to as a "classical result."[12]

Completing a Set

Sometimes collectibles must be bought blindly. There may be a total of n items in the set, but they are packaged and sold individually in such a way that the buyer can't tell at the time of purchase if he or she is getting one that is already owned or not. An example is prizes found inside cereal boxes.

Suppose there are n distinct items in a complete set. If the items are not numbered (or if we choose to ignore any numbers that may be present on them), we can call the first to be opened 1. If the second prize that is opened is the same as the first, it is called 1 again, but if it is different, we call it 2. We continue in this manner, assigning the next integer not yet used to any prize that is opened and not found to be a duplicate. An item that is a duplicate is assigned the same number as when it was first encountered.

Trying to complete a set could be frustrating. Even in the easy case of $n = 3$, we might find our sequence of purchases is assigned values like so: 1222112121121123. This indicates that the first 2 purchases were distinct items, but then the next 13 purchases were all one or the other of these. Finally, the 16th item purchased completed the set.

Sometimes a buyer is lucky. A set of three items could be completed in just three or four purchases. For three purchases, using the notation established, it can only happen in one way: 123. But for four purchases, the possibilities are: 1123, 1213, and 1223 (three ways). Of course, it could take more than four purchases to get all three items.

For five purchases, there are seven ways: 11123, 11213, 12113, 11223, 12123, 12213, and 12223.

For six purchases, there are 15 ways: 111123, 111213, 112113, 121113, 111223, 112123, 112213, 121123, 121213, 122113, 112223, 121223, 122123, 122213, and 122223.

Summarizing the results so far, we have:

1 way to complete the set with 3 purchases

3 ways to complete the set with 4 purchases

7 ways to complete the set with 5 purchases

15 ways to complete the set with 6 purchases

If the context in which you're seeing this problem has led you to conclude that it is not a coincidence that 1, 3, 7, and 15 are Stirling numbers of the second kind, you're

---

12 Treviño, Enrique, "A Short Proof of a Sum of Powers Formula," *American Mathematical Monthly*, Vol. 125, No. 7, August-September 2018, p. 659, available at http://campus.lakeforest.edu/trevino/SumOfConsecutiveMathBit. pdf. This paper gives a one-page combinatorial proof.

right! And it doesn't matter how many distinct prizes there are. The answer will still be a Stirling number of the second kind, no matter how large this number is. In general, $S(n, k)$ is the number of ways in which k distinct prizes can be collected in exactly n purchases.

Stirling Numbers of the First Kind

Now, to answer a rather obvious question you likely posed by this point.

Introducing Stirling numbers of the second kind before Stirling numbers of the first kind is really not that strange. That's actually the order in which Stirling presented them, although, as has already been noted, he was not egotistical enough to name either set of numbers after himself. The numbers of the first kind arose from a sort of inversion of the polynomial problem of expressing powers of x as "factorials," like so:[13]

$$\frac{1}{x^2} = \frac{1}{x(x+1)} + \frac{1}{x(x+1)(x+2)} + \frac{2}{x(x+1)(x+2)(x+3)} + \frac{6}{x(x+1)(x+2)(x+3)(x+4)} + \cdots$$

The numbers in the numerators, 1, 1, 2, 6, ..., are Stirling numbers of the first kind. The triangle that includes these numbers (running down the left-hand side) is shown below.

$$
\begin{array}{ccccccccc}
& & & & 1 & & & & \\
& & & 1 & & 1 & & & \\
& & 2 & & 3 & & 1 & & \\
& 6 & & 11 & & 6 & & 1 & \\
24 & & 50 & & 35 & & 10 & & 1
\end{array}
$$

(unsigned) Stirling numbers of the first kind

Entry n, k of this triangle is often denoted as $\begin{bmatrix} n \\ k \end{bmatrix}$. We have a simple rule for generating further entries: $\begin{bmatrix} n+1 \\ k \end{bmatrix} = \begin{bmatrix} n \\ k-1 \end{bmatrix} + n \begin{bmatrix} n \\ k \end{bmatrix}$. Like Stirling numbers of the second kind, these numbers arise in other problems. One of these resembles a problem previously encountered.

Back to the Table(s)

In Chapter 8, we saw that the number of ways n people can be seated at a round table, where rotations of a seating arrangement are considered the same is $(n - 1)!$. Stirling numbers of the first kind answer a related problem – in how many ways can n people be seated at k tables (with at least one person at each), if the tables are indistinguishable and rotations of the tables are considered the same? The answer is simply $\begin{bmatrix} n \\ k \end{bmatrix}$. This is

---

13 Stirling, James, *Methodus Differentialis: sive Tractatus de Summatione et Interpolatione Serierum Infinitarum*, G. Bowyer, London, 1730, p. 10.

a true generalization of the original problem, for when we let $k = 1$ (for the case of just one table), $\begin{bmatrix} n \\ 1 \end{bmatrix} = (n-1)!$, which is our previous answer.

There's also a nice connection between Stirling numbers of the first kind and the harmonic series:

$$\begin{bmatrix} n \\ 2 \end{bmatrix} = (n-1)!\left(1 + \frac{1}{2} + \frac{1}{3} + \frac{1}{4} + \cdots + \frac{1}{n-1}\right)$$

It's sometimes specified that the numbers given above are the *unsigned* Stirling numbers of the first kind, for there's another version in which many of the entries are negative. Obviously, the *signed* Stirling numbers of the first kind do not answer the same questions.

These numbers arise as coefficients of the polynomials produced by the falling factorials, like so:

$$x^{\underline{n}} = \begin{bmatrix} n \\ 0 \end{bmatrix} + \begin{bmatrix} n \\ 1 \end{bmatrix}x + \begin{bmatrix} n \\ 2 \end{bmatrix}x^2 + \cdots + \begin{bmatrix} n \\ k \end{bmatrix}x^n$$

For example, $x^{\underline{3}} = x(x-1)(x-2) = 2x - 3x^2 + 1x^3$. This gives us the row 2, -3, 1.

$$
\begin{array}{ccccccccc}
 & & & & 1 & & & & \\
 & & & -1 & & 1 & & & \\
 & & 2 & & -3 & & 1 & & \\
 & -6 & & 11 & & -6 & & 1 & \\
24 & & -50 & & 35 & & -10 & & 1
\end{array}
$$

Signed Stirling numbers of the first kind

More entries in this triangle may be generated by using the relation $\begin{bmatrix} n+1 \\ k \end{bmatrix} = \begin{bmatrix} n \\ k-1 \end{bmatrix} - n\begin{bmatrix} n \\ k \end{bmatrix}$.

Leibniz, Again

As in so many other areas, Leibniz was here first. His investigation led him to discover Stirling numbers sometime before 1700, which is 30 years before Stirling's publication of them.[14]

A New Perspective from Japan

As was mentioned before, Stirling's use of the numbers was not in combinatorics. The first recognition of their value in that field came from Japan, when Masanobu

---

14 Wilson, Robin, and John J. Watkins, editors, *Combinatorics: Ancient & Modern*, Oxford University Press, Oxford, 2013, p. 154 and p. 162.

Saka independently discovered the numbers while investigating the problem of splitting a set of n distinct objects into k nonempty disjoint subsets. This is how the numbers are typically introduced in texts today. Saka published his results in 1782 in *Sanpō-Gakkai* (The Sea of Learning on Mathematics). Saka also found the formula $\left\{ \begin{matrix} n+1 \\ k \end{matrix} \right\} = k \left\{ \begin{matrix} n \\ k \end{matrix} \right\} + \left\{ \begin{matrix} n \\ k-1 \end{matrix} \right\}$.[15]

Bell Numbers

In Pascal's triangle, the sums of the rows were always powers of two. Does Stirling's triangle give anything special for these sums? Recalling the definition of the Stirling numbers of the second kind, we see that $\sum_{k=0}^{n} S_n^{(k)}$ gives the number of ways a set with n elements can be partitioned into (any number of) disjoint nonempty subsets. These sums are known as **Bell numbers**, denoted by B_n. A table of the first few is provided.

n	B_n
1	1
2	2
3	5
4	15
5	52
6	203
7	877
8	4,140
9	21,147
10	115,975

The Bell numbers may be calculated directly without reference to Stirling numbers of the second kind by using the formula

$$B_n = \sum_{k=0}^{n-1} B_k \binom{n-1}{k}.$$

There are several more ways to find Bell numbers. Comtet's formula (1974) involves the ceiling function and e.[16]

$$B_n = \left\lceil \frac{1}{e} \sum_{k=1}^{2n} \frac{k^n}{k!} \right\rceil$$

15 Wilson, Robin, and John J. Watkins, editors, *Combinatorics: Ancient & Modern*, Oxford University Press, Oxford, 2013, p. 26.

16 http://mathworld.wolfram.com/BellNumber.html

Dobiński's formula is[17]

$$B_n = \frac{1}{e} \sum_{k \geq 0}^{\infty} \frac{k^n}{k!} \, (n \geq 0).$$

Yet another way to calculate these numbers is by using what has become known as **Bell's triangle**. To do so, start with a 1, just like in Pascal's triangle. Each new row then starts with the last number of the previous row and continues to the right by adding each previous number to the number above it.[18]

1				
1	**2**			
2	**3**	**5**		
5	**7**	**10**	**15**	
15	**20**	**27**	**37**	**52**
52...				

Start with 1
Start with 1, add 1 + 1 to get 2
Start with 2, add 2 + 1 = 3, 3 + 2 = 5
Start with 5, add 5 + 2 = 7, 7 + 3 = 10, 10 + 5 = 15
Start with 15, add 15 + 5 = 20, 20 + 7 = 27, 27 + 10 = 37, 37 + 15 = 52
Start with 52, ... etc...

The Bell numbers may then be read off either side of the triangle.

FIGURE 11.1
E. T. Bell (1883–1960).

Courtesy of the Archives, California Institute of Technology.

---

17 Wilf, Herbert S., *Generatingfunctionology*, Academic Press, San Diego, CA, 1990. (Second edition, 1994; Third edition, A K Peters/CRC Press, 2005). The Second edition may be downloaded for free from https://www.math.upenn.edu/~wilf/DownldGF.html. Page 21 of this second edition is cited here. Also see http://mathworld.wolfram.com/BellNumber.html

18 http://www.pballew.net/Bellno.html provided this clear explanation.

Bell numbers are named after Eric Temple Bell (Figure 11.1). In the world of mathematics, Bell is best known for his books. *Men of Mathematics* (1937) is still in print. I highly recommend it. Bell let his strong personality shine through the book, which makes it very entertaining. He's been criticized for being sexist and, given the title of his most popular book, it isn't hard to see why. But the book *was* written in 1937, and it did include some of the great women of mathematics. In addition to his mathematical books and papers, Bell wrote science fiction under the pseudonym John Taine.[19] Two of his book covers are reproduced below. Take a look and form an opinion as to whether these books are free of sexism or not (Figures 11.2 and 11.3).

FIGURE 11.2
A literary triumph!

Scanned from the collection of the author.

---

19 See http://www-groups.dcs.st-and.ac.uk/~history/Mathematicians/Bell.html for more.

FIGURE 11.3
Another literary triumph!

Scanned from the collection of the author.

I think they're horribly sexist – worse than his math books! On the cover of *The Greatest Adventure*, the man appears to be attempting to fight a giant dinosaur with a small knife, while the woman intelligently runs away. This makes men look stupid. And on the cover of *Seeds of Life*, an overconfident male is engaged in what is probably a dangerous chemical experiment, while the woman, looking appropriately nervous, faces the other direction to protect her eyes from a possible explosion. Again, Bell makes the woman look smarter. Maybe she remembered what happened to Solomon Lefschetz (a *male* chemist who turned to mathematics, after blowing his hands off in a laboratory).[20]

Bernoulli's Triangle

In Chapter 10, I mentioned how Jacob Bernoulli rediscovered Pascal's triangle. He also had a triangle of his own, which appeared in his *Ars Conjectandi* in 1713, although it

---

20 Krantz, Steven G., *Mathematical Apocrypha: Stories and Anecdotes of Mathematicians and the Mathematical*, Mathematical Association of America, 2002, p. 148.

wasn't named after him until 1987![21] It arose from Bernoulli's investigation of the problem of points, detailed in Chapter 12.

$$
\begin{array}{ccccccc}
 & & & 1 & & & \\
 & & 1 & & 2 & & \\
 & 1 & & 3 & & 4 & \\
1 & & 4 & & 7 & & 8 \\
1 & & 5 & & 11 & 15 & 16 \\
1 & 6 & 16 & 26 & 31 & 32 &
\end{array}
$$

Entry $B_{n,k}$ of this triangle is the sum of the first $k + 1$ entries of the nth row of Pascal's triangle. That is, $B_{n,k} = \sum_{i=0}^{k} \binom{n}{i}$.

It seems like every mathematician has a triangle, Pascal, Stirling, Bell, Bernoulli, ...

And Now for Something Completely Different?

If you are familiar with matrices, that's great. If not, for the purposes of this section it suffices to think of them as just rectangular grids of numbers. The **main diagonal** in the example below is indicated by fives.

$$
\begin{pmatrix}
5 & -1 & 20 \\
0 & 5 & 4 \\
-7 & 19 & 5
\end{pmatrix}
$$

A partial permutation matrix is a matrix that has all of its entries from the set $\{0,1\}$ and has no more than a single one in any row or column. The matrix need not be invertible, that is, it need not have a 1 in every row and column. An upper triangular matrix is one that does not have any nonzero entries below the main diagonal. So, combining these definitions we have **upper triangular partial permutation matrices**. Some examples follow.

1 by 1

$$(0)(1)$$

2 possibilities

2 by 2

$$
\begin{pmatrix} 0 & 0 \\ 0 & 0 \end{pmatrix}
\begin{pmatrix} 1 & 0 \\ 0 & 0 \end{pmatrix}
\begin{pmatrix} 0 & 1 \\ 0 & 0 \end{pmatrix}
\begin{pmatrix} 0 & 0 \\ 0 & 1 \end{pmatrix}
\begin{pmatrix} 1 & 0 \\ 0 & 1 \end{pmatrix}
$$

5 possibilities

---

21 See Edwards, A. W. F., "Pascal's Triangle – And Bernoulli's and Vieta's," *Mathematical Spectrum*, Vol. 20, No. 2, 1987–1988, pp. 33–37.

3 by 3

$$
\begin{pmatrix} 0 & 0 & 0 \\ 0 & 0 & 0 \\ 0 & 0 & 0 \end{pmatrix}
\begin{pmatrix} 1 & 0 & 0 \\ 0 & 0 & 0 \\ 0 & 0 & 0 \end{pmatrix}
\begin{pmatrix} 0 & 1 & 0 \\ 0 & 0 & 0 \\ 0 & 0 & 0 \end{pmatrix}
\begin{pmatrix} 0 & 0 & 1 \\ 0 & 0 & 0 \\ 0 & 0 & 0 \end{pmatrix}
\begin{pmatrix} 0 & 0 & 0 \\ 0 & 1 & 0 \\ 0 & 0 & 0 \end{pmatrix}
\begin{pmatrix} 0 & 0 & 0 \\ 0 & 0 & 1 \\ 0 & 0 & 0 \end{pmatrix}
$$

$$
\begin{pmatrix} 0 & 0 & 0 \\ 0 & 0 & 0 \\ 0 & 0 & 1 \end{pmatrix}
\begin{pmatrix} 1 & 0 & 0 \\ 0 & 1 & 0 \\ 0 & 0 & 0 \end{pmatrix}
\begin{pmatrix} 1 & 0 & 0 \\ 0 & 0 & 1 \\ 0 & 0 & 0 \end{pmatrix}
\begin{pmatrix} 1 & 0 & 0 \\ 0 & 0 & 0 \\ 0 & 0 & 1 \end{pmatrix}
\begin{pmatrix} 0 & 1 & 0 \\ 0 & 0 & 1 \\ 0 & 0 & 0 \end{pmatrix}
\begin{pmatrix} 0 & 1 & 0 \\ 0 & 0 & 0 \\ 0 & 0 & 1 \end{pmatrix}
$$

$$
\begin{pmatrix} 0 & 0 & 1 \\ 0 & 1 & 0 \\ 0 & 0 & 0 \end{pmatrix}
\begin{pmatrix} 0 & 0 & 0 \\ 0 & 1 & 0 \\ 0 & 0 & 1 \end{pmatrix}
\begin{pmatrix} 1 & 0 & 0 \\ 0 & 1 & 0 \\ 0 & 0 & 1 \end{pmatrix}
$$

15 possibilities

For this special type of matrix, the rank is the number of 1s that are present. Canadian mathematicians David Borwein, Stuart Rankin, and Lex Renner found that the number of such n by n matrices with rank r is $S(n + 1, n + 1 - r)$.[22] So, if we let P_n to denote the total number of n by n upper triangular partial permutation matrices, then $P_n = \sum_{r=0}^{n} S(n + 1, n + 1 - r)$. But this is just B_{n-1}. That is, $P_n = B_{n-1}$. Thus, Bell numbers arise in a context that seems quite different from the manner in which they were defined.

Let's put another wrinkle on the problem. Take a larger matrix, say 4 by 4.

$$
\begin{pmatrix} a & b & c & d \\ 0 & e & f & g \\ 0 & 0 & h & i \\ 0 & 0 & 0 & j \end{pmatrix}
$$

As before, we must have 0s below the main diagonal and no more than a single 1 in any row or column. But we now insist that the first diagonal above the main diagonal (entries b, f, and i) must consist solely of 0s. We denote this by $P(4, 1)$, indicating a 4 by 4 matrix with 1 extra row of 0s. Our original problem may be expressed in this notation as $P(4, 0)$, which has B_3 as the solution. If we insist on two diagonals of 0s above the main diagonal (b, f, i, c, and g all equal to 0), we denote the total by $P(4, 2)$. If we only allow 1s on the main diagonal we denote the total number of possibilities by $P(4, 3)$, as there are three diagonals of 0s above the main diagonal. A table of values for $P(n, k)$ is provided. It appeared in 2000 in my first published paper.[23]

---

22 Borwein, David, Stuart Rankin, and Lex Renner, "Enumeration of Injective Partial Transformations," *Discrete Mathematics*, Vol. 73, 1989, pp. 291–296.
23 Bauer, Craig, "Triangular Monoids and an Analog to the Derived Sequence of a Solvable Group," *International Journal of Algebra and Computation*, Vol. 10, No. 3, 2000, pp. 309–321.

		1	2	3	4	5	n 6	7	8	9
	0	2	5	15	52	203	877	4,140	21,147	115,975
	1		4	10	29	97	366	1,534	7,050	35,167
	2			8	20	58	189	693	2,823	12,622
k	3				16	40	116	378	1,357	5,389
	4					32	80	232	756	2,714
	5						64	160	464	1,512
	6							128	320	928
	7								256	640
	8									512

Further entries of **Bauer's Triangle** may be computed, but there is no known formula for predicting all of the values. The special case n less than or equal to $2k + 4$ has been solved, but the rest is an open problem. Notice the doubling that occurs on the main diagonal. It also holds true for the diagonal immediately above. On the diagonal above that we go from 15 to 29. I spent a lot of time double-checking that 29, trying to make it a 30, but it was very stubborn and would not change. The known formulas for predicting some of the values are not pretty (see below and recall from Chapter 4 that $|S|$ is one of the notations used to indicate the size of the set S). In addition to a more complete solution, a more elegant solution is desired!

For $n \leq 2k + 2$,

$$|P_{n,k}| = \sum_{r=0}^{n-k-1} 2^{n-2r} \left\{ \begin{matrix} n-k \\ n-k-r \end{matrix} \right\}$$

$$|P_{n+1,k+1}| = 2|P_{n,k}|$$

There's a simple proof by cases for the second statement. It relies on a pair of functions, ϕ_1 and ϕ_2, that map matrices to smaller matrices. A, B, and C can be single values, or square blocks of values of any dimension. This is also the case for the 0s. In other words, the proof scales up.

Case 1. If n is even, let ϕ_1: $\begin{pmatrix} A & 0 & B \\ 0 & e_1 & 0 \\ 0 & 0 & C \end{pmatrix} \rightarrow \begin{pmatrix} A & B \\ 0 & C \end{pmatrix}$.

Case 2. If n is odd, let ϕ_2: $\begin{pmatrix} A & 0 & 0 & B \\ 0 & e_1 & 0 & 0 \\ 0 & 0 & e_2 & 0 \\ 0 & 0 & 0 & C \end{pmatrix} \rightarrow \begin{pmatrix} A & 0 & B \\ 0 & e_2 & 0 \\ 0 & 0 & C \end{pmatrix}$.

Because both functions are 2:1, the domain space must have twice as many elements as the image space (range). Q.E.D.

For $n = 2k + 3$,

$$|P_{n,k}| = \sum_{r=0}^{n-k-2} 2^{n-2r} \left\{ \begin{matrix} n-k \\ n-k-r \end{matrix} \right\} + \sum_{r=0}^{k-1} 2^{2k-2r-1} \left\{ \begin{matrix} k+1 \\ k+1-r \end{matrix} \right\} + 1$$

For $n = 2k + 4$,

$$|P_{n,k}| = \sum_{r=0}^{k+2} 2^{n-2r} \left\{ \begin{matrix} k+4 \\ k+4-r \end{matrix} \right\} + 2 \sum_{r=0}^{k} 2^{n-4-2r} \left\{ \begin{matrix} k+2 \\ k+2-r \end{matrix} \right\} + 2 \sum_{r=0}^{k} 2^{n-4-2r} \left\{ \begin{matrix} k+1 \\ k+1-r \end{matrix} \right\}$$

$$+ 2 \sum_{r=0}^{k-1} 2^{n-6-2r}(k-r) \left\{ \begin{matrix} k+1 \\ k+1-r \end{matrix} \right\} + \sum_{r=0}^{k-2} (2^{n-6-2r} - 3 \cdot 2^{n-8-2r}) \left\{ \begin{matrix} k \\ k-r \end{matrix} \right\} + 1$$

This is not elegant. Mathematicians don't want to see equations like this. Another approach is clearly needed.

The special case where $k = 1$ (one row above the main diagonal must contain only 0s) leads to a surprise. We can split the total count into groups, by rank, to get **Bauer's Other Triangle**:

1	1	1								
2	1	2	1							
3	1	4	4	1						
4	1	7	13	7	1					
n **5**	1	12	37	35	11	1				
6	1	20	99	149	80	16	1			
7	1	33	255	581	480	161	22	1		
8	1	54	642	2,147	2,581	1,301	294	29	1	
9	1	88	1,593	7,664	12,951	9,247	3,087	498	37	1

decreasing rank →

The maximum possible rank for a matrix in a particular row is n. There is only one such matrix having this rank for each n. Hence, the column of 1s. The next column, representing the number of matrices of rank $n - 1$, is the sum of the first n Fibonacci numbers! Surprise! What do horny immortal rabbits (see Chapter 13) have to do with my triangle?

Triangular numbers also connect with the triangle above. The number of n by n matrices of rank 1 is $n + T_{n-2}$.

If we wish to generate more rows for the triangle above, there's a nice shortcut.

$$P(n,r) = P(n-1,r) + (n-r) \cdot P(n-1,r-1) + P(n-2,r-2)$$

A diagram makes this pattern clearer. The squares in Figure 11.4 represents three positions in the triangle. The values in these positions are to be multiplied by the numbers in the squares, and the result placed diectly below m. In this instance, m stands for multiplier, and its value varies. In the formula before the

FIGURE 11.4
A diagram for generating entries in the $k = 1$ triangle.

Image created by Craig P. Bauer.

diagram, m is expressed as $n - r$. With this simple rule, it's easy to generate a large number of rows. Some shadings of this $k = 1$ triangle, mod various values, are shown in color figures 22 through 25 **(See color insert)**. They're left-justified for a change of pace.

There are many more triangles associated with this combinatorics problem.

In fact, we have an infinite sequence of triangles, $k = 0$, $k = 1$, $k = 2$, ... corresponding to the various number of diagonals, above the main diagonal, that must consist solely of 0s, and they are all distinct. Comparing terms in a fixed location of the triangles always gives a decreasing (convergent) sequence. Joint work of Bauer and Joe Molinar (an undergraduate at the time) quickly led to the realization that the limit triangle as k approaches infinity is Pascal's Triangle.

Moving on to the next triangle in our infinite sequence we have

	0	1								
	1	1	1							
	2	1	2	1						
	3	1	3	3	1					
	4	1	5	8	5	1				
n	5	1	8	20	20	8	1			
	6	1	12	46	71	46	12	1		
	7	1	18	102	232	226	96	17	1	
	8	1	27	222	722	1,009	635	183	23	1
	9	1	40	475	2,177	4,238	3,752	1,585	323	30	1

decreasing rank $\rightarrow$

The $k = 2$ triangle.

Molinar conjectured a dot diagram (Figure 11.5) to generate more rows. In this instance, the multiplier is $m = n - r - 1$.

A proof eluded Bauer and Molinar, but soon another undergraduate Emily Sullivan (Figure 11.7), became interested in the problem. Sullivan conjectured a nicer dot diagram (Figure 11.6), where $m = n - r$.

So who was right, Molinar or Sullivan? Working together, Bauer and Sullivan were able to prove the second dot diagram always worked. And, using this

FIGURE 11.5
A conjectured diagram for generating entries in the $k = 2$ triangle.

Image created by Craig P. Bauer.

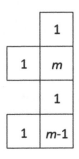

FIGURE 11.6
Another conjectured diagram for generating entries in the $k = 2$ triangle.

Image created by Craig P. Bauer.

fact, they were able to show that Molinar's conjecture was also true. So, both were right!

Looking at the diagrams for T_1 and T_2 (Figures 11.8 and 11.9), we are led to an obvious conjecture for T_3.

FIGURE 11.7
Emily Sullivan (1986–).

Courtesy of Emily Sullivan. In her senior year, Sullivan won two awards, one for being the best senior mathematics major and one for being the best female athlete. Today, she's an actuary. See https://twitter.com/Student_Actuary.

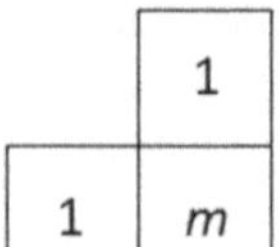

<table>
<tr><td></td><td>1</td></tr>
<tr><td>1</td><td>m</td></tr>
</table>

FIGURE 11.8
Dot diagram for T_1.

Image created by Craig P. Bauer.

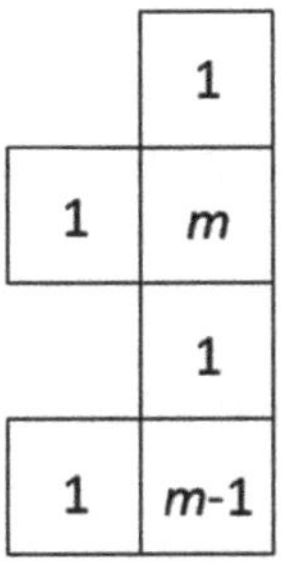

FIGURE 11.9
Dot diagram for T_2.

Image created by Craig P. Bauer.

Sadly, three copies of T_1 stacked up doesn't give us T_3. If anyone finds a formula for T_3, Bauer would be very happy to hear about it! The beginning of the T_3 triangle is provided below for anyone who wants to search for the pattern. The rightmost column gives the row sums.

0	1										1
1	1	1									2
2	1	2	1								4
3	1	3	3	1							8
4	1	4	6	4	1						16
5	1	6	13	13	6	1					40
6	1	9	28	40	28	9	1				116
7	1	13	58	117	117	58	13	1			378
8	1	18	113	321	451	321	113	18	1		1,357
9	1	25	217	849	1,638	1,613	815	206	24	1	5,389

n labels the rows; column 5 corresponds to n.

decreasing rank $\rightarrow$

The $k = 3$ triangle.

For the truly ambitious, T_4 is also given, along with the row sums.

0	1										1
1	1	1									2
2	1	2	1								4
3	1	3	3	1							8
4	1	4	6	4	1						16
5	1	5	10	10	5	1					32
6	1	7	19	26	19	7	1				80
7	1	10	37	68	68	37	10	1			232
8	1	14	71	175	234	175	71	14	1		756
9	1	19	131	434	772	772	434	131	19	1	2,714

decreasing rank $\rightarrow$

The $k = 4$ triangle.

Exercise Set 11

> If "Number rules the universe" as Pythagoras asserted, Number is merely our delegate to the throne, for we rule Number.
>
> E. T. Bell[24]

1. Flip back to Stirling's triangle at the beginning of this chapter and calculate the next row.

2. In how many ways can a set of 5 distinct objects be divided into three nonempty disjoint subsets?

3. In how many ways can a set of 6 distinct objects be divided into four nonempty disjoint subsets?

4. In how many ways can a set of 7 distinct objects be divided into four nonempty disjoint subsets?

5. In how many ways can a set of 8 distinct objects be divided into six nonempty disjoint subsets?

6. Calculate $\left\{ {5 \atop 3} \right\}$ using the non-recursive formula.

$$\left\{ {n \atop k} \right\} = \frac{1}{k!} \sum_{i=0}^{k} (-1)^i \binom{k}{i} (k-i)^n.$$

7. Calculate $\left\{ {7 \atop 4} \right\}$ using the non-recursive formula.

$$\left\{ {n \atop k} \right\} = \frac{1}{k!} \sum_{i=0}^{k} (-1)^i \binom{k}{i} (k-i)^n.$$

8. Prove that $\left\{ {n \atop 2} \right\} = 2^{n-1} - 1$, for $n > 0$.

9. Express x^5 as a sum of factorials.

10. Express x^6 as a sum of factorials.

11. How many ways are there to complete a set of 4 objects in 5 purchases?

12. How many ways are there to complete a set of 4 objects in 6 purchases?

13. How many ways are there to complete a set of 4 objects in 7 purchases?

14. Find a pattern in the triangle of Stirling numbers of the second kind not discussed in this chapter. You can search online for the pattern, but it might be more fun to explore on your own.

15. Look at the table of unsigned Stirling numbers of the first kind. Find a pattern for the row sums. What should the sum of the tenth row be?

16. Use the Bell's triangle provided in this chapter to calculate the next two Bell numbers. How could you check your answer?

---

24 Quoted in Eves, Howard W., *Mathematical Circles Revisited*, Prindle, Weber & Schmidt, Inc., Boston, MA, 1971, p. 14.

17. Find a connection between the second column of Bauer's Other Triangle, which begins 1, 2, 4, 7, 12, ..., and the Fibonacci numbers.

References/Further Reading

What we become depends on what we read after all of the professors have finished with us. The greatest university of all is a collection of books.
– Thomas Carlyle.

Ballew, Pat, "Bell Numbers," available online at www.pballew.net/Bellno.html.

Bauer, Craig, "Triangular Monoids and an Analog to the Derived Sequence of a Solvable Group," *International Journal of Algebra and Computation*, Vol. 10, No. 3, 2000, pp. 309–321.

Benjamin, Arthur T., Gregory O. Preston, and Jennifer J. Quinn, "A Stirling Encounter with Harmonic Numbers," *Mathematics Magazine*, Vol. 75, No. 2, 2002, pp. 95–103.

Borwein, David, Stewart Rankin, and Lex Renner, "Enumeration of Injective Partial Transformations," *Discrete Mathematics*, Vol. 73, No. 3, 1989, pp. 291–296.

Boyadzhiev, Khristo N., "Close Encounters with the Stirling Numbers of the Second Kind," *Mathematics Magazine*, Vol. 85, No. 4, October 2012, pp. 252–266.

Dickau, Robert M., "Bell Number Diagrams," *The Math Forum at NCTM*, available online at http://mathforum.org/advanced/robertd/bell.html.

Dickau, Robert M., "Stirling Numbers of the Second Kind," *The Math Forum at NCTM*, available online at http://mathforum.org/advanced/robertd/stirling2.html.

Edwards, Anthony William Fairbank "Pascal's Triangle – And Bernoulli's and Vieta's," *Mathematical Spectrum*, Vol. 20, No. 2, 1987–1988, pp. 33–37.

Eves, Howard W., *Mathematical Circles Revisited*, Prindle, Weber & Schmidt, Inc., Boston, MA, 1971.

Ginsburg, Jekuthiel, "Note on Stirling's Numbers," *The American Mathematical Monthly*, Vol. 35, No. 2, February 1928, pp. 77–80.

Glaisher, James Whitbread Lee., "Congruences Relating to the Sums of Product of the First *n* numbers and to Other Sums of Product," *Quarterly Journal of Mathematics*, Vol. 31, 1900, pp. 1–35.

Graham, Ronald L., Donald E. Knuth, and Oren Patashnik, *Concrete Mathematics*, Addison-Wesley, New York, 1994.

Krantz, Steven G., *Mathematical Apocrypha: Stories and Anecdotes of Mathematicians and the Mathematical*, Mathematical Association of America, 2002.

Mendelsohn, Nathan Saul, "Those Stirling Numbers Again," *Canadian Mathematical Bulletin*, Vol. 4, No. 2, 1961, pp. 149–151.

Nicholson, Peter, *Essays on the Combinatorial Analysis*, Longman, Hurst, Rees, Orme, and Brown, 1818, available online at https://books.google.com/books?id=PFdLAAAAMAAJ&printsec=frontcover&source=gbs_ge_summary_r&cad=0#v=onepage&q&f=false.

Nielsen, Niels, "Recherches sur les Polynomes et les Nombres de Stirling," *Annali di Matematica Pura ed Applicata*, Series 3, Vol. 10, No. 1, 1904, pp. 287–318.

Nielsen, Niels, *Handbuch der Theorie der Gammafunktion*, Teubner, Leipzig, 1906. This is a 326-page book that includes a chapter devoted to Stirling numbers. There are also sections on Stirling numbers in other chapters. Nielsen refers to "Die Stirlingschen Zahlen" here. This may be the first place they are named after Stirling. See pp. 66–69.

Reid, Constance, *The Search for E.T. Bell: Also Known as John Taine*, Cambridge University Press, Cambridge, UK, 1983. This is a biography.

Schlömilch, Oscar, "Recherches sur les coefficients des facultés analytiques," *Journal für die reine und angewandte Mathematik*, Vol. 44, 1852, pp. 344–355. Note: This journal was often referred to as *Crelle's Journal*, after an editor.

Setyobudhi, Yosua Jabby, *Stirling Numbers, Stirling Numbers for Complex Arguments*, LAP LAMBERT Academic Publishing, Saarbrücken, Germany, 2010.

Shallit, Jeffrey, "A Triangle for the Bell Numbers," in Hoggatt, Verner E. and Marjorie Bicknell Johnson, editors, *A Collection of Manuscripts Related to the Fibonacci Sequence*, 18th Anniversary Volume, Fibonacci Association, Santa Clara, CA, 1980, pp. 68–71, available online at www.fq.math.ca/Books/Collection/shallit.pdf.

Sloane, Neil J. A. and Simon Plouffe, *The Encyclopedia of Integer Sequences*, Academic Press, San Diego, CA, 1995.

Stirling, James, *Methodus differentialis: sive tractatus de summatione et interpolatione serierum infinitarum*, William Bowyer, London, 1730.

Treviño, Enrique, "A Short Proof of a Sum of Powers Formula," *American Mathematical Monthly*, Vol. 125, No. 7, August-September 2018, p. 659, available online at http://campus.lakeforest.edu/trevino/SumOfConsecutiveMathBit.pdf.

Weisstein, Eric W., "Bell Number," from *MathWorld*-A Wolfram Web Resource, available online at http://mathworld.wolfram.com/BellNumber.html.

Wilf, Herbert S., *Generatingfunctionology*, Academic Press, 1990. (Second edition, 1994; Third edition, A K Peters/CRC Press, 2005). The Second edition may be downloaded for free from www.math.upenn.edu/~wilf/DownldGF.html.

Wilson, Robin and John J. Watkins, editors, *Combinatorics: Ancient & Modern*, Oxford University Press, Oxford, 2013.

Young, Robert M., *Excursions in Calculus: An Interplay of the Continuous and the Discrete*, The Mathematical Association of America, 1992.

12

The Basics of Probability

...the theory of probabilities – that theory to which the most glorious objects of human research are indebted for the most glorious of illustrations

Edgar Allan Poe[1]

No matter how unlikely a thing is, if it happens, it happens.

Stanislaw Lem[2]

Disreputable Origins – Cardano

Mathematics can be applied to unravel the mysteries of nature, the very laws of the universe, but Gerolamo Cardano (1501–1576) applied it in another direction. His *Liber de Ludo Aleae* (The Book on Games of Chance), can be considered the first book on probability, but it's also a gambling manual for this was his motivation in developing the subject.[3] Cardano defined the odds in favor of an event as the ratio of the number of ways it can happen to the number of ways it can fail to happen. For example, the odds of rolling a six with a single die are 1:5, because there is only 1 way to get a 6 and 5 ways to *not* get a 6. Cardano calculated the odds of the various sums that can be obtained from rolling dice two and three at a time, as well as the odds connected with card games. He also defined the chance of an event, what mathematicians now call the probability, as the ratio of the number of ways it can happen to the *total* number of possible outcomes, both favorable and unfavorable.

The techniques of chapters 7 through 11 allow us to count the number of ways in which various events can occur, so to find the odds, or the probability, we just take the ratio of the relevant numbers. While odds are the usual way the chance of something happening is expressed in the context of gambling, mathematicians typically prefer probabilities. In forming either ratio, we need the various possible outcomes to be equiprobable. That is, no particular outcome can be any more likely than another. If this is not the case, we need to adjust for it.

Another result from combinatorics, the multiplication principle, also carries over to many probability problems. That is, if an event has probability p_1 of occurring and another event has probability p_2 of occurring, then the probability of both occurring

---

1 From his short story "Murders in the Rue Morgue."
2 From Stanislaw Lem's short story "Odds" (1978), translated by Michael Kandel.
3 *Liber de Ludo Aleae* remained unpublished during Cardano's lifetime. It first appeared in 1663 as part of a ten-volume edition of Cardano's writings. An English translation by S. H. Gould is included in Ore, Oystein, *Cardano, The Gambling Scholar*, Princeton University Press, Princeton, NJ, 1953. This translation also appeared by itself as Cardano, G., *The Book on Games of Chance*, translated by S. H. Gould, Holt, Rinehart and Winston, New York, 1961.

is often $p_1 \cdot p_2$. When this is the case, we say that the two events are **independent**. In some cases though, one event occurring makes another more or less likely to occur. The proper way to deal with probabilities for such **dependent** events is detailed in Chapter 21.

Disreputable Origins – Gombauld, Pascal, and Fermat

While Cardano's work was an important beginning, Blaise Pascal and Pierre de Fermat are generally credited for laying the foundation of the theory of probability. They did this in a series of letters exchanged in 1654.[4] Once again the work had its roots in a gambling problem.

The problem originated with a French nobleman, Antoine Gombauld, who liked to bet on dice games. For example, he bet on a six appearing, at least once, when a single die was rolled four times. This went well for him – he won more often than he lost. However, he moved on to betting on at least one pair of sixes when a pair of dice were rolled twenty-four times. This didn't go so well for him and he mentioned the fact to Blaise Pascal in a letter sent in 1654. Pascal, in turn, wrote to Fermat about the problem. The two mathematicians correctly calculated the probability that Gombauld would win at each game and found that the numbers matched his experiences.[5] The first game was in his favor, but the second was not. You are asked to calculate the exact probabilities in Exercises 1 and 2 at the end of this chapter.

Pascal and Fermat considered another gambling problem in their letters, the problem of points. This applies to a game played by two or more individuals that consist of multiple rounds, in each of which the winning player earns a point. The final winner is the first to reach some predetermined number of points. The problem arises when the game must be stopped before there's a winner. How should the stakes be divided if no player has reached the target number? If one player leads by a single point it doesn't seem fair to simply declare him the winner and let him take all of the money, for another player would have a chance of winning, if the game could be completed. Exactly how great is that other player's chance, for each possible number of points he is behind by? This problem dates back to at least 1494, when it saw print in Luca Pacioli's text *Summa de arithmetica, geometrica, proportioni et proportionalità* (Summary of arithmetic, geometry, proportions and proportionality).[6] His solution is presented in Exercise 3, where you're asked to determine whether it is fair or not. It's interesting to note that Pascal included this problem in his *Traite du Triangle Arithmetique*, written in the same year as his letters to Fermat. Pascal showed how the famous triangle could be used to solve the problem.[7] The Problem of Points led Pascal to develop a concept that came to be known as **expected value**. This is detailed later in this chapter.

---

4 Seven of these letters still exist and two more are known to have existed, but are now lost. See Wilson, Robin, and John J. Watkins, editors, *Combinatorics: Ancient & Modern*, Oxford University Press, Oxford, 2013, p. 147.

5 Burger, Edward B. and Michael Starbird, *The Heart of Mathematics*, third edition, Wiley, Hoboken, NJ, 2010, p. 587.

6 https://en.wikipedia.org/wiki/Problem_of_points

7 See the original, or Todhunter, Isaac, *A History of the Mathematical Theory of Probability from the Time of Pascal to that of Laplace*, Cambridge, 1865, reprinted by Chelsea in 1965, pp. 18–19.

There's a tendency among mathematicians to view everything through the lens of their subject. Pascal even attempted to quantify an emotional aspect of games of chance. He observed, "The excitement that a gambler feels when making a bet is equal to the amount he might win times the probability of winning it."[8]

Sometimes mathematics majors think their knowledge of probability gives them an advantage at casinos. Before going to one, they ought to look closer at the life of Gerolamo Cardano. This mathematician and gambler wrote the first book on probability, but apparently knowing more about the odds than anyone else wasn't enough to make him wealthy, or even to allow him to hold on to what he entered the games with. In 1533, he was forced to pawn his wife's jewelry and some of his furniture to pay gambling debts. Justifying the topic of his book, Cardano wrote:

> Even if gambling were altogether an evil, still, on account of the very large number of people who play, it would seem to be a natural evil. For that very reason it ought to be discussed by a medical doctor like one of the incurable diseases.[9]

Gambling is often more about psychology (being able to read your opponent and know when he's bluffing, for example) than mathematics. It has ruined many lives and can be a hard addiction to break for certain personality types. Gambling with others is bad enough, but challenging a casino is even worse. The only way to turn the odds in your favor at a casino is to cheat and if you do that you're taking a different kind of chance. Keep this in mind while learning about probability.

Example 1 – Full house
What is the probability of getting a full house in five card stud?

Solution: In Chapter 9, we saw that the number of ways to get a full house is 3,744. Since the total number of five card hands was calculated to be 2,598,960, the probability of a full house is $\frac{3,744}{2,598,960} \approx 0.00144$.

Probabilities are always between 0 and 1 inclusive. It should be clear that negative values are not possible, as both numerator and denominator represent the results of enumeration problems (the number of ways something can happen) and must therefore be nonnegative. Furthermore the numerator must be less than or equal to the denominator, as the denominator represents all possible outcomes. Hence, a probability cannot exceed 1.

If you wish to convert a probability to a "percent chance," multiply by 100. Thus, there is only about a 0.144% chance of getting a full house, much less than 1%.

Example 2 – Improving the odds?
If someone is thinking of a number between 1 and 9 inclusive, what is the probability of correctly guessing that number? What is the probability of the number being odd? Do you improve your chances of being correct by guessing an odd value?

Solution: Because there are 9 possible outcomes, but only 1 way to guess correctly, the probability of guessing correctly is 1/9. There are 5 odd numbers in the range

---

8 Rose, Nicholas J., *Mathematical Maxims and Minims*, Rome Press, Raleigh, NC, 1988.
9 from chapter 5 of Cardano's book, as quoted at http://www.columbia.edu/~hg17/ProbabilityWorkshop2012-2013/Excerpt%20from%20Hald.pdf, p. 38.

1 to 9, inclusive, so the probability that the number is odd is 5/9. However, we do not improve our chances by guessing an odd value, as each particular value is assumed to be equally probable and we may only guess one of them. If we are asked to guess the parity of the number (whether it is even or odd), we're better off guessing odd.

Example 3 – Guessing an integer

If someone is thinking of a nonnegative integer, what is the probability of guessing it correctly?

Solution: There is still only 1 way to guess correctly, but there are infinitely many possibilities, so the probability of guessing correctly is 0. Nevertheless, it is possible to guess correctly. Thus, an event with probability 0 can happen!

Example 4 – *"jeu de recontre"*

Of what possible value is knowing that the ratio of arrangements to derangements is close to e, as seen in Chapter 9? Quite a bit if you would like to play *jeu de recontre*.

This game existed in many variants under many names, but one version goes back to 18th-century France as *jeu de recontre*, and many mathematicians still use that name.[10] It's a two-player game where each player has an identical deck of cards. Within each deck all of the cards are distinct. The players can place their cards in any order they desire and the play consists of simultaneously turning their cards over one at a time. If at any point the cards turned over match, the player betting on a match wins. If the ends of the decks are reached without encountering a match, the other player wins.

Should you bet on or against there being a match? Well, the only way to avoid a match is for one deck to be a derangement of the other. Thus, the probability of no matches is D_n/A_n. This is approximately $1/e \approx 0.3679$. That is, there is about a 36.79% chance of no matches. Looked at from the other side, there's about a 63.21% of a match. Betting on a match is the smart thing to do!

This game attracted quite a bit of attention from mathematicians, including Nicolaus Bernoulli (1687–1759) and Pierre Rémond de Montmort (1678–1719).[11] It was actually an analysis of this game that led to the formula for derangements presented in Chapter 9. Because of tis connection, derangements are sometimes called **rencontres numbers**. De Montmort and Abraham de Moivre went on to apply derangements to other games played with cards and dice.[12] However, it was not until Euler took a crack at *jeu de recontre* that it was completely solved.[13]

---

10 Young, Robert M., *Excursions in Calculus: An Interplay of the Continuous and the Discrete*, The Mathematical Association of America, 1992, p. 32.

11 See Todhunter, Isaac, *A History of the Mathematical Theory of Probability from the Time of Pascal to that of Laplace*, Cambridge, 1865, reprinted by Chelsea in 1965, p. 91. These two men called the game *Treize* or *Jeu du Treize*. This version used decks of 13 cards each, hence the name.

12 Wilson, Robin, and John J. Watkins, editors, *Combinatorics: Ancient & Modern*, Oxford University Press, Oxford, 2013, p. 147.

13 Euler, Leonhard, "Calcul de la Probabilité dans le Jeu de Recontre," *Histoire de l'Académie Royale des Sciences et Belles Lettres*, 1753, pp. 255–270, reprinted in *Opera Omnia*, series prima, Vol. 7, starting on p. 11.

Euler initially proved it by applying, among other ideas, the hockey stick pattern of Pascal's triangle seen in Chapter 10.[14] However, 28 years later he returned to the problem, and found an easier proof.[15]

The Monty Hall Problem

As in combinatorics, there are problems in probability that seem to have several reasonable ways to get a solution, yet they all yield distinct values! Obviously, one or more must be wrong. The "Monty Hall Problem," which has garnered a great deal of media attention, provides a nice example. The problem was inspired by a TV game show, *Let's Make a Deal*, in which a contestant would get to pick one of three curtains and win whatever prize was hidden behind it. Typically, one curtain hid a car, another hid a goat, and the third had nothing behind it. If all this portion of the show consisted of was picking a curtain, there wouldn't be much mathematics to it. The probability of winning the car would simply be 1/3. What made it interesting was that after the contestant made his or her pick, the host, Monty Hall, would open a curtain that was not chosen and did not have a car behind it (because he knew which curtain hid the car, he could easily do this). Monty would then ask the contestant if he or she would like to switch to the other unopened curtain. Sometimes he would even offer a few hundred dollars to the person as an incentive to switch.

Should the contestant switch? Could this actually improve the chances of winning the car? This problem was posed to Marilyn vos Savant (Figure 12.1), author of the "Ask Marilyn" column in *Parade Magazine*. Marilyn was listed in the *Guinness Book of World Records* as having the highest recorded IQ, so it would seem that she could handle this basic mathematical question. In her September 9, 1990 column, she responded, "Yes, you should switch." She explained why she believed that switching increased the probability of winning the car from 1/3 to 2/3.

This caused a minor uproar which included some professors writing letters claiming she was wrong. To make matters worse, some of the letters were rude. One that Marilyn printed in the December 2, 1990 edition of her column came from a professor with a PhD in mechanical engineering. It read:

> You blew it, and you blew it big! Since you seem to have difficulty grasping the basic principle at work here, I'll explain. After the host reveals a goat, you now have a one-in-two chance of being correct. Whether you change your selection or not, the odds are the same. There is enough mathematical illiteracy in this country, and we don't need the world's highest IQ propagating more. Shame!

Marilyn responded by giving a different explanation as to why her answer was correct, but the rude letters saying she was wrong continued to flow in. One from a PhD professor of mathematics read:

---

14 Hinz, Andreas M., Sandi Klavžar, Uroš Milutinović, and Ciril Petr, *The Tower of Hanoi – Myths and Maths*, Birkhäuser, Heidelberg, Germany, 2013, p. 16.

15 Hinz, Andreas M., Sandi Klavžar, Uroš Milutinović, and Ciril Petr, *The Tower of Hanoi – Myths and Maths*, Birkhäuser, Heidelberg, Germany, 2013, p. 17, which cites Euler, Leonhard, "Solutio quaestionis curiosae ex doctrina combinationum," *Mémoires de l'Académie Impériale des Sciences de St. Pétersbourg*, Vol. 3, 1811, pp. 57–64.

FIGURE 12.1
Marilyn vos Savant (1946–).

Public domain.

> You are utterly incorrect about the game show question, and I hope this controversy
> will call some public attention to the serious national crisis in mathematical education.
> If you can admit your error, you will have contributed constructively towards the
> solution of a deplorable situation. How many irate mathematicians are needed to get
> you to change your mind?

Marilyn shared this letter and others in her February 17, 1991 column. She might have
mentioned that when a book was published in Germany in 1931 with the title *Hundert
Autoren Gegen Einstein* (A Hundred Authors against Einstein), Einstein responded, "If
I were wrong, then one would have been enough!"[16] Instead, she made a constructive
suggestion – that people actually carry out the experiment.[17] See what percentage of the
time you win by sticking to your initial guess and what percentage of the time you win
by switching. While mathematical arguments can be confusing, even to many mathematicians (!), it is hard to ignore the evidence of well-conducted experiments. When
Marilyn's fourth column addressing the topic appeared on July 7, 1991, it contained
letters of a different sort – they now said that she was correct.

As the paragraphs above indicate, there are several arguments to show that switching
is a wise move and that it does indeed double the probability of winning the car. An
argument involving a tree diagram is presented in Chapter 21. For now, a pair of
arguments not requiring an illustration are given.

The probability that the correct curtain was chosen is 1/3, so the probability that the
pick is wrong must be 2/3. That is, the probability that the car is behind one of the other

---

16 Hawking, Stephen, *A Brief History of Time*, Bantam Books, New York, 1988, p. 178.
17 On a smaller scale and not for keeps! With the help of a friend who hides a small item under one of three cups,
    it's easy to carry out the experiment as many times as is necessary to convince yourself whether switching
    truly doubles your chances of winning or not.

two curtains is 2/3. When Monty Hall reveals that one of the other curtains does not conceal the car, the probability of 2/3 is no longer split over two options. The curtain that he offers the contestant an option to switch to now has probability 2/3 of concealing the car.

There are a couple of real-world concerns that need to be taken into consideration. First, we assume that Monty always offers an opportunity to switch and does not only do so when the contestant actually picks the right curtain initially. That is, he is not just trying to save the show money! Second, we assume that when Monty has a choice of which curtain to open (e.g., when the contestant has picked correctly), he selects one of them at random. The curtains are numbered, but he does not simply open the first curtain without a prize each time.

Another way to look at this is by taking the problem to an extreme. Suppose there are 1,000 curtains, with a car behind just one of them. Pick one of the curtains. Now imagine that the host opens 998 of the other curtains, revealing that none of them concealed a car. There are now only two curtains that have not yet been opened – the one you selected and one other. Should you switch? The probability that your original guess was correct is 1/1,000, so the probability that the car is behind one of the other curtains is 999/1,000. It is clearly not behind any curtain that was opened, so the probability that it's behind the curtain you now have an opportunity to switch to is 999/1,000.

It's also easy to see that switching doubles your chances of winning by running a computer simulation. Marilyn suggested doing the experiment by hand, but computers weren't in every home then, as they are now.

While few readers will find themselves in the exact situation described above, the problem of "to switch or not to switch" does arise in other contexts. For example, when taking a multiple choice test and encountering a question you don't know the answer to, how many of you have heard that it is better to stick with your first guess? What was the person making this claim basing it on? Actual studies reveal that the opposite is true.

> ... evidence from research on testing ... indicates that changes from wrong to right on multiple-choice tests outnumber changes from right to wrong by a ratio of more than 2:1.[18]

Probability Becomes Respectable

Probability eventually rose from the gutter and ended up playing a role in nearly everything. For example, the Belgian mathematician Adolphe Quételet (1796–1874) applied it to social, economic, and biological problems,[19] and with the development of

---

18 I saw this quote on p. 42 of Rosenhouse, Jason, *The Monty Hall Problem: The Remarkable Story of Math's Most Contentious Brain Teaser*, Oxford University Press, Oxford, 2009, but he was quoting Granberg, Donald and Thad A. Brown, "The Monty Hall Dilemma," *Personality and Social Psychology Bulletin*, Vol. 21, No. 7, July 1995, pp. 711–723. Granberg and Brown, in turn, cite Geiger, Marshall A., "Changing Multiple-Choice Answers: Do Students Accurately Perceive Their Performance?" *Journal of Experimental Education*, Vol. 59, No. 3, 1991, pp. 250–257 and Mathews, C. O., "Erroneous First Impressions on Objective Tests," *Journal of Educational Psychology*, Vol. 20, No. 4, 1929, pp. 280–286.

19 Paulos, John Allen, *Beyond Numeracy: Ruminations of a Numbers Man*, Vintage Books, A Division of Random House, Inc., New York, 1992, p. 273.

quantum mechanics, probability was found to be essential in physics. So, in the spirit of respectability, let's consider some probability problems that have nothing to do with gambling.

Monte Carlo Methods

This method is only named after the famous casino town. It can be applied to a wide range of problems. The idea did arise from a card game, although in a refreshing change of pace, not one in which money typically changes hands. The discoverer, Stanislaw Ulam (Figure 12.2), explained:

> The first thoughts and attempts I made to practice [the Monte Carlo Method] were suggested by a question which occurred to me in 1946 as I was convalescing from an illness and playing solitaires. The question was what are the chances that a Canfield solitaire laid out with 52 cards will come out successfully? After spending a lot of time trying to estimate them by pure combinatorial calculations, I wondered whether a more practical method than "abstract thinking" might not be to lay it out say one hundred times and simply observe and count the number of successful plays. This was already possible to envisage with the beginning of the new era of fast computers, and I immediately thought of problems of neutron diffusion and other questions of mathematical physics, and more generally how to change processes described by certain differential equations into an equivalent form interpretable as a succession of random operations. Later... [in 1946, I] described the idea to John von Neumann, and we began to plan actual calculations.[20]

The probability of any poker hand in five card stud and the probability of any roll with any number of dice can be determined in this way as well. Simply deal the cards repeatedly and roll the dice repeatedly and see what percentage of the results

FIGURE 12.2
Stanislaw Ulam (1909–1984).

This is Ulam's badge photo from his time at Los Alamos (public domain).

---

20 Eckhardt, Roger, "Stan Ulam, John von Neumann, and the Monte Carlo Method," *Los Alamos Science*, Special Issue, 15, 1987, pp. 131–137, p. 131 quoted from here.

are as desired. But what makes the Monte Carlo method powerful is that it can be applied to a much wider range of problems, including some that are not posed as probability questions.

For example, this method can be applied to find the area under a curve, whether we can figure out how to integrate the function or not.

Example 5 – Estimating area

Estimate the area under $\sin(x)$ between $x = 0$ and $x = \pi$ without integrating or using a Riemann sum.

Solution: A sketch of the curve and a bounding rectangle (from $x = 0$ to $x = \pi$ and from $y = 0$ to $y = 1$) will make the solution clearer (Figure 12.3).

Imagine that this sketch is placed on a dart board far away from you. It's so distant that you can often get the dart within the bounding rectangle, but you have absolutely no control over *where* within that rectangle it hits. If you throw until you get 1,000 hits within the rectangle, the number of hits under the curve, say h, can be used to estimate the area under the curve. Simply calculate $(h/1{,}000)\cdot$(area of rectangle). In this example, the area of the rectangle is π, so the area under the curve will be approximately $h\cdot\pi/1{,}000$.

This may seem like a tedious method, but we can easily program a computer to carry out a simulation of the dart-throwing process. Have the computer generate a pair of random numbers, the first between 0 and π, and the second between 0 and 1. Call these numbers x and y, for they represent the location (x, y) where the simulated dart hits our rectangle. If $y > \sin(x)$, then the dart has landed above the curve, and if $y \leq \sin(x)$, then the dart has landed on or under the curve. Placing this in a loop and keeping track of how many darts land on or under the curve, we can very rapidly determine a value to multiply the area of the rectangle by to estimate the area under the curve. If the random number generator you're using is decent (none are truly random on traditional computers) and you do enough iterations, the estimate should be excellent.

To find the area we are after in this particular example, integrating would be easier, but the function $y = \sin(x)$ was only used for illustrative purposes. The Monte Carlo method works just as well for any function, whether we know how to integrate it or not.

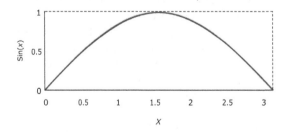

FIGURE 12.3
$\sin(x)$ from $x = 0$ to $x = \pi$.

Image created by Josh Gross.

Although she didn't ask for it to be run on a computer, the experiment that Marilyn suggested readers carry out to verify her solution to the Monty Hall problem was a Monte Carlo simulation.

Probability and Tom Clancy

Tom Clancy was an author well-known for his best-selling thrillers. In his novel *Debt of Honor*, he attempted to thrill his readers with the details of a probability calculation. The problem raises several important issues, so it's reproduced below.

> There were ten target points – missile silos, the intelligence data said, and it pleased the Colonel to be eliminating the hateful things, even though the price of that was the lives of other men. There were only three of them, and his bomber, like the others, carried only eight weapons. The total number of weapons carried for the mission was twenty-four, with two designated for each silo, and Zacharia's last four for the last target. Two bombs each. Every bomb had a 95 percent probability of hitting within four meters of the aim point, pretty good numbers really, except that this sort of mission had precisely no margin for error. Even the paper probability was less than half a percent chance of a double miss, but that number times ten targets meant a five percent chance that one would survive, and that could not be tolerated.[21]

The first issue is Clancy's use of the words "95 percent probability." This is not correct. The word "percent" should only be used with "chance" not "probability." Recall that a probability is always between 0 and 1, while a percent chance can range from 0 to 100. The mistake made here is a very common one. You may often here people speaking of "an 80 percent probability of rain," for example. To be technically correct, Clancy should have written "0.95 probability" or "95 percent chance."

But, jargon aside, is Clancy's mathematics correct? Near the end of the quoted passage we have "was less than half a percent chance of a double miss, but that number times ten targets meant a five percent chance that one would survive." For the moment, let's assume that the chance of a double miss is exactly half a percent (more on this soon!). Then, following Clancy's presentation, we multiply by 10 to find the percent chance (5% now) that one of the 10 targets is missed both times. This sounds reasonable, but let's take it to an extreme. Suppose there were 1,000 targets. In this case, half a percent chance times 1,000 targets gives a 500% chance that one is missed. This is ludicrous. We cannot have more than a 100% chance of anything! Also, with 1,000 targets it is possible that the bombers are extremely lucky and hit all of them. The percent chance must be a bit less than 100 here. Clancy made another common error here. The correct way to approach this problem is detailed below, but let's first look closely at the probability of a double miss for any particular silo.

Because the probability of each bomb hitting its designated target is independent of the probability of its partner hitting the target, we can calculate the probability of a double miss by multiplying the individual probabilities like so:

$$P(\text{Double Miss}) = (5/100)(5/100) = 1/400 = 0.0025$$

21 Clancy, Tom, *Debt of Honor*, G. P. Putnam & Sons, New York, 1994, p. 687.

Multiplying this probability by 100, we see that there's a 0.25% chance of a double miss. This is certainly less than half a percent, as Clancy claimed, so he's technically correct in this instance, although it seems likely that he would have written that it was a quarter percent chance, if he had known it.

Now, to find the probability that at least one of the ten silos survives, we cannot just multiply 0.0025 by 10, as was explained above. First we calculate the probability that exactly one silo survives. If one survives, with probability 0.0025, then the other nine must be destroyed, with probability $1 - 0.0025$ each. These are all independent events, so we have $(0.0025)(1 - 0.0025)^9$. But there are $_{10}C_1$ ways to select which one of the ten silos survives, so we must multiply the previous product by this factor as well. We end up with $(_{10}C_1)(0.0025)(1 - 0.0025)^9 \approx 0.0244430923$.

This is about 2.4%. So it appears that Clancy's final answer is wrong. Not surprising, given that the logic with which he approached the problem is flawed. But wait! Maybe what Clancy meant by "a five percent chance that one would survive" was that there was a 5% chance that *at least one* would survive. And, in any case, false premises can lead to true conclusions. Recall from Chapter 1 that FALSE → TRUE is TRUE. Perhaps Clancy got the final answer right by chance!

To find the probability that at least one survives we need to sum the probabilities for exactly one surviving, exactly two surviving, exactly three surviving, and so on all the way up to all ten surviving. In general, to calculate the probability of exactly n surviving, we have $_{10}C_n$ ways to choose which n survive and they each survive with probability 0.0025, while the remaining $10 - n$ silos are destroyed with probability $1 - 0.0025$ each. So, the probability of having exactly n survivors is $\binom{10}{n}(0.0025)^n(1 - 0.0025)^{10-n}$.

Because we want to sum these probabilities for all values of n between 1 and 10, inclusive, our answer is

$$\sum_{n=1}^{10}\binom{10}{n}(0.0025)^n(1 - 0.0025)^{10-n}.$$

$\approx 0.0244430923 + 0.0002756739734 + 0.000001842432571 + 0.000000008080844609$

$+ 0.0000000000243032921 + 0.00000000000000507587553$

$+ 0.00000000000000000726942432 + 0.0000000000000000000683216572$

$+ 0.00000000000000000000000380516052 + 0.000000000000000000000000000953674316$

As you can see, the probabilities that more than one target surviving are very small. We can now say with confidence that Clancy's final answer is incorrect.

Alternatively, we could calculate the probability above by taking one minus the complement. That is, one minus the probability that none survive or

$$1 - \binom{10}{0}(0.0025)^0(1 - 0.0025)^{10-0} = 1 - (0.9975)^{10} \approx 0.0247206168.$$

See how much easier this approach is? I first showed the longer method to make you truly appreciate the shortcut. Whenever you approach a problem in probability that seems difficult or tedious, you should consider calculating the complement instead. You may find that, as in the example above, it offers a much easier alternative.

Expected Value

As was mentioned earlier in this chapter, the Problem of Points – how to divide the stakes when a game ends prematurely – led Pascal to develop the idea of expected value. This is simply what the average result would be if the event were allowed to follow the course of chance again and again. A few examples will make this clear.

Example 6 – Rolling a die

If you roll a single die, the outcomes of 1, 2, 3, 4, 5, and 6 are all equally likely. They each have probability 1/6. So, the expected value is $E(X) = (1/6) \cdot (1) + (1/6) \cdot (2) + (1/6) \cdot (3) + (1/6) \cdot (4) + (1/6) \cdot (5) + (1/6) \cdot (6) = (1 + 2 + 3 + 4 + 5 + 6)/6 = 21/6 = 3.5$.

This is a result that will never occur! Yet, if you roll a die hundreds of times, you will find that the average of the results is very close to 3.5.

Note: physicists often notate the expected value of X as $\langle X \rangle$.

Example 7 – Roulette

In the American version of this game, there are 38 values that the ball can land on, 18 black, 18 red, and 2 green. If you correctly bet on black or red, you double your money. If you correctly bet on a specific value, you receive 35 times your bet back. Which bet has the better return, on average?

Solution:

Unlike the previous problem, the outcomes here (winning or losing) are not equally likely. If you bet on black or red, there are 18 winning values, so the probability of winning is 18/38 and the probability of losing is therefore 20/38. In the former you win $1 and in the latter you lose $1, so we have the expected value as a weighted average of these results, with the probabilities as the weights.

$E(X) = (Probability\ Win)(Amount\ Won) + (Probability\ Lose)(Amount\ Lost) = (18/38)(1) + (20/38)(-1) = -2/38 = -1/19 \approx -0.05263$. That is, you will lose about 5 cents, on average, every time you play. Notice that Amount Lost and the final answer are both negative numbers in this calculation, as in a debt.

If you bet on a specific number, there is only 1 way to win and 37 ways to lose, but the payout is, of course, much higher. For this kind of bet we have

$E(X) = (Probability\ Win)(Amount\ Won) + (Probability\ Lose)(Amount\ Lost) = (1/38)(35) + (37/38)(-1) = -2/38 = -1/19 \approx -0.05263$. Again, you will lose about 5 cents, on average, every time you play.

So, both bets are equally bad!

Example 8 – Betting on balls

> The lottery is society's way of taxing people who are bad at math.
> Ed Burger and Michael Starbird[22]

The Pennsylvania Lottery Pick 3 game consists of three containers, each containing ten ping pong balls bearing the single digits 0 through 9. The balls are bounced around

---

22 Burger, Edward B., and Michael Starbird, *The Heart of Mathematics*, third edition, Wiley, Hoboken, NJ, 2010, p. 835.

under the influence of blowing air and one ball is finally sucked up from each container to determine the winning three-digit number. The game pays 500 to 1 to players who correctly guess the result. It's easy to calculate the expected value of a $1 bet. Note that the amount won is only $499. It's necessary to subtract out the cost of the ticket from the winnings to find the net gain.

$$E(X) = (Probability\ Win)(Amount\ Won) + (Probability\ Lose)(Amount\ Lost)$$

$$E(X) = \left(\frac{1}{1,000}\right)(499) + \left(\frac{999}{1,000}\right)(-1)$$

$$E(X) = \frac{499}{1,000} - \frac{999}{1,000} = \frac{-500}{1,000} = -0.50.$$

This means that the expected outcome from a $1 bet in this game is a loss of 50 cents. This is impossible. You will either lose a dollar or win 499 dollars. While it doesn't realistically predict the outcome of a single event, it is a wonderful predictor of long-term behavior. If you play the lottery once a week for thirty years ($52 \times 30 = 1,560$ times in total), your outcome is likely close to the expected result of $(1,560)(-0.50) = -780$ dollars.

Nobody Expects...

Suppose you (or a computer) randomly picks numbers from the closed interval [0,1]. How many numbers would you expect need to be chosen, on average, to get a sum that exceeds 1? With calculus, it can be shown that the expected number is e. If you'd like to see the details, check out the reference.[23]

Is Belief in God Rational?

Pascal's philosophical work *Pensées* contains "Pascal's wager." It was an attempt to prove that believing in God is rational. It resembles an expected value problem.

> If God does not exist, one will lose nothing by believing in him, while if he does exist, one will lose everything by not believing.

There are several flaws in this argument. The existence of a God does not automatically imply any consequences for believing in him or not. Many religions claim that there are consequences, but others do not. Exactly what kind of God are we talking about here? If it's a God who places consequences on your personal beliefs concerning his existence, then which one is it? Some of these Gods are jealous and consider believing in *another*

---

23 Mackinnon, Nick, "Another Surprising Appearance of *e*," *Mathematical Gazette*, Vol. 74, No. 468, June 1990, pp. 167–169.

God to be as bad as, or worse than, believing in no God at all, so punishment could await you if you pick the wrong God!

Another flaw in Pascal's brief argument is his claim that "one will lose nothing by believing in him." Again, it depends on the religion. Maybe you lose one hour per week in which you are required to sit in church.[24] Perhaps you lose the pleasure of eating certain foods (pork, for example) or drinking certain beverages (alcohol for Muslims, coffee and tea for Mormons). It also depends on the person. If the religion promises eternal damnation for individuals who do not meet some very high standards, you might find yourself very stressed out over the possibility. This is a price that believers would pay that nonbelievers would not. In the next paragraph, I detail how Pascal's strong religious convictions arose and the actions he took as a consequence of his beliefs. Many would say that he paid a price for his beliefs.

Pascal's strong religious convictions arose before he came up with his wager. They may originate with a carriage accident that Pascal was lucky to survive. After his religious conversion he sold all of his possessions with the exception of a Bible and a few other books. He gave the money to the needy and then often had to beg for money himself to stay alive. He forced yet other hardships upon himself. E. T. Bell explained:

> Martyrlike, he wore an iron belt with points on the inside so that he was in constant discomfort. Whenever he felt happy, he would push the belt's points into his flesh to bring him to his senses.[25]

Practical Math

The 20th-century physicist Richard Feynman wasn't swayed by Pascal's argument, but he did, at an early age, try to apply the idea of an expected value to his life in a very practical way.

> When I was in high school, I had this notion that you could take the importance of the problem and multiply by your chance of solving it. You know how a technically minded kid is, he likes the idea of optimizing everything... anyway, if you can get the right combination of those factors, you don't spend your life getting nowhere with a profound problem, or solving lots of small problems that others could do just as well.[26]

Other (Job) Applications

There are many jobs available for those who can master the mathematics of probability:

> I was a recall coordinator. My job was to apply the formula. A new car built by my company leaves somewhere traveling at sixty miles per hour, the rear differential locks up, the car crashes and burns with everyone trapped inside. Now, should we

---

24 And that depends on you. Maybe you like sitting in church. There are many variables here!

25 Muir, Jane, *Of Men and Numbers: The Story of the Great Mathematicians*, Dover, New York, 1965, p. 98.

26 Feynman, Richard P., *The Pleasure of Finding Things Out*, Basic Books, Cambridge, MA, 2005, pp. 195–196.

initiate a recall? Take the number of vehicles in the field, a, multiply it by the probable rate of failure, b, then multiple the result by the average out of court settlement, c. a times b times c equals x. If x is less than the cost of the recall, we don't do one.

Tyler Durden[27]

Not all jobs that involve applying probabilities are so despicable, but like casinos, businesses must make sure that the odds are in their favor. Insurance companies provide an example. It doesn't matter if the insurance is for an automobile, a house, or simply an extended warranty on some product. The model provided in the example below can be adapted to any of these situations.

Example 9 – Life insurance
If you wish to insure your life for the next year, all the life insurance company has to know is the probability that you'll make it through the year (or not), and how much profit they would like to make, on average, on each policy. Let's say that you desire a $50,000 policy, have a 99% chance of surviving the year, and that the insurer desires a $100 profit. Then, from their perspective, $E(X)$ will be $100. And the calculation will take the form

$$100 = E(X) = (Probability\ You\ Die)(Premium - 50,000) + (Probability\ You\ Live)(Premium).$$

Plugging in the Die/Live probabilities, this becomes

$$100 = (0.01)(Premium - 50,000) + (0.99)(Premium).$$

Performing the steps needed to solve for the premium, we have

$$100 = (0.01)(Premium) - (0.01)(50,000) + (0.99)(Premium)$$

$$100 + (0.01)(50,000) = Premium$$

That is, $Premium = \$600$.

The insurer will sometimes lose this bet and your heirs will reap the benefit, but if they sell a large number of such policies, they will win nearly all of the bets and thereby come out ahead by an average of $100 per policy. E. T. Bell looked at the deal cynically.

> What you pay is based on the probability that you will die within the year. You bet that you will, the company bets that you won't. If you win, you lose – a modern version of the old saying that he who would save his life must lose it.[28]

Why would anyone who understands the mathematics take any of these losing bets? In some cases, you may not have a choice. A certain level of auto insurance may be required by law and a mortgage holder for your home may insist you have it insured. In other cases, not having insurance entails potential consequences scarier than the

---

27 From the film *Fight Club* (1999). See https://www.imdb.com/title/tt0137523/?ref_=fn_al_tt_1.
28 Bell, Eric Temple, *Mathematics, Queen and Servant of Science*, Copyright, 1951, British Edition 1952, reprinted 1954, 1958, 1961, Printed in Great Britain by Butler and Tanner Ltd., Frome and London, pp. 377-378.

expected financial loss of buying it, if it turns out you need it. You may buy life insurance so that you don't have to worry about how your family will keep up their standard of living, if you die sooner than you anticipate.

But when it comes to small stakes, as in extended warranties, you should just say no. You only have one life to gamble on (or not) with life insurance, but you will likely have hundreds of opportunities to gamble on extended warranties. You may occasionally wish you had purchased one, but the money you save when you do not will very likely more than cover such losses.

Not Worldwide!

As previous chapters showed, there's no shortage of mathematical results that were discovered independently by researchers all over the globe. However, not all math appears in this widespread manner. For example, while combinatorics developed independently in China, the Chinese did not take the next step to probability. Instead, they imported that knowledge from Europe[29] Nor are the Arabs known to have delved into probability.[30] It may be that the culture in Europe was more conducive to such development. Given the inspiration from games of chance that led to the study of probability in Europe, it seems likely that religious prohibitions played a role in preventing the subject from springing forth in some other parts of the world.

While probability has its origins in 17th-century Europe, it grew greatly in importance, like many other areas of mathematics, in the 20th century. It was only then that some of our modern notations developed. For example, using an E to denote expected value came about in 1901, when William Allen Whitworth used it in a script form.[31]

More to Come

Although this chapter is coming to an end, another important idea in probability (conditional probability) awaits you in Chapter 21.

Exercise Set 12

> Indeed "probability" has been called the most important concept in current science, "especially," as Russell remarked, "as nobody has the slightest idea what it means."
>
> E. T. Bell[32]

29 Wilson, Robin, and John J. Watkins, editors, *Combinatorics: Ancient & Modern*, Oxford University Press, Oxford, 2013, pp. 72–73.

30 Wilson, Robin, and John J. Watkins, editors, *Combinatorics: Ancient & Modern*, Oxford University Press, Oxford, 2013, p. 104.

31 Whitworth, William Allen, *Choice and Chance with One Thousand Exercises*, fifth edition, Deighton Bell, Cambridge, 1901.

32 Bell, Eric Temple, *Mathematics, Queen and Servant of Science*, Copyright, 1951, British Edition 1952, reprinted 1954, 1958, 1961, Printed in Great Britain by Butler and Tanner Ltd., Frome and London, p. 377.

1. If you bet on a six appearing, at least once, when a single die is rolled four times, what is the probability you will win?

2. If you bet on at least one pair of sixes when a pair of dice are rolled 24 times, what is the probability you will win?

3. Luca Pacioli's solution to the Problem of Points[33] was to divide the pot according to how many rounds each player had won. So if one player had four points and the other had six, 40% of the pot would go to the player in the weaker position and 60% to the player in the lead. The number of points needed to win did not factor into Pacioli's solution. Was Pacioli correct? Prove it or come up with an example where this is not the fairest distribution and show what would be fair.

4. How many times would you have to roll a pair of dice to have a greater than 50% chance of getting a sum of seven?

5. If someone flips a coin 15 times, what is the probability of getting heads at least once?

6. If a deck of cards is thoroughly shuffled and divided into three piles, face down, what's the probability, when the top card on each pile is turned over, that at least two will be of the same suit?

7. What percentage of million-digit numbers contains a 3?

8. If a die is tossed twice, what is the probability that the second result is larger than the first?

9. Suppose a trial is held in which each juror is "on the fence," having a 50% chance of voting for guilty and a 50% chance of voting for not guilty. What's the probability that the vote will be unanimous? There are about 154,000 jury trials in the United States each year.[34] If 10% of them fall into this category, and half of the defendants in these cases are innocent, how many of them should we expect to be convicted anyway?

10. [Theodore] Sturgeon's Law states "Ninety percent of everything is crap."[35] Assuming that this is true, if you read ten random novels, what is the probability that at least one of them is *not* crap?

11. The Weakest Link: suppose a chain is manufactured by a process that yields a 99% certainty of each link being able to hold a 200-pound load. If a 200-pound weight is attached to a length of chain consisting of 100 such links, what's the probability that it will hold? Ignore the weight of the chain!

12. If the probability of a typo at any given letter is 0.001, what's the probability that a 1,000-word essay (where the average word length is 4.5 letters) is typo free?

13. Suppose condoms are correctly used in attempts to prevent pregnancy. Given that they are 98% effective, after how many uses would the chance of pregnancy be greater than or equal to 50%?

---

33 Pacioli, Luca, *Summa de arithmetica, geometrica, proportioni et proportionalità*, Paganini, Venice, 1494.

34 http://iipdigital.usembassy.gov/st/english/publication/2009/07/20090706162635ebyessedo5.389911e-02.html

35 See Teresi, Dick, "People: Fuzz Busters," *OMNI*, July 1980, Vol. 2, No. 10, pp. 117-118, p. 118 specifically, for an account of how Sturgeon's law came about.

14. A study revealed the errors per medication in a hospital, per dose, amounted to 1.6%.[36] How many doses can be administered, given this error rate, until the probability of an error somewhere along the way exceeds 0.5?

15. In his book on the Monty Hall problem, Jason Rosenhouse wrote:

> I consulted a 2006 report from the U. S. Department of Transportation. It seems that the least reliable United States airlines misplace roughly twenty out of every one thousand bags, a rate of 2%.[37]

How many times would you have to check a bag with the least reliable airlines for the chance of at least one of them being misplaced to exceed 50%?

16. Consider the following lines from page 253 of *And the Band Played On* by Randy Shilts (St. Martin's Press, New York, 1987):

> By the end of December [1982], 1 in 333 single men over age fifteen in the Castro neighborhood already was diagnosed with AIDS. Factoring out heterosexual single men and the delay in reporting diagnoses, this meant that perhaps 1 in 100 gay men in this area already had AIDS. A person having twenty sexual contacts a year had 1 chance in 10 of making it with an AIDS sufferer.

Is it 1 chance in 10 or something else?

17. Another way to show that the rationals are a smaller infinity than the irrationals is to imagine rolling a ten-sided die to get the decimal digits of a "random" number. How likely is it that after some time the digits will begin to repeat, and then continue this pattern of repeats forever? This is what would have to happen for the number to be rational.

18. Consider a gambling game based on flipping a coin with a payoff when tails appears for the first time. If it happens on the first flip of the coin, the player wins $2. If it happens on the second flip, the player wins $4. If it happens on the third flip, the player wins $8. Continuing this pattern, if the first tails doesn't appear until the nth flip, then the player wins 2^n dollars. There is only one payoff. Once tails occurs, the player is paid the appropriate amount and the game ends. Calculate the expected value from the players' perspective. How much would you be willing to pay to play this game? If your result seems strange, know that this is referred to as the St. Petersburg Paradox. It was first described by Nicolas Bernoulli in 1713 in a letter to Pierre Raymond de Montmort.[38] Daniel Bernoulli resolved it in 1738 and published his results in the *Commentaries of the Imperial*

---

36 Panko, Raymond R., "Basic Error Rates," http://panko.shidler.hawaii.edu/HumanErr/Basic.htm. There are other good statistics dealing with errors provided here. The source for the example used above is Edmondson, Amy C., "Learning from Mistakes is Easier Said than Done: Group and Organizational Influences on the Detection and Correction of Human Error," *The Journal of Applied Behavioral Science*, Vol. 32, No. 1, March 1, 1996.

37 Rosenhouse, Jason, *The Monty Hall Problem: The Remarkable Story of Math's Most Contentious Brain Teaser*, Oxford University Press, Oxford, 2009, p. 71.

38 https://en.wikipedia.org/wiki/St._Petersburg_paradox

Academy of Science of Saint Petersburg, which led to the name this paradox received.[39]

19. When the Pennsylvania Lottery Pick 3 game began in 1971, numbers runners allowed their customers to bet on it with them, instead of the state, encouraging them by offering the better payoff of 600 to 1. What was the expected value of an illegal $1 bet?[40]

20. What's the probability of a bridge hand (13 cards) containing all of the spades? Note: this actually happened to someone.[41]

21. The book, *Sex: A Man's Guide*, states that prostate cancer "strikes one in nine men."[42] But then, on the next page, claims "If three or more relatives have gotten the disease, your chances are more than ten times greater."[43] What do you make of this?

22. The keys for 1988 Ford Escorts were made by cutting each of five positions on the blank key down to one of five different depths. How many possibilities does this allow? If the different keys were each made with equal probabilities, how many of the 380,000 Escorts manufactured that year could be opened with a particular key?[44]

23. The Games column in the May 1979 issue of *OMNI* included the following problem.

> Someone offers you the following bar bet: "You flip two coins, and I flip one. If you have more heads than I do, you win, otherwise, I win." Is this a good bet? Why, or why not?[45]

Give a mathematical argument in support of your answer.

24. Martin Hellman, famous for Diffie-Hellman key exchange in cryptography, is also a glider pilot with over 2,600 hours in the air. In his essay "Soaring, Cryptography and Nuclear Weapons" he described a glider maneuver that is 99.9% safe. He went on to claim that if a pilot does the maneuver 100 times "he stands roughly a 10% chance of being killed."[46] Assuming that death is the only possible result for the 0.1% of the time the maneuver isn't performed successfully, is his conclusion of a 10% chance of death correct?

---

39 Bernoulli, Daniel, "Specimen Theoriae Novae de Mensura Sortis," *Commentarii Academiae Scientiarum Imperialis Petropolitanae*, Tomus V (Commentaries of the Imperial Academy of Sciences of Petersburg, Vol. V), 1738, pp. 175–192, English translation by Louise Sommer, published as Bernoulli, Daniel, "Exposition of a New Theory on the Measurement of Risk," *Econometrica*, Journal of The Econometric Society, Vol. 22, No. 1, pp. 23–36. See https://en.wikipedia.org/wiki/St._Petersburg_paradox for more.

40 Krauss, Margaret, "Triple Six Fix: How Rigging the PA Lottery Inadvertently Contributed to its Success," June 19, 2015, http://wesa.fm/post/triple-six-fix-how-rigging-pa-lottery-inadvertently-contributed-its-success#stream/0

41 Kellogg, Chester E., "New Evidence (?) for 'Extra-Sensory Perception,'" *The Scientific Monthly*, Vol. 45, October 1937, pp. 331–343, specifically p. 336, and Browne, Owens Hand, "Comments and Criticisms: Causation. Chance, Determinism, and Freedom in Nature," *The Scientific Monthly*, Vol. 63, July 1946, p. 81.

42 Bechtel, Stefan, Laurence Roy Stains, and the editors of *Men's Health* books, *Sex: A Man's Guide*, Rodale Press, Emmaus, PA, 1996, p. 53.

43 Bechtel, Stefan, Laurence Roy Stains, and the editors of *Men's Health* books, *Sex: A Man's Guide*, Rodale Press, Emmaus, PA, 1996, p. 54.

44 Burger, Edward B., and Michael Starbird, *The Heart of Mathematics*, third edition, Wiley, Hoboken NJ, 2010, p. 642. The authors cite the April 1989 issue of the *Atlantic Monthly*.

45 Morris, Scot, "Games," *OMNI*, May 1979, Vol. 1, No. 8, pp. 144–145, p. 144 quoted here.

46 Hellman, Martin, "Soaring, Cryptography and Nuclear Weapons," October 21, 2008, available online at http://nuclearrisk.org/soaring_article.php.

25. The purpose of Hellman's essay, referenced in the exercise above, was to address the threat of nuclear war. Here's a paragraph from the essay that makes use of probability.

> On an annual basis, that makes relying on nuclear weapons a 99% safe maneuver. As with 99.9% safe maneuvers in soaring, that is not as safe as it sounds and is no cause for complacency. If we continue to rely on a strategy with a one percent failure rate per year, that adds up to about 10% in a decade and almost certain destruction within my grandchildren's lifetimes. Because the estimate was only accurate to an order of magnitude, the actual risk could be as much as three times greater or smaller. But even ⅓% per year adds up to roughly a 25% fatality rate for a child born today, and 3% per year would, with high probability, consign that child to an early, nuclear death.[47]

Is Hellman's math correct here? On a related note, there have been hundreds of what are codenamed "broken arrows." Some were never found.

26. In his report on the Challenger explosion, Richard Feynman included the following claim:

> Since there are two rockets on the Shuttle, these rocket failure rates must be doubled to get Shuttle failure rates from Solid Rocket Booster failure.[48]

Was Feynman correct?

27. What's the probability of scoring 70% or higher on a 10-question True/False test by guessing at the answer to each question?

28. What's the probability of scoring 70% or higher on a 10-question multiple choice test with four choices for each question by guessing at the answer to each question?

29. If a baseball player has a 30% of getting a hit at a given at bat, what's the probability that he gets a hit in exactly 4 out of his next 10 at bats?

30. If a baseball player has a 30% of getting a hit at a given at bat, what's the probability that he gets a hit in at least 4 out of his next 10 at bats?

31. **Benford's Law.** Write a program to generate a very long list of Fibonacci numbers. What percentage of these numbers begins with each of the digits one through nine?

32. **Benford's Law (again).** This is an exercise best explored by the whole class. Your task is to collect 100 numerical data points for something and see how many of these numbers begin with each of the digits one through nine. If your data is random, you might expect that the probability of beginning with any given digit is 1/9. After collecting your data, try to explain why it does or does not fit this expectation. What was the result for the data your classmates collected?

33. Sometimes when a poker night goes very late someone will remark, "We've been playing so long the same hands are starting to come around again." Suppose Joe is one of the players and the game is five card stud. How many hands should Joe expect to be dealt to make the probability that one of them has repeated exceed 0.5?

---

47 Hellman, Martin, "Soaring, Cryptography and Nuclear Weapons," October 21, 2008, available online at http://nuclearrisk.org/soaring_article.php.

48 Taken here from Feynman, Richard P., *The Pleasure of Finding Things Out*, Basic Books, Cambridge, MA, 2005, p. 154.

34. Suppose you have two well-shuffled decks of cards, with 52 cards in each deck, and you turn over the top card of each. What is the probability that at least one of these cards is the ace of spades?

35. If birthdays are uniformly distributed over 366 days (leap year is included), how many people must there be in a group to give a 100% chance that two share a birthday? What about for just a greater than 50% chance?

36. If there are 90 students in a class, what is the probability that at least 2 were born on the same day of the year (but not necessarily the same year)?

37. A coin is flipped repeatedly until the pattern Heads, Heads, Tails or Heads, Tails, Tails is seen. If HHT appears first, player 1 wins. If HTT appears first, player 2 wins. Who is more likely to win and by how much? Hint: one of the players really is more likely to win!

38. We saw in Example 3 that a probability of 0 doesn't mean that the event cannot occur. Does a probability of 1 mean that the event must occur? If not, give an example.

39. Suppose you are playing a variant of the Monty Hall game, in which there are 10 curtains, you pick 2, the host opens 4 and then offers you an opportunity to trade *both* of your curtains for *one* other? Should you do it?

40. If you roll five dice simultaneously, what is the probability that at least two show the same number?

41. Toss a fair coin until you get two heads in a row. What is the probability that there were n tosses in total?

42. Toss a fair coin until you get three heads in a row. What is the probability that there were n tosses in total?

43. Consider the following quote.

> I suppose we are dealing with the laws of statistical probability. A coin flipped in the air 100 times should turn 50 times heads and 50 times tails. The odds against the coin landing consecutively 50 times, on either heads or tails, are undoubtedly astronomical. It would probably be considered impossible.[49]

   a) What's the probability that a coin flipped 100 times lands heads exactly 50 times?

   b) What's the probability that somewhere in the 100 flips there's a string of 50 consecutive heads?

   c) What's the probability that somewhere on the 100 flips there's a string of 50 consecutive heads or 50 consecutive tails?

44. If a professor hands graded tests back randomly to a class of 10 students, what's the probability that at least one student is handed his or her own test?

45. If 50 single people living in separate units in an apartment complex all go out and get drunk and return to random apartments in the complex, what's the probability that at least one will return to his or her own apartment?

---

49 Lafferty, Lyndon E., *The Zodiac Killer Cover-Up, AKA The Silenced Badge*, Mandamus Publishing, Vallejo, CA, 2005, p. 227.

46. On June 7, 1989, a TV program aired with the title *Exploring Psychic Powers... Live*.[50] People who claimed to have various psychic powers were put to the test by James Randi, who had a long career as an illusionist, "The Amazing Randi," and a pair of professors.[51] Randi was familiar with tricks that fakers can use and helped to design tests that made such cheating impossible. A prize of $100,000 was awaited anyone who could beat the tests. The following paragraphs describe some of the tests. For each, calculate the probability that someone without any special powers could win simply by guessing. In instances where there's not enough information to give the probability, make some assumptions (be sure to state them!) and carry on.

   a) Can an astrologer determine someone's astrological sign by simply chatting with the person about topics other than birthdates? Joseph Meriwether claimed he could and was allowed to converse with 12 people having 12 different astrological signs. The $100,000 would be his if he could correctly match at least 10 people to the correct signs. He failed miserable, getting 0 correct.

   b) Can some people see auras? Barbara Martin claimed that she could and she was allowed to pick 10 people whose auras she claimed extended at least a foot over the tops of their heads. Her test involved screens numbered 1 through 10, behind *some* of which, some of these people stood. The idea was that she ought to be able to tell which screens hid people, because their auras would be visible over the top of the screens, which were less than a foot taller than the chosen people. To pass the test, she simply had to say which screens concealed people and which did not. She needed to make the correct assessment in every case, twice. That is, if she succeeded in the first attempt, a second trial would be conducted, in which she would also have to be 100% accurate. In the first trial, she claimed that every screen concealed a person, but only four actually did, so she missed the mark by quite a bit. There was no second trial.

   c) Dowsers claim to be able to locate water underground using only a Y-shaped twig or branch, or sometimes a metal rod or two. The idea is that the dowser can walk around and the twig or rod will shift to point down when he or she is over water. Drilling wells is not cheap, so if a dowser can guarantee that water is hit the first time, he could save his client a good deal of money. The dowser on this program, Forrest Bayes, was tested with 20 boxes. He was to indicate which of them contained 5-gallon water bottles. He claimed eight of them did. However, there was only water in 5 of the 20 boxes. It was not stated how accurate he needed to be to win the $100,000. Assume that 100% accuracy was required.

   d) In some films and TV programs, a psychic handles possessions of a killer or victim and get visions that lead to the person being located. This is called psychometry. Can anyone really do it? In the program being considered, Sharon McLaren-Straz was given the personal possessions of a dozen people – a watch and a keyring, in each case. She was to handle these objects

50 https://en.wikipedia.org/wiki/James_Randi#Exploring_Psychic_Powers..._Live_television_show You can watch the program at https://www.youtube.com/watch?v=KGsIQIVpbgY. Mathematician Arthur Benjamin is also in this program.

51 Dr. Stanley Krippner, parapsychologist, and Dr. Ray Hyman, a professor of psychology and a statistician.

and correctly pair the watches and keyrings. She needed to get at least 9 out of 12 right to win the $100,000, but she only correctly paired two sets.

e) Zener cards, (Figure 12.4), are well known and have been used since the 1930s to test for psychic abilities. They even made an appearance in the original *Ghostbusters* movie (1984).

FIGURE 12.4
The five kinds of Zener cards.

Image created by Mikhail Ryazanov. This file is licensed under the Creative Commons Attribution-Share Alike 3.0 Unported license.

The idea is that a person might be able to identify the images on the cards by only viewing the other sides, which are all the same, perhaps be reading the mind of a person viewing the design side. Because there are only five different images to guess from, and the same number of each in a deck, pure guessing would result, on average, in 50 correct identifications for a 250-card deck. Given such a deck, Valerie Swan was challenged to get at least 82 out of the 250 correct. This may not seem impressive, but the probability of exceeding chance to this degree is actually very small. Swan came nowhere near 82. She only got 50 right.

While the repeated failures did not surprise me at all, and I don't believe anyone possesses the abilities described above, I'm all for such tests. There are many results in math and physics that I wouldn't have believed that have, nevertheless, been shown to be true. If our gut feelings or intuition, whatever you want to call it, were always correct, there'd be no need for proof. There are many people out there with rare abilities – human calculators, contortionists, great athletes, etc. So, it's not inconceivable to me that other talents exist that have not yet been documented to my satisfaction.

Coincidences? The remaining problems present some startling coincidences. But given the population of the world, maybe we should expect such things to happen from time to time.

In reality, the most astonishingly incredible coincidence imaginable would be the complete absence of all coincidences.[52]

John Allen Paulos

---

52 Paulos, John Allen, *Beyond Numeracy: Ruminations of a Numbers Man*, Vintage Books, A Division of Random House, Inc., New York, 1992, p. 41.

47. February 29 only comes once every gour years, yet both Elizabeth Elchlinger of Parma, Ohio, and her son Michael were born on February 29. According to *Ripley's Believe it or Not!*, "The odds of a mother and son being born on that date are over two million to one."[53] Are these odds correct? Should we expect there to be a mother and son somewhere in the world who share this birthday?

48. E. H. Bisch and his wife had three children, all of whom were born on May 28 (1951, 1954, and 1958). According to *Ripley's Believe it or Not!*, "The odds against 3 children in the same family being born on the same date are 28,000,000 to 1."[54] Are these odds correct? Is the existence of three such children surprising?

49. In French/European-style roulette there are 18 red values, 18 black values, and just 1 green value. Back in January 1910, such a game at Monte Carlo resulted a win for red 27 times in a row.[55] What are the odds against this happening?

References/Further Reading

> Blessed is he that readeth.
> – Revelation 1:3, KJV

Aldrich, John, "Earliest Uses of Symbols in Probability and Statistics," April 18, 2014, available online at http://jeff560.tripod.com/stat.html.

Bell, Eric Temple, *Mathematics, Queen and Servant of Science*, Copyright, 1951, British Edition 1952, reprinted 1954, 1958, 1961, Printed in Great Britain by Butler and Tanner Ltd., Frome and London.

Bellhouse, David, "Decoding Cardano's Liber de Ludo Aleae," *Historia Mathematica*, Vol. 32, No. 2, May 2005, pp. 180–202, available online at www.sciencedirect.com/science/article/pii/S0315086004000400.

Burger, Edward B. and Michael Starbird, *The Heart of Mathematics*, third edition, Wiley, Hoboken, NJ, 2010.

Cardano, Gerolamo, *Liber De Ludo Aleae* remained unpublished during Cardano's lifetime. It first appeared in 1663 as part of a ten-volume edition of Cardano's writings. An English translation by S. H. Gould is included in Ore, Oystein, *Cardano, The Gambling Scholar*, Princeton University Press, Princeton, NJ, 1953. This translation also appeared by itself as Cardano, Gerolamo, *The Book on Games of Chance*, translated by S. H. Gould, Holt, Rinehart and Winston, New York, 1961.

Clancy, Tom, *Debt of Honor*, G. P. Putnam & Sons, New York, 1994. Clancy was a good writer, but not a good mathematician!.

Crockett, Zachary, "The Time Everyone 'Corrected' the World's Smartest Woman," *Priceonomics*, available online at https://priceonomics.com/the-time-everyone-corrected-the-worlds-smartest/.

Eckhardt, Roger, "Stan Ulam, John Von Neumann, and the Monte Carlo Method," *Los Alamos Science*, Special Issue, 15, 1987, pp. 131–137.

Edmondson, Amy C., "Learning from Mistakes Is Easier Said than Done: Group and Organizational Influences on the Detection and Correction of Human Error," *The Journal of Applied Behavioral Science*, Vol. 40, No. 1, March 1, 2004, pp. 66–90.

---

53 Packard, Mary and the editors of Ripley Entertainment Inc., *Ripley's Believe It or Not! Creepy Stuff*, Scholastic Inc., New York, 2001, p. 26.

54 Zimmerman, Howard, and Megan Miller, editors, *Ripley's Believe It or Not! Strange Coincidences*, Tom Doherty Associates, Inc., New York, 1990, p. 62.

55 *Ripley's Believe it or Not!*, 2nd Series, Pocket Books, New York, 1948, p. 69. A Simon and Schuster edition was previously published in 1931.

Euler, Leonhard, "Solutio quaestionis curiosae ex doctrina combinationum," *Mémoires de l'Académie Impériale des Sciences de St. Pétersbourg*, Vol. 3, 1811, pp. 57–64. See exercise 0.3 on p. 51 of Hinz et al.

Euler, Leonhard, "Calcul de la Probabilité dans le Jeu de Recontre," *Histoire de l'Académie Royale des Sciences et Belles Lettres*, 1753, pp. 255–270, reprinted in *Opera Omnia*, series prima, Vol. 7, starting on p. 11.

Feynman, Richard P., *The Pleasure of Finding Things Out*, Basic Books, Cambridge, MA, 2005.

Flannery, Sarah, *In Code: A Mathematical Journey*, Algonquin Books of Chapel Hill, Chapel Hill, NC, 2001. This autobiography contains very clear explanations of various ideas in mathematics, including the Monty Hall problem.

Franklin, James, *The Science of Conjecture: Evidence and Probability before Pascal*, Johns Hopkins University Press, Baltimore, MD, 2001.

Geiger, Marshall A., "Changing Multiple-Choice Answers: Do Students Accurately Perceive Their Performance?" *Journal of Experimental Education*, Vol. 59, No. 3, 1991, pp. 250–257.

Gorroochurn, Prakash, "Some Laws and Problems of Classical Probability and How Cardano Anticipated Them," *Chance*, Vol. 25, No. 4, 2012, pp. 13–20, available online at www.columbia.edu/~pg2113/index_files/Gorroochurn-Some%20Laws.pdf.

Granberg, Donald and Thad A. Brown, "The Monty Hall Dilemma," *Personality and Social Psychology Bulletin*, Vol. 21, No. 7, July 1995, pp. 711–723.

Hacking, Ian, *The Emergence of Probability: A Philosophical Study of Early Ideas about Probability, Induction and Statistical Inference*, Cambridge University Press, Cambridge, UK, 2006.

Hinz, Andreas M., Sandi Klavžar, Uroš Milutinović, and Ciril Petr, *The Tower of Hanoi – Myths and Maths*, Birkhäuser, Heidelberg, Germany, 2013.

Kac, Mark, *Enigmas of Chance*, HarperCollins, 1985. This is the autobiography of Mark Kac, a founder of modern probability theory. It's an accessible and entertaining read.

Kac, Mark and Stanislaw M. Ulam, *Mathematics and Logic*, Dover, New York, 1968.

Krauss, Margaret, "Triple Six Fix: How Rigging the PA Lottery Inadvertently Contributed to Its Success," WESA 90.5, Pittsburgh's NPR News Station, June 19, 2015, available online at http://wesa.fm/post/triple-six-fix-how-rigging-pa-lottery-inadvertently-contributed-its-success#stream/0.

Lucas, Stephen K. and Jason Rosenhouse, "Optimal Strategies for the Progressive Monty Hall Problem," *The Mathematical Gazette*, Vol. 93, No. 528, November 2009, pp. 410–419. This paper shows the connection between a variant of the Monty Hall problem and $1 - 1/e$. Also see pages 105–107 of Rosenhouse, Jason, *The Monty Hall Problem: The Remarkable Story of Math's Most Contentious Brain Teaser*, Oxford University Press, New York, 2009.

Mackinnon, Nick, "Another Surprising Appearance of e," *Mathematical Gazette*, Vol. 74, No. 468, June 1990, pp. 167–169.

Mathews, C. O., "Erroneous First Impressions on Objective Tests," *Journal of Educational Psychology*, Vol. 20, No. 4, 1929, pp. 280–286.

Maudlin, Tim, "Interpreting Probabilities: What's Interference Got to Do with It?," in Bricmont J., G. Ghirardi, D. Dürr, F. Petruccione, M. C. Galavotti, and N. Zanghi, editors, *Chance in Physics*, Springer Lecture Notes in Physics, Vol. 571, Springer, Berlin, 2001, pp. 283–288.

Maudlin, Tim, "What Could Be Objective about Probabilities?" *Studies in the History and Philosophy of Science, Part B: Studies in the History and Philosophy of Modern Physics*, Vol. 38, No. 2, June 2007, pp. 275–291.

Mueser, Peter R. and Donald Granberg, "The Monty Hall Dilemma Revisited: Understanding the Interaction of Problem Definition and Decision Making," May 1999, Experimental, 9906001, EconWPA, available online at http://ceadserv1.nku.edu/longa/classes/mat115_resources/docs/10.1.1.42.6708.pdf.

Muir, Jane, *Of Men and Numbers: The Story of the Great Mathematicians*, Dover, New York, 1965.

Pacioli, Luca, *Summa de arithmetica, geometrica, proportioni et proportionalità*, Paganini, Venice, 1494.

Panko, Raymond R., "Basic Error Rates," available online at http://panko.shidler.hawaii.edu/HumanErr/Basic.htm.

Paulos, John Allen, *Beyond Numeracy: Ruminations of a Numbers Man*, Vintage Books, A Division of Random House, Inc., New York, 1992.

Rabinovitch, Nachum L., *Probability and Statistical Inference in Ancient and Medieval Jewish Literature*, University of Toronto Press, Toronto, Canada, 1973.

Rose, Nicholas J., *Mathematical Maxims and Minims*, Rome Press, Raleigh, NC, 1988.

Rosenhouse, Jason, *The Monty Hall Problem: The Remarkable Story of Math's Most Contentious Brain Teaser*, Oxford University Press, New York, 2009.

Rosenthal, Jeffrey, *Struck by Lightning: The Curious World of Probabilities*, Joseph Henry Press, Washington, DC, 2006.

Todhunter, Isaac, *A History of the Mathematical Theory of Probability from the Time of Pascal to That of Laplace*, MacMillan and Co., Cambridge, UK, 1865, reprinted by Chelsea in 1965.

Vos Savant, Marilyn, *The Power of Logical Thinking: Easy Lessons in the Art of Reasoning ... and Hard Facts about its Absence in Our Lives*, St. Martin's Press, New York, 1996.

Whitworth, William Allen, *Choice and Chance with One Thousand Exercises*, fifth edition, Deighton Bell, Cambridge, 1901.

Wikipedia Contributors, "James Randi," *Wikipedia, The Free Encyclopedia*, available online at https://en.wikipedia.org/w/index.php?title=James_Randi&oldid=849916177. See the subsection titled "Exploring Psychic Powers. Live television show."

Wilson, Robin and John J. Watkins, editors, *Combinatorics: Ancient & Modern*, Oxford University Press, Oxford, 2013.

Young, Robert M., *Excursions in Calculus: An Interplay of the Continuous and the Discrete*, The Mathematical Association of America, 1992.

13

The Fibonacci Sequence

The Middle Ages, sometimes referred to as the Dark Ages, were indeed dark for mathematics. However, there was some light here and there, and the leading light was to be found in Italy, under the name Leonardo of Pisa. Today he is better known as Fibonacci (Figure 13.1), an abbreviated form of Filius Bonacci, meaning *son of Bonacci*.

Discrete Mathematics with Rabbits

In the third section of *Liber Abaci*, published in 1202, Fibonacci introduced and solved a problem that led to what we now call the Fibonacci sequence. The problem is inefficiently expressed from today's perspective, but it can be summarized as follows.[1]

> A man put a pair of rabbits in an enclosed space. How many pairs of rabbits are created from that pair in one year, if every month each pair begets a new pair which from the second month on becomes productive, and if no rabbits die?

FIGURE 13.1
Fibonacci (ca. 1170–ca. 1240).

Public domain. Actually, although this image is often purported to be Fibonacci, he (and all other mathematicians from this era) predates portraiture.

---

1 An English translation of the original is Fibonacci (author) and Laurence Sigler (translator), *Fibonacci's Liber Abaci: A Translation into Modern English of Leonardo Pisano's Book of Calculation*, Springer, 2002, pp. 404–405. The relevant passage is also available online at https://www.math.utah.edu/~beebe/software/java/fibonacci/liber-abaci.html.

Under these assumptions, if we start with one pair of immature rabbits, the population will grow as follows:

Month	Mature Pairs	Immature Pairs
1	0	1
2	1	0
3	1	1
4	2	1
5	3	2
6	5	3
7	8	5

Thus, the total number of rabbits at monthly intervals is given by the sequence 1, 1, 2, 3, 5, 8, 13, ...

Denoting the nth term by F_n, we have $F_1 = 1$, $F_2 = 1$, $F_3 = 2$, $F_4 = 3$, etc. Because each term is simply the sum of the two preceding terms, we may express the nth term by the formula $F_n = F_{n-1} + F_{n-2}$. Formulas that depend on previous values are called **recursive formulas**, **recurrence relations**, or **difference equations**.

An Old Approach

Approaching problems with recursion is a technique that goes back to the ancient world. Coincidentally, it was being explored in China at about the same time Fibonacci was active in Italy. Qin Jiushao[2] (ca. 1202–1261), for example, used a recursive method to approximate solutions to polynomials.[3] The technique is now known as Horner's method, or sometimes Ruffini-Horner, after mathematicians who carried out their work centuries later.

The Fibonacci sequence itself was presented by Virahanka in India sometime between 600 and 800 CE.[4] Also in India, Nārāyaṇa Paṇḍita's *Gaṇita Kaumudī* (Lotus Delight of Calculation), 1356, considered sequences that begin with a pair of 1s, like the Fibonacci sequence, but for which new terms are generated by summing any (fixed) number, n, of previous terms.[5] When $n = 2$, we get the Fibonacci sequence, but this is only a special case. For $n = 3$, the sequence runs 1, 1, 2, 4, 7, 13, 24, ... It should be clear that 24 was obtained from the sum $4 + 7 + 13$, but the third term in the sequence, 2, requires a bit of explanation. It should be the sum of the previous three

---

2 You'll sometimes see it spelled, especially in older books, as Ch'in Chiu-Shao or Ch'in Kiu-shao.

3 See http://www-groups.dcs.st-and.ac.uk/history/Biographies/Qin_Jiushao.html, https://en.wikipedia.org/wiki/Qin_Jiushao, and https://en.wikipedia.org/wiki/Horner%27s_method.

4 Singh, Parmanand, "The So-called Fibonacci Numbers in Ancient and medieval India," *Historia Mathematica*, Vol. 12, No. 3, 1985, pp. 229–244.

5 Wilson, Robin, and John J. Watkins, editors, *Combinatorics: Ancient & Modern*, Oxford University Press, Oxford, 2013, pp. 55-56. Also see Singh, Parmanand, "The So-called Fibonacci Numbers in Ancient and Medieval India," *Historia Mathematica*, Vol. 12, No. 3, 1985, pp. 229–244, pp. 237–238 cited here.

terms, but there are only two previous terms. In such cases, we just sum the terms that we actually have. This problem only arises at the beginning of the sequence and doesn't last long if n is small.

Waiting to Be Recognized

There are many beautiful formulas connected with Fibonacci numbers, but they don't follow close on the heels of Fibonacci's book. Partly responsible for the massive delay is the fact that the book was written long before the printing presses existed in Europe, and, as a consequence, it had to circulate in manuscript form. Even after printing hit Europe, centuries passed before *Liber Abaci* reached printed form in 1857.[6]

Indeed, it wasn't until 1876 that the sequence was named after Fibonacci. It was Édouard Lucas (1842–1891), a French mathematician, who finally did so.[7] Lucas's life is detailed in Chapter 14.

Fibonacci Numbers in Nature

Johannes Kepler (1571–1630) is famous for being the astronomer who first recognized that the orbits of the planets around the sun are better approximated as ellipses than as circles, but he also carried out some work with Fibonacci numbers. He noted, for example, that the number of petals on many flowers is a Fibonacci number.[8] These numbers also govern the arrangement of leaves. When leaves branch off of a central stem, at various angles, counting the leaves until arriving at one that is directly above the initial leaf often yields a Fibonacci number. The study of such things is known as phyllotaxis or phyllotaxy. A variety of example follow below.

Example 1 – Pinecones

Pinecones (Figure 13.2) exhibit spirals in two directions, as demonstrated in Figures 13.3 and 13.4.

The number of spirals in each direction is always a Fibonacci number, but depending on which direction you go, you get a different Fibonacci number. The example shown here manifests the numbers 8 and 13, but some other pinecones have 5 and 8.

---

6 Archibald, R. C., "Undergraduate Mathematics Clubs," *The American Mathematical Monthly*, Vol. 25, No. 5, May 1918, pp. 226–238, p. 235 cited here. The author gives the following in a footnote to this fact: "I1 liber Abbaci di Leonardo Pisano pubblicato da Baldassure Boncompagni, Roma, MDCCCLVII. For an analysis of this work see M. Cantor, Vorlesungen über Geschichte der Mathematik, Band II, 3. Auflage, Leipzig, Teubner, 1900, pp. 5–35."

7 Lucas, Édouard, "Sur la théorie des nombres premiers," Atti della Reale Accademia delle scienze di Torino, Vol. 11, May 1876, pp. 928–937. Lucas is stated as the author on p. 927, which leads into the piece. In a previous paper Lucas named the numbers after Lamé – see Édouard, "Note sur la triangle arithmétique de Pascal et sur la serie de Lamé," *Nouvelle correspondance mathématique*, tome 2, 1876, pp. 70–75.

8 Kepler, Johannes, The *Six-Cornered Snowflake: A New Year's Gift*, Paul Dry Books, Philadelphia, PA, 2010. This book reproduces the Latin with an English translation by Jacques Bromberg, pp. 65 and 67.

FIGURE 13.2
A pinecone viewed from the top.

Image modified by Josh Gross from https://pixabay.com/en/pinecones-pine-cone-nature-autumn-3026691/.

FIGURE 13.3
A direction with 8 spirals.

Image modified by Josh Gross from https://pixabay.
com/en/pinecones-pine-cone-nature-autumn-3026691/.

FIGURE 13.4
A direction with 13 spirals.

Image modified by Josh Gross from https://pixabay.
com/en/pinecones-pine-cone-nature-autumn-3026691/.

Example 2 – Pineapples

On a pineapple, you can observe spirals in three directions. In each case, the number of such spirals is a Fibonacci number. In Figure 13.5, they number 5, 8, and 13, from left to right.

FIGURE 13.5
The mathematics of pineapple spirals.

Image modified by Josh Gross from https://pxhere.com/en/photo/599758. The image is released free of copyrights under Creative Commons CC0.

For most pineapples the number of spirals in each direction is given by the numbers 8, 13, and 21. Some small ones offer 5, 8, and 13 instead, as shown above.[9] These numbers seem to have been first remarked on in print in 1933, but the author did not identify them as Fibonacci numbers.[10]

When artists portray pineapples, they often make them symmetric, stripping away the real mathematical beauty of the fruit. Even SpongeBob Squarepants's home fails in this respect.[11]

Example 3 – Sunflowers

Looking closely at the center of a sunflower, Fibonacci numbers arise in the number of spirals in each direction.

The usual numbers of spirals in a sunflower are 34 and 55, although 55 and 89, and even 89 and 144 have been found. The world record appears to be 144 and 233.[12]

Fibonacci numbers don't just appear in cones and flowers. They also arise in a lifeform directly dependent on flowers:

---

9 Onderdonk, Philip B., "Pineapples and Fibonacci Numbers," *The Fibonacci Quarterly*, December 1970, pp. 507–508.

10 Linford, M. B., "Fruit Quality Studies II, Eye Number and Eye Weight," *Pineapple Quarterly*, Vol. III, No. 4, December 1933, pp. 185–195.

11 The artist beyond SpongeBob addressed mathematical criticism expressed in an open YouTube video, https://www.youtube.com/watch?v=gBxeju8dMho. His response can be seen at http://kennypittenger.blogspot.com/2012/01/called-out.html.

12 Koshy, Thomas, *Fibonacci and Lucas Numbers with Applications*, John Wiley & Sons, Inc., New York, 2001, p. 19. Also see Pierce, John Crawford, "The Fibonacci Series," *The Scientific Monthly*, Vol. 73, October 1951, pp. 224–228 and O'Connell, Margaret K. and Daniel T. O'Connell, Letter to the Editor, *The Scientific Monthly*, Vol. 73, November 1951, p. 333.

FIGURE 13.6
Fibonacci numbers in the form of a sunflower's 34 and 55 spirals.

Image modified by Josh Gross from https://pixabay.com/en/flower-flowers-sunflower-yellow-194831/.

Example 4 – Bees!

Consider the family tree of a male bee. The males have mothers, but no fathers, while female bees have both. Thus the number of ancestors for a male bee any number of generations back will always be a Fibonacci number. See Figure 13.7 to convince yourself this is true.

So, numbers that were originally used to model rabbit populations ended up applying to bees! This connection was first put forth by W. Hope-Jones in 1921.[13] But he never dropped the name "Fibonacci" in his brief piece pointing out the connection (Figure 13.8).

The formula Hope-Jones gave generates Fibonacci numbers. For example, letting $n = 10$, we have

$$\frac{2^{12}}{\sqrt{5}}\left(\cos^{12}36° - \cos^{12}108°\right) = 144.$$

Thus, we see another neat connection – between Fibonacci numbers and trigonometry!

---

13 Hope-Jones, W., "The Bee and the Pentagon," *The Mathematical Gazette*, Vol. 10, No. 150, January 1921, p. 206.

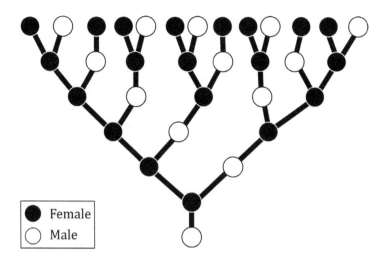

FIGURE 13.7
The Bee's Tree.

Image created by Josh Gross.

571. [I. 2. δ.] *The Bee and the Pentagon.*

In matters hexagonal the bee is our oldest teacher : but her connection with the regular pentagon is unsuspected by many naturalists, and very likely by herself.

The female bee has two parents, like man and other respectable animals ; but the male is a parthenogenetic product, and owns a mother only. From this it follows that the number of ancestors of n generations back possessed by the ordinary worker, or her sister the queen, is

$$\frac{2^{n+2}}{\sqrt{5}}\left(\cos^{n+2}36° - \cos^{n+2}108°\right).$$

W. Hope-Jones.

FIGURE 13.8
Where the bee-Fibonacci connection was first made.

Public domain.

More from Kepler – Fibonacci Ratios

Kepler's work with Fibonacci numbers wasn't limited to recognizing their role in nature. He also considered the ratios of consecutive Fibonacci numbers, $\lim_{n\to\infty}\frac{F_{n+1}}{F_n}$. This limit is especially interesting. It is not immediately clear that this will converge, but looking at some early values gives a nice surprise:

n	F_n	F_{n+1}	F_{n+1}/F_n
1	1	1	1
2	1	2	2
3	2	3	1.5
4	3	5	1.666666667
5	5	8	1.6
6	8	13	1.625
7	13	21	1.615384615
8	21	34	1.619047619
9	34	55	1.617647059
10	55	89	1.618181818
11	89	144	1.617977528
12	144	233	1.618055556
13	233	377	1.618025751
14	377	610	1.618037135
15	610	987	1.618032787
16	987	1,597	1.618034448
17	1,597	2,584	1.618033813
18	2,584	4,181	1.618034056

So, it appears that the sequence converges to a number whose first few digits are 1.61803. This fact was noticed by Johannes Kepler.[14] He referred to this number as the "Divine Proportion." Over the centuries it's gone by a few other names, including the Golden Ratio, the Golden Number, the Golden Section, and the Extreme and Mean Ratio. It's usually represented by the Greek letter φ and simply called "phi." While φ is now the more popular way of designating this number, it has often been denoted by another Greek letter, τ (tau). This usage arose from the Greek word for section (τομή), which begins with τ. An exact value is given by $\frac{1+\sqrt{5}}{2}$. More on this soon!

> Geometry has two great treasures: one is the theorem of Pythagoras; the other, the division of a line into extreme and mean ratio. The first we may compare to a measure of gold; the second we may name a precious jewel.
>
> Johannes Kepler[15]

Binet's Formula

Returning to the recursive formula $F_n = F_{n-1} + F_{n-2}$, we have an easy way to calculate F_n for small values of n, but we would not want to use it to calculate the millionth

---

14 Kepler, Johannes, The *Six-Cornered Snowflake: A New Year's Gift*, Paul Dry Books, Philadelphia, PA, 2010. This book reproduces the Latin with an English translation by Jacques Bromberg, p. 67.

15 Kepler, Johannes, *Mysterium Cosmographicum*, Tübingen, 1596.

Fibonacci number. By this method, we'd have to calculate all 999,999 smaller Fibonacci numbers first.

Fortunately, there's a way to explicitly solve for the nth term without calculating all previous terms.[16] Formulas that take this form are called non-recursive. A method by which such formulas can be found is detailed in Chapter 15. For now, simply observe the beauty of the result in this instance.

$$F_n = \frac{1}{\sqrt{5}} \left(\frac{1 + \sqrt{5}}{2} \right)^n - \frac{1}{\sqrt{5}} \left(\frac{1 - \sqrt{5}}{2} \right)^n$$

This formula was first found by Abraham de Moivre in 1718 and only *rediscovered* by Binet (Figure 13.9) in 1843.[17] Binet wasn't even a distant second. Other mathematicians between de Moivre and Binet who knew the result included Leonhard Euler and Daniel Bernoulli.[18] Nevertheless, it is generally known as "Binet's formula."

Using Binet's formula, it isn't hard to prove the ratio of Fibonacci numbers converges to φ, as Kepler had determined.[19]

FIGURE 13.9
Jacques Philippe Marie Binet (1786–1856).

Courtesy of Archives de l'Académie des sciences.

---

16 We actually already saw one example – the formula given to count a male bee's ancestors in each generation.

17 Young, Robert M., *Excursions in Calculus: An Interplay of the Continuous and the Discrete*, The Mathematical Association of America, 1992, p. 130.

18 http://mathworld.wolfram.com/BinetsFibonacciNumberFormula.html.

19 For an alternate proof, see Young, Robert M., *Excursions in Calculus: An Interplay of the Continuous and the Discrete*, The Mathematical Association of America, 1992, pp. 140–141.

$$\lim_{n\to\infty}\frac{F_{n+1}}{F_n}=\lim_{n\to\infty}\frac{\frac{1}{\sqrt5}\left(\frac{1+\sqrt5}{2}\right)^{n+1}-\frac{1}{\sqrt5}\left(\frac{1-\sqrt5}{2}\right)^{n+1}}{\frac{1}{\sqrt5}\left(\frac{1+\sqrt5}{2}\right)^{n}-\frac{1}{\sqrt5}\left(\frac{1-\sqrt5}{2}\right)^{n}}$$

$$=\frac{\lim_{n\to\infty}\left(\frac{1}{\sqrt5}\left(\frac{1+\sqrt5}{2}\right)^{n+1}-\frac{1}{\sqrt5}\left(\frac{1-\sqrt5}{2}\right)^{n+1}\right)}{\lim_{n\to\infty}\left(\frac{1}{\sqrt5}\left(\frac{1+\sqrt5}{2}\right)^{n}-\frac{1}{\sqrt5}\left(\frac{1-\sqrt5}{2}\right)^{n}\right)}$$

$$=\frac{\lim_{n\to\infty}\frac{1}{\sqrt5}\left(\frac{1+\sqrt5}{2}\right)^{n+1}-\lim_{n\to\infty}\frac{1}{\sqrt5}\left(\frac{1-\sqrt5}{2}\right)^{n+1}}{\lim_{n\to\infty}\frac{1}{\sqrt5}\left(\frac{1+\sqrt5}{2}\right)^{n}-\lim_{n\to\infty}\frac{1}{\sqrt5}\left(\frac{1-\sqrt5}{2}\right)^{n}}$$

But $\frac{1-\sqrt5}{2}\approx-0.618$, and any number less than 1 in absolute value heads to 0 as it's raised to larger and larger powers. Therefore, both the second limit in the numerator and the second limit in the denominator are 0. The above becomes

$$\frac{\lim_{n\to\infty}\frac{1}{\sqrt5}\left(\frac{1+\sqrt5}{2}\right)^{n+1}}{\lim_{n\to\infty}\frac{1}{\sqrt5}\left(\frac{1+\sqrt5}{2}\right)^{n}}=\lim_{n\to\infty}\frac{\frac{1}{\sqrt5}\left(\frac{1+\sqrt5}{2}\right)^{n+1}}{\frac{1}{\sqrt5}\left(\frac{1+\sqrt5}{2}\right)^{n}}=\lim_{n\to\infty}\frac{1+\sqrt5}{2}=\varphi.$$

Binet's formula can be evaluated for n other than natural numbers. This is another example of the interplay between the continuous and the discrete. A graph is given in Figure 13.10.

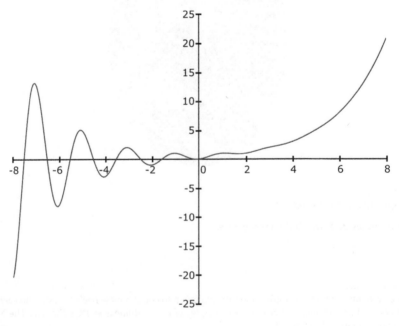

FIGURE 13.10
A continuous version of the Fibonacci numbers.

Image created by Josh Gross.

The behavior to the left of the y-axis might look strange, but consider what values F_n must take to preserve the rule $F_n = F_{n-1} + F_{n-2}$. Letting $n = 2$ in our recurrence relation, we have $F_2 = F_1 + F_0$. We know $F_1 = 1$ and $F_2 = 1$, so this gives us $1 = 1 + F_0$. Thus, $F_0 = 0$. Now, letting $n = 1$, we have $F_1 = F_0 + F_{-1}$, which gives us $1 = 0 + F_{-1}$. So, F_{-1} must be 1. Letting $n = 0$, we have $F_0 = F_{-1} + F_{-2}$, which gives us $0 = 1 + F_{-2}$. So, F_{-2} must be -1. We can keep going in this manner and the numbers will alternate between positive and negative, but in absolute value they'll be the same as the Fibonacci numbers that have already been seen.

There's much more that can be said about Fibonacci numbers, but before doing so, let's take a closer look at φ.

The Golden Ratio

Artists and mathematicians have a lot in common, including a love of the golden ratio.[20] In addition to this number's interesting mathematical properties, more of which will be detailed, it has a special aesthetic appeal.

Any rectangle with sides such that their ratio is φ is known as a **golden rectangle**. Such rectangles are generally considered to be the prettiest of all. They can be found in paintings (by Georges Seurat[21], George Bellows, Salvador Dali[22], Gustav Caillebotte[23], for example), architecture (a prominent example being Le Corbusier's use of the ratio in the United Nations Headquarters in New York City),[24] and even some music. Composers who made use of it include Claude Debussy and Béla Bartók.[25] In fact, the use of the golden ratio is so widespread in the art world that it has its own name in this context – **dynamic symmetry**.[26]

How far back do such applications go? It has often been claimed that a golden rectangle fits the Parthenon, in Athens Greece, for which construction began in 447 BCE. However, this claim has been disputed. What is not in dispute is that some ancient Greeks were aware of φ. Euclid himself referred to it as the "mean and extreme ratio." To see what he meant by this, consider the line segment in Figure 13.11, divided into pieces of length a and b. The division has been done in accordance with the Golden Ratio, if $(a + b)/a = a/b$. In this case, both fractions will equal φ.

20 Actually, mathematicians are artists. They simply work in different media from the people who are more widely viewed as artists and they are subject to different constraints.

21 Burger, Edward B. and Michael Starbird, *The Heart of Mathematics*, third edition, Wiley, Hoboken, NJ, 2010, p. 263.

22 Dali's "The Sacrament of the Last Supper" (owned by the National Gallery of Art, Washington, DC), for example.

23 Burger, Edward B. and Michael Starbird, *The Heart of Mathematics*, third edition, Wiley, Hoboken, NJ, 2010, p. 271, #7.

24 Burger, Edward B. and Michael Starbird, *The Heart of Mathematics*, third edition, Wiley, Hoboken, NJ, 2010, p. 263.

25 For Debussy, see Howat, Roy, *Debussy in Proportion: A Musical Analysis*, Cambridge University Press, Cambridge, 1983. For Bartók. see Kramer, Jonathan, "The Fibonacci Series in Twentieth-Century Music," *Journal of Music Theory*, Vol. 17, No. 1, Spring 1973, pp. 110–148.

26 Burger, Edward B. and Michael Starbird, *The Heart of Mathematics*, third edition, Wiley, Hoboken New Jersey, 2010, p. 263.

FIGURE 13.11
The mean and extreme ratio.

Image created by Josh Gross.

Despite the constant's antiquity, using the Greek letter φ to represent the Golden Ratio is a relatively recent phenomena. It was chosen by American electrical engineer and inventor Mark Barr in honor of Phidias, an ancient Greek sculptor who is believed by some to have made frequent intentional use of the ratio, although Barr himself is skeptical.[27] Barr's φ notation got a boost in 1914 when Theodore Andrea Cook used it in his book *The Curves of Life*.[28]

Luca Pacioli and the Divine Proportion

The Italian mathematician Luca Pacioli (Figure 13.12) has a few claims to fame. He's the first mathematician to not predate portraiture. What this means is that we have an image of him that was made by an artist who actually looked at him to create it. For all earlier mathematicians, the only likenesses we possess were created by artists who had never actually seen their subjects.

Pacioli also wrote several books, including *De Divina Proportione* (Milan, 1509). He might be better remembered in association with this book, which contained 60 illustrations showing the Golden Ratio, if he hadn't been nearly eclipsed by the creator of the illustrations, Leonardo da Vinci.

Golden rectangles can also be found in earlier paintings by Leonardo da Vinci, but there is debate over whether this was intentional or not. While da Vinci had to know about the ratio to create the illustrations for Pacioli's book, it is not clear for how long before this project he knew about it.

Psychological Studies

Gustav Fechner, a 19th-century German psychologist, measured thousands of man-made rectangles. Most were paintings, but windows, playing cards, and books were included.[29] It turned out the average proportions for these rectangles was close to the

27 Gardner, Martin, *The Second Scientific American Book of Mathematical Puzzles & Diversions*, The University of Chicago Press, Chicago, IL, 1961, 1987, p. 91. See Barr, Mark, "Parameters of Beauty," *Architecture* (N. Y.), Vol. 60, 1929, p. 325, reprinted as Barr, Mark "Parameters of beauty," *Think*, Vol. 10–11, International Business Machines Corporation, 1944, and https://en.wikipedia.org/wiki/Mark_Barr for Barr's skepticism.

28 Cook, Theodore Andrea, *The Curves of Life*, Constable and Company Ltd., London, 1914. See page 420. Available online at https://ia800204.us.archive.org/4/items/cu31924028937179/cu31924028937179.pdf

29 Gardner, Martin, *The Second Scientific American Book of Mathematical Puzzles & Diversions*, The University of Chicago Press, Chicago, IL, 1961, 1987, p. 98.

FIGURE 13.12
Luca Pacioli (ca. 1447–1517).

Public domain.

golden ratio, seemingly indicating a subconscious preference among many individuals for this ratio. Fechner tested this hypothesis by surveying people as to which of ten rectangles they found the most pleasing. Once again, the golden rectangle was the winner. However, in the decades that followed, other experimenters failed to replicate Fechner's results and his methods have been criticized.[30]

While we might never know exactly why the golden ratio pleases some people so much, this hasn't stopped animators from using it to make their creations more aesthetically pleasing and successful. One of many examples is Stephen Silver, an art teacher who has designed characters for Disney, Sony, and Nickelodeon. He describes the golden ratio as "the character designers' secret weapon."[31] Disney has even woven the Greek letter φ into its name logo. Take a close look at the final letter![32]

More on Phi

If your taste in art runs more in the direction of the abstract beauty of mathematics, consider the following:

---

30 Höge, Holger, "The Golden Section Hypothesis – Its Last Funeral," *Empirical Studies of the Arts*, Vol. 15, No. 2, 1997, pp. 233–255.

31 See https://www.goldennumber.net/golden-ratio-cartoon-character-design/ and https://www.youtube.com/watch?v=PEiaGGp5ddY&feature=youtu.be.

32 This and other examples can be seen at https://www.goldennumber.net/logo-design/.

$$\varphi = 1 + \cfrac{1}{1 + \cfrac{1}{1 + \cfrac{1}{1 + \cfrac{1}{1 + \cdots}}}}$$

This is actually easy to prove. Start by calling it x. We can't use φ, because that would be assuming what we are trying to prove. We have

$$x = 1 + \cfrac{1}{1 + \cfrac{1}{1 + \cfrac{1}{1 + \cfrac{1}{1 + \cdots}}}}$$

Notice that what we have under the top line is also x. Making this substitution gives

$$x = 1 + \frac{1}{x}$$

It's now easy to solve for x by first multiplying through by x to get

$$x^2 = x + 1$$

and then moving everything to the left-hand side

$$x^2 - x - 1 = 0$$

giving an equation that can be solved with the quadratic formula.

The solutions are $x = \frac{1+\sqrt{5}}{2}$ and $x = \frac{1-\sqrt{5}}{2}$, but we can quickly reject the second solution, as the continued fraction of all 1s is clearly not a negative number. Thus, we have $x = \frac{1+\sqrt{5}}{2} = \varphi$.

Another neat result involving φ concerns the sequence 1, φ, $\varphi + 1$, $2\varphi + 1$, $3\varphi + 2$, ... Like the Fibonacci sequence, each term is obtained from the sum of the two previous terms. But, if we look at the ratio of consecutive terms, it is always φ. While a Fibonacci-like sequence can be based off of any two initial values, this is the only such sequence where the ratio of consecutive terms is constant.[33] Because of this feature, the sequence can be expressed as 1, φ, φ^2, φ^3, φ^4...

Also, the floor function of Chapter 6, when applied to multiples of φ and $\varphi/(\varphi - 1)$, gives something interesting

$\lfloor \varphi \rfloor, \lfloor 2\varphi \rfloor, \lfloor 3\varphi \rfloor, \lfloor 4\varphi \rfloor, \lfloor 5\varphi \rfloor, \lfloor 6\varphi \rfloor, \lfloor 7\varphi \rfloor, \lfloor 8\varphi \rfloor, \lfloor 9\varphi \rfloor, \ldots$

$\quad = 1, 3, 4, 6, 8, 9, 11, 12, 14, \ldots$

$\lfloor \varphi/(\varphi - 1) \rfloor, \lfloor 2\varphi/(\varphi - 1) \rfloor, \lfloor 3\varphi/(\varphi - 1) \rfloor, \lfloor 4\varphi/(\varphi - 1) \rfloor, \lfloor 5\varphi/(\varphi - 1) \rfloor, \lfloor 6\varphi/(\varphi - 1) \rfloor, \ldots$

$\quad = 2, 5, 7, 10, 13, 15, \ldots$

---

33 Gardner, Martin, *The Second Scientific American Book of Mathematical Puzzles & Diversions*, The University of Chicago Press, Chicago, IL, 1961, 1987, p. 96.

Taken one at a time, you might not notice anything special about these integer sequences, but if you look at them together you'll see that they give the natural numbers. Each one appears exactly once![34]

Okay, this last result is a bit of a card trick. There's nothing special about φ in the above. *Any* irrational number greater than 1 can be used in its place to split the natural numbers into two sets! This result is referred to as Rayleigh's theorem or Beatty's theorem.[35]

For a fabulous geometric result connected with φ, we start with Figure 13.13.

This figure was designed by Roger Penrose in the 1970s to attack a tiling problem, that is, a problem in which specific shapes are arranged to completely cover the plane. We could tile the plane by simply using this rhombus repeatedly. It would be no harder or any more interesting than tiling a plane with triangles or rectangles. However, if we split the rhombus into two different shapes, as indicated by the shading, we are on our way to a much more interesting result.

Penrose called the shapes in Figure 13.14 kites and darts. He and John Horton Conway found a way to label the vertices using the letters H and T such that if a tiling was created in which shapes are only allowed to meet at vertices that are labeled in the same way, something strange results.[36] An example of this is shown in Figure 13.15.

What makes the above tiling strange (and probably what makes it beautiful, as well) is the fact that it is not periodic. If we could pick up an outline of the shapes, there's no

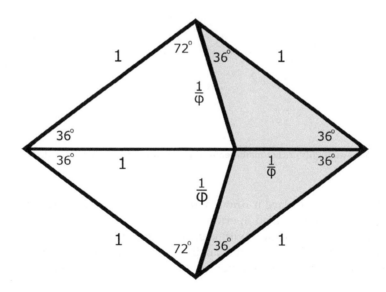

FIGURE 13.13
A shape with connections to φ.

Image created by Josh Gross.

34 I first saw this in Chamberland, Marc, *Single Digits: In Praise of Small Numbers*, Princeton University Press, Princeton, NJ, 2015, pp. 32–33, although this is not its first appearance.

35 See https://en.wikipedia.org/wiki/Beatty_sequence for more information, including a pair of proofs.

36 http://jwilson.coe.uga.edu/EMAT6680Fa05/Schultz/Penrose/Penrose_Main.html

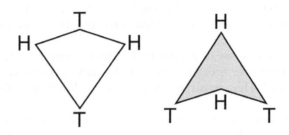

FIGURE 13.14
A "kite" and a "dart."

Image created by Josh Gross.

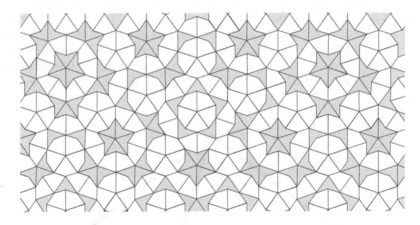

FIGURE 13.15
An aperiodic tiling.

Image from http://www.quadibloc.com/math/pen01.htm, courtesy of John Savard, redrawn by Josh Gross.

way we could shift it and place it down somewhere else so that it matched the outline once again. Such tilings are called aperiodic.

There was already a connection with φ in the design of the kite and dart shapes, but as a bonus, as the tilings become larger, using more kites and darts and covering more of the plane, the ratio of the number of kites used to the number of darts used converges to φ.

The quest to find a set of shapes (and rules for laying them) that force any tiling to be aperiodic did not begin with kites and darts. The first set consisted of 20,462 different shapes. This was reduced to 2 in stages, namely 104, 92, 6, 2.[37] Conway would like to see another reduction.

37 Roberts, Siobhan, *Genius at Play – The Curious Mind of John Horton Conway*, Bloomsbury, New York, 2015, p. 231.

And by the way, nobody knows that you can't get the number down to 1. Every now and then I've tried. But I haven't succeeded. Not yet.

John H. Conway[38]

Can't Spare a Square

Those of you who need applications may be delighted to know that toilet paper has been manufactured with a Penrose tiling pattern (Figure 13.16), and it wasn't just to make it pretty!

The non-repeating pattern means that the creases won't line up over any large areas as the paper is rolled. This means the roll will be rounder and the individual wraps unlikely to stick together.[39] But Sir Roger Penrose wasn't amused. He had patented the design, although the patent expired around the time the toilet paper came out. Pentaplex Ltd. was assigned the rights by Penrose and this British company shared his objections.[40] Penrose sued the manufacturer and a Pentaplex representative was quoted in the *London Times* concerning the indignity of the application in question.

> So often we read of very large companies riding roughshod over small businesses or individuals, but when it comes to the population of Great Britain being invited by a multinational [company] to wipe their bottoms on what appears to be the work of a knight of the realm without his permission, then a last stand must be made.[41]

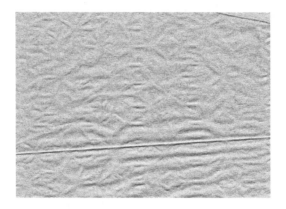

FIGURE 13.16
Toilet paper so good it ought to be illegal.

Image courtesy of Brian Hayes The chain of custody was from Marion Walter through Phil and Phylis Morrison to Hayes. See http://bit-player.org/2017/sir-roger-penroses-toilet-paper.

---

38 Roberts, Siobhan, *Genius at Play – The Curious Mind of John Horton Conway*, Bloomsbury, New York, 2015, p. 231.

39 The idea isn't as perfect as it sounds here. An exercise at the end of this chapter asks you to explain why.

40 Hayes, Brian, http://bit-player.org/

41 Quoted here from Hayes, Brian, http://bit-player.org/.

Like all of the other topics in this book, there's a lot more that could be said about aperiodic tilings. I hope you are inspired to seek out more on your own.

A Precise Value

While the constant π is better known and has been examined much more closely, with over 22 trillion digits of it now determined,[42] a very precise value for φ has been obtained. There's a program you can download that allows you to calculate over a million digits of φ on your own machine.[43] What you do with them all is up to you!

The Golden Spiral

There's a special curve known as the Golden Spiral. It can be approximated using nested Golden Rectangles, as shown below.

We start with a golden rectangle and then draw a line joining the longer sides, such that the golden rectangle is divided into a square and another, smaller, golden rectangle (Figure 13.17).

We then divide the new, smaller, golden rectangle into a square and a golden rectangle (Figure 13.18).

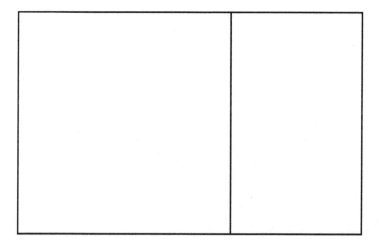

FIGURE 13.17
Getting started on the golden spiral.

Image created by Josh Gross.

42 Roeder, Oliver, "Even after 22 Trillion Digits, We're Still No Closer to the End of Pi," *FiveThirtyEight*, March 14, 2018, https://fivethirtyeight.com/features/even-after-22-trillion-digits-were-still-no-closer-to-the-end-of-pi/.

43 https://www.goldennumber.net/phi-million-places/.

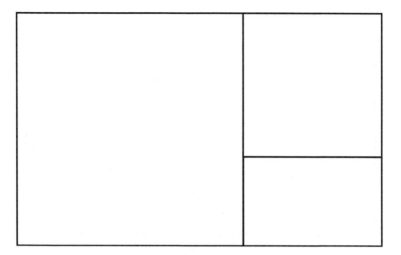

FIGURE 13.18
Dividing the smaller golden rectangle.

Image created by Josh Gross.

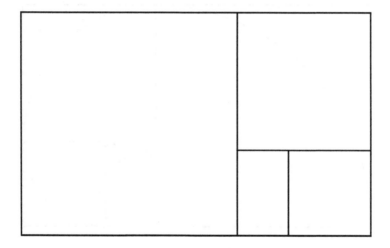

FIGURE 13.19
After another division.

Image created by Josh Gross.

Next, the golden rectangle in the lower right hand corner is split into a square and a golden rectangle (Figure 13.19).

We repeat this process, always dividing the new smallest golden rectangle into a square and an even smaller golden rectangle. Eventually resolution issues will cause us to stop. Figure 13.20 shows the result after seven more divisions.

To complete the spiral, simply fill in quarter circles, in each square, as in Figure 13.21.

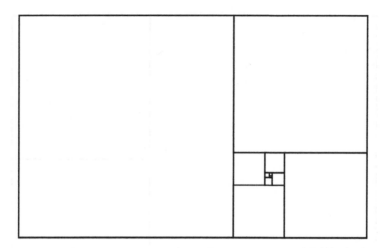

FIGURE 13.20
Continuing the process.

Image created by Josh Gross.

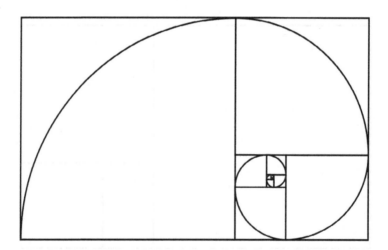

FIGURE 13.21
Sketching the spiral.

Image created by Josh Gross.

An approximation to the Golden Spiral

This approximation is sometimes called a Fibonacci spiral, because it can be obtained using Fibonacci numbers, as in Figure 13.22.

Some authors confuse matters by using the term Golden Spiral to refer to the approximation, thus failing to distinguish it from the true Golden Spiral, which does

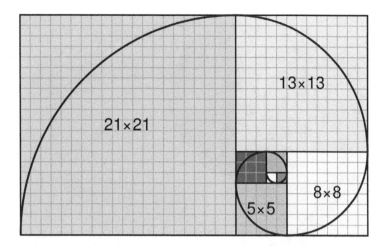

FIGURE 13.22
A Fibonacci spiral.

Image created by Jahobr. This file is made available under the Creative Commons CC0 1.0 Universal Public Domain Dedication.

not consist of exact quarter circle segments. But they can be forgiven, for the deviation is slight.[44]

The formula for the Golden Spiral, expressed in polar coordinates is $r = \varphi^{\frac{\theta^2}{\pi}}$. This is a special case of the logarithmic spiral, which has the general form $r = ae^{b\theta}$. Another special case, where the rate of growth is 0, is the circle.[45] The logarithmic spiral can also be expressed as $\theta = \frac{1}{b}\ln\left(\frac{r}{a}\right)$, which makes sense of its name.

Staying true to the rule of everything important having multiple names, a logarithmic spiral can be called an equiangular spiral (because every radius, drawn from the origin, hits the curve at the same angle), or even *Spira mirabilis* (the marvelous spiral), as Jacob Bernoulli christened it. Bernoulli was so fond of this curve that he asked for it to be carved on his tombstone, along with the inscription *Eadem numero mutate resurgo* (Though changed, I arise again exactly the same).[46] Such curves were first discovered by René Descartes, who discussed them with Mersenne in letters dated 1638.[47]

A Logarithmic Spiral in Nature

It's easy to construct an approximation of the Golden Spiral, as shown above, but why would a curve like this arise in nature? Well, there are many kinds of spirals, but

44 See https://en.wikipedia.org/wiki/Golden_spiral#/media/File:FakeRealLogSpiral.svg for a comparison.
45 In this case, the rate of growth is zero, See Maor, Eli, *e: The Story of a Number*, ninth printing with new material on pp. 183-186, Princeton University Press, Princeton, NJ, 199, p. 209.
46 Young, Robert M., *Excursions in Calculus: An Interplay of the Continuous and the Discrete*, The Mathematical Association of America, 1992, p. 153, which cites Moritz, Robert Edouard, *On Mathematics*, Dover, New York, 1958, p. 145. Previously issued as *Memorabilia Mathematica or the Philomath's Quotation-Book*.
47 Young, Robert M., *Excursions in Calculus: An Interplay of the Continuous and the Discrete*, The Mathematical Association of America, 1992, p. 148.

FIGURE 13.23
A cutaway view of a nautilus shell.

Image created by Chris 73 / Wikimedia Commons. This file is licensed under the Creative Commons Attribution-Share Alike 3.0 Unported license.

logarithmic spirals, of which the Golden Spiral is an example, are the only ones that do not change in shape as they spiral outward. This consistency may be why they arise in organisms whose growth is governed by simple rules. Perhaps the most interesting real-world example of a logarithmic spiral is the nautilus (Figure 13.23).

But there are patterns lurking beyond the shape of the spiral in this shell. In 1978, researchers discovered that the growth rate for a nautilus is one new chamber every lunar month. Because the manner in which the chambers grow indicates the number of days in a lunar month, and fossils showing this date back to 400 million years ago, this lowly creature has a story to tell about the history of our moon. And this story contradicted what astronomers had generally believed at the time. The evidence is that the moon was much closer to the earth and moving more rapidly than believed at such ancient dates.[48]

A Coincidental Application of Fibonacci Numbers

There's one neat application of φ that has nothing to do with nature. The fact that it works is a pure coincidence!

---

48 "Fossil Cuttlefish Draw Ancient Moon Nearer to Earth" *New Scientist*, October 19, 1978, p. 172. This reports on an article in *Nature*: Kahn, Peter G. K. and Stephen Pompea, "Nautiloid Growth Rhythms and Dynamical Evolution of the Earth–Moon System," *Nature*, Vol. 275, October 19, 1978, pp. 606–611. Also see "Undersea Calendar," *OMNI*, April 1979, Vol. 1, No. 7, p. 34.

Example 5 – Convert Kilometers to Miles or Miles to Kilometers[49]

So you don't have to flip back and forth, I'm writing out the first few Fibonacci numbers here: 1, 1, 2, 3, 5, 8, 13, 21, 34, 55, 89, 144, 233, 377, 610, 987, 1597, 2584, 4181, 6765, 10946,…

Many American highways have posted speed limits of 55 miles per hour. What is this in kilometers per hour? To answer this question, we locate 55 in our list of Fibonacci numbers and simply take the number after it. It's about 89 kilometers per hour. A more precise answer is 88.5139 km/hr, but the Fibonacci numbers got it right to the closest integer!

It's 611 kilometers from Annandale, Virginia[50] to Dayton, Ohio by plane, according to www.distancecalculator.net/. What is this in miles? Well, 611 is not a Fibonacci number, but it's very close to 610, which is. To convert from kilometers to miles, move to the next smallest Fibonacci number. Our answer is 377 km. Because we were actually a mile over 610, we might add 1 to our answer to get 378 km. A more precise answer is 379.658. The Fibonacci method didn't get us to the closest integer this time, but our error was still tiny.

Why does this work so well? It's because 1 mile = 1.60934 kilometers and this conversion factor is closely approximated by the ratio of consecutive Fibonacci numbers which, as we saw, converges to $\varphi \approx 1.618$. The Fibonacci method yields better results for some numbers than for others, because the ratio varies.

Some readers may object that the method isn't very useful because most numbers aren't Fibonacci numbers. It couldn't be used, for example, to convert the length of a marathon (26.2 miles) to kilometers. Or could it? Forgetting about the .2 part, we still seem to be in trouble; 26 is not a Fibonacci number. But, 26 = 21 + 5, and 21 and 5 are both Fibonacci numbers. We can then convert 21 to 34 and 5 to 8 to get a total of 42 kilometers. The precise answer is 26.2 miles = 42.16481. Our answer was correct to the closest kilometer, even after discarding the fractional mile.

But is this fair? You could argue that 26 was special, because it could be expressed as a sum of two Fibonacci numbers. This is true. Most numbers cannot be expressed as a sum of two Fibonacci numbers, but why limit ourselves to two? If we're willing to use more numbers, we can always apply this technique. This is a consequence of Zeckendorf's Theorem .

Zeckendorf's Theorem (1939) – Every positive integer can be uniquely expressed as a sum of non-consecutive Fibonacci numbers.[51]

Although Zeckendorf had made this discovery by 1939, he didn't publish it until 1972.[52] Meanwhile, in 1952, someone else put it into print.[53]

There are several ways to represent some numbers, if the "non-consecutive" constraint is removed. For example, 100 = 89 + 8 + 3 (this is the Zeckendorf representation), but we also have 100 = 89 + 8 + 2 + 1 (2 and 1 are consecutive Fibonacci numbers) and 100 = 55 + 34 + 8 + 3 (55 and 34 are consecutive Fibonacci numbers).

---

49 I first saw this at https://www.reddit.com/r/AskReddit/comments/4kz3di/whats_your_favourite_maths_fact/.

50 https://www.youtube.com/watch?v=l-GrF87b82Q.

51 Chamberland, Marc, *Single Digits: In Praise of Small Numbers*, Princeton University Press, Princeton, NJ, 2015, p. 6.

52 Kimberling, Clark, "Edouard Zeckendorf," *Fibonacci Quarterly*, Vol. 36, November 1998, pp. 416–418. Zeckendorf, Edouard, "Representation des nombres naturels par une somme de nombres de Fibonacci ou de nombres de Lucas," *Bulletin de la Societe Royale des Sciences de Liege*, Vol. 41, 1972, pp. 179–182.

53 Lekkerkerker, Cornelis Gerrit, "Voorstelling van natuurlijke getallen door een som van getallen van Fibonacci," *Simon Stevin*, Vol. 29, 1952-1953, pp. 190–195.

One thing that makes Zeckendorf's Theorem really neat is that it can be viewed as analogous to the fundamental theorem of arithmetic, which states that every integer (except 0, 1, and −1) has a unique representation as a product of primes (up to order and sign). Replacing "product" with "sum" and "primes" with "nonconsecutive Fibonacci numbers" gets us Zeckendorf's theorem.

More Neat Facts

Although Fibonacci numbers arise in nature, as seen in this chapter, they also have applications to the abstract world of mathematics. For example, Lucas used the Fibonacci-like sequence 1, 3, 4, 7, 11, ... (where each new term is the sum of the two previous values, but beginning with 1, 3, instead of 1, 1) to study the question of primality for Mersenne numbers.[54]

There are even more abstract applications. In 1900, when David Hilbert posed 23 problems that he felt it important for mathematics to address in the coming century, one of these (the 10th) concerned Diophantine equations. These are polynomial equations for which solutions where all of the variable are integers are sought. Hilbert posed the problem like so:

> Given a Diophantine equation with any number of unknown quantities and with rational integral numerical coefficients: To devise a process according to which it can be determined in a finite number of operations whether the equation is solvable in rational integers.

Hilbert's "rational integers" sounds redundant – what other kinds of integers are there? What he meant by this was what I've been referring to simply as integers: ...−3, −2, −1, 0, 1, 2, 3, ... Anyway, this problem was finally solved, in the negative, by Russian mathematician Yuri Matijasevich in 1970. That is, Matijasevich showed that no such process exists in general. Although the connection is far from obvious, the rate of growth for the Fibonacci numbers played a key role in the proof.[55]

A simpler result involving Fibonacci numbers arises from the rational expression $\frac{x}{1-x-x^2}$. Converting this into a polynomial of infinite degree by long division, we get $x + x^2 + 2x^3 + 3x^4 + 5x^5 + \ldots$, and recognize the coefficients as the famous sequence! Such magic falls under the umbrella of **generating functions**, a topic covered in Chapter 17.

There are numerous connections between the various topics covered in this text. A formula connecting Fibonacci numbers with combinations (as seen in Chapter 10) is given by

$$F_{n+1} = \sum_{k=0}^{\lfloor n/2 \rfloor} \binom{n-k}{k}$$

---

54 Young, Robert M., *Excursions in Calculus: An Interplay of the Continuous and the Discrete*, The Mathematical Association of America, 1992, p. 131, which cites Shanks, Daniel, *Solved and Unsolved Problems in Number Theory*, 3rd edition, Chelsea, New York, 1985 and Young, p. 135 #13, which cites Hardy, G. H. and E. M. Wright, *An Introduction to the Theory of Numbers*, 5th edition, Oxford University Press, Oxford, 1988, p. 223.

55 For a more accessible account of the proof, see Davis, Martin, "Hilbert's Tenth Problem Is Unsolvable," *The American Mathematical Monthly*, Vol. 80, No. 3, March 1973, pp. 233–269.

It's reproduced here just in case you're reading the chapters out of order!

The Fibonacci sequence arises in many places in math and science, far more than can be covered in a chapter or even a book. There's even a journal, *Fibonacci Quarterly*, devoted to it.[56] Actually, the journal is a lot broader than the title implies. Many of the articles could be introduced by the words "If you like the Fibonacci numbers, you might also like ..."

Exercise Set 13

> Making mistakes is a sign of creativity and strength; it is not a sign of weakness.
> Edward B. Burger and Michael Starbird[57]

1. Prove that $F_1 + F_2 + \cdots + F_n = F_{n+2} - 1$.

2. Prove that $F_1{}^2 + F_2{}^2 + \cdots + F_n{}^2 = F_n F_{n+1}$.

3. Form a conjecture concerning $(F_n)^2 + (F_{n+1})^2$.

4. Prove the conjecture you formed in the exercise above.

5. Form a conjecture concerning $(F_{n+1})^2 - (F_{n-1})^2$.

6. Prove the conjecture you formed in the exercise above.

7. Prove that $F_1 F_2 + F_2 F_3 + \cdots + F_{2n-1} F_{2n} = (F_{2n})^2$.

8. Prove or disprove: $\varphi^n = \varphi F_n + F_{n-1}, \ \forall \, n \in \mathbb{N}$.

9. Prove **Cassini's Identity (1680)**: $F_{n-1} \cdot F_{n+1} - F_n{}^2 = (-1)^n$.

10. Make a table of values for $\sqrt{\frac{F_{n+2}}{F_n}}$ starting with $n = 1$ and use this to guess the value of $\lim_{n \to \infty} \sqrt{\frac{F_{n+2}}{F_n}}$.[58]

11. Start a sequence with 0, 0, 1 and generate further terms by summing the previous three. This is called the Tribonacci sequence. Write out 15 terms, after the initial 0, 0, 1, and consider the quotients of consecutive terms. Do they approach a limiting value? If so, what is it?

12. Use the ratio test from calculus to show that $\sum_{n=1}^{\infty} \frac{1}{F_n}$ converges. This is known as the **Carl Series**.

13. Evaluate $\sqrt{1 + \sqrt{1 + \sqrt{1 + \cdots}}}$ exactly, in closed form, and completely simplified.

14. Evaluate $\sqrt{2 + \sqrt{2 + \sqrt{2 + \cdots}}}$ exactly, in closed form, and completely simplified.

15. Evaluate $\sqrt{2}^{\sqrt{2}^{\sqrt{2}^{\cdot^{\cdot^{\cdot}}}}}$ exactly, in closed form, and completely simplified.[59]

---

56 See the journal's homepage at https://www.fq.math.ca/. The current volume and the previous five volumes are only available to subscribers, but all earlier volumes going back to 1963 (volume 1) are available for free download.

57 Burger, Edward B. and Michael Starbird, *The Heart of Mathematics*, third edition, Wiley, Hoboken New Jersey, 2010, p. 215.

58 I first saw this in Burger, Edward B. and Michael Starbird, *The Heart of Mathematics*, third edition, Wiley, Hoboken, NJ, 2010, p. 75.

59 I first saw this in Chamberland, Marc, *Single Digits: In Praise of Small Numbers*, Princeton University Press, Princeton, NJ, 2015, p. 56.

16. When walking up a staircase, a person can take the steps one at time, two at a time, or any combination thereof. For example, you might take two on the first step, one on the second step, one on the third step, two on the fourth step, and so on. Approaching a staircase with n steps, how many different ways are there to ascend?

17. How many binary sequences (using only 0 and 1) are there of length n such that there is never a pair of consecutive 0s?

18. Use induction to prove that every positive integer can be expressed as a sum of non-consecutive Fibonacci numbers. Note: this is not quite Zeckendorf's theorem, because you are not asked to also prove that the representation is unique.

19. Go to www.fq.math.ca/and follow the link to back issues. Look at the papers until you come upon a result that you find interesting. Explain it in your own words.

References/Further Reading

No matter how busy you may think you are, you must find time for reading, or surrender yourself to self-chosen ignorance.

– Confucius

Archibald, Raymond C., "Undergraduate Mathematics Clubs," *The American Mathematical Monthly*, Vol. 25, No. 5, May 1918, pp. 226–238.

Archibald, Raymond C., "Notes on the Logarithmic Spiral, Golden Section and the Fibonacci Series," in Hambidge, Jay, editor, *Dynamic Symmetry: The Greek Vase*, Yale University Press, New Haven, CT, 1920.

Barr, Mark, "Parameters of Beauty," *Architecture: The AIA Journal*, Vol. 60, December, 1929, pp. 325–328, reprinted as Barr, Mark, "Parameters of beauty," *Think*, Vol. 10–11, International Business Machines Corporation, 1944.

Borissavliévitch, Miloutine and Louis Hautecœur, *The Golden Number and the Scientific Aesthetics of Architecture*, Alec Tiranti, London, 1958.

Brousseau, Alfred Brother, "Fibonacci Numbers and Geometry," *The Fibonacci Quarterly*, Vol. 10, No. 3, April 1972, pp. 303–318, 323, available online at www.fq.math.ca/10-3.html.

Brown, Jr., J. L., "Zeckendorf's Theorem and Some Applications," *The Fibonacci Quarterly*, Vol. 2, No. 3, October 1964, pp. 163–168.

Burger, Edward B. and Michael Starbird, *The Heart of Mathematics*, third edition, Wiley, Hoboken, NJ, 2010.

Chamberland, Marc, *Single Digits: In Praise of Small Numbers*, Princeton University Press, Princeton, NJ, 2015.

Chandra, Pravin and Eric W. Weisstein, "Fibonacci Number," from *MathWorld*-A Wolfram Web Resource, available online at http://mathworld.wolfram.com/FibonacciNumber.html.

Conway, John H. and Alex Ryba, "Fibonometry," *Mathematical Gazette*, Vol. 97, No. 540, 2013, pp. 494–495.

Conway, John H. and Alex Ryba, "The Extra Fibonacci Series and the Empire State Building," *Mathematical Intelligencer*, Vol. 38, No. 1, 2016, pp. 41–48.

Cook, Theodore Andrea, *The Curves of Life*, Constable and Company Ltd., London, 1914, available online at https://ia800204.us.archive.org/4/items/cu31924028937179/cu31924028937179.pdf.

Coxeter, Harold Scott MacDonald, "The Golden Section, Phyllotaxis and Wythoff's Game," *Scripta Mathematica*, Vol. 19, No. 2–3, June–September 1953, pp. 135–143.

Coxeter, Harold Scott MacDonald, "The Golden Section and Phyllotaxis," in *Introduction to Geometry*, Chapter 11, John Wiley and Sons, Inc., New York, 1961.

D' Agnese, Joseph (author) and John O'Brien (illustrator), *Blockhead: The Life of Fibonacci*, Henry Holt and Company, New York, 2010. This book is aimed at children.

Davis, Martin, "Hilbert's Tenth Problem Is Unsolvable," *The American Mathematical Monthly*, Vol. 80, No. 3, March 1973, pp. 233–269.

Devlin, Keith, *The Man of Numbers: Fibonacci's Arithmetic Revolution*, Walker Books, New York, 2011.

Dunlap, Richard A., *The Golden Ratio and Fibonacci Numbers*, World Scientific, London, 1997.

Fechner, Gustav, "Über die Frage des goldenen Schnittes (On the Question of the Golden section)," *Archiv für die zeichnenden Künste*, Vol. 11, 1865, pp. 100–112.

Fechner, Gustav, *Zur experimentalen ästhetik (On Experimental aesthetics)*, Hirzel, Leipzig, Germany, 1871.

Fechner, Gustav, *Vorschule der aesthetik (Preschool of Aesthetics)*, 2 volumes, Breitkopf & Härtel, Leipzig, Germany, 1876, available online at https://archive.org/details/vorschulederaest12fechuoft.

Fechner, Gustav, translated by, Niemann, Monika, Julia Quehl, Holger Höge and Carl von Ossietzky, "Various Attempts to Establish a Basic Form of Beauty: Experimental Aesthetics, Golden Section, and Square," *Empirical Studies of the Arts*, Vol. 15, No. 2, 1997, 115–130.

Fibonacci (author) and Laurence Sigler (translator), *Fibonacci's Liber Abaci: A Translation into Modern English of Leonardo Pisano's Book of Calculation*, Springer, 2002.

Fibonacci Quarterly (journal), http://www.fq.math.ca/. The current volume and the previous five volumes are only available to subscribers, but all earlier volumes going back to 1963 (volume 1) are available for free download.

"Fossil Cuttlefish Draw Ancient Moon Nearer to Earth," *New Scientist*, Vol. 80, No. 1125, October 19, 1978, p. 172.

Gardner, Martin, *The Second Scientific American Book of Mathematical Puzzles & Diversions*, The University of Chicago Press, Chicago, IL, 1961, 1987.

Green, Christopher D., "All That Glitters: A Review of Psychological Research on the Aesthetics of the Golden Section," *Perception*, Vol. 24, 1995, pp. 937–968.

Hardy, Godfrey Harold and Edward Maitland Wright, *An Introduction to the Theory of Numbers*, 5th edition, Oxford University Press, Oxford, 1988.

Hayes, Brian, "Sir Roger Penrose's Toilet Paper," December 24, 2017, available online at http://bit-player.org/. You'll have to scroll down a bit to get to the relevant piece.

Herz-Fischler, Roger, *A Mathematical History of Division in Extreme and Mean Ratio*, Wilfrid Laurier University Press, Waterloo, 1987.

Höge, Holger, "Fechner's Experimental Aesthetics and the Golden Section Hypothesis Today," *Empirical Studies of the Arts*, Vol. 13, No. 2, 1995, pp. 131–148.

Höge, Holger, "The Golden Section Hypothesis – A Funeral, but not the Last One ...," *Visual Arts Research*, Vol. 22, No. 1, 1996, pp. 79–89.

Höge, Holger, "The Golden Section Hypothesis – Its Last Funeral," *Empirical Studies of the Arts*, Vol. 15, No. 2, 1997, pp. 233–255.

Hoggatt, Verner E., *Fibonacci and Lucas Numbers*, Houghton Mifflin, Boston, MA, 1969.

Hope-Jones, W., "The Bee and the Pentagon," *The Mathematical Gazette*, Vol. 10, No. 150, January 1921, p. 206.

Howat, Roy, *Debussy in Proportion: A Musical Analysis*, Cambridge University Press, Cambridge, 1983.

Huntley, H. E., *The Divine Proportion: A Study in Mathematical Beauty*, Dover, New York, 1970.

Kahn, Peter G. K. and Stephen Pompea, "Nautiloid Growth Rhythms and Dynamical Evolution of the Earth–Moon System," *Nature*, Vol. 275, October 19, 1978, pp. 606–611.

Kepler, Johannes, *Mysterium Cosmographicum*, Tübingen, 1596.

Kepler, Johannes, *Strena Seu de Niva Sexangula*, Frankfurt am Main, 1611.

Kepler, Johannes, *A New Year's Gift, or on the Six-Cornered Snowflake*, Based on *Strena Seu De Niva Sexangula* and edited by Whyte, L.L., Oxford, 1966. This edition has both modernized Latin and an English translation by Colin Hardie.

Kepler, Johannes, *The Six-Cornered Snowflake: A New Year's Gift*, Paul Dry Books, Philadelphia, PA, 2010. This book reproduces the Latin with an English translation by Jacques Bromberg.

Kimberling, Clark, "Edouard Zeckendorf," *Fibonacci Quarterly*, Vol. 36, November 1998, pp. 416–418.

Koshy, Thomas, *Fibonacci and Lucas Numbers with Applications*, John Wiley & Sons, Inc., New York, 2001.

Kramer, Jonathan, "The Fibonacci Series in Twentieth-Century Music," *Journal of Music Theory*, Vol. 17, No. 1, Spring 1973, pp. 110–148.

Lekkerkerker, Cornelis Gerrit, "Voorstelling van natuurlijke getallen door een som van getallen van Fibonacci," *Simon Stevin*, Vol. 29, 1951, pp. 190–195. This paper presents Zeckendorf's theorem 20 years before Zeckendorf.

Linford, M. B., "Fruit Quality Studies II, Eye Number and Eye Weight," *Pineapple Quarterly*, Vol. III, No. 4, December 1933, pp. 185–195.

Livio, Mario, *The Golden Ratio: The Story of Phi, the World's Most Astonishing Number*, Broadway Books, New York, 2002.

Lucas, Édouard, "Note sur la triangle arithmétique de Pascal et sur la serie de Lamé," *Nouvelle correspondance mathématique*, tome 2, 1876, pp. 70–75.

Lucas, Édouard, "Sur la théorie des nombres premiers," *Atti della Reale Accademia delle scienze di Torino*, Vol. 11, May 1876, pp. 928–937. Lucas is stated as the author on p. 927, which leads into the piece.

Maor, Eli, *e: The Story of a Number*, ninth printing with new material on pp. 183–186, Princeton University Press, Princeton, NJ, 1998.

Markowsky, George, "Misconceptions about the Golden Ratio," *The College Mathematics Journal*, Vol. 23, No. 1, January 1992, pp. 2–19.

Meisner, Gary B., *GoldenNumber.net*, www.goldennumber.net/.

Meisner, Gary B. (author) and Rafael Araujo (artist), *The Golden Ratio: The Divine Beauty of Mathematics*, Race Point Publishing, Minneapolis, Minnesota, 2018.

Mitchison, Graeme J., "Phyllotaxis and the Fibonacci Series," *Science*, Vol. 196, No. 4287, April 15, 1977, pp. 270–275.

Moore, Richard E. M., "Mosaic Units: Patterns in Ancient Mosaics," *Fibonacci Quarterly*, Vol. 8, April 1970, pp. 281–310.

Moritz, Robert Edouard, *On Mathematics*, Dover, New York, 1958, p. 145. Previously issued as *Memorabilia Mathematica or the Philomath's Quotation-Book*.

Niven, Ivan, *An Introduction to the Theory of Numbers*, John Wiley & Sons, New York, 1991.

O'Connell, Margaret K. and Daniel T. O'Connell, "Letter to the Editor," *The Scientific Monthly*, Vol. 78, November 1951, p. 333.

Onderdonk, Philip B., "Pineapples and Fibonacci Numbers," *The Fibonacci Quarterly*, Vol. 8, No. 5, December 1970, pp. 507–508.

Pacioli, Luca, *Divina Proportione*, Paganini, Venice, 1509.

Phillips, Flip, J. Farley Norman, and Amanda M. Beers, "Fechner's Aesthetics Revisited," *Seeing and Perceiving*, Vol. 23, 2010, pp. 263–271.

Pierce, John Crawford, "The Fibonacci Series," *The Scientific Monthly*, Vol. 78, October 1951, pp. 224–228.

Posamentier, Alfred S. and Ingmar Lehmann, *The (Fabulous) Fibonacci Numbers*, Prometheus Books, Amherst, NY, 2007.

Posamentier, Alfred S. and Ingmar Lehmann, *The Glorious Golden Ratio*, Prometheus Books, Amherst, NY, 2012.

Roberts, Siobhan, *Genius at Play – The Curious Mind of John Horton Conway*, Bloomsbury, New York, 2015.

Roeder, Oliver, "Even after 22 Trillion Digits, We're Still No Closer to the End of Pi," *FiveThirtyEight*, March 14, 2018, available online at https://fivethirtyeight.com/features/even-after-22-trillion-digits-were-still-no-closer-to-the-end-of-pi/.

Runion, Garth E., *The Golden Section and Related Curiosa*, Glenview, IL, Scott, Foresman and Company, 1972.

Runion, Garth E., *The Golden Section*, Dale Seymour Publications, Palo Alto, CA, 1990.

Scholfield, P. H., *The Theory of Proportion in Architecture*, Cambridge University Press, Cambridge, UK, 1958.

Schultz, Kyle, "Penrose Tilings," The University of Georgia, available online at http://jwilson.coe.uga.edu/EMAT6680Fa05/Schultz/Penrose/Penrose_Main.html.

Shanks, Daniel, *Solved and Unsolved Problems in Number Theory*, 3rd edition, Chelsea, New York, 1985.

Silver, Stephen, "Art Lesson- How to Use the Golden Section for Character Design," available online at www.youtube.com/watch?v=PEiaGGp5ddY&feature=youtu.be.

Singh, Parmanand, "The So-called Fibonacci Numbers in Ancient and Medieval India," *Historia Mathematica*, Vol. 12, No. 3, 1985, pp. 229–244.

Sloane, Neil J. A., "Decimal Expansion of Golden Ratio Phi (or Tau) = (1 + sqrt(5))/2," *The On-Line Encyclopedia of Integer Sequences*, available online at https://oeis.org/A001622.

Sloane, Neil J. A., "Fibonacci Numbers: $F(n) = F(n-1) + F(n-2)$ with $F(0) = 0$ and $F(1) = 1$," *The On-Line Encyclopedia of Integer Sequences*, https://oeis.org/A000045.

Sobel, Dava, "Undersea Calendar," *OMNI*, April 1979, Vol. 1, No. 7, p. 34.

Thompson, D'Arcy Wentworth, *On Growth and Form*, Cambridge University Press, Cambridge, UK, 1917.

Vajda, Steven, *Fibonacci and Lucas Numbers and the Golden Section*, Halsted Press, New York, 1989.

Vorobyov, Nikolai Nikolayevich, *The Fibonacci Numbers*, D. C. Heath and Company, Boston, MA, 1963.

Weisstein, Eric W., "Binet's Fibonacci Number Formula," from *MathWorld*-A Wolfram Web Resource, available online at http://mathworld.wolfram.com/BinetsFibonacciNumberFormula.html.

Wilson, Robin and John J. Watkins, editors, *Combinatorics: Ancient & Modern*, Oxford University Press, Oxford, 2013.

Yarden, D., "A Bibliography of the Fibonacci Sequence," *Riveon Lematematika*, Vol. 22, January 1948, pp. 36–45.

Young, Robert M., *Excursions in Calculus: An Interplay of the Continuous and the Discrete*, The Mathematical Association of America, 1992.

Zeckendorf, Edouard, "Representation des nombres naturels par une somme de nombres de Fibonacci ou de nombres de Lucas," *Bulletin de la Societe Royale des Sciences de Liege*, Vol. 41, 1972, pp. 179–182.

Zeising, Adolf, *Aesthetische Forschungen (Aesthetic Inquiries)*, Medinger, Frankfurt am Main, 1855.

Zeising, Adolf, *Der goldene Schnitt*, Engelmann in Komm., Halle, Germany, 1884.

14

The Tower of Hanoi

What defines a major contribution to mathematics? Does one have to conquer a problem that has baffled mathematicians for over a century or create a new field of mathematics? Can the set of the most important mathematicians be defined as the recipients of math's major prizes, like the Fields Medal and the Wolf Prize?

Is there room on the list for those who have simply inspired the greats? Consider the case for W. W. Rouse Ball (Figure 14.1). This man only held an M.A., not a Ph.D., and, according to Mathematics Genealogy Project, only directed one dissertation.[1] His published writing was primarily on the history of mathematics. He also wrote *Mathematical Recreations and Essays*. As the title implies, it was on "recreational" mathematics, the sort an amateur might engage in as a hobby, and not the serious mathematics that's likely to gain the respect of one's peers. But this book, aimed not at specialists, but rather at a general audience, was wildly successful. It first appeared in 1892 and went through many editions. The 10th appeared in 1922. This edition is

FIGURE 14.1
W. W. Rouse Ball (1850–1925).

Public domain.

---

1 https://www.genealogy.math.ndsu.nodak.edu/id.php?id=27230 The student was Ernest Barnes, not a giant of mathematics.

special because a copy of it printed in 1928 was the book chosen by the winner of the Sherborne School's Christopher Morcom Science Prize, namely, Alan Turing.[2] This young man went on to become England's greatest cryptanalyst in World War II. His interest in the subject arose well before his war work and was very likely sparked by the final chapter of Ball's book. How highly should we rate the importance of inspiring a great mind?

When young people are deciding what subjects to pursue, it's not the latest research papers that capture their interest, but instead popularizations. Andrew Wiles's interest in Fermat's Last theorem sprang from a description of it in a book by E. T. Bell. Many other great mathematicians credit Bell for leading them to mathematics. Another great popularizer was Martin Gardner. His legacy is huge and there have even been conferences held in his honor.[3] Although he didn't even consider himself a mathematician, his superbly written essays and books inspired generations of young people to make the subject their own. An inspiration for a large number of the great cryptologists of the last fifty years is a historian, David Kahn. His best-known book, *The Codebreakers*, led many to careers at America's National Security Agency, and others, such as Martin Hellman and Whitfield Diffie, to greatness in the public side of the subject.

Ball died in 1925, a few years after the 10th edition of *Mathematical Recreations and Essays* had been published. The next edition was revised by H. S. M. Coxeter, a man who *did* do serious mathematics.[4] Coxeter was able to get Abraham Sinkov, a great cryptologist (now in the National Security Agency's Hall of Honor) to handle the chapter titled "Cryptographs and Ciphers."[5] These all-stars felt updating Ball's book important enough to devote some of their time to it, so again, how important should we consider the original effort by Ball? A 13th edition appeared in 1987 and is still in print. Few books stay in print for over 125 years!

In my opinion, it doesn't matter that this book focused on *recreational* mathematics. Once a bright young person is turned on to mathematics, he or she will quickly enough discover that there are many different kinds of mathematics that can be pursued, and may well pursue an area that has a serious impact on our world.

Speaking of which, Ball's book did contain one very serious problem, translated from the report given on it by Henri de Parville in 1884.[6] It is quoted here from the second edition:

> In the great temple at Benares, says he, beneath the dome which marks the centre of the world, rests a brass-plate in which are fixed three diamond needles, each a cubit high and as thick as the body of a bee. On one of these needles, at the creation, God placed sixty-four discs of pure gold, the largest disc resting on the brass plate, and the others getting smaller and smaller up to the top one. This is the Tower of Bramah. Day and night unceasingly the priests transfer the discs from one diamond needle to another according to the fixed and immutable laws of

2 Hodges, Andrew, *Alan Turing: The Enigma*, Simon and Schuster, New York, 1983, p. 56. Also see Christensen, Chris, "Alan Turing's First Cryptology Textbook and Sinkov's Revision of It," *Cryptologia*, Vol. 34, No. 1, January 2009, pp. 27–43.

3 https://en.wikipedia.org/wiki/Gathering_4_Gardner

4 http://www-history.mcs.st-andrews.ac.uk/Biographies/Ball.html

5 Christensen, Chris, "Alan Turing's First Cryptology Textbook and Sinkov's Revision of It," *Cryptologia*, Vol. 34, No. 1, January 2009, pp. 27–43.

6 de Parville, Henri, "La Tour d'Hanoï et la question du Tonkin," *La Nature*, Vol. 12, 1884, pp. 285–286.

Bramah, which require that the priest must not move more than one disc at a time and that he must place this disc on a needle so that there is no smaller disc below it. When the sixty-four discs shall have been thus transferred from the needle on which at the creation God placed them to one of the other needles, tower, temple, and Brahmins alike will crumble into dust, and with a thunder-clap the world will vanish.[7]

Naturally, interest in this problem persists to this day! It is generally referred to as the Tower of Hanoi, although Hanoi is not mentioned in the account reproduced above. Richard E. Korf used a computer to investigate a variant of this problem and found it convenient to make a distinction between disk and disc.

To avoid confusion, we use the term "disk" to refer to a magnetic disk, and the term "disc" to refer to a Towers of Hanoi disc.[8]

The results of his work will be seen later in this chapter, but his convention will be followed throughout. The only exceptions will be quoted material and citations.

So, how much time does the human race have left? It's hard to say exactly, given that we don't know when the priests began moving the discs, nor the rate at which they are moving them. Still, we can begin a mathematical investigation. It helps to have images of the Tower of Hanoi to look at. Though I have never been to Benares, I do have a smaller wooden version of the Tower (made for me by my great-grandfather), which is shown in Figure 14.2. As you can see, it has just seven discs.

FIGURE 14.2
The Tower of Hanoi.

Image created by Craig P. Bauer.

---

7 Ball, W. W. Rouse, *Mathematical Recreations and Essays*, second edition, Macmillan and Co., London, 1896, p. 92.

8 Korf, Richard E., "Linear-Time Disk-Based Implicit Graph Search," *Journal of the ACM*, Vol. 55, No. 6, December 2008, pp. 26:1–26:40; p. 26:3, footnote 2, quoted here.

In this smaller example, the mission is to move the tower of seven discs from the first peg to the third. The rules are the same as for the larger version:

1. Move one disc at a time.
2. Don't ever place a disc on top of a smaller disc.

What is the minimum number of moves needed to complete this smaller version? We could try various methods of solution, keeping track of the number of moves made, but simplifying things further might be a better idea. We can rapidly find the minimal solution for versions with even smaller numbers of discs. Often an approach like this gives insight into the general solution. That is, a pattern may be detected. For a single disc, we can obviously complete the task in one move. Two discs require three moves. Verify this for yourself! So far, we have

Number of Discs	Number of Moves Needed
1	1
2	3

One pattern for the minimum number of moves, as the number of discs grows, could be 1, 3, 5, 7, ... Algebraically, the minimum number of moves for n discs is then $2n - 1$. However, it is dangerous to base a conclusion on such a small data set. Let's examine the three-disc case. A step-by-step solution follows in Figures 14.3 through 14.10.

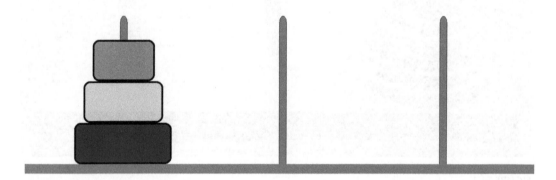

FIGURE 14.3
The starting position.

Image created by Garrett Ruths.

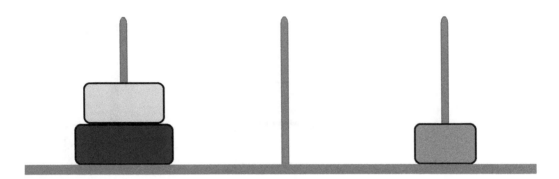

FIGURE 14.4
After move 1.

Image created by Garrett Ruths.

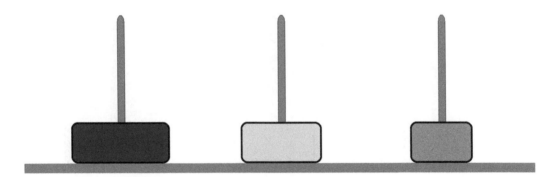

FIGURE 14.5
After move 2.

Image created by Garrett Ruths.

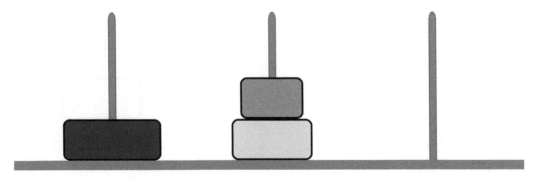

FIGURE 14.6
After move 3.

Image created by Garrett Ruths.

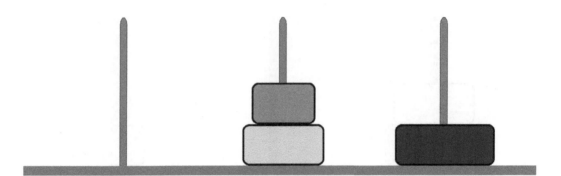

FIGURE 14.7
After move 4.

Image created by Garrett Ruths.

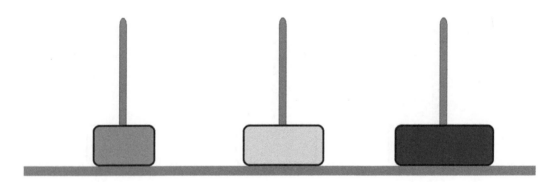

FIGURE 14.8
After move 5.

Image created by Garrett Ruths.

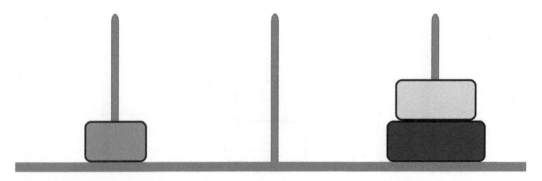

FIGURE 14.9
After move 6.

Image created by Garrett Ruths.

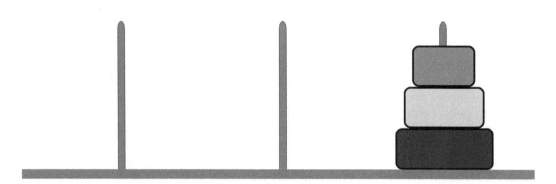

FIGURE 14.10
After move 7.

Image created by Garrett Ruths.

So, three discs require 7 moves.[9] Our table expands to:

Number of Discs	Number of Moves Needed
1	1
2	3
3	7

We see that the solution $2n - 1$ cannot work, as it incorrectly predicts the minimum number of moves required with three discs. This is a good time to stop reading and try to expand the table above on your own. You don't need to purchase or construct the Tower. There are many websites, some of which are listed in the References/Further Reading section at the end of this chapter, that allow you to manipulate a digital version of the Tower. See if you can find a better pattern for the minimum number of moves – one that continues! Once you've found it, continue reading and see if your answer is correct.

With four discs, experimentation shows that there's a solution in 15 moves. So, the sequence of the minimum number of moves begins 1, 3, 7, 15. As the number of discs grows, you may still be able to find solutions with a minimum number of moves, but you might well have less and less confidence that they are indeed minimal. For example, the next values in the sequence, for moving five and six discs, are 31 and 63, respectively. How confident can you be, after completing the six-disc task in 63 moves, that there's no solution in 62 moves? We clearly have an upper bound on the number of moves needed, and you may recognize a pattern in 1, 3, 7, 15, 31, 63, but we are far from a proof of minimality in regard to moving a six-disc tower.

So, let's look at this from another perspective. In order to move the Tower from the leftmost peg to the rightmost peg, we must move the largest disc. But the largest disc cannot be moved until there are no other discs on top of it. And it cannot be moved to the rightmost peg unless there are no discs there. Thus, just prior to moving the largest

---

9 If you do it in less, you're cheating!

disc to the desired position, it must be on the leftmost peg, while all the other discs are on the middle peg.

So, an optimal solution with seven discs involves somehow moving the top six discs to the second peg, followed by moving the largest disc to the third peg. But the number of moves required to move the top six discs to the middle peg is no different than the number of moves needed to move them to the third peg. The Tower is symmetric with respect to pegs. Letting M_n denote the minimum number of moves to transfer n discs, we have $M_7 = M_6 + 1 +$ the number of moves needed to get the six discs on the middle peg on top of the largest disc, which is now on the rightmost peg. But this last term is the same as the number of moves required to move six discs from one peg to any peg, namely, M_6 again. That is, it's now established that $M_7 = M_6 + 1 + M_6$. Simplifying a bit, $M_7 = 2M_6 + 1$. The solution has been reduced to a difference equation. We can now generate the answer to the minimum number of moves question for any number of discs. For example, $M_7 = 2 \cdot 63 + 1 = 127$. I solved this 7-disc version many times as a child and the world never ended.

A 10-disc version proved to be much more dangerous in an episode of the sci-fi series *Doctor Who*. With the last move, following the optimal algorithm, the doctor destroyed the world of the Celestial Toymaker.[10]

But what about the 64-disc version at Benares? If you are not keen to iterate by hand until getting to M_{64}, feel free to use a spreadsheet or a computer program to determine this value. Alternatively, a technique is developed in Chapter 15 that can be used to find the nth term of such a sequence non-recursively. That is, it can be determined without knowing any of the previous values. In this instance, the non-recursive formula is $M_n = 2^n - 1$. By coincidence, numbers of this form are known as Mersenne numbers. So, depending on the context, the M could stand for "Minimum number of moves" or "Mersenne." Both the recursive and non-recursive solutions appeared in a paper published in 1883.[11] Whichever approach you use, you'll find that the full-on Tower requires a minimum of 18,446,744,073,709,551,615 moves to complete.

As was mentioned before, we don't know when the Brahmins began, nor how quickly they're moving their discs, but if a move is made every second, the time required to complete a 64-disc Tower exceeds 500 billion years.

An Optimal Algorithm

In addition to knowing the minimum number of moves needed to solve the Tower, we'd also like to have an algorithm for carrying out the solution in this number of moves. It turns out that the optimal algorithm is very simple to describe.

If you have an odd number of discs, you will move the smallest disc during odd number moves (i.e., the first move, the third move, etc.) and it will always move one peg to the left. If it is already at the left-most peg, as it is when you start the puzzle, then move it like Pac-Man going through the side door, so that it ends up on the

---

10 *Doctor Who*, The Celestial Toymaker, Part 4: The Final Test, 1966. See https://en.wikipedia.org/wiki/The_Celestial_Toymaker for details.

11 de Longchamps, Gaston, "Variétés," *Journal de Mathématiques Spéciales*, Vol. 2, No. 2, 1883, pp. 286–287.

right-most peg. For even moves, simply make the only move possible that doesn't involve the smallest disc.

If the number of discs is even, the only change you need to make to the algorithm given above is to move the smallest disc to the right each time.

That's all there is to it!

This algorithm was first published by N. Claus de Siam in 1884.[12]

Despite having been discovered so long ago, the first rigorous proof that the number of moves this algorithm uses is indeed the minimum wasn't put forth until 1981. Previous authors made use of the fact that in an optimal solution the largest disc will only move once, without actually proving that this is the case.[13]

As was mentioned above, the References/Further Reading section at the end of this chapter lists several websites that allow you to manipulate a digital version of the Tower. Some of these also allow you to choose the number of discs and pegs, or to simply watch as the optimal solution is carried out automatically.

Another potential method of solution is to simply move discs at random and hope for the best. This random approach has actually been studied and a formula was found for the number of moves that this technique would require, on average. For an 8-disc version of the Tower, it's approximately 576,000.[14]

Given our previous calculation of 500 billion years to complete the 64-disc tower using the optimal algorithm, and that the universe is only estimated to be about 13.8 billion years old, it doesn't really matter when the Brahmins began moving discs. The movement of the last disc is still so far in the future that there's no need to worry now.

Still, it would be interesting to know more about what is going on at Benares. Tracing Henri de Parville's story back, we see that he got it from Professor N. Claus de Siam, mandarin of the Collège Li-Sou-Stian. This gentleman was reporting from Tonkin and had written about the Tower in 1883, calling it *La Tour D'Hanoi* (The Tower of Hanoi).[15]

It turns out that the letters in the name "N. Claus de Siam" can be anagrammed into "Lucas d'Amiens" and "Li-Sou-Stian" can become "Saint-Louis." It's no coincidence that the mathematician Édouard Lucas (Figure 14.11) was born in Amiens, France and was a professor at the Lycée Saint-Louis in Paris. Lucas's colorful pen name and the story of the Brahmins of Benares were simply made up to give what is basically just a mathematical game a neater context.

It's said that if you're a professor at a research institution, two of the worst things you can do for your career are to show an interest in teaching and to write anything at the popular level. Lucas did both. He was a "gifted and entertaining teacher"[16] and, like

12 Hinz, Andreas M., Sandi Klavžar, Uroš Milutinović, and Ciril Petr, *The Tower of Hanoi – Myths and Maths*, Birkhäuser, Heidelberg, Germany, 2013, p. 43, which cites N. Claus de Siam, "La Tour d'Hanoï, Jeu de calcul," *Science et Nature*, Vol. 1, 1884, pp. 127–128. Claus credited this algorithm to Raoul Olive, a nephew of Édouard Lucas. For this reason, it's been called "Olive's algorithm."

13 Hinz, Andreas M., Sandi Klavžar, Uroš Milutinović, and Ciril Petr, *The Tower of Hanoi – Myths and Maths*, Birkhäuser, Heidelberg, Germany, 2013, p. 44. A proof by induction is given on pp. 44–45. The 1981 proof is in Wood, Derick, "The Towers of Brahma and Hanoi Revisited," *Journal of Recreational Mathematics*, Vol. 14, No. 1, 1981, pp. 17–24.

14 Hinz, Andreas M., Sandi Klavžar, Uroš Milutinović, and Ciril Petr, *The Tower of Hanoi – Myths and Maths*, Birkhäuser, Heidelberg, Germany, 2013, p. 155.

15 Claus, N., La Tour d'Hanoï, Véritable casse-tête annamite, P. Bousrez, Tours, 1883.

16 Koshy, Thomas, *Fibonacci and Lucas Numbers with Applications*, John Wiley & Sons, Inc., New York, 2001, p. 6.

FIGURE 14.11
Édouard Lucas (1842–1891).

Public domain. Note: his full name was François Édouard Anatole Lucas.

Ball, a popularizer of mathematics. His great work in this direction was *Récréations mathématiques*, which ran to four volumes, published between 1882 and 1894, with the last two being posthumous. When a servant accidentally dropped plates at a banquet Lucas was attending, a shard badly cut the mathematician's cheek. Infection followed and Lucas died within a few days, at age 49.[17]

During his lifetime, Lucas's devotion to recreational mathematics and teaching seem to have cast a shadow over him, as far as the great mathematicians of the day were concerned. It is only fairly recently that he has gained the respect of mathematicians for his other work. *Récréations mathématiques* remains in print, but Lucas's collected mathematical papers has not yet been published.

You may recall from Chapter 13 that Lucas is the one responsible for giving the Fibonacci numbers their name. So, it's appropriate that there's a connection between these numbers and the Tower. To see this, count the number of distinct arrangements of discs that appear on the center peg, including the arrangement of no discs, when pursuing the optimal solution to the Tower. If n discs are used, this value will be the Fibonacci number F_{n+1}.[18]

---

17 Koshy, Thomas, *Fibonacci and Lucas Numbers with Applications*, John Wiley & Sons, Inc., New York, 2001, p. 6. "Obituary: Édouard Lucas," *Popular Science Monthly*, Vol. 40, January 1892, p. 432, available online at https:// en.wikisource.org/wiki/Popular_Science_Monthly/Volume_40/January_1892/Obituary_Notes. Also see Hinz, Andreas M., Sandi Klavžar, Uroš Milutinović, and Ciril Petr, *The Tower of Hanoi – Myths and Maths*, Birkhäuser, Heidelberg, Germany, 2013, pp. 12–13.

18 Hinz, Andreas M., Sandi Klavžar, Uroš Milutinović, and Ciril Petr, *The Tower of Hanoi – Myths and Maths*, Birkhäuser, Heidelberg, Germany, 2013, p. 129, exercise 2.6, which cites Bennish, Joseph, "The Tower of Hanoi Problem and Mathematical Thinking," *Missouri Journal of Mathematical Sciences*, Vol. 11, No. 3, Fall 1999, pp. 164–166.

An Open Problem

Henry Dudeney (Figure 14.12) is another example of a person who went by the beat of his own drum when it came to mathematics. He delighted in being a puzzle maker, and was one of the best England has seen.

In 1908, Dudeney put a twist on the Tower of Hanoi by adding a fourth peg.[19] He called the new challenge "The Reve's puzzle." This might seem to be much easier than the original. One can simply ignore one of the pegs in the middle and solve it just like the original Tower. However, if you use the full flexibility that the four pegs offer, trying to find an optimal solution, for an arbitrary number of discs, is extremely difficult.

In 1939, Bonnie M. Stewart, who was pursuing a Ph.D. at the University of Wisconsin, posed the problem in greater generality as a challenge to readers of *The American Mathematical Monthly*.[20] He asked for the minimum number of moves for *n* discs and *k* pegs, where *k* is any number greater than three. Two "solutions" saw print in the March 1941 issue of *The American Mathematical Monthly*, one from J. Sutherland Frame, and another from Stewart, the original poser of the problem.[21] However, as the editor

FIGURE 14.12
Henry Dudeney (1857–1930).

Public domain.

19 Hinz, Andreas M., Sandi Klavžar, Uroš Milutinović, and Ciril Petr, *The Tower of Hanoi – Myths and Maths*, Birkhäuser, Heidelberg, Germany, 2013, p. 46 and 165. In the footnote on p. 46, Hinz notes that Dudeney first put forth the four-peg version in some London newspapers in 1902 (Hinz doesn't specify which.). He then reproduced it in Dudeney, Henry E., *The Canterbury Puzzles and Other Curious Problems*, E. P. Dutton, New York, 1908.

20 Stewart, Bonnie M., "Problems and Solutions: Advanced Problem 3918," *The American Mathematical Monthly*, Vol. 46, No. 6, June–July 1939, p. 363.

21 Frame, J. Sutherland, "Problems and Solutions: Advanced Problems: Solutions: 3918," *The American Mathematical Monthly*, Vol. 48, No. 3, March 1941, pp. 216–217. Stewart, Bonnie M., "Problems and Solutions: Advanced Problems: Solutions: 3918," *The American Mathematical Monthly*, Vol. 48, No. 3, March 1941, pp. 217–219.

pointed out, both relied on an assumption that needed to be proven.[22] So, a solution had not really been obtained. The formula that the work of both Frame and Stewart leads to is now known as the Frame-Stewart conjecture.

In the special case of the Reve's puzzle with n discs the Frame-Stewart conjecture claims that the minimum number of moves needed is

$$R(n) = \sum_{i=1}^{k} i2^{i-1} - \left(\frac{k(k+1)}{2} - n\right)2^{k-1},$$

where k is the smallest integer such that $n \leq \frac{k(k+1)}{2}$.[23]

For example, to find the minimum number of moves for the 7-disc version, we first seek the smallest k such that $7 \leq \frac{k(k+1)}{2}$. We have $k = 4$, because $7 \leq \frac{4(4+1)}{2} = 10$, while $k = 3$ makes $\frac{k(k+1)}{2} = \frac{3(3+1)}{2} = 6$, which is less than 7.

Now that we know $k = 4$, we can evaluate

$$R(7) = \sum_{i=1}^{4} i2^{i-1} - \left(\frac{4(4+1)}{2} - 7\right)2^{4-1}$$

$$= 1 \cdot 2^{1-1} + 2 \cdot 2^{2-1} + 3 \cdot 2^{3-1} + 4 \cdot 2^{4-1} - \left(\frac{4(4+1)}{2} - 7\right)2^{4-1}$$

$$= 1 \cdot 2^0 + 2 \cdot 2^1 + 3 \cdot 2^2 + 4 \cdot 2^3 - \left(\frac{4(5)}{2} - 7\right)2^3$$

$$= 1 \cdot 1 + 2 \cdot 2 + 3 \cdot 4 + 4 \cdot 8 - (10 - 7)8$$

$$= 1 + 4 + 12 + 32 - (3)8$$

$$= 49 - 24$$

$$= 25.$$

This is far less than for the minimum number of moves needed for the same number of discs on the three-peg Tower.

As usual, researchers have managed to chip away at the problem of the Reve's puzzle without yet obtaining the complete solution. In 1999, Jens-P. Bode and Andreas M. Hinz verified that the conjecture is correct for 17 or fewer discs.[24] Richard E. Korf confirmed it through 24 discs in 2004[25] and Sebastian Strohhäcker took it through 25 discs a year

---

22 Dunkel, Otto, "Editorial Note," *The American Mathematical Monthly*, Vol. 48, No. 3, March 1941, p. 219. Also see Hinz, Andreas M., Sandi Klavžar, Uroš Milutinović, and Ciril Petr, *The Tower of Hanoi – Myths and Maths*, Birkhäuser, Heidelberg, Germany, 2013, pp. 175–177.

23 Rosen, Kenneth, *Handbook of Discrete and Combinatorial Mathematics*, CRC Press, Boca Raton, FL, 2000, p. 94.

24 Bode, Jens-P., and Andreas M. Hinz, "Results and Open Problems on the Tower of Hanoi," *Congressus Numerantium*, Vol. 139, 1999, pp. 113–122.

25 Korf, Richard E., "Best-First Frontier Search with Delayed Duplicate Detection," *Nineteenth National Conference on Artificial Intelligence* (AAAI-2004), The MIT Press, Cambridge, MA, 2004, pp. 650–657.

later.[26] Most recently, in 2007, Korf and Ariel Felner confirmed the claim through 30-disc.[27] The runtime for the 30-disc confirmation took over 28 days.[28]

If the formula for the Reve's puzzle is eventually proven to be correct for any number of discs, there's still a lot of work to do. Is the Frame-Stewart conjecture correct when the number of pegs is greater than four? If Donald Knuth is right, we may never know. In a 1979 letter to Martin Gardner, he commented, "I doubt if anyone will ever resolve the conjecture; it is truly difficult."[29]

Years later Knuth remarked:

> In the case of the 4-peg "Tower of Hanoi," there are many, many ways to achieve what we think is the minimum number of moves, but we have no good way to characterize all those solutions. So that's why I personally came to the conclusion that I was never going to be able to solve it, and I stopped working on it in 1972. But I spent a solid week working on it pretty hard.[30]

An Unexpected Connection

We can further analyze an n-disc, three-peg Tower. Begin by numbering the pegs, from left to right, 0, 1, and 2. Then any arrangement of the discs on the pegs, where no disc is on top of a smaller disc, may be represented by an n-tuple. The number in the first position gives the peg on which the largest disc sets, the number in the second position gives the peg on which the second largest disc sets, and so on. Finally the last number in the n-tuple gives the peg where the smallest disc sets.[31]

Using this representation, for a 7-disc Tower, the initial position is 0000000 and the final position, after all discs have been moved to the rightmost peg, is 2222222. The intermediate position shown in Figure 14.13 is represented by 0022210.

With each state of the Tower reduced to an n-tuple, all possible states can be written down. We can even spread them out on a piece of paper and draw a line from each state to every state that it can become by the movement of a single disc. There are many ways to make such sketches, depending on the relative positions of the various n-tuples. Some lovely arrangements, for $n = 1$, 2, and 3, are shown in Figures 14.4 and 14.5.

---

26 Strohhäcker, Sebastian, A Program to Find Distances in Tower of Hanoi and Related Graphs, Technical report, Technische Universität München, Garching, 2008.

27 Korf, Richard E. and Ariel Felner, "Recent Progress in Heuristic Search: A Case Study of the Four-Peg Towers of Hanoi Problem," in Veloso, Manuela M., editor, *Proceedings of the Twentieth International Joint Conference on Artificial Intelligence (IJCAI-07)*, AAAI Press, Menlo Park, CA, 2007, pp. 2324–2329. Also see Korf, Richard E., "Linear-Time Disk-Based Implicit Graph Search," *Journal of the ACM*, Vol. 55, No. 6, December 2008, pp. 26:1–26:40.

28 Korf, Richard E. and Ariel Felner, "Recent Progress in Heuristic Search: A Case Study of the Four-Peg Towers of Hanoi Problem," in Veloso, Manuela M., editor, *Proceedings of the Twentieth International Joint Conference on Artificial Intelligence (IJCAI-07)*, AAAI Press, Menlo Park, CA, 2007, pp. 2324–2329, p. 2329 cited here.

29 Lunnon, W. F., "The Reve's Puzzle," Correspondence, *The Computer Journal*, Vol. 29, No. 5, 1986, p. 478.

30 Knuth, Donald, "All Questions Answered," *Notices of the AMS*, Vol. 49, No. 3, March 2002, pp. 318–324, p. 321 quoted here.

31 This representation is shown in Hinz, Andreas M., Sandi Klavžar, Uroš Milutinović, and Ciril Petr, *The Tower of Hanoi – Myths and Maths*, Birkhäuser, Heidelberg, Germany, 2013, p. 72.

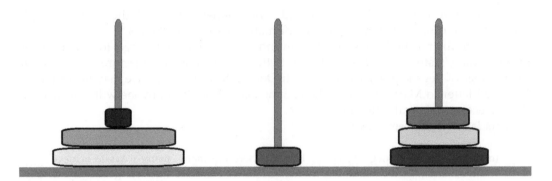

FIGURE 14.13
An intermediate state of the Tower, on the way to a solution.

Image created by Garret Ruths.

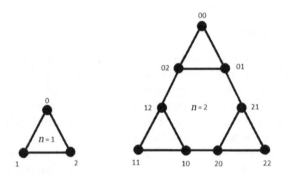

FIGURE 14.14
Hanoi graphs for one and two discs.

Images created by Garrett Ruths, based on Hinz, Andreas M., Sandi Klavžar, Uroš Milutinović, and Ciril Petr, *The Tower of Hanoi – Myths and Maths*, Birkhäuser, Heidelberg, Germany, 2013, p. 95.

The results resemble Sierpiński's triangle, seen in Chapter 10 when the entries of Pascal's triangle were shaded based on the results modulo 2. So we have an unexpected connection between mathematical topics! The connection is even more impressive when the graph is overlaid on A Sierpiński triangle, as shown in Figure 14.16.

The connection is no coincidence – the pattern does continue for larger numbers of discs.

The Tower of Hanoi was first represented by such a graph in a paper from 1944, although the trio of authors (Scorer, Grundy, and Smith) did not remark on the connection with Sierpiński's triangle.[32] The term **Hanoi graph** for such representations first appeared in 1986.[33] Many sources say that Ian Stewart, in 1989, was the first to

32 Scorer, R. S., Patrick Michael Grundy, and Cedric Austen Bardell Smith, "Some Binary Games," *The Mathematical Gazette*, Vol. 28, No. 280, July 1944, pp. 96–103.

33 Lu, Xuemiao, "Towers of Hanoi Graphs," *International Journal of Computer Mathematics*, Vol. 19, No. 1, 1986, pp. 23–38.

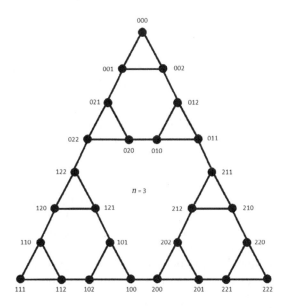

FIGURE 14.15
Hanoi graph for three discs.

Image created by Garrett Ruths, based on Hinz, Andreas M., Sandi Klavžar, Uroš Milutinović, and Ciril Petr, *The Tower of Hanoi – Myths and Maths*, Birkhäuser, Heidelberg, Germany, 2013, p. 95.

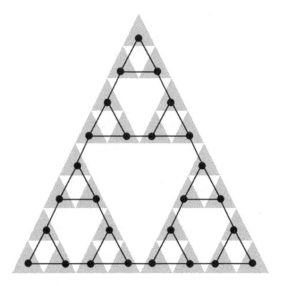

FIGURE 14.16
A Hanoi graph and Sierpiński's triangle.

Image created by Garrett Ruths, based on Hinz, Andreas M., Sandi Klavžar, Uroš Milutinović, and Ciril Petr, *The Tower of Hanoi – Myths and Maths*, Birkhäuser, Heidelberg, Germany, 2013, p. 162. A version of this image previously appeared in Hinz, Andreas M., and Andreas Schief, "The Average Distance on the Sierpiński Gasket," *Probability Theory and Related Fields*, Vol. 87, 1990, pp. 129–138, p. 134 cited here.

explicitly make the connection between Hanoi graphs and Sierpiński's triangle.[34] These include the book *The Tower of Hanoi – Myths and Maths* by Andreas Hinz, and colleagues, but in the introduction to this book, written by Ian Stewart, he says that "several others had already spotted the connection" before him and he only rediscovered it independently.[35] In an earlier paper, Stewart had written:

> The graph H_n for n-disc Hanoi was introduced by Scorer et al.,[36] rediscovered by Er,[37] re-rediscovered by Lu,[38] and (re)^{3}discovered by me.[39] It is in the nature of the topic that most people working on it don't know the literature, so (re)^{n}discovery for all $n \geq 4$ will occur with probability 1 in the long run.[40]

But having looked at the papers Stewart cites, I feel he deserves the credit. Scorer showed the graphs for $n = 2$ and $n = 3$, but didn't note their resemblance to Sierpiński's triangle. Er neither showed the graphs nor mentioned Sierpiński. Lu coined the term **Hanoi graph** and presented them for $n = 1$, 2, and 3, but like those who preceded him, didn't mention Sierpiński. Stewart did.

Theoretically, graphs for any number of discs can be made, but how far can we go before it becomes impractical? That is, how quickly does the number of states grow as the number of discs increases?

Well, for n discs there are $(1/2)(3^{n-1} + 1)$ legal states, so the number of possibilities grows exponentially.[41] With 40 discs, which is a much smaller stack than that used in Banares, there are 2,026,277,576,509,488,134 states. By comparison, the number of states of Rubik's cube is the smaller value 901,083,404,981,813,616.[42] If we consider illegal states, where a disc may rest on a smaller disc, as well as legal states, we have the nice formula $\frac{1}{2}(n+2)!$ for the number of possibilities with n discs.[43]

The graphs are pretty (when they can be drawn!), but what are they good for? Before answering, I'd like to point out that beauty is a sufficient justification! Few people ask what the practical applications of a beautiful painting or song are. These graphs are

---

34 For example, Romik, Dan, "Shortest Paths in the Tower of Hanoi Graph and Finite Automata," *SIAM Journal on Discrete Mathematics*, Vol. 20, No. 3, 2006, pp. 610–622, available online at https://arxiv.org/pdf/math/0310109.pdf, which cites Stewart, Ian, "Le lion, le lama et la laitue," *Pour la Science*, Vol. 142, 1989, pp. 102–107.

35 Hinz, Andreas M., Sandi Klavžar, Uroš Milutinović, and Ciril Petr, *The Tower of Hanoi – Myths and Maths*, Birkhäuser, Heidelberg, Germany, 2013, p. 120 for the claim that Stewart was first. Hinz cites Stewart, Ian, "Le lion, le lama et la laitue," *Pour la Science*, Vol. 142, 1989, pp. 102–107. Hinz, Andreas M., Sandi Klavžar, Uroš Milutinović, and Ciril Petr, *The Tower of Hanoi – Myths and Maths*, Birkhäuser, Heidelberg, Germany, 2013, see pp. v-vi for Stewart's statement that others spotted it before him.

36 Scorer, R. S., Patrick Michael Grundy, and Cedric Austen Bardell Smith, "Some Binary Games," *The Mathematical Gazette*, Vol. 28, No. 280, July 1944, pp. 96–103.

37 Er, M. C., "A General Algorithm for Finding a Shortest Path between Two N-Configurations," *Information Sciences*, Vol. 42, 1987, pp. 137–141.

38 Lu, Xuemiao, "Towers of Hanoi Graphs," *International Journal of Computer Mathematics*, Vol. 19, No. 1, 1986, pp. 23–38.

39 Stewart, Ian, "Le lion, le lama et la laitue," *Pour la Science*, Vol. 142, 1989, pp. 102–107.

40 Stewart, Ian, "Four Encounters with Sierpiński's Gasket," *Mathematical Intelligencer*, Winter 95, Vol. 17, No. 1, 1995, pp. 52–64, p. 58 cited here.

41 For a proof of this, see Hinz, Andreas M., Sandi Klavžar, Uroš Milutinović, and Ciril Petr, *The Tower of Hanoi – Myths and Maths*, Birkhäuser, Heidelberg, Germany, 2013, pp. 99-100.

42 Hinz, Andreas M., Sandi Klavžar, Uroš Milutinović, and Ciril Petr, *The Tower of Hanoi – Myths and Maths*, Birkhäuser, Heidelberg, Germany, 2013, p. 100.

43 Hinz, Andreas M., Sandi Klavžar, Uroš Milutinović, and Ciril Petr, *The Tower of Hanoi – Myths and Maths*, Birkhäuser, Heidelberg, Germany, 2013, p. 132.

beautiful, but as a bonus, they allow us to easily find a minimal sequence of moves to solve the Tower of Hanoi. Simply follow a shortest path (there may be more than one) from the point indicating the initial configuration to the endpoint. We can also quickly find such paths between any two legal states. You don't even need to be human to do this! A portion of the abstract for an academic paper from 2011 follows.

> We mapped all possible solutions to the Towers of Hanoi on a single graph and converted this into a maze for the ants to solve. We show that the ants are capable of solving the Towers of Hanoi, and are able to adapt when sections of the maze are blocked off and new sections installed.[44]

Not only did the ant colonies all manage to navigate the graph maze successfully, 93.3% found the *shortest* path within an hour.[45] Granted, it was only a three-disc version of the three-peg Tower, but I don't believe the ants could have solved even this small version of the problem without the graph maze.

Some graphs have unexpected features. As was mentioned above, there can be more than one optimal solution. Also, an optimal solution sometimes requires moving the largest disc twice.[46] No optimal solution involves moving the largest disc more than twice, at least not with the three-peg version.

The Reve's puzzle is different from Lucas's original puzzle in many ways. Not only is the minimum number of moves not known, the optimal paths between regular states can be quite strange. With four discs, there are states that have minimal paths between them in which the largest disc may be moved 1, 2, or 3 times. An example is 0233 → 3001.[47]

As more pegs and discs are added, the situation gets even weirder. No matter how large of a value you pick, there are versions of the puzzle in which the largest disc must be moved at least that many times in an optimal solution to transition between states that the puzzle can exist in! This can be formulated as a theorem, as follows.

Theorem Let n and p be the number of discs and pegs, respectively. If $n \geq p(p - 2)$, then $p - 1$ largest disc moves are needed for some solutions.[48]

If this hadn't been established mathematically and I was trying to solve a version of the Tower and found myself moving the largest disc for the twentieth time, I'd probably be certain that I wasn't following the optimal solution!

---

44 Reid, Chris R., David J. T. Sumpter, and Madeleine Beekman, "Optimisation in a Natural System: Argentine Ants Solve the Tower of Hanoi," *The Journal of Experimental Biology*, Vol. 214, 2011, pp. 50–58, available online at http://jeb.biologists.org/content/214/1/50.

45 Reid, Chris R., David J. T. Sumpter, and Madeleine Beekman, "Optimisation in a Natural System: Argentine Ants Solve the Tower of Hanoi," *The Journal of Experimental Biology*, Vol. 214, 2011, pp. 50–58, p. 54 cited here, available online at http://jeb.biologists.org/content/214/1/50.

46 Hinz, Andreas M., Sandi Klavžar, Uroš Milutinović, and Ciril Petr, *The Tower of Hanoi – Myths and Maths*, Birkhäuser, Heidelberg, Germany, 2013, p. 112 and p. 117.

47 Hinz, Andreas M., Sandi Klavžar, Uroš Milutinović, and Ciril Petr, *The Tower of Hanoi – Myths and Maths*, Birkhäuser, Heidelberg, Germany, 2013, p. 206. However, in the traditional task of moving the entire Tower from the leftmost peg to the rightmost peg, the optimal solution for the Reve's puzzle will involve moving the largest disc exactly once. See Hinz, Andreas M., Sandi Klavžar, Uroš Milutinović, and Ciril Petr, *The Tower of Hanoi – Myths and Maths*, Birkhäuser, Heidelberg, Germany, 2013, p. 175, which cites p. 119 of Bode, Jens-P., and Andreas M. Hinz, "Results and Open Problems on the Tower of Hanoi," *Congressus Numerantium*, Vol. 139, 1999, pp. 113–122, and provides a simple proof.

48 Hinz, Andreas M., Sandi Klavžar, Uroš Milutinović, and Ciril Petr, *The Tower of Hanoi – Myths and Maths*, Birkhäuser, Heidelberg, Germany, 2013, p. 206, which cites Aumann, S. Über die Anzahl der Stege auf Geodäten van Hanoi-Graphen, Diploma thesis, Ludwig-Maximilians-Universität, München, 2009, see Korollar 6.2.

There are many open problems concerning the number of largest disc moves in optimal solutions for various versions of the puzzle.[49] At this time, mathematicians have more questions than answers in this area!

The size of a graph G is defined to be the number of edges it has and is notated $||G||$. For Hanoi graphs, there's a formula to calculate this value directly from the number of pegs and discs. If $n \in N_0$ and $p \geq 3$, we have $||H_p^n|| = \frac{1}{4} \sum_{k=1}^{p} k(2p - k - 1) \left\{ \begin{matrix} n \\ k \end{matrix} \right\} p^k$. So, the Tower is connected to Stirling numbers of the second kind.[50]

Although it doesn't connect with other interesting topics, there is a simpler formula for $||H_p^n||$.[51] If $n \in N_0$ and $p \geq 3$, then $||H_p^n|| = \frac{p(p-1)}{4} (p^n - (p - 2)^n)$.

Still, none of this interesting mathematics has helped to solve the problem of the minimum number of moves for the Reve's puzzle. All we have is a conjecture. In the meanwhile, there are other variants.

Other Variants

From the beginning, there have been variants on the Tower of Hanoi. The Reve's puzzle with its four pegs is perhaps the most obvious, but it doesn't require much thought to come up with many more. In 1889, Lucas himself noted, "The number of problems which one can pose oneself on the *new* Tower of Hanoi is incalculable."[52]

Lucas's Five-Peg Puzzle (1889)[53] – This variant (Figure 14.17) has four pegs arranged to make a square, with a fifth peg in the center. Each outer peg initially has four discs of a particular color. That is, there are four towers altogether, each of a different color. The colors may be useful for planning moves, but there are no rules connected with them in this variant. Among the complete set of 16 discs, no two are the same size. The aim is to move the discs, following the usual rules, so that the final position consists of one large tower on the center peg, as shown below.

This particular game was offered for sale in Paris.[54]

49 See Hinz, Andreas M., Sandi Klavžar, Uroš Milutinović, and Ciril Petr, *The Tower of Hanoi – Myths and Maths*, Birkhäuser, Heidelberg, Germany, 2013, pp. 207–208.

50 Hinz, Andreas M., Sandi Klavžar, Uroš Milutinović, and Ciril Petr, *The Tower of Hanoi – Myths and Maths*, Birkhäuser, Heidelberg, Germany, 2013, p. 209, which cites Klavžar Sandi, Uroš Milutinović, and Ciril Petr, "Combinatorics of Topmost Discs of Multi-Peg Tower of Hanoi problem," *Ars Combinatoria*, Vol. 59, 2001, pp. 55–64, p. 62 cited here.

51 Hinz, Andreas M., Sandi Klavžar, Uroš Milutinović, and Ciril Petr, *The Tower of Hanoi – Myths and Maths*, Birkhäuser, Heidelberg, Germany, 2013, p. 192.

52 Original French: "Le nombre des problems que l'on peut se poser sur la *nouvelle* Tour d'Hanoï est incalculable." See Lucas, Édouard, "Nouveaux Jeux Scientifiques de M. Édouard Lucas," *La Nature*, Vol. 17, 1889, pp. 301–303, p. 303 quoted here. Translation taken from Hinz, Andreas M., Sandi Klavžar, Uroš Milutinović, and Ciril Petr, *The Tower of Hanoi – Myths and Maths*, Birkhäuser, Heidelberg, Germany, 2013, p. 211, footnote.

53 Hinz, Andreas M., Sandi Klavžar, Uroš Milutinović, and Ciril Petr, *The Tower of Hanoi – Myths and Maths*, Birkhäuser, Heidelberg, Germany, 2013, pp. 212–213. Originally proposed in Lucas, Édouard, "Nouveaux Jeux Scientifiques de M. Édouard Lucas," *La Nature*, Vol. 17, 1889, pp. 301–303. Also see Lucas, Édouard, Jeux Scientifiques, Première Série, N° 3, La Tour d'Hanoï, Chambon & Baye/Édouard Lucas, Paris, 1889.

54 Hinz, Andreas M., Sandi Klavžar, Uroš Milutinović, and Ciril Petr, *The Tower of Hanoi – Myths and Maths*, Birkhäuser, Heidelberg, Germany, 2013, p. 13.

FIGURE 14.17
A five-peg puzzle from Lucas.

Image from Lucas, Édouard, "Nouveaux Jeux Scientifiques de M. Édouard Lucas," *La Nature*, Vol. 17, 1889, pp. 301–303, p. 301 cited here (public domain).

Cyclic Tower of Hanoi (1944)[55] – It's easiest to understand this version, if you arrange the pegs in a circle. Then, the rule for this variant is that the discs can only be moved in one direction, say clock-wise, and only to the very next peg in each move. The goal is to move the tower from peg 0 to some other specific peg. The minimum number of moves needed to complete such tasks is not known.[56] Paul K. Stockmeyer's summary of this problem, which included his own efforts, ended with "In short, we have not been able to find any algorithm for this puzzle that is significantly better than a brute force breadth-first search of the state graph of the puzzle."[57]

Exchanging Discs Tower of Hanoi (1944)[58] – In this version, the task is to move the tower from one peg to another, but the only moves that are allowed are:

55 Hinz, Andreas M., Sandi Klavžar, Uroš Milutinović, and Ciril Petr, *The Tower of Hanoi – Myths and Maths*, Birkhäuser, Heidelberg, Germany, 2013, p. 256. Proposed in Scorer, R. S., Patrick Michael Grundy, and Cedric Austen Bardell Smith, "Some Binary Games," *The Mathematical Gazette*, Vol. 280, 1944, pp. 96–103. Lunnon, W. F., "The Reve's Puzzle," Correspondence, *The Computer Journal*, Vol. 29, No. 5, 1986, p. 478 references a later paper for the proposal.

56 Hinz, Andreas M., Sandi Klavžar, Uroš Milutinović, and Ciril Petr, *The Tower of Hanoi – Myths and Maths*, Birkhäuser, Heidelberg, Germany, 2013, p. 256.

57 Stockmeyer, Paul K., "Variations on the Four-Post Tower of Hanoi Puzzle," *Congressus Numerantium*, Vol. 102, 1994, pp. 3–12. Pages. 6–7 are relevant here.

58 Hinz, Andreas M., Sandi Klavžar, Uroš Milutinović, and Ciril Petr, *The Tower of Hanoi – Myths and Maths*, Birkhäuser, Heidelberg, Germany, 2013, pp. 214–215. See Scorer, R. S., Patrick Michael Grundy, and Cedric Austen Bardell Smith, "Some Binary Games," *The Mathematical Gazette*, Vol. 280, 1944, pp. 96–103 for the proposal and Stockmeyer, Paul K., C. Douglass Bateman, James W. Clark, Cyrus R. Eyster, Matthew T. Harrison, Nicholas A. Loehr, Patrick J. Rodriguez, and Joseph R. Simmons III, "Exchanging Disks in the Tower of Hanoi," *International Journal of Computer Mathematics*, Vol. 59, 1995, pp. 37–47 for the solution.

1. moving disc 1 from any peg to any peg
2. exchanging two top discs that are on different pegs (this counts as a single move), if there is no disc of intermediate size (i.e., the discs are consecutive, in terms of size)

The Tower of Antwerpen (1981/1982)[59] – This version has stacks of black, yellow, and red discs that differ only in their colors. The goal is to derange the colored towers. That is, rearrange them so that no tower occupies its original position. The discs are to be moved one at a time and a disc may be placed atop a disc of the same size, but not atop a smaller disc.

The Twin-Tower Problem (1991)[60] – This is a variant of The Tower of Antwerpen, with only two colored towers. The task is to interchange their positions, However, if the color of the towers are, say, red and blue, we have the additional restriction that a blue disc can be placed on a red disc of the same size, but a red disc cannot be placed on a blue disc of the same size.

The Star Tower of Hanoi (1994)[61] – In this variant there are four pegs. Three form the corners of a regular triangle and the fourth is located at this triangle's center. The tower starts on one of the outer three pegs and the task is to move it to a different outer peg.

The Little Tower of Antwerpen (2005)[62] – This is a variant of The Tower of Antwerpen, with only two colored towers. The task is to interchange their positions. This time, as in the original, a disc can always be placed atop a disc of the same size.

Linear Twin Hanoi (2007)[63] – This is a variant of The Little Tower of Antwerpen where a tower of one color is initially on peg 0 and a tower of another color is on peg 2. As before, the goal is to exchange the towers, but in this variant, no moves are allowed between pegs 0 and 2.

The Black and White Tower of Hanoi (2010)[64] – This variant uses four pegs and begins with two towers, one black and one white, of course. The discs are moved as usual, but

59 Hinz, Andreas M., Sandi Klavžar, Uroš Milutinović, and Ciril Petr, *The Tower of Hanoi – Myths and Maths*, Birkhäuser, Heidelberg, Germany, 2013, pp. 218–219. Hinz cites Wood, D., "The Towers of Brahma and Hanoi Revisited," *Journal of Recreational Mathematics*, Vol. 14, 1981–1982, pp. 17–24 for the proposal and Minsker, Steven, "The Towers of Antwerpen Problem," *Information Processing Letters*, Vol. 38, 1991, pp. 107–111 for the solution.

60 Hinz, Andreas M., Sandi Klavžar, Uroš Milutinović, and Ciril Petr, *The Tower of Hanoi – Myths and Maths*, Birkhäuser, Heidelberg, Germany, 2013, pp. 219–220. Hinz cites van Zanten, A. J., "An Optimal Algorithm for the Twin-Tower Problem," *Delft Progress Report*, Vol. 15, 1991, pp. 33–50 for the proposal.

61 Hinz, Andreas M., Sandi Klavžar, Uroš Milutinović, and Ciril Petr, *The Tower of Hanoi – Myths and Maths*, Birkhäuser, Heidelberg, Germany, 2013, pp. 251–254. Proposed in Stockmeyer, Paul K., "Variations on the Four-Post Tower of Hanoi Puzzle," *Congressus Numerantium*, Vol. 102, 1994, pp. 3–12. Hinz et al, p. 254 gives an open problem connected with this variant.

62 Hinz, Andreas M., Sandi Klavžar, Uroš Milutinović, and Ciril Petr, *The Tower of Hanoi – Myths and Maths*, Birkhäuser, Heidelberg, Germany, 2013, p. 219. Hinz cites Minsker, Steven, "The Little Towers of Antwerpen problem," *Information Processing Letters*, Vol. 94, 2005, pp. 197–201 for the proposal and solution.

63 Hinz, Andreas M., Sandi Klavžar, Uroš Milutinović, and Ciril Petr, *The Tower of Hanoi – Myths and Maths*, Birkhäuser, Heidelberg, Germany, 2013, p. 221. Hinz cites Minsker, Steven, "The Linear Twin Towers of Hanoi Problem," *ACM SIGCSE Bulletin*, Vol. 39, 2007, pp. 37–40 for the proposal and solution.

64 Hinz, Andreas M., Sandi Klavžar, Uroš Milutinović, and Ciril Petr, *The Tower of Hanoi – Myths and Maths*, Birkhäuser, Heidelberg, Germany, 2013, p. 215. Hinz cites Stockmeyer, Paul K., "New Variations on the Tower of Hanoi," *Congressus Numerantium*, Vol. 201, 2010, pp. 277–287 for the proposal and Hinz et al. gives the solution on p. 216.

a disc cannot be placed on top of a disc of the same size. The goal is to exchange the two towers.

Linear Tower of Hanoi, aka Three-in-a-Row Tower of Hanoi (1944)[65] – In this variant, moves between pegs 0 and 2 are not allowed.

Four-in-a-Row Tower of Hanoi (1994)[66] – Labeling the posts in this variant as 0, 1, 2, and 3, moves may only be made between neighboring posts. That is, $0 \leftrightarrow 1$, $1 \leftrightarrow 2$, and $2 \leftrightarrow 3$.

p-in-a-Row Tower of Hanoi, aka Linear Tower of Hanoi on p Pegs – This has nothing to do with men lining up at a wall or urinals. Instead, it's the obvious generalization of Three-in-a-Row Tower of Hanoi and Four-in-a-Row Tower of Hanoi. Moves can only be made between neighboring pegs. There's no Pac-Man-like movement between the left-most and rightmost pegs. In the excellent book on the Tower of Hanoi and its variants by Hinz and colleagues, the authors comment, "the case $p = 3$ is easy, the cases $p \geq 4$ seem notoriously difficult."[67]

There's even a variant called Monsters and Globes, in which the roles of the discs and the pegs are reversed. The difficulty should be equivalent to that of the original Tower of Hanoi, but for some reason most people find it harder, even those who already know how to solve the original![68]

In general, I didn't specify a certain number of discs in the variants detailed above. However, for a given variant there may be a number n, such that the puzzle is solvable for n discs, but not for $n + 1$ discs.[69] And the above is by no means an exhaustive list of Tower variants. I have, for example, left out variants that allow discs to be placed atop smaller discs in some cases. These include **The Bottleneck Tower of Hanoi (1981)** and **Sinner's Tower of Hanoi (2007)**.[70] I've also left out **The Tower of London (1982)**, which uses colored balls of uniform size instead of discs, and has variants of its own.[71] So,

---

65 Hinz, Andreas M., Sandi Klavžar, Uroš Milutinović, and Ciril Petr, *The Tower of Hanoi – Myths and Maths*, Birkhäuser, Heidelberg, Germany, 2013, pp. 242–244. See p. 99 of Scorer, R. S., Patrick Michael Grundy, and Cedric Austen Bardell Smith, "Some Binary Games," *The Mathematical Gazette*, Vol. 28, No. 280, July 1944, pp. 96–103 for the proposal. Also see Hering, H., "Genetische Mathematisierung am Beispiel einer Variante des Turm-von-Hanoi-Spiels," *Didaktik der Mathematik*, Vol. 4, 1978, pp. 307–317.

66 Hinz, Andreas M., Sandi Klavžar, Uroš Milutinović, and Ciril Petr, *The Tower of Hanoi – Myths and Maths*, Birkhäuser, Heidelberg, Germany, 2013, pp. 254–256. Also see Stockmeyer, Paul K., "Variations on the Four-Post Tower of Hanoi Puzzle," *Congressus Numerantium*, Vol. 102, 1994, pp. 3–12.

67 Hinz, Andreas M., Sandi Klavžar, Uroš Milutinović, and Ciril Petr, *The Tower of Hanoi – Myths and Maths*, Birkhäuser, Heidelberg, Germany, 2013, p. 255

68 Hinz, Andreas M., Sandi Klavžar, Uroš Milutinović, and Ciril Petr, *The Tower of Hanoi – Myths and Maths*, Birkhäuser, Heidelberg, Germany, 2013, p. 50. Also see p. 251 of Kotovsky, Kenneth, John R. Hayes, and Herbert A. Simon, "Why Are Some Problems Hard? Evidence from Tower of Hanoi," *Cognitive Psychology*, Vol. 17, 1985, pp. 248–294 and Hinz, Andreas M., "Graph Theory of Tower Tasks," *Behavioural Neurology*, Vol. 25, No. 1, 2012, pp. 13–22.

69 Hinz, Andreas M., Sandi Klavžar, Uroš Milutinović, and Ciril Petr, *The Tower of Hanoi – Myths and Maths*, Birkhäuser, Heidelberg, Germany, 2013, p. 258. Hinz et al. reference Leiss, Ernst L., "Solving the 'Towers of Hanoi' on Graphs," *Journal of Combinatorics, Information & System Sciences*, Vol. 8, No. 1, 1983, pp. 81–89 and Azriel, Dany, Noam Solomon, and Shay Solomon, "On an Infinite Family of Solvable Hanoi Graphs," *ACM Transactions on Algorithms*, Vol. 5, 2009, Article 13.

70 See Hinz, Andreas M., Sandi Klavžar, Uroš Milutinović, and Ciril Petr, *The Tower of Hanoi – Myths and Maths*, Birkhäuser, Heidelberg, Germany, 2013, pp. 222–226.

71 Hinz, Andreas M., Sandi Klavžar, Uroš Milutinović, and Ciril Petr, *The Tower of Hanoi – Myths and Maths*, Birkhäuser, Heidelberg, Germany, 2013, pp. 227–230, p. 227 cites Shallice, Timothy, "Specific Impairments of

even if all questions connected with the Reve's puzzle can be resolved, we still have a wealth of other Tower puzzles to consider!

Yet More Open Problems

There are many more open problems associated with the Tower of Hanoi and its variants. Rather than list them, I simply point the interested reader to pages 261–262 of *The Tower of Hanoi – Myths and Maths* by Andreas M. Hinz, and colleagues, which provide a wonderful summary.[72] This pointer will have to suffice for those interested in pursuing the topic further. The present chapter ends with just one more open problem, one concerning Hanoi graphs.

As the Hanoi graphs for the three-peg version become larger, with the addition of more discs, they can still be drawn in the plane. This means that there are ways to draw them so that the lines joining states that can be transitioned between in single moves never cross. However, the addition of a fourth peg changes matters. The four-peg version can only be drawn in the plane if there are two or fewer discs. And with more than four pegs, the graph cannot be drawn in the plane unless there are no discs at all! To summarize, H_3^n, H_4^1, H_4^2, and H_p^0 are planar, but all other Hanoi graphs are **non-planar**. That is, any depiction of them in the plane must include crossings.[73]

To make non-planar graphs easier to follow, they're often drawn with as few crossings as possible. This can be done by moving the points that need to be connected. The smallest number of crossings with which a graph can be drawn is called its **crossing number**. Finding these values for Hanoi graphs has turned out to be a very difficult problem. Hinz and colleagues note, "Not a single crossing number of non-planar Hanoi graphs for more than one disc is known explicitly!"[74] Figure 14.18 is a graph of the four-peg version of the Tower, with only three discs. It's shown with 72 crossings, but this is not known to be the minimum.

Can you do better or prove that this representation is the best possible?

Planning," *Philosophical Transactions of the Royal Society of London, Series B*, Vol. 298, 1982, pp. 199–209 for the proposal. Also see Hinz, Andreas M., Anton Kostov, Fabian Kneissl, Fatma Sürer, and Adrian Danek, "A Mathematical Model and a Computer Tool for the Tower of Hanoi and Tower of London Puzzles," *Information Sciences*, Vol. 179, 2009, pp. 2934–2947. Hinz et al. pp. 231–238 discuss Tower of London variants.

72 Hinz, Andreas M., Sandi Klavžar, Uroš Milutinović, and Ciril Petr, *The Tower of Hanoi – Myths and Maths*, Birkhäuser, Heidelberg, Germany, 2013, pp. 261–26.

73 Hinz, Andreas M., Sandi Klavžar, Uroš Milutinović, and Ciril Petr, *The Tower of Hanoi – Myths and Maths*, Birkhäuser, Heidelberg, Germany, 2013, p. 195, which cites Hinz, Andreas M. and Daniele Parisse, "On the Planarity of Hanoi Graphs," *Expositiones Mathematicae*, Vol. 20, 2002, pp. 263–268. See Theorem 2.

74 Hinz, Andreas M., Sandi Klavžar, Uroš Milutinović, and Ciril Petr, *The Tower of Hanoi – Myths and Maths*, Birkhäuser, Heidelberg, Germany, 2013, p. 195. This page also includes "Nobody has ever dared to approach the question of genera of Hanoi graphs." Also see p. 236. For some chipping away at these problems see Hinz et al. pp. 157–158 and their references: Klavžar, Sandi and Bojan Mohar, "Crossing Numbers of Sierpiński-Like Graphs," *Journal of Graph Theory*, Vol. 50, 2005, pp. 186–198; Schmid, R. S., Überschneidungszahlen und Geschlecht bei Hanoi- und Sierpiński-Graphen, Diploma thesis, Ludwig-Maximilians-Universität, München, 2010; Köhler, T., Überschneidungszahlen spezieller Hanoi- und Sierpiński-Graphen, Diploma thesis, Ludwig-Maximilians-Universität, München, 2011.

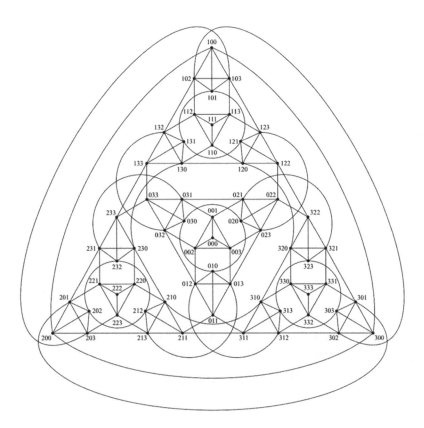

FIGURE 14.18
The four-peg, three-disc Tower of Hanoi graph, shown with 72 crossings.

Reprinted by permission from Springer Nature: Springer/Birkhäuser, *The Tower of Hanoi – Myths and Maths* by Andreas M. Hinz, Sandi Klavžar, Uroš Milutinović, and Ciril Petr, © 2013.

Exercise Set 14

> Man is a puzzle-solving animal.
>> Ronald A. Knox (1888–1957) Roman Catholic priest
>> and author of detective fiction[75]

> All the properties of the Tower of Hanoi problem follow trivially.
>> – R. J. Heard[76]

---

75 Quote found in Danesi, Marcel, *The Puzzle Instinct: The Meaning of Puzzles in Human Life*, Indiana University Press, Bloomington, IN, 2004, preface.
76 Heard, R. J., "The Tower of Hanoi as a Trivial Problem," *The Computer Journal*, Vol. 27, No. 1, 1984, p. 90.

1. Counting from the top, how do discs in odd positions move in the optimal solution to the three-peg Tower of Hanoi?

2. How many times does each disc move in the optimal solution to the three-peg Tower of Hanoi?

3. What is the minimum number of moves needed to solve the four-peg Tower of Hanoi with six discs?

4. What is the minimum number of moves needed to solve the four-peg Tower of Hanoi with eight discs?

5. Draw a picture of the three-peg Tower of Hanoi state represented by 1022012.

6. Draw a picture of the three-peg Tower of Hanoi state represented by 1010201.

7. Construct the state graph for the four-peg Tower of Hanoi with one disc. Make it planar.

8. Construct the state graph for the Linear Tower of Hanoi with two discs.

9. Calculate the value of $||H_3^3||$. Use the depiction of this graph (given in this chapter) to check your answer.

10. Calculate the value of $||H_3^5||$.

11. Calculate the value of $||H_4^5||$.

12. There are other ways to present the conjectured formula for the minimum number of moves needed to solve the Reve's puzzle.[77] Confirm that each of the following agrees with the formula I presented in the special case $n = 7$ discs.

    a. $s_n = s_{n-1} + 2^x$, where $x = \left\lfloor \frac{\sqrt{8n-7}-1}{2} \right\rfloor$ and $s_1 = 1$.

    b. A non-recursive formula is given by $s_n = 1 + \left[n - \frac{1}{2}x(x-1) - 1 \right]2^x$, where x is defined as above.

13. For the four-disc version of the Reve's puzzle, find the minimal solutions for $0233 \to 3001$ that move the largest disc 1, 2, and 3 times.

14. Give the smallest value of p for which there's a value for n guaranteeing that a puzzle with n discs and p pegs has a task requiring the largest disc to be moved at least 20 times.

References/Further Reading

The passion to possess books has never been more widespread than it is today; indeed, obsessive book collecting remains the only hobby to have a disease named after it.
                                                                                      – Nicholas A. Basbanes[78]

Allouche, Jean-Paul, "Note on the Cyclic Towers of Hanoi," *Theoretical Computer Science*, Vol. 123, No. 1, January 1994, pp. 3–7.

Arett, Danielle and Suzanne Dorée, "Coloring and Counting on the Tower of Hanoi Graphs," *Mathematics Magazine*, Vol. 83, June, 2010, pp. 200–209.

---

77 Weisstein, Eric W. "Tower of Hanoi." From *MathWorld*–A Wolfram Web Resource. http://mathworld. wolfram.com/TowerofHanoi.html.

78 Basbanes, Nicholas A., *A Gentle Madness: Bibliophiles, Bibliomanes, and the Eternal Passion for Books*, Henry Holt and Co., New York, 1995.

Azriel, Dany, Noam Solomon, and Shay Solomon, "On an Infinite Family of Solvable Hanoi Graphs," *ACM Transactions on Algorithms*, Vol. 5, November 2008, Article 13, pp. 13: 1–13:22.

Ball, W. W. Rouse, *Mathematical Recreations and Essays*, second edition, Macmillan and Co., London, 1896.

Barlow, Martin T. and Edwin A. Perkins, "Brownian Motion on the Sierpiński Gasket," *Probability Theory and Related Fields*, Vol. 79, No. 4, November 1988, pp. 543–623.

Bennish, Joseph, "The Tower of Hanoi Problem and Mathematical Thinking," *Missouri Journal of Mathematical Sciences*, Vol. 11, No. 3, Fall 1999, pp. 164–166.

Berlekamp, Elwyn R., John H. Conway, and Richard K. Guy, *Winning Ways for Your Mathematical Plays*, Vol. 4, Second edition, A K Peters, Wellesley, MA, 2004.

Bode, Jens-P., and Andreas M. Hinz, "Results and open problems on the Tower of Hanoi," *Congressus Numerantium*, Vol. 139, 1999, pp. 113–122.

Bogomolny, Alexander, "Sierpiński Gasket and Tower of Hanoi," Interactive Mathematics Miscellany and Puzzles, available online at www.cut-the-knot.org/triangle/Hanoi.shtml.

Bogomolny, Alexander, "Tower of Hanoi" from Interactive Mathematics Miscellany and Puzzles, https://www.cut-the-knot.org/recurrence/hanoi.shtml. This website has an applet for playing Tower of Hanoi.

Brousseau, Brother Alfred, "Tower of Hanoi with More Pegs," *Journal of Recreational Mathematics*, Vol. 8, No. 3, 1976, pp. 169–176.

Calkin, Neil and Herbert S. Wilf, "Recounting the Rationals," *The American Mathematical Monthly*, Vol. 107, No. 4, April 2000, pp. 360–363.

Chang, Shu-Chiuan, Lung-Chi Chen, and Wei-Shih Yang, "Spanning Trees on the Sierpiński Gasket," *Journal of Statistical Physics*, Vol. 126, No. 3, February 2007, pp. 649–667.

Chen, Xiao and Jian Shen, "On the Frame-Stewart Conjecture about the Towers of Hanoi," *SIAM Journal on Computing*, Vol. 33, No. 3, 2004, pp. 584–589.

Chen, Xiaomin, Bin Tian, and Lei Wang, "Santa Claus' Towers of Hanoi," *Graphs and Combinatorics*, Vol. 23, Supplement 1, 2007, pp. 153–167.

Christensen, Chris, "Alan Turing's First Cryptology Textbook and Sinkov's Revision of It," *Cryptologia*, Vol. 34, No. 1, January 2009, pp. 27–43.

Claus de Siam, N., *La Tour d'Hanoï, Véritable casse-tête annamite*, P. Bousrez, Tours, 1883.

Claus de Siam, N., "La Tour d'Hanoï, Jeu de calcul," *Science et Nature*, Vol. 1, 1884, pp. 127–128.

Crowe, D. W., "The *n*-dimensional Cube and the Tower of Hanoi," *The American Mathematical Monthly*, Vol. 63, No. 1, January 1956, pp. 29–30.

Cull, Paul and E. F. Ecklund, Jr., "On the Towers of Hanoi and Generalized Towers of Hanoi Problems," in Hoffman, Frederick et al., editors, *Proceedings of the Thirteenth Southeastern Conference on Combinatorics, Graph Theory and Computing*, Boca Raton, FL, 1982; *Congressus Numerantium*, Vol. 35, 1982, pp. 229–238.

Cull, Paul and Colin Gerety, "Is Towers of Hanoi Really Hard?," *Congressus Numerantium*, Vol. 47, 1985, pp. 237–242.

Cull, Paul and Ingrid Nelson, "Error-Correcting Codes on the Towers of Hanoi Graphs," *Discrete Mathematics*, Vol. 208/209, October 28, 1999, pp. 157–175.

de Longchamps, Gaston, "Variétés," *Journal de Mathématiques Spéciales*, Vol. 2, No. 2, 1883, pp. 286–287.

de Parville, Henri, "La Tour d'Hanoï et la question du Tonkin," *La Nature*, Vol. 12, 1884, pp. 285–286.

Dewdney, Alexander K., "Computer Recreations: Ying and Yang: Recursion and Iteration, the Tower of Hanoi and the Chinese Rings," *Scientific American*, Vol. 251, No. 5, November 1984, pp. 19–20, 22, 27–28. This article includes a solution to the Tower of Hanoi based on binary numbers.

Dudeney, Henry E., *The Canterbury Puzzles and Other Curious Problems*, Dutton, E. P., New York, 1908.

Dunkel, Otto, "Editorial Note," *The American Mathematical Monthly*, Vol. 48, No. 3, March 1941, p. 219.

Er, M. C., "An Analysis of the Generalized Towers of Hanoi Problem," *BIT Numerical Mathematics*, Vol. 23, No. 4, December 1983, pp. 429–435.

Er, M. C., "The Towers of Hanoi and Binary Numerals," *Journal of Information and Optimization Sciences*,", Vol. 6, No. 2, 1985, pp. 147–152.

Er, M. C., "A General Algorithm for Finding a Shortest Path between Two n-Configurations," *Information Sciences*, Vol. 42, No. 2, July 1987, pp. 137–141.

Frame, J. Sutherland, "Problems and Solutions: Advanced Problems: Solutions: 3918," *The American Mathematical Monthly*, Vol. 48, No. 3, March 1941, pp. 216–217.

Gardner, Martin, "Mathematical Games: About the Remarkable Similarity between the Icosian Game and the Towers of Hanoi," *Scientific American*, Vol. 196, May, 1957, pp. 150–156.

Gardner, Martin, "The Icosian Game and the Tower of Hanoi," in *Hexaflexagons and Other Mathematical Diversions: The First Scientific American Book of Puzzles and Games*, Simon and Schuster, New York, 1959, pp. 55–62.

Glaisher, James Whitbread Lee., "On the Residue of a Binomial-Theorem Coefficient with Respect to a Prime Modulus," *The Quarterly Journal of Pure and Applied Mathematics*, Vol. 30, 1899, pp. 150–156.

Heard, R. J., "The Tower of Hanoi as a Trivial Problem," *The Computer Journal*, Vol. 27, 1984, p. 90.

Hinz, Andreas M., "The Tower of Hanoi," *L'Enseignement Mathématique*, Vol. 35, 1989, pp. 289–321.

Hinz, Andreas M., "Pascal's Triangle and the Tower of Hanoi," *The American Mathematical Monthly*, Vol. 99, No. 6, June–July 1992, pp. 538–544.

Hinz, Andreas M., "Graph Theory of Tower Tasks," *Behavioural Neurology*, Vol. 25, No. 1, 2012, pp. 13–22.

Hinz, Andreas M., Sandi Klavžar, Uroš Milutinović, Daniele Parisse, and Ciril Petr, "Metric Properties of the Tower of Hanoi Graphs and Stern's Diatomic Sequence," *European Journal of Combinatorics*, Vol. 26, No. 5, July 2005, pp. 693–708.

Hinz, Andreas M., Sandi Klavžar, Uroš Milutinović, and Ciril Petr, *The Tower of Hanoi – Myths and Maths*, Birkhäuser, Heidelberg, Germany, 2013. This book is the best single source on the topic. The companion website to the book is http://tohbook.info/. I learned of the second edition, 2018, the day I finished the book you're holding!.

Hinz, Andreas M., Sandi Klavžar, and Sara Sabrina Zemljič, "Sierpiński Graphs as Spanning Subgraphs of Hanoi Graphs," *Central European Journal of Mathematics*, Vol. 11, No. 6, 2013, pp. 1153–1157.

Hinz, Andreas M. and Daniele Parisse, "On the Planarity of Hanoi Graphs," *Expositiones Mathematicae*, Vol. 20, No. 3, 2002, pp. 263–268.

Hinz, Andreas M. and Andreas Schief, "The Average Distance on the Sierpiński Gasket," *Probability Theory and Related Fields*, Vol. 87, No. 1, March 1990, pp. 129–138.

Hodges, Andrew, *Alan Turing: The Enigma*, Simon and Schuster, New York, 1983.

Kai, Yang and Xu Chuan, "The Preliminary Probe of 4-Peg Hanoi Tower," *Acta Scientarium Naturulium Universitatis Pekinensis*, Vol. 40, 2004, pp. 99–106.

Klavžar, Sandi, Uroš Milutinović, and Ciril Petr, "Combinatorics of Topmost Discs of Multi-Peg Tower of Hanoi Problem," *Ars Combinatoria*, Vol. 59, 2001, pp. 55–64.

Klavžar, Sandi and Uroš Milutinović, "Simple Explicit Formulas for the Frame-Stewart Numbers," *Annals of Combinatorics*, Vol. 6, No. 2, November 2002, pp. 157–167.

Klavžar, Sandi, Uroš Milutinović, and Ciril Petr, "On the Frame–Stewart Algorithm for the Multi-Peg Tower of Hanoi Problem," *Discrete Applied Mathematics*, Vol. 120, No. 1–3, August 15, 2002, pp. 141–157.

Klavžar, Sandi, Uroš Milutinović, and Ciril Petr, "Hanoi graphs and some classical numbers," *Expositiones Mathematicae*, Vol. 23, No. 4, December 2005, pp. 371–378.

Klavžar, Sandi and Bojan Mohar, "Crossing Numbers of Sierpiński-like Graphs," *Journal of Graph Theory*, Vol. 50, No. 3, November 2005, pp. 186–198.

Knuth, Donald E., "Computer Science and Mathematics: How a New Discipline Presently Interacts with an Old One, and What We May Expect in the Future," *American Scientist*, Vol. 61, No. 6, November–December 1973, pp. 707–713.

Knuth, Donald E., "All Questions Answered," *Notices of the American Mathematical Society*, Vol. 49, No. 3, March 2002, pp. 318–324.

Korf, Richard E., "Best-First Frontier Search with Delayed Duplicate Detection," *Nineteenth National Conference on Artificial Intelligence (AAAI-2004)*, The MIT Press, Cambridge, MA, 2004, pp. 650–657.

Korf, Richard E., "Linear-Time Disk-Based Implicit Graph Search," *Journal of the ACM*, Vol. 55, No. 6, December 2008, pp. 26:1–26:40.

Korf, Richard E. and Ariel Felner, "Recent Progress in Heuristic Search: A Case Study of the Four-Peg Towers of Hanoi Problem," in Veloso, Manuela M., editor, *Proceedings of the Twentieth International Joint Conference on Artificial Intelligence (IJCAI-07)*, AAAI Press, Menlo Park, CA, 2007, pp. 2324–2329.

Koshy, Thomas, *Fibonacci and Lucas Numbers with Applications*, John Wiley & Sons, New York, 2001.

Leiss, Ernst L., "Solving the 'Towers of Hanoi' on Graphs," *Journal of Combinatorics, Information & System Sciences*, Vol. 8, No. 1, 1983, pp. 81–89.

Lucas, Édouard, "Nouveaux Jeux Scientifiques de M. Édouard Lucas," *La Nature*, Vol. 17, 1889, pp. 301–303.

Lucas, Édouard, *Récréations Mathématiques III*, Gauthier-Villars et fils, Paris, 1893. pp. 55–59 represent Lucas's first writing on the Tower of Hanoi under his real name. It includes a variant with more pegs.

Lucas, Édouard and Jeux, *Scientifiques, Première Série, N° 3, La Tour d'Hanoï*, Chambon & Baye/ Édouard Lucas, Paris, 1889.

Lunnon, W. F., "The Reve's Puzzle," *Correspondence, The Computer Journal*, Vol. 29, No. 5, 1986, p. 478.

Majumdar, A. A. K., "A Note on the Iterative Algorithm for the Reve's Puzzle," *The Computer Journal*, Vol. 37, No. 5, January 1994, pp. 463–464.

Majumdar, A. A. K., "The Generalized Four-Peg Tower of Hanoi Problem," *Optimization*, Vol. 29, No. 4, 1994, pp. 349–360.

Matsuura, Akihiro, "Exact Analysis of the Recurrence Relations Generalized from the Tower of Hanoi," in *ANALCO '08 Proceedings of the Meeting on Analytic Algorithmics and Combinatorics, SIAM Proceedings in Applied Mathematics*, Vol. 129, 2008, pp. 228–233.

Minsker, Steven, "Another Brief Recursion Excursion to Hanoi," *ACM SIGCSE Bulletin*, Vol. 40, No. 4, December 2008, pp. 35–37.

O'Connor, John J. and Edmund F. Robertson, "Walter William Rouse Ball," *MacTutor Archive*, January 2015, available online at www.history.mcs.st-andrews.ac.uk/Biographies/Ball.html.

"Obituary: Édouard Lucas," *Popular Science Monthly*, Vol. 40, January 1892, p. 432, available online at https://en.wikisource.org/wiki/Popular_Science_Monthly/Volume_40/January_1892/ Obituary_Notes.

Poole, David G., "The Towers and Triangles of Professor Claus (Or, Pascal Knows Hanoi)," *Mathematics Magazine*, Vol. 67, No. 5, 1994, pp. 323–344.

Quint, J.-F., "Harmonic Analysis on the Pascal Graph," *Journal of Functional Analysis*, Vol. 256, No. 10, May 15, 2009, pp. 3409–3460.

Reid, Chris R., J. David, T. Sumpter, and Madeleine Beekman, "Optimisation in a Natural System: Argentine Ants Solve the Tower of Hanoi," *The Journal of Experimental Biology*, Vol. 214, 2011, pp. 50–58, available online at http://jeb.biologists.org/content/214/1/50.

Romik, Dan, "Shortest Paths in the Tower of Hanoi Graph and Finite Automata," *SIAM Journal on Discrete Mathematics*, Vol. 20, No. 3, August 2006, pp. 610–622.

Rosen, Kenneth, editor, *Handbook of Discrete and Combinatorial Mathematics*, CRC Press, Boca Raton, FL, 2000.

Rukhin, Andrey, *On the Generalized Tower of Hanoi Problem: An Introduction to Cluster Spaces*, Master thesis, University of Maryland, College Park, MD, 2004.

Scheinerman, Edward R., *Mathematics: A Discrete Introduction*, Second edition, Thomson Brooks/ Cole, Belmont, CA, 2006.

Scorer, R. S., Patrick Michael Grundy, and Cedric Austen Bardell Smith, "Some Binary Games," *The Mathematical Gazette*, Vol. 28, No. 280, July 1944, pp. 96–103.

Sloane, Neil J. A., "Reve's Puzzle: Number of Moves Needed to Solve the Towers of Hanoi Puzzle with 4 Pegs and *n* Disks, according to the Frame-Stewart Algorithm," *The On-Line Encyclopedia of Integer Sequences*, available online at http://oeis.org/A007664.

Stewart, Bonnie M., "Problems and Solutions: Advanced Problem 3918," *The American Mathematical Monthly*, Vol. 46, No. 6, June–July 1939, p. 363.

Stewart, Bonnie M., "Problems and Solutions: Advanced Problems: Solutions: 3918," *The American Mathematical Monthly*, Vol. 48, No. 3, March 1941, pp. 217–219.

Stewart, Ian, "Le lion, le lama et la laitue," *Pour la Science*, Vol. 142, 1989, pp. 102–107.

Stewart, Ian, "Four Encounters with Sierpiński's Gasket," *Mathematical Intelligencer*, Winter 95, Vol. 17, No. 1, 1995, pp. 52–64.

Stewart, Ian, "Sierpiński's Pathological Curve and its Modern Incarnations," *Wiadomości Matematyczne*, Vol. 48, No. 2, 2012, pp. 239–246.

Stockmeyer, Paul K., "Variations on the Four-Post Tower of Hanoi Puzzle," *Congressus Numerantium*, Vol. 102, 1994, pp. 3–12.

Stockmeyer, Paul K., The Tower of Hanoi: A Bibliography, Version 2.2, September 12, 2005, Corrected October 22, 2005, available online at http://zylla.wipos.p.lodz.pl/games/biblio2.pdf.

Stockmeyer, Paul K., "The Tower of Hanoi for Humans," in Beineke, Jennifer and Jason Rosenhouse, editors, *The Mathematics of Various Entertaining Subjects: Research in Games, Graphs, Counting, and Complexity*, Vol. 2, Princeton University Press, Princeton, NJ, 2017, pp. 52–70.

Stockmeyer, Paul K., C. Douglass Bateman, James W. Clark, Cyrus R. Eyster, Matthew T. Harrison, Nicholas A. Loehr, Patrick J. Rodriguez, and Joseph R. Simmons III, "Exchanging Disks in the Tower of Hanoi," *International Journal of Computer Mathematics*, Vol. 59, 1995, pp. 37–47.

Strohhäcker, Sebastian, *A Program to Find Distances in Tower of Hanoi and Related Graphs*, Technical report, Technische Universität München, Garching, 2008.

Szegedy, Mario, "In How Many Steps the k Peg Version of the Towers of Hanoi Game Can Be Solved?" in Meinel, Christoph and Sophie Tison, editors, *Proceedings of the 16th Annual Symposium on Theoretical Aspects of Computer Science (STACS '99)*, Lecture Notes in Computer Science Vol. 1563, Springer, Berlin, 1999, pp. 356–361.

"Towers of Hanoi Animation," *Towers of Hanoi*, available online at http://towersofhanoi.info/Animate.aspx. This website allows you to watch solutions being carried out for various numbers of discs and pegs, or try to carry them out yourself.

Van Zanten, A. J., "The Complexity of an Optimal Algorithm for the Generalized Tower of Hanoi Problem," *International Journal of Computer Mathematics*, Vol. 36, No. 1–2, 1990, pp. 1–8.

Walsh, Timothy R., "The Towers of Hanoi Revisited: Moving the Rings by Counting the Moves," *Information Processing Letters*, Vol. 15, No. 2, September 6, 1982, pp. 64–67.

Walsh, Timothy R., "A Case for Iteration," *Congressus Numerantium*, Vol. 40, 1983, pp. 409–417.

Walsh, Timothy R., "Iteration Strikes Back – At the Cyclic Towers of Hanoi," *Information Processing Letters*, Vol. 16, No. 2, February 1983, pp. 91–93.

Weisstein, Eric W., "Tower of Hanoi." From MathWorld–A Wolfram Web Resource, available online at http://mathworld.wolfram.com/TowerofHanoi.html.

Wexelblat, Richard L., "Editorial," *ACM SIGPLAN Notices*, Vol. 20, 1985, pp. 1.

Wikipedia Contributors, "Gathering 4 Gardner," *Wikipedia, The Free Encyclopedia*, available online at https://en.wikipedia.org/w/index.php?title=Gathering_4_Gardner&oldid=844247679.

Williams, Lauren Kelly, "Tower of Hanoi," *Department of Mathematics and Information Technology, Mercyhurst University*, 2016, available online at http://math.mercyhurst.edu/~lwilliams/Applets/TowerOfHanoi.html. This is one of many online versions of the Tower of Hanoi.

Wood, Derick, "The Towers of Brahma and Hanoi Revisited," *Journal of Recreational Mathematics*, Vol. 14, No. 1, 1981, pp. 17–24.

Xue-Miao, Lu, "Towers of Hanoi Graphs," *International Journal of Computer Mathematics*, Vol. 19, No. 1, 1986, pp. 23–38.

Xue-Miao, Lu, "Towers of Hanoi Problem with Arbitrary $k \geq 3$ Pegs," *International Journal of Computer Mathematics*, Vol. 24, No. 1, 1988, pp. 39–54.

15

Population Models

Difference equations are the discrete version of differential equations. The Fibonacci sequence provided one example and the Tower of Hanoi another. In this chapter we'll see how to solve some of the easier classes of difference equations, as well as observe some surprising results.

> Differential equations describe processes that change smoothly over time, but differential equations are hard to compute. Simpler equations – "difference equations" – can be used for processes that jump from state to state. Fortunately, many animal populations do what they do in neat one-year intervals. Changes year to year are often more important than changes on a continuum. Unlike people, many insects, for example, stick to a single breeding season, so their generations do not overlap. To guess next spring's gypsy moth population or next winter's measles epidemic, an ecologist might only need to know the corresponding figure for this year. A year-by-year facsimile produces no more than a shadow of a system's intricacies, but in many real applications the shadow gives all the information a scientist needs. – James Gleick[1]

Populations may be modeled by difference equations of various degrees of sophistication. Fibonacci's early attempt for rabbits is a very simple example. A simple model for humans is given by $x_n = x_{n-1} + rx_{n-1}$. This means that the population in year n is the previous year's population, year $n - 1$, plus some new people, rx_{n-1}. The number of new people depends on the previous year's population. The value of r is the fraction of people from the previous year's population who gave birth. Hence, $0 \leq r \leq 1$. We call r the annual rate of growth. The equation can be rewritten as $x_n = (1 + r)\, x_{n-1}$.

Some sample values are provided, on the following page, for $x_0 = 500$, an initial population, and $r = 0.02$. We simply multiply each year's population by 1.02 to get the population in the next year. The fractional parts in the table are not meaningful in this context, but become extremely important in other contexts.

We could continue the table as long as we like, but this is not an efficient manner for finding out what the population will be after 1,000 years! The method above is recursive. That means that one (or more) previous values must be known in order to find the next value. We can, however, change this. Observe:

$$x_1 = 1.02\, x_0$$
$$x_2 = 1.02\, x_1 = 1.02\, (1.02\, x_0) = (1.02)^2 x_0$$
$$x_3 = 1.02\, x_2 = 1.02\, (1.02)^2 x_0 = (1.02)^3 x_0$$

The pattern is clear: $x_n = (1.02)^n\, x_0$. This can be proven by induction. Using this non-recursive formula, we can find the population at any time, without any previous values except for the starting value.

---

1 Gleick, James, *Chaos: Making a New Science*, Penguin Books, New York, 1988, p. 61.

n	Population
0	500
1	510
2	520.2
3	530.604
4	541.21608
5	552.0404016
6	563.0812096
7	574.3428338
8	585.8296905
9	597.5462843
10	609.4972100

For example, $x_{1000} = (1.02)^{1000} x_0 = (1.02)^{1000} 500$ (using our initial value)

$$\approx 1.99 \times 10^{11}$$

This difference equation ignored deaths and other factors that affect populations. If we desire more accurate long-term predictions, we must improve the model. Factoring in deaths is not hard. Our constant r is simply adjusted to be the difference between the birth rate and the death rate.

Malthusian Growth

Thomas Malthus (Figure 15.1) is usually given credit for the population model investigated above. It's sometimes referred to as "Malthusian Growth" in his honor. A few passages from his 1798 piece "An Essay on the Principle of Population" are reproduced below with my comments.

> Assuming then my postulata as granted, I say, that the power of population is indefinitely greater than the power in the earth to produce subsistence for man.[2]
> Population, when unchecked, increases in a geometrical ratio. Subsistence increases only in an arithmetical ratio. A slight acquaintance with numbers will shew the immensity of the first power in comparison of the second.[3]

In our example above, the value for the geometric ratio was 1.02. Malthus never states the value (which varies from group to group), but simply provides examples of how long it has taken various populations to double. He also claims that food production increases more slowly, at a linear rate.

---

2 Malthus, Thomas, *An Essay on the Principle of Population*, J. Johnson, London, 1798, p. 4.
3 Malthus, Thomas, *An Essay on the Principle of Population*, J. Johnson, London, 1798, p. 4.

FIGURE 15.1
Thomas Malthus (1766–1834).

Public domain.

By that law of our nature which makes food necessary to the life of man, the effects of these two unequal powers must be kept equal.[4]

This implies a strong and constantly operating check on population from the difficulty of subsistence. This difficulty must fall somewhere and must necessarily be severely felt by a large portion of mankind.[5]

This is the scary part. If increases in food production cannot keep pace with increases in population, deaths must result.

The germs of existence contained in this spot of earth, with ample food, and ample room to expand in, would fill millions of worlds in the course of a few thousand years.[6]

Although Malthus doesn't get into it, the passage above indicates that expanding out to other planets for food cultivation and/or population relocation (e.g., terraforming Mars) can't provide a long-term solution.

In the United States of America, where the means of subsistence have been more ample, the manners of the people more pure, and consequently the checks to early marriages fewer, than in any of the modern states of Europe, the population has been found to double itself in twenty-five years.[7]

This ratio of increase, though short of the utmost power of population, yet as the result of actual experience, we will take as our rule, and say, that population, when unchecked, goes on doubling itself every twenty-five years or increases in a geometrical ratio.[8]

---

4 Malthus, Thomas, *An Essay on the Principle of Population*, J. Johnson, London, 1798, p. 5.

5 Malthus, Thomas, *An Essay on the Principle of Population*, J. Johnson, London, 1798, p. 5.

6 Malthus, Thomas, *An Essay on the Principle of Population*, J. Johnson, London, 1798, p. 5.

7 Malthus, Thomas, *An Essay on the Principle of Population*, J. Johnson, London, 1798, p. 7.

8 Malthus, Thomas, *An Essay on the Principle of Population*, J. Johnson, London, 1798, p. 7.

Returning to our example where the growth rate was 2%, we can easily calculate how long it will take the population to double. Our formula is $x_n = (1.02)^n x_0$, with $x_0 = 500$. We want to know when $x_n = 1000$, so we simply solve $1000 = (1.02)^n 500$. Dividing both sides by 500 gives $2 = (1.02)^n$. We can then take the natural log of both sides to get $\ln(2) = \ln[(1.02)^n]$. Recalling the rules of logarithms, we write $\ln(2) = n \cdot \ln(1.02)$ and then solve for n as $\ln(2)/\ln(1.02) \approx 35$ years.

This is a decade longer than Malthus's example. We can use the formula $x_n = (1 + r)^n x_0$ and an assumed 25-year doubling time to find the value of r in Malthus's model. It doesn't matter what starting population x_0 is used. We simply find the value of r that leads to twice that amount, $2x_0$, in 25 years. We have $2x_0 = (1 + r)^{25} x_0$, so dividing through by x_0 gives $2 = (1 + r)^{25}$. This implies $2^{1/25} = 1 + r$, which gives $r = 2^{1/25} - 1 \approx 0.0281$. That is, a growth rate of about 2.81%.

Malthus goes into detail as to the effect this doubling of the population every 25 years will have:

> If I allow that by the best possible policy, by breaking up more land and by great encouragements to agriculture, the produce of this Island may be doubled in the first twenty-five years, I think it will be allowing as much as any person can well demand.[9]
>
> In the next twenty-five years, it is impossible to suppose that the produce could be quadrupled. It would be contrary to all our knowledge of the qualities of land. The very utmost that we can conceive, is, that the increase in the second twenty-five years might equal the present produce. Let us then take this for our rule, though certainly far beyond the truth, and allow that, by great exertion, the whole produce of the Island might be increased every twenty-five years, by a quantity of subsistence equal to what it at present produces. The most enthusiastic speculator cannot suppose a greater increase than this. In a few centuries it would make every acre of land in the Island like a garden.[10]
>
> Yet this ratio of increase is evidently arithmetical.[11]
>
> It may be fairly said, therefore, that the means of subsistence increase in an arithmetical ratio. Let us now bring the effects of these two ratios together.[12]
>
> The population of the Island is computed to be about seven millions, and we will suppose the present produce equal to the support of such a number. In the first twenty-five years the population would be fourteen millions, and the food being also doubled, the means of subsistence would be equal to this increase. In the next twenty-five years the population would be twenty-eight millions, and the means of subsistence only equal to the support of twenty-one millions. In the next period, the population would be fifty-six millions, and the means of subsistence just sufficient for half that number. And at the conclusion of the 1st century the population would be one hundred and twelve millions and the means of subsistence only equal to the support of thirty-five millions, which would leave a population of seventy-seven millions totally unprovided for.[13]

Malthus never provides a difference equation (or a differential equation, if you prefer a continuous model), but in the following paragraph he illustrates the idea numerically.

9 Malthus, Thomas, *An Essay on the Principle of Population*, J. Johnson, London, 1798, p. 7.

10 Malthus, Thomas, *An Essay on the Principle of Population*, J. Johnson, London, 1798, p. 7.

11 Malthus, Thomas, *An Essay on the Principle of Population*, J. Johnson, London, 1798, p. 7.

12 Malthus, Thomas, *An Essay on the Principle of Population*, J. Johnson, London, 1798, p. 7.

13 Malthus, Thomas, *An Essay on the Principle of Population*, J. Johnson, London, 1798, p. 7.

> Taking the population of the world at any number, a thousand millions, for instance, the human species would increase in the ratio of – 1, 2, 4, 8, 16, 32, 64, 128, 256, 512, etc. and subsistence as – 1, 2, 3, 4, 5, 6, 7, 8, 9, 10, etc. In two centuries and a quarter, the population would be to the means of subsistence as 512 to 10: in three centuries as 4096 to 13, and in two thousand years the difference would be almost incalculable, though the produce in that time would have increased to an immense extent.[14]

In Malthus's view, there was no real solution to this serious problem.

> It is, undoubtedly, a most disheartening reflection that the great obstacle in the way to any extraordinary improvement in society is of a nature that we can never hope to overcome. The perpetual tendency in the race of man to increase beyond the means of subsistence is one of the general laws of animated nature which we can have no reason to expect will change. [...] But if we proceed without a thorough knowledge and accurate comprehension of the nature, extent, and magnitude of the difficulties we have to encounter, or if we unwisely direct our efforts towards an object in which we cannot hope for success, we shall not only exhaust our strength in fruitless exertions and remain at as great a distance as ever from the summit of our wishes, but we shall be perpetually crushed by the recoil of this rock of Sisyphus.[15]

According to Malthus, attempts to make cultivation keep pace with population, although in vain, lead to innovation and progress for mankind. For this reason, Malthus believed that the mismatch between the growth of populations and the growth of the food supply was part of God's plan:

> To furnish the most unremitted excitements of this kind, and to urge man to further the gracious designs of Providence by the full cultivation of the earth, it has been ordained that population should increase much faster than food.[16]
>
> Had population and food increased in the same ratio, it is probable that man might never have emerged from the savage state.[17]
>
> But it is impossible that this law can operate, and produce the effects apparently intended by the Supreme Being, without occasioning partial evil."[18]
>
> The partial pain, therefore, that is inflicted by the supreme Creator, while he is forming numberless beings to a capacity of the highest enjoyments, is but as the dust of the balance in comparison of the happiness that is communicated, and we have every reason to think that there is no more evil in the world than what is absolutely necessary as one of the ingredients in the mighty process.[19]

Malthus's writing is economical rather than mathematical. Also, he ties in religion strongly, as the above passages demonstrate.

The Logistic Model

In the difference equation $x_n = (1 + r)^n x_0$, if the constant is positive, the population will grow without bound over the long run, which is not realistic in a habitat with limited

---

14 Malthus, Thomas, *An Essay on the Principle of Population*, J. Johnson, London, 1798, p. 8.
15 Malthus, Thomas, *An Essay on the Principle of Population*, J. Johnson, London, 1798, pp. 108–109.
16 Malthus, Thomas, *An Essay on the Principle of Population*, J. Johnson, London, 1798, p. 114.
17 Malthus, Thomas, *An Essay on the Principle of Population*, J. Johnson, London, 1798, p. 115.
18 Malthus, Thomas, *An Essay on the Principle of Population*, J. Johnson, London, 1798, p. 115.
19 Malthus, Thomas, *An Essay on the Principle of Population*, J. Johnson, London, 1798, p. 123.

resources. In reality, most species have an equilibrium population, often referred to as the **carrying capacity**. If the population for some reason rises above this value, the death rate will increase and the population will decrease again.

In 1838, 40 years after Malthus's frequently cited essay, the Belgian mathematician Pierre Francois Verhulst (Figure 15.2) provided a better mathematical model of population growth that accounted for the above effect. His model was expressed as a differential equation.[20] Despite being a great improvement, it apparently didn't catch on. It was rediscovered in 1920 by Raymond Pearl and Lowell J. Reed, both of Johns Hopkins.[21] These men improved upon it a bit, expressing it as:

$$\frac{dP}{dt} = rP\left(1 - \frac{P}{K}\right)$$

In this version, the carrying capacity is represented by K. Nevertheless, word still failed to reach many people interested in such things and it was re-rediscovered by Alfred J. Lotka in 1925.

The differential equation is referred to as the logistic equation or the logistic model, or even the Verhulst model, in honor of its original discoverer.

If P is tiny compared to K, then $\left(1 - \frac{P}{K}\right)$ is very close to 1 and the above differential equation is nearly $\frac{dP}{dt} = rP$. That is, we have exponential growth. However, if P is close to the value of K, the value of $\left(1 - \frac{P}{K}\right)$ has a significant damping effect on the change in population. This damping becomes stronger as P gets closer to K.

FIGURE 15.2
Pierre Francois Verhulst (1804–1849).

Public domain.

---

20 Verhulst, Pierre-François, "Notice sur la loi que la population poursuit dans son accroissement," *Correspondance mathématique et physique*, Vol. 10, 1838, pp. 113–121.

21 Pearl, Raymond and Lowell J. Reed, "On the Rate of Growth of the Population of the United States since 1790 and Its Mathematical Representation," *Proceedings of the National Academy of Sciences*, Vol. 6, No. 6, June 15, 1920, pp. 275–288.

Today, the equation is very well-known. It's used all over the place for a wide range of populations, as well as modeling the growth of tumors, and, in chemistry, the concentration of reactants and products in autocatalytic reactions.[22]

Finding the solution to Verhulst's differential equation can be done via separation of variables, and is left as an exercise. Graphing it gives a shape like the one shown in Figure 15.3.

It's important to understand that this new population model is still just a model. Too much trust has been put in it at times. For example, in 1924, George Udny Yule, a British statistician and Fellow of the Royal Society, and Raymond Pearl, a rediscoverer of the Verhulst model, independently applied it to conclude that the limiting value for the population of the United States was about 199 million. The report on this in *The Scientific Monthly* included "According to this our country is already more than half filled up and the danger is lest it should fill up too fast and with the wrong sort of folks."[23]

The population of the United States, as of this writing, is over 330 million.[24]

The graph in Figure 15.4, still showing a limiting value of about 199 million people for the United States, appeared in the January 1935 issue of *The Scientific Monthly*. To be fair, the author of the article that contained it, Warren D. Smith, expressed skepticism as to its predictive power. He also expressed skepticism that the world population could ever reach 7.689 billion, a potential value found by Albrecht Penck. Smith wrote, "It is of course quite improbable that the world population will ever reach the stupendous total estimated in the above paragraphs."[25] We are now over 7.7 billion.[26]

Predictive power aside, the version of the logistic equation described above belongs to the world of continuous mathematics. The discrete version of the logistic equation came later and it's far more interesting. It may be written as $P_{t+1} = P_t(a - bP_t)$, where a and

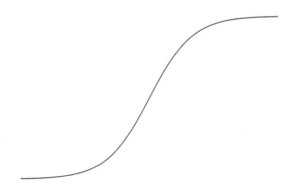

FIGURE 15.3
The solution to the logistic equation.

Public domain.

---

22 https://en.wikipedia.org/wiki/Logistic_function

23 Slosson, Edwin E., "The Progress of Science," *The Scientific Monthly*, Vol. 19, November 1924, pp. 553–560, p. 559 cited here.

24 https://www.census.gov/popclock/

25 Smith, Warren D., "World Population," *The Scientific Monthly*, Vol. 40, January 1935, pp. 33–43, p. 41 cited here.

26 https://www.worldometers.info/world-population/.

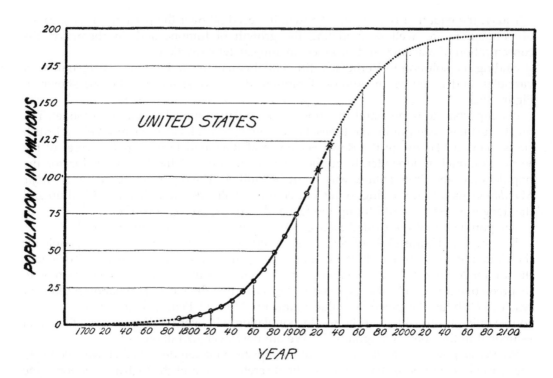

FIGURE 15.4
A population model for the United States published in 1935.

Smith, Warren D., "World Population," *The Scientific Monthly,* **Vol. 40, January 1935, pp. 33–43, graph on p. 35 (public domain).**

b are constants. Making the substitution $X_t = bP_t/a$ converts the equation to the form $X_{t+1} = aX_t(1 - X_t)$.

In either form it's known as the **Logistic Difference Equation** or the **Logistic Map**. This model cannot be solved (in closed form) in its greatest generality.[27] The only values for which exact solutions are known are $a = -2$, $a = 2$, and $a = 4$.[28] However, we can investigate the behavior easily with a computer. This was done in the 1970s by Robert May, a physicist turned applied mathematician. In 1976, he published a survey paper on the topic of simple difference equations with complicated behavior. In the introduction to this paper, May wrote:

> The review ends with an evangelical plea for the introduction of these difference equations into elementary mathematics courses, so that students' intuition may be enriched by seeing the wild things that simple nonlinear equations can do.[29]

27 http://mathworld.wolfram.com/LogisticMap.html
28 http://mathworld.wolfram.com/LogisticMap.html
29 May, Robert M., "Simple Mathematical Models with very Complicated Dynamics," *Nature*, Vol. 261, June 10, 1976, pp. 459–467, p. 459 cited here.

Instead of immediately revealing the wild behavior, the paragraphs below slowly explore the equations, so that you can get a small taste of the process of discovery that the original researchers went through. There would have been nothing to prevent you from being the one to make these discoveries, if you could only have been there first with a modest computer.

We start the investigation of $X_{t+1} = aX_t(1 - X_t)$ by picking a value for a and an initial value X_0, and performing a sufficient number of iterations to make the long-term behavior clear. For example, picking $a = 2.5$, and $X_0 = 0.5$, gives $X_1 = 0.625$, $X_2 = 0.585938$, $X_3 = 0.606537$, etc. Continuing to iterate, we find that the value tends to a limiting value of 0.6.

Trying the slightly larger value $a = 2.6$ (and using the same starting value $X_0 = 0.5$), our iterations lead us to 0.615385. This isn't very exciting yet! At first, increasing the value of the constant a simply leads to slightly larger convergent values upon iteration of the difference equation. However, something interesting happens once we pass 3.

At $a = 3.1$, the values produced by iterating the logistic difference equation do not converge. Instead, we eventually have the alternating values 0.764567 and 0.558014. A range of values for a yield such cycles of length two. As a grows, the distance between the values it cycles between grow as well.

Let's now iterate using $a = 3.5$. After many iterations, we find ourselves stuck in a loop again, but this time it doesn't consist of two values. We loop between 0.500884, 0.874997, 0.38282, and 0.826941. That is, we have a cycle of length four. We may refer to this as a 4-cycle or say that we have period four.

We can look for a value of a between 3.1 and 3.5 with the expectation of finding a 3-cycle, but we won't succeed. The change from a 2-cycle to a 4-cycle is abrupt. Looking beyond $a = 3.5$, we can find values that yield periods of length 8, 16, 32, etc. We can obtain all periods that are powers of two! However, the range of values that yield each period gets smaller and smaller. It's a very narrow range of values for a that yield period 2^{100}, but it does exist. After the powers of 2, the behavior gets even stranger. At $a = 3.8284$ we have period three.

By experimenting with various values for a, we can pinpoint the points where the long-term behavior transitions from period 2 to 4, from 4 to 8, etc. These transitional points are referred to as bifurcation points. An important discovery concerning their locations was made by Mitchell Feigenbaum (Figure 15.5).

Feigenbaum's Constant

The police in the small town of Los Alamos, New Mexico, worried briefly in 1974 about a man seen prowling in the dark, night after night, the red glow of his cigarette floating along the back streets. He would pace for hours, heading nowhere in the starlight that hammers down through the thin air of the mesas. The police were not the only ones to wonder. At the national laboratory some physicists had learned that their newest colleague was experimenting with twenty-six-hour days, which meant that his waking schedule would slowly roll in and out of phase with theirs. This bordered on strange, even for the Theoretical Division. – James Gleick[30]

---

30 Gleick, James, *Chaos: Making a New Science*, Penguin Books, New York, 1987, p.1.

FIGURE 15.5
Mitchell Feigenbaum.

Courtesy of The Rockefeller University.

As you might guess, the man described above is Mitchell Feigenbaum. We'll first look at his discovery and then find out how well his twenty-six-hour days experiment worked out. To help us understand the results being described, a graph is presented (Figure 15.6) with the value of a on the x-axis and the corresponding cycle values on the y-axis.

As the graph is traced out from left to right we see it repeatedly split in two. These splits are called bifurcation points. Feigenbaum's discovery concerns the horizontal distance between bifurcations. A few of these distances are labeled in Figure 15.7.

Feigenbaum discovered that the ratio of the distances between bifurcation points, $\frac{L_n}{L_{n+1}}$, approaches a limiting value. That is, $\lim_{n \to \infty} \frac{L_n}{L_{n+1}} = 4.6692016091 \ldots$. This constant, represented by the lowercase Greek letter δ (delta), doesn't just arise for this one particular function. Other very simple difference equations yield similar graphs and the same value pops up for any quadratic that exhibits the period doubling phenomenon. The other degree polynomials, cubic, quartic, quintic, etc., each have their own constant. The existence of δ shows that the graphs are self-similar and gives their scale. Hence, we have a connection with fractals, which we will be seen more fully in Chapter 18.

Note the chaotic behavior as we move to the right. This indicates values of a that never become periodic. One can literally iterate forever and never return to a previously obtained value. The moral is that complicated (chaotic) behavior can arise from *simple* discrete mathematics.

A Small Proof

The 1975 paper *Period Three Implies Chaos* by Tien-Yien Li and James A. Yorke provided a proof of a surprising result. If a continuous function, such as the one iterated above,

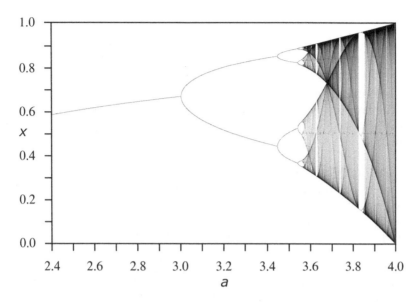

FIGURE 15.6
A diagram showing bifurcations and the chaotic region.

Public domain.

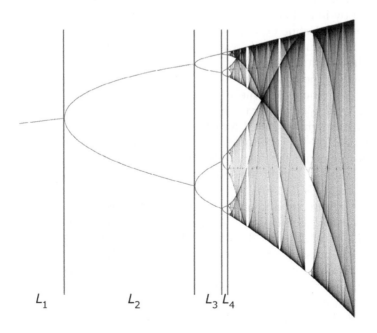

FIGURE 15.7
How Feigenbaum's constant arises.

Adapted from Figure 15.6 by Josh Gross.

has a cycle of period three, then it also has a cycle of period k for *every* natural number k, as well as chaotic behavior for some initial condition. Only the first half of this claim will be verified here. A few lemmas are needed first.

Lemma 0: Let $G: I \to R$ be continuous, where I is an interval. For any compact interval $I_1 \subseteq G(I)$ there is a compact interval $Q \subseteq I$ such that $G(Q) = I_1$.
(For purposes here, it's okay if you think of "compact" as meaning closed. The real definition can be found in an Advanced Calculus / Elementary Analysis or Topology text or online.)

Proof: Let $I_1 = [G(p), G(q)]$, where $p, q \in I$. If $p < q$, let r be the last point of $[p, q]$ where $G(r) = G(p)$ and let s be the first point after r where $G(s) = G(q)$. Then $G([r, s]) = I_1$. Similar reasoning applies when $p > q$.
   This is exactly how the proof of this lemma is presented by Li and Yorke, but I would suggest you draw a picture to help convince yourself that it's true.

Lemma 1: Let $F: J \to J$ be continuous and let $\{I_n\}_{n=0}^{\infty}$ be a sequence of compact intervals with $I_n \subseteq J$ and $I_{n+1} \subset F(I_n)$ for all n. (Note: $\{I_n\}_{n=0}^{\infty} = I_1, I_2, I_3, \ldots$)
   Then there is a sequence of compact intervals Q_n such that $Q_{n+1} \subset Q_n \subset I_0$ and $F^n(Q_n) = I_n$ for $n \geq 0$. For any $x \in Q = \cap\, Q_n$ we have $F^n(x) \in I_n$ for all n.
   (Note: We have another sequence $Q_1, Q_2, Q_3, \ldots$ of closed intervals and we define Q to be the points that are common to all; i.e., the intersection of all of the Q_i, $i = 1, 2, \ldots$)

Proof: We need to show that the sequence Q_n exists and has the properties claimed above. A proof by induction will be used.
   Base Case ($n = 0$): Define $Q_0 = I_0$. Then we certainly have $Q_0 \subset I_0$. It remains to show that $F^0(Q_0) = I_0$. Because $Q_0 = I_0$, we have $F^0(Q_0) = F^0(I_0)$. The notation F^0 means "do not perform the function F," so $F^0(I_0) = I_0$. Piecing this together, we have $F^0(Q_0) = I_0$, as desired, completing the base case.
   Induction Assumption: Assume $F^{n-1}(Q_{n-1}) = I_{n-1}$. Now take F of both sides to get $F\big(F^{n-1}(Q_{n-1})\big) = F(I_{n-1})$, which may be rewritten as $F^n(Q_{n-1}) = F(I_{n-1})$. We will use this equation later. It is lightly shaded for easy reference.
   We were given $I_{n+1} \subset F(I_n)$ for all n (see the statement of lemma 1), so $I_n \subset F(I_{n-1})$ (we simply decremented the subscripts – the meaning of the statement is unchanged). Again the equation is lightly shaded for easy reference. Piecing together the lightly shaded equations gives $I_n \subset F(I_{n-1}) = F^n(Q_{n-1})$
   We will now get prepared to apply lemma 0 to this last equation. This means making sure the conditions of lemma 0 are satisfied. Let $G = F^n$. F is continuous (we were given this fact), so F^n is continuous (the composition of continuous functions is continuous). Because $G = F^n$, G must be continuous. We have $I_n \subset F(I_{n-1}) = G(Q_{n-1})$, which, ignoring the middle expression, says $I_n \subset G(Q_{n-1})$. Q_{n-1} is an interval and I_n is a compact (closed) interval with $I_n \subset G(Q_{n-1})$, so the conditions of lemma 0 are satisfied with I_n playing the role of I_1 and Q_{n-1} playing the role of I. Hence, lemma 0 tells us that there is a compact (closed) interval $Q_n \subset Q_{n-1}$ such that $G(Q_n) = I_n$. Because $G = F^n$, this may be rewritten as $F^n(Q_n) = I_n$, completing the induction proof.
   Note: Because lemma 0 was used in the proof of lemma 1, it was necessary to prove them in this order.

Lemma 2: Let $G: J \to R$ be continuous. Let $I \subseteq J$ be a compact interval. Assume $I \subseteq G(I)$. Then there is a point $p \in I$ such that $G(p) = p$.

This is one of a large group of "fixed point" theorems. It tells us that under the given conditions, there is a point p that stays fixed (gets sent to itself) under the function G.

Proof: As with lemma 0, a picture makes the proof tremendously easier to see, so please sketch one out as you read. The text of the proof is, following Li and Yorke:

Let $I = [\beta_0, \beta_1]$. Choose α_i ($i = 0, 1$) in I such that $G(\alpha_i) = \beta_i$. It follows $\alpha_0 - G(\alpha_0) \geq 0$ and $\alpha_1 - G(\alpha_1) \leq 0$ and so continuity implies $G(\beta) - \beta$ must be 0 for some $\beta \in I$.

Theorem 1: Let J be an interval and let $F: J \to J$ be continuous. Assume there is a point $a \in J$ for which the points $b = F(a)$, $c = F^2(a)$ and $d = F^3(a)$, satisfy $d \leq a < b < c$ (or $d \geq a > b > c$). Then for every $k = 1, 2, \dots$ there is a periodic point in J having period k.

Proof: We will assume $d \leq a < b < c$. The proof for the case $d \geq a > b > c$ is similar.

Define $K = [a, b]$ and $L = [b, c]$. Warning: don't confuse K with k. We let k represent some positive integer. We will show F has a point of period k no matter what the value of k is. For $k > 1$, let $\{I_n\}$ be the sequence of intervals $I_n = L$ for $n = 0, 1, 2, \dots k - 2$ and $I_{k-1} = K$ and define I_n to be periodic inductively, $I_{n+k} = I_n$ for $n = 0, 1, 2, \dots$

Example: If $k = 5$, we have

$$
\begin{array}{ll}
I_0 = L = [b, c] & I_5 = I_0 = L = [b, c] \\
I_1 = L = [b, c] & I_6 = I_1 = L = [b, c] \\
I_2 = L = [b, c] & I_7 = I_2 = L = [b, c] \\
I_3 = L = [b, c] & I_8 = I_3 = L = [b, c] \\
I_4 = K = [a, b] & I_9 = I_4 = K = [a, b]
\end{array}
$$

(I added 5 to the subscripts in the first column to get the second column.)

In the special case $k = 1$, we define $I_n = L = [b, c]$ for all n.

Note that the I_n above satisfy the requirements of lemma 1. Consider $F(I_0)$. $F(b) = c$ and $F(c) = d \leq a$, so $F(I_0) \supset [a, c] \supset [b, c] = I_1$. That is, $I_1 \subseteq F(I_0)$, as lemma 1 requires. Similarly $F(I_3) \supset [a, c] \supset [a, b] = I_4$, so $I_4 \subseteq F(I_3)$. The last distinct case that needs to be considered is $F(I_4)$. We have $F(a) = b$ and $F(b) = c$, so $[b, c] \subseteq F(I_4)$. Thus, $I_5 \subseteq F(I_4)$. Hence, the conditions of lemma 1 are satisfied.

Now let Q_n be the sets in the proof of lemma 1. We have $Q_{n+1} \subset Q_n \subset I_0$ for all $n \geq 0$ and $Q_0 = I_0$, so $Q_k \subset Q_0$. I've lightly shaded this last inclusion for future reference. We also have $F^k(Q_k) = I_k$ (by lemma 1) and $I_k = L = I_0$ (by definition) and $I_0 = Q_0$, so transitivity gives us $F^k(Q_k) = Q_0$. The two lightly shaded equations now give $Q_k \subset F^k(Q_k)$. Q_k is a compact (closed) interval and F is continuous, which implies F^k is continuous, so letting $G = F^k$ gives us $Q_k \subset G(Q_k)$, which satisfies the conditions of lemma 2. Hence, G has a fixed point in Q_k. Let's denote this point by p_k. $G(p_k) = p_k \Rightarrow F^k(p_k) = p_k$.

We claim that p_k is a point of period k for F. It only remains to verify that p_k doesn't have some period smaller than k. The example for $k = 5$ will serve to illuminate the argument. First note that $p_k \in Q_k \subset Q_0 = I_0 = L = [b, c]$. Also, $F(p_k) \in F(Q_k) \subset F(Q_1) = I_1 = L = [b, c]$, where the last inclusion follows from the fact that $Q_k \subset Q_1$. The same argument shows $F^k(p_k) \subset F^k(Q_k) = I_k$

To summarize, for the example with $k = 5$, we get

$$p_k \in I_0 = L = [b, c]$$

$$F(p_k) \in I_1 = L = [b, c]$$

$$F^2(p_k) \in I_2 = L = [b, c]$$

$$F^3(p_k) \in I_3 = L = [b, c]$$

$$F^4(p_k) \in I_4 = K = [a, b]$$

The values p_k, $F(p_k)$, $F^2(p_k)$, $F^3(p_k)$, and $F^4(p_k)$ cannot all be distinct if the period of p_k under F is less than k (= 5 in this example). That is, $F^4(p_k)$ would have to repeat some previous value. All previous values were in $L = [b, c]$, so $F^4(p_k)$ would have to be in L. However $F^4(p_k) \in K = [a, b]$. Thus the only possibility is $F^4(p_k) = b$. That is, $F^{k-1}(p_k) = b$. This implies $F(F^{k-1}(p_k)) = F(b)$, which simplifies to $F^k(p_k) = c$. Again taking F of both sides gives $F^{k+1}(p_k) = F(c) = d$. Recall that $d \le a < b$. So, $F^{k+1}(p_k) < b$, if p_k has a period *less than* k. The inequality is lightly shaded, because it will be needed shortly. On the other hand, we showed $F^k(p_k) = p_k$, so taking F of both sides gives $F^{k+1}(p_k) = F(p_k)$. $p_k \in Q_k \subset Q_1$, so $F(p_k) \subset F(Q_1) = I_1$, where the last equality follows from the statement of lemma 1.

That is, $F^{k+1}(p_k) \subset I_1 = L = [b, c]$, which contradicts the lightly shaded inequality above. Because the period of p_k is now seen to be k under F, and k is an arbitrary positive integer, the proof is complete. **Q.E.D.**

Notice that although a 3-cycle will satisfy the conditions of theorem 1, the condition is actually more general.

A proof that the presence of period three implies all other periods appeared before the famous "Period Three Implies Chaos" paper, but it was in Russian, written by Oleksandr Sharkovsky.[31] While the results of Sharkovsky, Li, and Yorke deepen our understanding of the discrete logistic equation, it is still not completely understood.[32]

Now, as to Feigenbaum's personal attempt at twenty-six-hour days, "his experiment in personal quasi-periodicity came to an end when he decided he could no longer bear waking to the setting sun, as had to happen every few days."[33]

---

31 The name is sometimes written Sharkovskii, in the Latin alphabet. The reference is Sharkovsky, A. N., "Сосуществование циклов непрерывного преобразования прямой в себя," *Ukrainskij matematicheskij zhurnal*, Vol. 16, No. 1, 1964, pp. 61–71, available online at http://umj.imath.kiev.ua/archiv/1964/01/umj_1964_01_9991_18320.pdf. For an English translation see Sharkovsky, A. N., "Co-existence of Cycles of a Continuous Mapping of the Line into Itself," *International Journal of Bifurcation and Chaos*, Vol. 5, No. 5, October 1995, pp. 1263–1273. You can also find a proof, in English, in Devaney, Robert L., *An Introduction to Chaotic Dynamical Systems*, 2nd ed., Addison-Wesley, Redwood City, CA, 1989, pp. 60–62. Some web references are https://en.wikipedia.org/wiki/Sharkovskii%27s_theorem and http://www-groups.dcs.st-and.ac.uk/history/Printonly/Sharkovsky.html.

32 Young, Robert M., *Excursions in Calculus: An Interplay of the Continuous and the Discrete*, The Mathematical Association of America, 1992, p. 182.

33 Gleick, James, *Chaos: Making a New Science*, Penguin Books, New York, 1987, p. 2.

Connections

By the way, the bifurcation diagram above has a connection to the Mandelbrot set, which is, in turn, connected to the Fibonacci numbers. These connections are detailed in Chapter 18.

Speaking of Fibonacci his simplistic model of rabbit populations yielded the difference equation $F_n = F_{n-1} + F_{n-2}$. We may rewrite it as $F_n - F_{n-1} - F_{n-2} = 0$ and form the **characteristic equation** (aka **characteristic polynomial**) $\lambda^2 - \lambda - 1 = 0$. The roots, found by using the quadratic formula, are $\frac{1 \pm \sqrt{5}}{2}$. Do these values look familiar? We've begun a process through which the formula

$$F_n = \frac{1}{\sqrt{5}}\left(\frac{1 + \sqrt{5}}{2}\right)^n - \frac{1}{\sqrt{5}}\left(\frac{1 - \sqrt{5}}{2}\right)^n$$

can be derived. The formula was first found by Abraham de Moivre in 1718, but the proof didn't arrive until Nicolas Bernoulli discovered it ten years later.

Because the characteristic polynomial had the distinct real roots $\frac{1 + \sqrt{5}}{2}$ and $\frac{1 - \sqrt{5}}{2}$, the **general solution** to the difference equation it represents is given by

$$F_n = A\left(\frac{1 + \sqrt{5}}{2}\right)^n + B\left(\frac{1 - \sqrt{5}}{2}\right)^n.$$

Initial conditions allow us to solve for A and B to get a **particular solution**.

$$F_0 = 0 \Rightarrow A + B = 0.$$

$$F_1 = 1 \Rightarrow A\left(\frac{1 + \sqrt{5}}{2}\right) + B\left(\frac{1 + \sqrt{5}}{2}\right) = 1.$$

Substituting $-A$ in for B in the second equation and simplifying gives $2A\left(\frac{\sqrt{5}}{2}\right) = 1$, so $A = \frac{1}{\sqrt{5}}$. It then quickly follows that $B = \frac{-1}{\sqrt{5}}$. The particular solution to the Fibonacci difference equation is $F_n = \frac{1}{\sqrt{5}}\left(\frac{1+\sqrt{5}}{2}\right)^n - \frac{1}{\sqrt{5}}\left(\frac{1-\sqrt{5}}{2}\right)^n$, as claimed.

Now that I've verified a Fibonacci formula from a previous chapter, I'm presenting yet another that is even more surprising.[34]

$$F_n = i^{n-1}\frac{\sin(nz)}{\sin(z)}, \text{ where } z = \frac{\pi}{2} + i\ln\left(\frac{1 + \sqrt{5}}{2}\right)$$

So, if you were surprised by the formula that connected Fibonacci numbers (positive integers) to the Golden Ratio (an irrational number), how do you feel about this new formula that expresses these integers in terms of a combination of $i = \sqrt{-1}$, a trigonometric function, π, and φ? The presence of the natural logarithm also brings e into this equation.

---

34 I found this formula in Schroeder, Manfred R., *Number Theory in Science and Communication*, Springer, Berlin, 1984. p. 65.

Let's now return to solving difference equations through the use of the characteristic equation. In general, if λ is a root of the characteristic equation, then $x_n = \lambda^n$ is a solution to the difference equation. Also, if x_n and y_n are solutions, then $Ax_n + By_n$ is also a solution for all constants A and B. These constants may be determined if initial conditions are present, as was the case for the difference equation describing the Fibonacci numbers. We had $F_0 = 0$ and $F_1 = 1$.

Let's look at some examples.

Example 1

Consider the difference equation $x_n = x_{n-1} + 2x_{n-2}$. This may be rewritten as $x_n - x_{n-1} - 2x_{n-2} = 0$.

We may then form the characteristic equation $\lambda^2 - \lambda - 2 = 0$.

This factors easily as $(\lambda - 2)(\lambda + 1) = 0$, which implies $\lambda = 2$ or $\lambda = -1$.

Hence, the general solution is of the form $x_n = A \cdot 2^n + B \cdot (-1)^n$.

If we had initial conditions, we could solve for A and B and obtain a particular solution.

The next example shows what to do in the case of the two roots being the same.

Example 2

Consider the difference equation $x_n = 6x_{n-1} + 9x_{n-2}$. This may be rewritten as $x_n - 6x_{n-1} + 9x_{n-2} = 0$.

The characteristic equation is $\lambda^2 - 6\lambda + 9 = 0$.

This factors as $(\lambda - 3)(\lambda - 3) = 0$, which gives $\lambda = 3$ as a double root.

In this situation, we raise 3 to the nth power twice, but multiply the second instance by n. That is, the general solution is of the form $x_n = A \cdot 3^n + Bn \cdot 3^n$.

In general, if λ is a root of multiplicity m, it contributes the following terms to our general solution:

$$A_1 \cdot \lambda^n + A_2 n \cdot \lambda^n + A_3 n^2 \cdot \lambda^n + \cdots + A_m n^{m-1} \cdot \lambda^n.$$

Sometimes the roots turn out to be complex numbers. What to do in this situation will be addressed in Chapter 17.

More Improvements to Come

The population models described in this chapter isolate the populations from outside influences, but this is not natural. In the real world people immigrate and emigrate. This will be considered in the next chapter, which begins with applying the difference equations from this chapter to financial problems. As the problems become more complex, new features have to be added and these new features are then made use of in population modeling, as well.

Open Problem

It is not known if Feigenbaum's constant is an algebraic or a transcendental number.

Exercise Set 15

> I have always believed in getting things done – in meeting challenges head-on and solving problems before they get out of hand.
>
> Arnold Schwarzenegger[35]

1. How long will it take for a population that grows by 3% per year to double?
2. How long will it take for a population that grows by 6% per year to double?
3. What is the annual growth rate of a population that doubles in 20 years?
4. What is the annual growth rate of a population that doubles in 40 years?
5. What is the annual growth rate of a population that grows from 8 million to 10 million in 15 years?
6. What is the annual growth rate of a population that grows from 8 million to 10 million in 30 years?
7. The bit of dialog reproduced below is from the 1994 movie *The Puppet Masters*. It concerns rapidly multiplying alien invaders.

   ANDREW NIVENS: How long does it take the creature to prepare for division?
   GRAVES: 12 hours.
   　　　　 So, let's assume we started with a thousand of them.
   　　　　 We'll be conservative.
   　　　　 That's one day, 4,000.
   　　　　 Two days, 16,000.
   　　　　 In two weeks there would be ...
   ANDREW NIVENS: More than 250 billion.[36]

   If the premises that the invaders double their population every 12 hours and that their initial population was 1,000 are both correct, would there really be 250 billion of them in two weeks?
8. Answer the following brain teaser from the April 1979 issue of *OMNI*, without resorting to a calculator or using any formulas.

   An amoeba divides and reproduces itself every minute. Two amoebas in a test tube can fill it to capacity in 2 hours. How long would it take 1 amoeba to fill another test tube of equal capacity?[37]

9. Nine digit social security numbers can only distinguish 1 billion Americans. When will this no longer suffice, given the present population of 330 million and an annual growth rate of 0.6%?[38] Assume growth is exponential – don't use the logistic model for this problem.

---

35 Schwarzenegger, Arnold with Charles Gaines, *Arnold's Fitness for Kids ages birth-5: a guide to health, exercise, and nutrition*, Doubleday, New York, 1993, p. 1.

36 Taken from https://www.springfieldspringfield.co.uk/movie_script.php?movie=the-puppet-masters. More information on this movie can be found at https://www.imdb.com/title/tt0111003/?ref_=fn_al_tt_1.

37 Morris, Scot, "Games," *OMNI*, April 1979, Vol. 1, No. 7, pp. 144–145. See p. 144, problem 12.

38 Growth rate taken from http://www.worldometers.info/world-population/us-population/.

10. Use separation of variables to solve for P in $\frac{dP}{dt} = rP\left(1 - \frac{P}{K}\right)$.

11. Verify that the substitution $X_t = bP_t/a$ converts $P_{t+1} = P_t(a - bP_t)$ to the form $X_{t+1} = aX_t(1 - X_t)$.

12. Letting $a = 3.52$ in $X_{t+1} = aX_t(1 - X_t)$, what period results?

13. Find a value for a in $X_{t+1} = aX_t(1 - X_t)$ that leads to period eight.

14. What periods can you find for the function $f(x) = 4x(1 - x)$?

15. Look up the definition for the second Feigenbaum constant and explain it clearly in your own words. Illustrations can be included as part of your explanation.

16. What are the relative costs of software and hardware and how have they changed over time? Has the cost of software followed a logistic curve, when plotted as a percentage of the whole, as some claim? Explore the literature yourself to decide how the graph should be drawn. Support your conclusion as thoroughly as possible. A pair of references are:

Boehm, Barry, "Software and its Impact," *Datamation*, May 1973, p. 49.

Ceruzzi, Paul E., *A History of Modern Computing*, The MIT Press, Cambridge, Massachusetts, 1998, p 82.

17. Consider Figure 15.8 below.

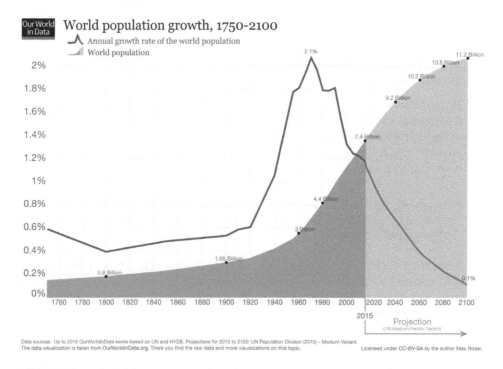

FIGURE 15.8
A graph of the world's population, including a projection into the future.

Max Roser and Esteban Ortiz-Ospina (2018) – "World Population Growth." Published online at OurWorldInData.org. Retrieved from https://ourworldindata.org/world-population-growth.

What sort of model is used to predict the future world population here? What is the justification for the model? Do you think it's the correct model? Give evidence to support your answer for the final question.

18. Find the general solution to the difference equation $x_n = 10x_{n-1} - 21x_{n-2}$.

19. Find the general solution to the difference equation $x_n = 4x_{n-1} + 12x_{n-2}$.

20. Find the general solution to the difference equation $x_n = -4x_{n-1} + 32x_{n-2}$.

21. Find the general solution to the difference equation $x_n = -8x_{n-1} - 7x_{n-2}$.

22. Find the general solution to the difference equation $x_n = 10x_{n-1} - 25x_{n-2}$.

23. Find the specific solution to the difference equation $x_n = 11x_{n-1} - 18x_{n-2}$ with initial conditions $x_0 = 1$ and $x_1 = 16$.

24. Find the specific solution to the difference equation $x_n = 7x_{n-1} - 12x_{n-2}$ with initial conditions $x_0 = -2$ and $x_1 = -4$.

25. Find the specific solution to the difference equation $x_n = -6x_{n-1} - 8x_{n-2}$ with initial conditions $x_0 = 0$ and $x_1 = 1$.

26. What is the general solution for a difference equation whose characteristic polynomial has the roots $1, -1, 2, -2$?

27. What is the general solution for a difference equation whose characteristic polynomial has the roots $1, 1, 2, 3$?

28. What is the general solution for a difference equation whose characteristic polynomial has the roots $6, 6, 6$?

29. What is the general solution for a difference equation whose characteristic polynomial has the roots $1, 2, 2, 3, 3, 3$?

References/Further Reading

RTFM
  – Anonymous.

Ausloos, Marcel and Michel Dirickx, editors, *The Logistic Map and the Route to Chaos – From the Beginnings to Modern Applications,* Springer, Berlin, 2006.

Bonar, James, *Malthus and His Work,* Macmillan and Co., London, 1885.

Bowen, Ezra, "Malthus, a Revaluation," *The Scientific Monthly,* Vol. 30, May 1930, pp. 465–471. This article offers a bit of historic background on Malthus's essay, as well as reactions to it. Of interest is how it motivated Wallace and Darwin in their conception of evolution (see p. 468).

Devaney, Robert L., *An Introduction to Chaotic Dynamical Systems,* 2nd edition, Addison-Wesley, Redwood City, CA, 1989.

Devaney, Robert L., *Chaos, Fractals, and Dynamics: Computer Experiments in Mathematics,* Addison-Wesley, Menlo Park, CA, 2000.

Feigenbaum, Mitchell J., "Quantitative Universality for a Class of Nonlinear Transformations," *Journal of Statistical Physics,* Vol. 19, 1978, pp. 25–52.

Feigenbaum, Mitchell J., "The Universal Metric Properties of Nonlinear Transformations," *Journal of Statistical Physics,* Vol. 21, 1979, pp. 669–706.

Feigenbaum, Mitchell J., "Universal Behavior in Nonlinear Systems," *Los Alamos Science,* Vol. 1, 1981, pp. 4–27.

Gleick, James, *Chaos: Making a New Science,* Penguin Books, New York, 1987.

Gurel, Okan and Otto E. Rössler, editors, "Bifurcation Theory and Applications in Scientific Disciplines," *Annals of the New York Academy of Sciences*, Vol. 316, 1979, pp. 1–708.

The International Society of Malthus, http://desip.igc.org/malthus/.

Li, Tien-Yien, and James A. Yorke, "Period Three Implies Chaos," *The American Mathematical Monthly*, Vol. 82, No. 10, December 1975, pp. 985–992.

Malthus, Thomas, *An Essay on the Principle of Population*, J. Johnson, London, 1798.

May, Robert M., "Simple Mathematical Models with Very Complicated Dynamics," *Nature*, Vol 261, June 10, 1976, pp. 459–467.

Pearl, Raymond and Lowell J. Reed, "On the Rate of Growth of the Population of the United States since 1790 and its Mathematical Representation," *Proceedings of the National Academy of Sciences*, Vol. 6, No. 6, June 15, 1920, pp. 275–288.

Schroeder, Manfred R., *Number Theory in Science and Communication*, Springer Verlag, Berlin, 1984.

Verhulst, Pierre-François, "Notice sur la loi que la population poursuit dans son accroissement," *Correspondance mathématique et physique*, Vol. 10, 1838, pp. 113–121.

Verhulst, Pierre-François, *Traité élémentaire des fonctions elliptiques: ouvrage destiné à faire suite aux traités élémentaires de calcul intégral*, Hayez, Bruxelles, 1841.

Verhulst, Pierre-François, "Recherches mathématiques sur la loi d'accroissement de la population," (Mathematical Researches into the Law of Population Growth Increase)," *Nouveaux Mémoires de l'Académie Royale des Sciences et Belles-Lettres de Bruxelles*, Vol. 18, 1845, pp. 1–42.

Verhulst, Pierre-François, "Deuxième Mémoire sur la Loi D'accroissement de la Population," *Mémoires de l'Académie Royale des Sciences, des Lettres et des Beaux-Arts de*, Vol. 20, 1847, pp. 1–32.

Weisstein, Eric W., "Logistic Map," from *MathWorld*-A Wolfram Web Resource, available online at http://mathworld.wolfram.com/LogisticMap.html.

Wikipedia contributors, "Sharkovskii's Theorem," *Wikipedia, The Free Encyclopedia*, available online at https://en.wikipedia.org/w/index.php?title=Sharkovskii%27s_theorem&oldid=843619012.

Young, Robert M., *Excursions in Calculus: An Interplay of the Continuous and the Discrete*, The Mathematical Association of America, 1992.

16

Financial Mathematics (and More)

Difference equations of the form $X_{n+1} = aX_n$ are easy to solve. We saw examples of these in the previous chapter, but they also arise in the context of financial mathematics.

Example 1 – Annually compounded interest
Suppose $500 is deposited in a bank account that earns 2% interest compounded annually. How long will it take until the initial deposit is doubled?

Solution
This problem is perfectly modeled by the difference equation $X_{n+1} = aX_n$, with $a = 1.02$. As in the population model from Chapter 15, the solution is given by $X_n = (1.02)^n X_0$. Now, let's see how long it would take to double our money at this rate. We start with $x_0 = 500$ and we want to know when $x_n = 1,000$, so we simply solve $1,000 = (1.02)^n\, 500$. Dividing both sides by 500 gives $2 = (1.02)^n$. We can then take the natural log of both sides to get $ln(2) = ln[(1.02)^n]$. Recalling a rules of logarithms, we write $ln(2) = n \cdot ln(1.02)$, which gives us $n = ln(2)/ln(1.02) \approx 35$ years.

More Frequent Compoundings

Interest on money in a bank account may be compounded more often. For example, an account paying 2% interest compounded biannually would pay half of the 2% every six months. That is, it pays 1% every six months

Time (in years)	Amount
0.0	500
0.5	505
1.0	510.05
1.5	515.1505
2.0	520.302005
2.5	525.5050251
3.0	530.7600753
3.5	536.0676761
4.0	541.4283528
4.5	546.8426363
5.0	552.3110627

I have not rounded off to the closest cent, as a bank would. The amount in the account at the end of each year is slightly more than when interest was compounded annually. This is because we're earning interest on the interest more often. In general (for

n compoundings per year) the formula is $A = P(1 + r/n)^{nt}$, where P is the principal (initial deposit), r is the interest rate, and t is the number of years. This equation should make perfect sense. r/n is the fraction of the interest credited at each compounding and nt is the total number of compoundings. You may think of this as the number of iterations of the difference equation.

Biannual compounding serves well to illustrate the idea of more frequent compoundings, but it's unusual in the real world. Most accounts compound the interest monthly.

Example 2 – Monthly compounding
How long would it take to double your money if the interest rate is 2% and interest is compounded monthly?

Solution
To answer this question we solve $2P = P(1 + 0.02/12)^{12t}$. We don't need to know the initial deposit, P. A deposit of $1 will double just as quickly as a deposit of $500. This is because the Ps drop out of our equation leaving $2 = (1 + 0.02/12)^{12t}$. We again use logs to solve for the variable: $ln(2) = ln\left[(1 + 0.02/12)^{12t}\right]$ becomes $ln(2) = 12t \cdot ln\,(1 + 0.02/12)$, which gives us $t = ln(2)/[12 \cdot ln(1 + 0.02/12)] \approx 34.686$ years. So the initial deposit is doubled months earlier than when interest is compounded annually.

Continuously Compounded Interest

Can we improve things dramatically by compounding infinitely often? Let's see:

$$A = \lim_{n \to \infty} P\left(1 + \frac{r}{n}\right)^{nt}$$

A substitution will make this easier to solve. Let $x = n/r$. It follows that $r/n = 1/x$ and $n = xr$. The above equation becomes

$$\lim_{x \to \infty} P\left(1 + \frac{1}{x}\right)^{xrt}$$

Because n is gone, we change the limit from n approaching infinity to x approaching infinity. This makes sense, as $n = xr$ and r is fixed.

P, the initial deposit, has nothing to do with x, so we may pull it outside of the limit.

$$P \lim_{x \to \infty} \left(1 + \frac{1}{x}\right)^{xrt}$$

Similarly, the exponents r and t can be taken outside of the limit.

$$P\left[\lim_{x \to \infty} \left(1 + \frac{1}{x}\right)^{x}\right]^{rt}$$

Does anyone recognize the value of the limit? It's our old friend e. So our formula becomes

$$A = Pe^{rt}.$$

An approximate value for e (100 digits after the decimal place) is

```
2. 71828 18284 59045 23536
   02874 71352 66249 77572
   47093 69995 95749 66967
   62772 40766 30353 54759
   45713 82178 52516 64274
```

A book by Edward Teller, who is known as the father of the atomic bomb, gives a trick for remembering a few of the digits given above.

> For those familiar with American history, the first nine digits [of e] after the decimal point can be remembered by e = 2.7(Andrew Jackson)^2, or e = 2.718281828..., because Andrew Jackson was elected President of the United States in 1828.[1]

Teller didn't mention it, but Jackson was the seventh U.S. President, so it is appropriate that the digits representing his year of election appear right after the 7. If you wish to remember six more digits, simply recall the angles of the famous right triangle 45, 90, 45.

Leonhard Euler, a man whose name appears in over half of the chapters in this book, is the one who introduced the notation e to represent this number, but it was Jacob Bernoulli, in 1690, who first discovered it, while investigating the problem of compound interest detailed above.[2] But for those of you who are not big fans of applied mathematics, let it be known that there's a great more to this special number.

The Number e and (Non-financial) Mathematics

While Euler was born too late (1707) to be the one to discover e, he contributed much more than mere notation. For instance, in 1737, he proved that e is irrational.[3] For comparison, the irrationality of π, a much older and more widely known number, wasn't established until the 1760s (by Johann Heinrich Lambert).[4]

Recall from Chapter 4 that a number is said to be transcendental if it can never be the root of a polynomial with integer coefficients. Both e and π are transcendental, but once again the proof came first for the younger number. The transcendence of e was first chipped away at by Joseph Liouville, who showed that e cannot be the root of a quadratic with integer

---

1 Teller, Edward, Wendy Teller, and Wilson Talley, *Conversations on the Dark Secrets of Physics*, Plenum Press, New York, 1991, p. 87, footnote.

2 Bernoulli, Jacob, "Quæstiones nonnullæ de usuris, cum solutione problematis de sorte alearum, propositi in Ephem. Gall. A. 1685" (Some questions about interest, with a solution of a problem about games of chance, proposed in *Ephemerides Eruditorum Gallicanæ* [*Journal des Savants*] in 1685), *Acta eruditorum*, May 1690, pp. 219–223.

3 His proof saw print seven years later as Euler, Leonhard, "De fractionibus continuis dissertation" (A dissertation on continued fractions), *Commentarii academiae scientiarum Petropolitanae*, Vol. 9, 1744, pp. 98–137. Also see Maor, Eli, *e: The Story of a Number*, ninth printing with new material on pp. 183–186, Princeton University Press, Princeton, NJ, 1998, p. 192. A proof is given in Appendix 2, pp. 201–203.

4 Lambert, Johann Heinrich, "Mémoire sur quelques propriétés remarquables des quantités transcendantes circulaires et logarithmétiques," *Histoire Acad. roy. des sciences et belles lettres*, Berlin, 1768 (but found in 1761), pp. 265–322, available online at www.kuttaka.org/~JHL/L1768b.pdf.

coefficients.[5] In 1873, Charles Hermite (1822–1901) showed that it cannot be the root of any polynomial with integer coefficient, thus providing a complete proof. This proof, however, ran to over 30 pages. To be fair, Hermite's lengthy paper did offer *two* proofs.[6]

Hermite's paper served as the launching pad for Ferdinand von Lindemann's 1882 proof that π is transcendental, with *e* playing a key role in the proof.[7] This, by the way, answered the question of whether or not it was possible to square the circle, a problem that had persisted since the time of the ancient Greeks. The answer is no. The construction cannot be carried out with straightedge and compass in a finite number of steps. But despite the importance of the result, Lindemann was in for some criticism. Recall Leopold Kronecker's reaction, quoted in Chapter 0:

> What good your beautiful proof on [the transcendence of] π: Why investigate such problems, given that irrational numbers do not even exist?
>
> Leopold Kronecker

This extreme position may remind you of the intuitionist mentioned in Chapter 2. Indeed, Kronecker is often referred to as a "pre-intuitionist," holding views similar to Brouwer's before Brouwer was born. As was mentioned in Chapter 0, Kronecker's most famous quote on the matter of how mathematics should be constructed is "God made the integers, all else is the work of man." Stephen Hawking later used a portion of this quote as the title of an anthology of mathematical papers by some of the greats.[8]

And the smack-talk continued. Many years later, the *Dictionary of Scientific Biography* referred to Lindemann as "a mediocre mathematician."[9]

We can use *e* to take an unromantic look at love and marriage.[10] The first step is to estimate how many girlfriends/boyfriends you expect to have up to the age at which you want to get married. Let this number be *n*. An important assumption, which may not be true (!), is that you cannot ever "go back." That is, once you move on to someone else, you cannot decide a past partner was superior and then (successfully) propose to that person. So, if you pass on person *k*, it is in the hope that one of your future *n* – *k* girlfriends/boyfriends will be a better match. There's no strategy that guarantees the best outcome. That is, you may end up having

---

5 Maor, Eli, *e: The Story of a Number*, ninth printing with new material on pp. 183–186, Princeton University Press, Princeton, NJ, 1998, p. 192.

6 Maor, Eli, *e: The Story of a Number*, ninth printing with new material on pp. 183–186, Princeton University Press, Princeton, NJ, 1998, pp. 192–193. On p. 196 Maor gives the references "See David Eugene Smith, *A Source Book in Mathematics* (1929; rpt. Dover, New York, 1959), pp. 99–106 [for original proof, I presume]. For Hilbert's simplified version of Hermite's proof, see Simmons, *Calculus with Analytic Geometry*, pp. 737–739." A shorter proof also appears in Jacobsen, Nathan, *Basic Algebra I*, second edition, W. H. Freeman and Company, New York, 1985, pp. 277–286 give proofs of the transcendence of both *e* and π.

7 Ferdinand von Lindemann, "Über die Zahl π," *Mathematische Annalen*, Vol. 20, No. 2, June 1882, pp 213–225. Also see Maor, Eli, *e: The Story of a Number*, ninth printing with new material on pp. 183–186, Princeton University Press, Princeton, NJ, 1998, p. 193.

8 Hawking, Stephen, editor, *God Created The Integers: The Mathematical Breakthroughs that Changed History*, Running Press, Philadelphia, PA, 2007. While Hawking is credited for the editing and commentary, a good case has been made for him actually having very little to do with it. See Stillwell, John, "Review of *God Created the Integers*," *American Mathematical Monthly*, Vol. 114, No. 3, March 2007, pp. 267–271.

9 Maor, Eli, *e: The Story of a Number*, ninth printing with new material on pp. 183–186, Princeton University Press, Princeton, NJ, 1998, p. 194, citing Gillispie, Charles Coulston, editor, *Dictionary of Scientific Biography*, Charles Scribner's Sons, New York, 1972.

10 What follows is adapted from Paulos, John Allen, *Beyond Numeracy: Ruminations of a Numbers Man*, Vintage Books, A Division of Random House, Inc., New York, 1992, p. 64.

regrets no matter what strategy you follow, but the best possible strategy involves *e*. To follow this strategy, first calculate n/e and round it off to the closest integer. Call this number *k*. For the sake of illustration, I'll suppose your *n* is 20. Then n/e rounds to 7, so you have $k = 7$. What do you do with this magic number? This is the number of potential mates that you *pass* on. That is, you do not propose to any of the first 7, no matter what! I did say that this is an unromantic approach, right? Then, as you encounter further possibilities, perhaps as many as $n - k$, you propose to the first one who is superior to all of the first *k*. The probability of this strategy rewarding you with the best possible mate of the *n* possibilities you would encounter is only about 37%, or more exactly $1/e$. Of course, if everyone followed this strategy, you might have the bad luck of meeting the right person after your value of *k* has been passed, but before hers or his has! That leads us to another assumption that this model makes use of – that the answer will always be yes. Overconfidence, pass it on!

When records are first collected, they're frequently broken, whether it be for sports, weather-related phenomena, or nearly anything else. But as more and more years pass, record-breaking becomes rarer and rarer. Exceptions occur in cases where outside influences make it ever-easier to break records, such as better performance-enhancing drugs becoming available to athletes. Ignoring such unnatural influences, the number, *n*, of times you would expect a record to be broken, and thus reset, over *t* years, for large values of *t*, is connected to *e* via the formula $e = \sqrt[n]{t}$. To solve for the expected number of record-breaking events in *t* years, we first raise both sides of this equation to the nth power getting $e^n = t$, and then take the natural logarithm of both side to get $n = ln(t)$. John Allen Paulos notes that *t* "must be sufficiently – i.e., humongously – large for the approximations to be accurate."[11]

Back to Financial Math

From the perspective of the investor the more often interest is compounded, the better the result. So, just how good is continuously compounded interest?

Example 3 – Doubling your money with continuously compounded interest
If interest is compounded at 2%, we saw that it takes about 35 years for the initial deposit to double when interest is compounded annually and 34.686 years when its compounded monthly. How long does it take to double with continuously compounded interest?

Solution
Using Pe^{rt} at 2%, we have $2P = Pe^{0.02t}$. The *P*s drop out giving $2 = e^{0.02t}$. Taking the natural log of both sides gives $ln(2) = 0.02t$. Thus, $t = ln(2)/0.02 \approx 34.65$. So, even when compounding infinitely often, 2% interest sucks.

Repeated Deposits *or* Who Wants to Be a Millionaire?

Most of us don't simply open a bank account with an initial deposit and then leave it alone for years. If we're trying to save money for retirement, we should make regular deposits.

---

11 Paulos, John Allen, *Beyond Numeracy: Ruminations of a Numbers Man*, Vintage Books, A Division of Random House, Inc., New York, 1992, p. 65.

How large do you think the monthly deposits would have to be, at say 4% interest compounded monthly, if we wish to have \$1,000,000 at the end of 30 years?

Our difference equation would be $x_{n+1} = (1 + .04/12)x_n + m$, where n represents time in months and m represents the monthly deposits. Notice that the amount earned due to interest, $(1 + 0.04/12)x_n$, depends on the previous month's balance, while the increase due to deposits, m, does not.

While we can (and will!) solve this specific problem, let's approach it more generally using a difference equation of the form $x_{n+1} = ax_n + b$. One way to see the solution is to write the first few terms, simplifying as you go, and look for a pattern.

Let x_0 denote the initial value.

Then, $x_1 = ax_0 + b$
$$x_2 = ax_1 + b = a(ax_0 + b) + b = a^2x_0 + ab + b$$
$$x_3 = ax_3 + b = a(a^2x_0 + ab + b) + b = a^3x_0 + a^2b + ab + b$$
$$x_4 = ax_4 + b = a(a^3x_0 + a^2b + ab + b) + b = a^4x_0 + a^3b + a^2b + ab + b$$

The pattern, which may be verified by induction, is:

$$x_n = a^nx_0 + a^{n-1}b + a^{n-2}b + \cdots + ab + b = a^nx_0 + b(a^{n-1} + a^{n-2} + \cdots + a + 1)$$

To further compact this result, we perform a common trick.

Let $\quad S = a^{n-1} + a^{n-2} + \cdots + a + 1.$
Then $aS = a^n + a^{n-1} + \cdots + a^2 + a.$
and $\quad S - aS = 1 - a^n$

Hence, $(1 - a)S = 1 - a^n \Rightarrow S = (1 - a^n)/(1 - a)$, which is precisely the formula for the sum of a finite geometric series. As you see, this is easily derived if forgotten. As an exercise, you're asked to show that an infinite geometric sum, $S = 1 + a + a^2 + a^3 + \cdots$ simplifies to $\frac{1}{1-a}$, provided that $|a| < 1$.

We now have the solution

$$x_n = a^nx_0 + b\left(\frac{1 - a^n}{1 - a}\right)$$

For the problem that was initially posed, we have $n = 360$ (30 years times 12 compoundings per year), $a = 1 + 0.04/12$, and b representing the monthly deposits necessary to make $x_{360} = \$1,000,000$. We'll assume $x_0 = 0$. That is, we aren't starting off with any special initial deposit; we simply make our monthly deposits. We have

$$1,000,000 = \left(1 + \frac{0.04}{12}\right)^{360}(0) + b\left(\frac{1 - \left(1 + \frac{0.04}{12}\right)^{360}}{1 - \left(1 + \frac{0.04}{12}\right)}\right).$$

Because x_0 was 0, the above simplifies to

$$1,000,000 = b\left(\frac{1 - \left(1 + \frac{0.04}{12}\right)^{360}}{1 - \left(1 + \frac{0.04}{12}\right)}\right).$$

It's then a simple matter to solve for b.

$$b = \frac{1,000,000}{\left(\frac{1 - \left(1 + \frac{0.04}{12}\right)^{360}}{1 - \left(1 + \frac{0.04}{12}\right)}\right)} \approx 1,440.82.$$

So, monthly deposits of $1,440.82 will get us to our goal of $1,000,000 in 30 years (if we can consistently earn 4% interest compounded monthly and somehow avoid taxes!).

A Mathematical Aside[12]

The approach used to show that $a^{n-1} + a^{n-2} + \cdots + a + 1 = (1 - a^n)/(1 - a)$, comes in handy in many situations. For example, consider the infinite series $1 + 2(1/2) + 3(1/2)^2 + 4(1/2)^3 + 5(1/2)^4 + \cdots$ While it does involve powers of $1/2$, the integers these powers are multiplied by prevent it from being a geometric series. We cannot apply the formula derived earlier. But watch what happens when we perform the same steps that led to that (useless here) formula.

Let	$S =$	$1 +$	$2(1/2) +$	$3(1/2)^2 +$	$4(1/2)^3 +$	$5(1/2)^4 + \cdots$
Then	$(1/2)S =$		$1(1/2) +$	$2(1/2)^2 +$	$3(1/2)^3 +$	$4(1/2)^4 + \cdots$

and $S - (1/2)S = 1 + \quad (1/2) + \quad (1/2)^2 + \quad (1/2)^3 + \quad (1/2)^4 + \cdots$

The right-hand side of this last equation is a geometric sum, which we can easily evaluate. We get $S - (1/2)S = 2$. Hence, $(1/2)S = 2 \Rightarrow S = 4$.

Loans and Mortgages

The equation $x_{n+1} = ax_n + b$ isn't just relevant to savings accounts. It may also be applied to loans, whether they be college loans, car loans, or mortgages. For example, suppose you have a 30 years mortgage for $200,000 at 5% interest compounded monthly. What will your monthly payments be?

Again, the solution to our equation is $x_n = a^n x_0 + b\left(\frac{1 - a^n}{1 - a}\right)$. We start by recognizing that $x_0 = 200,000$ (what we start off owing), $a = 1 + \frac{0.05}{12}$, and that we wish to solve for b. We

---

12 The problem (and solution) described in this section appears in Hathout, Dean, *Wearing Gauss's Jersey*, CRC Press, Boca Raton, FL, 2013, pp. 25–26. This is not the first appearance of the problem, but I reference it this way, because it's a wonderful book. You should read it!

also need to realize that $x_{360} = 0$. This simply means that the amount owed will be 0 at the end of the 30-year term. We have

$$0 = \left(1 + \frac{0.05}{12}\right)^{360} (200,000) + b \left(\frac{1 - \left(1 + \frac{0.05}{12}\right)^{360}}{1 - \left(1 + \frac{0.05}{12}\right)}\right).$$

Although there's a bit more algebra than in the previous example, it's still just algebra. Working carefully, we obtain

$$b = \frac{-\left(1 + \frac{0.05}{12}\right)^{360} (200,000)}{\left(\frac{1 - \left(1 + \frac{0.05}{12}\right)^{360}}{1 - \left(1 + \frac{0.05}{12}\right)}\right)}.$$

And then (carefully again) plugging into a calculator, we get $b \approx -1{,}073.64$.

The negative sign means that the monthly payments *decrease* the amount owed on the loan, as we would hope! Monthly payments are $1,073.64.

Now, suppose interest rates drop to 4.5%. What would the monthly payments be on the same $200,000 loan? We must calculate

$$b = \frac{-\left(1 + \frac{0.045}{12}\right)^{360} (200,000)}{\left(\frac{1 - \left(1 + \frac{0.045}{12}\right)^{360}}{1 - \left(1 + \frac{0.045}{12}\right)}\right)}.$$

This time we get monthly payments of $1,013.37. So an extra $60.27 would be available each month to spend (or save) as desired.

Alternatively, we can ask how much more could have been borrowed if we wish to make the original payments of $1,073.64. To determine this, we return to our formula

$$x_n = a^n x_0 + b \left(\frac{1 - a^n}{1 - a}\right).$$

This time, we have $a = 1 + \frac{0.045}{12}$ and $b = -1{,}073.64$ and we know $x_{360} = 0$, but we need to find x_0. Plugging in known values gives

$$0 = \left(1 + \frac{0.045}{12}\right)^{360} x_0 - 1{,}073.64 \left(\frac{1 - \left(1 + \frac{0.045}{12}\right)^{360}}{1 - \left(1 + \frac{0.045}{12}\right)}\right).$$

Solving for x_0 gives

$$x_0 = \frac{1{,}073.64 \left(\frac{1 - \left(1 + \frac{0.045}{12}\right)^{360}}{1 - \left(1 + \frac{0.045}{12}\right)}\right)}{\left(1 + \frac{0.045}{12}\right)^{360}} \approx 211{,}894.83.$$

So, an interest rate that is only half a percent lower (in this case) allows the borrower an extra $11,894.83 for the same monthly payments. This is why it is important to negotiate the best interest rate you possibly can when obtaining a mortgage.

Once you get a mortgage, you should strongly consider paying a little extra each month, if you can afford it. To see the effect of such "prepaying" we continue the previous example (a loan at 4.5% with required monthly payments of $1,073.64), but we now assume the borrower can make one extra payment per year. To simplify matters, we'll assume he divides up the extra payment into twelve pieces and pays an extra $1,073.64/12 \approx 89.47$ each month. How much sooner will the loan be paid off in this instance? Naively, we might guess that an extra payment per year for 30 years would retire the loan 30 months early. But if we could intuit answers in this manner, we wouldn't need math! Let's look at our equation again, using payments of $1,073.64 + 89.47 = 1,163.11$.

$$0 = \left(1 + \frac{0.045}{12}\right)^n (211,894.83) - 1,163.11 \left(\frac{1 - \left(1 + \frac{0.045}{12}\right)^n}{1 - \left(1 + \frac{0.045}{12}\right)}\right)$$

This time, we need to solve for n. We start by distributing the $1,163.11$ on the right-hand side.

$$0 = \left(1 + \frac{0.045}{12}\right)^n (211,894.83) - \frac{1,163.11}{1 - \left(1 + \frac{0.045}{12}\right)} + \frac{1,163.11\left(1 + \frac{0.045}{12}\right)^n}{1 - \left(1 + \frac{0.045}{12}\right)}$$

We then move the term without an n in it to the left-hand side.

$$\frac{1,163.11}{1 - \left(1 + \frac{0.045}{12}\right)} = \left(1 + \frac{0.045}{12}\right)^n (211,894.83) + \frac{1,163.11\left(1 + \frac{0.045}{12}\right)^n}{1 - \left(1 + \frac{0.045}{12}\right)}$$

Factoring out a $(1 + 0.045/12)^n$ from the right-hand side gives

$$\frac{1,163.11}{1 - \left(1 + \frac{0.045}{12}\right)} = \left(1 + \frac{0.045}{12}\right)^n \left[211,894.83 + \frac{1,163.11}{1 - \left(1 + \frac{0.045}{12}\right)}\right]$$

We then isolate the piece raised to the nth power by dividing both sides by the bracketed quantity.

$$\frac{\frac{1,163.11}{1 - \left(1 + \frac{0.045}{12}\right)}}{211,894.83 + \frac{1,163.11}{1 - \left(1 + \frac{0.045}{12}\right)}} = \left(1 + \frac{0.045}{12}\right)^n$$

We now switch sides and take a decimal approximation, because the expression is a bit messy.

$$\left(1 + \frac{0.045}{12}\right)^n = \frac{\frac{1,163.11}{1 - \left(1 + \frac{0.045}{12}\right)}}{211,894.83 + \frac{1,163.11}{1 - \left(1 + \frac{0.045}{12}\right)}} \approx 3.156298919$$

To solve for the exponent, we take the ln of both sides.

$$ln\left(\left(1+\frac{0.045}{12}\right)^{n}\right) \approx \ln(3.156298919)$$

$$n \cdot ln\left(1+\frac{0.045}{12}\right) \approx \ln(3.156298919)$$

$$n \approx \frac{ln(3.156298919)}{ln\left(1+\frac{0.045}{12}\right)} \approx 307.08$$

Rounding this off to 307, we see the loan is paid off about 53 months before the original 360 month term. We save about 4.4 years off the original loan!

Not Just Math

In order to make smart financial decisions, you don't just need to know some mathematics – you also have to be street smart. Mostly this means learning from experience, not just your own but also that of others, and not trusting everyone!

1. **Avoid variable rate mortgages!**

   When a mortgage is called "fixed rate" it means that the interest rate will never change. However, there are also "variable rate" mortgages. These can be tempting, because the initial rate is often lower than a typical fixed rate. If someone is trying to encourage you to get this sort of mortgage and you express concern that the rate could increase, you may be told that it won't necessarily increase and that it could actually decrease. Don't bet on it! Some takers of such mortgages saw the rate rise so high that it got to the point that the interest that was tacked on each month exceeded the monthly payments they were making. Imagine that, paying your mortgage only to owe more than before!

2. **Trust no one.**

   Unfortunately, people providing financial services sometimes lie to those they should be serving. Others avoid flat-out lies, but behave unethically. There's plenty of behavior they can engage in that is not in your best interest. To give you a sense of how common this is, consider the phrase "churning the account." This consists of a broker executing trades simply to generate commissions, rather than to improve your position.[13] While there are laws against it, the fact that this activity has a name (and that it was necessary to pass laws against it) shows that it happens. So, don't just go along with whatever a financial advisor might suggest. Be suspicious and think for yourself!

3. **Remember the regulators.**

   I won't go into detail here, but I once had a dispute over $1,500 with a bank. The bank was supposed to wire the money, but it never arrived and they kept insisting

---

13 See https://en.wikipedia.org/wiki/Churning_(finance), for example.

the problem was with the bank it was supposed to be wired to. Despite repeated calls from me, my bank didn't make things right. The money was simply gone. Finally I contacted someone at the FDIC, an organization that regulates banks. I was told that I had to put my complaint in writing and mail it to them. I did so and very soon after the time it took for the letter to arrive at its destination, I received a call from my bank letting me know that the money was back in my account. The caller offered some excuse that I hadn't heard before in any of my conversations with bank employees, but the real reason the money was back was because the FDIC had a lot more pull than I did. Keep regulators in mind whenever you have a problem, whether it be with a financial institution, a landlord, an employer, etc.

e Everywhere!

While establishing the irrationality and transcendence of *e* required some work, the results can't be said to be surprising, given that almost all real numbers fall into these categories. In the following sections a few of the results that truly make *e* special are presented.

e in Pascal's Triangle

The (arithmetic) mean of a set of numbers is simply their average. The geometric mean is calculated differently. If we have n numbers, their geometric mean is the nth root of their product. In the table below, the geometric mean for each of the first few rows of Pascal's triangle is calculated, as is the nth root of each geometric mean, where n is the row of the triangle in question. The top row is taken to be row 0, which leaves the first value undefined.

Row							Geometric Mean	(Geometric Mean)$^{1/n}$
1							1	Undefined
1	1						$\sqrt{1} = 1$	1
1	2	1					$\sqrt[3]{2} \approx 1.260$	$\sqrt{1.260} \approx 1.122$
1	3	3	1				$\sqrt[4]{9} \approx 1.732$	$\sqrt[3]{1.732} \approx 1.201$
1	4	6	4	1			$\sqrt[5]{96} \approx 2.491$	$\sqrt[4]{2.491} \approx 1.256$
1	5	10	10	5	1		$\sqrt[6]{2500} \approx 3.684$	$\sqrt[5]{3.684} \approx 1.298$
1	6	15	20	15	6	1	$\sqrt[7]{162000} \approx 5.549$	$\sqrt[6]{5.549} \approx 1.331$

The first few values of a sequence can often be misleading, but the entries in the rightmost column of the table above show an increasing sequence. The question is does it grow forever or converge to a limiting value? Skipping ahead, row 10 has a geometric mean whose 10th root is about 1.413. Not recognizing this number, we jump ahead again to the 100th row to get about 1.605. This can be quickly calculated with WolframAlpha, as can the larger values that follow.[14] Jumping again to the 1,000th and 5,000th rows, we get 1.642 and 1.647. So, it looks like the sequence is converging to 1.64

---

14 www.wolframalpha.com/input/?i=((prod_(k%3D0)%5E100+(binom(100,k))+)%5E(1%2F101))%5E(1%2F100).

something. Maybe! The sequence is converging so slowly, we can't be sure of much at all at this point. Finally, evaluating the 6,400th row we get 1.647. This is about the limit of what WolframAlpha can do for us. But, for someone who lives and breathes math, it might be enough. This last number is close to 1.6487, which is the beginning of $\sqrt{e}$. As is so often the case in this book, this is not a coincidence. The sequence of nth roots of the geometric means of the entries in the rows of Pascal's triangle, does indeed converge to $\sqrt{e}$.[15]

The above provides a good example of the value of mathematical proof. Sequences don't always converge rapidly. If you have many digits after the decimal place matching the suspected limit after just a few values, you may be full of confidence that you are correct, but how confident can you be in the case of something like what we have here?

Graphing e^x

Of course, e is of great importance in calculus and differential equations because the function $f(x) = e^x$ is its own derivative. Another neat feature of this function is that it doesn't pass through any algebraic points other than $(0, 1)$.[16] Heinrich Dörrie put the matter into perspective.

> Since algebraic points are omnipresent in densely concentrated quantities within the plane, the exponential curve accomplishes the remarkably difficult feat of winding between all these points without touching any of them. The same is, naturally, also true of the logarithmic curve $y = \ln x$.[17]

Elementary Functions?

Slight twists on the function $f(x) = e^x$ also offer interesting results.

In Chapter 8 we saw that $f(n) = n!$ can't be expressed as an elementary function. Similarly, there is no elementary function whose derivative is $\frac{e^{-x}}{x}$.[18] Another example is e^{-x^2}, but in this case there is some beautiful mathematics connected with it. The formula below was found by the great Indian mathematician Srinivasa Ramanujan (1887–1920).[19] It includes a continued fraction.

---

15 I first saw this as an exercise in Young, Robert M., *Excursions in Calculus: An Interplay of the Continuous and the Discrete*, The Mathematical Association of America, 1992, p. 201. For a proof see https://tcgrubb.wordpress.com/2017/01/29/asymptotic-growth-rate-of-geometric-mean-of-binomial-coefficients/.

16 Maor, Eli, *e: The Story of a Number*, ninth printing with new material on pp. 183–186, Princeton University Press, Princeton, NJ, 1998, pp. 195–196 note 5.

17 Dörrie, Heinrich, translated by David Antin, *100 Great Problems of Elementary Mathematics: Their History and Solution*, Dover, New York, 1965, p. 136, available online at http://pyrkov-professor.ru/Portals/0/Media teka/Obzor/dorrie_h_100_great_problems_of_elementary_mathematics.pdf. See Maor, Eli, *e: The Story of a Number*, ninth printing with new material on pp. 183–186, Princeton University Press, Princeton, NJ, 1998, p. 192 for Lambert's result that implies this.

18 Maor, Eli, *e: The Story of a Number*, ninth printing with new material on pp. 183–186, Princeton University Press, Princeton, NJ, 1998, p. 208.

19 It can be found in Clawson, Calvin C., *Mathematical Mysteries: The Beauty and Magic of Numbers*, Springer, Boston, MA, 1996, see Chapter 10.

$$\int_0^a e^{-x^2} dx = \frac{1}{2}\pi^{1/2} - \cfrac{e^{-a^2}}{2a + \cfrac{1}{a + \cfrac{2}{2a + \cfrac{3}{a + \cfrac{4}{2a + \ldots}}}}}$$

Another Continued Fraction

Here's another pretty pattern[20] involving e.

$$\frac{1}{e-1} = \cfrac{1}{1 + \cfrac{2}{2 + \cfrac{3}{3 + \cfrac{4}{4 + \ldots}}}}$$

Sometimes irrational numbers are said to have no pattern behind them. Well, if you express them in base 10, that's true, but if you express these numbers by other means, there are sometimes very simple patterns to be found.

The Most Efficient Base (aka Radix Economy)

Let's compare the efficiency of representing numbers in various bases. Because radix is another word for base, this is referred to as radix economy. Initially, the discussion will be limited to base 10 (because we're used to it), base 2 (because computers are used to it), and base 3 (because, … well you'll soon see!).

Take the decimal (base 10) number 763, for example. It requires three positions (figures) to be noted. Because this base has 10 possibilities for any given position, we could consider the efficiency to be $3 \cdot 10 = 30$. Expressing 763 as a sum of powers of 2, we have $1 \cdot 2^9 + 1 \cdot 2^7 + 1 \cdot 2^6 + 1 \cdot 2^5 + 1 \cdot 2^4 + 1 \cdot 2^3 + 1 \cdot 2^1 + 1 \cdot 2^0$, so its binary (base 2) representation is 1011111011. Thus, 10 figures are required, but each has only two possibilities, so our efficiency in this base is $10 \cdot 2 = 20$, which is much better. Expressing 763 as a sum of powers of 3, we have $1 \cdot 3^6 + 1 \cdot 3^3 + 2 \cdot 3^1 + 1 \cdot 3^0$. Hence, in ternary (base 3), the number is represented as 1001021. Seven figures are required and each has three possibilities, so our efficiency is $7 \cdot 3 = 21$. Of the bases considered, base 2 is the most efficient.

Now, let's consider the base 10 number 1,729. Because it has four digits, its efficiency is $4 \cdot 10 = 40$. Expressing this number as a sum of powers of 2, we have $1 \cdot 2^{10} + 1 \cdot 2^9 + 1 \cdot 2^7 + 1 \cdot 2^6 + 1 \cdot 2^0$, so its base 2 representation is 11011000001. This uses 11 bits, so the number's base 2 efficiency is $11 \cdot 2 = 22$. Finally, expressing it in terms of powers of 3,

20 Clawson, Calvin C., *Mathematical Mysteries: The Beauty and Magic of Numbers*, Springer, Boston, MA, 1996, see Chapter 10.

we have $2 \cdot 3^6 + 1 \cdot 3^5 + 1 \cdot 3^3 + 1 \cdot 3^0$. This is written as 2101001, which has 7 figures, so its efficiency is $7 \cdot 3 = 21$. This is the smallest efficiency for this number, so in this instance base 3 is the winner.

We could look at more examples and we'd see that both base 2 and base 3 have victories, but base 10 *never* wins. Of course, to run more tests, we don't really need the complete representation of the test numbers in each base. We only need to know how many symbols are required, so it suffices to know the highest power of the base that does not exceed the given number. That highest power will be $\lfloor \log_b n \rfloor$. Once again, the floor function allows us to express what we need tersely. But the base b representation includes powers of b from n down to 0, so the total number of symbols is $\lfloor \log_b n + 1 \rfloor$. To get the efficiency of the number, we must multiply this by the base. Introducing the notation $E(b, n)$ for the efficiency of n in base b, we have $E(b,n) = b \cdot \lfloor \log_b n + 1 \rfloor$.

With this formula, programs can be written to test which base is best for huge ranges of numbers. Such programs quickly reveal 8,487 numbers for which base 2 gives a more efficient representation than base 3.[21] The first is 1 and the last is 65,535, which is the Mersenne number 2^{16}–1. There are also numbers for which the bases 2 and 3 are equally efficient. However, base 3 is more efficient than base 2 for *infinitely* many values.[22] Moreover, 2 is nearly the only base that ever beats 3. For just a pair of numbers, namely, 3 and 4, base 5 beats it.[23] There are no other bases, besides 2 and 5, that ever beat 3.[24]

What happens if we broaden the numbers we are willing to accept as bases to include all of the reals that exceed 1? We have $E(b,n) = b \cdot \lfloor \log_b n + 1 \rfloor \approx b \cdot \log_b n = b \cdot \frac{\ln n}{\ln b} = \ln n \cdot \frac{b}{\ln b}$. Examining $\ln n \cdot \frac{b}{\ln b}$, we find that it achieves its minimum value when $b = e$. It doesn't matter what value is chosen for n, because $\ln n$ is just a constant multiplier. Hence, e is the most efficient base. The most efficient *integer* base is 3, because 3 is much closer to e than 2 is.

A practical application of this is providing three options at each step in phone menus to allow callers to navigate the system in the most efficient manner (minimizing the total number of options they must listen to).

For a dissenting view see http://hummusandmagnets.tumblr.com/post/48664858476/the-most-efficient-radix-is-not-e.

If you're uncomfortable with the idea of an irrational number serving as the base of a number system, take a look at quarter-imaginary numeral system proposes by Donald E. Knuth in 1960.[25] It uses the imaginary number $2i$ as the base. It makes base e seem very reasonable by comparison!

---

21 I saw this in Hayes, Brian, "Third Base," *American Scientist*, Vol. 89, No. 6, 2001, pp. 490–494, p. 491, Figure 2 caption cited here, and Hayes, Brian, "Third Base," in *Group Theory in the Bedroom and Other Mathematical Diversions*, pp. 179–200. Hill and Wang, New York, 2008, and also confirmed it for myself using Microsoft Excel.

22 Hayes, Brian, "Third Base," *American Scientist*, Vol. 89, No. 6, 2001, pp. 490–494, p. 491, Figure 2 caption cited here.

23 Brian Hayes indicated that there is just a single value for which base 5 is the most efficient. I contacted him and discovered that he was calculating the efficiency of a base somewhat differently than I do here. I had misunderstood his approach, but nevertheless obtained an identical answer for the number of times base 2 beats base 3 and only disagreed on one value concerning base 5!

24 The number of times that 2 beats 3 was given by Hayes. I found the largest such number with a quick program written in Microsoft Excel.

25 Knuth, Donald E., "An Imaginary Number System," *Communications of the ACM*, Vol. 3, No. 4, April 1960, pp. 245–247.

Back to Population Models

In Chapter 15, difference equations were used to model populations. The new equations developed in this chapter can also be applied in that direction.

Example 4 – Horrific years in Ireland

At the start of 1847, the population of Ireland was around 8 million, but about 250,000 left the country that year because of the potato blight that caused the Great Famine.[26] It's difficult to estimate the difference between the birth rate and the death rate during this time,[27] but this number is needed, so the value of 1.6%, given by James Stewart, in his popular calculus text, will be used here.[28] If the birth and death rates remained the same, along with the rate of emigration, what would the fate of Ireland have been?

Solution

The difference equation modeling Ireland's population is of the form $x_{n+1} = ax_n + b$, the same as in the "Who wants to be a millionaire?" and mortgage problems. If we start the clock ticking at the beginning of 1847, we have $x_0 = 8{,}000{,}000$, representing the initial population of Ireland. The value of b is $-250{,}000$, the number of people leaving the country that year, and in each future year, if we assume the numbers don't change. Finally, the value of a is 1.016, representing an annual increase of 1.6% in the population due to there being more births than deaths. As was seen previously, the solution to a difference equation of this form is

$$x_n = a^n x_0 + b\left(\frac{1 - a^n}{1 - a}\right).$$

Plugging in the values given above, it becomes

$$x_n = 1.016^n (8{,}000{,}000) - 250{,}000 \left(\frac{1 - 1.016^n}{-0.016}\right).$$

Predicting the population in any future year is easy with a calculator. The year 1848 corresponds to $n = 1$, which is

$$x_1 = 8{,}128{,}000 - 250{,}000 = 7{,}878{,}000.$$

The population of Ireland has decreased.
Looking ahead to 1857, $n = 10$, gives

$$x_{10} = 1.016^{10} (8{,}000{,}000) - 250{,}000 \left(\frac{1 - 1.016^{10}}{-0.016}\right)$$

26 www.mapspictures.com/ireland/history/ireland_population.php.

27 See Boyle Phelim P. and Cormac Ó. Gráda, "Fertility trends, excess mortality, and the Great Irish Famine," *Demography*, Vol. 23, No. 4, November 1986, pp. 543–562 to get a better understanding of this challenge.

28 Stewart, James, *Single Variable Calculus: Concepts and Contexts*, fourth edition, Cengage Learning, Brooks/Cole, Belmont, CA, 2009, p. 528.

$$\approx 9,376,204 - 2,687899 = 6,688,305.$$

The decrease in the population is continuing. The constant exodus of 250,000 people per year will continue to lower the population. Is there a point at which the population hits zero? This can be investigated by solving for n in

$$0 = 1.016^n (8,000,000) - 250,000 \left(\frac{1 - 1.016^n}{-0.016} \right).$$

The steps are shown below, but a computer algebra system makes quick work of this problem.

$$0 = 1.016^n (8,000,000) + 15,625,000(1 - 1.016^n)$$

$$0 = 1.016^n (8,000,000) + 15,625,000 - 1.016^n (15,625,000)$$

$$0 = 1.016^n (-7,625,000) + 15,625,000$$

$$1.016^n (7,625,000) = 15,625,000$$

$$1.016^n = \frac{15,625,000}{7,625,000}$$

$$ln(1.016^n) = ln\left(\frac{15,625,000}{7,625,000} \right)$$

$$n \cdot ln(1.016) = ln\left(\frac{15,625,000}{7,625,000} \right)$$

$$n = \frac{ln\left(\frac{15,625,000}{7,625,000} \right)}{ln(1.016)}$$

$$n \approx 45.2$$

So, if things had continued exactly as they were going in 1847, Ireland would have been empty by early 1892. Obviously, this isn't the way things turned out!

Open Problems

In general, determining whether a number is transcendental or algebraic is very difficult. Hilbert stressed the importance of resolving this difficulty by making it the seventh problem in his famous list from 1900. He believed numbers of the form α^β, with both α and β algebraic and β irrational (and $\alpha \neq 0$ or 1) would always be transcendental. As an example, he offered the number $2^{\sqrt{2}}$. It ought to be transcendental. As a different sort of example he also mentioned e^π. This does not fit the form Hilbert described, because e is transcendental, but his conjecture, if true, could still be used to determine its nature, like so: Suppose e^π is algebraic. Then it could play the role of α in Hilbert's conjecture, with $i = \sqrt{-1}$, an irrational algebraic number, in the role of β. Then $\alpha^\beta = (e^\pi)^i = e^{\pi i}$ would be transcendental. But we have $e^{\pi i} = -1$, which is an algebraic number. Thus, we have a contradiction. Therefore our supposition that e^{π} is algebraic must be wrong and we can conclude that e^{π} is transcendental.[29]

So, we have an easy proof by contradiction that e^{π} is transcendental, if Hilbert's conjecture can be verified. Well, the Russian mathematician, Alexandr Osipovich Gelfond (1906–1968), found another way. In 1929, he proved e^{π} is transcendental without confirming Hilbert's conjecture.[30] A year later, Rodion Kuzmin established $2^{\sqrt{2}}$ as transcendental, again without Hilbert's conjecture.[31] Thus, the greater challenge of finding a general technique to solve such problems remained unconquered, for a time.

Finally, in 1934, Gelfond and a German mathematician, Theodor Schneider, independently proved Hilbert's conjecture.[32] But while the result allows us to state the nature of a great many numbers, it is far from a complete answer when it comes to questions of transcendence. For example, it's still an open question whether or not the following are transcendental: π^e, π^{π}, e^e. Won't someone please find out?

There's also an open problem concerning powers of 2 expressed in base 3. Both 2^2 and 2^8 are written in base 3 without any 2s. Their base 3 representations are 11 and 100111, respectively. Ronald Graham and Paul Erdös conjectured that these are the only powers of 2 with this property. That is, the base 3 representations of all other powers of 2 contain at least one 2.[33] A search for a counterexample carried out by Ilan Vardi, with some help from a computer, found none up to $2^{6,973,568,802}$.[34] Whether or not there's a counterexample somewhere beyond this ridiculously huge value remains unknown.

Exercise Set 16

> Knowledge and productivity are like compound interest. The more you know, the more you learn; the more you learn, the more you can do; the more you can do, the more the opportunity. I don't want to give you a rate, but it is a very high rate. Given two people with exactly the same ability, the one person who manages day in and day out to get in one more hour of thinking will be tremendously more productive over a lifetime.
>
> Richard Hamming

1. How long will it take for money invested at 6%, compounded annually to double?
2. How long will it take for money invested at 6%, compounded monthly to double?

---

29 If you don't like proof by contradiction, you can rewrite e^{π} as i^{-2i}, which fits Hilbert's form. See Maor, Eli, *e: The Story of a Number*, ninth printing with new material on pp. 183–186, Princeton University Press, Princeton, NJ, 1998, p. 194 and 178.

30 Gelfond, Alexandr, "Sur les nombres transcendants," *Comptes rendus de l'Académie des Sciences*, Paris, Vol. 189, 1929, pp. 1224–1226.

31 Kuzmin, Rodion A., "Sur une nouvelle classe de nombres transcendants," *Bulletin de l'Académie des Sciences de l'URSS*, Leningrad, Series 7, Vol. 3, No. 6, 1930, pp. 585–597.

32 Gelfond, Alexandr, "Sur le septièe problème de D. Hilbert," *Doklady Akademiya Nauk*, Vol. 2, 1934, pp. 1–3 in Russian, pp. 4–6 in French. Schneider, Theodor, "Transzendenzuntersuchungen periodischer Funktionen I.," *Journal für die reine und angewandte Mathematik*, Vol. 172, 1935, pp. 65–69.

33 Erdös, Paul and Ronald L. Graham, "Old and New Problems and Results in Combinatorial Number Theory," *L'Enseignement Mathématique*, Université de Genève, Geneva, 1980.

34 Vardi, Ilan, "The Digits of 2^n in Base Three," in *Computational Recreations in Mathematica*, Addison-Wesley, Reading, MA, 1991, pp. 20–25.

3. How long will it take for money invested at 6%, compounded continuously to double?

4. What interest rate, compounded monthly, is required for an initial deposit to double in exactly 10 years?

5. What interest rate, compounded continuously, is required for an initial deposit to double in exactly 10 years?

6. What annual interest rate is equivalent to 3% compounded continuously?

7. What continuously compounded interest rate is equivalent to 7% compounded annually?

8. Claude Shannon invested in Teledyne and got "an annual compound return of 27% over twenty-five years."[35] Using these numbers, what would an initial investment of $1,000 be by the end?

9. What are the monthly payments on a 30-year mortgage for $150,000, if the interest rate is 5%, compounded monthly?

10. If the interest rate in problem 9 above is lowered to 4.5%, how much lower would the monthly payments be? How much more could be borrowed for the same monthly payments?

11. If monthly payments of $500 are made, how long will it take to pay off a $10,000 loan at 4% interest compounded monthly?

12. How much quicker will the loan in problem 11 above be paid off if payments of $550 are made each month?

13. If a person has $100,000 in the bank earning 2% interest, compounded monthly, for how long may he withdraw $800 per month before the account is empty?

14. In 1923, Germany was experiencing hyperinflation. At the start of that year, a loaf of bread cost ¼ Reichsmark, but in November it was priced at 80 billion Reichsmarks.[36] If we model this as a continuously compounded interest problem, what is the value of r?

15. East Germany's population shrank from 18.9 million in 1949 to 16.1 million in 1990.[37] If the rate of decrease remained fixed, when would the population dip below 10 million? Don't break this one down by looking at birth rates, death rates, emigration, or immigration.

16. The quotation below details hyperinflation in China.

By the end of 1941 the Shanghai wholesale price index stood at 15.98. By December 1945 it had reached 177,088, and by the end of 1947 it was 16,759,000. In December 1948 the index had risen to 36,788,000,000, and in April 1949 it was at 151,733,000,000,000.[38]

Interpreting "end of 1941" as January 1, 1942 and "in April 1949" as April 15, 1949, determine the value of r over this interval, assuming continuous compounding.

35 Soni, Jimmy and Rob Goodman, *A Mind at Play: How Claude Shannon Invented the Information Age*, Simon & Schuster, New York, 2017, p. 239.

36 Llewellyn, Jennifer, and Steve Thompson, "The 1923 hyperinflation," *Alpha History*, 2014, accessed June 12, 2018, http://alphahistory.com/weimarrepublic/1923-hyperinflation/.

37 www.populstat.info/Europe/germanec.htm.

38 Ebeling, Richard M., "The Great Chinese Inflation," *FEE, Foundation for Economic Education*, July 5, 2010, https://fee.org/articles/the-great-chinese-inflation/.

You may ignore the data between the endpoints. They are only included for completeness.

17. Prove that $S = 1 + a + a^2 + a^3 + \cdots$ simplifies to $\frac{1}{1-a}$, provided that $|a| < 1$.

18. According to Ibn Khallikan (1211–1282), the Indian King Shirham wanted to reward Grand Vizier Sissa ben Dahir for inventing the game of chess and asked him to name his reward. The King was surprised by the seemingly modest suggestion:

   Majesty, I would be happy if you were to give me a grain of wheat to place on the first square of the chessboard, and two grains of wheat to place on the second square, and four grains of wheat to place on the third, and eight grains of wheat to place on the fourth, and so on for the sixty-four squares.[39]

   How many grains were requested, in all?

19. The following brain teaser was posed to John von Neumann.

   Two bicyclists are 20 miles apart and head toward each other at 10 miles per hour each. At the same time a fly traveling at a steady 15 miles per hour starts from the front wheel of the northbound bicycle. It lands on the front wheel of the southbound bicycle, and then instantly turns around and flies back, and after next landing instantly flies north again. Question: What total distance did the fly cover before it was crushed between the two front wheels?[40]

   Now, there's an easy way to find the distance, and there's a hard way to find it. The hard way involves summing an infinite series (which you can do). Von Neumann, when asked, immediately gave the correct answer. The person who had posed the problem disappointedly said, "Oh, you've heard the trick before." Von Neumann responded, "What trick? I simply summed the infinite series."[41]

20. Suppose a ball is dropped from a height of 10 feet and each bounce takes it up to 40% of the previous bounce's height. What is the total back and forth distance traveled?

21. If you could fold a piece of paper of standard thickness, 0.1 millimeters, as many times as you pleased, how many folds would it take until it was over 1 kilometer thick?

22. How thick would a piece of paper of standard thickness, 0.1 millimeters, be if it were possible to fold it 100 times?

23. Show that if $2^n - 1$ is prime, then $N = 2^{n-1}(2^n - 1)$ is a perfect number. Flip back to Chapter 10, if you've forgotten what perfect numbers are. Hint: When a problem appears in a textbook, the context usually provides a clue as to how it can be solved. There's almost always some new tool in that particular chapter that can be applied to the problem. In this case it's the formula for the sum of a finite geometric sequence. First list all of the factors of N in two groups, depending on

39 Pickover, Clifford A., *The Math Book: From Pythagoras to the 57th Dimension*, Sterling, New York, 2009, p. 102. Note: The story exists in many versions. This one, the oldest known, is from 1256.

40 MacRae, Norman, *John von Neumann: The Scientific Genius Who Pioneered the Modern Computer, Game Theory, Nuclear Deterrence, and Much More*, Pantheon Books, New York, 1992, p. 10.

41 MacRae, Norman, *John von Neumann: The Scientific Genius Who Pioneered the Modern Computer, Game Theory, Nuclear Deterrence, and Much More*, Pantheon Books, New York, 1992, p. 10.

whether the factor is a multiple of the prime $2^n - 1$, or not, and then apply the formula to sum each group.

24. What is the converse of the statement you were asked to prove in the exercise above? Euler proved that the converse is true, if we restrict ourselves to even numbers. It is not known if the converse holds in general. That is, there might be an odd perfect number that isn't of the form stated above. If so, it's known that it would have to be very large! See Chapter 10 for more.

25. Be warned! This problem, from the October 1978 issue of *OMNI*, is a brain teaser.[42] You need to think creatively to solve it! Here it is: Fill in the missing term, indicated by –, in the sequence below.

    10, 11, 12, 13, 14, 15, 16, 17, 20, 22, 24, 31, 100, – , 10000, 1111111111111111

26. Consider the geometric series $S = 9/10 + 9/10^2 + 9/10^3 + \cdots$ It's not hard to see that $S = 0.9999...$, but when we use the formula for an infinite geometric series, $S = a/(1 - r)$ we get $S = (9/10)/(1 - 0.1) = 0.9/0.9 = 1$. What gives? Is there an error in one approach or the other or is $0.9999... = 1$?

27. Make a table of values for $\frac{\sqrt[n]{n!}}{n}$ and use it to predict the value of $\lim_{n\to\infty} \frac{\sqrt[n]{n!}}{n}$.

28. Make a table of values for $n\sin(2\pi e n!)$ and use it to predict the value of $\lim_{n\to\infty} n\sin(2\pi e n!)$. The proof of this quickly leads to a proof that e is irrational, because if it were rational, the limit would be zero.

29. In a game played with a single die, the players take turns rolling the die until someone gets a 6. That person is declared the winner. Obviously the person who rolls first has an advantage. What is the probability he or she will win?

References/Further Reading

I insist on a lot of time being spent, almost every day, to just sit and think. That is very uncommon in American business. I read and think. So I do more reading and thinking, and make less impulse decisions than most people in business. I do it because I like this kind of life.

– Warren Buffett (3rd richest person in the world).

I really had a lot of dreams when I was a kid, and I think a great deal of that grew out of the fact that I had a chance to read a lot.

– Bill Gates (2nd richest person in the world).

I'm going to go do this crazy thing. I'm going to start this company selling books online.

– Jeff Bezos (richest person in the world)[43].

---

42 Morris, Scot, "Games," *OMNI*, Vol. 1, No. 1, October 1978, pp. 174–175, problem 10. The answer is given on p. 140 of this issue.

43 Well, these were the rankings when I completed the writing of the book. Now that I'm examining the copyedited manuscript, things have changed a bit. There will probably be more changes soon, but I expect you will always be able to find quotes about the importance of books from whoever makes the top 3!

Bergman, George, "A Number System with an Irrational Base," *Mathematics Magazine*, Vol. 31, No. 2, November–December 1957, pp. 98–110. In this paper, the base described is the Golden Ratio.

Bernoulli, Jacob, "Quæstiones nonnullæ de usuris, cum solutione problematis de sorte alearum, propositi in Ephem. Gall. A. 1685" (Some Questions about Interest, with a Solution of a Problem about Games of Chance, Proposed in *Ephemerides Eruditorum Gallicanæ* [*Journal des Savants*] in 1685), *Acta eruditorum*, May 1690, pp. 219–223.

Boyle, Phelim P. and Cormac Ó. Gráda, "Fertility Trends, Excess Mortality, and the Great Irish Famine," *Demography*, Vol. 23, No. 4, November 1986, pp. 543–562.

Calinger, Ronald, editor, *Classics of Mathematics*, Moore Publishing Company, Oak Park, IL, 1982, pp. 653–677 reproduce Hilbert's 1900 lecture, after a short introduction. Hilbert's seventh problem is on p. 667.

Clawson, Calvin C., *Mathematical Mysteries: The Beauty and Magic of Numbers*, Springer, Boston, MA, 1996.

Dörrie, Heinrich, translated by David Antin, *100 Great Problems of Elementary Mathematics: Their History and Solution*, Dover, New York, 1965, available online at http://pyrkov-professor.ru/Portals/0/Mediateka/Obzor/dorrie_h_100_great_problems_of_elementary_mathematics.pdf.

Ebeling, Richard M., "The Great Chinese Inflation," *FEE, Foundation for Economic Education*, July 5, 2010, available online at https://fee.org/articles/the-great-chinese-inflation/.

Engineering Research Associates, Inc., *High-Speed Computing Devices*, McGraw-Hill, New York, 1950.

Gillispie, Charles Coulston, editor, *Dictionary of Scientific Biography*, Charles Scribner's Sons, New York, 1972.

Hathout, Dean, *Wearing Gauss's Jersey*, CRC Press, Boca Raton, FL, 2013.

Hawking, Stephen, editor, *God Created the Integers: The Mathematical Breakthroughs That Changed History, Running Press*, Philadelphia, PA, 2007. While Hawking is credited for the editing and commentary, a good case has been made for him actually having very little to do with it. See Stillwell, John, "Review of *God Created the Integers*," *American Mathematical Monthly*, Vol. 114, No. 3, March 2007, pp. 267-271.

Hayes, Brian, "Third Base," *American Scientist*, Vol. 89, No. 6, 2001, pp. 490–494.

Hayes, Brian., "Third Base," in *Group Theory in the Bedroom and Other Mathematical Diversions*, Hill and Wang, New York, 2008, pp. 179–200.

Jacobsen, Nathan, *Basic Algebra I*, second edition, W. H. Freeman and Company, New York, 1985.

Knuth, Donald, *Surreal Numbers: How Two Ex-Students Turned on to Pure Mathematics and Found Total Happiness*, Addison-Wesley, Reading, MA, 1974.

Lambert, Johann Heinrich, "Mémoire sur quelques propriétés remarquables des quantités transcendantes circulaires et logarithmétiques," *Histoire Acad. roy. des sciences et belles lettres*, Berlin, Année 1768 (but found in 1761), pp. 265–322, available online at www.kuttaka.org/~JHL/L1768b.pdf.

Llewellyn, Jennifer, and Steve Thompson, "The 1923 Hyperinflation," *Alpha History*, 2014, accessed June 12, 2018, available online at http://alphahistory.com/weimarrepublic/1923-hyperinflation/.

MacRae, Norman, *John von Neumann: The Scientific Genius Who Pioneered the Modern Computer, Game Theory, Nuclear Deterrence, and Much More*, Pantheon Books, New York, 1992.

Maor, Eli, *e: The Story of a Number*, ninth printing with new material on pp. 183–186, Princeton University Press, Princeton, NJ, 1998.

"The Most Efficient Radix Is Not e," *Hummus and Magnets*, April 23, 2013, available online at http://hummusandmagnets.tumblr.com/post/48664858476/the-most-efficient-radix-is-not-e This gives a dissenting view concerning the efficiency of *e* as a base.

Ornstein, Leonard, "Hierarchic Heuristics: Their Relevance-to Economic Pattern-Recognition and High-Speed Data-Processing," 1969, available online at https://pdfs.semanticscholar.org/0441/d73255b1d4a924aea3edf59807616e4e3539.pdf. This is relevant for its discussion of the efficiencies of different bases.

Paulos, John Allen, *Beyond Numeracy: Ruminations of a Numbers Man*, Vintage Books, A Division of Random House, Inc., New York, 1992.

Pickover, Clifford A., *The Math Book: From Pythagoras to the 57th Dimension*, Sterling, New York, 2009, p. 102.

Roy, Ranjan, "The Discovery of the Series Formula for π by Leibniz, Gregory and Nilakantha," *Mathematics Magazine*, Vol. 63, No. 5, December 1990, pp. 291–306.

Simmons, George F., *Calculus with Analytic Geometry*, McGraw-Hill, New York, 1985. See pp. 737–739 for Hilbert's simplified version of Hermite's proof.

Tubbs, Robert, "Hilbert's Seventh Problem: Solutions and Extensions," available online at http://euclid.colorado.edu/~tubbs/courses/courses.html.

Von Lindemann, Ferdinand, "Über die Zahl π," *Mathematische Annalen*, Vol. 20, No. 2, June 1882, pp. 213–225.

Wiener, Norbert, *Cybernetics or Control and Communication in the Animal and the Machine*, The M.I.T. Press, Cambridge, MA, 1948, second edition 1961. Wiener argues that 2 is the most efficient base.

Wikipedia contributors, "Churning (Finance)," *Wikipedia, The Free Encyclopedia*, available online at https://en.wikipedia.org/w/index.php?title=Churning_(finance)&oldid=797485194.

Young, Robert M., *Excursions in Calculus: An Interplay of the Continuous and the Discrete*, The Mathematical Association of America, 1992.

17

More Difference Equations

> There is no guarantee that shortcuts exist for predicting the consequences of a recursive rule.
>
> William Poundstone[1]

Recursion has been used to solve a wide variety of mathematical problems from ancient times to the present. In this chapter, diverse applications, many connected to previously covered topics, are examined, but please keep in mind the fact that finding a recursion isn't always useful. In some cases no techniques exist for shortcutting it and the iterations can only presently be computed one at a time. On the other hand, that's sometimes good enough!

Example 1 – Square roots[2]

Well over 2,000 years ago, the Babylonians had an algorithm for approximating the square root of a number, a.[3] The algorithm begins with a guess, $x_0 > 0$, and then, generates a sequence of ever-better approximations, via the difference equation

$$x_{n+1} = \frac{1}{2}\left(x_n + \frac{a}{x_n}\right).$$

For example, to approximate $\sqrt{2}$, we could begin with the (poor) guess $x_0 = 1$. The first few iterations are as follows.

$$x_1 = \frac{1}{2}\left(1 + \frac{2}{1}\right) = \frac{3}{2}$$

$$x_2 = \frac{1}{2}\left(\frac{3}{2} + \frac{2}{\frac{3}{2}}\right) = \frac{17}{12}$$

$$x_3 = \frac{1}{2}\left(\frac{17}{12} + \frac{2}{\frac{17}{12}}\right) = \frac{577}{408}$$

This last value is approximately 1.414216, which is very close to $\sqrt{2} \approx 1.414214$.

The ancient Greeks also knew of this method and it is often credited to Heron of Alexandria (ca. 10 CE–ca. 70 CE), even though the Babylonians were aware of it centuries earlier.

---

1 Poundstone, William, *The Recursive Universe: Cosmic Complexity and the Limits of Scientific Knowledge*, Willow Morrow and Company, Inc, New York, 1985, p. 127.

2 Young, Robert M., *Excursions in Calculus: An Interplay of the Continuous and the Discrete*, The Mathematical Association of America, 1992, pp. 19–20.

3 Van der Waerden, Bartel L., *Science Awakening*, 3rd edition, Wolters-Noordhoff, Groningen, 1971, p. 45.

Example 2 – A surprising result?[4]

Speaking of square roots, let's consider a difference equation that involves one: $x_{n+1} = \sqrt{1 + x_n}$. If we give this the initial condition $x_0 = 0$, the first few values are $x_1 = \sqrt{1}$, $x_2 = \sqrt{1 + \sqrt{1}}$, $x_3 = \sqrt{1 + \sqrt{1 + \sqrt{1}}}$, ... As we continue, we get nested roots of 1 forever. Does this look familiar? It was seen in Chapter 13 (Exercise 13), and it converges to φ, the Golden Ratio.

Newton's Method

Newton's method is a technique used to approximate the roots of a differentiable function, $f(x)$. We start with a guess, x_0 and then calculate the tangent line to the curve at that point. Where this tangent line hits the x-axis is our next estimate. We then repeat the process to get third, fourth, fifth, etc. estimates. Figure 17.1 below shows an example where this approach works well.

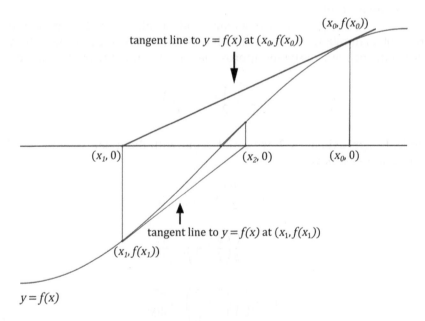

FIGURE 17.1

A geometrical depiction of Newton's method.

Image created by Josh Gross.

This is a geometric examination of the algorithm, but we may also look at it algebraically.

---

4 Young, Robert M., *Excursions in Calculus: An Interplay of the Continuous and the Discrete*, The Mathematical Association of America, 1992, pp. 171–172.

Suppose the initial guess is x_0. Then we have the point $(x_0, f(x_0))$ on the curve. The slope of the tangent line at that point will be $f'(x_0)$. Knowing a point and the slope, we may use the point-slope form of the equation of a straight line to express the tangent line.

$$y - y_0 = m(x - x_0) \Rightarrow y - f(x_0) = f'(x_0)(x - x_0),$$

which may be rewritten as

$$y = f(x_0) + f'(x_0)(x - x_0).$$

We are interested in where this curve hits the x-axis. That is, we want to find the x value for which $y = 0$. We have

$$0 = f(x_0) + f'(x_0)(x - x_0).$$

We solve for x like so:

$$-f(x_0) = f'(x_0)(x - x_0)$$

$$\frac{-f(x_0)}{f'(x_0)} = x - x_0$$

$$x_0 - \frac{f(x_0)}{f'(x_0)} = x$$

$$x = x_0 - \frac{f(x_0)}{f'(x_0)}$$

Because this point is our new estimate, we label it x_1.

$$x_1 = x_0 - \frac{f(x_0)}{f'(x_0)}$$

Repeating this process, treating each new x value as the guess for the next iteration, will often take us closer and closer to a root. The general term is

$$x_{n+1} = x_n - \frac{f(x_n)}{f'(x_n)}$$

However, this process doesn't always work. If our initial guess is a max or min of the function, we're in trouble, because $f'(x_0) = 0$ in this case and we cannot divide by 0.

Example 3 – Estimating a root of $f(x) = \frac{1}{2}x^3 - 2x + 1$ using Newton's method
We don't need a graph of the function to apply Newton's method, but one is given (Figure 17.2) to shed some light on what's happening.

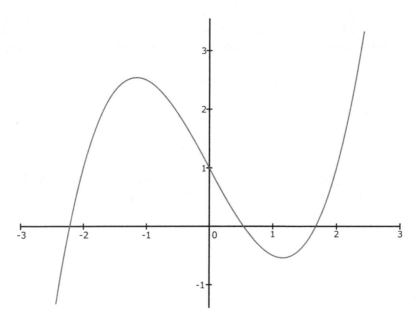

FIGURE 17.2
A graph of $f(x) = \frac{1}{2}x^3 - 2x + 1$.

Image created by Josh Gross.

There are three roots for this polynomial. Newton's method will usually converge to the root closest to the initial guess. The initial guess doesn't have to be great, so let's start with $x_0 = 0$. The first iteration gives

$$x_1 = x_0 - \frac{f(x_0)}{f'(x_0)} = 0 - \frac{f(0)}{f'(0)} = -\frac{\frac{1}{2}(0)^3 - 2(0) + 1}{\frac{3}{2}(0)^2 - 2} = -\frac{1}{-2} = \frac{1}{2}.$$

From the graph, you can see that this does indeed take us nearer to the closest root. The next iteration is

$$x_2 = x_1 - \frac{f(x_1)}{f'(x_1)} = \frac{1}{2} - \frac{f\left(\frac{1}{2}\right)}{f'\left(\frac{1}{2}\right)} = \frac{1}{2} - \frac{\frac{1}{2}\left(\frac{1}{2}\right)^3 - 2\left(\frac{1}{2}\right) + 1}{\frac{3}{2}\left(\frac{1}{2}\right)^2 - 2} = \frac{1}{2} - \frac{\frac{1}{16}}{\frac{-13}{8}} = \frac{7}{13}.$$

So, with a lousy guess and just two iterations we obtained $\frac{7}{13} \approx 0.538$, which is close to the true value of about 0.539.

Although this method is named after Isaac Newton, he doesn't deserve all of the credit. Newton only looked at a special case of the method and his approach was so different (he didn't use calculus, for example) that you might not even recognize what he did as what we today call Newton's method![5] The method was simplified by Joseph Raphson and is therefore occasionally referred to as the Newton–Raphson method, but at this stage it still

---

5 https://en.wikipedia.org/wiki/Newton%27s_method.

did not make use of calculus.[6] It was only in 1740 that Thomas Simpson, after whom Simpson's rule in calculus is named, published it in a form very much like what was presented here.[7] Newton's method is implemented in a more colorful way in Chapter 18.

Example 4 – Derangements

Derangements appeared in Chapters 9 and 12, but these both came before the difference equations chapters, so a lovely formula was postponed until now. It is motivated below by an example, counting the number of derangements on five objects. For convenience, let **a**, **b**, **c**, **d**, and **e** denote the objects. To count the number of derangements on these objects, two cases are considered:

Case 1: a switches positions with some other letter

> There are four choices as to which letter **a** can switch positions with. For the sake of illustration, suppose that letter is **b**. Then, our derangement looks like this:

> **a**, **b**, **c**, **d**, **e** (original)
> **b**, **a**, ?, ?, ? (deranged)

> The positions indicated by question marks can be anything, as long as none of **c**, **d**, and **e** retain their original position. Because that's the only restriction, and **c**, **d**, and **e** are, collectively, three objects, there are D_3 possibilities. By symmetry, there are also D_3 derangements when **a** switches positions with any of the other three letters. Thus, altogether we have $4D_3$ possibilities in this case.

Case 2: a and **b** don't switch positions

> In this case, **a** is sent to something that doesn't go to **a**. It's easy to see that every derangement falls into exactly one of our two cases. In the present case, there are four choices where **a** can go, because it can go to anything other than **a**. Please note that it can even go to **b**, but if it does, then **b** cannot go to **a**.

> **a**, **b**, **c**, **d**, **e** (original)
> **b**, ?, ?, ?, ? (deranged)

> Now, in the second row, representing the deranged values, we'd like to know the number of ways in which the question marks can be filled in with **a**, **c**, **d**, and **e**, subject to the following constraints.

> **a** cannot appear under **b**
> **c** cannot appear under **c**
> **d** cannot appear under **d**
> **e** cannot appear under **e**

> The last three constraints are simply part of the definition of a derangement, but the first is a bit different. It follows from the fact that we are in Case 2, where **a** and **b** don't switch positions. A little trick will make things easier. To help remember that **a** cannot appear under **b**, when writing **a**, draw it very poorly so it looks more like a **b**. Our constraints then take the form

---

6 https://en.wikipedia.org/wiki/Newton%27s_method.

7 Simpson, Thomas, *Essays on Several Curious and Useful Subjects in Speculative and Mix'd Mathematicks*, 1740.

"**b**" cannot appear under **b**

c cannot appear under **c**

d cannot appear under **d**

e cannot appear under **e**

And we know exactly how many possibilities satisfy these slightly rewritten constraints, namely, D_4, the number of derangements on four objects (**b**, **c**, **d**, and **e**).

So, the total number of possibilities for this case is $4D_4$.

Combining the two case, we have $D_5 = 4D_3 + 4D_4$. While the argument above was specifically for the case $n = 5$, it easily adapts to any larger value. In general, the result is

$$D_n = (n-1)(D_{n-1} + D_{n-2}), \text{ for } n \geq 2.[8]$$

Pierre Rémond de Montmort, in the context of his investigation of the card game, *Treize*,[9] gave this formula, without a proof, in 1708.[10] He also found the formula $D_n = n!\left(1 - \frac{1}{1!} + \frac{1}{2!} - \frac{1}{3!} + \cdots + \frac{(-1)^n}{n!}\right)$, as was related in a footnote in Chapter 9.

Example 5 – Euler's primes

In Chapter 2, the function $f(x) = x^2 + x + 41$ served as an example that not all patterns continue. Plugging in the integers 0 through 39, we always get a prime out, but at 40 we suddenly get a composite number.

It's easy to show that we can have similarly misleading results arise from a difference equation, because we can recast the function $f(x)$ in this form.[11]

$$x_{n+1} = x_n + 2(n+1)$$

Given the initial condition $x_0 = 41$, we have

$$x_1 = x_0 + 2(0+1) = 41 + 2 = 43$$

$$x_2 = x_1 + 2(1+1) = 43 + 4 = 47$$

$$x_3 = x_2 + 2(2+1) = 47 + 6 = 53$$

etc.

We get the exact same sequence as when we evaluated $f(x)$.

We can also roll backwards and find values for $x_{-1}, x_{-2}, x_{-3}, \cdots$ To begin doing so, let $n = -1$ in $x_{n+1} = x_n + 2(n+1)$ to get $x_0 = x_{-1} + 2(-1+1) = x_{-1}$. We know, $x_0 = 41$, so we

---

8 For a different proof, see Vilenkin, N. Ya, *Combinatorics*, Academic Press, New York and London, 1971, p. 53. Page 49 tells us how many matches can be expected, depending on how many objects are being rearranged. There are connections to ESP experiments. See Exercise 46 in Chapter 12.

9 This is *jeu de recontre* played with 13-card decks.

10 Wilson, Robin, and John J. Watkins, editors, *Combinatorics: Ancient & Modern*, Oxford University Press, Oxford, 2013, p. 160, which cites de Montmort, Pierre Rémond, *Essay d'Analyse sur les Jeux de Hazard*, Paris, 1708; second edition, J. Quillau, Paris, 1713, reprinted by Chelsea, 1980.

11 I first saw this in Young, Robert M., *Excursions in Calculus: An Interplay of the Continuous and the Discrete*, The Mathematical Association of America, 1992, p. 64.

also have $x_{-1} = 41$. Similarly, we can find x_{-2} by letting $n = -2$ in $x_{n+1} = x_n + 2(n+1)$. This gives $x_{-1} = x_{-2} + 2(-2+1) = x_{-2} - 2$. Using the just established fact $x_{-1} = 41$, this becomes $41 = x_{-2} - 2$. Hence, $x_{-2} = 43$. Continuing like this, we values of x_n for $n = -1$, $-2, -3,...$ match the values of x_n for $n = 0, 1, 2,...$. That is, $x_{-n} = x_{n-1}$ for all natural numbers n.

Thus, x_n gives a prime for the consecutive integers $-40, -39, -38,..., 37, 38, 39$. This is 80 values in a row![12]

Example 6 – Euler's pentagonal number theorem[13]

Euler wanted a formula for $\sigma(n)$, the sum of the positive divisors of n. In Chapter 10, we saw that numbers for which $\sigma(n) = 2n$ are perfect, but there remain many open questions about such numbers. Could a formula help answer some of them? Well, it depends on what form the formula takes. Euler was quickly able to see that $\sigma(n) = n + 1$ when n is prime and that $\sigma(ab) = \sigma(a) \cdot \sigma(b)$ when a and b are relatively prime. Verification of these facts is left as an exercise. With these two pieces of information, and the fundamental theorem of arithmetic, we are close to a formula. The fundamental theorem of arithmetic tells us that every integer, except 0, 1, and -1, can be expressed uniquely as a product of primes, up to order and sign. So, assuming n is not $-1, 0$, or 1, we can write $n = p_1^{a_1} p_2^{a_2} p_3^{a_3} \cdots p_k^{a_k}$, where the p_i are distinct primes and k and the a_i are positive integers. Because distinct primes are relatively prime, we have

$$\sigma(n) = \sigma\left(p_1^{a_1} p_2^{a_2} p_3^{a_3} \cdots p_k^{a_k}\right) = \sigma\left(p_1^{a_1}\right)\sigma\left(p_2^{a_2}\right)\sigma\left(p_3^{a_3}\right) \cdots \sigma\left(p_k^{a_k}\right)$$

All we need to get a final answer is a nice formula for evaluating σ for powers or primes. It's not hard to see (this detail is also left as an exercise) that $\sigma(p^a) = \frac{p^{a+1}-1}{p-1}$. Combining our formulas, we have $\sigma(n) = \sigma\left(p_1^{a_1} p_2^{a_2} p_3^{a_3} \cdots p_k^{a_k}\right) = \frac{p_1^{a_1+1}-1}{p_1-1} \cdot \frac{p_2^{a_2+1}-1}{p_2-1} \cdot \frac{p_3^{a_3+1}-1}{p_3-1} \cdots \frac{p_k^{a_k+1}-1}{p_k-1}$. But this formula is far from ideal, because it requires us to know the complete factorization of n, if we are to use it, and factoring can be extremely hard when the prime factors are large. Fortunately, Euler came up with another formula, although it is a strange one! Here it is:

$$\begin{aligned} \sigma(n) = \ & \sigma(n-1) \ + \sigma(n-2) \ - \sigma(n-5) \ - \sigma(n-7) \\ & + \ \sigma(n-12) + \sigma(n-15) - \sigma(n-22) - \sigma(n-26) \\ & + \ \sigma(n-35) + \sigma(n-40) - \sigma(n-51) - \sigma(n-57) \\ & + \ \sigma(n-70) + \sigma(n-77) - \sigma(n-92) - \sigma(n-100) + \cdots \end{aligned}$$

Odd as this formula is, it has yet to be beaten. To this day, nobody has found a more efficient method for determining $\sigma(n)$ than this recursive equation.[14]

---

12 This is presented, with a different notation, in Young, Robert M., *Excursions in Calculus: An Interplay of the Continuous and the Discrete*, The Mathematical Association of America, 1992, p. 64.

13 The following example is adapted from Young, Robert M., *Excursions in Calculus: An Interplay of the Continuous and the Discrete*, The Mathematical Association of America, 1992, pp. 357–368, which, in turn, cites Polya, George, *Mathematics and Plausible Reasoning, Volume 1: Induction and Analogy in Mathematics*, Princeton University Press, Princeton, NJ, 1954, pp. 91–98. Note: there's much more to Euler's Pentagonal Number Theorem than is shared here.

14 Wilson, Robin, and John J. Watkins, editors, *Combinatorics: Ancient & Modern*, Oxford University Press, Oxford, 2013, p. 208.

Euler specified that the recursion is to stop before hitting $\sigma(k)$ for negative values of k and that $\sigma(0)$ is to be taken as n whenever it is encountered. How odd that $\sigma(0)$ would not have a unique value!

Euler didn't have a proof that his formula always worked. He showed that it held for a great many values, and he could have stopped there, but he was no man of mystery. He was happy to share where his formula came from. That the signs of the terms alternate, a pair of positives, a pair of negatives, is easy to see, but what about the numbers? Euler revealed the pattern so that readers could easily find which values appeared next in the recursion. He did this by listing the numbers he had used, along with the differences between them:

Numbers   1 2 5 7 12 15 22 26 35 40 51 57 70 77 92 100
Differences   1 3 2 5   3   7   4   9   5   11 6   13 7   15 8

The differences might look patternless at first, but let's look separately at the values in the odd and even positions:

Differences in odd positions 1 2 3 4 5 6 7 8
Differences in even positions 3 5 7 9 11 13 15

The pattern is now crystal clear, the next difference will be in an even position, and is therefore 17, making the next value in our recursion $+ \sigma(n - 117)$.

While it's now easy to generate further terms, so that the formula can be applied to larger values of n, it's still unclear why Euler would have taken this approach, as opposed to any one of a million other possible approaches, to the problem. For the answer to that, you will have to wait for (or skip ahead to) Chapter 22.

As a closing note on the topic, for now, I point out that this is the result, described in Chapter 2, that Euler was unable to prove and attempted to convince the reader of by showing that it held for the integer values 1 through 20, as well as 101 and 301.

Euler had better luck with other difference equations. In fact, he was the one who came up with the characteristic equation method detailed in Chapter 15. But how does it work when the roots are complex numbers?

Complex Roots

Previously, we considered the difference equation $x_n - x_{n-1} - 2x_{n-2} = 0$ and found that the general solution is of the form $x_n = A \cdot 2^n + B \cdot (-1)^n$. If we're given initial conditions $x_0 = 1$ and $x_1 = 3$, we may solve for A and B. We get:

$$x_0 = 1 \Rightarrow 1 = A \cdot 2^0 + B(-1)^0 \Rightarrow 1 = A + B$$

$$x_1 = 3 \Rightarrow 3 = A \cdot 2^1 + B(-1)^1 \Rightarrow 3 = 2A - B$$

Adding the two equations together gives $4 = 3A$, which shows $A = 4/3$. It then follows easily from either equation that $B = -1/3$. Thus, our particular solution, subject to the given initial conditions is $x_n = (4/3) 2^n + (-1/3)(-1)^n$.

We found the general solution in Chapter 15 by calculating the roots of the characteristic equation, as above. In this example the roots were distinct, but we also investigated the case of a repeated root. In both cases though, the roots were real. This will not always be the case. We now look at how to handle complex roots.

Example 7 – Find the general solution to $x_{n+1} = x_n - x_{n-1}$ with initial conditions $x_0 = 1$ and $x_{-1} = 1$.

Solution
The characteristic equation is $\lambda^2 - \lambda + 1 = 0$, which has the complex roots $\frac{1}{2} \pm \frac{\sqrt{3}}{2}i$.

Continuing as we did when the roots were real, we get $x_n = A\left(\frac{1}{2} + \frac{\sqrt{3}}{2}i\right)^n + B\left(\frac{1}{2} - \frac{\sqrt{3}}{2}i\right)^n$.

Although this is a correct answer, it's not as nice as desired. It would be a pain to evaluate for large n. Fortunately there are helpful formulas that can be applied to convert our answer to a more useful form. To see how the first formula works, a review of conversion between rectangular and polar coordinates is needed.

In general, a number expressed in rectangular form as $a + bi$, or simply (a, b), can be expressed in polar coordinates by the pair (r, θ), where $r = \sqrt{a^2 + b^2}$ and $\theta = \arctan\left(\frac{b}{a}\right)$.

To convert from the polar coordinate pair (r, θ) back to rectangular coordinates, we calculate $x = r\cos(\theta)$ and $y = r\sin(\theta)$. These equations then give us

$$x + yi = r\cos(\theta) + r\sin(\theta)i = r\left(\cos(\theta) + i\sin(\theta)\right)$$

The expression $\cos(\theta) + i\sin(\theta)$ appears often enough in mathematics to warrant an abbreviation. It can be written as $\text{cis}(\theta)$.

Returning to our answer $x_n = A\left(\frac{1}{2} + \frac{\sqrt{3}}{2}i\right)^n + B\left(\frac{1}{2} - \frac{\sqrt{3}}{2}i\right)^n$ and using the formula above, we can rewrite it in the form $x_n = A(r(\cos(\theta) + i\sin(\theta)))^n + B(r(\cos(\theta) - i\sin(\theta)))^n$.

In this particular example, we have $r = \sqrt{\left(\frac{1}{2}\right)^2 + \left(\frac{\sqrt{3}}{2}\right)^2} = 1$ (for both roots), $\theta = \arctan\left(\frac{\frac{\sqrt{3}}{2}}{\frac{1}{2}}\right) = \arctan(\sqrt{3}) = \frac{\pi}{3}$ for the first root, and $\theta = \arctan\left(\frac{\frac{-\sqrt{3}}{2}}{\frac{1}{2}}\right) = \arctan(-\sqrt{3}) = \frac{-\pi}{3}$ for the second root.

So, $x_n = A(r(\cos(\theta) + i\sin(\theta)))^n + B(r(\cos(\theta) - i\sin(\theta)))^n$ becomes

$$x_n = A\left(\cos\left(\frac{\pi}{3}\right) + i\sin\left(\frac{\pi}{3}\right)\right)^n + B\left(\cos\left(\frac{-\pi}{3}\right) - i\sin\left(\frac{-\pi}{3}\right)\right)^n.$$

This may not seem much better than our original answer, because we still have a hassle evaluating it for large n. However, we have one more result that can be applied.

De Moivre's theorem: $(\cos(x) + i\sin(x))^n = \cos(nx) + i\sin(nx)$

This theorem brought the tools of trigonometry to bear on problems in analysis, but it was a long time in coming. Abraham de Moivre found it in 1722, but a proof had to wait for Euler, who finally provided it sometime in 1748 or 1749.[15] Despite this long wait, the

---

15 For historical details, see www.history.mcs.st-and.ac.uk/Biographies/De_Moivre.html and https://en.wiki
pedia.org/wiki/Abraham_de_Moivre.

proof, follows pretty easily by induction, if you can recall the trig identities you saw back in high school.

Using this result, we get

$$x_n = A\left(\cos\left(n\frac{\pi}{3}\right) + i\sin\left(n\frac{\pi}{3}\right)\right) + B\left(\cos\left(n\frac{-\pi}{3}\right) + i\sin\left(n\frac{-\pi}{3}\right)\right).$$

Recalling that $\cos(-x) = \cos(x)$ and $\sin(-x) = -\sin(x)$, we can rewrite the above as

$$x_n = A\left(\cos\left(n\frac{\pi}{3}\right) + i\sin\left(n\frac{\pi}{3}\right)\right) + B\left(\cos\left(n\frac{\pi}{3}\right) - i\sin\left(n\frac{\pi}{3}\right)\right).$$

Collecting like terms then gives

$$x_n = (A + B)\cos\left(n\frac{\pi}{3}\right) + (A - B)i\sin\left(n\frac{\pi}{3}\right).$$

A and B were unknown constants, so combinations of them are unknown constants that may as well be relabeled as A and B. That is, we can write

$$x_n = A\cos\left(n\frac{\pi}{3}\right) + Bi\sin\left(n\frac{\pi}{3}\right),$$

where it's understood that the constants here are not the same as the ones in our previous equation.

We finally have an answer in a form that is easy to evaluate for large values of n. Turning to our initial conditions $x_0 = 1$ and $x_{-1} = 1$, we can solve for A and B.

$$x_0 = 1 \Rightarrow 1 = A\cos(0) + Bi\sin(0) \Rightarrow 1 = A$$

$$x_{-1} = 1 \Rightarrow 1 = \cos\left(\frac{-\pi}{3}\right) + Bi\sin\left(\frac{-\pi}{3}\right) \Rightarrow 1 = \frac{1}{2} + Bi\left(\frac{-\sqrt{3}}{2}\right)$$

$$\Rightarrow \frac{1}{2} = Bi\left(\frac{-\sqrt{3}}{2}\right) \Rightarrow \frac{i}{2} = -B\left(\frac{-\sqrt{3}}{2}\right) = B\left(\frac{\sqrt{3}}{2}\right) \text{ (multiplying both sides by } i\text{)}$$

$$\Rightarrow \frac{i}{2}\left(\frac{2}{\sqrt{3}}\right) = B \Rightarrow B = \frac{i}{\sqrt{3}}$$

So, our answer becomes

$$x_n = \cos\left(n\frac{\pi}{3}\right) + \frac{i}{\sqrt{3}}i\sin\left(n\frac{\pi}{3}\right),$$

which simplifies to

$$x_n = \cos\left(n\frac{\pi}{3}\right) - \frac{1}{\sqrt{3}}\sin\left(n\frac{\pi}{3}\right).$$

It seems like there's a formula for everything. If there's one that predicts the time of your death, would you want to know it? De Moivre didn't have much of a choice. Late in his life he couldn't help but notice that he slept 15 minutes longer each night than the night before. It was trivial to observe that once a night's sleep grew to 24 hours, there would be no waking up! On the day that the sequence indicated that would occur, de Moivre did indeed die![16]

Improving on Cantor[17]

In Chapter 4, it was demonstrated, ala Georg Cantor, that there are the same number of rational numbers as natural numbers. To do so, a zig-zag path was woven through the lattice points of the first quadrant of the Cartesian coordinate system. For some points the ratio of the x and y values reduced to a rational number already encountered. These were skipped. It was a very easy proof, but the skipping is a little annoying. And it can be avoided by making use of a very simple difference equation.

Let the initial conditions be $f_0 = 0$ and $f_1 = 1$.

Now define the difference equation piece-wise, based on whether the given term is even or odd, like so

$$f_{2n} = f_n$$

$$f_{2n+1} = f_n + f_{n+1}$$

The result is called **Stern's diatomic sequence** and it begins 0, 1, 1, 2, 1, 3, 2, 3, 1, 4, 3, 5, 2, 5, 3, 4, ... It doesn't look very interesting and there's no obvious connection to the rationals, but looking closer, you may notice that consecutive terms are always relatively prime. Generating more terms will not yield an exception. Of course, even if we generate a million terms, this does not prove the pattern continues forever. It can be proven by other means, though. Now, consider a new sequence defined by

$$g_n = \frac{f_n}{f_{n+1}}$$

The first few terms are 0, 1, 1/2, 2, 1/3, 3/2, 2/3, 3, 1/4, 4/3, 3/5, 5/2, 2/5, 5/3, 3/4, ...

This sequence contains every nonnegative rational number exactly once. So, we have a one-to-one correspondence between the natural numbers and the rationals given by

$$n \leftrightarrow g_n$$

While verifying all of the claims made about the sequence requires some work, isn't the result prettier than the correspondence given by Cantor?

---

16 http://www.history.mcs.st-and.ac.uk/Biographies/De_Moivre.html.

17 I first saw this in Chamberland, Marc, *Single Digits: In Praise of Small Numbers*, Princeton University Press, Princeton, NJ, 2015, pp. 9–10.

The Gingerbreadman (1984)[18]

Consider the following system of difference equations, in which the absolute value function appears.

$$x_{n+1} = 1 - y_n + |x_n|$$

$$y_{n+1} = x_n$$

If we start out with $x_n = 0$ and $y_n = 0$, then the iterations cycle around the six values (0, 0), (1, 0), (2, 1), (2, 2), (1, 2) and (0, 1).

For another set of initial conditions, $x_n = 1$ and $y_n = 1$, we go nowhere! That is, (1, 1) is a fixed point. It's actually the only fixed point.[19]

There are much more interesting initial conditions. Robert Devaney, the discoverer of this set of equations, iterated the point (0.001, –0.001) 50,000 times and noted that it "seems to fill a region in the plane that resembles a gingerbreadman or, as some would say, a rabbit's head."[20] This result is shown in Figure 17.3 (after 70,000 iterations). The cycle of six values mentioned above forms the vertices of the hexagon at the center of the figure.

If someone performed enough iterations, would the body, with the exception of the six hexagonal holes, eventually get filled in completely? Devaney answered this question as follows:

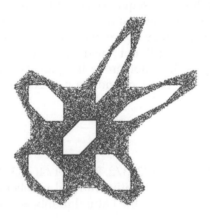

FIGURE 17.3
The Gingerbreadman.

Image created by Josh Gross using 70,000 iterations.

18 Devaney, Robert L., "A Piecewise Linear Model for the Zones of Instability of an Area Preserving Map," *Physica D: Nonlinear Phenomena*, Vol. 10, No. 3, March 1984, pp. 387–393.

19 Devaney, Robert L., "A Piecewise Linear Model for the Zones of Instability of an Area Preserving Map," *Physica D: Nonlinear Phenomena*, Vol. 10, No. 3, March 1984, pp. 387–393, p. 387 cited here.

20 Devaney, Robert L., "The Gingerbreadman," *Algorithm*, Vol. 3, January 1992, pp. 15–16, p. 15 quoted here.

THE COMPUTER LIES
Despite appearances to the contrary, the orbit of (0.001, -0.00l) is *not* a dense orbit as I have just defined it. In fact, this orbit is really a very long cycle with period 7509. Amazingly, every point in the plane with rational coordinates lies on a cycle that has some (possibly very large) period. The point can be proved right here.[21]

Devaney went on to offer a quick proof that whenever the initial conditions are rational numbers, the result is a cycle. He defined G to be the points in the gingerbreadman excluding the boundary and the boundary of the six hexagonal regions, and offered another surprise.

It turns out that the dynamics of the iteration within G is chaotic, in a sense: It can be proved that there is a point (x^*, y^*) the orbit of which comes arbitrarily close to any point in G! In other words, if you choose any small rectangular region in G, no matter how small, the orbit of this mysterious point (x^*, y^*) eventually enters the rectangle.[22]

Some exercises at the end of this chapter ask you to further explore the dynamics of this simple system of difference equations.

OEIS

In earlier chapters, a few problems were attacked by examining simpler cases and looking for patterns. Examples include the number of distinct seating arrangements for *n* people at a round table and the minimum number of moves needed to solve the *n*-disc Tower of Hanoi. In both cases a sequence of integers was found that suggested a formula. But what can be done when the sequence that is found is unfamiliar and no formula can be readily conjectured? In such instances, it's a good idea to turn to *The On-Line Encyclopedia of Integer Sequences* at https://oeis.org/. This website contains over 330,000 integer sequences, along with formulas, references (both print and online), and comments. Simply enter the terms you've determined and a list of sequences that contain those values as consecutive terms is displayed. You may well find that one of them is exactly what you need. Two sequences contained in OEIS follow.

Chess Games[23]

After 0 moves, there's only one possible configuration a chess board can be in – the same position as when the pieces were first set on the board. But once the pieces start moving, the number of possible games quickly grows. There are 20 possible games after the first move, 400 after the second, 8,902 after the third, ... Only 11 more terms are known. Here's the present state of knowledge concerning the sequence for the number of possible chess games after *n* moves.

---

21 Devaney, Robert L., "The Gingerbreadman," *Algorithm*, Vol. 3, January 1992, pp. 15–16, p. 16 quoted here.
22 Devaney, Robert L., "The Gingerbreadman," *Algorithm*, Vol. 3, January 1992, pp. 15–16, p. 15 quoted here.
23 http://oeis.org/A048987.

1, 20, 400, 8902, 197281, 4865609, 119060324, 3195901860, 84998978956, 2439530234167, 69352859712417, 2097651003696806, 62854969236701747, 1981066775000396239, 61885021521585529237, ...

This sequence was initially computed for the Encyclopedia by Ken Thompson in the 1990s, but others have added terms since then.

There are many other sequences for which only a few terms are known. An example is Hadamard's maximal determinant problem.

Hadamard's Maximal Determinant Problem

What is the largest determinant of any $n \times n$ matrix with entries that are 0 and 1? Here are the terms of the sequence that are presently known, starting with $n = 1$.

1, 1, 1, 2, 3, 5, 9, 32, 56, 144, 320, 1458, 3645, 9477, 25515, 131072, 327680, 1114112, 3411968, 19531250, 56640625, ...

The next term is unknown.[24] That is, it's not known what the largest possible determinant is for a 22 × 22 matrix made up of 0s and 1s. We could easily write a computer program to try all possible combinations. Unfortunately, this number is quite large, as there are $22^2 = 484$ entries, each of which may be 0 or 1. The total number of matrices is therefore $2^{484} \approx 4.99 \times 10^{145}$. We know the sequence is nondecreasing, as we can always take an $n - 1 \times n - 1$ matrix with maximum determinant and add a row on the right and a column on the bottom with all 0s except for the last entry, which is made a 1. This $n \times n$ matrix will have the same determinant as the smaller matrix that is embedded in it. Can you show that the sequence is strictly increasing, after the first few values? If you cannot solve a problem completely, don't ignore it! Chip away at it a little bit. You or someone else may chip away a little more later on and eventually an insight may be gained and the original problem solved.

To see other "Sequences That Need Extending" go to https://oeis.org/more.html. A link on this page will take you to a list of over 19,000 such sequences. Can you help with one of them?

The On-Line Encyclopedia of Integer Sequences was created by Neil Sloane (Figure 17.4). The first incarnation was a book, published in 1973.[25] A new book, with over twice as many sequences, coauthored with Simon Plouffe, came out in 1995.[26] With the rapid growth of the database that followed, thanks to reader contributions, the move to an online format made sense.

In addition to his work on the Encyclopedia, Sloane has published in many areas, including combinatorics, error-correcting codes, sphere packing, rock climbing, and cryptography![27]

---

24 http://oeis.org/A003432.

25 Sloane, Neil J. A., *A Handbook of Integer Sequences*, Academic Press, New York and London, 1973.

26 Sloane, Neil J. A. and Simon Plouffe, *The Encyclopedia of Integer Sequences*, Academic Press, San Diego, CA, 1995.

27 See http://neilsloane.com/ and https://en.wikipedia.org/wiki/Neil_Sloane.

FIGURE 17.4
Neil Sloane (1939–).

Courtesy of Neil J. A. Sloane.

Generating Functions

> A generating function is a clothesline on which we hang up a sequence of numbers for display.
>
> Herbert S. Wilf[28]

Given a finite sequence $a_0, a_1, a_2, ..., a_{n-1}, a_n,$ the generating function for it is the polynomial

$$G(x) = a_0 + a_1x + a_2x^2 + \cdots + a_{n-1}x^{n-1} + a_nx^n = \sum_{i=0}^{n} a_ix^i$$

If we are dealing with an infinite sequence $a_0, a_1, a_2, ..., a_n, ...$ the generating function is given by

$$G(x) = a_0 + a_1x + a_2x^2 + \cdots + a_nx^n + \cdots = \sum_{i=0}^{\infty} a_ix^i$$

You should be aware that z is often used in place of x and the generating function $G(z)$ may be referred to as the z transform of the sequence $\{a_n\}$.

Example 8 – Power sets

Although the phrase "generating function" wasn't used at the time, an example was seen in Chapter 4, when the following was shown as a means of generating the elements of the power set of $S = \{a, b, c\}$:

---

28 Wilf, Herbert S., *Generatingfunctionology*, Second edition, Academic Press, 1994, p. 1.

$$(x + a)(x + b)(x + c) = x^3 + (a + b + c)x^2 + (bc + ac + ab)x + abc.$$

Recall: The constant term is abc, corresponding to the set $\{a, b, c\}$. The linear term has coefficient $bc + ac + ab$, corresponding to the sets $\{b, c\}$, $\{a, c\}$, and $\{a, b\}$. The quadratic term has coefficient $(a + b + c)$, corresponding to the sets $\{a\}$, $\{b\}$, and $\{c\}$. Finally, the cubic term has coefficient 1, which does not involve any of the element of our set S, so it corresponds to the empty set, $\varnothing$.

An alternative multiplication to achieve the same sort of result was also given:

$$(1 + ax)(1 + bx)(1 + cx) = 1 + (a + b + c)x + (bc + ac + ab)x^2 + (abc)x^3.$$

As was mentioned in Chapter 4, these results generalize to n-element sets.

It's now time to put a twist on these results. Suppose that instead of generating the elements of a power set, we merely want to know how many elements there are of any given size?

To answer this new question, simply let $a = b = c = 1$ in the alternative multiplication given above. This gives

$$(1 + x)^3 = 1 + 3x + 3x^2 + x^3.$$

In general, we get

$$(1 + x)^n = \binom{n}{0} + \binom{n}{1}x + \binom{n}{2}x^2 + \cdots + \binom{n}{n}x^n.$$

There are other combinatorial problems that can be solved with this sort of approach.[29]

Notice that letting $x = 1$, in the above equation shows that the sum of a row of Pascal's triangle is 2^n. There are other ways in which generating functions connect with Pascal's triangle.

Example 9 – Pascal's triangle

Consider the first few columns of the left-justified Pascal's triangle shown below.

$$
\begin{array}{ccccc}
1 & & & & \\
1 & 1 & & & \\
1 & 2 & 1 & & \\
1 & 3 & 3 & 1 & \\
1 & 4 & 6 & 4 & 1 \\
\vdots & \vdots & \vdots & \vdots & \vdots \\
\binom{n}{0} & \binom{n}{1} & \binom{n}{2} & \binom{n}{3} & \binom{n}{4}
\end{array}
$$

---

29 Wilson, Robin, and John J. Watkins, editors, *Combinatorics: Ancient & Modern*, Oxford University Press, Oxford, 2013, p. 287.

The generating functions for these columns are $\frac{1}{1-x}$, $\frac{x}{(1-x)^2}$, $\frac{x^2}{(1-x)^3}$, $\frac{x^3}{(1-x)^4}$, and $\frac{x^4}{(1-x)^5}$. This pattern continues.[30]

Example 10 – An infinite geometric series.

The simple steps below show how the closed form of the generating function for an infinite geometric series can be found.

$$G(x) = 1 + ax + a^2x^2 + a^3x^3 + \cdots$$

$$G(x) - 1 = ax + a^2x^2 + a^3x^3 + \cdots$$

$$G(x) - 1 = (ax)\left(1 + ax + a^2x^2 + \cdots\right)$$

$$\frac{G(x) - 1}{ax} = 1 + ax + a^2x^2 + \cdots$$

$$\frac{G(x) - 1}{ax} = G(x)$$

$$G(x) - 1 = ax \cdot G(x)$$

$$G(x) - ax \cdot G(x) = 1$$

$$G(x) \cdot (1 - ax) = 1$$

$$G(x) = \frac{1}{1 - ax}$$

Example 11 – The Fibonacci sequence

Let

$$G(x) = F_0 + F_1x + F_2x^2 + \cdots + F_nx^n + \cdots,$$

where F_n is the nth Fibonacci number.

We can convert this to a closed form by beginning with the equation $F_n = F_{n-1} + F_{n-2}$. Multiplying both sides of this difference equation by x^n gives $F_nx^n = F_{n-1}x^n + F_{n-2}x^n$.

This last equation is true for $n = 2, 3, 4, \ldots$, and we may imagine that we have written out all of these instances. Summing them yields

$$\sum_{n=2}^{\infty} F_nx^n = \sum_{n=2}^{\infty} F_{n-1}x^n + \sum_{n=2}^{\infty} F_{n-2}x^n.$$

Notice that

---

30 Hoggatt, Jr., Verner Emil, "A New Angle on Pascal's Triangle," *The Fibonacci Quarterly*, Vol. 6, No. 4, October 1968, pp. 221–234.

$$\sum_{n=2}^{\infty} F_n x^n = G(x) - F_0 - F_1 x$$

$$\sum_{n=2}^{\infty} F_{n-1} x^n = x(G(x) - F_0)$$

$$\sum_{n=2}^{\infty} F_{n-2} x^n = x^2 G(x)$$

Making these substitutions, we have

$$G(x) - F_0 - F_1 x = x(G(x) - F_0) + x^2 G(x)$$

$$G(x) - F_0 - F_1 x = xG(x) - xF_0 + x^2 G(x)$$

$$G(x) - xG(x) - x^2 G(x) = F_1 x - xF_0 + F_0$$

$$(1 - x - x^2) G(x) = F_1 x - xF_0 + F_0$$

$$G(x) = \frac{F_1 x - xF_0 + F_0}{1 - x - x^2}$$

But $F_0 = 0$ and $F_1 = 1$, so the above simplifies to

$$G(x) = \frac{x}{1 - x - x^2}.$$

This approach can be used for other difference equations.

There's an even simpler expression that contains all of the Fibonacci numbers.[31]

$$\frac{1}{89} = 0.01 + 0.001 + 0.0002 + 0.00003 + 0.000005 + 0.0000008 + 0.00000013$$
$$+ 0.000000021 + 0.0000000034 + \cdots$$

While this is not a generating function, it arises easily from the one above. We have

$$G(x) = \frac{x}{1 - x - x^2} = x + x^2 + 2x^3 + 3x^4 + 5x^5 + 8x^6 + 13x^7 + 21x^8 + \cdots$$

Letting $x = 1/10$, this becomes

---

31 I first saw this at www.reddit.com/r/AskReddit/comments/4kz3di/whats_your_favourite_maths_fact/.

$$\frac{\frac{1}{10}}{1 - \frac{1}{10} - \left(\frac{1}{10}\right)^2} = 0.1 + 0.01 + 0.002 + 0.0003 + 0.00005 + 0.000008 + 0.0000013$$
$$+ 0.00000021 + 0.000000034 + \cdots,$$

which simplifies to

$$\frac{10}{89} = 0.1 + 0.01 + 0.002 + 0.0003 + 0.00005 + 0.000008 + 0.0000013 + 0.00000021$$
$$+ 0.000000034 + \cdots,$$

Finally, dividing both sides by 10 gives the claimed result.

$$\frac{1}{89} = 0.01 + 0.001 + 0.0002 + 0.00003 + 0.000005 + 0.0000008 + 0.00000013$$
$$+ 0.000000021 + 0.0000000034 + \cdots$$

And the fraction is easy to remember, because 89 is itself a Fibonacci number!

Here are a few sequences and their generating functions. Some were described above and others are new.

Sequence	Generating Function
$1, 1, 1, 1, 1, \ldots$	$\frac{1}{1-x}$
$1, 2, 3, 4, 5, \ldots$	$\frac{1}{(1-x)^2}$
$1^2, 2^2, 3^2, 4^2, 5^2, \ldots$	$\frac{1+x}{(1-x)^3}$
$1^3, 2^3, 3^3, 4^3, 5^3, \ldots$	$\frac{1+4x+x^2}{(1-x)^4}$
$1, 1/2, 1/3, 1/4, 1/5, \ldots$	$\ln \frac{1}{1-x}$
$H_1, H_2, H_3, H_4, H_5, \ldots$ where $H_n = \sum_{i=1}^{n} \frac{1}{i}$ (Harmonic series)	$\frac{1}{1-x} \ln \left(\frac{1}{1-x}\right)$

Abraham de Moivre is responsible for originating the idea of generating functions. Other mathematicians, including James Stirling, Leonhard Euler, and Pierre Simon Laplace further developed the approach, making it more powerful.[32]

One of the developments was a new kind of generating function that's better suited to certain types of sequences. It's called the **exponential generating function**. Given an infinite sequence $a_0, a_1, a_2, \ldots, a_n, \ldots$ the exponential generating function is the polynomial

---

32 Knuth, Donald E., *The Art of Computer Programming*, Vol. 1, Fundamental Algorithms, 3rd edition, Addison-Wesley, New York, 1997, p. 87.

$$G(x) = \sum_{i=0}^{\infty} a_i \frac{x^i}{i!}$$

We may replace infinity by n in the case where the sequence is finite.
   Here are a few sequences and their (exponential) generating functions.

Sequence	Generating Function
$1, 1, 1, 1, 1, \ldots$	e^x
$D_1, D_2, D_3, D_4, D_5, \ldots$ (Derangements)	$\frac{e^{-x}}{1-x}$
Stirling Numbers of the second Kind	$\frac{1}{k!} (e^x - 1)^k$

Exercise Set 17

Hofstadter's Law: It always takes longer than you expect, even when you take into account Hofstadter's Law.[33]

1. Use the Babylonian method detailed in Example 1 to estimate $\sqrt{10}$ using the initial guess $x_0 = 3$. Calculate x_1 through x_3. How many digits after the decimal point does x_3 correctly give?

2. Use the Babylonian method detailed in Example 1 to estimate $\sqrt{5}$ using the initial guess $x_0 = 2$. Calculate x_1 through x_3. How many digits after the decimal point does x_3 correctly give?

3. Use the Babylonian method detailed in Example 1 to estimate $\sqrt{42}$ using the initial guess $x_0 = 6$. Calculate x_1 through x_3. How many digits after the decimal point does x_3 correctly give?

4. Go back to the difference equation given in Example 2 and, for each of the following initial conditions, iterate far enough to guess what the sequence x_0, x_1, x_2, ... converges to.

   a. $x_0 = 10$

   b. $x_0 = 5$

   c. $x_0 = 42$

5. Given $f(x) = -x^2 + 2x + 4$, apply Newton's method with initial guess $x_0 = 0$. What is the value of x_2?

6. Given $f(x) = -x^2 + 2x + 4$, apply Newton's method with initial guess $x_0 = 2$. What is the value of x_2? Why does this seem to head to a different value than the result of Exercise 5?

---

33 Hofstadter, Douglas R., *Gödel, Escher, Bach: An Eternal Golden Braid*. Vintage Books Edition, New York, September 1980, p. 152.

7. Use Newton's method to approximate $\sqrt[3]{10}$ with initial guess $x_0 = 2$. Only calculate x_1 and x_2. Compare x_2 to the exact answer. How many digits after the decimal point are correct?

8. Given $f(x) = x^3 - 2x + 2$, apply Newton's method with initial guess $x_0 = 0$. What is the value of x_{100}?

9. Show graphically how Newton's method can fail, without getting caught in a cycle.

10. Show that the Babylonian method detailed in Example 1 is a special case of Newton's method.

11. Given $D_1 = 0$ and $D_2 = 1$, repeatedly apply $D_n = (n-1)(D_{n-1} + D_{n-2})$ to calculate the number of derangements on 7 items.

12. Prove $D_n = n \cdot D_{n-1} + (-1)^n$.

    Although Exercises 13, 14, and 15 concern $\sigma(n)$, you do not want to approach these problems with Euler's pentagonal theorem!

13. Prove that $\sigma(n) = n + 1$ when n is prime.

14. Show that $\sigma(ab) = \sigma(a) \cdot \sigma(b)$ when a and b are relatively prime.

15. Show that for any prime p and any positive integer a, $\sigma(p^a) = \frac{p^{a+1}-1}{p-1}$. Hint: Write out all of the divisors of p^a and then use the formula for a finite geometric sum to add them together.

16. Find the general solution to $x_{n+1} = -2x_n - 2x_{n-1}$

17. Find the specific solution to $x_{n+1} = -2x_n - 2x_{n-1}$ with initial conditions $x_0 = 1$ and $x_1 = 3$.

18. Find the general solution to $x_{n+1} = -4x_n - 8x_{n-1}$.

19. Find the specific solution to $x_{n+1} = -4x_n - 8x_{n-1}$ with initial conditions $x_0 = 0$ and $x_1 = 2$.

20. Give a proof by induction of De Moivre's theorem, $(\cos(x) + i\sin(x))^n = \cos(nx) + i\sin(nx)$. Hint: In the induction step, you'll need to make use of a trig identity that you've probably seen before, but forgotten.

21. Why isn't the formula $e^{0\pi i} = 1$, as impressive as Euler's? It involves the same numbers, but the author of the book I found it in described it as "vacuous."[34]

22. For the gingerbreadman equations

$$x_{n+1} = 1 - y_n + |x_n|$$

$$y_{n+1} = x_n$$

what is the period of the point (1, 5)?

23. What happens if you iterate a point that lies inside the hexagon on the chest of the gingerbreadman?

24. What happens if you iterate the point $(-1, -1)$, which lies inside the hexagon on the head of the gingerbreadman?

25. What happens if you iterate a point other than $(-1, -1)$ from among those that lie inside the hexagon on the head of the gingerbreadman?

---

34 Paulos, John Allen, *Beyond Numeracy*, Vintage Books, New York, 1992, p. 117.

26. Go to https://oeis.org/more.html and find the link that takes you to the list of about 16,000 "Sequences That Need Extending." Examine these sequences until you find one that intrigues you. Explain it in your own words. How would you begin to attack this problem?

27. Find the generating function of the sequence defined by $a_0 = 0$, $a_1 = 1$, and $a_n + 2a_{n-1} - 15a_{n-2} = 0$ for $n > 1$.

28. Find the generating function of the sequence defined by $a_0 = 1$, $a_1 = 1$, and $a_n + 6a_{n-1} + 16a_{n-2} = 0$ for $n > 1$.

29. Find the generating function of the sequence defined by $a_0 = 1$, $a_1 = 2$, and $a_n + 15a_{n-1} - 50a_{n-2} = 0$ for $n > 1$.

30. Find the generating function of the sequence defined by $a_0 = 0$, $a_1 = 2$, and $a_n + 6a_{n-1} + 27a_{n-2} = 0$ for $n > 1$.

31. Find the generating function of the sequence defined by $a_0 = 1$, $a_1 = 0$, and $a_n - 13a_{n-1} - 30a_{n-2} = 0$ for $n > 1$.

32. $\frac{1}{1-x^2}$ is the generating function for what sequence?

33. $\frac{x}{(1-x)^3}$ is the generating function for what sequence?

34. Use the binomial theorem to prove that

$$(1 + x)^n = \binom{n}{0} + \binom{n}{1}x + \binom{n}{2}x^2 + \cdots + \binom{n}{n}x^n.$$

35. Consider difference equation $p_n = p_{n-1} + (-1)^n \frac{4}{2n+1}$ with initial condition $p_0 = 4$. What happens when it's iterated?[35]

36. The April 1979 issue of *OMNI* featured "The World's hardest IQ Test."[36] Eight out of the 56 questions consisted of sequences of integers for which the test taker was asked to find the next term. Five choices were provided for each, but I'm making it a bit harder by not sharing them here! Some of the answers can be found in the Online Encyclopedia of Integer Sequences. However, it's more fun to see if you can get them without resorting to the Internet.

    a. 5, 8, 17, 24, 37, ?
    b. 1, 3, 4, 8, 15, 27, 50, ?
    c. 7, 10, 5, 9, 3, 8, 2, ?
    d. 5, 7, 3, 1, 4, 5, 9, 5, ?
    e. 8, 10, 7, 12, 5, 16, 3, ?
    f. 11, 13, 17, 25, 32, 37, 47, 58, 71, ?
    g. 2, 3, 6, 10, 17, 28, ?
    h. 6, 6, 4, 8, 4, 12, 4, 16, ?

37. In July 1981, *OMNI* challenged readers with "The *OMNI*-Mensa I.Q. Test"[37] As with the previous IQ test, this one featured "find the next term" problems. They are reproduced below. The answers were given in the August 1981 issue. The IQ

35 Taken from Poundstone, William, *The Recursive Universe: Cosmic Complexity and the Limits of Scientific Knowledge*, Willow Morrow and Company, Inc, New York, 1985, pp. 123–124.

36 Morris, Scot, "The World's Hardest IQ Test," *OMNI*, April 1979, Vol. 1, No. 7, pp. 116–120.

37 Morris, Scot, "The *OMNI*-Mensa I. Q. Test," *OMNI*, Vol. 3, No. 10, July 1981, pp. 96–98.

test was given to 88 members of Mensa and this follow-up piece revealed how many of them correctly answered each question.[38]

    a. "What number follows logically in this series?
        2, 3, 5, 9, 17 _____"
    b. "Find the number that logically completes this series.
        1, 2, 6, 12, 36 _____"
    c. "What number logically comes next in this series?"
        7, 12, 27, 72 _____"
    d. "What number follows logically in this series?
        9, 12, 21, 48 _____ (a) 69 (b) 70 (c) 129 (d) 144"
    e. "What is the next number in this series?
        21, 20, 18, 15, 11 _____"
    Each of these questions contained an error! Can you spot it?

38. **Cobweb diagrams** are a useful tool for investigating the iteration of difference equations. Do an online search to learn about these and give an example. In addition to the diagram, give an explanation in your own words as to what is happening in it.

References/Further Reading

I have never been able to resist a book about books.
        – Anne Fadiman, *Ex Libris: Confessions of a Common Reader.*

Bender, Edward A., Jay R. Goldman, and Gian-Carlo Rota, "Enumerative Uses of Generating Functions," *Indiana University Mathematics Journal*, Vol. 20, No. 8, 1971, pp. 753–764.

Calkin, Neil and Herbert S. Wilf, "Recounting the Rationals," *The American Mathematical Monthly*, Vol. 107, No. 4, April 2000, pp. 360–363. This is a Stern sequence reference.

Chamberland, Marc, *Single Digits: In Praise of Small Numbers*, Princeton University Press, Princeton, NJ, 2015.

Comtet, Louis, *Advanced Combinatorics: The Art of Finite and Infinite Expansions*, D. Reidel Publ. Co., Boston, MA, 1974.

de Montmort, Pierre Rémond, *Essay d'Analyse sur les Jeux de Hazard*, Paris, 1708; 2nd edition, J. Quillau, Paris, 1713, reprinted by Chelsea, 1980.

Deuflhard, Peter, "A Short History of Newton's Method," *Documenta Mathematica*, *Optimization Stories*, Extra Volume ISMP, 2012, pp. 25–30, available online at https://emis.math.unistra.fr/journals/DMJDMV/vol-ismp/13_deuflhard-peter.pdf.

Devaney, Robert L., "A Piecewise Linear Model for the Zones of Instability of an Area Preserving Map," *Physica D: Nonlinear Phenomena*, Vol. 10, No. 3, March 1984, pp. 387–393.

Devaney, Robert L., "Fractal Patterns Arising in Chaotic Dynamical Systems," in Peitgen, Heinz-Otto and Dietmar Saupe, editors, *The Science of Fractal Images*, Springer-Verlag, New York, 1988, pp. 137–168. See especially "A Chaotic Gingerbreadman," §3.2.3, pp. 149–150.

Devaney, Robert L., "The Gingerbreadman," *Algorithm*, Vol. 3, January 1992, pp. 15–16.

Dijkstra, Edsger W., *Selected Writings on Computing: A Personal Perspective*, Springer, New York, 1982, pp. 215–232. This is a Stern sequence reference.

---

38 Morris, Scot, "IQ Test Answers," *OMNI*, Vol. 3, No. 11, August 1981, pp. 88 and 122.

Doubilet, Peter, Gian-Carlo Rota, and Richard Stanley, "On the Foundations of Combinatorial Theory VI: The Idea of Generating Function," *Proceedings of the Sixth Berkeley Symposium on Statistics and Probability*, Vol. 2, 1972, pp. 267–318.

Gauvrit, Nicolas, Jean-Paul Delahaye, and Hector Zenil, "Sloane's Gap: Do Mathematical and Social Factors Explain the Distribution of Numbers in the OEIS?" 2013.

Gibbons, Jeremy, David Lester, and Richard Bird, "Functional Pearl: Enumerating the Rationals," *Journal of Functional Programming*, Vol. 16, No. 3, 2006, pp. 281–291. This is a Stern sequence reference.

Goulden, Ian P. and David M. Jackson, *Combinatorial Enumeration*, John Wiley and Sons, New York, 1983.

Hayman, Walter, "A Generalisation of Stirling's Formula," *Journal für die reine und angewandte Mathematik*, Vol. 196, 1956, pp. 67–95.

Hoggatt, Jr., Verner E., "A New Angle on Pascal's Triangle," *The Fibonacci Quarterly*, Vol. 6, No. 4, October 1968, pp. 221–234.

Klarner, David A., "Some Results Concerning Polyominoes," *Fibonacci Quarterly*, Vol. 3, No. 1, February 1965, pp. 9–20.

Liversidge, Anthony, "Interview, Neil Sloane," *OMNI*, Vol. 11, No. 12, September 1989, pp. 78–80, 82–83, 96–97, 99–101.

McEliece, Robert J., Robert B. Ash, and Carol Ash, *Introduction to Discrete Mathematics*, Random House, New York, 1989.

Mu, Dr., "Cowculations: Gingerbread Man," *Quantum*, January–February, 1998, pp. 55–57.

Northshield, Sam, "Stern's Diatomic Sequence," *American Mathematical Monthly*, Vol. 117, No. 7, August 2010, pp. 581–598.

O'Connor, John J. and Edmund F. Robertson, "Abraham de Moivre," *MacTutor archive*, June 2004, available online at www.history.mcs.st-and.ac.uk/Biographies/De_Moivre.html.

Polya, George, *Mathematics and Plausible Reasoning, Volume 1: Induction and Analogy in Mathematics*, Princeton University Press, Princeton, NJ, 1954.

Pólya, George, "On Picture-Writing," *The American Mathematical Monthly*, Vol. 63, No. 10, December 1956, pp. 689–697. This article details some applications of generating functions.

Poundstone, William, *The Recursive Universe: Cosmic Complexity and the Limits of Scientific Knowledge*, Willow Morrow and Company, Inc, New York, 1985.

Rosen, Kenneth H., editor, *Handbook of Discrete and Combinatorial Mathematics*, CRC Press, Boca Raton, FL, 2000.

Schneider, Ivo, "Der Mathematiker Abraham de Moivre," *Archive for History of Exact Sciences*, Vol. 5, No. 3–4, January 1968, pp. 177–317.

Simpson, Thomas, *Essays on Several Curious and Useful Subjects in Speculative and Mix'd Mathematicks*, 1740.

Sloane, Neil J. A., *A Handbook of Integer Sequences*, Academic Press, New York and London, 1973.

Sloane, Neil J. A., "Stern's Diatomic Series (Or Stern-Brocot sequence)," *The On-Line Encyclopedia of Integer Sequences*, available online at http://oeis.org/A002487.

Sloane, Neil J. A. and Simon Plouffe, *The Encyclopedia of Integer Sequences*, Academic Press, San Diego, CA, 1995.

"Sloane's Gap," *Numberphile*, available online at www.youtube.com/watch?v=_YysNM2JoFo.

Stanley, Richard P., "Generating Functions," in Rota, Gian-Carlo, editor, *MAA Studies in Combinatorics*, Mathematical Association of America, Washington, DC, 1978, pp. 100–141.

Stanley, Richard P., *Enumerative Combinatorics*, Wadsworth, Monterey, CA, 1986.

Van der Waerden, Bartel Leendert, *Science Awakening*, 3rd edition, Wolters-Noordhoff, Groningen, 1971.

Wikipedia contributors, "Newton's Method," *Wikipedia, The Free Encyclopedia*, available online at https://en.wikipedia.org/w/index.php?title=Newton%27s_method&oldid=848618978.

Wilf, Herbert S., *Algorithms and Complexity*, Prentice Hall, Englewood Cliffs, NJ, 1986.

Wilf, Herbert S., *Generatingfunctionology*, Academic Press, 1990. (Second edition, 1994; Third edition, A K Peters/CRC Press, 2005). The Second edition may be downloaded for free from www. math.upenn.edu/~wilf/DownldGF.html.

Wilson, Robin and John J. Watkins, editors, *Combinatorics: Ancient & Modern*, Oxford University Press, Oxford, 2013.

Young, Robert M., *Excursions In Calculus: An Interplay of the Continuous and the Discrete*, The Mathematical Association of America, 1992.

18

Chaos Theory and Fractals

It's sometimes claimed that the term "Butterfly effect," which is used in chaos theory, arose from the 1952 Ray Bradbury (Figure 18.1) short story, *A Sound of Thunder*.[1] The idea is that a system that is highly dependent on initial conditions, such as the weather, may be radically different, in the long term, if a minor change, like the flapping of a butterfly's wings, is introduced or eliminated.

In Ray Bradbury's story, a butterfly is accidentally killed by a time traveler 60 million years before his birth. Upon returning to his own time, he finds that the result of a recent presidential election has changed as a consequence.

Sensitive Dependence on Initial Conditions

While Bradbury's story is a more poetic introduction to the topic of the butterfly effect, the historic origin of the concept, without the colorful name, arose much earlier.

Henri Poincaré (pwan ka ray) (Figure 18.2) was a towering intellect who is considered by some to be "the last universalist," that is, the last person to understand the whole of mathematics and physics, as it was known at his time.[2] Today, these subjects are far too broad and deep for anyone to have a firm grasp of it all. The name Poincaré is not nearly as widely recognized as Einstein, but Poincaré put forth $E = mc^2$ before his more famous colleague, although he did give it in the less catchy form $m = \frac{E}{c^2}$.[3]

Poincaré is also considered to be the father of chaos theory. The idea resulted from a challenge posed in 1889 by Oscar II, King of Sweden and Norway, at the suggestion of Sweden's top mathematician, Gösta Mittag-Leffler. The King asked for a solution to the n-body problem. This concerns the motions of n bodies in space, subject to the force of gravity. For just two bodies, say a planet and a sun, Johannes Kepler had found the solution of the planet pursuing an elliptical orbit about the sun, but when a third body is introduced, the problem becomes significantly harder. Two planets orbiting a sun pull on each other, so their orbits are no longer perfect ellipses. Similarly, a moon orbiting

---

1 It first appeared in the June 28, 1952 issue of *Collier's* magazine. It also appeared in the Ray Bradbury short story collections *The Golden Apples of the Sun*, Doubleday, New York, 1953, and *R is for Rocket*, Doubleday, New York, 1962. Yet another appearance is in *The Stories of Ray Bradbury*, Alfred A. Knopf, New York, 1980.

2 For example, see Ginoux, Jean-Marc and Christian Gerini, *Henri Poincaré: A Biography through the Daily Papers*, World Scientific, Singapore, 2013. Ian Stewart refers to him as "one of the last to range over almost the entire mathematical landscape of his day." in Stewart, Ian, *Significant Figures: The Lives and Work of Great Mathematicians*, Basic Books, New York, p. 198.

3 Poincaré, Henri, *Arch. Néerland. Sci.*, Vol. 2, No. 5, pp. 252–278, 1900. For a discussion of the context of Poincaré's discovery, as well as what others knew and when they knew it, see Auffray, Jean-Paul, "Dual Origin of *E=mc²*," 2006, available online at https://arxiv.org/abs/physics/0608289.

FIGURE 18.1
Ray Bradbury (1920–2012).

Public domain.

FIGURE 18.2
Henri Poincaré (1854–1912).

Public domain.

a planet that, in turn, orbits a sun makes for a much more complex system. Finding a solution for an arbitrarily large number of bodies seemed unlikely, but if it could be obtained it would answer important questions. For example, is our solar system stable or can large (unpleasant) changes occur?

Poincaré won the prize, even though he had only addressed a special case of the three-body problem.[4] His prize essay was to be published in *Acta Mathematica*, but a problem arose. Lars Edvard Phragmén, in helping Gösta Mittag-Leffler edit the journal, contacted Poincaré concerning some mathematics in the paper that he couldn't follow. Poincaré looked at this portion of the paper again and realized that he had made a mistake.[5]

Naturally, Poincaré corrected his mathematics, but this led him to recognize a more serious difficulty. He learned that small changes in the initial conditions eventually caused huge differences in the state of the system. If the initial conditions weren't known exactly, no long-term predictions were possible. He wrote,

> When one tries to depict the figure formed by these two curves [representing the orbits] and their infinity of intersections, each of which corresponds to a doubly asymptotic solution, these instructions form a kind of net, web or infinitely tight mesh; neither of the two curves can ever intersect itself, but must fold back on itself in a very complex way in order to intersect all the links of the mesh infinitely often.
>
> One is struck by the complexity of this figure that I am not even attempting to draw.[6]

Computers would eventually produce the figures humans cannot and the latter would dub them **strange attractors**, but that time was still far in the future. For now, what matters is that an error led to the discovery of sensitive dependence on initial conditions and the birth of chaos theory.

Charles Fort vs. the Astronomers

Charles Fort discussed the *n*-body problem in his book *New Lands*.[7] Some quotations follow. He began by comparing astronomers to astrologers. Is this fair? Let's see!

> ...no particular differences can be noted between astrologers and astronomers, and that both represent engulfment in Dark Ages. Lord Bacon pointed out that astrologers had squirmed into prestige and emolument by shooting at marks, disregarding their misses, and recording their hits with unseemly advertisement.

Well, the comparison might be fair if Fort went on to provide evidence of astronomers presenting an imbalanced accounting of their predictions. He first repeated his claim, in different language.

---

4 The special case was that of two bodies of equal mass orbiting each other at opposite points on a circle, with the third body being so light as to have almost no effect on them. See Stewart, Ian, *Significant Figures: The Lives and Work of Great Mathematicians*, Basic Books, New York, p. 198.

5 "The correspondence between Phragmén and Poincaré, as well as the printed copy of the original memoir annotated with Poincaré's corrections, are preserved in the Institut Mittag-Leffler so it is possible to see exactly the extent of the error." – email to the author from June Barrow-Green, May 14, 2018.

6 Quoted here from Barrow-Green, June, *Poincaré and the Three Body Problem*, Vol. 2, American Mathematical Society, Providence, RI, 1997, p. 162. See Mandelbrot, Benoit, *The Fractal Geometry of Nature*, Updated and Augmented, W. H. Freeman and Company, New York, NY, 1983, p. 414 for a slightly different translation. Or see Poincaré, Henri, *New Methods in Celestial Mechanics*, 1892–III, pp. 389–390.

7 Fort, Charles, *New Lands*, Boni & Liveright, New York, 1923. There's a nice annotated edition available online at http://www.resologist.net/landsei.htm.

...our data are oppressed by a tyranny of false announcements; that there never has
been an astronomic discovery other than the observational or the accidental.

Eventually he gets to the heart of the matter. It was found that Uranus's orbit failed to
match what was predicted by the astronomer's mathematical modeling of our solar
system. This isn't very surprising, given that the n-body problem is unsolved. Nevertheless, it was taken as evidence of there being a previously undetected planet beyond
the orbit of Uranus that must be exerting a gravitational influence on that body. This is
the planet that, when finally discovered, was named Neptune.

> In *The Story of the Heavens*, Sir Robert Ball's opinion of the discovery of Neptune is that
> it is a triumph unparalleled in the annals of science. He lavishes–the great astronomer
> Leverrier, buried for months in profound meditations–the dramatic moment–Leverrier
> rises from his calculations and points to the sky – "Lo!" there a new planet is found.

But Fort points out that Leverrier wasn't the only astronomer trying to mathematically
solve the problem.

> According to Leverrier, there was a planet external to Uranus; according to Hansen,
> there were two; according to Airy, "doubtful if there were one."
>   One planet was found–so calculated Leverrier, in his profound meditations. Suppose
> two had been found–confirmation of the brilliant computations by Hansen. None–the
> opinion of the great astronomer, Sir George Airy.
>   Leverrier calculated that the hypothetical planet was at a distance from the sun,
> within the limits of 35 and 37.9 times this earth's distance from the sun. The new
> planet was found in a position said to be 30 times this earth's distance from the
> sun. The discrepancy was so great that, in the United States, astronomers refused
> to accept that Neptune had been discovered by means of calculation: see such
> publications as the *American Journal of Science*, of the period. Upon August 29, 1849,
> Dr. Babinet read, to the French Academy a paper in which he showed that, by
> observations of three years, the revolution of Neptune would have to be placed at
> 165 years. Between the limits of 207 and 233 years was the period that Leverrier
> had calculated. Simultaneously, in England, Adams had calculated. Upon Septem-
> ber 2, 1846, after he had, for at least a month, been charting the stars in the region
> toward which Adams had pointed, Prof. Challis wrote to Sir George Airy that this
> work would occupy his time for three more months. This indicates the extent of
> the region toward which Adams had pointed.

Having given the context of the discovery, Fort noted how it was stripped away from
the scientific histories.

> So it is our expression that hosts of astronomers calculate, and calculation-mad,
> calculate and calculate and calculate, and that when one of them does point within
> 600,000,000 miles (by conventional measurements) of something that is found, he is
> the Leverrier of the text books; that the others are the Prof. Chases not of the text
> books.

The resemblance to the astrologers is clear.

> ...the disproportionality of balancing one Leverrier against hundreds of Chases...

Fort then explained why the astronomers have so much trouble (it's a hard problem!) and showed how history was repeating itself.

> ...We begin with the next to the simplest problem in celestial mechanics: that is the formulation of the inter-actions of the sun and the moon and this earth. In the highest mathematics, final, sacred mathematics, can this next to the simplest problem in so-called mathematical astronomy be solved?"
>
> It cannot be solved.
>
> Every now and then, somebody announces that he has solved the Problem of the Three Bodies, but it is always an incomplete, or impressionistic demonstration, compounded of abstractions, and ignoring the conditions of bodies in space. Over and over we shall find vacancy under supposed achievements; elaborate structures that are pretensions without foundation. Here we learn that astronomers can not formulate the inter-actions of three bodies in space, but calculate anyway, and publish what they call the formula of a planet that is inter-acting with a thousand other bodies. They explain. It will be one of our most lasting impressions of astronomers: they explain and explain and explain. The astronomers explain that, though in finer terms, the mutual effects of three planets can not be determined, so dominant is the power of the sun that all other effects are negligible.
>
> Before the discovery of Uranus, there was no way by which the miracles of the astro-magicians could be tested. They said that their formulas worked out, and external inquiry was panic-stricken at the mention of a formula. But Uranus was discovered, and the magicians were called upon to calculate his path. They did calculate, and, if Uranus had moved in a regular path, I do not mean to say that astronomers or college boys have no mathematics by which to determine anything so simple.
>
> They computed the orbit of Uranus.
>
> He went somewhere else.
>
> They explained. They computed some more. They went on explaining and computing, year in and year out, and the planet Uranus kept on going somewhere else. Then they conceived of a powerful perturbing force beyond Uranus–so then that at the distance of Uranus the sun is not so dominant–in which case the effects of Saturn upon Uranus and Uranus upon Saturn are not so negligible–on through complexes of inter-actions that infinitely intensify by cumulativeness into a black outlook for the whole brilliant system. The palæo-astronomers calculated, and for more than fifty years pointed variously at the sky. Finally two of them, of course agreeing upon the general background of Uranus, pointed within distances that are conventionally supposed to have been about six hundred millions of miles of Neptune, and now it is religiously, if not insolently, said that the discovery of Neptune was not accidental –
>
> That the test of that which is not accidental is ability to do it again –
>
> That it is within the power of anybody, who does not know a hyperbola from a cosine, to find out whether the astronomers are led by a cloud of rubbish by day and a pillar of bosh by night –
>
> If, by the magic of his mathematics, any astronomer could have pointed to the position of Neptune, let him point to the planet past Neptune. According to the same reasoning by which a planet past Uranus was supposed to be, a Trans-Neptunian planet may be supposed to be. Neptune shows perturbations similar to those of Uranus.
>
> According to Prof. Todd there is such a planet, and it revolves around the sun once in 375 years. There are two according to Prof. Forbes: one revolving once in 1,000 years, and the other once in 5,000 years. See Macpherson's *A Century's Progress in Astronomy*. It exists according to Dr. Eric Doolittle, and revolves once in 283 years, (*Sci. Amer.*, 122-641). According to Mr. Hind it revolves once in 1,600 years, (*Smithson. Miscell. Cols.*, 20-20).

Fort closed with

> Let three bodies inter-act. There is no calculus by which their inter-actions can be formulated. But there are a thousand inter-acting bodies in this solar system–or supposed solar system–and we find that the higher prestige in our existence is built upon the tangled assertions that there are magicians who can compute in a thousand quantities, though they cannot compute in three.

The three-body problem has still not been solved.[8] So, it's no surprise that attempts to account for the motions of *all* of the bodies in the solar system, aren't working out either. The reason extra bodies pose such a great complication is due to the fact that the force involved (gravitation) is nonlinear. It obeys an inverse square law. Nonlinear equations are where chaos is likely to appear.

Lessons Learned?

I've devoted a fair amount of space to this for several reasons. It's important that we view science and mathematics from a detached perspective. Loving these subjects doesn't make them more accurate, nor does it make their practitioners infallible. In fact, much of the progress that has been made came by rejecting what the greats of the past had done.

We went from having the Sun orbit the Earth, a fact that seemed completely obvious, to having all of the planets orbit the Sun, in perfect *circles*. Eventually Kepler saw that this too was wrong and changed the orbits to ellipses. Newton lent his authority to the matter by showing that an ellipse is the path implied by his law of universal gravitation. This was considered a mathematical proof, but again it wasn't right. There were perturbations that couldn't be accounted for no matter how the scientists looked at the matter. E.T. Bell explained how the next version of astronomical truth came about.

> According to the Newtonian law of gravitation the orbit of a single (unperturbed) planet is an ellipse with the sun at one focus. In Einstein's theory the orbit is almost an ellipse, but the path does not close after one revolution of the planet. At each revolution the orbit advances by a small amount in the direction of motion of the planet. Although small, the advance is easily measurable in a century. Einstein's theory predicted for the planet Mercury an advance of 43 seconds of arc per century in addition to the advance of 532 seconds per century due to the perturbations of the other planets as calculated on the Newtonian theory. The observed advance is 574 seconds per century. Einstein's correction is only 1 second off the observed discrepancy, well within the limits of observation.[9]

Do you think this is the final word? If so, how lucky of you to be born shortly after the matter was finally settled for good. I think one of the lessons that an honest appraisal of

---

8 There is also the one-body problem, which is being worked on by many single mathematicians, physicists, and astronomers.

9 Bell, Eric Temple, *Mathematics, Queen and Servant of Science*, Copyright, 1951, British Edition 1952, reprinted 1954, 1958, I961, Printed in Great Britain by Butler and Tanner Ltd., Frome and London, p. 216.

scientific history teaches us is to not be arrogant. We make progress by questioning the greats of the past, as well as ourselves.

On a lighter note, *Science Made Stupid* gives the following definition:

three-body problem the problem faced by a triple murderer in hiding the evidence[10]

Predicting the Weather (Or Not)

> In making his seminal discoveries about chaos, Lorenz used one of the most effective methods of investigation in history: discovery by accident.
>
> Edward B. Burger and Michael Starbird[11]

As with Poincaré, an error propelled M.I.T. meteorologist Edward Lorenz (Figure 18.3) into the mathematics of chaos theory. He had been working on weather prediction. One of his models, developed in 1960, consisted of a system of equations in 12 variables, the values of which he entered into a computer to see how the system evolved. Lorenz's computer had an accuracy of six digits after the decimal place, but when he printed the results, he had the machine round to just three. This was to become important in 1961, when Lorenz made his error. He later explained exactly what happened.

FIGURE 18.3
Edward Lorenz (1917–2008).

Courtesy of Kerry Emanuel.

---

10 Weller, Tom, *Science Made Stupid*, Houghton Mifflin Company, Boston, MA, 1985, p. 75.
11 Burger, Edward B. and Michael Starbird, *The Heart of Mathematics*, third edition, Wiley, Hoboken, NJ, 2010, p. 559.

At one point I decided to repeat some of the computations in order to examine what was happening in greater detail. I stopped the computer, typed in a line of numbers that it had printed out a while earlier, and set it running again. I went down the hall for a cup of coffee and returned after about an hour, during which time the computer had simulated about two months of weather. The numbers being printed were nothing like the old ones. I immediately suspected a weak vacuum tube or some other computer trouble, which was not uncommon, but before calling for service I decided to see just where the mistake had occurred, knowing that this could speed up the servicing process. Instead of a sudden break, I found that the new values at first repeated the old ones, but soon afterward differed by one and then several units in the last decimal place, and then began to differ in the next to the last place and then in the place before that. In fact, the differences more or less steadily doubled in size every four days or so, until all resemblance with the original output disappeared somewhere in the second month. This was enough to tell me what had happened: the numbers I had typed in were not the exact original numbers, but were the rounded-off values that had appeared in the original printout. The initial round-off errors were the culprits; they were steadily amplifying until they dominated the solution. In today's terminology, there was chaos.[12]

The image shown in Figure 18.4 overlays Lorenz's original result and the rerun result with the rounded off initial values.

Lorenz, who has referred to himself as "a professional meteorologist and an amateur mathematician,"[13] was not familiar with Poincaré's work and did not expect this sensitive dependence for his equations.[14] He quickly recognized its importance, though. If his model was reasonable, his discovery implied that it's impossible to make accurate long-term forecasts.

Lorenz referred to the divergent results caused by slightly altered initial conditions as the butterfly effect, although he later stated that he had not been thinking of the Ray Bradbury story.[15]

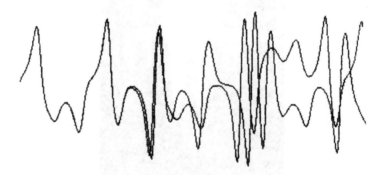

FIGURE 18.4
Lorenz's data.

Courtesy of Ian Stewart.

12 Lorenz, Edward N., *The Essence of Chaos*, University of Washington Press, Seattle, WA, 1993, pp. 134–136.
13 Lorenz, Edward N., *The Essence of Chaos*, University of Washington Press, Seattle, WA, 1993, p. 132.
14 Lorenz, Edward N., *The Essence of Chaos*, University of Washington Press, Seattle, WA, 1993, p. 133.
15 Lorenz, Edward N., *The Essence of Chaos*, University of Washington Press, Seattle, WA, 1993, p. 15.

There were opportunities to discover the butterfly effect well before both Lorenz and Poincaré. For example, Verhulst's model, seen in Chapter 15 has sensitive dependence on both the value of the constant and the initial population, but it was Lorenz's efforts that brought the result to the attention of other researchers in a big way. He went on to find a simpler system of equations that exhibited the same sensitive dependence on initial conditions. It is given below.[16]

$$\frac{dx}{dt} = -\sigma x + \sigma y$$

$$\frac{dy}{dt} = -xz + rx - y$$

$$\frac{dz}{dt} = xy - bz.$$

Lorenz published these equations in 1963 in a paper titled "Deterministic Nonperiodic Flow."[17] Some of Lorenz's graphs illustrated the nonperiodic behavior he had found, where cycles do not overlap, but instead produce odd designs. An example of one of these, viewed from a different angle than presented in his paper, appears as color figure 26 **(See color insert)**. It is a three-dimensional object, depicted in two dimensions. The lines do not intersect, despite what the image seems to indicate.

This is the sort of result that Poincaré had found, concerning which he remarked, "One is struck by the complexity of this figure that I am not even attempting to draw."[18] With the passage of time, technology made renderings possible. In 1971, they were dubbed **strange attractors** in a paper by David Ruelle and Floris Takens.[19] The one in color figure 26 **(See color insert)**, the Lorenz attractor, came to be called the **butterfly attractor**. It's shape resembles a butterfly's wings, circling us back to the Ray Bradbury story. Reflecting on his famous paper years later, Lorenz wrote, "I never guessed at the time that this attractor would for a while, much later, become the feature of the paper that would draw the most attention."[20]

Such attractors, once called "bad curves,"[21] are now known to be completely normal, arising in a wide range of disciplines.

---

16 See Lorenz, Edward N., *The Essence of Chaos*, University of Washington Press, Seattle, WA, 1993, p. 188, or the original paper, Lorenz, Edward N., "Deterministic Nonperiodic Flow," *Journal of Atmospheric Sciences*, Vol. 20, March 1963, pp. 130–141, p. 135 in particular.

17 Lorenz, Edward N., "Deterministic Nonperiodic Flow," *Journal of Atmospheric Sciences*, Vol. 20, March 1963, pp. 130–141.

18 Quoted here from Barrow-Green, June, *Poincaré and the Three Body Problem*, Volume 2, American Mathematical Society, Providence, RI, 1997, p. 162. See Mandelbrot, Benoit, *The Fractal Geometry of Nature*, Updated and Augmented, W. H. Freeman and Company, New York, 1983, p. 414 for a slightly different translation. Or see Poincaré, Henri, *New Methods in Celestial Mechanics*, 1892–III, pp. 389–390.

19 Ruelle, David and Floris Takens, "On the Nature of Turbulence," *Communications in Mathematical Physics*, Vol. 20, 1971, pp. 167–192.

20 Lorenz, Edward N., *The Essence of Chaos*, University of Washington Press, Seattle, WA, 1993, p. 141.

21 Lorenz, Edward N., *The Essence of Chaos*, University of Washington Press, Seattle, WA, 1993, p. 123, citing Cartwright, Mary L. and John Edensor Littlewood, "On Non-Linear Differential Equations of the Second Order: I. the Equation $\ddot{y} - k(1 - y^2)\dot{y} + y = b\lambda k\cos(\lambda t + a)$, k Large," *Journal of the London Mathematical Society*, Vol. s1–20, No. 3, 1945, pp. 180–189.

Lorenz's professional life seems to be characterized by surprises. In his book, *The Essence of Chaos*, he related one surprise that didn't involve any work on his end.

> A few years after chaos had burst into prominence, I received a reprint bearing the curious title "Lorenz Knots are Prime."[22] I had not realized that I had any knots, let alone prime ones, which I had never heard of.[23]

He soon learned that the author "was referring to the butterfly-shaped attractor, and he had shown that some of the closed solution curves that lay in it were knotted."[24]

A Catchy Name!

In 1975, Tien-Yien Li and James A. Yorke (Figure 18.5) introduced the word "Chaos" in the context of systems with sensitive dependence on initial conditions.[25] And so, "chaos

FIGURE 18.5
James A. Yorke (1941–).

Courtesy of Jim Yorke.

22 Williams, R. F., "Lorenz Knots are Prime," *Ergodic Theory and Dynamical Systems*, Vol. 4, No. 1, March 1984, pp. 147–163.
23 Lorenz, Edward N., *The Essence of Chaos*, University of Washington Press, Seattle, WA, 1993, p. 146.
24 Lorenz, Edward N., *The Essence of Chaos*, University of Washington Press, Seattle, WA, 1993, p. 146.
25 Li, Tien-Yien and James A. Yorke, "Period Three Implies Chaos," *The American Mathematical Monthly*, Vol. 82, No. 10, December 1975, pp. 985–992.

theory" came to be a specialization for some mathematicians and scientists. However, these words are not universally loved, and some researchers prefer to say they study "complex dynamical systems."

In the true spirit of science, Yorke has been unafraid to challenge authority. When the U.S. government used an aerial photograph to support their claim of a sparse crowd at a Vietnam War protest held in Washington DC, Yorke didn't accept the story unquestioningly. He looked at the picture and, in particular, the shadow cast by the Washington monument. After applying some trigonometry, Yorke had proof that the photograph had been taken after the protest had ended. Some of the protestors were still there, but the crowd was greatly diminished from what it was at the height of the rally, and the government's attempt to mislead the people was exposed![26]

Laplace and Determinism

Why is sensitive dependence on initial conditions such a big deal? Well, in its own way, it was as much a dream-killer as Gödel's incompleteness theorem. The dream it killed was that of Pierre Simon Laplace (1749–1827) and many other physicists. Laplace had summed it up like so.

> We may regard the present state of the universe as the effect of its past and the cause of its future. An intellect which at a certain moment would know all forces that set nature in motion, and all positions of all items of which nature is composed, if this intellect were also vast enough to submit these data to analysis, it would embrace in a single formula the movements of the greatest bodies of the universe and those of the tiniest atom; for such an intellect nothing would be uncertain and the future just like the past would be present before its eyes.[27]

This describes what came to be known as the clock-work universe. Everything was predetermined by the initial conditions, just like a grandfather clock, whose gears could be relied upon to turn in a predictable way.

A vast intellect attempting to predict the future would have to know the initial conditions with extreme precision, because of sensitive dependence on initial conditions. But the Heisenberg uncertainty principle, put forth by Werner Heisenberg in 1927, showed that some uncertainty in these values must always be present. The more precisely we measure a particle's position, the less certain we can be concerning its momentum, and vice versa. More formally, the product of our uncertainties in regard to position and momentum is bounded below by a constant.

The combination of uncertain initial conditions and systems that are sensitively dependent on said conditions was a one-two knockout punch for the clock-work universe.

---

26 Gleick, James, *Chaos: Making a New Science*, Viking, New York, NY, 1987, p. 65.

27 Laplace, Pierre Simon, *A Philosophical Essay on Probabilities*, translated into English from the original French 6th ed. by Truscott, Frederick Wilson and Frederick Lincoln Emory, Dover, New York, 1951, p.4.

A Gallery of Monsters

> These new functions, violating laws deemed perfect, were looked upon as signs of anarchy
> and chaos which mocked the order and harmony previous generations had sought.
>
> Morris Kline[28]

The overthrow of determinism was not the only problem mathematicians faced. There was also a growing collection of bizarre counterexamples to seemingly reasonable mathematical conjectures. It got so bad that Poincaré complained, "In the old days when people invented a new function they had something useful in mind. Now, they invent them deliberately just to invalidate our ancestors' reasoning, and that is all they are ever going to get out of them."[29] Little did he know that these functions would be found to have deep connections with a field that he created, chaos theory. In the following sections, some of these functions are examined. Poincaré referred to them as a "gallery of monsters."

Cantor's Ternary Set (1874)

Some of Georg Cantor's work was detailed in Chapter 4, but there's much more. The odd set detailed here was presented by Cantor in 1883,[30] although its initial discovery goes back to Henry John Stephen Smith in 1874.[31]

We begin with a line segment (the top of the image shown in Figure 18.6) and then remove the middle third. This gives us the pair of horizontal lines occupying the second position in Figure 18.6. Now, from each of those two shorter lines, we remove the middle third. We're left with four short line segments. Next, remove the middle third of each of these. We get eight even shorter line segments. Continue repeating this process. At every step, the number of (disconnected) line segments doubles, but their lengths are reduced to one-third of their previous lengths.

The limit of this process is the set we are interested in. It consists of infinitely many "infinitely tiny" pieces. To find the total length of these pieces, we look at the complement, that is, the pieces that have been removed, and subtract this from the total original length. If we start with a line segment of length 1, then the lengths of the segments removed at each stage is given by the sequence $\frac{1}{3}, 2\left(\frac{1}{3}\right)^2, 4\left(\frac{1}{3}\right)^3, 8\left(\frac{1}{3}\right)^4, \cdots$ For example, in the fourth round of deletions, there are 8 segments removed, and each is of length $\left(\frac{1}{3}\right)^4$. So, the total length removed is $\frac{1}{3} + 2\left(\frac{1}{3}\right)^2 + 4\left(\frac{1}{3}\right)^3 + 8\left(\frac{1}{3}\right)^4 + \cdots$. Following a bit of algebra, this will be easy to

28 Young, Robert M., *Excursions in Calculus: An Interplay of the Continuous and the Discrete*, The Mathematical Association of America, 1992, p. 325, quoting from Morris Kline, *Mathematical Thought from Ancient to Modern Times*, Oxford University Press, New York, 1972, p. 973.

29 http://www-history.mcs.st-andrews.ac.uk/Quotations/Poincare.html

30 Cantor, Georg, "Über unendliche, lineare Punktmannichfaltigkeiten" (On infinite, linear point-manifolds), *Mathematische Annalen*, Vol. 21, No. 4, 1883, pp. 545–591.

31 Smith, Henry J. S., "On the integration of discontinuous functions," *Proceedings of the London Mathematical Society*, First series, Vol. 6, 1874, pp. 140–153. Or see Mandelbrot, Benoit, *The Fractal Geometry of Nature*, Updated and Augmented, W. H. Freeman and Company, New York, 1983, p. 409 for details of Smith's work.

FIGURE 18.6
Iterating on the way to the Cantor ternary set.

Image created by Craig P. Bauer.

evaluate. First, factor out $\frac{1}{3}$ to get $\frac{1}{3}\left(1 + 2\left(\frac{1}{3}\right)^1 + 4\left(\frac{1}{3}\right)^2 + 8\left(\frac{1}{3}\right)^3 + \cdots\right)$. Next, express the even numbers as powers of 2, like so $\frac{1}{3}\left(1 + 2^1\left(\frac{1}{3}\right)^1 + 2^2\left(\frac{1}{3}\right)^2 + 2^3\left(\frac{1}{3}\right)^3 + \cdots\right)$. Now, because the powers match in each term we can combine them to get $\frac{1}{3}\left(1 + \left(\frac{2}{3}\right)^1 + \left(\frac{2}{3}\right)^2 + \left(\frac{2}{3}\right)^3 + \cdots\right)$. The parentheses can now be seen to contain an infinite geometric series with common ratio $\frac{2}{3}$. Using the formula from Chapter 16, this simplifies to $\frac{1}{3}\left(\frac{1}{1-\frac{2}{3}}\right) = \frac{1}{3}\left(\frac{1}{\frac{1}{3}}\right) = \frac{1}{3}\left(\frac{3}{1}\right) = 1$. That is, the sum of the lengths removed at each iteration equals that of the entire original interval. So, why do we have infinitely many pieces left in the limit when constructing the Cantor set?

Peano Space-Filling Curve (1890)[32]

The next monstrous result was put forth by Giuseppe Peano in 1890 and it's exactly what it sounds like, a curve that completely fills in an area, even though it has no thickness. Peano's four-page paper didn't include any illustrations, but illustrations make the curve much easier to understand, so they're used here.[33] Think of it as the limit of a sequence of curves, the first three terms of which are shown graphically in Figure 18.7.

It's probably not immediately clear to you from these terms, how the next term in this graphical sequence should be drawn. Another pair of images (Figure 18.8) will help.

The left image of Figure 18.8 shows the first term, arises from drawing a line that passes through the centers of the nine squares indicated by the dotted lines. In the image on the right, showing the second term, this shape appears (shrunk down) in the bottom left corner, hitting the centers of one-ninth of the total number of rectangles in the image. The line is then continued to pass through the centers of the nine squares

32 Peano, Giuseppe, "Sur une courbe, qui remplit toute une aire plane," *Mathematische Annalen*, Vol. 36, No. 1, 1890, pp. 157–160.

33 For the first illustration, see Moore, Eliakim Hastings, "On Certain Crinkly Curves," *Transactions of the American Mathematical Society*, Vol. 1, No. 1, January 1900, pp. 72–90.

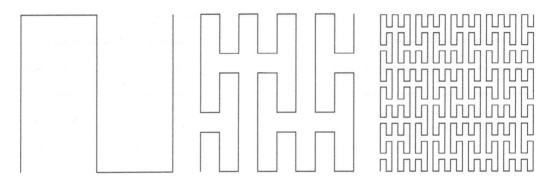

FIGURE 18.7
The first three iterations of a Peano curve.

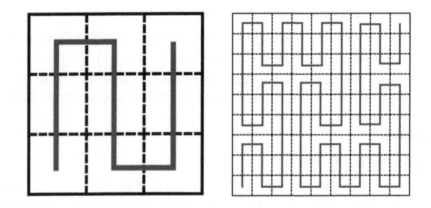

FIGURE 18.8
The first two iterations, with the grids used to generate them.

above the original image. The path it follows is a reflection of the original term. This process is repeated for the next nine squares (in the upper left corner of the second iteration). The curve then continues through three 9 by 9 squares going from the top middle of the image to the bottom middle of the image. Finally, the curve passes through three more 9 by 9 squares going up the right hand side to complete the second iteration.

For the third term, the second image would be shrunk down to occupy the left and bottom thirds of a square, and the process described above is repeated.

Taking the limit of this sequence, the large square is filled, by a line. In this monstrous result, a line seems to have area! What inspired Peano to create such a crazy curve? It was Cantor's proof that a line segment contains the same number of points as appear in a square (see Chapter 4).

Everything had come unstrung! It is difficult to put into words the effect that Peano's result had on the mathematical world. It seemed that everything was in ruins, that all the basic mathematical concepts had lost their meaning.

N. Ya Vilenkin[34]

The Koch Curve (1904)

I turn away with fright and horror from this lamentable plague of functions which do not have derivatives.

Charles Hermite[35]

The next figure to be examined starts off like Cantor's ternary set in that we remove the middle third of a line segment. However, this time we replace it with two sides of a triangle, and the result is that what was a straight line segment now has three sharp corners. As Figure 18.9 shows, this process is repeated for each new straight line

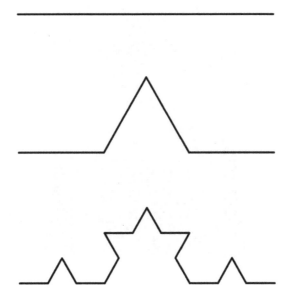

FIGURE 18.9
Initial iterations leading to Koch's curve.

Public domain.

34 Quoted here from Mandelbrot, Benoit, *The Fractal Geometry of Nature*, Updated and Augmented, W. H. Freeman and Company, New York, 1983, pp. 58–59.

35 Excerpted from a May 20, 1893 letter from Charles Hermite to Thomas Stieltjes. See Hermite, Charles and Thomas Stieltjes, *Correspondance d'Hermite et de Stieltjes*, edited by Baillaud, B., and H. Bourget, Vol. II, Gauthier-Villars, Paris, 1905, p. 318, https://archive.org/details/correspondanced01bourgoog.

segment. Continuing indefinitely, the limiting value of this sequence of curves is called Koch's curve.

This strange curve was presented by Helge von Koch (Figure 18.10), a Swedish mathematician, in 1904.[36] His purpose was to create a continuous curve that didn't have a tangent line at any of its points. Just as the absolute value function has a sharp corner at $x = 0$ that prevents it from having a tangent line (or a derivative) at that point, so Koch's curve has sharp corners *everywhere* and thus cannot have a tangent line at *any* point.

Koch's curve was not the first of this type. Karl Weierstrass, for example, found a continuous curve with no tangents decades before Koch, but Koch's example is easier to construct and understand. Weierstrass's example was

$$W(x) = \sum_{k=0}^{\infty} a^k \cos(b^k \pi x)$$

for $0 < a < 1$, b an odd integer, and $ab > 1 + 3\pi/2$.[37]

FIGURE 18.10
Helge von Koch (1870–1924).

Public domain.

---

36 Koch, Helge von, "Sur une courbe continue sans tangente, obtenue par une construction géométrique élémentaire," (On a continuous curve without tangents, constructible from elementary geometry), Arkiv för matematik, astronomi och fysik, Vol. 1, 1904, pp. 681–704.

37 Weierstrass presented this to the Berlin Academy in 1872. It was first published by du Bois-Reymond in 1875. However, Riemann apparently had a similar result, which he used in lectures, as early as 1861. Embracing all of this and Koch's work, at least temporally, was Bernard Bolzano, who had an example in about 1830, which didn't see print until 1922. These historical details were found in Thim, Johan, *Continuous Nowhere Differentiable Functions*, Master Thesis, Department of Mathematics, Luleå University of Technology, December 2003 (see the abstract).

Weierstrass, in 1857, defended such pure mathematics as follows.

> To the question of whether it is really possible to extract something useful from the abstract theories that modern mathematics seems to favor, one could answer that it was only on the basis of pure speculation that Greek mathematicians derived the properties of conic sections, long before one could guess that they represent the planets' orbits.[38]

Turning back to Koch's simpler example, if instead of starting with a line segment, we begin with an equilateral triangle, the process generates a pretty image known as Koch's Snowflake. It's shown in Figure 18.11, with the center filled in.

As an exercise, you are asked to show that this figure has an infinite perimeter and a finite area.

The three-dimensional version of the Koch snowflake is formed in the obvious way, start with a tetrahedron and then place smaller tetrahedrons on each face, repeatedly. However, we get something unexpected in the limit – a cube. The sides are perfectly smooth![39]

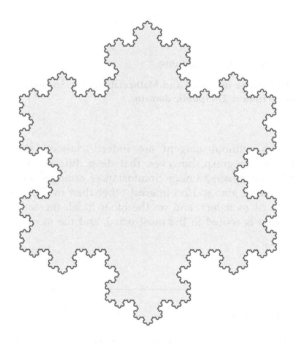

FIGURE 18.11
Koch's Snowflake.

Public domain.

---

38 Mandelbrot, Benoit, *The Fractal Geometry of Nature*, Updated and Augmented, W. H. Freeman and Company, New York, 1983, p. 421. Mandelbrot's source for this quote was Hilbert, David, *Gesammelte Abhandlungen*, Vol. 3, Springer, Berlin, 1932, pp. 337–338.

39 Morris, Scot, "Games," *OMNI*, Vol. 11, No. 2, November 1988, pp. 124–125.

The Koch curve isn't as artificial as you might think. Such "crinkly curves," as they were once called, arise naturally, as the following snippet from an article in the 1933 volume of *The Scientific Monthly* notes.

> It has even been suggested that possibly we deal in nature frequently with such crinkly curves. If we consider the barometric pressure as a function of the time, we get a zigzag curve like Fig. 7, and if we enlarge the curve, then, within the limits of accuracy of the recording instrument, we still have the same kind of zigzag curve. Have we really any guarantee that, however much we enlarge the scale, our curve will ever reduce to a decent smooth curve?[40]

The accompanying Figure 7 is reproduced here as Figure 18.12.

A deeper understanding of these crinkly curves was still decades away.

FIGURE 18.12
A graph of barometric pressure as a function of time.

Kempner, Aubrey J., "The Paradox in Nature and Mathematics," *The Scientific Monthly*, **Vol. 37, September 1933, pp. 207–217, image from p. 214 (public domain).**

> Examples of curves without tangent are indeed classical since Riemann and Weierstrass. Anyone can grasp, however, that deep differences exist between, on the one hand, a fact established under circumstances arranged for the enjoyment of the mind, with no other aim and no interest other than to show its possibility, an exhibit in a gallery of monsters, and on the other hand, the same fact as encountered in a theory that is rooted in the most usual, and the most essential problems of analysis.[41]
>
> Jacques Hadamard, 1912

Sierpiński's Triangle (1915)

The Polish mathematician Wacław Sierpiński pursued a path similar to Koch, in creating his first famous figure. He began with a simple shape consisting of three straight line segments, as shown in Figure 18.13.

---

40 Kempner, Aubrey J., "The Paradox in Nature and Mathematics," *The Scientific Monthly*, Vol. 37, September 1933, pp. 207–217, p. 214 quoted here.
41 Hadamard, Jacques, "L'œuvre mathématique de Poincaré," *Acta Mathematica*, Vol. 38, 1912, pp. 203–287, quoted portion from a translation given in Mandelbrot, Benoit, *The Fractal Geometry of Nature*, Updated and Augmented, W. H. Freeman and Company, New York, 1983, p. 415.

FIGURE 18.13
A simple shape to get Started.

Each straight piece was then replaced by a scaled-down version of the entire image. The replacements were positioned so that they stay within a bounding triangle and do not overlap or cross each other. A few iterations are shown in Figure 18.14.

FIGURE 18.14
Sierpiński's triangle, as it was first derived.

Thus, the triangle seen in Chapters 10 and 14 debuted as a curve.[42] It's sometimes referred to as Sierpiński's gasket,[43] but as it did not originally arise from removing sections from a two-dimensional figure, I find calling it Sierpiński's triangle more appropriate (and more aesthetically pleasing).

However, if you prefer, the triangle can be obtained by starting with a filled-in triangle and removing an inverted triangle from its center, leaving three triangles behind, and then repeating the removal process with the new triangles, ad infinitum. The first few iterations of this process are shown in Figure 18.15.

FIGURE 18.15
A more "gasket-like" approach to realizing Sierpiński's triangle.

Public domain.

Sierpiński's Carpet (1916)[44]

In a paper published a year after the introduction of his famous triangle, Sierpiński used the "removal method," but not with a triangle. He performed the operations described above on a square. To begin, imagine a square divided into nine smaller squares, as in the technique used to get started on the Peano curve. But all we do in this instance is remove the center square. Then, for each of the remaining eight squares, divide them up into nine squares and remove the center squares. Continuing this process indefinitely, we get what has become known as **Sierpiński's carpet**. An illustration of the first few iterations of the process is given in Figure 18.16.

This object, in the limit, has an infinite boundary, but no area.

---

42 Sierpiński, Wacław, "Sur une courbe dont tout point est un point de ramification," *Comptes Rendus*, Paris, Vol. 160, 1915, pp. 302–305. Reprinted in Sierpiński, Wacław, 1974–, *Oeuvres Choisis*, Vol. II, Ed. S. Hartman et al. Warsaw: Éditions scientifiques, pp. 99–106, available online at http://plouffe.fr/simon/math/Sierpinski%20Oeuvres%20Choisies%20II.pdf.

43 Mandelbrot, Benoit, *The Fractal Geometry of Nature*, Updated and Augmented, W. H. Freeman and Company, New York, 1983, p. 463 mentions that Mandelbrot coined the term "gasket" in this context in English, without a French equivalent.

44 Sierpiński, Wacław, "Sur une courbe cantorienne qui contient une image biunivoque et continue de toute courbe donnée," Comptes Rendus, Paris, Vol. 162, 1916, pp. 629–632. Reprinted in Sierpiński 1974–, II, 107–119. This gives the carpet. Reprinted in Sierpiński, Wacław, 1974–, *Oeuvres Choisis*, Vol. II, Ed. S. Hartman et al. Warsaw: Éditions scientifiques, pp. 107–119, available online at http://plouffe.fr/simon/math/Sierpinski%20Oeuvres%20Choisies%20II.pdf.

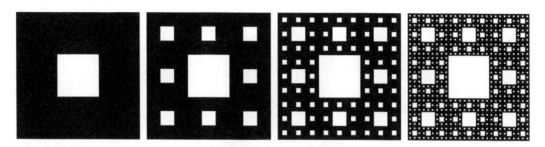

FIGURE 18.16
Sierpiński Carpet.

Image created by Cody Spath.

The Menger Sponge (1926)

The three-dimensional version of the Sierpiński carpet was first described (although not yet accompanied by an illustration) in 1926.[45] It's called the Menger sponge (Figure 18.17), after its discoverer, Austrian mathematician Karl Menger (Figure 18.18). It has infinite surface area, but no volume.

The iterations are carried out in the same manner as for the two-dimensional case, but are performed on all six faces[46] and the removed areas cut all the way through the cube.

FIGURE 18.17
Menger sponge (after four iterations).

Image created by Niabot. This file is licensed under the Creative Commons Attribution 3.0 Unported license.

---

45 Menger, Karl, "Allgemeine Räume und Cartesische Räume. I.", *Communications to the Amsterdam Academy of Sciences*,1926, English translation in Edgar, Gerald A., editor, *Classics on Fractals*, Addison-Wesley Publishing Company, Reading, MA, 1993, pp. 103–116.
46 Actually, three faces are enough, as long as you select three that can be seen at the same time. Cutting through the cube will produce the same effect as if the operation had also been performed from the opposite sides.

Nonmathematical Challenges

Sierpiński's mathematical works have gained him immortality, but his life was far
from easy.

Waclaw Sierpiński (Figure 18.19) attended school in Poland during a period of
Russian occupation. The Russians forced their language and culture upon the Poles,
while discouraging learning. However, life was to become far worse years later. In 1939,
the Nazis overran Poland. In a lecture, Sierpiński related some of the horrific events
from the years that followed.

> In July 1941 one of my oldest students Stanislaw Ruziewicz was murdered. He was
> a retired professor of Jan Kazimierz University in Lvov, the last rector of Foreign
> Trade Academy in Lvov, an outstanding mathematician and an excellent teacher.
>
> In 1943 one of my most distinguished students Stanislaw Saks was murdered. He was
> an assistant professor at Warsaw University, one of the leading experts in the world in
> the theory of the integral, the author of its monograph, which was published in Polish,
> French and in English.
>
> In 1942 another student of mine Adolf Lindenbaum was murdered. He was the
> assistant professor of Warsaw University, a distinguished author in the field of the set
> theory.
>
> It does not end the war losses of Polish mathematics. Besides the above mentioned
> more friends of mine were murdered. They are professor Hoborski, professor
> A. Łomnicki, professor Wl. Stozek, professor K. Bartel, dr A. Rajchman, dr Juliusz
> Schauder, dr Herman Auerbach, dr Jacob. Some of my friends died in connection
> with war events: Samuel Dickstein-nestor of Polish mathematicians, professor of
> honour of Warsaw University, who devoted at least two thirds of the century to

FIGURE 18.19
Waclaw Sierpiński (1882–1969).

Public domain.

scientific and pedagogic work, professor Witold Wilkosz, professor Stefan Kempisty, professor Antoni Przeborski, dr Stefan Kwietniewski, professor Stanislaw Zaremba.

I must add that one of our mathematicians is still in Germany in the concentration camp and the whereabouts of some other assistant professors is still not known.

Thus more than half of the mathematicians, who lectured in our academic schools were killed. It was a great loss for Polish mathematics, which was developing favourably in some fields such as set theory and topology and it took the lead in the world mathematics. The only hope to develop Polish mathematics is the new talents among our young people. We hope they will devote their talents and their work to studying mathematics.

Except the lamented personel losses Polish mathematics suffered in consequence of German barbarity during the war it also suffered material losses. They burned down Warsaw University Library which contained several thousand volumes, magazines, mathematical books and thousands of reprints of mathematical works by different authors.

Nearly all the editions of Fundamenta Mathematicae (32 volumes) and ten volumes of Mathematical Monograph were completely burned down. Private libraries of all the four professors of mathematics from Warsaw University and also quite a number of manuscripts of their works and handbooks written during the war were burnt down, too.[47]

The Nazis burned Sierpiński's house down in 1944.

---

47 Rotkiewicz, A., "W. Sierpiński's Works on the Theory of Numbers," *Rendiconti del Circolo Matematico di Palermo*, Vol. 21, No. 1–2, January 1972, pp. 5–24, pp. 14–15 quoted here, available online at https://slideheaven.com/w-sierpiskis-works-on-the-theory-of-numbers.html.

Somehow, despite severe oppression at different points in his life from Communists and Nazis, Sierpiński managed to write 724 papers and 50 books.[48] This made him one of the most prolific mathematicians. Perhaps mathematics is part of what got him through some of his hardships. It can provide an escape of sorts.

All of the objects described in the "Gallery of Monsters" sections were considered pure mathematics, but that was to change.

Benoit Mandelbrot – A Scientific Maverick

The title of maverick was not one forced on Mandelbrot (Figure 18.20) by others. Rather, he was proud to claim the designation himself.[49] In his book, *The Fractal Geometry of Nature*, Mandelbrot quoted William James on other scientific mavericks. A fuller version of the quoted material follows below.

> "The great field for new discoveries," said a scientific friend to me the other day, "is always the Unclassified Residuum." Round about the accredited and orderly facts of every science there ever floats a sort of dust-cloud of exceptional observations, of occurrences minute and irregular, and seldom met with, which it always proves more easy to ignore to than to attend to. The ideal of every science is that of a closed and completed system of truth. The charm of most sciences to their more passive disciples consists in their appearing, in fact, to wear just this ideal form. Each one of our

FIGURE 18.20
Benoit Mandelbrot (1924–2010).

Image created by Rama. This file is licensed under the Creative Commons Attribution-Share Alike 2.0 France license.

---

48 http://www-groups.dcs.st-and.ac.uk/~history/Mathematicians/Sierpinski.html
49 See Mandelbrot, Benoit, *The Fractal Geometry of Nature*, Updated and Augmented, W. H. Freeman and Company, New York, 1983, pp. 391–392. Also, his posthumously published autobiography was titled *The Fractalist: Memoir of a Scientific Maverick.*

various *ologies* seems to offer a definite head of classification for every possible phenomenon of the sort which it professes to cover; and so far from free is most men's fancy, that, when a consistent and organized scheme of the sort has once been comprehended and assimilated, a different scheme is unimaginable. No alternative, whether to whole or parts, can any longer be conceived as possible. Phenomena unclassifiable within the system are therefore paradoxical absurdities, and must be held untrue. When, moreover, as so often happens, the reports of them are vague and indirect; when they come as mere marvels and oddities rather than as things of serious moment, – one neglects or denies them with the best of scientific consciences. Only the born geniuses let themselves be worried and fascinated by these outstanding exceptions, and get no peace till they are brought within the fold. Your Galileos, Galvanis, Fresnels, Purkinjes, and Darwins are always getting confounded and troubled by insignificant things. Any one will renovate his science who will steadily look after the irregular phenomena. And when the science is renewed, its new formulas often have more of the voice of the exceptions in them than of what were supposed to be the rules.[50]

Mandelbrot clearly saw himself as kin to the scientists James singled out and followed his quote of James with

This Essay, whose ambition is indeed to renew the Geometry of Nature, relies upon many puzzles so unclassified that they are only published when the censors nod.[51]

There were occasionally mathematicians who described what were usually referred to as monsters in far more positive terms. For example, N. Ya Vilenkin included the Koch curve, Sierpiński carpet, Menger sponge, Peano curve, and more in a chapter of his book, *Stories about Sets*, titled "Remarkable Functions and Curves, or a Stroll through a Mathematical Art Museum."[52] However, it was Benoit Mandelbrot who tamed the monsters by examining them both mathematically and in the context of physical phenomena that were also not understood. His work unified the unexplained topics in a way that offered a foothold for the ascent to comprehension. The key was **fractals**.

Mathematics is the art of giving the same name to different things.

Henri Poincaré[53]

This is exactly what Mandelbrot did when he coined the term fractal in 1975, although he resisted giving the term a precise definition until 1977.

In a later edition of the book in which he gave the definition, Mandelbrot noted, "Although the term *fractal* is defined in Chapter 3, I continue to believe that one would

---

50 Mandelbrot quoted portions of this in Mandelbrot, Benoit, *The Fractal Geometry of Nature*, Updated and Augmented, W. H. Freeman and Company, New York, 1983, p. 28. The fuller quote given here is taken from James, William, *The Will to Believe: And Other Essays in Popular Philosophy*, Longmans Green and Co., New York, 1897, pp. 299–300.

51 Mandelbrot, Benoit, *The Fractal Geometry of Nature*, Updated and Augmented, W. H. Freeman and Company, New York, 1983, p. 28. Charles Fort also took it as his mission to draw attention to the "Unclassified Residuum," but he was not a scientist and was unable to initiate a revolution in scientific thought through his actions.

52 Vilenkin, N. Ya, *Stories about Sets*, Academic Press, New York and London, 1968. This is an English translation of the original Russian edition from 1965.

53 Poincaré, Henri, translated by George Bruce Halsted, *The Foundations of Science*, The Science Press, New York and Garrison, New York, 1913, p. 375.

do better without a definition (my 1975 Essay included none)."[54] So, please don't get hung up on the technical details of the definition that follows!

Mandelbrot used two different measures of dimension to define a fractal:

> A **fractal** is by definition a set for which the Hausdorff Besicovitch dimension strictly exceeds the topological dimension.[55]

There is no need to define the Hausdorff Besicovitch dimension here. It suffices to understand that it needn't be an integer, whereas the topological dimension must be. That is, fractals have something to do with objects that have fractional dimension.

Another, less formal (and less precise), definition of a fractal is an object that has self-similarity at different scales. As an example, consider the Sierpiński triangle. If we have three such triangles, we may place two of them side by side, so that their bases form a straight line, and then place the third on top, with its bottom left and bottom right corners resting on the peaks of the first two triangles. The result is simply a larger version of the original triangle.

So, for the Sierpiński triangle, we have a larger version produced by 3 copies, each of which is ½ as large (in terms of the length of a side) as the new creation. Mandelbrot used these numbers, 3 and ½, to calculate its fractal dimension, D. The relevant formula, which can be applied to many more such objects, is $D = \log(C)/\log(1/M)$, where C is the number of scaled down copies needed to make a larger copy and M is the size of the smaller copy in comparison to the larger. For the Sierpiński triangle $C = 3$ and $M = ½$, so we have $D = \log(3)/\log(2) \approx 1.585$. Note: D is also known as the similarity dimension.

The fractal dimension formula arises from taking something that is discrete and making it continuous. To see this, consider some nonfractal examples.

Four copies of a square can be assembled to make a larger square (Figure 18.21). The original squares have length ½, compared to the larger square. That is, $C = 4$ and $M = ½$.

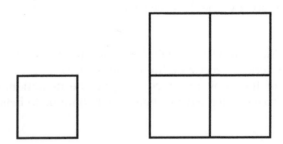

FIGURE 18.21
Combining squares to make a larger square.

Image created by Craig P. Bauer.

---

54 Mandelbrot, Benoit, *The Fractal Geometry of Nature*, Updated and Augmented, W. H. Freeman and Company, New York, 1983, p. 361.
55 Mandelbrot, Benoit, *The Fractal Geometry of Nature*, Updated and Augmented, W. H. Freeman and Company, New York, 1983, p.15. Page 31 has "I propose that curves for which the fractal dimension exceeds the topological dimension 1 be called *fractal curves*." See p. 4 for the origin of the word fractal.

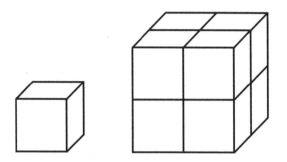

FIGURE 18.22
Combining cubes to make a larger cube.

Image created by Craig P. Bauer.

Eight copies of a cube can be assembled to make a larger cube (Figure 18.22). The original cubes have length ½, compared to the larger cube. We have, $C = 8$ and $M = ½$. With the pair of data points derived above, we can make a small table.

Object	Dimension, D	Scaling Factor, M	Number of Copies, C
Square	2	½	4
Cube	3	½	8

For both of our data points, the relationship $C = (1/M)^D$ is satisfied. While it's tricky to continue with an example in four dimensions, it isn't hard to add a one-dimensional example to our table. You are asked to do this in the exercises. It also satisfies $C = (1/M)^D$.

To solve for D in the equation above, we begin by taking the logarithm of both sides, $\log(C) = \log((1/M)^D)$. This gives, $\log(C) = D \cdot \log(1/M)$, which allows us to solve for D. We get $D = \log(C)/\log(1/M)$.

While the examples of the square and the cube have integer dimensions, the formula derived above is a generalization that can be used on examples where the value of D turns out to be a fraction, as we saw in the case of Sierpiński's triangle. The formula helps to unify the strange geometric objects shown earlier in this chapter. In the exercise set, you're asked to use it to calculate their fractal dimensions.

There are also exercises that ask you to find connections between different fractals. There are many examples of this. One connection is given by a Peano curve Mandelbrot found that fills the Koch snowflake.[56] A bigger connection is that between fractals and strange attractors. Not only are strange attractors fractals, when the intersection of

56 Mandelbrot, Benoit, *The Fractal Geometry of Nature,* Updated and Augmented, W. H. Freeman and Company, New York, 1983, p. 68.

a strange attractor is taken with the right manifold, a Cantor set results.[57] So, chaos theory is seen to connect with fractal geometry.[58]

All of this unification is what is meant when mathematicians say that Mandelbrot tamed the monsters. Connections were also found to many other topics not thought of as part of the gallery of monsters. Indeed, the subject has easy connections with nearly every chapter of this book. Only a few examples can be detailed here, but you're encouraged to explore more on your own.

Mandelbrot explained how fractals tied in to the ancient debate, detailed in Chapter 0, on the nature of the universe – continuous or discrete?

> Note that the Cantor dust may be seen as defusing an ancient paradox: it is divisible without end but is not continuous.[59]

The Greeks had thought that infinite divisibility implied continuity and this was used as an argument in favor of a continuous universe. Cantor dusts show that this argument is not correct.

A connection with Chapter 13, in which tilings were considered, is given by a fractal known as a dragon curve. It's actually an example of a Peano curve.[60] These objects have infinite perimeter and finite area and they can be fit together perfectly to tile the plane, as indicated in color figure 27 **(See color insert)**. There are other ways to tile the plane with this unusual shape.[61]

Chapter 15 included a graph of the bifurcations of the discrete logistic equation. Edward Lorenz described a connection between this graph and a Cantor set.

> The values of A [this is how Lorenz denoted the constant in the equation] for which the behavior is periodic form an infinite number of finite intervals, while the values for which it is chaotic lie between 3.57 and 4.0 and form a Cantor set whose dimension appears to be 1.0. The Cantor set resembles the one that would be formed by taking an interval of unit length, removing the middle fourth, then removing the middle ninth of the resulting pieces, then the middle sixteenth of the four pieces, etc. Here the sum of the removed pieces is only ½.[62]

Some applications of fractals are presented at the end of this chapter, but for now, so that you can see that such work did have tangible benefits, for at least two people, consider the size of the prize described in the article reproduced in Figure 18.23.

---

57 Lorenz, Edward N., *The Essence of Chaos*, University of Washington Press, Seattle, WA, 1993, p. 212 uses such an intersection as part of the definition: "**Strange attractor.** An attractor with a fractal structure; one whose intersection with a suitable manifold is a Cantor set." Also see Mandelbrot, Benoit, *The Fractal Geometry of Nature*, Updated and Augmented, W. H. Freeman and Company, New York, 1983, p. 197. Mandelbrot called them fractal attractors. On p. 198 Mandelbrot expresses his personal dislike for the word "strange" in this context. Given that such fractal attractors arise all over the place, can they really be considered strange? They're more like the norm!

58 Lorenz, Edward N., *The Essence of Chaos*, University of Washington Press, Seattle, WA, 1993, p. 176. Pages 176–178 detail how to measure the fractal dimension of strange attractors and give connections with the Cantor set.

59 Mandelbrot, Benoit, *The Fractal Geometry of Nature*, Updated and Augmented, W. H. Freeman and Company, New York, 1983, p. 406. Also see p. 81.

60 Mandelbrot, Benoit, *The Fractal Geometry of Nature*, Updated and Augmented, W. H. Freeman and Company, New York, 1983, pp. 66–67.

61 Examples are given at https://en.wikipedia.org/wiki/Dragon_curve.

62 Lorenz, Edward N., *The Essence of Chaos*, University of Washington Press, Seattle, WA, 1993, p. 194.

March 2003 **FOCUS**

Benoit Mandelbrot and James Yorke Win Japan Prize

Benoit B. Mandelbrot, 78, Sterling Professor of Mathematical Sciences, Yale University; and James A. Yorke, 61, Distinguished University Professor of Mathematics and Physics, University of Maryland, have won the Japan Prize. They will share the $412,000 prize money from the Science and Technology Foundation of Japan.

The Japan Prize recognizes "original and outstanding achievements that contribute to the progress of science and technology and the promotion of peace and prosperity of mankind."

Mandelbrot has been called the "father of fractals." In 1993, when he won the Wolf Prize for Physics, Mandelbrot was cited for "having changed our view of nature." Physicist Michael V. Berry has written that "fractal geometry is one of those concepts which at first sight invites disbelief but on second thought becomes natural that one wonders why it has only recently been developed." John Wheeler (the Princeton physicist) wrote that "no

Benoit B. Mandelbrot

one will be considered scientifically literate tomorrow who is not familiar with fractals."

"Fifty years ago" said Mandelbrot, "when I began to study complexity for its own sake, I was very lonely," Mandelbrot said. "Today, it is the theme of this great prize and I am utterly delighted to be chosen as a recipient. Early on, I became a wanderer-by-choice between the disciplines, and between theory and applications. Electing to live as a constant maverick, I allowed my interests to move in and out of mathematics, in and out of physics, of economics, or diverse other fields of physical and social sciences, and even music and art. I showed that very simple formulas can generate objects that exhibit an extraordinary wealth of structure. Lately, I have also been very active in college and high school education. I feel extraordinarily privileged that my professional life has continued long enough to allow me to merge every one of my activities into a reasonable beginning of a science of roughness."

James Yorke has been called "Dr. Chaos," that is, the one who found the universal mechanism underlying nonlinear phenomena, naming it chaos.

Early in his academic career, James Yorke quickly earned a reputation as a man

with an unpredictable mind. "He thinks very unconventionally," says Edward Ott, a professor of physics and electrical engineering at the University of Maryland, who has collaborated with

James Yorke

Yorke. "When I'm talking to him in the hall sometimes, he'll say something that seems completely bizarre to me. Then I'll go away and I'll think, 'Hey, that was very good.'"

In 1975, Yorke published the math paper that made him famous. It was called "Period Three Implies Chaos," and it gave a name to the emerging new field that was thereafter known as chaos theory.

In his best-selling 1987 book, *Chaos: Making a New Science*, author James Gleick summed up the revolutionary effect of Yorke's paper: "Yorke had offered more than mathematical result. He had sent a message to physicists: Chaos is ubiquitous; it is stable; it is structured."

More information about the award and this year's winners can be found at http://www.japanprize.jp/English.htm.

FIGURE 18.23
A benefit of research.

Fractals in Nature

Science having long suffered from Euclid's barrenness in models for the unsmooth patterns of nature, the fact that fractals release us from unquestionable inappropriateness was reason to rejoice.

Benoit Mandelbrot[63]

63 Mandelbrot, Benoit, *The Fractal Geometry of Nature*, Updated and Augmented, W. H. Freeman and Company, New York, 1983, pp. 226–227.

In Chapter 13, Fibonacci numbers were defined and followed by details of various places they appear in nature. Similarly, once fractals were recognized, mathematicians began seeing them everywhere. The next few sections point out some examples.

Fractal Food!

Broccoli is often cited as an example of a fractal. The prettiest variety seems to be Romanesco Broccoli (Figure 18.24).

Some nonedible plants also take fractal forms.

FIGURE 18.24
Romanesco Broccoli.

Courtesy of Ian Stewart. This is just way too pretty to eat.

The Fern

A simple rule that generates the artificial fern seen in Figure 18.25, on the right, was published by Michael Barnsley in his 1988 book *Fractals Everywhere*.[64]

---

64 Barnsley, Michael, *Fractals Everywhere*, Academic Press, Boston, MA, 1988.

FIGURE 18.25
An actual fern (left). A computer-simulated fern (right).

Pratt, Anne, *The ferns of Great Britain, and their allies the club-mosses, pepperworts, and horsetails*, Printed for the Society for Promoting Christian Knowledge, London, 1855. Public domain. (left)

Image created by António Miguel de Campos. Public domain. (right)

Biological Fractals

Living fractals are not limited to plants. They also appear in animals, including us, in the branchings of our lungs. The fractal form of our lungs (Figure 18.26) serves to maximize the all-important surface-area-to-volume ratio mentioned in Chapter 6, making respiration easier. Viewed upside-down, the lungs resemble a tree. Heliotropy does not mean "growing towards the sun," but rather growing in a way that maximizes the amount of sunlight that falls on the leaves. Of course, trees also breathe, so their forms are well-adapted in a pair of ways.

Circulatory systems are also fractals.[65] Writing about fractals in our bodies got Mandelbrot excited enough to express himself in all capitals with "LEBESGUE-OSGOOD MONSTERS ARE THE VERY SUBSTANCE OF OUR FLESH!"[66]

With all of the living fractals, it's interesting that, before the term fractal even existed, a botanist discovered one in the realm of *physics*.

---

65 Mandelbrot, Benoit, *The Fractal Geometry of Nature*, Updated and Augmented, W. H. Freeman and Company, New York, 1983, pp. 149–150.

66 Mandelbrot, Benoit, *The Fractal Geometry of Nature*, Updated and Augmented, W. H. Freeman and Company, New York, 1983, pp. 149–150. Also see p. 159. Okay, I'll admit that the use of all caps was because it was a section header and they all took this form, but Mandelbrot didn't have to use an exclamation mark!

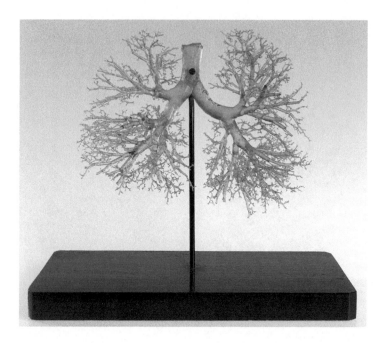

FIGURE 18.26
A cast of human lungs.

Image from Centre for Research Collections University of Edinburgh. This file is licensed under the Creative Commons Attribution-Share Alike 2.0 Generic license. Also see Mandelbrot, Benoit, *The Fractal Geometry of Nature*, Updated and Augmented, W. H. Freeman and Company, New York, 1983, pp. 150 and 157–159.

Brownian motion

Back in 1827, British botanist Robert Brown noticed a strange phenomenon. Under the magnification provided by his microscope, he observed that pollen particles suspended in water seem to move about randomly, and no matter how long he watched, they never came to a stop. An example of what a particle's path could look like is given in Figure 18.27.

What could be causing this weird behavior? In 1905, the correct solution was given by Albert Einstein.[67] He realized that water molecules, too small for Brown to see, but numbering in the tens of millions, were bumping into the pollen particle, from various angles, causing it to move about. Such collisions occur with great frequency. If the position of the pollen particle is plotted at one second intervals, and successive points connected with straight lines, we get a sketch like the one shown above. However, if the position is noted every 1/100 of a second then each straight line

---

67 Einstein, Albert, "Über die von der molekularkinetischen Theorie der Wärme geforderte Bewegung von in ruhenden Flüssigkeiten suspendierten Teilchen," *Annalen der Physik*, Vol. 17, No. 8, 1905, pp. 549–560. Others gave the correct explanation before Einstein, but were more or less ignored. See Heyl, Paul R., *The Scientific Monthly*, June 1934, pp. 493–500, p. 496 relevant here.

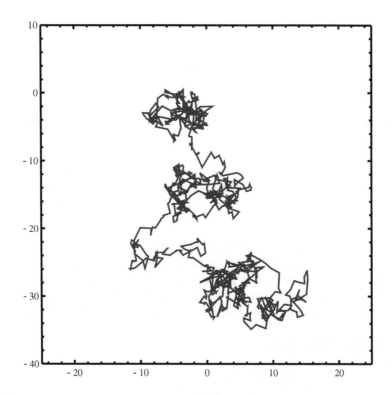

FIGURE 18.27
Brownian motion.

Public domain.

segment in the present sketch would come to resemble the entire image. That is, zooming in, we find that pieces of this particle's path resemble the entire path. It should be noted that the path is not *identical* at different scales. Mandelbrot explained that "different parts of the trail of Brownian motion can never be superposed on each other... Nevertheless, the parts can be made to be superposable in a statistical sense."[68]

Brownian motion turns out to be a naturally occurring space-filling curve. In the unbounded trail produced by Brownian motion, the particle comes arbitrarily close to each point in the plane infinitely often. Interestingly, the probability of any particular point being hit exactly is zero.[69] In any case, the curve traced out in Brownian motion has dimensions $D = 2$ and $D_T = 1$.[70] Because $D > D_T$, one of the definitions for a fractal is satisfied. Jean Perrin compared this natural curve to

---

68 Mandelbrot, Benoit, *The Fractal Geometry of Nature*, Updated and Augmented, W. H. Freeman and Company, New York, 1983, p. 18.

69 Mandelbrot, Benoit, *The Fractal Geometry of Nature*, Updated and Augmented, W. H. Freeman and Company, New York, 1983, p. 235.

70 Mandelbrot, Benoit, *The Fractal Geometry of Nature*, Updated and Augmented, W. H. Freeman and Company, New York, 1983, p. 15. See p. 237 for the details.

the artificial continuous, but nondifferentiable, curves long before Mandelbrot began his work.[71]

Attempts have been made to use Brownian motion to model stock prices.[72]

While fractals have not yet resolved the mystery of the market, they have closed the case on another mystery.

The Border Mystery

> Richardson was a very interesting and original character who seldom thought on the same lines as did his contemporaries, and often was not understood by them.
>
> G. I. Taylor[73]

Lewis Fry Richardson (Figure 18.28) had a broad background, having received his B.A. in physics, mathematics, chemistry, biology, and zoology.[74] Like Lorenz, Richardson investigated weather prediction, but he began his work earlier, during World War I, without a computer, and missed making Lorenz's discoveries. Still, in 1922, parodying Jonathan Swift, he gave a description that ties weather to the world of fractals.

> Big whorls have little whorls
> That feed on their velocity,
> And little whorls have lesser whorls
> And so on to viscosity. – Lewis F. Richardson[75]

In 1926 Richardson even briefly mentioned the Weierstrass function in connection with wind.[76]

During World War I, Richardson also drove an ambulance, but, as a Quaker and a conscientious objector, avoided combat.[77] Afterwards, he set out to understand why

---

71 Mandelbrot, Benoit, *The Fractal Geometry of Nature*, Updated and Augmented, W. H. Freeman and Company, New York, 1983, p. 232.

72 Mandelbrot, Benoit, *The Fractal Geometry of Nature*, Updated and Augmented, W. H. Freeman and Company, New York, 1983, p. 394 discusses the earliest such effort.

73 Quoted in Mandelbrot, Benoit, *The Fractal Geometry of Nature*, Updated and Augmented, W. H. Freeman and Company, New York, 1983, p. 401.

74 Mandelbrot, Benoit, *The Fractal Geometry of Nature*, Updated and Augmented, W. H. Freeman and Company, New York, 1983, p. 401.

75 Richardson, Lewis Fry, *Weather Prediction by Numerical Process*, Cambridge University Press, Cambridge, UK, 1922, p. 66. The original lines by Swift were
   "So, Nat'ralists observ a Flea.
   Hath smaller Fleas that on him prey,
   And these have smaller Fleas to bit 'em.
   And so proceed ad infinitum."
See Swift, Jonathan, *On Poetry, a Rhapsody*, 1733.

76 This is discussed in Mandelbrot, Benoit, *The Fractal Geometry of Nature*, Updated and Augmented, W. H. Freeman and Company, New York, 1983, pp. 402–403.

77 Lorenz, Edward N., *The Essence of Chaos*, University of Washington Press, Seattle, WA, 1993, p. 51 and Mandelbrot, Benoit, *The Fractal Geometry of Nature*, Updated and Augmented, W. H. Freeman and Company, New York, 1983, p. 401.

FIGURE 18.28
Lewis Fry Richardson (1881–1953).

Public domain.

nations go to war. His results could be considered inconclusive. Nothing correlated strongly with war breaking out. The best was a somewhat weak correlation with nations having religious differences. However, in carrying out his research, a new mystery arose. Richardson thought sharing a border might correlate with going to war, so he looked up the lengths of various common borders to see if wars were more frequent when these lengths were longer. Mandelbrot came upon this work and detailed it in his own writing.

> Richardson's search in encyclopedias reveals notable differences in the lengths of the common land frontiers claimed by Spain and Portugal (987 versus 1214 km), and by the Netherlands and Belgium (380 versus 449 km).[78]

It seems like a joke – the smaller nation gives a larger value for the border, as if it has an inferiority complex and has to exaggerate its size! Richardson figured out what was happening. The length depends on the size of the measuring stick used. A smaller stick captures more of the zigs and zigs in the border, yielding a greater total length. Letting the stick length be represented by ε, Mandelbrot noted, "the 20%

---

78 Mandelbrot, Benoit, *The Fractal Geometry of Nature*, Updated and Augmented, W. H. Freeman and Company, New York, 1983, p.33. Also see p. 27 and p. 29, where Mandelbrot states, "I knew of Richardson as a great scientist whose originality mixed with eccentricity." For the original see Richardson, Lewis Fry, "The Problem of Contiguity: An appendix to Statistics of Deadly Quarrels," *General systems Yearbook*, Vol. 6, 1961, pp. 139–187. Mandelbrot, p. 28 mentions other people who recognized the difficulty in measuring coastlines.

differences between these claims can be accounted for by assuming that the ε's differ by a factor of 2, which is not unlikely."[79] Richardson found that as ε heads to zero, the length can grow without bound. Of course, this won't happen if the border is, as in much of the western United States, a straight line. It occurs when the border follows a river or mountain range. It's also seen in the measurement of coastlines. Mandelbrot, having come upon Richardson's result in his wide reading, made it famous in a paper he published in *Science* in 1967 titled "How Long Is the Coast of Britain? Statistical Self-Similarity and Fractional Dimension."[80] The answer is ∞. Another way to approach the question is to give the fractal dimension of the border or coastline, which will be a number between 1 and 2, along with the length using a specified measuring stick size. The same sort of argument can be applied to measuring the course of rivers.[81]

Fractals in Art

Fractals are common in today's art, but they also appeared long before Mandelbrot coined the term in 1975. A great example was produced around 1830–1832 by the Japanese artist Katsushika Hokusai.

Just when you thought it was safe to go in the water ... fractal waves! (Figure 18.29)

While some artistic license was used in the waves depicted by Hokusai, there are many examples of fractals in nature. In fact, fractal geometry forms a better model for nature than Euclidean geometry. As Mandelbrot remarked, "Clouds are not spheres, mountains are not cones, coastlines are not circles, and bark is not smooth, nor does lightning travel in a straight line."[82] The ubiquity of fractals inspired one mathematician to write a short poem.

> Bounded by the fractal coast and the fractal sea,
> I embrace the math that embraces me.[83]

Mathematics is truly everywhere. I often bump into it when I think I'm investigating something else. Reading a book in German (to improve my skills with the language), I came across the following in the section on Alexander Humboldt:

> Vom Kraterrand des Riesenvulkans sah Humboldt die Vegetationsgürtel, die den Berg umgaben, und ein Gesetz der von ihm geschaffenen Pflanzengeographie offenbarte sich ihm. Er erkannte, daß die Vegetationsgürtel eines in den Tropen gelegenen Berges

---

79 Mandelbrot, Benoit, *The Fractal Geometry of Nature*, Updated and Augmented, W. H. Freeman and Company, New York, 1983, p. 33.

80 Mandelbrot, Benoit, "How Long Is the Coast of Britain? Statistical Self-Similarity and Fractional Dimension," *Science*, Vol. 156, No. 3775, pp. 636–638, 1967.

81 Mandelbrot, Benoit, *The Fractal Geometry of Nature*, Updated and Augmented, W. H. Freeman and Company, New York, 1983, p. 225.

82 Mandelbrot, Benoit, *The Fractal Geometry of Nature*, Updated and Augmented, W. H. Freeman and Company, New York, 1983, p. 1.

83 Bauer, Craig, "Exploring Mathematics with the Computer," *PRIMUS*, Vol. 20, No. 1, 2009, pp. 50–89.

FIGURE 18.29
The Great Wave at Kanagawa (Kanagawa Oki Nami Ura), from a Series of Thirty-six Views of Mount Fuji.

Public domain.

vom Fuß des Berges bis dahin, wo der ewige Schnee beginnt, den Vegetationszonen der Erde entsprechen, d.h. den vom Äquator bis zu den Polen hin klimatisch unterscheidbaren Pflanzengruppen.[84]

A translation reads:

From the crater edge of the giant volcano, Humboldt saw the vegetation belts that surrounded the mountain and a law of plant geography revealed itself to him. He recognized that the vegetation belts of a mountain in the tropics, from the foot of the mountain to where the eternal snow begins, correspond to the vegetation zones of the Earth, i.e. the plant groups climatically distinguishable from the equator to the poles.

Humboldt made this observation during a visit to the Canary Island Tenerife. We can describe his discovery as a realization that the earth may be viewed as having the fractal property of self-similarity on different scales. Of course, Humboldt could not have described the earth as fractal-like, as the term had yet to be coined.

Diamonds are distributed fractally through the Earth's crust, as are the quakes that shake it.[85] Zooming out, arguments have been made in favor of a fractal

---

84 Spann, Meno, and C. R. Goedsche, *Deutsche Denker und Forscher*, Appleton-Century-Crofts, New York, 1954, p. 128.
85 Mandelbrot, Benoit, *The Fractal Geometry of Nature*, Updated and Augmented, W. H. Freeman and Company, New York, 1983, p. 94 (diamonds) and p. 461 (earthquakes).

universe, in terms of the distribution of stars and galaxies.[86] And fractal telescope arrays are used by today's astronomers to further our knowledge of the universe.[87]

At the other extreme, that of the microcosm, the influence of fractals has also been felt. For example, the spectra of some organic molecules have a stunning resemblance to Cantor dust.[88] Also, Richard Feynman, who pops up in several chapters of this book, found fractal trees useful in visualizing and modeling high energy particle collisions.[89] Quantum particles appear to follow continuous nondifferentiable paths, like those in Brownian motion.[90]

The irony of monsters from the world of pure mathematics being found in nature was not lost on the American physicist Freeman Dyson. In a review of Mandelbrot's *The Fractal Geometry of Nature*, he wrote,

> A great revolution of ideas separates the classical mathematics of the 19th century from the modern mathematics of the 20th. Classical mathematics had its roots in the regular geometric structures of Euclid and the continuously evolving dynamics of Newton. Modern mathematics began with Cantor's set theory and Peano's space-filling curve. Historically, the revolution was forced by the discovery of mathematical structures that did not fit the patterns of Euclid and Newton. These new structures were regarded by contemporary mathematicians as "pathological." They were described as a "gallery of monsters," kin to the cubist painting and atonal music that were upsetting established standards of taste in the arts at about the same time. The mathematicians who created the monsters regarded them as important in showing that the world of pure mathematics contains a richness of possibilities going far beyond the simple structures that they saw in nature. Twentieth-century mathematics flowered in the belief that it had transcended completely the limitations imposed by its natural origins.
>
> Now, as Mandelbrot points out with one example after another, we see that nature has played a joke on the mathematicians. The 19th-century mathematicians may have been lacking in imagination, but nature was not. The same pathological structures that the mathematicians invented to break loose from 19th-century naturalism turn out to be inherent in familiar objects all around us in nature.[91]

---

86 Mandelbrot, Benoit, *The Fractal Geometry of Nature*, Updated and Augmented, W. H. Freeman and Company, New York, 1983, pp. 91–92.

87 Mandelbrot, Benoit, *The Fractal Geometry of Nature*, Updated and Augmented, W. H. Freeman and Company, New York, 1983, p. 93.

88 Mandelbrot, Benoit, *The Fractal Geometry of Nature*, Updated and Augmented, W. H. Freeman and Company, New York, 1983, p. 81.

89 Mandelbrot, Benoit, *The Fractal Geometry of Nature*, Updated and Augmented, W. H. Freeman and Company, New York, 1983, p. 154, which cites Feynman, Richard P., in Casalbuoni, Roberto, Gabor Domokos, and Susan Kovesi-Domokos, editors, *Proceedings of the Third Workshop on Current Problems in High Energy Particle Theory*, Johns Hopkins University Press, Baltimore, MD, 1979.

90 Mandelbrot, Benoit, *The Fractal Geometry of Nature*, Updated and Augmented, W. H. Freeman and Company, New York, 1983, p. 239, which provides three references.

91 Mandelbrot, Benoit, *The Fractal Geometry of Nature*, Updated and Augmented, W. H. Freeman and Company, New York, 1983, pp. 3–4, which quotes from Dyson, Freeman, "Characterizing Irregularity," *Science*, New Series, Vol. 200, No. 4342, May 12, 1978, pp. 677–678.

The Chaos Game (1988)[92]

Dice go back to ancient Egypt, and may be much older. Yet, in 1988, Michael Barnsley found a brand new game to play with one. Place three points on a plane so that they form an equilateral triangle. Label them a, b, and c. Now start at any of these points and roll a die. If the result is a 1 or a 2, move halfway to a and place another point there. If the result is a 3 or a 4, move halfway to b and place a point there. If the result is a 5 or 6, move halfway to c and place a point there. Now repeat. Continue moving to the midpoint of the last point you placed and the vertex indicated by the die roll. You are now playing the chaos game. The game will go faster if you play it on a computer and the image shown in Figure 18.30 will result.

Sierpiński's triangle arises again! How many different ways are there to generate this fractal?

FIGURE 18.30
The result of the chaos game.

Image created by Josh Gross.

Modified Chaos Games

We can usually find new results by putting small twists on previous work. For example, if we play the chaos game with a larger number of points, we have something known as the

92 Barnsley, Michael F., *Fractals Everywhere*, Academic Press, Boston, MA, 1988.

modified chaos game. We may also color the points based on how many times they are hit. This will reveal more patterns lurking amid the chaos. Using five points (forming a regular pentagon) and a particular coloring scheme, we get the image shown on the cover of this book. So, now you know!

Some very pretty examples of color schemes may be found in Field, Michael, and Martin Golubitsky, *Symmetry in Chaos: A Search for Pattern in Mathematics, Art and Nature*, Oxford University Press, Oxford, 1992. See page 166 for example.

Some of the exercises at the end of this chapter ask you to investigate the modified chaos game with more points, as well as take a closer look at Figure 18.30.

The points one moves to in a modified chaos game needn't be the vertices of a regular polygon, but if you would like them to be, a mathematical question arises: How can we determine the coordinates for each point? It turns out that using complex numbers makes this problem very easy.

The great German mathematician Carl Friedrich Gauss (1777–1855) graphed complex numbers, of the form $a + bi$, by plotting them as ordered pairs (a, b). He believed that "just as objective an existence can be assigned to imaginary as to negative quantities."[93] We will make use of this technique along with a result due to Euler.

Using polar coordinates allows us to find what are known as the nth roots of unity. That is, solutions to $x^n = 1$ for any particular n. We represent (r, θ) by $re^{\theta i}$. This is convenient, because multiplying complex numbers expressed in polar coordinates amounts to multiplying the r values and adding the angles. The nth roots of unity are spaced evenly around the unit circle, starting at 1, so they are given by $e^{2\pi k i/n}$ for $k = 0, 1, 2, ..., n - 1$.

For example, the third roots of unity can be found by letting $n = 3$, while k goes from 0 to 2. That is, e^{0i}, $e^{2\pi i/3}$, and $e^{4\pi i/3}$. If you don't like polar coordinates, you may use Euler's formula $e^{ix} = \cos(x) + i\sin(x)$ to switch back to rectangular coordinates and plot the points as Gauss did.[94]

$$e^{0i} = \cos(0) + i\sin(0) = 1 + 0i$$

$$e^{2\pi i/3} = \cos(2\pi/3) + i\sin(2\pi/3) = -0.5 + 0.866i$$

$$e^{4\pi i/3} = \cos(4\pi/3) + i\sin(4\pi/3) = -0.5 - 0.866i$$

Plotting (1, 0), (−0.5, 0.866), and (−0.5, −0.866) gives three points evenly spaced on the unit circle. These can be taken as the corners of an equilateral triangle. You may rescale and shift to make the unit circle as large as the monitor allows.

---

93 Muir, Jane, *Of Men and Numbers: The Story of the Great Mathematicians*, Dover, NY, 1965, p. 163. Years before Gauss, two other, less well-known, men plotted complex numbers in the same way. They were Caspar Wessel, a Norwegian surveyor, and Jean Robert Argand, a French accountant and bookstore manager. See Maor, Eli, *e: The Story of a Number*, ninth printing with new material on pp. 183–186, Princeton University Press, Princeton, NJ, 1998, p. 166.

94 Euler gave this formula in *Introductio in analysin infinitorum* (1748). Letting $x = \pi$ yields:
$e^{i\pi} = \cos \pi + i \sin \pi = -1$, giving the famous result $e^{i\pi} + 1 = 0$.

Back to Basics

The simplest method for getting a fractal from a random process doesn't require complex numbers or even a die. All you need is a coin and a pair of points labeled 0 and 1. Start at 0 and flip the coin repeatedly. Every time you get heads, move 2/3 of the way from where you are to 1 and mark the point. When you get tails, move 2/3 of the way from where you are to 0 and mark the point. This process generates the Cantor set.

The Mandelbrot Set[95]

> The Mandelbrot Set is an intellectual triumph that spans mathematics and art.
> Edward B. Burger and Michael Starbird[96]

To see this complex figure, which arises from some very simple mathematics, turn to color figure 28 **(See color insert)**. Instructions for generating it follow here.

Pick any point c in the complex plane and calculate: $Z_1 = c^2 + c$, $Z_2 = Z_1^2 + c$, $Z_3 = Z_2^2 + c, ..., Z_{n+1} = Z_n^2 + c, ...$[97]

If Z_n remains bounded as n approaches infinity, then point c is in the Mandelbrot set, and is colored black in color figure 28 **(See color insert)**. The boundary of the black area is a fractal with dimension 2. It's considered to be "almost space-filling."[98] Note that objects with fractional dimension are fractals, but there are also fractals with whole number dimension, of which this is an example. This is part of why Mandelbrot was so reluctant to give a definition for fractal.

The different colors used for various points represent the number of iterations it takes those points until their distances from the origin exceed $\sqrt{5}$. Any point for which this happens will not be in the Mandelbrot set.[99] The idea of using colors in this manner came from John H. Hubbard of Cornell University.[100]

There are a great number of beautiful pictures that can be produced by zooming in on various sections of the Mandelbrot set. See color figures 29 and 30 **(See color insert)**. In these close-ups, the points in the Mandelbrot set are yellow. Such artistic choices can be made by the programmer. When Mandelbrot first generated images of his famous set, they were nowhere near as beautiful as those here. He claimed to have first done so in late 1979, when computer graphics were much more primitive than they

---

95 At this point it should come as no surprise that there were others, before Mandelbrot, for whom the set could be named. For details see Horgan, John, "Mandelbrot Set-To," *Scientific American*, Vol. 262, No. 4, April 1990, pp. 30, 32–34. Nearly 20 years later, this piece was posted online with a title change: Horgan, John, "Who Discovered the Mandelbrot Set?" *Scientific American*, March 13, 2009, https://www.scientifica merican.com/article/mandelbrot-set-1990-horgan/.

96 Burger, Edward B. and Michael Starbird, *The Heart of Mathematics*, third edition, Wiley, Hoboken, NJ, 2010, p. 459.

97 Or you can start with $Z_0 = 0$ and iterate $Z_{n+1} = Z_n^2 + c$. The same values will be generated.

98 Stewart, Ian, *Significant Figures: The Lives and Work of Great Mathematicians*, Basic Books, NY, p. 266.

99 http://www.math.utah.edu/~alfeld/math/mandelbrot/mandelbrot.html

100 Young, Robert M., *Excursions in Calculus: An Interplay of the Continuous and the Discrete*, The Mathematical Association of America, 1992, p. 175 footnote

are today.[101] He first published a variant of the set shown here in late 1980.[102] He put forth the version detailed here in 1982.[103]

There are other ways to look at it. For example, complex numbers can be eliminated by replacing z with $x + yi$ and c with $a + bi$ and simplifying. The result is

$$x_{n+1} = x_n^2 + y_n^2 + a,$$

$$y_{n+1} = 2x_n y_n + b.$$

This is somewhat similar in appearance to the equations that generate the gingerbreadman.

A Lovely Pattern

Zooming in slightly on a portion of the Mandelbrot set (Figure 18.31) allows a lovely pattern to be illustrated.

The image shown in Figure 18.31 clearly shows some of the "bulbs" that surround the main portion of the Mandelbrot set. Looking at the largest bulb, the top portion resembles lightning, although it branches out from a central point in three directions. We can label the branches, 1, 2, 3, as we move in a counterclockwise direction, beginning with the first branch after the one that connects to the main body of the Mandelbrot set. This is shown in Figure 18.32 for clarity.

The shortest branch (other than the one connecting to the main portion of the set) is branch 1. That is, the shortest branch is number 1 of 3. Hence, we refer to the 1/3 bulb. This method assigns every bulb a unique fraction strictly between 0 and 1.

Amazingly, the fractions representing bulbs include *every* fraction between 0 and 1, and they are all in reduced form. That is, there's a 1/3 bulb, but no 2/6 bulb. Also, every fraction being represented means, for example, that there's a 75,689/104,729 bulb. I wouldn't want to count them, but this bulb has 104,729 branches and branch 75,689 is the shortest. These fractions are all represented *in numerical order* by bulbs positioned around the perimeter of the Mandelbrot set.

Returning to Figure 18.31, the large bulb on the left corresponds to the fraction 2/5. Now that I've identified the 1/3 bulb and the 2/5 bulb, we can determine the value of the largest bulb between these two by adding fractions the way we wanted to when we first encountered them in elementary school. That is, 1/3 + 2/5 = 3/8. Just add the tops and add the bottoms! The largest bulb between the 1/3 bulb and the 2/5 bulb is the 3/8 bulb. This special kind of addition, known as **Farey addition**, always works in this

---

101 Horgan, John, "Mandelbrot Set-To," *Scientific American*, Vol. 262, No. 4, April 1990, pp. 30, 32–34. Nearly 20 years later, this piece was posted online with a title change: Horgan, John, "Who Discovered the Mandelbrot Set?" *Scientific American*, March 13, 2009, https://www.scientificamerican.com/article/mandelbrot-set-1990-horgan/.

102 Mandelbrot, Benoit, "Fractal aspects of the iteration of $z \rightarrow \lambda z(1-z)$ for complex λ and z," *Annals of the New York Academy of Sciences*, Vol. 357, No. 1, December 26, 1980, pp. 249–259.

103 Horgan, John, "Mandelbrot Set-To," *Scientific American*, Vol. 262, No. 4, April 1990, pp. 30, 32–34 p. 33 cited here. Nearly 20 years later, this piece was posted online with a title change: Horgan, John, "Who Discovered the Mandelbrot Set?" *Scientific American*, March 13, 2009, https://www.scientificamerican.com/article/mandelbrot-set-1990-horgan/.

FIGURE 18.31
A view of the Mandelbrot set emphasizing some of the bulbs.

Cropped version of an image created by Wolfgang Beyer with the program *Ultra Fractal 3*. This file is licensed under the Creative Commons Attribution-Share Alike 3.0 Unported license.

context. The largest bulb between the 2/5 bulb and the 3/8 bulb is the 5/13 bulb. Now, look closely at the numbers in the numerators and denominators of the sequence of fractions I'm generating. They are all Fibonacci numbers!

Farey addition is named after John Farey, Sr. (1766–1826), a British geologist. It was found to be useful long before the Mandelbrot set was discovered. It even comes in handy for today's sports fans. In baseball, a batter who goes 2 for 5 in one game, and 1 for 3 in the next, has gone 2/5 + 1/3 = 3/8, or 3 for 8 over the course of both games.

Although the above might seem playful, the Mandelbrot set is an object of very serious mathematical study and has led to at least two Fields Medals.[104]

Not Just for Textbooks...

In 1991, the Mandelbrot set appeared in the form of a crop circle 10 miles south of Cambridge, England. When Mandelbrot heard about this, he said he was very pleased

104 Stewart, Ian, *Significant Figures" The Lives and Work of Great Mathematicians*, Basic Books, New York, p. 266.

FIGURE 18.32
Labeling a bulb's branches.

Cropped and altered version of an image created by Wolfgang Beyer with the program *Ultra Fractal 3*. This file is licensed under the Creative Commons Attribution-Share Alike 3.0 Unported license.

to hear of the theory "taking root."[105] Another professor observed that the crop circle representation wasn't perfect. Therefore, he concluded, it couldn't have been made by a mathematician. Maybe, he speculated, it was someone from the engineering department.

Mathematical Connections

Almost all of the topics covered in this book connect with each other. The few that don't probably actually do, just in ways we haven't been clever enough to discover yet.

Figure 18.33 shows how the Mandelbrot set exhibits period doubling with a ratio that converges to Feigenbaum's constant, bringing us back to Chapter 15. You are led to make the connection a bit stronger in Exercise 39 at the end of this chapter.

The sharpness of the image of the Mandelbrot set is determined by how many iterations are performed. In 1991, David Boll was attempting to show that the neck (where the heart-shaped piece meets the largest circular piece) of the Mandelbrot set is

---

105 Davis, Beth, editor, *Ciphers in the Crops: The Fractal and Geometric Circles of 1991*, Gateway Books, Bath, UK, 1992, p.10.

FIGURE 18.33
The Mandelbrot set and a bifurcation diagram.

Image created by Georg-Johann Lay. Public domain.

ε	Number of iterations
1	3
0.1	33
0.01	315
0.001	3143
0.0001	31,417
0.00001	314,160
0.000001	3,141,593
0.0000001	31,415,928

infinitely thin, when he made a cool discovery. The table above, giving the number of iterations required to get the thickness less than various ε values, makes the connection with π crystal clear.[106]

In 2001, Aaron Klebanoff proved that the pattern continues forever.[107]

---

106 Boll, David, "Pi, and the Mandelbrot Set." Archived at https://web.archive.org/web/20140317091957/ https://home.comcast.net/~davejanelle/mandel.html.

107 Klebanoff, Aaron, "π in the Mandelbrot Set," *Fractals, Complex Geometry, Patterns, and Scaling in Nature and Society*, Vol. 9, No. 4, December 2001, available online at http://www.pi-e.de/PDF/mandel.pdf.

One measures a circle beginning anywhere.

Charles Fort

Julia Sets

Equations very similar to the one used to generate the Mandelbrot set were investigated by Gaston Julia long before Mandelbrot. However, without computers to uncover their great beauty, they were largely forgotten by mathematicians, with the notable exception of Mandelbrot (!), until the late 1970s, when Mandelbrot's work brought attention back to them.

Earlier in this chapter, mention was made of Richardson's attempts to determine the causes of war and the probabilities of their occurring in the future. Gaston Julia (Figure 18.34) saw the horrors of war first hand. In World War I, he was nearly blinded and suffered the loss of his nose, an event that was followed by painful operations. Afterwards, he hid his severe disfigurement under a piece of leather. Julia had already made a reputation in mathematics prior to the war and, through all of his miseries, he continued his work.[108]

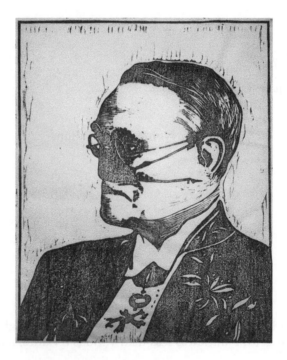

FIGURE 18.34
Gaston Julia (1893–1978).

Gaston Julia, woodblock print by Morgen Bell, 2008.

---

108 http://www.educ.fc.ul.pt/icm/icm2003/icm14/Julia.htm and https://www.britannica.com/biography/Gaston-Maurice-Julia

Julia began his work on the sets that would be named after him with a prize-winning paper published in 1918.[109] Weighing in at 199 pages, one could call it a book.

A Julia set is generated in a manner almost identical to that used for the Mandelbrot set. The only difference is that the same $+ c$ is used every time. That is, we fix c and then, for each Z_0 in the complex plane, we calculate: $Z_1 = Z_0^2 + c$, $Z_2 = Z_1^2 + c$, $Z_3 = Z_2^2 + c$, ..., $Z_{n+1} = Z_n^2 + c$, ... If the iterations beginning with Z_0 stay bounded (do not move arbitrarily far from the origin), then we say Z_0 is in the Julia set for the point c. While there's just one Mandelbrot set, there's a Julia set for each complex number. Three examples are given in Figures 18.35–18.37. In these images, points in the Julia set are colored black, while those outside the Julia set are white. Depending on the value of c, a wide variety of images may result. Some Julia sets can be used like the inkblots in a Rorschach test. What do you see in the following?

FIGURE 18.35
Julia set for $c = -0.123 + 0.745i$.

Image created by Matthew Jones.

---

109 Julia, Gaston, "Mémoire sur l'itération des fonctions rationnelles," *Journal de mathématiques pures et appliquées*," Vol. 8, No. 1, 1918, pp. 47–246. It won the Grand Prix from the French Academy of Sciences.

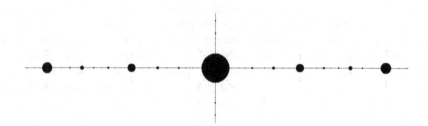

FIGURE 18.36
Julia set for $c = -1.7549$.

Image created by Matthew Jones.

FIGURE 18.37
Julia set for $c = -1$.

Image created by Matthew Jones.

Not all Julia sets are so exciting. For example, the plain-old circle is a Julia set. As an exercise, you're asked to find a value of c that demonstrates this. On the plus side, this means there's a very easy connection between at least one Julia set and π.

Answers to Julia–Rorshach test:

$c = -0.123 + 0.745i$ is "the rabbit."

$c = -1.7549$ is "the airplane."

$c = -1$ is "the basilica" (including its reflection in the water).[110]

Kathryn A. Lindsey (Figure 18.38) took a different approach to Julia sets. Instead of looking at a Julia set and giving it a name based on what it resembled, she drew pictures and then found Julia sets that matched them.[111]

Figures 18.39–18.42 are some of her examples.

---

110 Mandelbrot called this name "a mathematician's wild extrapolation." He preferred calling it "San Marco Dragon." See Mandelbrot, Benoit, *The Fractal Geometry of Nature*, Updated and Augmented, W. H. Freeman and Company, New York, 1983, p. 185.

111 Lindsey, Kathryn A., "Shapes of Polynomial Julia Sets," *Ergodic Theory and Dynamical Systems*, Vol. 35, No. 6, September 2015, pp. 1913–1924.

FIGURE 18.38
Kathryn A. Lindsey (1984–).

Courtesy of Kathryn A. Lindsey.

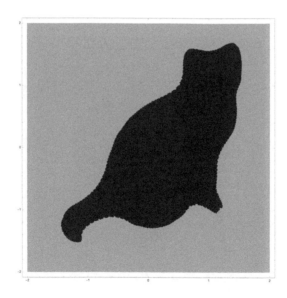

FIGURE 18.39
"The cat" Julia set.

Courtesy of Kathryn A. Lindsey.

FIGURE 18.40
"The butterfly" Julia set.

Courtesy of Kathryn A. Lindsey.

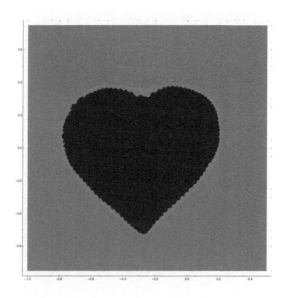

FIGURE 18.41
"The heart" Julia set.

Courtesy of Kathryn A. Lindsey.

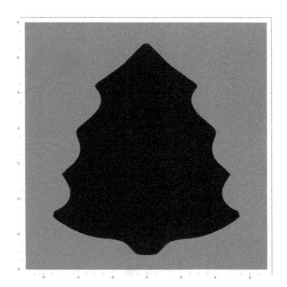

FIGURE 18.42
"The Christmas tree" Julia set.

Courtesy of Kathryn A. Lindsey.

In a nutshell, what Lindsey showed was that for any closed curve, there's a polynomial whose Julia set matches it as closely as desired. That is, an arbitrarily small value can be chosen (any $\varepsilon > 0$) and the border of the Julia set will never be more than that value away from the closed curve. She also showed that shapes with holes (also in the form of closed curves) can be obtained as Julia sets as well, but for these polynomials do not suffice. It is rational functions that do the trick.

In another direction, Julia sets can be made more attractive by using a wider range of colors, as was done for the Mandelbrot set. Examples are given in color figures 31 and 32 **(See color insert)**.

More Connections

There are much deeper connections between Julia sets and the Mandelbrot set than the similar ways in which they are generated and their common fractal appearance. The Mandelbrot set is a sort of roadmap to the Julia sets, a way of understanding all of them at once. To see the connections, we must look at Julia sets more closely.

Some Julia sets are connected. That is, there's a path between any two points in the Julia sets that stay entirely within the Julia set. Other Julia sets are disconnected (Figure 18.43).

It turns out that the Julia set for c is connected if and only if c is in the Mandelbrot set. An equivalent way of saying this is, the Julia set for c is connected if and only if $Z_0 = 0$ is in that Julia set. That is, if and only if $Z_0 = 0$ is bounded under iteration of $Z_{n+1} = Z_n^2 + c$.

Robert DeVaney made a nice video illustrating some of this. It shows the Julia sets corresponding to a point moving through the Mandelbrot set. The video can now be

FIGURE 18.43
(Disconnected) Julia set for $c = 0.22710406638167324 - 0.5316489047745387i$.

Image created by Josh Gross.

found on YouTube.[112] While the graphics are not up to today's standards, the narration provided by Robert Devaney makes it worth viewing. Newer, more sophisticated animations are also now available online.[113]

The Mandelbrot set and the infinite number of Julia sets are just the beginning. Similar images can be made by looking at other difference equations, such as $Z_{n+1} = Z_n^3 + c$. We can play these fractal image games with any polynomial. And if you want to see what happens with trig functions, logarithmic functions, exponential functions, etc., you can get an idea by using their Taylor polynomial approximations. Go explore!

Newton's Method

Computers didn't just breathe new life into Gaston Julia's work, they reached all the way back to the 1600s and Isaac Newton.

Newton's method, previously seen in Chapter 17, is a technique used to approximate the roots of a differentiable function, $f(x)$. We start with a guess, x_0 and then calculate new values via the difference equation

$$x_{n+1} = x_n - \frac{f(x_n)}{f'(x_n)}.$$

---

112 At https://www.youtube.com/watch?v=2r9Zrbgtr_g, for example.
113 See https://www.youtube.com/watch?v=DWASyGtvatY, for example.

Repeating this process, treating each new value as the guess for the next iteration, will often take us closer and closer to a root. However, this process doesn't always work. If our initial guess, x_0, is a max or min of the function, we're in trouble, because $f'(x_n) = 0$, causing division by zero in the difference equation.

With the computer we may very rapidly carry out this process using millions of points in the complex plane as our initial guesses. If each root is assigned a color, we may then color each point representing a guess with the color of the root the method converges to for that value. We use black if the method does not converge for the fixed number of iterations that is performed. Color figure 33 **(See color insert)** shows the results for $x^4 = 1$ on the interval from −3 to 3 for both the real and imaginary components of $a + bi$.

There are several more complicated coloring schemes that result in prettier images. For example, we may use shades of the given colors to indicate how quickly points converge to the roots those colors represent.

The example used here was $x^4 = 1$, but there appear to be an infinite number of functions that lead to interesting images when Newton's method is applied in the manner described above. John H. Hubbard even found a function for which Newton's method produces an image of the Mandelbrot set![114]

Fractal images can be shared with friends and family, who may well appreciate their beauty, whether they're interested in the mathematics behind them or not. But the study of such things began long before such images could be generated. Arthur Cayley (1821–1895), for example, examined Newton's method in the complex plane back in the 19th century.[115]

Pascal's triangle was seen to produce fractals in Chapter 10, so when a new triangle (Stirling's) was examined in Chapter 11, it was natural to carry out the same steps and, indeed, fractals arose again. In what new directions can we take the idea of fractals associated with Newton's methods? We can, of course, look at different equations, as was indicated above, but a bigger change would be to use a different root approximation algorithm. There are several, Newton's is just the best known.

A picture generated using the "iteration method"[116] of root approximation for $x^3 - 1 = 0$ is given as color figure 34 **(See color insert)**. As above, points colored black failed to converge after the fixed number of iterations used in the program. They may converge for a larger number of iterations. Because there's a lot more black than in the image for Newton's method, we can tell at a glance that this method doesn't succeed quickly for as wide a range of initial values.

Clifford A. Pickover, a prolific and entertaining author, devoted Chapter 15 of his book *Computers, Pattern, Chaos and Beauty: Graphics from an Unseen World* to root approximation algorithms and their associated fractal images.[117]

---

114 Horgan, John, "Mandelbrot Set-To," *Scientific American*, Vol. 262, No. 4, April 1990, pp. 30, 32–34 p. 34 cited here. Nearly 20 years later, this piece was posted online with a title change: Horgan, John, "Who Discovered the Mandelbrot Set?" *Scientific American*, March 13, 2009, https://www.scientificamerican.com/article/mandelbrot-set-1990-horgan/.

115 Young, Robert M., *Excursions in Calculus: An Interplay of the Continuous and the Discrete*, The Mathematical Association of America, 1992, p. 172.

116 Details of this method can be found at https://web.archive.org/web/20020203000033/http://www.ping.be/math/root.htm.

117 Pickover, Clifford A., *Computers, Pattern, Chaos and Beauty: Graphics from an Unseen World*, St. Martin's Press, New York, 1990.

Image Compression

While the ultimate role that fractals and chaos theory will play in the science of the future is yet to be determined, some useful applications have already been found. One of these is image compression. This is part of a larger area known as data compression. Being able to reduce the size of files, whether they be image, sound, video, or text files, can effectively increase storage space, allow larger files to be attached to emails, and decrease download time. In order to better understand the advantages of the image compression application, a few definitions are needed.

Lossless compression refers to methods that reduce the size of a file in a way that can be undone without any loss of data. That is, the algorithm takes a file of some specific size, reduces it to a smaller size, and when the file is decompressed it is *identical* to the original.

Lossy compression refers to methods in which data will be lost when the file is compressed. That is, the decompressed file will resemble the original, but will not be identical to it.

The advantage of lossy compression is that it often allows for much greater compression. And sometimes we really don't care about the lost data. For example, audio files can be compressed by eliminating the frequencies that humans can't hear. While this is a loss, it's not one that's likely to be noticed. A possible exception is offered by The Beatles' *Sgt. Pepper's Lonely Hearts Club Band* album. It includes a dog whistle at the end of *A Day in the Life*.[118] The intent was that some fans would notice their dogs reacting every time the album ended, but not know why.

Fractal data compression algorithms fall in the category of lossy compression, but they offer another advantage for image files besides a high degree of compression. These algorithms work by finding portions of the image that are self-similar at different scales and saving representations of them in the form of equations. When the file is decompressed, it can be decompressed to any size. If the desired size is larger than the original, values can be interpolated in the equations, between what were the original pixels. For a normal file, stored as color values for a two-dimensional array of pixels, zooming simply enlarges the pixels. The effect, known as pixilation, is to lose the illusion of continuity in the image. It becomes visibly discrete and, as one further enlarges it, the pixels grow bigger and bigger. Hence, a regular image that isn't of very high resolution may look terrible if it's enlarged to poster size. An image that was stored as a set of equations, however, may look fine. The interpolated values are completely fictitious, as they did not exist in the original image, but they will be in the style of the original. For example, if the image is of a person's face, slightly different skin tones will be seen, as is normal. They will not be the exact variations that the subject's face exhibits, but they will look natural.

Please see the references for the exact details of how fractal data compression works. One of the pioneers in this area is Michael Barnsley, whose simulated fern and chaos game were described earlier in this chapter.

---

118 Arnold, Brandy, "VIDEO: A Beatles Song That Only Dogs Can Hear!" *The Dogington Post*, July 20, 2016, http://www.dogingtonpost.com/video-a-beatles-song-that-only-dogs-can-hear/.

More Applications

Some three-dimensional images also have connections with fractals, namely holograms. If one takes only a small piece of the hologram and discards the rest, the entire image is still present! Admittedly, the resolution is decreased, but this is still impressive.[119]

Traditional antennae are optimized for a particular frequency. An adjustable antenna allows the user to change this frequency, but depending on the application, adjustment is not always practical. Antennae with self-similar designs, that is, fractal antennae, can offer better performance at a variety of frequencies, without adjustment. They can also be made more compact.[120] This is separate from arrays of antennae laid out in fractal patterns. Fractal telescope arrays were mentioned very briefly earlier in this chapter.

Open Problem

An easily stated open problem is "Is the Mandelbrot set pathwise-connected?" At the risk of oversimplifying, this is basically asking whether, given any two points in the Mandelbrot set, it is always possible to draw a path from one to the other that never leaves the set. It has been proven that the Mandelbrot set is connected,[121] but being pathwise-connected, although sounding similar, is actually a stronger requirement than being connected. That is, there are sets that are connected, but not pathwise-connected.

Exercise Set 18

> There are no solved problems; there are only problems that are more or less solved.
>
> Henri Poincaré[122]

1. As of this writing there are various "predictions" (distance from the sun, period, etc.) for a new planet X in our solar system. What are they? Are they in close agreement or do they vary widely? Make use of as many sources as you can and compare the results.

2. What does the latest evidence suggest in regard to the stability (or lack thereof) of our solar system?

3. Consider the following system of difference equations in polar coordinates, known as the standard map, put forth by Russian physicist Boris Chirikov.

$$r_{n+1} = r_n + a \cdot \sin(\theta_n)$$

---

119 Brown, Ronald, *LASERS Tools of Modern Technology*, Doubleday, Garden City, NY, 1968, p. 108.

120 Wikipedia contributors, "Fractal antenna," *Wikipedia, The Free Encyclopedia*, https://en.wikipedia.org/w/index.php?title=Fractal_antenna&oldid=837695170.

121 Douady, Adrien and John Hubbard, "Itération des polynômes quadratiques complexes," *Comptes Rendus*, Paris, Vol. 294, 1982, pp. 123–126. This is also the paper in which the set was named after Mandelbrot.

122 Taken here from Young, Robert M., *Excursions in Calculus: An Interplay of the Continuous and the Discrete*, The Mathematical Association of America, 1992, p. 94.

$$\theta_{n+1} = \theta_n + r_{n+1}$$

Can you find a fixed point?

4. Investigate the standard map with a computer program. Relate whatever you discover.

5. What is the fractal dimension of Cantor's ternary set?

6. How many gaps are there in Cantor's ternary set after the nth iteration?

7. If we start out with a line segment of length one and form a Cantor set by repeatedly removing the middle fourth of each line segment, what length is removed at each iteration and what length is left after each iteration? What is the final result?

8. Consider the variant of the Cantor set for which 1/4 of each line segment is removed at each step. What is the dimension of this object?

9. If we start out with a line segment of length one and form a Cantor set by repeatedly removing the middle third of each line segment, what length is removed at each iteration and what length is left after each iteration? What is the limiting result of this process?

10. What is the fractal dimension of Koch's curve?

11. How many straight line segments are there in the nth iteration of Koch's snowflake?

12. How many corners are there in the nth iteration of Koch's snowflake?

13. Prove that Koch's snowflake has an infinite perimeter and a finite area.

14. How many solid (filled in) triangles are there in the nth iteration of Sierpiński's triangle?

15. What is the area of Sierpiński's triangle after the nth iteration? What is its final area?

16. Removing the first square from the center of Sierpiński's carpet takes away 1/9 of the total area. Continuing on through the endless iterations, what percentage is removed in total?

17. What is the fractal dimension of Sierpiński's carpet?

18. What is the fractal dimension of the object generated by the process shown in Figure 18.44?

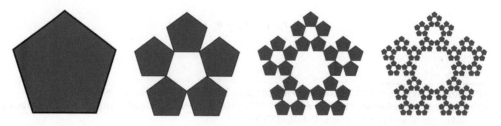

FIGURE 18.44
Sierpiński Pentagon.

Image created by Josh Gross.

19. What is the fractal dimension of the Menger sponge?

20. If you start to generate a Menger sponge from a unit cube, how much volume is removed at each iteration?

21. Can you intersect a line with a Menger sponge to get a Cantor set? If so, how? If not, why not?

22. Start with a solid (filled-in) square. Divide it into four squares of equal size and discard the square in the upper left. Now do the same for each of the three remaining squares. If this process is repeated forever, what will the resulting image look like?

23. Earlier in this chapter the following table was constructed

Object	Dimension, *D*	Scaling Factor, *M*	Number of Copies, *C*
Square	2	½	4
Cube	3	½	8

What values should be used if an entry is to be added for "Line"?

24. Make another new entry for the table in the exercise above, this time for "Triangle." Does it fit in or does it contradict the formula the other lines led to?

25. Dragon curves can be used to tile the plane. Find another example of a fractal that offers a tiling.

26. Give a pair of examples of fractals that arise in nature that were not included in this chapter.

27. Give a pair of examples of fractals in art that were not included in this chapter.

28. What are the fourth roots of unity? Express them in both polar and rectangular coordinates.

29. What are the fifth roots of unity? Express them in both polar and rectangular coordinates.

30. What are the sixth roots of unity? Express them in both polar and rectangular coordinates.

31. Write a program to investigate what happens if you play the modified chaos game with four points, located at the corners of a square.

32. Write a program to investigate what happens if you play the modified chaos game with six points, located at the corners of a regular hexagon.

33. Repeat the problem above, but now color code the pixels that are hit by how often they are hit. The exact details of this step (i.e., how often pixels need to be hit to change colors) are left to you. Is the result prettier?

34. Prove that iterating the chaos game can never result in a point being placed inside the central inverted triangle.

35. Describe a doubling phenomenon that arises in the modified chaos game played on a regular pentagon.

36. For the equation that generates the Mandelbrot set, prove that replacing z with $x + yi$ and c with $a + bi$, and simplifying, gives

$$x_{n+1} = x_n^2 + y_n^2 + a,$$

$$y_{n+1} = 2x_n y_n + b.$$

37. Does $Z_{n+1} = Z_n{}^2 + (1 + i)$ have a fixed point? Either find it or explain why there isn't one.

38. Investigate the periodicity of the Mandelbrot sets for real values of c, both positive and negative. Connect your result with another topic from this book.

39. Start with the logistic equation, $X_{n+1} = AX_n(1 - X_n)$ and substitute $c = \frac{A}{2} - \frac{A^2}{4}$ and $Z = \frac{A(1-2X)}{2}$. What is the result after simplifying?

40. Prove that the result of Farey addition is always somewhere between the two numbers that were added together.

41. What shape is the Julia set for the constant 0? Hint: this offers an easy connection between a Julia set and π.

42. What shape is the Julia set for the constant $3/5 + (4/5)i$?

43. Is the Julia set for the constant i connected? Don't just answer "yes" or "no." Give a proof.

44. Prove or disprove: all Julia sets are bounded.

45. Prove or disprove: every Julia set has at least one fixed point.

46. Write a program to produce a fractal image, as was done for Newton's method, but use Halley's method for root approximation instead. Which function you investigate is up to you, but if the result isn't interesting please try again. The difference equation for Halley's method is

$$x_{n+1} = x_n - \frac{2f(x_n)f'(x_n)}{2[f'(x_n)]^2 - f(x_n)''(x_n)}$$

47. Explain the error in the following quote from a character in Michael Crichton's novel *Timeline*.

    We compress it by using a lossless fractal algorithm.[123]

48. Explain the error in the following quote from Dr. Ian Malcolm, the mathematician in Michael Crichton's novel *Jurassic Park*.

    Physics has had great success at describing certain kinds of behavior: planets in orbit, spacecraft going to the moon, pendulums and springs and rolling balls, that sort of thing. The regular movement of objects. These are described by what are called linear equations, and mathematicians can solve those equations easily. We've been doing it for hundreds of years.[124]

---

123 Crichton, Michael, *Timeline*, paperback edition, Ballantine Books, New York, 1999, p. 137.
124 Crichton, Michael, *Jurassic Park*, Ballantine Books, New York, paperback edition, 1991, pp. 73–74.

References/Further Reading

> You don't have to burn books to destroy a culture. Just get people to stop reading them.
> – Ray Bradbury.

Abrahams, Marc, editor, *The Best of Annals of Improbable Research*, W.H. Freeman & Company, 1998. This is science humor. In one paper, the authors apologize to France for excessive rainfall caused by a butterfly in Lausanne, Switzerland.

Bader, Michael, *Space-Filling Curves, an Introduction with Applications in Scientific Computing*, Springer, Berlin, Germany, 2013.

Bai-lin, Hao, *Chaos*, World Scientific, Singapore, 1984. This is a collection of reprinted articles.

Bai-lin, Hao, *Chaos II*, World Scientific, Singapore, 1990. This is a collection of reprinted articles.

Barnsley, Michael, "A Better Way to Compress Images," *BYTE*, January 1988.

Barnsley, Michael, *Fractals Everywhere*, Academic Press, Boston, MA, 1988.

Barnsley, Michael and Arnaud E. Jacquin, "Application of Recurrent Iterated Function Systems to Images," *SPIE*, Vol. 1001, 1988, pp. 122–131.

Barnsley, Michael F. and Lyman P. Hurd, *Fractal Image Compression*, A. K. Peters, Wellesley, MA, 1993.

Berger, J. M. and Benoit B. Mandelbrot, "A New Model for the Clustering of Errors on Telephone Circuits," *IBM Journal of Research and Development*, Vol. 7, 1963, pp. 224–236.

Blumenthal, Leonard M. and Karl Menger, *Studies in Geometry*, W. H. Freeman and Company, San Francisco, CA, 1970.

Boll, David, "Pi, and the Mandelbrot Set." Archived at https://web.archive.org/web/20140317091957/https://home.comcast.net/~davejanelle/mandel.html. This web reference includes a link to the proof.

Bradbury, Ray, "A Sound of Thunder," *Collier's*, June 28, 1952. This short story also appeared in the Ray Bradbury collections *The Golden Apples of the Sun*, Doubleday, New York, 1953, and *R is for Rocket*, Doubleday, New York, 1962. Yet another appearance is in *The Stories of Ray Bradbury*, Alfred A. Knopf, New York, 1980.

Burger, Edward B. and Michael Starbird, *The Heart of Mathematics*, third edition, Wiley, Hoboken, NJ, 2010.

Chaos: A New Science, VHS video, 52 min color, New Dimensions Media Inc., Jacksonville, IL, 1988. Producers, David Barlow, Rex Cowan; photography, Chris Goodger, Robin Riseley, Simon French; editor, Phil McDonald. Cast: John Castle. An Independent Communications Associates Words Edge Films Production for Channel 4. This documentary features many mathematical all-stars. Tim Palmer, Edward Lorenz, Robert Devaney, Heinz-Otto Peitgen, David Rand, Mitchell Feigenbaum, Benoit Mandelbrot, Michael Barnsley, Harry Swinney, Richard Cohen, James Ramsey, and Ian Stewart all make appearances!.

Davis, Chandler, and Donald E. Knuth, "Number Representations and Dragon Curves," *Journal of Recreational Mathematics*, Vol. 3, 1970, pp. 66–81 and 133–149.

Devaney, Robert L., *An Introduction to Chaotic Dynamical Systems*, 2nd edition, Addison-Wesley, Redwood City, CA, 1989.

Devaney, Robert L., *Chaos, Fractals, and Dynamics: Computer Experiments in Mathematics*, Addison-Wesley, Menlo Park, CA, 1990.

Dewdney, Alan K., "Computer Recreations: A Computer Microscope Zooms in for a Look at the Most Complex Object in Mathematics," *Scientific American*, Vol. 253, No. 2, August 1985, pp. 16–21, 24.

Douady, Adrien, and John H. Hubbard, "Itération Des Polynomes Quadratiques Complexes," *Comptes Rendus*, Paris, Vol. 294, 1982, pp. 123–126.

Einstein, Albert, "Über die von der molekularkinetischen Theorie der Wärme geforderte Bewegung von in ruhenden Flüssigkeiten suspendierten Teilchen," *Annalen der Physik*, Vol. 17, No. 8, 1905, pp. 549–560.

Falconer, Kenneth, *Fractal Geometry*, Wiley, New York, 1990.

Field, Michael and Martin Golubitsky, *Symmetry in Chaos: A Search for Pattern in Mathematics, Art and Nature*, Oxford University Press, Oxford, 1992.

Flake, Gary William, *The Computational Beauty of Nature: Computer Explorations of Fractals, Chaos, Complex Systems, and Adaptation*, MIT Press, Cambridge, MA, 1998.

Fort, Charles, *New Lands*, Boni & Liveright, New York, 1923. This work is also available online at www.sacred-texts.com/fort/lands/index.htm.

Gardner, Martin, "Mathematical Games: In Which "Monster" Curves Force Redefinition of the Word "Curve"," *Scientific American*, Vol. 235, December 1976, pp. 124–133.

Glass, Leon, and Michael C. Mackey, *From Clocks to Chaos*, Princeton University Press, Princeton, NJ, 1988.

Gleick, James, *Chaos: Making a New Science*, Viking, New York, 1987.

Guckenheimer, John, "A Strange, Strange Attractor," in Marsden, Jerrold E. and Marjorie McCracken, editors, *The Hopf Bifurcation and its Applications*, Springer-Verlag, New York, 1976, pp. 368–381.

Hofstadter, Douglas R., *Gödel, Escher, Bach: An Eternal Golden Braid*, Basic Books, New York, 1979.

Hofstadter, Douglas R., "Strange Attractors: Mathematical Patterns Delicately Poised between Order and Chaos," *Scientific American*, Vol. 245, November 1981, pp. 16–29.

Holden, Arun V., editor, *Chaos*, Princeton University Press, Princeton, NJ, 1986.

Jacquin, Arnaud E., "A Novel Fractal Block-coding Technique for Digital Images," *ICASSP '90*, 1990.

"Jonathan Coulton + Mandelbrot Set HD," available online at www.youtube.com/watch?v=ZDU40eUcTj0. A long zoom into the Mandelbrot set, accompanied by a fun song by Jonathan Coulton.

Kaplan, James L. and James A. Yorke, "Chaotic Behavior of Multidimensional Difference Equations," in Peitgen, Heinz-Otto, and Hans-Otto Walther, editors, *Functional Differential Equations and Approximation of Fixed Points*, Lecture Notes in Mathematics, Vol. 730, Springer-Verlag, Berlin, 1979, pp. 204–227.

Keim, Brandon, "The Human Genome in 3 Dimensions," *Wired*, October 8, 2009, available online at www.wired.com/2009/10/fractal-genome/. This piece reports on the discovery of the connection between the human genome and space-filling curves.

Klebanoff, Aaron, "π in the Mandelbrot Set," *Fractals, Complex Geometry, Patterns, and Scaling in Nature and Society*, Vol. 9, No. 4, December 2001, available online at http://www.pi-e.de/PDF/mandel.pdf.

Kline, S. A., "On Curves of Fractional Dimensions," *Journal of the London Mathematical Society*, Vol. 20, 1945, pp. 79–86.

Lamb, Evelyn, "Fractal Kitties Illustrate the Endless Possibilities for Julia Sets," *Scientific American*, September 26, 2012, available online at https://blogs.scientificamerican.com/observations/fractal-kitties-illustrate-the-endless-possibilities-for-julia-sets/.

Lieberman-Aiden, Erez, Nynke L. van Berkum, Louise Williams, Maxim Imakaev, Tobias Ragoczy, Agnes Telling, Ido Amit, Bryan R. Lajoie, Peter J. Sabo, Michael O. Dorschner, Richard Sandstrom, Bradley Bernstein, M. A. Bender, Mark Groudine, Andreas Gnirke, John Stamatoyannopoulos, Leonid A. Mirny, Eric S. Lander, and Job Dekker, "Comprehensive Mapping of Long-Range Interactions Reveals Folding Principles of the Human Genome," *Science*, Vol. 326, No. 5950, October 9, 2009, pp. 289–293. This details the connection between the human genome and space-filling curves.

Lindsey, Kathryn A., "Shapes of Polynomial Julia Sets," *Ergodic Theory and Dynamical Systems*, Vol. 35, No. 6, September 2015, pp. 1913–1924.

Lorenz, Edward N., "Deterministic Nonperiodic Flow," *Journal of Atmospheric Sciences*, Vol. 20, March 1963, pp. 130–141.

Lorenz, Edward N., *The Essence of Chaos*, University of Washington Press, Seattle, WA, 1993.

Mandelbrot, Benoit, "The Stable Paretian Income Distribution, When the Apparent Exponent Is Near Two," *International Economic Review*, Vol. 4, 1963, pp. 111–115.

Mandelbrot, Benoit, "The Variation of Certain Speculative Prices," *Journal of Business*, Vol. 36, 1963, pp. 394–419.

Mandelbrot, Benoit, "How Long Is the Coast of Britain? Statistical Self-similarity and Fractional Dimension," *Science*, Vol. 156, No. 3775, May 5 1967, pp. 636–638.

Mandelbrot, Benoit, *Les Objets Fractals: Forme, Hasard et Dimension*, Flammarion, Paris, France, 1975.

Mandelbrot, Benoit, *Fractals: Form, Chance, and Dimension*, W. H. Freeman & Co, San Francisco, CA, 1977. This is an updated English translation of *Les Objets Fractals: Forme, Hasard et Dimension*.

Mandelbrot, Benoit, "The Fractal Geometry of Trees and Other Natural Phenomena," in Miles, Roger E. and Jean Serra, editors, *Buffon Bicentenary Symposium on Geometrical Probability*, Lecture Notes in Biomathematics, Vol 23, Springer, New York, 1978, pp. 235–249.

Mandelbrot, Benoit, "Comment on Bifurcation Theory and Fractals," in Gurel, Okan, and Otto E. Rössler, editors, *Bifurcation Theory and Applications*, Annals of the New York Academy of Sciences, Vol. 316, 1979, pp. 463–464.

Mandelbrot, Benoit, "Fractal Aspects of the Iteration $z \rightarrow \lambda z(1 - z)$ for Complex λ and z," in Helleman, Robert H. G., editor, *Non-linear Dynamics*, Annals of the New York Academy of Sciences, Vol. 357, 1980, pp. 249–259.

Mandelbrot, Benoit, "Scalebound or Scaling Shapes: A Useful Distinction in the Visual Arts and in the Natural Sciences," *Leonardo*, Vol. 14, No. 1, Winter 1981, pp. 45–47.

Mandelbrot, Benoit, "On the Quadratic Mapping $z \rightarrow z^2 - \mu$ for Complex μ and z: The Fractal Structure of its M Set, and Scaling," *Physica D: Nonlinear Phenomena*, Vol. 7, No. 1–3, May 1983, pp. 224–239.

Mandelbrot, Benoit, *The Fractal Geometry of Nature*, Updated and Augmented, W. H. Freeman and Company, New York, 1983. This is an update of *Fractals: Form, Chance, and Dimension*.

From the Foreword: "My first scientific publication came out on April 30, 1951. Over the years it had seemed to many that each of my investigations was aimed in a different direction. But this apparent disorder was misleading: it hid a strong unity of purpose, which the present Essay, like its two predecessors, is intended to reveal. Against odds, most of my works turn out to have been the birth pangs of a new scientific discipline."

Mandelbrot, Benoit, *The Fractalist: Memoir of a Scientific Maverick*, Pantheon Books, New York, 2012.

Mandelbrot, Benoit and John W. Van Ness, "Fractional Brownian Motions, Fractional Noises and Applications," *SIAM Review*, Vol. 10, 1968, pp. 422–437.

McMahon, Thomas A., "The Mechanical Design of Trees," *Scientific American*, Vol. 233, No. 1, July 1975, pp. 92–102.

McMahon, Thomas A. and Richard E. Kronauer, "Tree Structures: Deducing the Principle of Mechanical Design," *Journal of Theoretical Biology*, Vol. 59, No. 2, July 7, 1976, pp. 443–466.

Menger, Karl, "What Is Dimension," *American Mathematical Monthly*, Vol. 50, No. 1, January 1943, pp. 2–7.

Moon, Francis C., *Chaotic Vibrations*, Wiley, New York, 1987.

Moore, Eliakim Hastings, "On Certain Crinkly Curves," *Transactions of the American Mathematical Society*, Vol. 1, No. 1, January 1900, pp. 72–90.

Osgood, William F., "A Jordan Curve of Positive Area," *Transactions of the American Mathematical Society*, Vol. 4, No. 1, January 1903, pp. 107–112.

Peitgen, Heinz-Otto and Peter H. Richter, *The Beauty of Fractals*, Springer, Berlin, 1986.

Peitgen, Heinz-Otto and Dietmar Saupe, editors, *The Science of Fractal Images*, Springer, New York, 1988.

Peterson, Ivars, *Newton's Clock: Chaos in the Solar System*, MacMillan, New York, 1993.

Pi and the Mandelbrot set, https://web.archive.org/web/20150421145609/https://people.math.osu.edu/edgar.2/piand.html.

Pickover, Clifford A., *Computers, Pattern, Chaos and Beauty: Graphics from an Unseen World*, St. Martin's Press, New York, 1990. Pickover has written many great books on topics in computer graphics. Chapter 15 of this one addresses root approximation algorithms and their associated fractal images.

Prigogine, Ilya and Isabelle Stengers, *Order Out of Chaos*, Bantam, New York, 1984.

Richardson, Lewis Fry, *Arms and Insecurity: A Mathematical Study of the Causes and Origins of War*, Rashevsky, Nicolas and Ernesto Trucco, editors, Boxwood Press, Pacific Grove, CA, 1960.

Richardson, Lewis Fry, *Statistics of Deadly Quarrels*, Wright, Quincy and C. C. Lienau, editors, Boxwood Press, Pacific Grove, CA, 1960.

Richardson, Lewis Fry, "The Problem of Contiguity: An Appendix to Statistics of Deadly Quarrels," *General Systems Yearbook*, Vol. 6, 1961, pp. 139–187.

Rose, N. J., "The Pascal Triangle and Sierpiński's Tree," in *Mathematical Calendar*, Rome Press, Raleigh, NC, 1981.

Ruelle, David, "Strange Attractors as a Mathematical Explanation of Turbulence," in Rosenblatt, Murray, and Charles W. Van Atta, editors, *Statistical Models and Turbulence*, Lecture Notes in Physics, Vol. 12, Springer, New York, 1972, pp. 292–299.

Ruelle, David, "Strange Attractors," *The Mathematical Intelligencer*, Vol. 2, No. 3, 1980, pp. 126–137.

Ruelle, David, *Chaotic Evolution and Strange Attractors*, Cambridge University Press, Cambridge, 1989.

Ruelle, David, *Chance and Chaos*, Princeton University Press, Princeton, NJ, 1991.

Ruelle, David and F. Takens, "On the Nature of Turbulence," *Communications in Mathematical Physics*, Vol. 20, 1971, pp. 167–192. Mandelbrot noted that strange attractors developed independently of fractals and "the theory of strange attractors took off for earnest with the study of turbulence in Ruelle & Takens 1971."[125].

Schuster, Heinz Georg, *Deterministic Chaos*, Physik-Verlag, Weinheim, 1984.

Shu-Yu, Zhang, *Bibliography on Chaos*, World Scientific, Singapore, 1991. This list contains 7,469 items, of which 303 are books.

Smale, Stephen, "How I Got Started in Dynamical Systems," in Smale, Stephen, editor, *The Mathematics of Time*, Springer, New York, 1980, pp. 147–151.

Sparrow, Colin, *The Lorenz Equations: Bifurcations, Chaos, and Strange Attractors*, Springer, New York, 1982.

Stewart, George R., *Storm*, Random House, New York, 1941. This novel, which appeared before Ray Bradbury's short story, *A Sound of Thunder*, includes a comment about a man's sneeze in China being able to cause snow in New York.

Stewart, Ian, "Four Encounters with Sierpiński's Gasket," *Mathematical Intelligencer*, Winter 95, Vol. 17, No. 1, 1995, pp. 52–64.

Stewart, Ian, *Does God Play Dice? the New Mathematics of Chaos*, second edition, Wiley-Blackwell, Hoboken, NJ, 2002.

Strichartz, Robert S., *Differential Equations on Fractals*, Princeton University Press, Princeton, NJ, 2006.

Strogatz, Steven H., *Nonlinear Dynamics and Chaos with Applications to Physics, Biology, Chemistry, and Engineering*, Addison Wesley, Reading, MA, 1994.

"Time and Punishment," *The Simpsons*, Season 6, Episode 6 (Treehouse of Horror 5), October 30, 1994. This segment is a parody of the Ray Bradbury short story "A Sound of Thunder."

Ueda, Yoshisuke, "Strange Attractors and the Origin of Chaos," in Ueda, Yoshisuke, editor, *The Road to Chaos*, Aerial Press, Santa Cruz, CA, 1992, pp. 185–216.

Waite, Jon and Mark Beaumont, *An Introduction to Block Based Fractal Image Coding*, preprint, British Telecom Research Station, Ipswich, 1991.

Whyburn, Gordon Thomas, "Topological Characterization of the Sierpiński Curve," *Fundamenta Mathematicae*, Vol. 45, 1958, pp. 320–324.

Wikibooks Contributors, "Pictures of Julia and Mandelbrot Sets/History," *Wikibooks, The Free Textbook Project*, 2010, available online at https://en.wikibooks.org/w/index.php?title=Pictures_of_Julia_and_Mandelbrot_Sets/History&oldid=1998543.

Williams, R. F., "Lorenz Knots are Prime," *Ergodic Theory and Dynamical Systems*, Vol. 4, No. 1, March 1984, pp. 147–163.

Note: I want to change the way you look at the world. Understanding the mathematics behind natural phenomena doesn't rob them of their mystery, but rather adds extra layers of beauty that relatively few are aware of and able to appreciate.

---

125 Mandelbrot, Benoit, *The Fractal Geometry of Nature*, Updated and Augmented, W. H. Freeman and Company, New York, 1983, p. 193.

19

Cellular Automata

FIGURE 19.1
Stanislaw Ulam (1909–1984)

Public domain.

Back in Chapter 12, you were introduced to Stanislaw Ulam (Figure 19.1) in connection with the Monte Carlo method. He's also a key figure in the history of cellular automata, but before detailing his impact on this field, another problem associated with him is examined. It's a problem that sounds simple, but has stumped the mathematical world's best minds.

Ulam's Conjecture

The conjecture concerns a very simple function:

$$f(n) = \begin{cases} \dfrac{n}{2} & \text{if } n \text{ is even} \\ 3n + 1 & \text{if } n \text{ is odd} \end{cases}$$

If you don't like piece-wise defined functions, it can be rewritten as

$$f(n) = \frac{1}{2} n \cos^2\left(\frac{\pi}{2} n\right) + (3n + 1)\sin^2\left(\frac{\pi}{2} n\right).$$

This form is not nearly as nice to work with for our initial investigations, but it will be revisited at the end of this section, and seen to lead to something intriguing.

In any case, the interesting behavior occurs when the function is iterated. Starting with $n = 10$, we have $f(10) = 5$, because 10 is even. Plugging this new value back into f, we get $f(5) = 16$, according to the case n odd. Plugging each new result into f a few more times we get 8, 4, 2, 1, 4, 2, 1, 4, 2, 1, ... We are stuck in a cycle.

This is all very simple and there might not seem like much more to say, because every positive integer for which this function is iterated seems to eventually end up stuck in this very same 4, 2, 1 cycle. The rub is that no one has been able to find a proof. Perhaps there's some number that hasn't been tried yet that yields a different long-term behavior. If so, it would have to be very large. Mathematicians and computer scientists have been writing programs for years to test as many values as possible and it's now established that any counterexample would have to exceed 20×2^{58}.[1] It's also been proven that a cycle other than 4, 2, 1 would have to consist of billions of terms.[2]

The many names given to this problem indicate how much attention it has received over the decades. It became known as Ulam's conjecture, after Ulam introduced the problem to Los Alamos.[3] Its other names include the Collatz conjecture, Hasse's algorithm, the Syracuse problem, Kakutani's problem, and the Hailstone problem. The numbers it generates are sometimes called Hailstone sequences. And sometimes it's simply referred to as the $3n + 1$ problem.

While most of the names pay tribute to people who worked on the problem, the "Hailstone" option actually gives a sense of what happens as we iterate. The numbers tend to move up and down (in terms of size) repeatedly, like hailstones forming in the upper atmosphere. Finally the numbers fall down, 4, 2, 1, hitting the ground.

It's believed that the first person to examine the problem was Lothar Collatz in the early 1930s.[4] But despite it spreading by a combination of word and mouth and rediscovery, in the decades that followed, nothing was put into print on the topic until the 1970s.[5]

Paul Erdös said, "Mathematics is not yet ready for such problems."[6] He offered $500 for its solution.[7]

There are many ways to investigate the problem with a computer. One is to plot the trajectories of a large number of small values on the Cartesian coordinate system. For example, the trajectory beginning with 9 is 9, 28, 14, 7, 22, 11, 34, 17, 52, 26, 13, 40, 20, 10, 5, 16, 8, 4, 2, 1, so all of the following points would be plotted: (9, 9), (9, 28), (9, 14), (9, 7), (9, 22), (9, 11), (9, 34), (9, 17), (9, 52), (9, 26), (9, 13), (9, 40), (9, 20), (9, 10), (9, 5), (9, 16), (9, 8), (9, 4), (9, 2),

1 Lagarias, Jeffrey C., editor, *The Ultimate Challenge: The 3x + 1 Problem*, American Mathematical Society, Providence, RI, 2010, p. 16.

2 Chamberland, Marc, *Single Digits: In Praise of Small Numbers*, Princeton University Press, Princeton, NJ, 2015, p. 71. Lagarias, Jeffrey C., editor, *The Ultimate Challenge: The 3x + 1 Problem*, American Mathematical Society, Providence, RI, 2010, p. 5.

3 Poundstone, William, *The Recursive Universe: Cosmic Complexity and the Limits of Scientific Knowledge*, Willow Morrow and Company, Inc, New York, 1985, p. 122.

4 Lagarias, Jeffrey C., editor, *The Ultimate Challenge: The 3x + 1 Problem*, American Mathematical Society, Providence, RI, 2010, p. 5.

5 Lagarias, Jeffrey C., editor, *The Ultimate Challenge: The 3x + 1 Problem*, American Mathematical Society, Providence, RI, 2010, p. 5.

6 Lagarias, Jeffrey C., "The 3x + 1 Problem and Its Generalizations," *The American Mathematical Monthly*, Vol. 92, No. 1, January 1985, pp. 3–23.

7 http://en.wikipedia.org/wiki/Collatz_conjecture

(9, 1). Figure 19.2 includes the points produced from all starting values between 1 and 10,000. The graph was compressed both horizontally and vertically, so several trajectories are represented by each vertical column of pixels. Also, the top of the graph is not shown. That is, some trajectories attain values higher than the maximum shown on the *y*-axis.

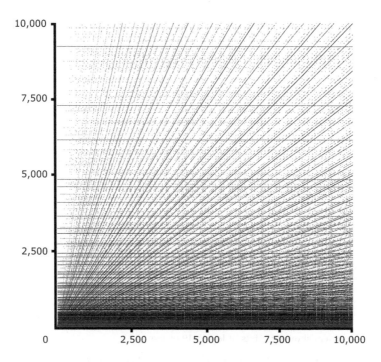

FIGURE 19.2
Does this graph provide any insight into the 3*n* + 1 problem?

Image created by Josh Gross.

The 3*n* + 1 problem connects with several chapters in this book. Obviously, the chapters on difference equations fall into this category, but Chapters 20 and 21 are also relevant, as Figure 19.3 shows.

To show a connection with another previous chapter, we return to the non-piece-wise form of the function,

$$f(n) = \frac{1}{2}n\cos^2\left(\frac{\pi}{2}n\right) + (3n + 1)\sin^2\left(\frac{\pi}{2}n\right).$$

In this form, it isn't necessary to decide if *n* is even or odd; you simply plug it in and the trig functions can be thought of as making the decision for you. In a similarly thoughtless way, you can plug in non-integers and get results. You can even plug in complex numbers. In that case, the function would normally be written using *z* as the variable. Recall that we typically use *n* to indicate an integer argument, *x* to indicate a real number, and *z* to indicate a complex number. Our new complex valued function,

$$f(z) = \frac{1}{2}z\cos^2\left(\frac{\pi}{2}z\right) + (3z + 1)\sin^2\left(\frac{\pi}{2}z\right),$$

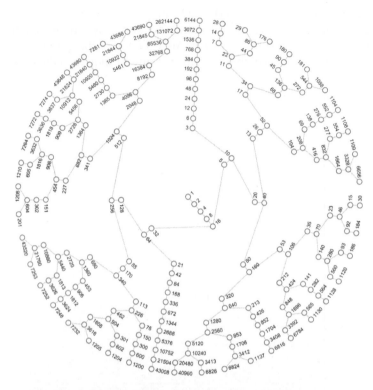

FIGURE 19.3

Graph of the $3n + 1$ problem showing orbits of length 19 or less.

Collatz Graph by Jason Davies, https://www.jasondavies.com/collatz-graph/.

is very interesting when iterated. As with other complex functions we've iterated, initial values that converge or get stuck in cycles are plotted in black and other points are plotted in various colors depending on how quickly they diverge. The result, for a portion of the complex plane, is shown in color figure 35 **(See color insert)**. Portions of this image bear a strong resemblance to the Mandelbrot set, but such visualizations for the $3n + 1$ problem came much later, in 1999.[8] It's exciting when an old problem can be put in a new light.

Despite all of the above, John H. Conway, in a paper published in March 2013, wrote, "the Collatz $3n + 1$ Conjecture is itself very likely to be unsettleable."[9] What he meant by this was that he did not believe a proof could be found, because a proof does not exist.

Is Conway right or will some new mathematical tool, to be developed in the decades to come and not imaginable now, lead to a proof? Only time will tell! It's now time to turn to another problem that attracted Ulam's attention, and is the focus of this chapter.

---

8 Letherman, Simon, Dierk Schleicher, and Reg Wood, "The $(3n + 1)$-Problem and Holomorphic Dynamics," *Experimental Mathematics*. Vol. 8, No. 3, 1999, pp. 241–252.

9 Conway, John H. "On Unsettleable Arithmetical Problems," *The American Mathematical Monthly*, Vol. 120, No. 3, March 2013, pp. 192–198. Also see Roberts, Siobhan, *Genius at Play – The Curious Mind of John Horton Conway*, Bloomsbury, New York, 2015, pp. 377–379.

Tessellation Structures

As often happens when a field is in its infancy, different researchers will attach different names to the same concepts. So, if you look in older books you may see the sort of things explored here called **tessellation structures** or **iterative circuit computers**.[10] Stanislaw Ulam used an even longer name, **recursively defined geometric objects**.[11] Today, they're primarily known as **cellular automata**.

One of Ulam's examples can be easily investigated on a piece of graph paper.[12] Start out with just one square filled in, as in Figure 19.4.

This single filled in square can be thought of as a "new" square. Now iterate. In each iteration, a square is filled in, if the following are all satisfied.

1) It borders exactly one new square (where "new" means a square filled in at the previous iteration), whether it is to the east, west, north, or south.

2) It does not touch any "old" squares (i.e., squares that were filled in sometime prior to the previous iteration). Here "touch" includes cells in the four cardinal directions, as well as cells that contact the given cell diagonally, at a corner. One exception is allowed. It can touch its grandparent square – the one that led to the creation of the new square that is, in turn, giving rise to the square currently being considered.

3) It does not touch any other square that rules 1 and 2 allow to be filled in, unless they arise from having bordered the same new square, according to rule 1.

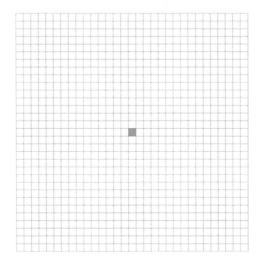

FIGURE 19.4
Time = 0.

Image created by Craig P. Bauer.

10 Burks, Arthur W., editor, *Essays on Cellular Automata*, University of Illinois Press, Urbana, 1970, p.xi.

11 Poundstone, William, *The Recursive Universe: Cosmic Complexity and the Limits of Scientific Knowledge*, Willow Morrow and Company, Inc, New York, 1985, p. 14.

12 This is adapted from Poundstone, William, *The Recursive Universe: Cosmic Complexity and the Limits of Scientific Knowledge*, Willow Morrow and Company, Inc, New York, 1985, pp. 133–136.

The result after the first iteration is shown in Figure 19.5. The results for further iterations are shown in Figures 19.6 to 19.17.

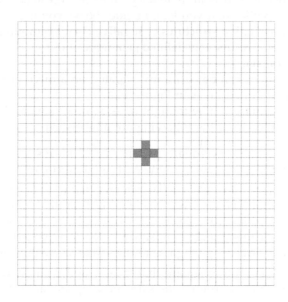

FIGURE 19.5
Time = 1 (first generation).

Image created by Craig P. Bauer.

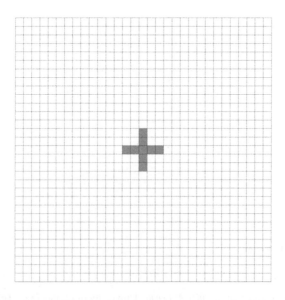

FIGURE 19.6
Time = 2.

Image created by Craig P. Bauer.

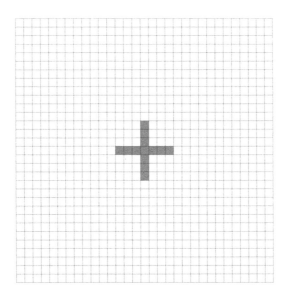

FIGURE 19.7
Time = 3.

Image created by Craig P. Bauer.

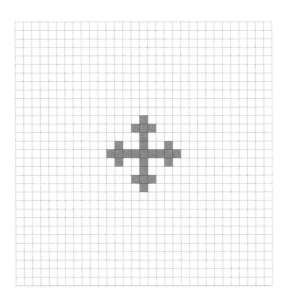

FIGURE 19.8
Time = 4.

Image created by Craig P. Bauer.

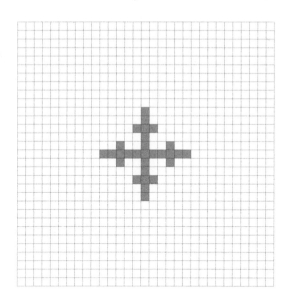

FIGURE 19.9
Time = 5.

Image created by Craig P. Bauer.

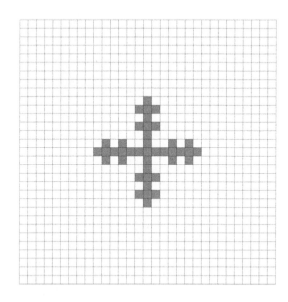

FIGURE 19.10
Time = 6.

Image created by Craig P. Bauer.

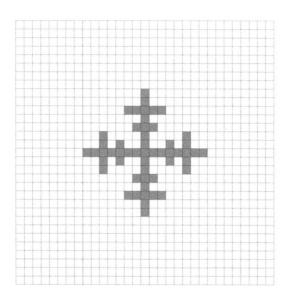

FIGURE 19.11
Time = 7.

Image created by Craig P. Bauer.

FIGURE 19.12
Time = 8.

Image created by Craig P. Bauer.

FIGURE 19.13
Time = 9.

Image created by Craig P. Bauer.

FIGURE 19.14
Time = 10.

Image created by Craig P. Bauer.

FIGURE 19.15
Time = 11.

Image created by Craig P. Bauer.

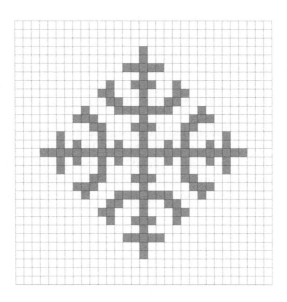

FIGURE 19.16
Time = 12.

Image created by Craig P. Bauer.

FIGURE 19.17
Time = 13.

Image created by Craig P. Bauer.

This just one example of what Ulam investigated in the 1940s. He had access to early computers at Los Alamos, so the problems weren't as tedious as they are on graph paper. He varied the rules so that cells could turn on or off and investigated other initial conditions. For example, if we start with two squares filled in, some distance apart, what happens when the structures these seeds generate approach each other?[13] He even used grids made up of triangles, and hexagons.[14] In yet another direction, he investigated three-dimensional cellular automata.[15]

Many of these examples illustrated how simple rules can give rise to complex results. This is a phenomenon we saw in Chapter 18, although at the time Ulam was studying cellular automata, the term "fractal" had not yet been coined.

Ulam was not the only one investigating such structures, but a huge contribution to the field was to be made by a man originally working on a different topic.

---

13 Poundstone, William, *The Recursive Universe: Cosmic Complexity and the Limits of Scientific Knowledge*, Willow Morrow and Company, Inc, New York, 1985, pp. 135–136.

14 Poundstone, William, *The Recursive Universe: Cosmic Complexity and the Limits of Scientific Knowledge*, Willow Morrow and Company, Inc, New York, 1985, p. 15.

15 Poundstone, William, *The Recursive Universe: Cosmic Complexity and the Limits of Scientific Knowledge*, Willow Morrow and Company, Inc, New York, 1985, p. 135. For much more detail, see Ulam's papers in Burks, Arthur W., editor, *Essays on Cellular Automata*, University of Illinois Press, Urbana and Chicago, 1970.

John von Neumann and Self-Reproducing Machines

There aren't many mathematicians who have been the topic of feature films, but I believe a good one could be made about John von Neumann. He was an interesting person throughout his entire life, beginning with his childhood in Hungary.

> At the age of six, he was able to exchange jokes with his father in classical Greek. The Neumann family sometimes entertained guests with demonstrations of Johnny's ability to memorize phone books. A guest would select a page and column of the phonebook at random. Young Johnny read the column over a few times, then handed the book back to the guest. He could answer any question put to him (who has number such and such?) or recite names, addresses, and numbers in order.[16]

Von Neumann made an even bigger impression at the Technische Hochschule in Zürich. George Pólya, one of his instructors, later remarked,

> Johnny was the only student I was ever afraid of. If in the course of a lecture I stated an unsolved problem, the chances were he'd come to me as soon as the lecture was over, with the complete solution in a few scribbles on a slip of paper.[17]

In 1926, Neumann earned a doctorate in mathematics from the University of Budapest for a dissertation on set theory. He taught in Germany and did postdoctoral work at Göttingen, under David Hilbert. In 1930, he came to America to take a position at Princeton University. A few years later he accepted a position at Princeton's Institute for Advanced Studies. Having scared Pólya as a student, von Neumann now scared students of his own.

> His fluid line of thought was difficult for those less gifted to follow. He was notorious for dashing out equations on a small portion of the available blackboard and erasing expressions before students could copy them.[18]

Neumann spent summers in Europe and kept positions in Germany until 1933, when he resigned because of the Nazis. He was a non-practicing (agnostic) Jew.

On a lighter note, Figure 19.18 shows von Neumann with his wife Klara and their dog Inverse. The dog would sit when told to stand and go when told to come.[19]

In the 1950s, von Neumann was investigating the possibility of a machine making a duplicate of itself. Could it be done? If so, one machine would become two, and each would be capable of duplicating itself, yielding four. Next there would be eight, then sixteen. If enough material were available, the exponential growth would soon result in as many copies of the machine as we might desire.

He also asked if a machine could make a machine more complex than itself.

---

16 Poundstone, William, *Prisoner's Dilemma*, Doubleday, New York, 1992, p. 12.
17 Halmos, Paul R., "The Legend of John von Neumann," *American Mathematical Monthly*, Vol. 80, No. 4, April 1973, pp. 382–394.
18 Poundstone, William, *Prisoner's Dilemma*, Doubleday, New York, 1992, p. 17.
19 Stueben, Michael with Diane Sandford, *Twenty Years before the Blackboard*, The Mathematical Association of America, 1998, p. 131.

FIGURE 19.18
Klara, Inverse, and John von Neumann (1903–1957).

Alan Richards photographer. From the Shelby White and Leon Levy Archives Center, Institute for Advanced Study, Princeton, NJ, USA.

Von Neumann's first attempt at a self-replicating machine involved the original floating in a lake with various parts, so that they could be reached more easily.[20] His social skills saved him from this probably too-challenging problem. Generally, it's good to let people know what you're working on. Someone might have a valuable insight into your problem. Such was the case in this instance. When von Neumann discussed the problem with Stanislaw Ulam, the latter suggested that he pursue it more abstractly with cellular automata, instead of actual physical parts.[21] This was von Neumann's first exposure to cellular automata, but he was, as we've seen, a quick study. He later remarked,

> By axiomatizing [self-reproductive automata with cellular automata], one has thrown half of the problem out the window, and it may be the more important half. One has resigned oneself not to explain how these parts are made up of real things, specifically, how these parts are made up of actual elementary particles, or even of higher chemical molecules. […] The question we hope to answer now […] is: what are the basic

20 Poundstone, William, *The Recursive Universe: Cosmic Complexity and the Limits of Scientific Knowledge*, Willow Morrow and Company, Inc, New York, 1985, pp. 182–183, 185

21 https://www.rennard.org/alife/english/acintrogb01.html and Poundstone, William, *The Recursive Universe: Cosmic Complexity and the Limits of Scientific Knowledge*, Willow Morrow and Company, Inc, New York, 1985, p. 15.

principles which underlie the organization of these elementary parts in living organisms? My discussion will be limited to that point of view.[22]

Von Neumann's final two-dimensional model was much more complex than Ulam's OFF or ON cells. Each cell could be OFF or in any one of 28 other states, representing machine components.[23] And he estimated that there were about 200,000 of these cells altogether.[24] Part of the reason he only gave an estimate was due to the fact that his proof was an *existence* proof. He did not have a working model in full detail.

Although far from simple, von Neumann's creation was a universal replicator; it could make a copy of anything, including itself.[25] Ulam had found much simpler automata that could reproduce, but the manner in which they did so was more like crystal growth than biological reproduction.[26]

In contrast, the logic of the method found by von Neumann was seen to be nearly identical to reproduction in living cells, when the dynamics of the latter was finally worked out. Portions of von Neumann's machine correspond to DNA, ribosomes, and enzymes.[27]

Von Neumann also answered his second question in the affirmative – a machine can make a machine more complex than itself.[28] Hence, the evolutionary process is not limited to traditional biological lifeforms. This brings us back to the brief discussion in Chapter 0 about the difficulty of defining life. If machines can reproduce and evolve, does life really have to be carbon-based?

Impressive as von Neumann's results were, it's not worth looking at them in greater detail here, because the same sort of thing is much more easily explained via a game developed in the 1970s. As with so many other results in math and science, a simpler proof (or explanation) was found later. We improve upon the masters that came before us.

Von Neumann was a great lover of the nightlife and famous for throwing large parties. He posed a strong contrast to the stereotypical introverted mathematician. In his prime, he seems to have been a man of limitless energy. While his aggressive outgoing personality likely helped his career, some manifestations of it are not to be admired. For example, he is said to have been able to drink a quart of rye whisky in an hour, *and then drive*. In possibly related news, he wrecked a car nearly every year.[29] We don't have proof that alcohol caused these wrecks and there's actually an account that singing was to

---

22 https://www.rennard.org/alife/english/acintrogb01.html

23 Poundstone, William, *The Recursive Universe: Cosmic Complexity and the Limits of Scientific Knowledge*, Willow Morrow and Company, Inc, New York, 1985, p. 15 and p. 185. The description continues on pp. 186–189.

24 Poundstone, William, *The Recursive Universe: Cosmic Complexity and the Limits of Scientific Knowledge*, Willow Morrow and Company, Inc, New York, 1985, p. 16 and p. 195.

25 https://www.rennard.org/alife/english/acintrogb01.html and Poundstone, William, *The Recursive Universe: Cosmic Complexity and the Limits of Scientific Knowledge*, Willow Morrow and Company, Inc, New York, 1985, p. 16.

26 https://www.rennard.org/alife/english/acintrogb01.html and Poundstone, William, *The Recursive Universe: Cosmic Complexity and the Limits of Scientific Knowledge*, Willow Morrow and Company, Inc, New York, 1985, p. 136

27 Poundstone, William, *The Recursive Universe: Cosmic Complexity and the Limits of Scientific Knowledge*, Willow Morrow and Company, Inc, New York, 1985, p. 18. Page 190 points out a difference between von Neumann's model and the manner in which organic life replicates.

28 Poundstone, William, *The Recursive Universe: Cosmic Complexity and the Limits of Scientific Knowledge*, Willow Morrow and Company, Inc, New York, 1985, p. 17 and p. 195.

29 Poundstone, William, *The Recursive Universe: Cosmic Complexity and the Limits of Scientific Knowledge*, Willow Morrow and Company, Inc, New York, 1985, p. 178. Poundstone, William, *Prisoner's Dilemma*, Doubleday, New York, 1992, p. 25.

blame! Apparently von Neumann would sway back and forth while singing, and turn the wheel as he did so. Or perhaps the accidents were a consequence of his aggressiveness.

Von Neumann was extremely aggressive when it came to issues connected with America's nuclear arsenal. He played an important role in the development of the atomic bomb at Los Alamos, including calculating the height at which it should be exploded for maximum damage.[30] In 1946, he watched atomic tests on Bikini Atoll. He pushed strongly for development of the more powerful hydrogen bomb, even though there was no immediate threat. In fact, *he* was to become a threatening figure. In 1950, he advocated launching an unprovoked nuclear attack on all of Russia's major cities and military targets.[31] He thought this should be done as soon as possible.

> If you say why not bomb them tomorrow, I say why not today? If you say today at 5 o'clock, I say why not one o'clock?[32]
>
>   John von Neumann

When Linus Pauling spoke up about the increased risk of cancer due to fallout from open-air testing of nuclear weapons, von Neumann opposed him claiming that human tolerance for the fallout was much greater than actual exposures. Von Neumann even opposed further study of the problem as "contrary to the interests of the United States."[33]

Various studies have now estimated the number of cancer deaths in America directly attributable to fallout from nuclear testing as ranging from the tens of thousands to well over half a million.[34] Von Neumann may have been one of them. By the time he was diagnosed with a malignant bone cancer, it was already spreading to other parts of his body.[35] After a lifetime as an agnostic, on his deathbed, he asked for a Catholic priest. Steve Joshua Heims described von Neumann during his final days.

> ...religion did not prevent him from suffering; even his mind, the amulet on which he had always been able to rely, was becoming less dependable. Then came complete psychological breakdown; panic, screams of uncontrollable terror every night. His friend Edward Teller said, "I think that von Neumann suffered more when his mind would no longer function, than I have ever seen any human being suffer."[36]

Von Neumann's write-up of his self-reproducing machine was not finished at the time of his death, but a colleague completed it in von Neumann's style.[37]

---

30 Hodges, Andrew, *Alan Turing: The Enigma*, Simon & Schuster, New York, 1983, p. 312.

31 Poundstone, William, *The Recursive Universe: Cosmic Complexity and the Limits of Scientific Knowledge*, Willow Morrow and Company, Inc, New York, 1985, p. 181.

32 Blair, Jr., Clay, "The Passing of a Great Mind," *Life*, February 25, 1957, p. 96.

33 Masani, Pesi R., *Norbert Wiener 1894–1964*, Volume 5 of Vita Mathematica, Birkhäuser, Basel, 1990, p. 312.

34 See www.newscientist.com/article/dn1993-nuclear-test-fall-out-killed-thousands-in-us/ for one of the lower estimates and https://qz.com/1163140/us-nuclear-tests-killed-american-civilians-on-a-scale-comparable-to-hiroshima-and-nagasaki/ for one of the higher estimates.

35 Poundstone, William, *The Recursive Universe: Cosmic Complexity and the Limits of Scientific Knowledge*, Willow Morrow and Company, Inc, New York, 1985, p. 181 and Hodges, Andrew, *Alan Turing: The Enigma*, Simon & Schuster, New York, 1983, p. 363.

36 Heims, Steve Joshua, *John von Neumann and Norbert Wiener: From Mathematics to the Technologies of Life and Death*, Cambridge University Press, Cambridge, MA, 1980, p. 370

37 Poundstone, William, *The Recursive Universe: Cosmic Complexity and the Limits of Scientific Knowledge*, Willow Morrow and Company, Inc, New York, 1985, p. 182. Also, Roberts, Siobhan, *Genius at Play –The Curious Mind of John Horton Conway*, Bloomsbury, New York, 2015, p. 113.

Self-reproducing machines are now called von Neumann machines in recognition of his work. They penetrated pop culture in the form of the monoliths in the science fiction series begun with *2001: A Space Odyssey*. The potential value isn't limited to science fiction though. Robert A. Frosch, NASA's fifth administrator, suggested von Neumann machines be used for space exploration and colonization. His proposed version was "a machine that, either totally automatically or with minimal human intervention and guidance, can use solar energy and local materials on the earth or the moon or an asteroid or elsewhere in the solar system to build a replica of itself."[38]

Martin Gardner – The Great Popularizer

In March 1970, Martin Gardner received a 12-foot-long letter discussing various mathematical ideas.[39] Something like this might normally be dismissed as the work of a crackpot, but this particular communication came from University of Cambridge mathematician John H. Conway.

Nine feet down, the letter contained a description of a cellular automata that was destined to capture far wider interest than any that had come before.[40] It had three things in its favor:

1. It was incredibly simple. There were only 2 states, not the 29 of the von Neumann machine.

2. Martin Gardner wrote about it in the October 1970 installment of his column in *Scientific American*.[41] Gardner was one of the all-time great popularizers of mathematics. He had the ability to present the automata to a wide audience in an entertaining manner.

3. It was 1970. Computers weren't in many homes, but more and more people had access to them at work. Some readers could run their own simulations and the number of them capable of doing so would grow rapidly in the years to come.

And the catchy name didn't hurt either. It wasn't typically called a cellular automaton, an iterative circuit computer, or a tessellation structure. It was **The Game of Life** or just **Life**.

In this game every square cell has eight potential neighbors – the cells that only touch it at a corner count as neighbors, if they are occupied. The rules are as follows

1. **Survivals.** If a cell has two or three neighbors, it makes it to the next generation.

2. **Deaths.** If a cell has four or more neighbors, it dies from overpopulation. Also, a cell with zero neighbors or just one dies from isolation.

3. **Births.** If an empty cell has exactly three neighbors, it becomes occupied.[42]

---

38 Teresi, Dick, "Frosch Drops a Bomb," *OMNI*, Vol. 2, No. 5, February 1980, p. 119.
39 Roberts, Siobhan, *Genius at Play – The Curious Mind of John Horton Conway*, Bloomsbury, New York, 2015, p. 110.
40 Roberts, Siobhan, *Genius at Play – The Curious Mind of John Horton Conway*, Bloomsbury, New York, 2015, p. 110.
41 Gardner, Martin, "Mathematical Games: The Fantastic Combinations of John Conway's New Solitaire Game 'Life,'" *Scientific American*, Vol. 223, October 1970, pp. 120–123.
42 Adapted from Gardner, Martin, "Mathematical Games: The Fantastic Combinations of John Conway's New Solitaire Game 'Life,'" *Scientific American*, Vol. 223, October 1970, pp. 120–123.

It should be noted that all of the changes happen at once. We don't change the state of cells based on other changes to be made in the same round. Each cell is updated as if there was nothing else going on at that stage.

Conway's simple rules led to a wide range of behavior.

There are configurations that die out (Figures 19.19a–c):

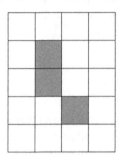 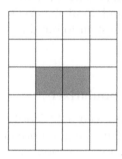 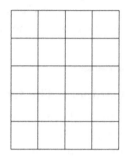

FIGURE 19.19
(a) Initial value; (b) first generation; (c) second generation.

Image created by Craig P. Bauer.

There are stable configurations known as "still lifes" (Figures 19.20a–b):

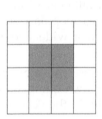 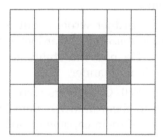

FIGURE 19.20
(a) The "block." (b). The "beehive."

Image created by Craig P. Bauer.

There are configurations that cycle with period two (Figure 19.21):

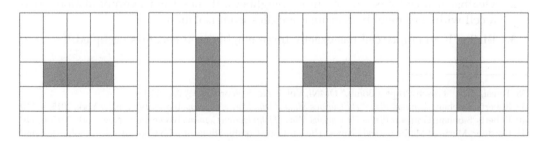

FIGURE 19.21
The "blinker."

Image created by Craig P. Bauer.

There are configurations that cycle with period three (Figure 19.22):

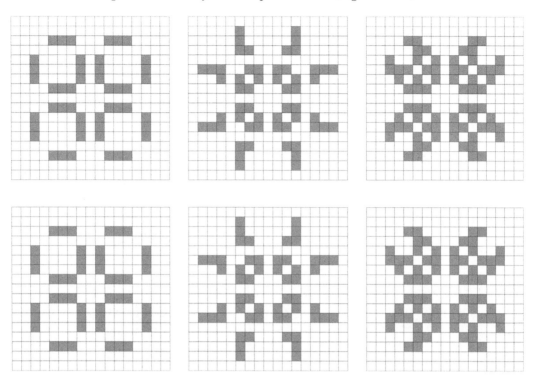

FIGURE 19.22
The "Pulsar."

Image created by Craig P. Bauer.

There are configurations with larger periods (Figure 19.23). The lifeform that resembles a leaning 8 has period eight.[43] Most initial conditions eventually end in some combination of stable objects and oscillators,[44] but there are lifeforms that move across the grid (Figure 19.24).

Every four iterations, the Glider moves one square in a diagonal direction. Which of the four diagonal directions it moves in depends on its orientation. It will move forever, if it doesn't encounter any obstacles.

Conway measured the velocity of objects like the glider by dividing the number of generations needed to return the object to its original configuration, but translated to a new position, by the number of cells it has been translated by. The glider takes four generations to advance one cell (diagonally), so its speed is ¼. With this definition of velocity, Conway defined the speed of light to be 1. Thus, the glider moves at ¼ the speed of light. While the speed of light would normally be the limit, Conway concluded

---

43 See http://conwaylife.com/w/index.php?title=Figure_eight for more.
44 Gardner, Martin, "Mathematical Games: The Fantastic Combinations of John Conway's New Solitaire Game 'Life,'" *Scientific American*, Vol. 223, October 1970.

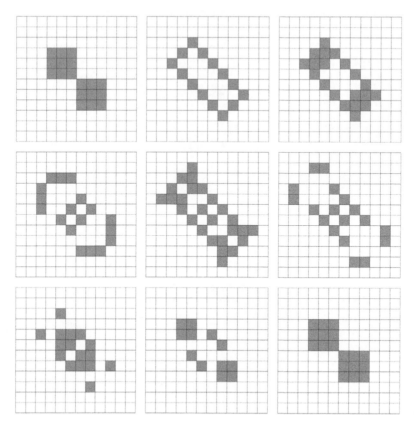

FIGURE 19.23
"Figure eight."

Image created by Craig P. Bauer.

that a lifeform moving in an unobstructed direction cannot exceed ½ the speed of light.[45] He challenged Martin Gardner's readers to find an example of a lifeform that attains that limit.[46]

There are many objects on the move in the Game of Life. The glider is merely the simplest. The general term for such movers is spaceship. While the Glider class moves diagonally, other spaceships are restricted to either horizontal or vertical movement.[47] But they do move faster, at the limit of ½ light speed.[48] There are also "overweight spaceships"

45 Poundstone, William, *The Recursive Universe: Cosmic Complexity and the Limits of Scientific Knowledge*, Willow Morrow and Company, Inc, New York, 1985, pp. 79–80.
46 Gardner, Martin, "Mathematical Games: The Fantastic Combinations of John Conway's New Solitaire Game 'Life'," *Scientific American*, Vol. 223, October 1970.
47 Poundstone, William, *The Recursive Universe: Cosmic Complexity and the Limits of Scientific Knowledge*, Willow Morrow and Company, Inc, New York, 1985, p. 80.
48 Poundstone, William, *The Recursive Universe: Cosmic Complexity and the Limits of Scientific Knowledge*, Willow Morrow and Company, Inc, New York, 1985, p. 80.

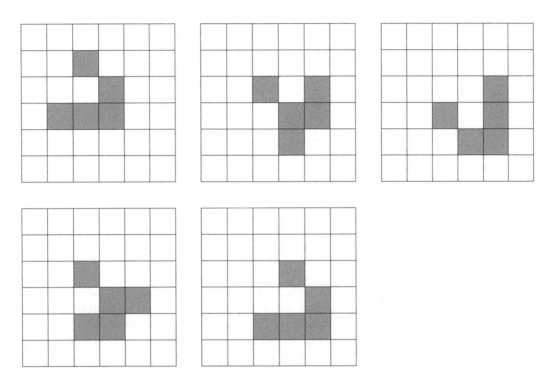

FIGURE 19.24
The "Glider."

Image created by Craig P. Bauer.

that cannot travel alone, but rather must move as part of a "flotilla."[49] A "lightweight spaceship" is shown in Figure 19.25.

Life was not something that Conway stumbled upon. He put in a lot of time considering various rules. It was actually years of experimentation.[50] Some mathematicians thoroughly examine everything that has previously been done on the research topics they choose, while others fear being contaminated by the thought of their predecessors and ending up stuck in the same ruts. Conway was somewhere in the middle in this instance. He knew about von Neumann's work, but not Ulam's.[51] when Ulam saw Gardner's article on Life, he mailed Conway reprints of some of his own related work. Because of political differences, Conway didn't think he'd like Ulam.[52] Conway was a radical, both politically and socially, and had been arrested at a Ban the

49 Poundstone, William, *The Recursive Universe: Cosmic Complexity and the Limits of Scientific Knowledge*, Willow Morrow and Company, Inc, New York, 1985, p. 86.

50 Roberts, Siobhan, *Genius at Play – The Curious Mind of John Horton Conway*, Bloomsbury, New York, 2015, p. 104.

51 Roberts, Siobhan, *Genius at Play – The Curious Mind of John Horton Conway*, Bloomsbury, New York, 2015, p. 146.

52 Roberts, Siobhan, *Genius at Play – The Curious Mind of John Horton Conway*, Bloomsbury, New York, 2015, p. 146.

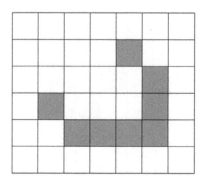

FIGURE 19.25
A "lightweight spaceship."

Image created by Craig P. Bauer.

Bomb protest and spent 11 days in jail.[53] By contrast, Ulam had helped create the hydrogen bomb. Nevertheless, to his surprise, Conway ended up finding Ulam "thoroughly congenial."[54] They actually did have some things in common. It probably didn't come up, but neither man was religious.[55] This, however, is a trait that most 20th-century research mathematicians share.

Conway's goals for Life were:

1. No initial pattern could be easily proven to grow without limit, in terms of the number of live cells.
2. Some initial patterns should *appear* to grow without limit.
3. Some simple initial patterns should grow and change for many generations before finally dying out completely, arriving at a stable form, or entering an endless oscillation.[56]

Conway also wanted the rules to be as simple as possible and the future of various configurations difficult to predict. He succeeded on all counts and then went on to make a prediction.

Conway believed that there was no initial configuration that would lead to unlimited growth. In other words, there was a sequence of numbers, a_n, such that if one started with n live cells, no matter how they were positioned, endless iteration would never result in more than a_n live cells in any generation. As an incentive for Gardener's readers to confirm his conjecture (or find a counterexample), Conway offered a $50 prize to the

---

53 Roberts, Siobhan, *Genius at Play – The Curious Mind of John Horton Conway*, Bloomsbury, New York, 2015, p. 36 and 40.
54 Roberts, Siobhan, *Genius at Play – The Curious Mind of John Horton Conway*, Bloomsbury, New York, 2015, p. 146
55 Roberts, Siobhan, *Genius at Play – The Curious Mind of John Horton Conway*, Bloomsbury, New York, 2015, p. 36 and pp. 104–105.
56 Adapted from Gardner, Martin, "Mathematical Games: The Fantastic Combinations of John Conway's New Solitaire Game 'Life,'" *Scientific American*, Vol. 223, October 1970.

first person who could prove him right or wrong. He even gave a possible lead to those who doubted his claim. A way to generate unlimited growth would be to find a form that "shoots" objects out of itself. Conway called this a gun. If the objects it produced were gliders, they would continuously move away, leaving room for the following gliders to exit unobstructed. Another possibility he gave was a "puffer train," something that moved like a glider, but left behind a trail of "smoke."[57] But, again, his intuition was that these did not exist. The prize was only offered for a proof or disproof before the end of the year, and the offer appeared in the October issue of *Scientific American*, so the clock was ticking!

Elsewhere in this book, I've praised popularizers of mathematics. Conway was perhaps returning a favor in instigating this column in a general interest periodical, and offering a $50 prize. Back when Conway was in high school, he had read W. W. Rouse Ball's *Mathematical Recreations and Essays*.[58] Later on, in 1956, he read Gardner's first column in *Scientific American* and became interested in flexagons.[59] How much did popularizers contribute to his initial interest in mathematics?

There were other open problems presented in this debut article for the Game of Life. One concerned the "R pentomino" shown in Figure 19.26.

Gardner wrote, "Its fate is not yet known. Conway has tracked it for 460 moves." It would have been amusing if a reader found that it stabilized after 461 moves, but this wasn't the case. When a solution did come, in late 1970, it confirmed that the pattern did indeed settle down, as Conway thought all patterns must, but it required 1,103 moves. Along the way it ejected six gliders that managed to survive.[60] A glider gun would have to eject gliders endlessly to beat Conway's conjecture in opposition to unlimited growth.

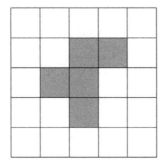

FIGURE 19.26
The "R pentomino."

Image created by Craig P. Bauer.

---

57 Gardner, Martin, "Mathematical Games: The Fantastic Combinations of John Conway's New Solitaire Game 'Life,'" *Scientific American*, Vol. 223, October 1970.

58 Roberts, Siobhan, *Genius at Play – The Curious Mind of John Horton Conway*, Bloomsbury, New York, 2015, p. 13, pp. 33–34.

59 Roberts, Siobhan, *Genius at Play – The Curious Mind of John Horton Conway*, Bloomsbury, New York, 2015, p. 33.

60 Poundstone, William, *The Recursive Universe: Cosmic Complexity and the Limits of Scientific Knowledge*, Willow Morrow and Company, Inc, New York, 1985, pp. 35–36. The final steady state is detailed on p. 36 and shown on p. 37.

Long-lived forms, like the R pentomino, became known as "methuselahs," after the Biblical figure with a purported lifespan of 969 years.

In addition to the gliders, the R pentomino generated five examples of still lifes. This might sound like a lot, especially considering the R pentomino's modest size, but it turned out that there are infinitely many still lifes, so it only provided the tiniest glimpse of Life's possibilities.[61]

Gardner announced the resolution of the R pentomino problem in the January 1971 installment of his column, giving the credit to the first solution he received, from researchers at Case Western Reserve University.[62] A list of others who obtained this solution a bit too late was given and it included Leonard Adleman, who would become the A in RSA public key cryptography.

An even more impressive methusaleh (Figure 19.27) was discovered by 1971.[63]

Dubbed "Acorn" this seven-cell configuration has a lifespan of 5,206 iterations, prior to stabilization.[64]

While some still lifes and oscillators were presented above, a complete list was never considered, because there are infinitely many of each.[65] Still, a few more are interesting enough that they should be shared. For example, the "eater" is Life's version of a black hole (Figure 19.28).[66] Moving objects that pass into an eater's domain are usually either totally or partially destroyed, while the eater itself, after the encounter, returns to its original form.[67] The eater's ability to repair itself brings to mind Von Neumann's self-reproducing automata.

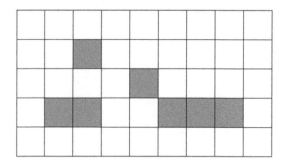

FIGURE 19.27
"Acorn."

Image created by Craig P. Bauer.

---

61 Poundstone, William, *The Recursive Universe: Cosmic Complexity and the Limits of Scientific Knowledge*, Willow Morrow and Company, Inc, New York, 1985, p. 36.

62 They were Gary Filipski and Brad Morgan.

63 It was published in the June 1971 issue of *Lifeline*, but it is unclear when exactly it was discovered. See www. conwaylife.com/wiki/Lifeline_Volume_2.

64 See Poundstone, William, *The Recursive Universe: Cosmic Complexity and the Limits of Scientific Knowledge*, Willow Morrow and Company, Inc, New York, 1985, p. 104 for the history and http://conwaylife.appspot. com/pattern/acorn for the full run.

65 Poundstone, William, *The Recursive Universe: Cosmic Complexity and the Limits of Scientific Knowledge*, Willow Morrow and Company, Inc, New York, 1985, p. 40.

66 It was discovered by William Gosper, Jr.

67 Poundstone, William, *The Recursive Universe: Cosmic Complexity and the Limits of Scientific Knowledge*, Willow Morrow and Company, Inc, New York, 1985, p. 38. Eaters are not indestructible – see Poundstone, p. 40.

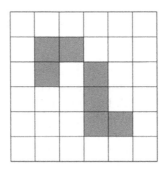

FIGURE 19.28
The "eater."

Image created by Craig P. Bauer.

But the eater is not undefeatable. A Glider can come at it from behind and destroy it.[68]
Turning away from such violence and getting back to creation, it was observed that most patterns in Life can arise from a variety of initial conditions.[69] But there are also "Garden-of-Eden" patterns, named so by John W. Tukey,[70] that can only be artificially created. They cannot arise in any way other than as initial conditions. This was first proven by Alvy Ray Smith III shortly after Martin Gardner wrote about the game.[71] Smith used a technique developed by Edward F. Moore that had been described informally in a paper published in Scientific American back in 1964.[72] However, Smith couldn't provide an example. His was an existence proof only. The first example was put forth by Roger Banks and Steve Ward in 1971.[73]

Unlimited Growth?

In the February 1971 installment of his *Scientific American* column, Gardner announced the winner of the $50 prize for resolving the question of unlimited growth.[74]

---

68 See Poundstone, William, *The Recursive Universe: Cosmic Complexity and the Limits of Scientific Knowledge*, Willow Morrow and Company, Inc, New York, 1985, p. 224, for an illustration of the proper angle of attack.

69 Poundstone, William, *The Recursive Universe: Cosmic Complexity and the Limits of Scientific Knowledge*, Willow Morrow and Company, Inc, New York, 1985, p. 48.

70 Gardner, Martin, "Mathematical Games," *Scientific American*, Vol. 224, No. 2, February 1971, pp. 112–117, p. 113 cited here.

71 Gardner, Martin, "Mathematical Games," *Scientific American*, Vol. 224, No. 2, February 1971, pp. 112–117.

72 Moore, Edward F., "Mathematics in the Biological Sciences," *Scientific American*, Vol. 211, No. 3, September 1964, pp. 148–154, 156–158, 160, 162, 164.

73 Poundstone, William, *The Recursive Universe: Cosmic Complexity and the Limits of Scientific Knowledge*, Willow Morrow and Company, Inc, New York, 1985, pp. 49–50. *Lifeline* Vol. 3, September 1971, *Lifeline* Vol. 4, December 1971.

74 Gardner, Martin, "Mathematical Games," *Scientific American*, Vol. 224, No. 2, February 1971, pp. 112–117, p. 115 announces the result and p. 113 shows it.

It turned out that Conway's intuition was wrong. Glider Guns and Puffer Trains *both* exist.[75] The winners were a group of MIT students at Marvin Minsky's Artificial Intelligence Lab.[76] They had constructed a glider gun that sends a glider out every 30 generations (Figure 19.29).

The design is not as complicated as might have been imagined, and it turns out that it can arise naturally. If 13 gliders collide in a specific way, a glider gun results![77] This contrasts strongly with the real world. Imagine crashing 13 cars together with the result that an auto factory is created and an endless string of cars roll forth from the collision.

In general, the results of collision are very difficult to predict. With just two gliders, there are 73 distinct outcomes possible, depending on exactly how they hit.[78] Comparisons have been made with collisions of subatomic particles, which also have a variety of outcomes and often cannot be predicted with certainty.[79]

In addition to Glider Guns, it turned out that Puffer Trains also exist. There's even the "Space Rake," which is a Glider Gun and a Puffer Train all in one![80]

While Conway was wrong, he may have taken consolation in the fact that the disproof of his conjecture was carried out in the manner he had suggested, by a glider gun. Also, his offer was only $50. On another occasion, while lecturing, he intended to offer a much larger prize of $1,000, but he got confused and said $10,000. And the prize was claimed! Conway mailed off the check for $10,000, but the winner, Colin Mallows, heard about

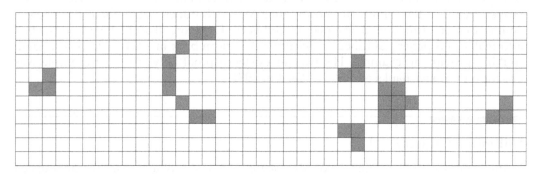

FIGURE 19.29
The first Glider Gun.

Image created by Craig P. Bauer.

---

75 For details of Puffer Trains, see Poundstone, William, *The Recursive Universe: Cosmic Complexity and the Limits of Scientific Knowledge*, Willow Morrow and Company, Inc, New York, 1985, pp. 111–114.

76 Robert April, Michael Beeler, R. William Gosper, Jr., Richard Howell, Rich Schroeppel, and Michael Speciner.

77 Poundstone, William, *The Recursive Universe: Cosmic Complexity and the Limits of Scientific Knowledge*, Willow Morrow and Company, Inc, New York, 1985, pp. 105–107. Also see Berlekamp, Elwyn, John Conway, and Richard Guy, *Winning Ways (for Your Mathematical Plays)*, Vol. 2, Academic Press, New York, 1982, pp. 836–837. The collision takes 67 generations to yield the Glider Gun.

78 Poundstone, William, *The Recursive Universe: Cosmic Complexity and the Limits of Scientific Knowledge*, Willow Morrow and Company, Inc, New York, 1985, pp. 80–81.

79 Poundstone, William, *The Recursive Universe: Cosmic Complexity and the Limits of Scientific Knowledge*, Willow Morrow and Company, Inc, New York, 1985, p. 84.

80 Poundstone, William, *The Recursive Universe: Cosmic Complexity and the Limits of Scientific Knowledge*, Willow Morrow and Company, Inc, New York, 1985, p. 113. There's a picture on p. 115.

the error and informed Conway that he was okay with $1,000. He cashed the smaller check and framed the larger one.[81]

Glider guns turned out to be much more important in the Game of Life than the small prize indicated. If they aren't interfered with, gliders can only move in one direction, but with other forms on the grid, a glider can be bounced back and forth like a ball in the old Pong video game. It's called the shuttling glider.[82] We can think of this whole system as a lifeform. It's a sophisticated example of an oscillator. Making the path the shuttling glider is passed along longer gives us an oscillator with a period as large as desired.

But the importance of gliders goes beyond setting records for periods, as will be seen in just a moment. First, it should be noted that Gardner's revisiting the Game of Life in his February 1971 column,[83] wasn't enough to meet the public's demand. Once word got out about this fascinating game, players wanted more. One of the players, Robert T. Wainwright, became an editor and publisher, putting out a newsletter titled *Lifeline* from 1971 to 1973.[84] The can all now be read online.[85]

The first issue of *Lifeline* appeared in March 1971, the same month in which a big announcement was made in Gardner's column.

> John Horton Conway and R. William Gosper, Jr. (who with the aid of his colleagues at the Massachusetts Institute of Technology found the glider gun described last month), have independently succeeded in "universalizing" the cellular space of Conway's "life" game. By using gliders as pulses they were able to simulate a Turing machine and create a universal calculator.[86]

Basically, the gliders played the role of electrons in a more traditional computer. It took some ingenuity, but logic gate lifeforms were created and universality was achieved.[87] Thus, the Game of Life could carry out any computation that a traditional computer could. It would not be anywhere near as efficient, but the sky was the limit for Life enthusiasts.

Conway, though, didn't seem to have any interest in publishing his proof.[88] He had also failed to publish his dissertation.[89] Perhaps he shared the perspective of Claude Shannon, who remarked, "After I had found the answers it was always painful to write

81 Roberts, Siobhan, *Genius at Play – The Curious Mind of John Horton Conway*, Bloomsbury, New York, 2015, pp. 282–283, 299–302, 305. Other instances of prize offers made by Conway are on p. 381 and p. 384. Also see http://mathworld.wolfram.com/Hofstadter-Conway10000-DollarSequence.html and http://oeis.org/A004001.

82 Poundstone, William, *The Recursive Universe: Cosmic Complexity and the Limits of Scientific Knowledge*, Willow Morrow and Company, Inc, New York, 1985, pp. 44–46. There are also non-glider shuttles. See pp. 88–89.

83 Gardner, Martin, "Mathematical Games," *Scientific American*, Vol. 224, No. 2, February 1971, pp. 112–117.

84 Gardner, Martin, "Mathematical Games," *Scientific American*, Vol. 224, No. 2, February 1971, pp. 112–117, p. 115 cited here. Also, Poundstone, William, *The Recursive Universe: Cosmic Complexity and the Limits of Scientific Knowledge*, Willow Morrow and Company, Inc, New York, 1985, p. 7.

85 http://conwaylife.com/wiki/Category:Lifeline_issues

86 Gardner, Martin, "Mathematical Games," *Scientific American*, Vol. 224, No. 3, March 1971, pp. 106–109, p. 109 cited here.

87 For a more detailed discussion of the mechanics of this, see Poundstone, William, *The Recursive Universe: Cosmic Complexity and the Limits of Scientific Knowledge*, Willow Morrow and Company, Inc, New York, 1985.

88 Roberts, Siobhan, *Genius at Play – The Curious Mind of John Horton Conway*, Bloomsbury, New York, 2015, pp. 144–145. It didn't appear until 1982, in Berlekamp, Elwyn, John Conway, and Richard Guy, *Winning Ways (for Your Mathematical Plays)*, Vol. 2, Academic Press, New York, 1982.

89 Roberts, Siobhan, *Genius at Play – The Curious Mind of John Horton Conway*, Bloomsbury, New York, 2015, p. 56.

them up or to publish them (which is how you get the acclaim)."[90] On the other hand, Conway did publish a lot of other material, including several books.

Others did publish on Life, as well as speak about it at conferences. In 2010, at the 9th "Gathering for Gardner," aka G4G9,[91] a conference held in the popularizer's honor, Tom Rokicki spoke about a Life program that had recently been created by Adam P. Goucher to calculate and display the digits of π.[92] Goucher warmed up for this Life programming task by first getting the cellular automata to calculate and display the endless digits of the Golden Ratio.[93]

Conway had predicted that the π calculator was possible back in 1970,[94] but not everyone could see the beauty and value in the game. In 1974, as Gardner's columns continued to attract young people to mathematics and the Game of Life was attracting them to computer programming, *Time* magazine published a piece whining about the alleged "millions of dollars in valuable computer time may have already been wasted by the game's growing horde of fanatics."[95]

As was seen in Chapter 12, mathematical analysis of other games, from the world of gambling, led to the creation of probability. Who can say where study of the Game of Life will lead? The game has been around for nearly 50 years, but gambling existed far longer before it "paid off" mathematically. Shame on *Time* for being so short-sighted!

Self-Reproducing Life?

When Conway's proof finally saw print in 1982, it went further. After showing that lifeforms can act as universal computers, as claimed back in 1971,[96] he went on to look at self-reproducing lifeforms, improving on what von Neumann did with 29 states by only using 2. However, he did not attempt to estimate the minimum size of such lifeforms. Like von Neumann, he had an existence proof.

Not only did Conway establish the existence of self-reproducing lifeforms, he also showed that some can move.[97] The chapter was written in an enthusiastic tone that used almost as many exclamation points as a silver age comic book written by Stan Lee. An excerpt that introduced another big result follows.[98]

90 Soni, Jimmy and Rob Goodman, *A Mind at Play: How Claude Shannon Invented the Information Age*, Simon & Schuster, New York, 2017, p. 60.

91 http://www.gathering4gardner.org/g4g9-2010-recap/

92 Roberts, Siobhan, *Genius at Play – The Curious Mind of John Horton Conway*, Bloomsbury, New York, 2015, p. 166.

93 http://pentadecathlon.com/lifeNews/2011/01/phi_and_pi_calculators.html

94 Roberts, Siobhan, *Genius at Play – The Curious Mind of John Horton Conway*, Bloomsbury, New York, 2015, p. 166

95 "The Game of Life," *Time*, January 21, 1974. Also see Roberts, Siobhan, *Genius at Play – The Curious Mind of John Horton Conway*, Bloomsbury, New York, 2015, p. 147.

96 Stephen Wolfram didn't consider it a complete proof. While he does not doubt that Life is universal, he wrote, "the fact remains that a complete and rigorous proof of universality has apparently still never been given for the Game of Life." See Wolfram, Stephen, *A New Kind of Science*, Wolfram Media, Inc., Champaign, IL, 2002, p. 1117.

97 Berlekamp, Elwyn, John Conway, and Richard Guy, *Winning Ways (for Your Mathematical Plays)*, Vol. 2, Academic Press, New York, 1982, pp. 848–849. Also see Poundstone, William, *The Recursive Universe: Cosmic Complexity and the Limits of Scientific Knowledge*, Willow Morrow and Company, Inc, New York, 1985, p. 224.

98 The quotation is from Berlekamp, Elwyn, John Conway, and Richard Guy, *Winning Ways (for Your Mathematical Plays)*, Vol. 2, Academic Press, New York, 1982, pp. 829–830.

LIFE'S PROBLEMS ARE HARD!

The questions we posed about the ultimate destiny of Life configurations may not seem very mathematical. After all, Life's but a game! Surely there aren't any difficult mathematical problems.

Well, yes there are! Indeed we can prove the astonishing fact that *every* sufficiently well-stated problem can be reduced to a question about Life! Those apparently trivial problems about Life histories can be arbitrarily difficult!

This requires some explanation. Conway showed how a computer constructed in the Game of Life could be programmed to self-destruct (i.e., cause itself to die out, leaving no cells behind) upon finding a solution to a problem. Why is this important? Well, the solution sought could be to a famous problem like Fermat's last theorem, which was actually just a conjecture at the time. It claimed that $x^n + y^n = z^n$ has no solutions in the positive integers, if n is an integer greater than 2. If the computer is programmed to self-destruct upon finding a solution, all we have to do, to determine if a solution exists or not, is predict the fate of the lifeform doing the computing. If it eventually dies out, there's a solution. If it persists forever, no solution exists. Therefore, if there's a general method for determining whether or not a given lifeform dies out, we could use it to solve important problems.[99]

But we know that some mathematical statements are undecidable (see Chapter 4), so we've just established that the fate of some lifeforms is unpredictable.[100] As Conway wrote, "LIFE'S PROBLEMS ARE HARD!"

Putting all of this together gives mobile self-reproducing lifeforms that can grow in unpredictably ways and, as a consequence of collisions, can mutate. Does this sound a little bit like the history of life on Earth? Conway believed similar results are possible in Life.

It's probable, given a large enough Life space, initially in a random state, that after a long time, intelligent self-reproducing animals will emerge and populate parts of the space.[101]

Looking at it from the other perspective:

If a two-state cellular automaton can produce such varied and esoteric phenomena from these simple laws, how much more so in our own universe?[102]

As was seen in Chapter 18, simple rules can lead to complex results. The rules or laws of physics may be very simple.

Analogies with real life processes are impossible to resist. If a primordial broth of amino-acids is large enough, and there is sufficient time, self-replicating moving automata may result from transition rules built into the structure of matter and the laws of nature. There is even the very remote possibility that space-time itself is

---

99 This is also explained in Poundstone, William, *The Recursive Universe: Cosmic Complexity and the Limits of Scientific Knowledge*, Willow Morrow and Company, Inc, New York, 1985, pp. 197–198.

100 Berlekamp, Elwyn, John Conway, and Richard Guy, *Winning Ways (for Your Mathematical Plays)*, Vol. 2, Academic Press, New York 1982, p. 847.

101 Berlekamp, Elwyn, John Conway, and Richard Guy, *Winning Ways (for Your Mathematical Plays)*, Vol. 2, Academic Press, New York, 1982, p. 849.

102 Berlekamp, Elwyn, John Conway, and Richard Guy, *Winning Ways (for Your Mathematical Plays)*, Vol. 2, Academic Press, New York, 1982, p. 849.

granular, composed of discrete units, and that the universe, as Edward Fredkin of M.I.T. and others have suggested, is a cellular automaton run by an enormous computer. If so, what we call motion may be only simulated motion. A moving particle in the ultimate microlevel may be essentially the same as one of our gliders, appearing to move on the macrolevel, whereas actually there is only an alteration of states of basic space-time cells in obedience to transition rules that have yet to be discovered.[103]

A major lesson from this chapter, as well as the previous chapter, is that complex results can arise from simple rules with simple and/or random initial conditions. This has implications for evolution. Creationists try to argue that life is "too complex" to have arisen in the manner that scientists have identified. This chapter and the previous chapter show that complexity doesn't have to be born from complexity, nor does it need divine intervention to manifest itself.

Variants

Many questions about the Game of Life have gone unsolved, but in the meanwhile researchers have pursued variants of the game. The original is played on an infinite plane, but a program can easily be modified to wrap around, when an object goes off a finite screen. Identifying the right edge with the left edge and the top edge with the bottom edge yields a torus (donut) as the playing field.[104]

In "3–4 Life," a cell is on in the next generation if and only if it has 3 or 4 neighbors in the current generation.[105] Other variants include Life without Death, Larger than Life, Real-Life (aka RealLife), and HighLife.[106]

Afterlife

So, what did Conway do after Life? Before answering that it should be mentioned that 1969, which Conway usually rounds to 1970, was his best year professionally.[107] In this year, in addition to Life, he made important contributions to the classification of finite simple groups (the tremendous collaborative proof mentioned in Chapter 2) and discovered surreal numbers. Both are too off-topic to delve into here, but I will note that Donald Knuth was inspired to write a novelette dealing with surreal numbers in which Conway (Figure 19.30) played the role of God (no image available).[108]

---

103 Berlekamp, Elwyn, John Conway, and Richard Guy, *Winning Ways (for Your Mathematical Plays)*, Vol. 2, Academic Press, New York, 1982, p. 849.

104 Poundstone, William, *The Recursive Universe: Cosmic Complexity and the Limits of Scientific Knowledge*, Willow Morrow and Company, Inc, New York, 1985, p. 234

105 See Poundstone, William, *The Recursive Universe: Cosmic Complexity and the Limits of Scientific Knowledge*, Willow Morrow and Company, Inc, New York, 1985, pp. 138–141 for more.

106 Roberts, Siobhan, *Genius at Play – The Curious Mind of John Horton Conway*, Bloomsbury, New York, 2015, p. 132.

107 Roberts, Siobhan, *Genius at Play – The Curious Mind of John Horton Conway*, Bloomsbury, New York, 2015, p. 171.

108 Knuth, Donald, *Surreal Numbers: How Two Ex-Students Turned on to Pure Mathematics and Found Total Happiness*, Addison-Wesley, Reading, MA, 1974.

FIGURE 19.30
John Horton Conway (1937–).

Image created by Thane Plambeck. This file is licensed under the Creative Commons Attribution 2.0 Generic license.

With a packed resume, Conway ventured to California Institute of Technology for a stint as a visiting professor in 1972.[109] While there he bought and drove a Lincoln Continental, despite not having a California driver's license. When he got pulled over by the police he pretended to be his graduate student, Robert Curtis, who did have a license.[110]

> However, to do good work in math, you have to be somewhat irresponsible.
>
> John Conway[111]

Jumping ahead a decade and back into Conway's office at the University of Cambridge, we seem to have "Anarchy in the U.K."

> Conway is incredibly untidy. The tables in his room at the Department of Pure Mathematics and Mathematical Statistics in Cambridge are heaped high with papers, books, unanswered letters, notes, models, charts, tables, diagrams, dead cups of coffee and an amazing assortment of bric-a-brac, which has overflowed most of the floor and all of the chairs, so that it is hard to take more than a pace or two into the room and impossible to sit down[112]

In 1985, Conway was invited to give a talk at Princeton University. Really, he was being eyed up for a position there, which he immediately accepted when it was offered.[113] He

---

109 This is the year given in Roberts, Siobhan, *Genius at Play – The Curious Mind of John Horton Conway*, Bloomsbury, New York, 2015, p. 220, but according to Conway's Curriculum Vitae, available at www.math.princeton.edu/sites/default/files/2017-03/ConwayCV.pdf, it was 1971. I checked with CalTech archives on May 29, 2018, and learned that Siobhan was correct.
110 Roberts, Siobhan, *Genius at Play – The Curious Mind of John Horton Conway*, Bloomsbury, New York, 2015, p. 221.
111 Roberts, Siobhan, *Genius at Play – The Curious Mind of John Horton Conway*, Bloomsbury, New York, 2015, p. 287.
112 Guy, Richard K., "John Horton Conway: Mathematical Magus," *The Two-Year College Mathematics Journal*, Vol. 13, No. 5, November, 1982, pp. 290–299, p. 292 quoted here.
113 Roberts, Siobhan, *Genius at Play – The Curious Mind of John Horton Conway*, Bloomsbury, New York, 2015, p. 266.

became the "John von Neumann Distinguished Professor in Applied and Computational Mathematics."[114] It was a new location, but Conway's piece of it took on the feel of his old stomping grounds. A reporter described the state of Conway's office at Princeton in June 1993, as follows.

> I entered the doorway and was astonished at how messy it was. The floor was covered in garbage, paper, food wrappings, etc., and it was actually kind of smelly, and lots of weird stuff, paper polyhedral and so forth, was hanging from the ceiling.[115]

According to Siobhan Roberts, Conway's biographer, his offices "always overflow with unopened mail, and he seldom reads any of his copious e-mail, either."[116] Other faculty and students weren't the only ones to notice the problematic state of Conway's office. The fire marshal declared it a hazard.[117] Conway solved the problem by essentially abandoning his office and hanging out in the math department's third-floor common room instead.[118]

Roberts noted that Conway's mother-in-law does his laundry and his finances are a mess.[119] It's easy to believe that the laundry would reach a state similar to his office without the help of another, and that the finances are in such a state. His personal appearance is consistent with all of the above. He gets his hair cut about once a year[120] and wears T-Shirts almost exclusively. He'd often lecture barefoot.[121]

Still, Conway couldn't very accurately be described as barefoot and carefree. As an undergraduate he was upset enough to make a suicide attempt, perhaps over his less than desired performance on Part II of the Tripos exam. Decades later when describing the attempt, he couldn't clearly recall his motivation, although it seemed important at the time.[122] Time has a way of putting things in perspective.

Decades later, Conway's life once again took a bad turn. He was in a second marriage that was falling apart and a physical problem was added to his emotional stress, when he had a heart attack running to catch a train. Naturally, these events depressed him greatly and, in the spring of 1993, he made another suicide attempt.[123]

Despite his idiosyncrasies / eccentricities, Conway's managed to get a lot done. While some might attribute his failures to laziness, those who know him better describe him as a workaholic.[124] Things that go askew are more due to his lack of organization and his absentmindedness. On January 6, 2008, at the Joint Mathematics Meetings of the American Mathematical Society and the Mathematical Association of America, held in San Diego, California, a large crowd eagerly awaited a prestigious invited address that

---

114 Roberts, Siobhan, *Genius at Play – The Curious Mind of John Horton Conway*, Bloomsbury, New York, 2015, p. 272.
115 The reporter was John Horgan. Quoted here from Roberts, Siobhan, *Genius at Play – The Curious Mind of John Horton Conway*, Bloomsbury, New York, 2015, p. 308.
116 Roberts, Siobhan, *Genius at Play – The Curious Mind of John Horton Conway*, Bloomsbury, New York, 2015, p. xxiii.
117 Roberts, Siobhan, *Genius at Play – The Curious Mind of John Horton Conway*, Bloomsbury, New York, 2015, p. 380.
118 Roberts, Siobhan, *Genius at Play – The Curious Mind of John Horton Conway*, Bloomsbury, New York, 2015, p. xiii.
119 Roberts, Siobhan, *Genius at Play – The Curious Mind of John Horton Conway*, Bloomsbury, New York, 2015, p. 108.
120 Roberts, Siobhan, *Genius at Play – The Curious Mind of John Horton Conway*, Bloomsbury, New York, 2015, p. 367.
121 Roberts, Siobhan, *Genius at Play – The Curious Mind of John Horton Conway*, Bloomsbury, New York, 2015, p. 75 and 97.
122 Roberts, Siobhan, *Genius at Play – The Curious Mind of John Horton Conway*, Bloomsbury, New York, 2015, p. 37.
123 Roberts, Siobhan, *Genius at Play – The Curious Mind of John Horton Conway*, Bloomsbury, New York, 2015, p. 311.
124 Roberts, Siobhan, *Genius at Play – The Curious Mind of John Horton Conway*, Bloomsbury, New York, 2015, p. 186 states, "Berlekamp and Guy both painted a picture of Conway as a workaholic."

Conway was to give. However, Conway was, at that moment, in Princeton, New Jersey, thinking that the conference was the following week.[125]

Like others who rose to the top through a combination of natural talent and hard work, Conway thinks nearly anyone else could do the same.

> I tend to believe that anybody could do, roughly speaking, what I can do. I think it's pretty easy. And I often teach people who don't think they can do such and such, and they can in the end. Now, on the other hand, it's a very common opinion that special gifts come at birth or something. And all I can say is that I don't really want to believe it.[126]

Along with this high opinion of the potential of others, Conway has a very high opinion of himself. His biographer called him an egomaniac. She also described him more specifically as an "insecure egotist," noting, "He very much cares what other people think."[127]

While the quirks can make life harder for Conway and some of the people around him he wouldn't be Conway without them. They permeate his lectures and books. Too many mathematicians think the sole purpose of delivering a lecture or writing a paper or book is to convey information. They make no effort to convey it in an entertaining way. Conway is a nice exception. Who else would title a book *The Sensual (Quadratic) Form*? As another small example, his book *On Numbers and Games* ends with the following.

> Theorem 100. *This is the last theorem in this book.*   *(The Proof is obvious.)*[128]

The next big steps in the field of cellular automata were made by a man who is Conway's opposite in many ways.

Stephen Wolfram and *A New Kind of Science*

Superficially, John Conway and Stephen Wolfram (Figure 19.31) might appear to have a lot in common. They were both born in England, earned Ph.D.s, employed in Princeton, New Jersey, as professors, earned worldwide reputations, and widely considered to have giant egos.

But the differences are just as numerous. For example, Wolfram is quiet. You might even say shy or introverted. Conway is extroverted and has a reputation for making conversations one-sided, but when I met Wolfram, he asked about *my* research. Wolfram is highly organized, and is still with his first wife, in contrast to Conway's three ex-wives.[129] And it seems like Wolfram really likes to write! His 2002 book, *A New Kind of Science*, is over 1,200 pages long.[130] There are no coauthors.

---

125 I was in the audience and only learned what happened a decade later when reading *Genius at Play*. See p. 341.

126 Roberts, Siobhan, *Genius at Play – The Curious Mind of John Horton Conway*, Bloomsbury, New York, 2015, p. 369.

127 Roberts, Siobhan, *Genius at Play – The Curious Mind of John Horton Conway*, Bloomsbury, New York, 2015, p. xi and p. xii.

128 Conway, John H., *On Numbers and Games*, second edition, A. K. Peters, Ltd., Natick, MA, p. 224.

129 Donald E. Knuth is another example of the extremely organized personality type. See Roberts, Siobhan, *Genius at Play – The Curious Mind of John Horton Conway*, Bloomsbury, New York, 2015, p. 175, for an illustration of this. It's interesting that both organized and disorganized personality types can rise to the top. This has also been seen in cryptology. Both the sloppy Herbert O. Yardley and the precise William F. Friedman are in the National Security Agency's Hall of Honor.

130 Wolfram, Stephen, *A New Kind of Science*, Wolfram Media, Inc., Champaign, IL, 2002.

FIGURE 19.31
Stephen Wolfram (1959–).

Wolfram was one of the huge number of young people investigating the Game of Life in the early 1970s.[131] He went to Eton and Oxford, and, in 1979, he earned his Ph.D. in theoretical physics from Caltech, despite being only 20 years old.[132] Looking back on these years, he commented:

> I had the good fortune never to have to go to classes or anything like that, because I learned what I needed to know from books. From everything I've seen, courses are just a waste of time, and one can learn more things a lot more quickly just by reading about them.[133]

Switching coasts, he arrived in Princeton, New Jersey, years before Conway, but not at the same location. Conway was at Princeton University, while Wolfram was hired by Princeton's Institute for Advanced Studies. The Institute is one of the very best places in the world to be employed as a professor of mathematics or physics. There is no teaching requirement. Working there, you may simply pursue your own research. For Wolfram, this wasn't good enough. He left the Institute and founded his own institute, Wolfram Research.[134] He had developed the computer algebra system Mathematica and now continued to push forward with the new endeavor, in parallel with his continued work on cellular automata.

Wolfram's background in physics is likely an important part of why he approached the topic of cellular automata in a manner different than what would be the typical mathematician's approach.

> "The obsession with proof," Wolfram declares, has kept mathematicians from discovering the vast new realms of phenomena accessible to computers. Even the most

---

131 Roberts, Siobhan, *Genius at Play – The Curious Mind of John Horton Conway*, Bloomsbury, New York, 2015, pp. 151–153.

132 http://www.wolframscience.com/events/bio.html

133 Regis, Jr., Edward, "Einstein's Sanctum," *OMNI*, September 1984, Vol. 6, No. 12, pp. 88–90, 93, 142, and 144. Wolfram is described and interviewed on pages 142 and 144. The quote appears on page 144.

134 The Mathworld website, which is a very useful for reference for mathematicians and computer scientists, was developed by Eric Weisstein, under the support of Wolfram Research.

intrepid mathematical experimentalists are for the most part "not going far enough," he says. "They're taking existing questions in mathematics and investigating those. They are adding a few little curlicues to the top of a gigantic structure."[135]

A New Kind of Science details the discoveries of Wolfram's nearly three-decade-long investigation of cellular automata. An online copy is available for free viewing, following registration. Such an immense work cannot be done justice in a few pages. The pages that follow simply convey some small portions of it. The book includes results of both Wolfram and earlier researchers.

There's no need, if interesting results are desired, to make the cellular automata more complicated. One of the classes that Wolfram presented is even simpler than life, in that it's only one-dimensional. Conway suggested calling it LineLife.[136] It begins with a row of cells, each of which can be ON or OFF. In the examples that follow, the ON cells are indicated in gray. Each position in the next generation is determined by its state and the states of its immediate neighbors to the left and right. Because each of these cells can exist in two states, the number of possibilities for all three cells is, by the multiplication principle 2·2·2 = 8. What to do in each of the eight cases can be easily illustrated by a diagram like the one given in Figure 19.32.

For example, the third "T" from the left indicates that if the pixel in question is OFF (white) and its neighbors are both ON (gray), then the pixel in question will be OFF in the next generation. This is indicated by the white cell at the bottom of the T.

Associating the new values with 0 and 1 (for OFF and ON) for each of the cases, and reading the numbers from left to right, the rule can be tersely represented as 01011010. Converting this form base 2 to base 10, it becomes 90. Hence, it is referred to as Rule 90.

Starting with a row of OFF cells with a single ON cell in the center and iterating, we would get something that looks like the Game of Life played on a line. However, a much more interesting result arises from displaying the successive generations simultaneously, placing each new generation (an individual line) underneath the previous generation. Doing so for Rule 90, with the initial condition stated above, yields a familiar object (Figure 19.33).

The resemblance is stronger if more iterations are carried out (Figure 19.34).

Another rule is shown in Figure 19.35.

Because the row of single cells is 00011110, this is called Rule 30. Its behavior is quite different (Figure 19.36).

FIGURE 19.32
A diagrammatic rule for a cellular automaton.

Image created by Craig P. Bauer.

---

135 Horgan, John, "The Death of Proof," *Scientific American*, Vol. 269, No. 4, October 1993, pp. 92–95, 98–103, p. 100 quoted here.
136 Roberts, Siobhan, *Genius at Play – The Curious Mind of John Horton Conway*, Bloomsbury, New York, 2015, p. 153.

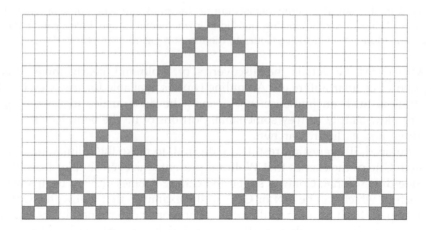

FIGURE 19.33
The first 16 rows of Rule 90 (15 generations).

Image created by Craig P. Bauer.

FIGURE 19.34
The first 512 rows of Rule 90 (511 generations).

Image created by Craig P. Bauer.

FIGURE 19.35
Rule 30.

Image created by Craig P. Bauer.

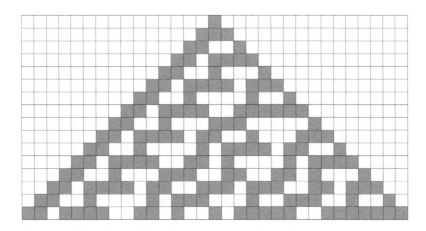

FIGURE 19.36
The first 16 rows of Rule 30 (15 generations).

Image created by Craig P. Bauer.

Some patterns are apparent, such as the state of the cells on the two diagonals leading off from the sole ON cell in generation zero. However, there is no better way to predict some of the cells than to iterate. That is, Rule 30 is chaotic.[137] For example, the center column, 1, 1, 0, 1, 1, 1, 0, 0, 1, 1, 0, 0, 0, 1, 0, 1, …, is essentially unpredictable.[138] Wolfram used this rule in *Mathematica* as the random number generator.[139]

Looking at a one-dimensional cellular automaton more generally, we see that there are 8 different cases for which we need to decide whether the rule will require a given a cell in the next generation to be ON or OFF. Because we have 2 options for each of the 8 cases, there are a total of $2^8 = 256$ possible rules. And there's quite a bit of variety in the results. Wolfram examined all of them and found that 88 of them are fundamentally inequivalent.[140] You're asked to investigate a few in the exercises at the end of this chapter.

Rule 110 turned out to be universal. That is, with the cells set to appropriate patterns for the initial row (generation 0), it can run whatever program is desired, acting as a universal computer.[141]

More complicated rules can be made. For example, instead of two states (ON and OFF), the cells can have many states. These states are easy to represent on a computer screen by making use of several colors. Some of the results are pretty, but they aren't more interesting mathematically. The tremendous complexity that arises from cellular automata can be achieved with simple rules and making matters more complicated doesn't seem to return anything of greater value.

---

137 Wolfram, Stephen, *A New Kind of Science*, Wolfram Media, Inc., Champaign, IL, 2002, p. 871.
138 https://oeis.org/A051023
139 Wolfram, Stephen, *A New Kind of Science*, Wolfram Media, Inc., Champaign, IL, 2002, p. 317.
140 Wolfram, Stephen, *A New Kind of Science*, Wolfram Media, Inc., Champaign, IL, 2002, p. 57.
141 See Roberts, Siobhan, *Genius at Play – The Curious Mind of John Horton Conway*, Bloomsbury, New York, 2015, p. 153 for the context and history of this discovery.

Chapter 9 of Wolfram's book is titled *Fundamental Physics*. It investigates the role simple rules like those of cellular automata might play in determining the physics of a discrete universe. The chapter concludes with

> And all of this supports my strong belief that in the end it will turn out that every detail of our universe does indeed follow rules that can be represented by a very simple program – and that everything we see will ultimately emerge from just running this program.[142]

Applications and the Future of Cellular Automata

For thousands of years, philosophers and scientists have been aware of the concept of **emergence**. It refers to how complex structure arises from simple systems, in which the initial conditions may be random. The system can be physical, as in the formation of a snowflake, or biological, like the organization of a colony of ants. Despite the antiquity of the concept, using the word emergence in this context, only goes back to the 19th century,[143] and research in this area has exploded in recent decades with the rise of chaos theory, fractals, and cellular automata.

In this chapter's examples, the initial conditions were highly organized; having a single ON cell in a row, for example. However, the diverse results that Wolfram and others observed also arise from random initial conditions, so the work is relevant to problems in emergence, whether or not cellular automata turn out to underlie our very reality. How important will posterity ultimately judge cellular automata to be? Opinions vary widely, with one extreme represented by Wolfram.

> In a 1995 telephone interview, Wolfram assured former *Scientific American* correspondent John Horgan that his new science would be the biggest thing since Newton's discoveries. When Horgan asked why Newton, why not Einstein? Wolfram replied that his ideas were even bigger than Einstein's.[144]

Are We Simulated?

Suppose that the universe *is* some sort of cellular automata. We're a part of the universe, so does that mean we're simulations? If so, who is running the simulation?

Science fiction authors and filmmakers have been exploring issues like this for decades. The most widely known example today is *The Matrix* film franchise, but this was far from the first. Investigations into the two questions "What is reality?" and "What is human?" make up a large portion of the writings of Philip K. Dick. Films based on his works include

---

142 Wolfram, Stephen, *A New Kind of Science*, Wolfram Media, Inc., Champaign, IL, 2002, p. 545
143 Lewes, George Henry, *Problems of Life and Mind*, Trübner, London, 1875.
144 Naiditch, David, "Divine Secrets of the Ya-Ya Universe, Stephen Wolfram: A New Kind of Science – or a Not-So-New Kind of Computer Program?" *Skeptic*, Vol. 10, No. 2, 2003. Also see Roberts, Siobhan, *Genius at Play – The Curious Mind of John Horton Conway*, Bloomsbury, New York, 2015, pp. 155–156.

Bladerunner (1982), from the book *Do Androids Dream of Electric Sheep?*, and *Total Recall* (1990), from the short story "We Can Remember It for You Wholesale."

For decades, one of the biggest names in science fiction was Isaac Asimov (1920–1992). But he was much more than a science fiction author. He was a renaissance man who bridged the worlds of the sciences and the humanities. He had a Ph.D. in chemistry and the more than 500 books that he wrote or edited included a huge number on popular science. He also wrote about history, literature, religion, and much more.[145] The Isaac Asimov Memorial Debate is an annual event and in 2016 the question was "Is the Universe a Simulation?"[146] The event was hosted by Neil deGrasse Tyson, who at the start of the program seemed to be snickering at the idea. At the end, the participants were asked to give their opinions as to the percent chance that the universe is a simulation. One passed on this question. The answers of the others were: effectively 0% (Lisa Randall), 1% (James Gates), 17% (Max Tegmark[147]), and 42% (David Chalmers). Although Gates put the chance at only 1%, he was intrigued by the presence of error-correcting codes at a fundamental level in particle physics. He said, this brought him to the "very stark realization" that he "could no longer say that people like Max were crazy."

After this Tyson said,

> I think the likelihood may be very high. ... I'm elsewhere on YouTube saying this. ... It is *easy* for me to imagine that *everything* in our lives is just the creation of some other entity for their entertainment. It is *easy* for me to think that. So, whatever the likelihood is, 0 percent, 1 percent, 17, 42, no answer, I'm saying the day we learn that it is true, I will be the only one in the room saying "I'm not surprised."

Elsewhere, Tyson said he thinks there's a 50/50 chance we're living in a simulation.[148] One of the best lines from the debate was delivered by Tyson, summing up the perspective of Max Tegmark, "Everything is mathematics. And if everything is mathematics, then everything is programmable."

A highly cited paper on this topic by Nick Bostrom is titled "Are We Living in a Computer Simulation."[149] Bostrom believes that there's a 20% chance that the answer is yes, although he does qualify that it's subjective.[150]

Others put the chances much higher. For example, in 2016 Elon Musk said, "There's a one in billions chance we're in base reality."[151] To understand what Musk means by "base reality," I turn back to science fiction. In the film *The Thirteenth Floor* (1999),

---

145 Roberts, Siobhan, *Genius at Play – The Curious mind of John Horton Conway*, Bloomsbury, New York, 2015, p. 429 notes that Isaac Asimov inspired Freeman Dyson, who influenced Conway. On a personal note, as a high school student, I found Asimov's popular science works inspirational and had the pleasure of hearing him lecture twice. INSPIRATION Pass It On. www.passiton.com/

146 https://www.youtube.com/watch?v=wgSZA3NPpBs

147 See Tegmark, Max, *Our Mathematical Universe: My Quest for the Ultimate Nature of Reality*, Vintage, 2014.

148 Gohd, Chelsea, "Are We Living in a Computer Simulation? Elon Musk Thinks So." *Futurism*, April 13, 2017, available online at https://futurism.com/are-we-living-in-a-computer-simulation-elon-musk-thinks-so/.

149 Bostrom, Nick, "Are We Living in a Computer Simulation?" *The Philosophical Quarterly*, Vol. 53, No. 211, April 2003, pp. 243–255.

150 https://getpocket.com/explore/item/this-cartoon-explains-why-elon-musk-thinks-we-re-characters-in-a-computer-simulation-he-might-be-1329513053

151 Gohd, Chelsea, "Are We Living in a Computer Simulation? Elon Musk Thinks So." *Futurism*, April 13, 2017, available online at https://futurism.com/are-we-living-in-a-computer-simulation-elon-musk-thinks-so/.

a simulation creates a simulation of its own. So, the question might not be as simple as simulation versus reality. There could be simulations within simulations, as deep as one cares to go, nested like Doron Zeilberger's lemmas. The question could be how many levels from the base reality is the present simulation. Readers who find such things intellectually intriguing can find a list of more films in this subsubgenre online.[152]

Musk explained the reason for his near certainty that we're living in a computer simulation.

> The strongest argument for us being in a simulation, probably being in a simulation, is the following: 40 years ago, we had Pong, two rectangles and a dot…That is what games were. Now, 40 years later, we have photorealistic 3D simulations with millions of people playing simultaneously, and it's getting better every year. And soon we'll have virtual reality, augmented reality. If you assume any rate of improvement at all, the games will become indistinguishable from reality.[153]

Ideas like this entered the mind of at least one cellular automata researcher early on. Edward Fredkin (1934–) started his investigations in the 1950s and was an early adopter of the theory of our universe as computer, or "digital physics" as he called it. Fredkin is an interesting character in a number of ways. He managed to become a full professor at MIT, despite the fact that he didn't even have a bachelor's degree.[154] He learned to program computers while in the Air Force, after flunking out of CalTech in his sophomore year.[155] Fortunately, there have been many instances, over the centuries, of universities being more interested in what a potential faculty member knows than how he or she learned it.

> [Fredkin] began discussing with [Marvin] Minsky the possibility of becoming a visiting professor at MIT. The idea of bringing a college dropout onto the faculty, Minsky recalls, was not as outlandish as it now sounds; computer science had become an academic discipline so suddenly that many of its leading lights possessed meager formal credentials.[156]

After barely a year of teaching at MIT, Fredkin was made a full professor.[157] In addition to rubbing elbows with the big names at MIT, a sabbatical starting in the fall of 1974 allowed Fredkin to learn from Richard Feynman, then at CalTech. In exchange for what Feynman could teach him about physics, Fredkin helped Feynman with computer science.[158] While Feynman is famous for viewing the world in his own unique way,

---

152 See https://letterboxd.com/chrbutler/list/simulated-reality/.

153 Gohd, Chelsea, "Are We Living in a Computer Simulation? Elon Musk Thinks So." *Futurism*, April 13, 2017, available online at https://futurism.com/are-we-living-in-a-computer-simulation-elon-musk-thinks-so/.

154 Wright, Robert, "Did the Universe Just Happen?" *Atlantic Monthly*, Vol., 261, No. 4, April 1988, pp. 29–44, www.theatlantic.com/past/docs/issues/88apr/wright.htm.

155 Wright, Robert, "Did the Universe Just Happen?" *Atlantic Monthly*, Vol., 261, No. 4, April 1988, pp. 29–44, www.theatlantic.com/past/docs/issues/88apr/wright.htm.

156 Wright, Robert, "Did the Universe Just Happen?" *Atlantic Monthly*, Vol., 261, No. 4, April 1988, pp. 29–44, www.theatlantic.com/past/docs/issues/88apr/wright.htm.

157 Wright, Robert, "Did the Universe Just Happen?" *Atlantic Monthly*, Vol., 261, No. 4, April 1988, pp. 29–44, www.theatlantic.com/past/docs/issues/88apr/wright.htm.

158 Wright, Robert, "Did the Universe Just Happen?" *Atlantic Monthly*, Vol., 261, No. 4, April 1988, pp. 29–44, www.theatlantic.com/past/docs/issues/88apr/wright.htm.

Fredkin managed to win others over to his perspective. One of his MIT colleagues, Tomasso Toffoli, published a paper in *Physica D* titled "Cellular automata as an alternative to (rather than an approximation of) differential equations in modeling physics."[159] The universe can be approached with continuous or discrete models, but as Toffoli points out in his paper, "it becomes obvious that the conceptual development of mathematical physics must have been strongly influenced by the nature of the available computational tools."[160] Prior to the existence of modern computers, the discrete approach simply wasn't practical. So, a path has been pursued for hundreds of years, not because it was the best path, but because it was the most accessible for nearly all of that time.

There are many philosophical or even theological conclusions one could attempt to draw from the perspective of the universe as a sort of cellular automata.

> Fredkin is at pains to clarify his position: "I guess what I'm saying is – I don't have any religious belief. I don't believe that there is a God. I don't believe in Christianity or Judaism or anything like that, okay? I'm not an atheist, I'm not an agnostic, I'm just in a simple state. I don't know what there is or might be. But what I can say is that it seems likely to me that this particular universe we have is a consequence of something I would call intelligent." Does he mean that there's something out there that wanted to get the answer to a question? "Yeah." Something that set up the universe to see what would happen? "In some way, yes."[161]

Back down on Earth, in 1997, a prize of $100,000, established by Fredkin for the creators of the first computer program that could beat a current world chess champion was awarded. Deep Blue had defeated Gary Kasparov.[162] The large prize makes one wonder if Fredkin was rooting for the machines, like Claude Shannon (see Chapter 1).

Amit Hagar traced similar ideas back hundreds of years.

> The thesis that ultimate reality is computational is not new. Leibniz already envisioned the world as an automaton in his "Monadology," Charles Babbage thought natural laws were like programs run by his analytical engine, and recently, under the PCTT, dynamical evolutions are regarded as computations.[163]

If reality could be shown to be discrete (see Chapter 0), this would be a big point in favor of the simulation hypothesis.

> If you have a crazy idea and it works out, then suddenly you are the great genius. Einstein had a crazy idea and it worked out. Newton had a crazy idea and it was criticized as occultism. Most people who have crazy ideas just have crazy ideas. On

159 Toffoli, Tommaso, "Cellular Automata as an Alternative to (Rather than an Approximation of) Differential Equations in Modeling Physics," *Physica D: Nonlinear Phenomena*, Vol. 10, No. 1–2, January 1984, pp. 117–127.

160 Toffoli, Tommaso, "Cellular Automata as an Alternative to (Rather than an Approximation of) Differential Equations in Modeling Physics," *Physica D: Nonlinear Phenomena*, Vol. 10, No. 1–2, January 1984, pp. 117–127, p. 117 quoted here.

161 Wright, Robert, "Did the Universe Just Happen?" *Atlantic Monthly*, Vol., 261, No. 4, April 1988, pp. 29–44, www.theatlantic.com/past/docs/issues/88apr/wright.htm.

162 https://www.chess.com/chessopedia/view/fredkin-prize

163 Hagar, Amit, *Discrete or Continuous?: The Quest for Fundamental Length in Modern Physics*, Cambridge University Press, Cambridge, UK, 2014, p. 40.

the other hand, the obvious idea turns out not to be true. You know: obviously, the world is flat.

– John H. Conway[164]

Open Problems

While the previous section could be considered a far-fetched extremely challenging open problem, there are many more reasonable open problems associated with cellular automata. Examples follow.

One of the first questions any mathematician looking into the Game of Life would ask is, amazingly, still open. Are there oscillators of all periods? Nobody knows! Periods that have yet to be found are 19, 23, 38 and 41.[165]

Stephen Wolfram produced a 45-page document listing open problems.[166] They range widely in apparent difficulty level. Some are of a historical nature. An example is:

> What discoveries about cellular automata were made in the 1950s and early 1960s in the context of cryptographic work? There is good evidence that extensive work was done, but it appears still to be classified. Is it in fact? Is it also secret in the former Soviet Union?[167]

And, of course, there's the open problem that began this chapter.

> The $3x + 1$ problem remains unsolved, and a solution remains unapproachable at present.
>
> Jeffrey C. Lagarias[168]

Exercise Set 19

> In mathematics you don't understand things. You just get used to them.
> John von Neumann[169]

> I'm confused at various times. In fact, I'm confused at all times. It's a permanent state.
> John H. Conway[170]

164 Roberts, Siobhan, *Genius at Play – The Curious Mind of John Horton Conway*, Bloomsbury, New York, 2015, p. 166.
165 http://conwaylife.com/wiki/Omniperiodic
166 Wolfram, Stephen, A New Kind of Science, Open Problems and Projects, Incomplete Preliminary version, June 26, 2003, www.wolframscience.com/openproblems/NKSOpenProblems.pdf.
167 Wolfram, Stephen, A New Kind of Science, Open Problems and Projects, Incomplete Preliminary version, June 26, 2003, www.wolframscience.com/openproblems/NKSOpenProblems.pdf, p. 7.
168 Lagarias, Jeffrey C., editor, *The Ultimate Challenge: The 3x + 1 Problem*, American Mathematical Society, Providence, RI, 2010, p. 16.
169 Zukav, Gary, *The Dancing Wu Li Masters*. Bantam Books, p. 208, footnote.
170 Roberts, Siobhan, *Genius at Play – The Curious Mind of John Horton Conway*, Bloomsbury, New York, 2015, p. 150.

1. Iterate

$$f(n) = \begin{cases} \dfrac{n}{2} & \text{if } n \text{ is even} \\ 3n+1 & \text{if } n \text{ is odd} \end{cases}$$

for the following initial values. How long does it take to reach 1 in each case?

   a) 12

   b) 15

   c) 69

   d) 100

2. Construct a diagram showing the paths of the first 10 integers to 1 under iteration of

$$f(n) = \begin{cases} \dfrac{n}{2} & \text{if } n \text{ is even} \\ 3n+1 & \text{if } n \text{ is odd} \end{cases}$$

3. What does the form shown in Figure 19.37 evolve into, following the rules of the Game of Life?

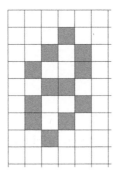

FIGURE 19.37
What happens to this "8"?

Image created by Craig P. Bauer.

4. In which direction does the Lightweight Spaceship shown in this chapter move?

5. What is the period of the object shown in Figure 19.38 in Life? Based on its appearances as it iterates, what would you call it?

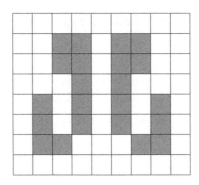

FIGURE 19.38
What is the fate of this symmetric form?

Image created by Craig P. Bauer.

Once you've answered these questions, try to find the object online. What is it called there? Is your name better or worse. Why?

6. What is the fate of diagonal lines, like the one shown in Figure 19.39 in Life? Answer this question in general, not just for the specific length shown.

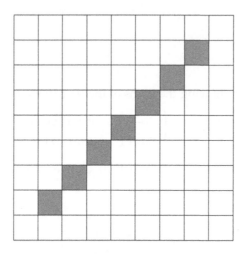

FIGURE 19.39
A diagonal line.

Image created by Craig P. Bauer.

7. What is the fate of the lifeform shown in Figure 19.40?

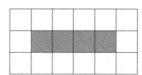

FIGURE 19.40
A four cell line.

Image created by Craig P. Bauer.

8. What is the fate of the lifeform shown in Figure 19.41?

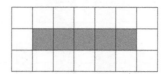

FIGURE 19.41
A five cell line.

Image created by Craig P. Bauer.

9. What is the fate of the lifeform shown in Figure 19.42?

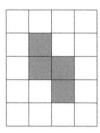

FIGURE 19.42
What is my fate?

Image created by Craig P. Bauer.

10. What is the fate of the lifeform shown in Figure 19.43?

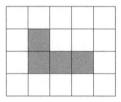

FIGURE 19.43
What is my fate?

Image created by Craig P. Bauer.

11. What is the period of the lifeform shown in Figure 19.44?

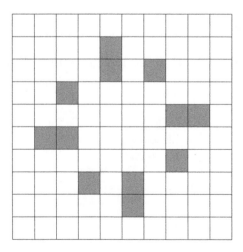

FIGURE 19.44
What is my period?

Image created by Craig P. Bauer.

12. What do the lifeforms in Figures 19.45 through 19.47 all lead to?

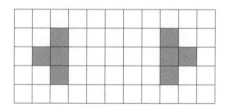

FIGURE 19.45
Where do I go?

Image created by Craig P. Bauer.

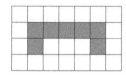

FIGURE 19.46
Where do I go?

Image created by Craig P. Bauer.

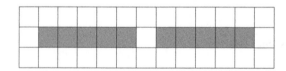

FIGURE 19.47
Where do I go?

Image created by Craig P. Bauer.

13. What are the periods of the lifeforms in Figures 19.48 through 19.50?

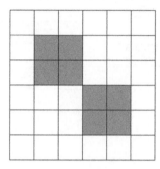

FIGURE 19.48
"Beacon."

Image created by Craig P. Bauer.

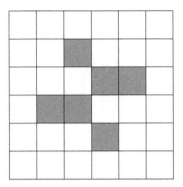

FIGURE 19.49
"Clock."

Image created by Craig P. Bauer.

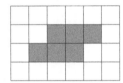

FIGURE 19.50
"Toad."

Image created by Craig P. Bauer.

14. What is the period of the Galaxy shown in Figure 19.51?

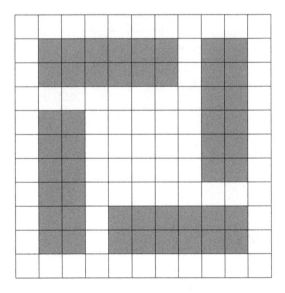

FIGURE 19.51
"Galaxy."

Image created by Craig P. Bauer.

15. Two similar looking lifeforms are shown in Figures 19.52 and 19.53. It takes 173 generations for pi to settle down.[171] What does H do?

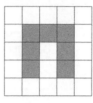

FIGURE 19.52
"Pi."
Image created by Craig P. Bauer.

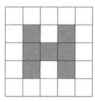

FIGURE 19.53
"H."
Image created by Craig P. Bauer.

16. What happens to the pinwheel (Figure 19.54) as it's iterated?[172]

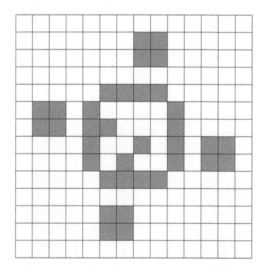

FIGURE 19.54
The "pinwheel."
Image created by Craig P. Bauer.

---

171 Gardner, Martin, "Mathematical Games: The Fantastic Combinations of John Conway's New Solitaire Game 'Life,'" *Scientific American*, Vol. 223, October 1970, p. 122.

172 Gardner, Martin, "Mathematical Games: The Fantastic Combinations of John Conway's New Solitaire Game 'Life,'" *Scientific American*, Vol. 223, October 1970, p. 122.

17. Which of the lifeforms shown in Figures 19.55 through 19.57 are stable?

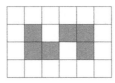

FIGURE 19.55
"Snake."

Image created by Craig P. Bauer.

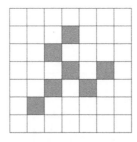

FIGURE 19.56
"Fork."

Image created by Craig P. Bauer.

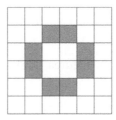

FIGURE 19.57
"Pond."

Image created by Craig P. Bauer.

18. Which of the lifeforms shown in Figures 19.58 through 19.60 are stable?

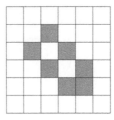

FIGURE 19.58
"Long boat."

Image created by Craig P. Bauer.

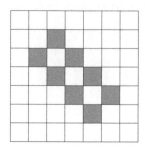

FIGURE 19.59
"Long barge."

Image created by Craig P. Bauer.

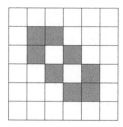

FIGURE 19.60
"Long ship."

Image created by Craig P. Bauer.

19. Which of the lifeforms shown in Figures 19.61 and 19.62 are stable?

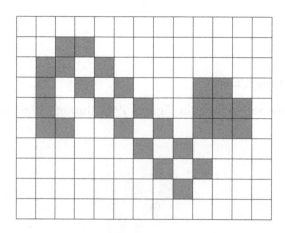

FIGURE 19.61
"Long cat."

Image created by Craig P. Bauer.

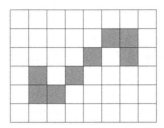

FIGURE 19.62
"Long snake."

Image created by Craig P. Bauer.

20. Create a glider that has both horizontal and vertical symmetry or prove that it's impossible.

21. Create a glider that looks the same when it is rotated to the right by 90 degrees or prove that it's impossible.

22. Could there be a glider that has bilateral symmetry, along a single axis or diagonal?

23. Find a description of Edward Fredkin's automata online or in print. Clearly state the rules he made use of and provide an example of a form that self-replicates under these rules. This last step should be especially easy.

24. Which of Stephen Wolfram's rules is indicated by the image below[173]?

25. Which of Stephen Wolfram's rules is indicated by the image below[174]?

26. Create a pictorial representation, like those in Exercises 24 and 25 above to represent Stephen Wolfram's Rule 166.

27. Create a pictorial representation, like those in Exercises 24 and 25 above to represent Stephen Wolfram's Rule 173.

28. Starting with a row of 31 cells with all of them OFF except for the one in the middle, iterate Stephen Wolfram's Rule 110 to generate 15 more rows.

29. Starting with a row of 31 cells with all of them OFF except for the one in the middle, iterate Stephen Wolfram's Rule 250 to generate 15 more rows.

---

173 Image created by Craig P. Bauer.
174 Image created by Craig P. Bauer.

30. Starting with a row of 31 cells with all of them OFF except for the one in the middle, iterate Stephen Wolfram's Rule 254 to generate 15 more rows.

31. Starting with a row of 31 cells with all of them OFF except for the one in the middle, iterate Stephen Wolfram's Rule 36 to generate 15 more rows.

32. Starting with a row of 31 cells with all of them OFF except for the one in the middle, iterate Stephen Wolfram's Rule 46 to generate 15 more rows.

33. Starting with a row of 31 cells with all of them OFF except for the one in the middle, iterate Stephen Wolfram's Rule 198 to generate 15 more rows.

34. What do you think the probability of our universe being a simulation is? Justify your answer.

35. Go to https://nickbostrom.com/. Find a paper that looks interesting and read it. Summarize it in your own words.

▬▬▬▬▬

References/Further Reading

> I own about five thousand books. I have books about most areas of science, as well as about mathematics, history and philosophy of science, and various areas of art and practical work that relate to science I have been interested in. I have collected a fairly large number of older books – often quite obscure in their times – that are relevant to my various interests. Among individual books my favorites are reference books and original books about ideas. My current favorite reference books are *Encyclopedia Britannica*, *EDM* (*Encyclopedic Dictionary of Mathematics*), and *DSB* (*Dictionary of Scientific Biography*).
> – Stephen Wolfram[175]

Abt, Clark C., *Serious Games: The Art and Science of Games That Simulate Life*, The Viking Press, New York, 1970.

Adamatzky, Andrew, editor, *Game of Life Cellular Automata*, Springer, London, 2010.

Adamatzky, Andrew, "Game of Life on Phyllosilicates: Gliders, Oscillators and Still Life," *Physics Letters A*, Vol. 377, No. 25–27, October 1, 2013, pp. 1597–1605, available online at https://arxiv.org/pdf/1306.0253.pdf.

Arbib, Michael A., "Simple Self-Reproducing Universal Automata," *Information and Control*, Vol. 9, 1966, pp. 177–189.

Bays, Carter, "The Discovery of Glider Guns in a Game of Life for the Triangular Tessellation," *Journal of Cellular Automata*, Vol. 2, No. 4, 2007, pp. 345–350, available online at www.old citypublishing.com/FullText/JCAfulltext/JCA2.4fulltext/JCAv2n4p345-350Bays.pdf.

Beer, Randall D., "The Cognitive Domain of a Glider in the Game of Life," *Artificial Life*, Vol. 20, No. 2, Spring 2014, pp. 183–206.

Beer, Randall D., "Characterizing Autopoiesis in the Game of Life," *Artificial Life*, Vol. 21, No. 1, Winter 2015, pp. 1–19.

Berlekamp, Elwyn, John Conway, and Richard Guy, *Winning Ways (For Your Mathematical Plays)*, Vol 2, Academic Press, New York, 1982. The dedication to this book reads, "To Martin Gardner who has brought more mathematics to more millions than anyone else." See Chapter 25, "What is Life?".

Bosch, Robert and Michael Trick, "Constraint Programming and Hybrid Formulations for Three Life Designs," *Annals of Operations Research*, Vol. 130, No. 1–4, August 2004, pp. 41–56.

---

175 http://www.stephenwolfram.com/qanda/lifeandtimes.html

Bostrom, Nick, "Are We Living in a Computer Simulation?" *The Philosophical Quarterly*, Vol. 53, No. 211, April 2003, pp. 243–255.

Bostrom, Nick, "Nick Bostrom's Homepage," available online at https://nickbostrom.com/.

Burks, Arthur W., editor, *Essays on Cellular Automata*, University of Illinois Press, Urbana and Chicago, 1970.

Callahan, Paul, "Patterns, Programs, and Links for Conway's Game of Life," available online at http://radicaleye.com/lifepage/.

Campbell, Jeremy, *Grammatical Man: Information Theory, Entropy, Language, and Life*, Simon & Schuster, New York, 1982.

Chamberland, Marc, "A Continuous Extension of the $3x + 1$ Problem to the Real Line," *Dynamics of Continuous, Discrete and Impulsive Systems*, Vol. 2, No. 4, 1996, pp. 495–509.

Chu, Geoffrey and Peter J. Stuckey, "A Complete Solution to the Maximum Density Still Life Problem," *Artificial Intelligence*, Vol. 184–185, June 2012, pp. 1–16.

Codd, Edgar F., *Cellular Automata*, Academic Press, New York and London, 1968.

Conway, John H. and Francis Y. C. Fung, *The Sensual (Quadratic) Form*, MAA, Washington, DC, 1997.

Cook, Matthew, "Universality in Elementary Cellular Automata," *Complex Systems*, Vol. 15, No. 1, 2004, pp. 1–40.

de Oliveira, Gina Maira Barbosa and Cardoso Siqueira Sandra Regina, "Parameter Characterization of Two-Dimensional Cellular Automata Rule Space," *Physica D: Nonlinear Phenomena*, Vol. 217, No. 1, May 2006, pp. 1–6.

Dunn, Ashley, "Predictably, the Game of Life Is Not Predictable," *New York Times*, July 16, 1997, available online at https://archive.nytimes.com/www.nytimes.com/library/cyber/surf/071697mind.html.

Elran, Yossi, "Retrolife and the Pawns Neighbors," *The College Mathematics Journal*, Vol. 43, No. 2, March 2012, pp. 147–151.

Evans, Kellie Michelle, "Larger than Life: Digital Creatures in a Family of Two-Dimensional Cellular Automata," *Discrete Mathematics and Theoretical Computer Science Proceedings*, AA (DM-CCG), 2001, pp. 177–192, available online at www.emis.ams.org/journals/DMTCS/pdfpapers/dmAA0113.pdf.

Farmer, Doyne, Tommaso Toffoli, and Stephen Wolfram, *Cellular Automata*, North-Holland Physics Publishing, New York, 1984.

Fredkin, Edward, "Digital Mechanics: An Informational Process Based on Reversible Universal Cellular Automata," *Physica D: Nonlinear Phenomena*, Vol. 45, No. 1–3, September 1990, pp. 254–270.

Gardner, Martin, "Mathematical Games, the Fantastic Combinations of John Conway's New Solitaire Game 'Life'," *Scientific American*, Vol. 223, October 1970, pp. 120–123.

Gardner, Martin, "Mathematical Games," *Scientific American*, Vol. 224, No. 2, February 1971, pp. 112–117.

Gardner, Martin, "Mathematical Games," *Scientific American*, Vol. 224, No. 3, March 1971, pp. 106–109.

Gardner, Martin, *Wheels, Life and Other Mathematical Amusements*, W. H. Freeman, New York and San Francisco, 1983. Gardner's writings on the Game of Life are collected here.

Gohd, Chelsea, "Are We Living in a Computer Simulation? Elon Musk Thinks So," *Futurism*, April 13, 2017, available online at https://futurism.com/are-we-living-in-a-computer-simulation-elon-musk-thinks-so/.

Gotts, Nicholas M., "Ramifying Feedback Networks, Cross-Scale Interactions, and Emergent Quasi Individuals in Conway's Game of Life," *Artificial Life*, Vol. 15, No. 3, Summer 2009, pp. 351–375.

Guy, Richard K., "John Horton Conway: Mathematical Magus," *Two-Year College Mathematics Journal*, Vol. 13, No. 5, November 1982, pp. 290–298.

Guy, Richard K., "Conway's Prime Producing Machine," *Mathematics Magazine*, Vol. 56, No. 1, January 1983, pp. 26–33.

Guy, Richard K., "Don't Try to Solve These Problems," *The American Mathematical Monthly*, Vol. 90, No. 1, January 1983, pp. 35–38, 39–41. The $3n + 1$ problem is included.

Hargittai, Istvan, "John Conway – Mathematician of Symmetry and Everything Else," *Mathematical Intelligencer*, Vol. 23, No. 2, 2001, pp. 6–14.

Heims, Steve J., *John Von Neumann and Norbert Wiener: From Mathematics to the Technologies of Life and Death*, M.I.T. Press, Cambridge, MA, and London, 1980.

Horgan, John, "The Death of Proof," *Scientific American*, Vol. 269, No. 4, October 1993, pp. 92–95, 98–103.

Johnston, Nathaniel, "The Collatz Conjecture as a Fractal," June 20, 2009, available online at www.njohnston.ca/2009/06/the-collatz-conjecture-as-a-fractal/.

Johnston, Nathaniel, *ConwayLife.com – A Community for Conway's Game of Life and Related Cellular Automata*, www.conwaylife.com/.

"Kaleidoscope of 3D Life," updated December 8, 2000, www.ibiblio.org/e-notes/Life/Game.htm.

Khovanova, Tanya, "Conway's Recipe for Success," *Tanya Khovanova's Math Blog*, March 6, 2010, http://blog.tanyakhovanova.com/?p=214.

Khovanova, Tanya, "The Greatest Mathematician Alive," *Tanya Khovanova's Math Blog*, March 20, 2010, http://blog.tanyakhovanova.com/?p=218.

Khovanova, Tanya, "The Sexual Side of Life," *Tanya Khovanova's Math Blog*, July 25, 2010, https://blog.tanyakhovanova.com/2010/07/the-sexual-side-of-life/.

Knuth, Donald E., *Surreal Numbers: How Two Ex-Students Turned on to Pure Mathematics and Found Total Happiness*, Addison-Wesley, Reading, MA, 1974.

Kolata, Gina, "Scientist at Work: John H. Conway; At Home in the Elusive World of Mathematics," *New York Times*, October 12, 1993, www.nytimes.com/1993/10/12/science/scientist-at-work-john-h-conway-at-home-in-the-elusive-world-of-mathematics.html.

Lagarias, Jeffrey C., editor, *The Ultimate Challenge: The 3x + 1 Problem*, American Mathematical Society, Providence, RI, 2010.

Letherman, Simon, Dierk Schleicher, and Reg Wood, "The $3n + 1$ problem and Holomorphic Dynamics," *Experimental Mathematics*, Vol. 8, No. 3, 1999, pp. 241–251.

Lewes, George Henry, *Problems of Life and Mind*, Trübner, London, 1875.

Lippy, Tod, "John Conway," *Esopus*, No. 6, "Process", 2006, pp. 33–52.

Mallows, Colin, "Conway's Challenge Sequence," *American Mathematical Monthly*, Vol. 98, No. 1, January 1991, pp. 5–20.

Mallows, Colin, "Conway's Challenge, Mallows Replies," *American Mathematical Monthly*, Vol. 99, No. 6, June–July 1992, pp. 563–564.

Margolus, Norman and Tomasso Toffoli, *Cellular Automata Machines*, MIT Press, Cambridge, MA, 1987.

Meinhardt, Hans, *The Algorithmic Beauty of Sea Shells*, Springer, Berlin, 1995.

Moore, Edward F., "Machine Models of Self-reproduction," in Bellman, Richard, editor, *Proceedings of Symposia in Applied Mathematics*, Vol 14, American Mathematical Society, Providence, RI, 1962, pp. 17–34.

Moore, Edward F., "Mathematics in the Biological Sciences," *Scientific American*, Vol. 211, No. 3, September 1964, pp. 148–154, 156–158, 160, 162, 164.

Naiditch, David, "Divine Secrets of the Ya-Ya Universe, Stephen Wolfram: A New Kind of Science – Or a Not-So-New Kind of Computer Program?" *Skeptic*, Vol. 10, No. 2, 2003, pp. 30–49.

Poundstone, William, *The Recursive Universe: Cosmic Complexity and the Limits of Scientific Knowledge*, Willow Morrow and Company, Inc, New York, 1985.

Poundstone, William, *Prisoner's Dilemma*, Doubleday, New York, 1992.

Radenski, Atanas, "Using MapReduce Streaming for Distributed Life Simulation on the Cloud," in Pietro, Liò, Orazio Miglino, Giuseppe Nicosia, Stefano Nolfi and Mario Pavone, editors, *Advances in Artificial Life, ECAL 2013, Proceedings of the Twelfth European Conference on the Synthesis and Simulation of Living Systems*, MIT Press, Cambridge, MA, September 2013, pp. 284–291, available online at https://pdfs.semanticscholar.org/d06c/6c3f7e8e3ff972a955a9bfb21c7722a3c51b.pdf.

Rangel-Mondragón, Jaime "A Catalog of Cellular Automata," available online at http://library.wolfram.com/infocenter/MathSource/505.

Rendell, Paul, "A Universal Turing Machine in Conway's Game of Life," in Smari, Waleed W. and John P McIntire, editors, *Proceedings of the 2011 International Conference on High Performance Computing & Simulation (HPCS 2011)*, Istanbul, Turkey, July 4–8, 2011, IEEE, pp. 764–772, available online at http://uncomp.uwe.ac.uk/CAAA2011/Program_files/764-772.pdf.

Rennard, Jean-Philippe, "Implementation of Logical Functions in the Game of Life," in Adamatzky, Andrew, editor, *Collision-Based Computing*, Springer, London, 2002, pp. 491–512, available online at https://arxiv.org/ftp/cs/papers/0406/0406009.pdf and www.rennard. org/alife/CollisionBasedRennard.pdf.

Roberts, Siobhan, "Character Assassination," available online at www.youtube.com/watch? v=hslrD0Y837c. This is a talk given by the author of *Genius at Play – The Curious Mind of John Horton Conway* on March 21, 2014.

Roberts, Siobhan, *Genius at Play – The Curious Mind of John Horton Conway*, Bloomsbury, New York, 2015.

Rufford, Nick, "Numbers Man Hits $10,000 Jackpot," *Sunday Times*, London, September 11, 1988.

Schleicher, Dierk, "Interview with John Horton Conway," *Notices of the American Mathematical Society*, Vol. 60, No. 5, May 2013, pp. 567–575.

Seife, Charles, "Mathemagician: Impressions of Conway," *Sciences* May–June 1994, pp. 12–15. A version of this article appears online at www.users.cloud9.net/~cgseife/conway.html.

Shannon, Claude E. and John McCarthy, editors, *Automata Studies*, Princeton University Press, Princeton, NJ, 1956. This was the book that inspired Conway to pursue cellular automata.[176].

Silva, Jorge-Nuno, "Breakfast with John Horton Conway," *Newsletter of the European Mathematical Society*, No. 57, September 2005, pp. 32–34.

Simons Foundation, "Science Lives: John Conway," Interview by Alex Ryba, filming by George Csicsery, May 12, 2011, https://www.simonsfoundation.org/2014/04/04/john-conway/.

Sloane, Neil J. A., Sequence A051023, "Middle Column of Rule-30 1-D Cellular Automaton, from a Lone 1 Cell," *The On-Line Encyclopedia of Integer Sequences*, available online at https://oeis. org/A051023.

Smith, Alvy Ray, "Cellular Automata Theory. Tech. Report No. 2," Digital Systems Lab., Stanford Electronics Labs., Stanford University, 1969.

Summers, Jason, "Game of Life Status Page," available online at http://entropymine.com/jason/life/ status.html. This page has a great summary of what is known about the periods of various lifeforms.

Taubes, Gary, "John Horton Conway: A Mathematical Madness in Cambridge," *Discover*, August 1984, pp. 41–50.

Tegmark, Max, *Our Mathematical Universe: My Quest for the Ultimate Nature of Reality*, Vintage, New York, 2014.

Thomas, Rachel, "Games, Life and the Game of Life," *Plus Magazine*, May 1, 2002, available online at http://plus.maths.org/content/games-life-and-game-life.

Toffoli, Tommaso, "Cellular Automata as an Alternative to (Rather than an Approximation of) Differential Equations in Modeling Physics," *Physica D: Nonlinear Phenomena*, Vol. 10, No. 1–2, January 1984, pp. 117–127.

Ulam, Stanislaw M., *Adventures of a Mathematician*, Charles Scribner's Sons, New York, 1976.

Von Neumann, John, *Theory of Self-Reproducing Automata*, edited and completed by Arthur W. Burks, University of Illinois Press, Urbana and Chicago, 1966.

Wainwright, Robert T., editor, *Lifeline: A Quarterly Newsletter for Enthusiasts of John Conway's Game of Life*, Nos. 1–11, 1971–1973, available online at http://conwaylife.com/wiki/Category: Lifeline_issues.

Weinberg, Steven, "Is the Universe a Computer?" *New York Review of Books*, October 24, 2002.

Wikipedia Contributors, "Collatz Conjecture," *Wikipedia, The Free Encyclopedia*, available online at https://en.wikipedia.org/w/index.php?title=Collatz_conjecture&oldid=849382377.

---

176 Roberts, Siobhan, *Genius at Play – The Curious Mind of John Horton Conway*, Bloomsbury USA, 2015, p. 111.

Willson, Stephen J., "Cellular Automata Can Generate Fractals," *Discrete Applied Mathematics*, Vol. 8, No. 1, April 1984, pp. 91–99.

Wirsching, Günther J., *The Dynamical System Generated by the 3n+1 Function*, Lecture Notes in Mathematics, Vol 1681, Springer, Berlin, 1998.

Wolfram, Stephen, "Universality and Complexity in Cellular Automata," *Physica D*, Vol. 10, 1984, pp. 1–35.

Wolfram, Stephen, "Statistical Mechanics of Cellular Automata," *Reviews of Modern Physics*, Vol. 55, No. 3, July–September 1983, pp. 601–644.

Wolfram, Stephen, *Theory and Applications of Cellular Automata*, World Scientific, Singapore, 1986.

Wolfram, Stephen, *A New Kind of Science*, Wolfram Media, Inc., Champaign, IL, 2002.

Wright, Robert, "Did the Universe Just Happen?" *Atlantic Monthly*, Vol. 261, No. 4, April 1988, pp. 29–44, available online at www.theatlantic.com/past/docs/issues/88apr/wright.htm.

20

Graph Theory

Men are never more ingenious than in inventing games.

Leonhard Euler, 1715[1]

The Bridges of Königsberg (1736)[2]

While probability arose from games of chance (gambling), graph theory traces its roots to a friendlier sort of game, more of a challenge really, in which no money changed hands. Leonhard Euler explained the challenge in a 1736 paper, as follows:[3]

> The problem, which I am told is widely known, is as follows: In Königsberg in Prussia, there is an island A, called the *Kneiphof*; the river which surrounds it is divided into two branches, as can be seen in the Figure, and these branches are crossed by seven bridges [...] Concerning these bridges, it was asked whether anyone could arrange a route in such a way that he would cross each bridge once and only once. I was told that some people had asserted that this was impossible, while others were in doubt; but nobody would actually assert that it could be done.

For convenience, a map is reproduced in Figure 20.1, in place of Euler's Figure. You are invited to try to solve the puzzle by finding a path that will cross each bridge exactly once. Is such a path possible?

Explanations of Euler's solution often begin with the diagram shown in Figure 20.2.

Point A represents the center island, Kneiphof. It can be reduced to a point because it doesn't matter where in this island the stroller is. He or she may move about on it freely. All that matters is which bridges are crossed. The lines a, b, c, d, and e represent the five bridges that have an end in Kneiphof. The other ends, the points B, C, and D, represent the areas of Königsberg to the south, north, and east of Kneiphof, respectively. Like Kneiphof, these portions of the city have been reduced to points.

If there is a path, it begins somewhere and ends somewhere (possibly the same point at which it began), but of the points A, B, C, and D, at least two of them cannot be

---

1 Chamberland, Marc, *Single Digits: in praise of small numbers*, Princeton University Press, Princeton, NJ, 2015, p. 187.

2 Today, the city is known as Kaliningrad, Russia.

3 The original paper is Euler, Leonhard, "Solutio Problematis ad Geometriam situs Pertinentis" (The solution of a problem relating to the geometry of position), *Commentarii Academiae Scientiarum Imperialis Petropolitanae*, Vol. 8, 1736 (so dated, although it did not appear until 1741), pp. 128–140. The paper was published again in *Commentarri Academiae Scientiarum Imperialis Petropolitanae*, Vol. 8, edition nova, Bononiae, 1752, pp. 116–126 and also appears in *Opera Omnia, Series 1*, Vol. 7, pp. 1–10 and (in an English translation quoted from here) in Biggs, Norman L., E. Keith Lloyd, and Robin J. Wilson, *Graph Theory, 1736–1936*, Clarendon Press, Oxford, 1976 (reissued in paperback in 1998).

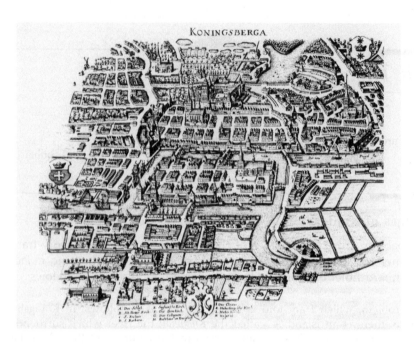

FIGURE 20.1
The Bridges of Königsberg.

Public domain.

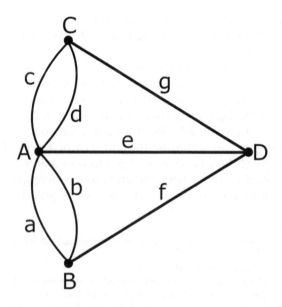

FIGURE 20.2
An abstract view of the problem.

Image created by Josh Gross.

starting or ending points. Thus, in a successful path, there are at least two points that are entered via a bridge (because we did not start there) and then departed via a different bridge (because we did not end there). If such a point is visited a second time, it must be entered and departed from via a new pair of bridges. In a larger problem of this type, involving more bridges, it's possible that this point would be passed through yet again via another pair of bridges. In any case, the number of bridges connected to this point would have to be even. But looking at the diagram above, we see that no point has an even number of bridges attached to it. Hence, we have a contradiction and the path cannot exist.

Euler addressed the general case, with any arrangement of landmasses and bridges, and correctly concluded,

> If there are more than two areas to which an odd number of bridges lead, then such a journey is impossible.
>
> If, however, the number of bridges is odd for exactly two areas, then the journey is possible only if it starts in either of these two areas.
>
> If, finally, there are no areas to which an odd number of bridges lead, then the required journey can be accomplished starting from any area.[4]

Diagrams like the one used in this proof come in handy in a variety of disciplines and, as a consequence, the terminology varies widely. The **points** (representing the land masses in this instance) can also be called **nodes** or **vertices**. The **lines** connecting the points are often called **edges** or **links**. The number of edges meeting at a given vertex, which plays such an important role in the proof here, is referred to as the **degree** or **valency** of that vertex. And the diagram itself is a **graph** or **network**.

Euler's contributions are remembered in today's terminology. We refer to a route that passes through every edge of a graph exactly once as an **Eulerian path** (aka **Eulerian trail** and **Euler walk**). If it happens that the path ends at the same vertex it began at, then we have an **Eulerian circuit** or an **Eulerian cycle**. An Eulerian circuit exists if and only if when a list is made of all of the graph's edges, each vertex appears as an endpoint of an edge an even number of times. When a circuit exists, it's easy to find. Begin by finding a path that begins and ends at the same point. Don't worry if there are edges that weren't traversed! If that happens, find an edge that connects with a point in your path and make a new path using it and other previously unused edges. You'll be able to make it so that it returns to the point on the original path that you began at. This new path can now be incorporated into your original path as a sort of "detour." You may have to add several such detours, but eventually every edge will be used and the result will be a circuit.

As Euler's initial application indicates, graph theory doesn't have anything to do with the x- and y- axes. Rather it deals with networks. Another example of a graph is shown in Figure 20.3. Only the manner in which the points are connected matters. This graph may be redrawn like the one shown in Figure 20.4. Because the connections are unchanged, these two graphs are considered to be identical. Typically, one attempts to draw a graph in the simplest possible manner, taking advantage of any symmetries that may be exploited.

---

4 From the English translation of Euler's paper.

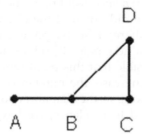

FIGURE 20.3
An example of a graph.

Image created by Craig P. Bauer.

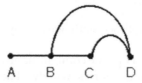

FIGURE 20.4
A graph that is mathematically identical to Figure 20.3.

Image created by Craig P. Bauer.

An edge may connect a point to itself. This is called a **loop** (or **self-loop** or **sling**). Multiple edges may connect the same vertices. If no loops or multiple edges are present, we have a **simple** (or **simplicial**) **graph**.

The graph shown in Figure 20.5 has a loop at A and two edges connecting C and D (such edges are said to be **parallel**). Loops count twice for valency. In Figure 20.4, the valency of A is 1, B has valency 3, C has valency 2, and D has valency 2. The sum of the valencies is 8. In Figure 20.5, every vertex has valency 3.

Although Euler's proof above was presented in a way that is typical for today's texts, the abstract diagram that was used didn't actually appear in Euler's paper. This useful

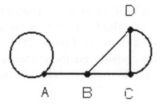

FIGURE 20.5
A graph with a loop.

Image created by Craig P. Bauer.

reference seems to have appeared for the first time in 1892 in W. W. Rouse Ball's famous book on recreational mathematics.[5] Besides the graph, there are some other improvements in the modern presentations (the introduction of the concept of degree, for example), but the basic approach is equivalent to that of Euler.

It's one matter to show that a solution exists for a particular system, and another to actually find it. Euler indicated how such a solution could be found.[6]

> When it has been determined that such a journey can be made, one still has to find how it should be arranged. For this I use the following rule: let those pairs of bridges which lead from one area to another be mentally removed, thereby considerably reducing the number of bridges; it is then an easy task to construct the required route across the remaining bridges, and the bridges which have been removed will not significantly alter the route found, as will become clear after a little thought. I do not therefore think it worthwhile to give any further details concerning the finding of the routes.

This is far from a detailed algorithm for finding a path when one exists. In fact, Euler had not really proven that a path must exist in this case. A complete proof didn't appear until 1873, when Carl Hierholzer's work on the topic, apparently carried out without knowledge of Euler's paper, saw print posthumously.[7]

Still, it isn't surprising that Euler didn't devote more ink to the bridges of Konigsberg problem, given a comment he made about it in a letter to an Italian mathematician, Giovanni Marinoni.

> This question is so banal, but seemed to me worthy of attention in that [neither] geometry, nor algebra, nor even the art of counting was sufficient to solve it.[8]

Euler also wrote

> Thus, you see. Most noble Sir, how this type of solution bears little relationship to mathematics, and I do not understand why you expect a mathematician to produce it, rather than anyone else, for the solution is based on reason alone, and its discovery does not depend on any mathematical principle.[9]

So, a banal question led to the birth of a new area of mathematics!

---

5 Ball, W. W. Rouse, *Mathematical Recreations and Problems of Past and Present Times* (later entitled *Mathematical Recreations and Essays*), Macmillan, 1892. The statement that this seems to be the diagram's first appearance is from, Wilson, Robin, and John J. Watkins, editors, *Combinatorics: Ancient & Modern*, Oxford University Press, Oxford, 2013, p. 188.

6 From the English translation of Euler's paper.

7 Hierholzer, Carl, "Uber die Moglichkeit, einen Lineanzug ohne Wiederholung und ohne Unterbrechung zu umfahren," *Mathematische Annalen*, Vol. 6, No. 1, 1873, pp. 30–32. A translation appears on pp. 11–12 of Biggs, Norman L., E. Keith Lloyd, and Robin J. Wilson, *Graph Theory, 1736–1936*, Clarendon Press, Oxford, 1976 (reissued in paperback in 1998).

8 See Hopkins, Brian, and Robin Wilson, "The Truth about Königsberg," *College Mathematics Journal*, Vol. 35, No. 3, May 2004, pp. 198–207, quote on p. 202, or Sachs, Horst, Michael Stiebitz, and Robin J. Wilson, "An historical note: Euler's Konigsberg letters," *Journal of Graph Theory*, Vol. 12, No. 1, Spring 1988, pp. 133–139, quote on p. 136.

9 Sachs, Horst, Michael Stiebitz, and Robin J. Wilson, "An historical note: Euler's Königsberg letters," *Journal of Graph Theory*, Vol. 12, No. 1, Spring 1988, pp. 133–139, quote on p. 136.

More Games

It took a long time for graph theory to leave the "game stage." A few other puzzles that were investigated early on are detailed in the following sections.

Tracing Complete Graphs (1809)

In 1809, Louis Poinsot, a French mathematician, posed the general problem of using a thread to trace out all possible connections between a set of n points, returning at the end to the starting point.[10]

If it could be done for small values of n, the results would appear as shown in Figure 20.6. These graphs, from left to right, are denoted as K_1, K_2, K_3, K_4, and K_5.

FIGURE 20.6
The first few complete graphs.

Image created by Craig P. Bauer.

Today, such graphs are known as **complete graphs**, because they contain all possible edges. A complete graph with n vertices is denoted K_n. While such graphs exist, they cannot always be traced out in the manner described by Poinsot. In fact, only the odd values for n have solutions. But, as n increases, the number of solutions for these odd cases grows dramatically.

Four Stroke Minimum (1844)[11]

When an object can't be traced with a single line (aka stroke), a natural question to ask is what is the minimum number of strokes that are needed? What do you think the answer is for the simple figure shown in Figure 20.7?

This problem is left as an exercise, but if you aren't sure whether your answer is correct or not, you can compare it with what's given on p. 187 of Wilson, Robin and John J. Watkins, editors, *Combinatorics: Ancient & Modern*, Oxford University Press, Oxford, 2013.

---

10 Poinsot, Louis, "Mémoire sur les polygones et les polyedres," *Journal de l'École polytechnique*. Vol. 4, 1809–1810, pp. 4, 16–48. See Wilson, Robin, and John J. Watkins, editors, *Combinatorics: Ancient & Modern*, Oxford University Press, Oxford, 2013, p. 186 for a discussion of this paper.

11 Clausen, Thomas, "De linearum tertii ordinis proprietatibus," *Astronomische Nachrichten*, Vol. 21, No. 14, Whole No. 494, 1844, pp. 209–216.

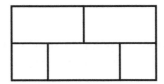

FIGURE 20.7
How many non-overlapping lines are needed to trace this?

Image created by Craig P. Bauer.

The Birth of Topology? (1847)[12]

If you're ready for a problem that looks a lot tougher than the bridges of Königsberg, but can actually be done with a single line, consider Figure 20.8.

This example appeared in Johann Benedict Listing's *Vorstudien zur Topologie* (Preliminary studies on topology), 1847, which has the claim to fame of being the first place the word topology appeared in print, albeit in its German version.[13] Graph theory can be thought of as a sort of special case of topology, a more general field, so it's not surprising that the disciplines have common roots. Topology will not be discussed in this text. The focus here is graph theory, which tends to be less abstract. The focus of graph theory is often on applications.

FIGURE 20.8
A symmetrical challenge.

Image redrawn by Josh Gross.

---

12 Listing, Johann Benedict, "Vorstudien zur Topologie," *Gottinger Studien, Erste Abteilung: Mathematische und naturwissenschaftliche Abhandlungen*, 1847, pp. 811–875. It was also published as a stand-alone book, Gottingen, 1848.

13 It previously appeared in a letter Listing wrote in 1836. Source: Wilson, Robin, and John J. Watkins, editors, *Combinatorics: Ancient & Modern*, Oxford University Press, Oxford, 2013, p. 187. Richeson, David, *Euler's Gem: The Polyhedron Formula and the Birth of Topology*, Princeton University Press, Princeton, NJ, 2012, p. 157 notes that the word had a botanical meaning before this.

The Icosian Game (1857)[14]

While Euler took a game away from the public by proving that there's no solution, the Irish mathematician William Rowan Hamilton (Figure 20.9) gave them a new one (Figure 20.10).

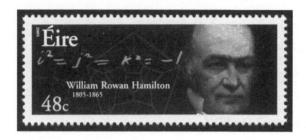

FIGURE 20.9
Sir William Rowan Hamilton (1805–1865).

Public domain.

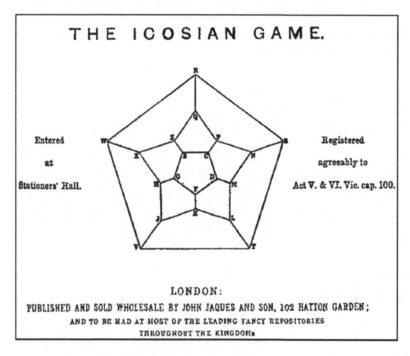

FIGURE 20.10
A sketch of the Hamilton's Icosian Game, from the instructions.

Hamilton, W. R., Instructions for the Icosian Game, John Jaques and Son, London, 1859 (public domain).

---

14 https://en.wikipedia.org/wiki/Icosian_game

The game board was a piece of wood with 20 holes (Figure 20.11), corresponding to the vertices of a dodecahedron – not an icosahedron as one might expect from the name![15] The holes represented cities, whose names began with the 20 English consonants, from Brussels to Zanzibar. The game came with a variety of challenges. One of these was to place pegs bearing the numbers 1 through 20 in all of the holes representing the cities by only progressing along the lines and never returning to a city once it has been visited.

This might sound like an example of the type of problem Euler addressed, but it is not. An Eulerian path passes through every edge exactly once. The problem Hamilton posed does not require every edge to be traversed. It is the vertices that must all be hit in this instance. A smaller example will help illustrate the difference. The graph of a cube is shown in Figure 20.12.

There's no Euler path for this graph, because all eight of its vertices are odd. However, starting with A and traversing from vertex to vertex alphabetically, all the way to H, every vertex is hit exactly once.

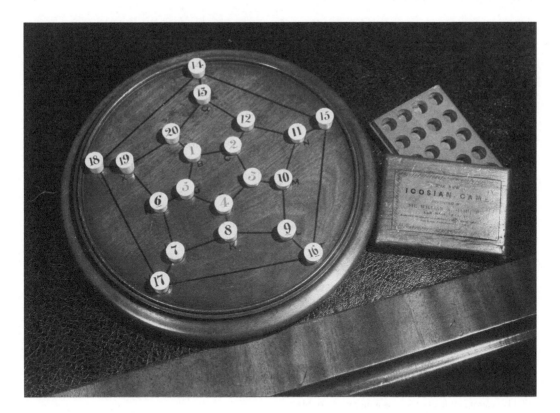

FIGURE 20.11
The Icosian game board and pieces.

By permission of the Royal Irish Academy © RIA.

---

15 Hamilton had looked at cycles of faces on an icosahedron, which correspond to cycles of vertices on a dodecahedron, so there is a connection after all! Source: Wilson, Robin, and John J. Watkins, editors, *Combinatorics: Ancient & Modern*, Oxford University Press, Oxford, 2013, p. 191.

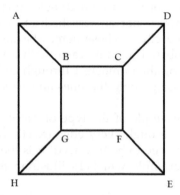

FIGURE 20.12
The graph of a cube.

Image created by Craig P. Bauer.

Because we have a truly different problem, there's terminology to distinguish it. We refer to a **Hamiltonian path**, and if it loops back to its starting point, as it does in the example above, if we pass from H back to A as a last step, a **Hamiltonian circuit** or a **Hamiltonian cycle**.[16]

Hamilton, it should be noted, was much more than a mathematician. Karl Weierstrass remarked, "A mathematician who is not also something of a poet will never be a complete mathematician." While this was just a colorful way of saying that a great mathematician must have the creative spark and sense of beauty that poets are known to possess, Hamilton was literally a poet. He counted Wordsworth[17] and Coleridge[18] among his friends.[19]

And he had acquired other skills. E. T. Bell described Hamilton's Uncle James as "an inhumanly accomplished linguist – Greek, Latin, Hebrew, Sanskrit, Chaldee, Pali, and heaven knows what other heathen dialects came to his tongue as readily as the more civilized languages of Continental Europe and Ireland."[20] At the age of 3, William Rowan Hamilton was already showing signs of his great intelligence and his father decided that the boy should learn the languages his uncle knew. As a result, he was turned into "one of the most shocking examples of a linguistic monstrosity in history."[21] By the age of 13, Hamilton had mastered 13 languages.

As a child, he had a very even temper. This may have changed as he aged, for there's a story in which he challenged someone to "mortal combat" for calling him a liar.[22]

---

16 Despite Hamilton's name being attached to these terms, Revd Thomas Penyngton Kirkman investigated the topic a few months before Hamilton, and in greater generality. Source: Wilson, Robin, and John J. Watkins, editors, *Combinatorics: Ancient & Modern*, Oxford University Press, Oxford, 2013, p. 191.
17 *I Wandered Lonely as a Cloud* is a nice poem of his.
18 Best known for *The Rime of the Ancient Mariner*.
19 Turnbull, Herbert Westren, *The Great Mathematicians*, Barnes & Noble Books, New York, NY, 1993, pp.129–130. This is a reprint of the original 1929 edition.
20 Bell, Eric Temple, *Men of Mathematics*, Dover Publications, New York, NY, 1937, p. 341. The racist tone is Bell's, not mine!
21 Bell, Eric Temple, *Men of Mathematics*, Dover Publications, New York, NY, 1937, p. 341. I don't get why this bothers Bell so much. I think it's cool.
22 Actual mortal combat, not the video game. See Bell, Eric Temple, *Men of Mathematics*, Dover Publications, New York, NY, 1937, p. 342.

When Hamilton was 12, he met Zerah Colburn (1804–1839), the American calculating boy. The two competed and Hamilton lost, but he benefited greatly from the meeting, as this is what marked the origin of his interest in mathematics.

Hamilton found an error in a proof by Laplace and, by the age of 17, had done original work in optics. When Hamilton was a 21-year-old undergraduate at Trinity College in Dublin, he was made a professor of astronomy there. He was knighted in 1835.

Hamilton's best-known mathematical result, from 1843, is an extension of the complex numbers known as the quaternions. It consists of ordered quadruples of real numbers $(\alpha, \beta, \gamma, \delta)$ with $1 = (1, 0, 0, 0)$, $i = (0, 1, 0, 0)$, $j = (0, 0, 1, 0)$, and $k = (0, 0, 0, 1)$. Furthermore, $i^2 = j^2 = k^2 = -1 = ijk$. The quaternions can be seen on the postage stamp picture of Hamilton reproduced in Figure 20.9. Hamilton also coined the terms *vector*, *scalar*, and *tensor*.[23]

While Hamilton was a very successful mathematician (and linguist), his Icosian game didn't catch on. To give you an idea of how slow sales must have been, there are at present only four copies of the original game known to still exist.[24] Hamilton sold the rights to the game to the manufacturer for £ 25.[25]

At this point, you're probably expecting some easily stated rule for which graphs have Hamiltonian cycles and which do not. However, no such rule is known; it's an open problem![26] As is usually the case, some partial results have been obtained.[27]

Back to Euler – Polyhedra (1750)[28]

Euler's work in graph theory wasn't limited to getting the party started. He made another huge contribution to the field with a problem originating with solid objects (not in the plane!). Letting F = the number of faces, V = the number of vertices, and E = the number of edges, he discovered the relationship $F + V = E + 2$. For example, a cube has 6 faces, 8 vertices, and 12 edges, and sure enough $6 + 8 = 12 + 2$. The result wasn't actually new; Descartes used this result over a century before Euler, but he didn't have a proof and he never published his formula.[29] Actually, Euler's proof doesn't meet

---

23 Gallian, Joseph, *Contemporary Abstract Algebra, Second Edition*, D. C. Heath and Company, Lexington, MA, 1990, p. 425.

24 https://www.puzzlemuseum.com/month/picm02/200207icosian.htm

25 https://www.puzzlemuseum.com/month/picm02/200207icosian.htm

26 Wilson, Robin, and John J. Watkins, editors, *Combinatorics: Ancient & Modern*, Oxford University Press, Oxford, 2013, p. 191.

27 For example, Kirkman, Thomas P., "On the representation of polyhedra," *Philosophical Transactions of the Royal Society of London*, Vol. 146, 1856, pp. 413–418 established that there are never Hamiltonian cycles on graphs arising from polyhedra with faces consisting of an even number of sides and having a total number of vertices that is odd. He tried to prove more than this, but made errors. See Wilson, Robin, and John J. Watkins, editors, *Combinatorics: Ancient & Modern*, Oxford University Press, Oxford, 2013, p. 190.

28 It wasn't in a journal that Euler put this result forth, but rather in a 1750 letter to Christian Goldbach. Wilson, Robin, and John J. Watkins, editors, *Combinatorics: Ancient & Modern*, Oxford University Press, Oxford, 2013, p. 192 includes a picture of part of the letter. Euler offered a proof in 1751.

29 Bell, Eric Temple, *Mathematics, Queen and Servant of Science*, Copyright, 1951, British Edition 1952, reprinted 1954, 1958, 1961, Printed in Great Britain by Butler and Tanner, Ltd., Frome and London, p. 153. For a more complete analysis of Descartes's work on this topic, see Richeson, David, *Euler's Gem: The Polyhedron Formula and the Birth of Topology*, Princeton University Press, Princeton, NJ, 2012, pp. 81–86.

today's rigorous standards, but it can be patched.[30] The first complete proof was put forth by Adrien-Marie Legendre in 1794, and a much nicer proof came later, in 1813, from Augustin-Louis Cauchy.[31] Both proofs were completely different from Euler's attempt. A proof that owes its initial step to Cauchy is given here.

Proof

Imagine that the polyhedron is set on a very hot surface. As it begins to melt, the bottom face, the one in contact with the hot surface, spreads out evenly so that all of its edges remain straight lines. The polyhedron gradually flattens out. As an example, viewed from above, a cube would look like Figure 20.13.

This reduces the polyhedron to a **planar** graph. This is a special kind of graph that can be drawn in the plane without any lines crossing where there are not vertices.

Edges and vertices may be counted as before and the enclosed regions correspond to faces. However, one face has been "lost." This is the face the polyhedron was resting on when it began to melt. To make up for this lost face, we can count the outside of the figure as a face.

All of the polyhedra for which Euler's theorem is valid, may be converted to connected planar graphs in this manner. So, we can prove the theorem by showing that it holds on all such graphs. Thus, we reduce a three-dimensional headache to a two-dimensional problem. To show that the claim holds for planar graphs, we proceed by induction on the number of vertices.

The base case is 1 vertex. In this instance we have $F = 1$, $V = 1$, and $E = 0$, so $F + V = E + 2$ holds. Don't worry that this doesn't correspond to a polyhedron! You may simply think of it as a degenerate case. Anyway, it's okay if the claim holds for graphs that don't correspond to polyhedra, as long as every polyhedron we're

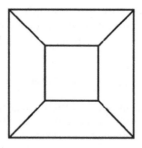

FIGURE 20.13
A Melting Cube.

Image created by Craig P. Bauer.

---

30 For Euler's proof, see Euler, Leonhard, "Elementa doctrinae solidorum," (Elements of the doctrine of solids), and "Demonstratio nonnullarum insignium proprietatum, quibus solida hedris planis inclusa sunt praedita," (Proof of some of the properties of solid bodies enclosed by planes), *Novi Commentarii Academiae Scientiarum Imperialis Petropolitanae*, Vol. 4, 1758, pp. 109–140 and pp. 140–160.

31 See Legendre, Adrien-Marie, *Élements de Géométrie*, first edition, Firmin Didot, Paris, 1794 and Cauchy, Augustin-Louis, "Recherches sur les polyèdres – premier mémoire," *Journal de l'Ecole Polytechnique*, Vol. 9, 1813, pp. 68–86.

interested in has a graph we can verify the claim for, the extra graphs don't do any harm!

For the induction step, we assume the claim holds for all connected planar graphs with n vertices. Now suppose we have a graph with $n + 1$ vertices. For this larger graph, we can shrink one of its edges until the vertices that are its two endpoints merge, becoming a single vertex. This will reduce both the number of edges and vertices by 1. The resulting graph must satisfy $F + V = E + 2$, by our induction assumption. Because the larger graph had one more vertex and one more edge, $F + V = E + 2$ is satisfied for the larger graph as well. Thus, the claim is proven by induction.

Regardless of who really deserves credit for this result, it's now known as Euler's polyhedral formula. Even if it should be named after someone else, Euler's receiving credit doesn't make up for the many times history has unintentionally slighted him. It's often been claimed, only half in jest, that "objects in mathematics are named after the first person after Euler to discover them."[32]

Although the relationship $F + V = E + 2$ holds for all planar graphs, it doesn't hold for any nonplanar graphs. This is a useful fact, for the failure of $F + V = E + 2$ can serve to prove that a given graph cannot be drawn without crossings.

Exceptions?

Applying Euler's formula to another solid figure, a washer (shown in Figure 20.14 for convenience), something strange happens.

The washer has 4 faces (top, bottom, inside, and outside), 0 vertices (the edges never meet each other), and 4 edges, so when we try to use the formula $F + V = E + 2$, we get $4 + 0 \neq 4 + 2$.

Is the problem the hole or the fact that the graph is not connected, or the combination of the two? Or is it the curved faces pose a problem? In the exercises at the end of this chapter, you're asked to determine whether Euler's formula holds for various solids. Investigation of this problem led to the development of topology. Rather than giving the solution, I encourage you to read *Euler's Gem: The Polyhedron Formula and the Birth of Topology* by David Richeson,[33] which gives a much fuller account than space allows for in the present book.

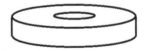

FIGURE 20.14
A washer.

Image created by Craig P. Bauer.

---

32 See Richeson, David, *Euler's Gem: The Polyhedron Formula and the Birth of Topology*, Princeton University Press, Princeton, NJ, 2012, pp. 85–86, for example.

33 Richeson, David, *Euler's Gem: The Polyhedron Formula and the Birth of Topology*, Princeton University Press, Princeton, NJ, 2012.

Special Graphs

The discussion thus far has already included examples of two special kinds of graphs, complete graphs and planar graphs. These two types go together only briefly.

K_4, when it was first introduced in this chapter, was drawn in a manner that caused two of its edges to cross. So, you might conclude that it is not planar. However, we can redraw K_4, as shown in Figure 20.15, without a crossing, so it actually is planar.

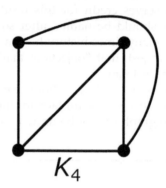

FIGURE 20.15
A graph of K_4 with no crossings.

Image created by Craig P. Bauer.

Whether a graph is planar or not doesn't depend on how it is drawn, but rather on how it could be drawn. Emboldened by success with K_4, we might expect to be able to redraw K_5 so that it too is seen to be planar. This, however, is impossible.

Theorem: K_5 Is nonplanar

Proof
It will be convenient to label the five points as a, b, c, d, and e. Because all points must be connected, there will be a triangle formed by the edges a–b, b–c, and c–a (Figure 20.16).

Now, the points d and e remain to be positioned. They must both be inside the triangle (case 1), or they must both be outside the triangle (case 2), for if one was inside and one was outside, a crossing would be needed to connect d and e with an edge.

Case 1. d and e both inside triangle abc.
Wherever d lies inside triangle abc, there is essentially one way to connect d to a, b, and c, as is shown in Figure 20.17.

However, when one attempts to connect d, three triangular regions will be created inside abc. The point e must lie in one of these regions, but no matter which it is, there will be a point, either a, b, or c, that e cannot be connected to without a crossing.

Case 2. d and e both outside triangle abc.
As in Case 1, we begin by connecting d to a, b, and c. No matter how this is done, two triangular regions will be created outside of abc. One such example is shown in Figure 20.18.

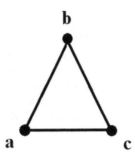

FIGURE 20.16
A necessary triangle.

Image created by Craig P. Bauer.

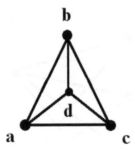

FIGURE 20.17
Connecting d from within.

Image created by Craig P. Bauer.

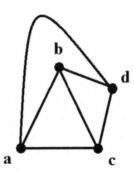

FIGURE 20.18
Connecting d from without.

Image created by Craig P. Bauer.

If e is placed inside one of these new triangular regions, there will be a point that e cannot be connected to without a crossing. However, if e is positioned exterior to all triangular regions, there will also be a point that e cannot be connected to without a crossing.

Because each case necessitates a crossing to complete the connections, the above shows that K_5 cannot be drawn without a crossing. That is, K_5 is nonplanar.

There are other proofs. One popular proof relies on Euler's theorem.[34]

Complete graphs have a history that goes back far beyond graph theory itself. For example, K_{16} appears in a manuscript by Ramon Llull from 1280 CE.[35]

Bipartite Graphs

If we split the vertices into two groups (often drawn so each group lies on a line) and include all edges joining a vertex on one side to a vertex on the other side, then we have a **complete bipartite graph**. We may not include edges joining vertices within the same group. An example is shown in Figure 20.19. It's denoted by $K_{4,2}$, because there are 4 vertices on one side and 2 on the other.

FIGURE 20.19
An example of a complete bipartite graph.

Image created by Craig P. Bauer.

The next graph (Figure 20.20) does not contain all the edges joining vertices from each group. It is still a bipartite graph, but it is not a *complete* bipartite graph.

Not only does this graph fail to be complete, it lacks another property that's been taken for granted up to this point. No matter what point you start at, there are other points that you can never reach, no matter what path you follow. If it is possible to reach a given point from any other point, we say the graph is **connected**. The example above is not, but it can be split into subgraphs that are. These subgraphs are referred to as (connected) **components** of the graph.

---

34 See Richeson, David, *Euler's Gem: The Polyhedron Formula and the Birth of Topology*, Princeton University Press, Princeton, NJ, 2012, p. 124, for example.

35 There's a reproduction of it in Wilson, Robin, and John J. Watkins, editors, *Combinatorics: Ancient & Modern*, Oxford University Press, Oxford, 2013, p. 15.

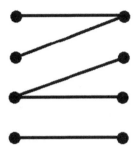

FIGURE 20.20
An bipartite graph that isn't complete.

Image created by Craig P. Bauer.

Like complete graphs, complete bipartite graphs predate graph theory. The example shown in Figure 20.21 is from Athanasius Kircher's *Ars Magna Sciendi*, published in 1669.

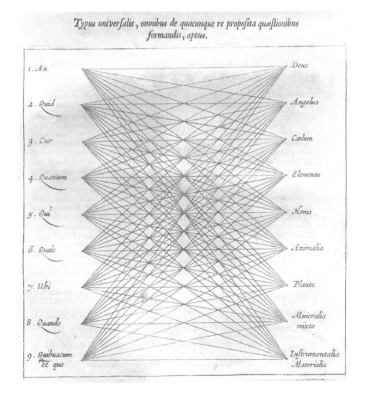

FIGURE 20.21
A page from *Ars Magna Sciendi* showing a complete bipartite graph.

Courtesy of the The Erwin Tomash Library on the History of Computing, http://www.cbi.umn.edu/ hostedpublications/Tomash/Images%20web%20site/Image%20files/K%20Images/pages/Kircher.Ars% 20magna%20sciendi.1669.combination%20diagrams.htm.

Some pretty patterns emerge if we place the vertices evenly around the perimeter of a circle (Figure 20.22). Can you identify which points would be on each side, if these graphs were drawn like the previous examples?

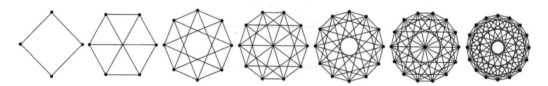

FIGURE 20.22
More bipartite graphs.

Image created by Josh Gross.

FIGURE 20.23
The bipartite graph $K_{18,18}$.

Image created by Josh Gross.

Postmodernist novelist Umberto Eco actually incorporated $K_{18,18}$ (Figure 20.23) into his novel *Foucalt's Pendulum* in an important way.[36] While this graph probably reached more people than most bipartite graphs considered important by mathematicians, it's a far cry from the best-known bipartite graph. This honor goes to a graph associated with (surprise) another puzzle.

The "gas-water-electricity problem" or the "utilities problem," as it is often known, is old enough that there is no certainty concerning its origin, but it is known to have been presented in 1913 by Henry Dudeney (1857–1930), who wrote that it was "as old as the hills."

---

36 Eco, Umberto, translated by William Weaver, *Foucault's Pendulum*, Harcourt Brace Jovanovich, San Diego, CA, 1989, p. 473.

He also noted, "It is much older than electric lighting, or even gas, but the new dress brings it up to date."[37] Dudeney was famous for various logical and mathematical puzzles and games of his own creation. Such work served to popularize mathematics and inspire bright young minds to pursue the subject, just as Édouard Lucas had before him, as W. W. Rouse Ball was doing contemporaneously, and as Martin Gardner's writings would decades later.

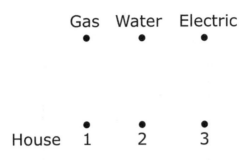

FIGURE 20.24
The Three Utilities Problem.

Image created by Josh Gross.

The utilities problem consists of three homes, each of which needs to be connected to the three utilities, gas, water, and electric. The challenge is to draw the connections such that none of the lines cross. In the lingo of graph theory, you are asked to draw $K_{3,3}$ so that it is seen to be planar. Can you do it? Is it even possible? Try it for yourself first using Figure 20.24 and then read on to see if your answer is correct.

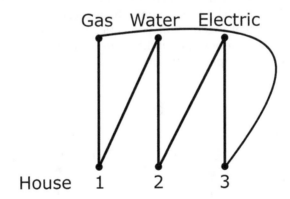

FIGURE 20.25
Some of the connections that must be made.

Image created Josh Gross.

You can start investigating by making the connections shown in Figure 20.25. It makes no difference how you draw the lines. The important part is that you connect House 1 to both Gas and Water, House 2 to both Water and Electric, and House 3 to both Electric and Gas.

---

37 Dudeney, Henry E., "Perplexities," *The Strand Magazine*, Vol. 46, July 1913, p. 110.

No matter how you make these connections, you'll get a cycle. You can think of it as going from Gas → House 1 → Water → House 2 → Electric → House 3 → Gas. Because all of these connections must be present, we must have a cycle with six edges. That is, we have a hexagon. It is redrawn in Figure 20.26, to make the argument that follows clearer.

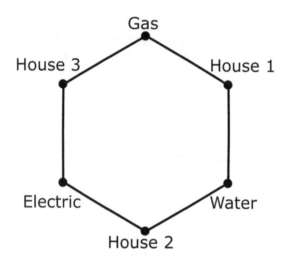

FIGURE 20.26
A necessary hexagon.

Image created by Josh Gross.

There are three more connections that need to be made: House 1 → Electric, House 2 → Gas, and House 3 → Water. Each of these connections must be made inside or outside of the hexagon. By the Dirichlet pigeon-hole principle, at least two must be inside, or at least two must be outside. But both possibilities produce crossings. The illustrations in Figure 20.27a and b give examples of this.

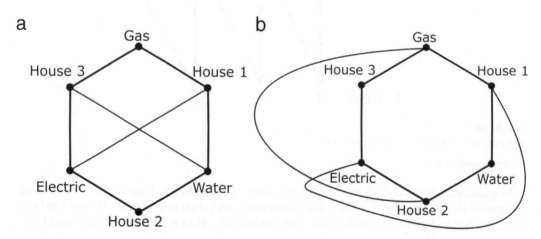

FIGURES 20.27
(a) A failed attempt. (b) Another failed attempt.

Image created by Josh Gross.

If you're not completely convinced, you can try other combinations of edges and see what happens when they are both inside or both outside. It won't take long to try all of them and they will all lead to crossings.

While there's no mathematical solution, one can come up with what might be called a "wise-ass" solution. Dudeney offered one of these by letting a pipe run through one of the homes.[38]

Having confirmed that the two graphs K_5 and $K_{3,3}$ are not planar, we are now in a position to determine whether any graph is planar or not. This is because of Kuratowski's theorem.

Kuratowski's theorem (1930) A finite graph is planar if and only if it does not contain K_5 or $K_{3,3}$ as a subgraph.[39]

While this result is named after the Polish mathematician Kazimierz Kuratowski (1896–1980), it was also found (independently) by two American mathematicians, Orrin Frink and Paul A. Smith, in the same year, although they never published their proof.[40]

A major landmark was reached a few years after this theorem. In 1936, 200 years after Euler essentially created the field of graph theory, the first book devoted to the subject appeared. Up until this time, it was only papers and portions of books that addressed graph theory. This historic book, *Theorie der endlichen und unendlichen Graphen* (Theory of Finite and Infinite Graphs), was authored by Dénes König (Figure 20.28) and became the standard reference on the subject for 30 years.[41] The world missed out on more contributions König might have made because, as a Hungarian Jew, he chose to commit suicide in 1944 in order to avoid a possibly worse fate.[42]

If we attempt the utilities problem on the surface of a torus (donut), it can be solved. That is, $K_{3,3}$ is planar on a torus. Thus, on different surfaces, the answer to the question of planarity can vary. It turns out that for any particular surface it can be determined whether or not a given graph is planar by seeing if it contains any of the forbidden subgraphs for that surface. This result was a consequence of work done by Neil Robertson and Paul Seymour in the 1980s. Unfortunately, the number

---

38 Dudeney, Henry E., "Perplexities," *The Strand Magazine*, Vol. 46, August 1913, p. 221.

39 Kuratowski, Casimir, "Sur le probleme des courbes gauches en topologie," *Fundamenta Mathematicae*, Vol. 15, No. 1, 1930, pp. 271–283.

40 Wilson, Robin, and John J. Watkins, editors, *Combinatorics: Ancient & Modern*, Oxford University Press, Oxford, 2013, p. 333 and https://en.wikipedia.org/wiki/Kuratowski%27s_theorem, which cites Frink, Orrin and Paul A. Smith, "Irreducible Non-Planar Graphs," *Bulletin of the AMS*, Vol. 36, No. 3, 1930, p. 214

41 Konig, Dénes, *Theorie der endlichen und unendlichen Graphen*, Akademische Verlag, 1936. An English translation made by Richard McCoart with commentary by William Thomas "Bill" Tutte is König, Dénes, *Theory of Finite and Infinite Graphs*. Birkhäuser, Boston, MA, 1990. See Wilson, Robin, and John J. Watkins, editors, *Combinatorics: Ancient & Modern*, Oxford University Press, Oxford, 2013, p. 322 for a discussion of the importance of this book.

42 https://en.wikipedia.org/wiki/D%C3%A9nes_K%C5%91nig and http://www-history.mcs.st-andrews.ac.uk/Biographies/Konig_Denes.html

FIGURE 20.28
Dénes König (1884–1944).

Photo of König is courtesy of János Bolyai Mathematical Society.

of forbidden subgraphs can be large. While there are just two for the plane, the torus has hundreds. As a small consolation, the number is always finite.[43]

For nonplanar graphs, it is often desirable to draw them with as few crossings as possible. This minimum is called the **crossing number** of the graph and it is often extremely difficult to determine. An example of this was seen in Chapter 14. But this is sometimes far from a pure math problem with no application. It can actually be a matter of life and death.

43 Wilson, Robin, and John J. Watkins, editors, *Combinatorics: Ancient & Modern*, Oxford University Press, Oxford, 2013, p. 339, which cites a survey on the topic: Robertson, Neal and P. D. Seymour, "Graph Minors – A Survey," in Anderson, Ian, editor, *Surveys in Combinatorics*, London Mathematical Society Lecture Note Series, Vol. 103, Cambridge University Press, 1985, pp. 153–171.

Consider railroad crossings. These are dangerous points of intersection and it has long been recognized that designing a graph that minimizes these saves lives.[44] Of course, bridges and tunnels can eliminate the danger of crossings, but these are expensive to build. Ideally, the optimal design would minimize both lives lost and total cost.

Less serious, but still of financial concern is the design of circuit boards. Manufacturing boards with no crossings is cheaper.

The next section addresses a problem that may not at first glance seem to have anything to do with graph theory, but as it turns out graph theory was the right tool.

The Four-Color Map Theorem (1976)

When coloring regions on a map, whether they be counties, states, or countries, it seems that four colors are always enough to distinguish neighbors. An example of this is shown in color figure 36 **(See color insert)**. No neighboring states share the same color. If they did, it would be potentially confusing. The two might be taken for a single larger state. It's okay, however, for states that only touch at a point (like a corner) to be the same color.

It's sometimes claimed that mapmakers were aware of the sufficiency of four colors for hundreds of years, but there is little evidence of this.[45] Old maps often make use of far more than four colors. Of course, mapmaking is part art and artists are not all minimalists.

In any case, as far as can currently be documented, the four-color conjecture first hit paper in a letter that Augustus De Morgan addressed to Sir William Rowan Hamilton on October 23, 1852. It was not De Morgan who had come up with the conjecture, but rather one of his students, Frederick Guthrie, who thought it was an established fact and simply wanted to know the reason for it.[46] De Morgan only relayed the query to Hamilton. He wrote,

> A student of mine asked me to-day to give him a reason for a fact which I did not know was a fact, and do not yet. He says that if a figure be anyhow divided, and the compartments differently colored, so that figures with any portion of common boundary *line* are differently colored – four colors may be wanted, but not more. Query cannot a necessity for five or more be invented? ... The more I think of it, the more evident it seems. If you retort with some very simple case which makes me out a stupid animal, I think I must do as the Sphynx did.[47]

---

44 See, for example, Steinberg, S. Sidney, "Our Highways, Arteries of the Nation," *The Scientific Monthly*, Vol. 41, November 1935, pp. 458–461, p. 459 is especially relevant here. It states, "About 1,500 persons are killed annually and three times as many injured in railroad grade crossing accidents." At the time, there were about 240,000 crossings, of which 30,000 were considered to be "extremely dangerous."

45 Richeson, David, *Euler's Gem: The Polyhedron Formula and the Birth of Topology*, Princeton University Press, Princeton, NJ, 2012, p. 131, which cites May, Kenneth O., "The Origin of the Four-Color Conjecture," *ISIS*, Vol. 56, No. 3, Autumn 1965, pp. 346–348.

46 "The student was later identified as Frederick Guthrie, who claimed that the problem was due to his brother Francis; the latter had formulated it while colouring the counties of a map of England." From Wilson, Robin, and John J. Watkins, editors, *Combinatorics: Ancient & Modern*, Oxford University Press, Oxford, 2013, pp. 198–199.

47 Graves, Robert Perceval, *Life of Sir William Rowan Hamilton*, 1889, three volumes, Hodges, Figgis, & Co., Dublin, Vol. 3, p. 423.

However, Hamilton was not to be lured into working on the problem – he replied, "I am not likely to attempt your quaternion of color very soon."[48] De Morgan went on to present the problem to more mathematicians, but none could offer a proof.

It's important to clarify what counts as a region on a map in the context of this problem. In particular, a region must be connected to all areas considered to be part of that region. Or, as David Richeson put it, "we do not allow disconnected 'imperialistic' nations."[49] Given this constraint, it's still easy to show that *some* maps require four colors (see Figure 20.29), and De Morgan was able to do this, but showing that four colors are enough for *any* map (that is, a fifth color is never needed) turned out to be extremely difficult!

Although Hamilton was not among them, several big names were attracted to the problem and some, including Arthur Cayley,[50] Clarence Newton Reynolds, Jr., Henri Lebesgue, George D. Birkhoff, and Oswald Veblen were able to chip away at it.[51] A few proofs were announced, but they all turned out to have flaws. Alfred Bray Kempe's failed attempt from 1879 was the closest (it took 11 years for the error to be spotted!) and it remained important even after collapsing. This is because Kempe showed how graph theory is relevant to the problem. By placing a point in each location to be colored and connecting the points to neighboring regions, a graph is produced. The original problem is now equivalent to coloring the vertices of this graph so that no vertices joined by an edge share the same color.[52]

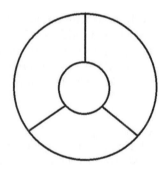

FIGURE 20.29
A map that requires four colors.

Image created by Craig P. Bauer.

48 Quoted from Richeson, David, *Euler's Gem: The Polyhedron Formula and the Birth of Topology*, Princeton University Press, Princeton, NJ, 2012, p. 132.

49 Richeson, David, *Euler's Gem: The Polyhedron Formula and the Birth of Topology*, Princeton University Press, Princeton, NJ, 2012, p. 131.

50 Cayley, Arthur, "On the colouring of maps," *Proceedings of the Royal Geographical Society and Monthly Record of Geography*, New Series, Vol. 1, No. 4, April 1879, pp. 259–261.

51 See Wilson, Robin, and John J. Watkins, editors, *Combinatorics: Ancient & Modern*, Oxford University Press, Oxford, 2013, pp. 334–335 and Richeson, David, *Euler's Gem: The Polyhedron Formula and the Birth of Topology*, Princeton University Press, Princeton, NJ, 2012, p. 141.

52 For discussion of Kempe's work, see Sipka, Timothy, "Alfred Bray Kempe's "Proof" of the Four-Color Theorem," *Math Horizons*, Vol. 10, No. 2, November 2002, pp. 21–23, 26, and Wilson, Robin, and John J. Watkins, editors, *Combinatorics: Ancient & Modern*, Oxford University Press, Oxford, 2013, p. 340.

Percy John Heawood, who was the one to find the flaw in Kempe's proof, nevertheless managed to use Kempe's work as a launching pad to successfully prove that *five* colors are always sufficient.[53]

The best-known person, although not the best *mathematician*, to consider the problem was Lewis Carroll, famed author of *Alice in Wonderland*. Carroll was actually a pseudonym used by mathematician Charles Dodgson. This playful author never claimed to have found a proof, but he did manage to have fun by turning the conjecture into a two-player game.[54]

In 1922, Philip Franklin managed a proof for the special case of 25 or fewer regions.[55] Others were able to raise this bound. By 1940, the claim was established for maps with 35 regions or less.[56] Still, there was no grand proof.

In the meanwhile, similar problems that actually sounded more difficult were solved. Coloring regions on a globe is no different than on a map. If four colors suffice for one, then fours colors will suffice for the other. But what about coloring regions on a torus? How many colors are sufficient, and sometimes necessary? Back in 1890, Heawood showed that the magic number for the torus is 7.[57]

> If you don't like the question, just answer the one you want to! Woo!
>
> David Lee Roth[58]

Mathematicians were following this advice long before Roth was born. That's how the tough question about coloring a map received an answer about coloring a torus. Heawood answered other questions that weren't asked. He found that for tori with g holes, the minimum number of colors sufficient for any division into connected regions is

$$\left\lfloor \frac{7 + \sqrt{1 + 48g}}{2} \right\rfloor.$$

Notice how this formula makes use of the floor function from Chapter 6.

Kempe had shown that the four-color map conjecture is equivalent to looking at colorings of vertices in graphs that correspond to maps, and mathematicians began looking at such colorings for arbitrary graphs – including ones that don't have any corresponding map! Again, answers were given to a question different from the one originally asked.

The minimum number of colors needed for the vertices of a particular graph G, such that no vertices joined by an edge share the same color, is called the graph's **chromatic number**. It is typically denoted by $\chi(G)$. An example of a theorem in this direction is:

53 Heawood, Percy John, "Map-colour theorem," *Quarterly Journal of Pure and Applied Mathematics*, Vol. 24, 1890, pp. 332–338.

54 Wilson, Robin, and John J. Watkins, editors, *Combinatorics: Ancient & Modern*, Oxford University Press, Oxford, 2013, p. 201. Also see Richeson, David, *Euler's Gem: The Polyhedron Formula and the Birth of Topology*, Princeton University Press, Princeton, NJ, 2012, p. 130.

55 Franklin, Philip, "The Four Color Problem," *American Journal of Mathematics*, Vol. 44, 1922, pp. 225–236.

56 Bell, Eric Temple, *Mathematics, Queen and Servant of Science*, Copyright, 1951, British Edition 1952, reprinted 1954, 1958, I961, Printed in Great Britain by Butler and Tanner Ltd., Frome and London, p. 152.

57 Heawood, Percy John, "Map-Colour Theorem," *Quarterly Journal of Pure and Applied Mathematics*, Vol. 24, 1890, pp. 332–338.

58 Quoted by Henry Rollins on the spoken word album Rollins, Henry, *Spoken Word Guy*, 2010, Disc 01, Track 04.

Theorem If G is a connected graph, but not a complete graph or a cycle of odd length, then $\chi(G)$ is less than or equal to the maximum vertex degree.[59]

Mathematicians also looked at the **chromatic index** for various graphs. This is the minimum number of colors needed to color the *edges*, so that any two edges that share a vertex will be different colors. For this problem, the maximum vertex degree, k, provides a minimum value, and one more than this value is always sufficient.[60] Exactly which graphs require k colors and which require $k + 1$ colors remains an open problem.[61] Claude Shannon found a connection between electrical relays and edge-coloring back in 1949.[62]

Finally in 1976, a pair of professors at University of Illinois at Urbana-Champaign found a complete proof.[63] They were Kenneth Appel (Figure 20.30), an American, and Wolfgang Haken (Figure 20.31), an immigrant from Germany. Their proof was not what the world expected. They built upon work done by German mathematician Heinrich Heesch and wrote a computer program to investigate the huge number of cases that the proof had been reduced to. Their program took 1,200 hours, generated a stack of printouts four feet tall, and, if bug-free, concluded the proof.[64] Such a "proof by computer" is far from satisfying for many mathematicians. As John Allen Paulos put it,

> It's clear that the proof of the four-color theorem is not elegant, compelling, or natural. It certainly doesn't belong in what mathematician Paul Erdös calls God's Book, a collection of ideal proofs which do have these properties. Still, it is an impressive resolution of an old problem.[65]

Some simplifications have been made in the proof since its original publication, but the result is still nowhere near being comprehensible to a mathematician working without technology.[66]

1977 was a big year for graph theory. In addition to Appel and Haken's proof seeing print, *Journal of Graph Theory*, the first journal devoted exclusively to the subject, hit the newsstands (well, big university libraries, anyway).

---

59 Brooks, Robert L., "On Colouring the Nodes of a Network," *Proceedings of the Cambridge Philosophical Society*, Vol. 37, No. 2, April 1941, pp. 194–197.

60 See Wilson, Robin, and John J. Watkins, editors, *Combinatorics: Ancient & Modern*, Oxford University Press, Oxford, 2013, p. 341 and Vizing, Vadim G., "On an Estimate of the Chromatic Class of a *p*-Graph," *Diskret. Analiz*, Vol. 3, 1964, pp. 25–30.

61 Wilson, Robin, and John J. Watkins, editors, *Combinatorics: Ancient & Modern*, Oxford University Press, Oxford, 2013, p. 341.

62 Shannon, Claude E., "A theorem on coloring the lines of a network," *Journal of Mathematics and Physics*, Vol. 28, 1949, pp. 148–151.

63 See Appel, Kenneth and Wolfgang Haken, "Every Planar Map is Four Colorable," Research Announcements, *Bulletin of the American Mathematical Society*, Vol. 82, No. 5, September 1976, pp. 711–712, Appel, Kenneth and Wolfgang Haken, "Every Planar Map is Four Colorable Part I: Discharging," *Illinois Journal of Mathematics*, Vol. 21, No. 3, 1977, pp. 429–490, and Appel, Kenneth, Wolfgang Haken, and John Koch, "Every Planar Map is Four Colorable Part II: Reducibility," *Illinois Journal of Mathematics*, Vol. 21, No. 3, 1977, pp. 491–567.

64 See Richeson, David, *Euler's Gem: The Polyhedron Formula and the Birth of Topology*, Princeton University Press, Princeton, NJ, 2012, pp. 142–143 and Wilson, Robin, and John J. Watkins, editors, *Combinatorics: Ancient & Modern*, Oxford University Press, Oxford, 2013, p. 336.

65 Paulos, John Allen, *Beyond Numeracy: Ruminations of a Numbers Man*, Vintage Books, A Division of Random House, Inc., New York, NY, 1992, p. 43.

66 See Robertson, Neil, Daniel Sanders, Paul Seymour, and Robin Thomas, "The Four-Colour Theorem," *Journal of Combinatorial Theory*, Series B, Vol. 70, No. 1, May 1997, pp. 2–44 and Thomas, Robin "An Update on the Four-Colour Theorem," *Notices of the American Mathematical Society*, Vol. 45, No. 7, August 1998, pp. 848–859.

FIGURE 20.30
Kenneth Appel (1932–2013).

Courtesy of the University of Illinois Archives. Kenneth Appel, ca. 1985, found in RS 39/2/26, Box 2.

In 2002, Robin Wilson put out a popular math book, *Four Colors Suffice: How the Map Problem was Solved*. It reated the story of the four-color map conjecture and its ultimate proof.[67] I heard Wilson lecture on the topic and appreciated how he mixed in some humor (often hard to do in a math lecture). First off, he wore a dress shirt in which the sleeves, collar, etc. looked like they had been stitched together from random shirts in his size. The connection was that four colors were used and no neighboring regions were of the same color. Then as a kind Brit lecturing in America, he offered to remove the "u" from "Colour" in the title of his talk – "Four Colours Suffice," although he did insist, in the interest of consistency, in also removing the "u" from "Four."

Matchings and Factorings

In my discrete mathematics class, I always pair the students up to work on the homework. Okay, some semesters there's an odd number of students, so "pairing" doesn't work, but for the sake of this discussion, let's assume that the number of students is even. Still, there are other problems. Some students don't want to work together, others

---

67 Wilson, Robin, *Four Colors Suffice: How the Map Problem was Solved*, Penguin, London, 2002.

FIGURE 20.31
Wolfgang Haken (1928-).

Courtesy of the University of Illinois Archives. Wolfgang Haken, undated, found in RS 39/1/11, Haken, Wolfgang.

might want to but their schedules are incompatible, and yet others are badly mismatched in terms of ability and/or work ethic. Can graph theory help find a solution? Yes, if there is one!

The first step in attacking this problem is to form a graph where the vertices are the students in the class and the edges indicate which ones could work together. A solution exists if there's a subset of the edges such that every vertex belongs to exactly one edge. If this is the case, we say that we have a **perfect matching**.

Perfect matching problems arise in other contexts, such as when trying to assign a group of employees to various tasks. The employees' capabilities will vary and while some may be able to handle several different tasks, others might only be able to handle unique tasks. A perfect matching would ensure that every task is successfully assigned. A similar problem arises in a company looking to hire a number of people to fill different roles. The number of positions each applicant is qualified to fill will vary. In general, this is referred to as the assignment problem.

Unlike my initial example, the graph for the assignment problem will always be bipartite, as employees (or applicants) can be paired with tasks (or positions), but employees are not paired with other employees, nor are tasks paired with each other.

There's a test to see if a given assignment problem has a solution. It's given by the following theorem due to British mathematician Philip Hall.[68]

Hall's Theorem (1935) – Every position can be filled if and only if each set of n positions has at least n applicants.

A perfect matching is also called a 1-factor, because it divides the vertices up into disjoint graphs with 1 edge each, so that every vertex has degree 1. Similarly, we have 2-factors (every vertex has degree 2), 3-factors, etc. The study of 1-factors originated as a spin-off from investigations of the four-color map conjecture.[69]

Weighted Graphs and Shortest Paths

It's often useful to associate numerical values with edges of graphs. If this is done, we have a weighted graph. We are still free to draw it however we like. For example, an edge assigned a weight of 2 can be longer than another edge that's assigned a weight of 10. The values can correspond to real physical distances, financial costs, or anything else. An example is given in Figure 20.32.

The shortest path between two nodes in a graph is often of interest. In Chapter 14 we saw how a shortest path in a Hanoi graph corresponds to an optimal solution. For weighted graphs, the shortest path is not necessarily the route with the least number of edges connecting the nodes, but rather the route that minimizes the sum of the weights of the edges connecting the nodes. If the weights of the edges are all the same, as they can be considered to be in an unweighted graph, these two definitions of "shortest" coincide. So, an algorithm that minimizes the sum of the weights for a path connecting a pair of points will yield the minimal path in any graph, weighted or not. Fortunately, we have such an algorithm. It is presented below, but if you don't understand it at first,

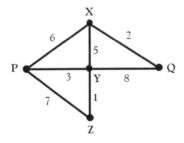

FIGURE 20.32
A weighted graph.

Image created by Craig P. Bauer.

68 Hall, Philip, "On Representations of Subsets," *Journal of the London Mathematical Society*, Vol. 10, 1935, pp. 26–30.
69 Wilson, Robin, and John J. Watkins, editors, *Combinatorics: Ancient & Modern*, Oxford University Press, Oxford, 2013, p. 343.

you should skip ahead to the example and then take another look at the algorithm. Examples almost always make algorithms clearer.

Also, be aware that there are many different ways to present this algorithm and the number of steps varies widely. I use two steps in the explanation below, just as Edsger Dijkstra did in his original paper, but I differ from him in using levels of indentation to make Step 1 clearer.

Dijkstra's algorithm (1959)[70]

When applying Dijkstra's algorithm to find the shortest path from P to Q, the nodes are divided into three disjoint sets, which change over the course of the algorithm. The sets are defined as follows.

> **Set A** consists of the nodes for which the shortest path from P is known.
> **Set B** consists of nodes not in A, but connected to nodes in A by single edges.
> **Set C** consists of nodes not in A or B. These are nodes that have been rejected or not yet considered.

Similarly, the edges are split into three disjoint sets, which also change over the course of the algorithm. They are defined as follows.

> **Set I** consists of the edges that occur in the shortest path from P to some other node in A.
> **Set II** consists of the edges from which the next edge to be added to Set I will be selected.
> **Set III** consists of edges not in Sets I or II. These are edges that have been rejected or not yet considered.

Initially, all nodes belong to Set C and all edges belong to Set III. Because the path is to begin at node P, the first move is to put P in Set A. Following that the following two steps are repeated until Q is moved to Set A.

> **Step 1:** Consider all edges that connect the node just added to Set A with other nodes in Sets B and C.
>
> > If a particular node, call it R, is in Set B, determine if the edge being considered offers a shorter path from P to R than the edge in Set II.
> > > If it does, move it to Set II and reject the edge that it beat. That is, move the defeated edge to Set III.
> > > If it does not, reject the edge being considered.
> > If node R is not in Set B, then move it to Set B and move the corresponding edge to Set II.

---

70 Dijkstra, Edsger W., "A Note on Two Problems in Connexion With Graphs," *Numerische Mathematik*, Vol. 1, 1959, pp. 269–271. It's only three pages long, but as of this writing Dijkstra's paper has earned 25,012 citations on Google Scholar. Note: The algorithm was published in 1959, but Dijkstsa came up with it in 1956. See https://en.wikipedia.org/wiki/Dijkstra%27s_algorithm.

Step 2: For each node in Set B, there's just one way to connect it to P, if we only allow the use of edges in Set I and a single edge from Set II. Thus, under these constraints, each node in B has a fixed distance from P. Take the node for which this distance is the smallest and move it to Set A. Also, move the corresponding edge from Set II to Set I.

Example

Dijkstra's algorithm is now applied to find the shortest path from P to Q in the weighted graph shown in figure 20.32. Of course, this path could be found more rapidly by simply looking at the graph. This example is merely intended to show different aspects of the algorithm, and not to reach the answer in the quickest possible time. For much larger problems, it really is a more practical means of solution than visual inspection. Also, the algorithm can be easily coded up and run on a computer, whereas the technique of visual inspection currently cannot.

We begin with all nodes in Set C and all edges in Set III. Because we want a path from P to Q, we move P to Set A. We now repeat Steps 1 and 2 until Q is moved to Set A. For clarity, every iteration of these two steps will be called a "round."

Round 1: Step 1: The node that was just added to Set A was node P, so we consider edges that connect other nodes to node P. These are the edges with weights 6, 3, and 7. They connect P to nodes X, Y, and Z, respectively. These nodes are all in Set C, so in this step we move them to Set B and their corresponding edges are moved to Set II. Our sets are now A = {P}, B = {X, Y, Z}, C = {Q}, I = {}, II = {3, 6, 7}, and III = {1, 2, 5, 8}.

Step 2: The nodes X, Y, and Z can be connected to P by edges in Set II of length 6, 3, and 7, respectively. The smallest of these is 3, so we move node Y to Set A and the edge with weight 3 to Set I. Our sets are now A = {P, Y}, B = {X, Z}, C = {Q}, I = {3}, II = {6, 7}, and III = {1, 2, 5, 8}.

Round 2: Step 1: The node that was just added to Set A was node Y, so we consider edges that connect other nodes to node Y. These are the edges with weights 5, 8, and 1. They connect Y to nodes X, Q, and Z, respectively. We consider these three nodes individually, beginning with X. Node X is in Set B and the edge connecting it to node Y has weight 5, giving the path from P to X a total length of 8. This is not a shorter path from P to X than the edge belonging to Set II, which has weight 6, so we reject the edge with weight 5 and it stays in Set III. Now consider node Q. Node Q is in Set C, so we move it to Set B and move its corresponding edge, which has weight 8, to Set II. Finally, we consider node Z. This node is in Set B and the edge connecting it to node Y has weight 1, giving the path from P to Z a total length of 4. This is a shorter path from P to Z than the edge belonging to Set II, which has weight 7, so we move the edge of weight 7 to Set III and the edge of weight 1 to Set II. Our sets are now A = {P, Y}, B = {X, Z, Q}, C = {}, I = {3}, II = {1, 6, 8}, and III = {2, 5, 7}.

Step 2: The nodes X, Z, and Q can be connected to P by edges in Set II of length 6, 1, and 8, respectively, for total lengths from P to each node of 6, 4, and 11. The smallest of these is 4, so we move node Z to Set A and the edge with weight 1 to Set I. Our sets are now A = {P, Y, Z}, B = {X, Q}, C = {}, I = {1, 3}, II = {6, 8}, and III = {2, 5, 7}.

Round 3: Step 1: The node that was just added to Set A was node Z, so we consider edges that connect other nodes (from Sets B and C) to node Z. There aren't any, so we move to Step 2.

Step 2: The nodes X and Q can be connected to P by edges in Sets I and II with total path lengths 6 (using the edge of length 6) and 11 (using the edges of length 3 and 8). The path from P to X is shorter than the path from P to Q, so we move X to Set A and the corresponding edge, of length 6, to Set I. Our sets are now A = {P, Y, Z, X}, B = {Q}, C = {}, I = {1, 3, 6}, II = {8}, III = {2, 5, 7}.

Round 4: Step 1: The node that was just added to Set A was node X, so we consider edges that connect other nodes (from Sets B and C) to node X. The only possibility is node Q, which is in Set B, and has corresponding edge 2 in Set III. The total path length from P to Q, using edge 6 in Set I and the edge of length 2, is 8. We must compare this to the path that uses the edge in Set II, which has length 8. For this latter path, the total distance is 3 + 8 = 11, which is greater than the first option considered. So, the edge of length 8 is rejected (and moved to Set III), while Q is moved to Set A and its corresponding edge, of length 2, is moved to Set I. Our sets are now A = {P, Y, Z, X, Q}, B={}, C = {}, I = {1, 2, 3, 6}, II = {}, and III = {5, 7, 8}.

Step 2: There are no nodes in Set B, so there is nothing to do for Step 2.

Q has been moved to Set A, so we're done. The edges in Set I give a subgraph (Figure 20.33) of the original weighted graph.

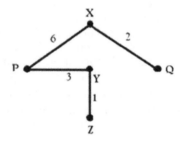

FIGURE 20.33
The result of Dijkstra's algorithm.

Image created by Craig P. Bauer.

In this subgraph, the shortest (and only) path from P to Q is along the edges of length 6 and 2. In carrying out the algorithm, the shortest path from P to Y and the shortest path from P to Z were also generated. This will not always be the case, but we will always get the shortest path from the desired starting point to the desired ending point.

Other Algorithms (or Lack thereof) in Graph Theory

The problem faced by a mail carrier who needs to traverse every road in a given area to make his deliveries, and wishes to do so in the least possible time, is easily translated to the language of graph theory.[71] Assuming that the least time solution is the same as the least distance solution, an optimal route will traverse every street (i.e., every edge of the graph) exactly once. Thus, if there's an Eulerian circuit for the graph, it provides the best possible solution. If there is no such circuit, then one or more edges must be traversed repeatedly. This challenge is sometimes called the **Chinese postman problem**, in honor

---

71 A snowplow driver faces the same problem.

of Kwan Mei-Ko, a Chinese mathematician who was the first to investigate and publish on the problem, in 1960.[72] A polynomial time algorithm that determines the optimal solution was published by Jack Edmonds (of the University of Waterloo in Canada) and Ellis L. Johnson (of the IBM Watson Research Center in America) in 1973.[73]

Now, changing the problem a bit, suppose the mission is to visit a specific location in each of 100 cities. Any route may be taken and, once again, the goal is to find the quickest route. Many roads will not be traversed. This is not how mail gets delivered, but there are people whose jobs entail such travel. Hence, this new challenge is known as the **traveling salesman problem**. Despite the fact that the problem goes back to at least 1831,[74] there's still no algorithm known that can efficiently solve it in general! Here, "efficiently" means in polynomial time or better. It may well be that no such algorithm exists.

Colorful People

The most famous individual to work in graph theory since Euler was Paul Erdös (pronounced "air-dish") (Figure 20.34).

The Atlantic Monthly described Erdös as "a socially helpless Hungarian who has thought about more mathematical problems than anyone else in history." Still, he was far from anti-social. In fact, he coauthored papers with over 500 different people. The following pages detail his unique path through life and some of his mathematical work.

During World War I, Erdös's father was captured by the Russians. He spent six years in captivity in Siberia, returning home in 1920. He learned to read English while a P.O.W., but he didn't know the pronunciation, so he simply guessed at it. He passed his strange accent on to his son.

Despite a quota on the number of Jews allowed to enter Hungarian Universities, Erdös was accepted in 1930. As an 18-year-old student, he found a new proof for an old result, namely that for a natural number $n > 1$ there is always a prime between n and $2n$. This result was initially posed as a conjecture in 1845 by Joseph Bertrand, who confirmed it for all n from 2 to 3,000,000.[75] A proof that it always works eluded Bertrand. The first (complicated) proof came from Pafnuty Chebyshev in 1852. Erdös's proof was simple. It was the subject of his first paper. Eventually, he would write over 1,500 more mathematical papers.

While the "prime between n and $2n$" theorem is firmly established, a related problem, known as Legendre's Conjecture remains open. It claims that there is always a prime between n^2 and $(n+1)^2$.[76]

---

72 Kwan, Mei-Ko, "Graphic Programming Using Odd or Even Points," *Acta Mathematica Sinica* (in Chinese), Vol. 10, 1960, pp. 263–266. For an English translation, see Kwan, Mei-Ko, "Graphic Programming Using Odd or Even Points," *Chinese Mathematics*, Vol. 1, 1962, pp. 273–277.

73 Edmonds Jack and Ellis L. Johnson, "Matching, Euler Tours and the Chinese Postman," *Mathematical Programming*, Vol. 5, No. 1, December 1973, pp. 88–124.

74 Voigt, B. F., *Der Handlungsreisende, wie er sein soll und was er zu thun hat, um Auftrage zu erhalten und einer glucklichen Erfolgs in seinen Geschaften gewiss zu sein – von einemalten Commis-Voyageur* (The travelling salesman, how he must be and what he should do in order to get commissions and be sure of the happy success in his business – by an old *commis-voyageur*), Ilmenau, 1831, reprinted by Schramm, 1981.

75 Young, Robert M., *Excursions in Calculus: An Interplay of the Continuous and the Discrete*, The Mathematical Association of America, 1992, p. 34.

76 Chamberland, Marc, *Single Digits: in Praise of Small Numbers*, Princeton University Press, Princeton, NJ, 2015, p. 36. Also see https://en.wikipedia.org/wiki/Legendre%27s_conjecture.

FIGURE 20.34
Paul Erdös (1913–1996).

Photo of Paul Erdös by George Csicsery taken in 1989 at Random Graphs in Poznan, Poland. © Zala Films. All Rights Reserved.

Erdös received a doctorate from the University Pázmány Péter (in Budapest) in 1934. Increasing anti-Semitism motivated him to leave Hungary for Manchester, England, where he carried out postdoctoral work. After Hitler came to power, Erdös traveled to America and began a fellowship at Princeton. Unfortunately he didn't meet Princeton's standards. He had hoped for a one-year extension on his fellowship, but was only offered six months. Princeton found him "… uncouth and unconventional …"[77]

In 1941, the police picked up Erdös and two other mathematicians near a military radio transmitter on Long Island. They had been discussing mathematics and did not notice the NO TRESPASSING sign. They were released, but the FBI opened a file on Erdös.

Erdös had a chance to land a position in the Manhattan project. However, the answers he gave to the interview questions weren't the ones needed to get the job. For example, Erdös admitted that he would wish to return to Budapest after the war. It could be that Erdös didn't want to work at Los Alamos and was simply amusing himself.

---

77 http://www-groups.dcs.st-and.ac.uk/~history/Mathematicians/Erdos.html

In 1943, Erdös worked part-time at Purdue University. He didn't hear from his family between 1941 and 1945, when he learned that his father died of a heart attack three years earlier. One of his cousins had been sent to Auschwitz and survived, while four of his aunts and uncles were murdered. Erdös returned to Hungary briefly in 1948 to see his remaining family members. He then began to travel frequently between England and America.

In 1949, Erdös, along with Atle Selberg, found another simple proof of an old result concerning primes; this time for the famous prime number theorem. Replacing a complicated proof with one that's more elegant is always considered an important achievement. The prime number theorem concerns the number of primes less than or equal to n. This is usually denoted by $\pi(n)$ and has nothing to do with 3.1415 .... The theorem can be stated tersely as $\pi(n) \sim \frac{n}{\ln(n)}$. Recall from Chapter 8 that the symbol $\sim$ is read as "asymptotically approaches" or "is asymptotically equal to" and means

$$\lim_{n \to \infty} \frac{\pi(n)}{\frac{n}{\ln(n)}} = 1.$$

This result had been conjectured by Carl Friedrich Gauss around 1800, but the first proofs came, independently, from Jacques-Salomon Hadamard and Charles Jean de la Vallée-Poussin in 1896.[78] The theorem offers insight into how many primes we can expect between n and $2n$. We have

$$\pi(2n) - \pi(n) \approx \frac{2n}{\ln(2n)} - \frac{n}{\ln(n)} \approx \frac{n}{\ln(n)}.$$

Because $\frac{n}{\ln(n)}$ grows rapidly with n, for large enough n, there should always be a prime between n and $2n$.[79] But, of course, "should" is not a word we like to see in proofs. This is just an argument in favor of there always being a prime between n and $2n$, not a proof. Erdös's teenage proof of the result was along different lines.

During the early 1950s (the McCarthy era in America), Erdös aroused suspicion from government authorities. U.S. immigration decided to ask him some questions following his return from a conference in Amsterdam. For "What do you think of Marx?" Erdös answered, "I'm not competent to judge, but no doubt he was a great man." This was followed by questions as to whether he would ever return to Hungary. Erdös said:

> I'm not planning to visit Hungary now because I don't know whether they would let me back out. I'm planning to go only to England and Holland. Of course, my mother is there and I have many friends there.[80]

Erdös was finally told that he would not be allowed back to the United States, if he left again. Files indicate that the official reasons were a correspondence he had with a Chinese mathematician in America, who decided to return to China and Erdös's FBI record. Despite the threat, Erdös chose to leave the country again. He explained later,

---

78 Chamberland, Marc, *Single Digits: in Praise of Small Numbers*, Princeton University Press, Princeton, NJ, 2015, p. 35.

79 I first saw this simple argument in Chamberland, Marc, *Single Digits: in Praise of Small Numbers*, Princeton University Press, Princeton, NJ, 2015, p. 35–36.

80 http://www-groups.dcs.st-and.ac.uk/~history/Mathematicians/Erdos.html

Since I don't let Sam and Joe tell me where I am traveling, I chose freedom. I still feel I acted in the best traditions of America: You don't let the government push you around.[81]

Erdös spent a lot of time in Israel in the following years. He began to refer to the "U.S. Iron Curtain" that had barred him entry. He was allowed to return to America with a visa in 1963.[82] Erdös then said, "Sam finally admitted me because he thinks I am too old to overthrow him."[83]

By this time, however, Erdös had become a perpetual traveler moving from one university to another, and from the home of one mathematician to another saying, "Another roof, another proof." He had no family, no property, and no fixed address. Despite this he was able to make a living through mathematics. He would travel from place to place, giving lectures for small fees and staying with mathematicians. He carried all of his possessions in a medium-sized suitcase and said, "Property is a nuisance." He frequently visited up to 15 places in a single month. He was like a mathematical rock star on a permanent world tour. He described typical days in his life as follows:

There was a Saturday meeting in Winnipeg on number theory and computing. On Saturday evening we had a dinner in a Hungarian restaurant, a farewell dinner for the speaker. Then on Sunday morning I flew to Toronto. I was met at the airport, and we went to Waterloo to a picnic. In the evening I was taken back to Toronto, and I flew to London where I lectured at 11 o'clock at Imperial College.[84]

Erdös was well cared for on his travels. He never cooked anything and never drove a car. He continued to put in 19-hour days doing math into his 70s. He greeted people, not with "Hello" or "How are you doing?" but rather phrases like "Consider the following problem..."[85] He referred to children as "epsilons," as ε is used in mathematics to denote a small quantity. Some of this slang has caught on with other mathematicians. American mathematician Ralph P. Boas, Jr. dedicated a book "to my epsilons."[86]

At one conference, Erdös met a mathematician who he learned was from Vancouver. Erdös said to him, "Oh, then you must know my good friend Elliot Mendelson." The man replied, "I *am* your good friend Elliot Mendelson."[87] His memory for mathematics was perfect.

Erdös won many prizes including the Wolf Prize of $50,000 in 1983. He kept $750, established a scholarship fund, and gave the rest away. He offered cash prizes for the solutions to unsolved problems. Prizes ranged from $5 to $10,000 depending on how difficult he felt the problem was. The check shown in Chapter 7 was an Erdös prize check, awarded over a decade after Erdös's death. His longtime friend Ronald Graham continues to oversee such awards.

---

81 Schechter, Bruce, *My Brain Is Open: The Mathematical Journeys of Paul Erdös*, Simon & Schuster, New York, NY, 1998, p. 165.

82 http://www-groups.dcs.st-and.ac.uk/~history/Mathematicians/Erdos.html.

83 Schechter, Bruce, *My Brain Is Open: The Mathematical Journeys of Paul Erdös*, Simon & Schuster, New York, NY, 1998, p.167.

84 Gallian, Joseph A., *Contemporary Abstract Algebra*, Second Edition, D. C. Heath and Company, Lexington, MA, 1990, p. 426.

85 Gallian, Joseph A., *Contemporary Abstract Algebra*, Second Edition, D. C. Heath and Company, Lexington, MA, 1990, p. 427.

86 Boas, Jr., Ralph P., *A Primer of Real Functions*, The Mathematical Association of America, 1994 reprinting.

87 Schechter, Bruce, *My Brain Is Open: The Mathematical Journeys of Paul Erdös*, Simon & Schuster, New York, NY, 1998, p. 178.

In 1976, Stanislaw Ulam gave this description of Erdös:

> ... Erdös is somewhat below medium height, an extremely nervous and agitated person.... His eyes indicated he was always thinking about mathematics, a process interrupted only by his rather pessimistic statements on world affairs, politics, or human affairs in general, which he viewed darkly.... His peculiarities are so numerous it is impossible to describe them all.[88]

Erdös vs. the Facists

> "Throughout his life he would fearlessly defy "fascist" authorities of every stripe, be they armed thugs, mindless university bureaucrats, the U.S. Immigration Service, the Hungarian secret police, the FBI, Los Angeles traffic cops, or the SF himself. Although Erdös hated fascists, he loved the word and applied it liberally to anything he didn't like. Mel Henriksen, who wrote one joint paper with Erdös, contributes this story: "Paul and I once went to a colleague's apartment where some kittens had been born. Paul picked one up, but returned it gently to the box when the kitten scratched. 'Fascist cat!' he exclaimed. The lady of the house questioned how he could call such a little kitten a fascist. 'If you were a mouse,' he said, 'you'd know.'"[89]

We've already seen several examples of Erdös's unique slang, his spectacular vernacular, as it is. Here are a few more:[90]

Erdösese	English
Supreme Fascist	God
Captured	Married
Liberated	Divorced
Recaptured	Remarried
Trivial being	A nonmathematician
Fascist	Anyone or anything annoying
Preaching	Giving a math lecture
To die	To stop doing math
Sam and Joe show	International news
On the long wavelength	Communist

Why are you a physicist? Why aren't you a mathematician?

Paul Erdös

Apparently thinking that anyone with a great mind ought to be pursuing mathematics, Erdös tried to make mathematical converts of people he met. He sometimes succeeded.

---

88 http://www-groups.dcs.st-and.ac.uk/~history/Mathematicians/Erdos.html.

89 This paragraph, with an error fixed, is from https://web.archive.org/web/20020203021833/http://www.paulerdos.com/tales/tale_01.html.

90 From http://www.paulerdos.com/5.html.

His greatest success may well have been William Thomas "Bill" Tutte. This success led Cedric A. B. Smith to pose the question "Did Erdös save western civilization?"[91]

It seems unlikely; Erdös wasn't involved in the Manhattan project, nor did he serve the military in another capacity during World War II. Yet he had posed a mathematical conjecture that attracted the attention of Tutte, who was a student of chemistry at the time. This problem allowed Tutte's mathematical ability to shine, which in turn allowed an excellent letter of recommendation to be written for him, which led to his employment at Bletchley Park, where Nazi codes and ciphers were attacked. Tutte was responsible for breaking Lorenz ciphers used by the Nazi high command. The intelligence gathered from the military secrets contained in these ciphers helped the Allies win World War II. And so, western civilization was saved. So, by a very indirect route, Erdös's influence was felt. His conjecture, as it turns out, was wrong, but it was a good mistake. If he had gotten it right, it might not have intrigued Tutte and he might have remained a chemist.[92]

While there were certainly other men and women at Bletchley Park to whom the world owes a debt, Tutte's contributions were among the greatest.

> There were three heroes at Bletchley Park: Alan Turing, who broke the naval Enigma that helped Britain not lose the war in 1941; Bill Tutte, who broke the Lorenz system, which helped shorten the war; and Tommy Flowers, the father of the computer. Britain was extraordinarily lucky to have these three great men in one place during the darkest times of the Second World War.[93]

Tutte also did important work in graph theory.

Erdös was a great poser and solver of problems, but he was much more. He introduced new methods to mathematics and is even responsible for the creation of an important new field, which although it sounds pure is now of great practical importance. It's described in the following section.

Random Graphs

If you start with a collection of vertices and then select pairs of them at random to join with edges, how many selections would you expect to have to make before ending up with a graph that's connected? It's problems like this that Paul Erdös and Alfréd Rényi took on in a series of eight papers beginning in 1959.[94] The answer is surprisingly small. For n vertices, n random edges are likely to produce a connected graph.

---

91 Smith, Cedric A. B., "Did Erdös Save Western Civilization?" in Graham, Ronald L., Jaroslav Nešetřil, and Steve Butler, editors, *The Mathematics of Paul Erdös I*, second edition, Springer, 2013, pp. 81–92.

92 Like Alan Turing, Tutte read W. W. Rouse Ball's *Mathematical Recreations and Essays*, while a student. So, although Erdös was certainly a key influence on his career, Tutte had also benefited from exposure to one of the great popularizers of mathematics.

93 Roberts, Captain Jerry. *Lorenz: Breaking Hitler's Top Secret Code at Bletchley Park*. The History Press, Stroud, Gloucestershire, 2017, p. 18.

94 The most important were Erdös, Paul and Alfréd Rényi, "On Random Graphs I," *Publicationes Mathematicae (Debrecen)*, Vol. 6, 1959, pp. 290–297, Erdös, Paul and Alfréd Rényi, "On the Evolution of Random Graphs," *Publications of the Mathematical Institute of the Hungarian Academy of Sciences*, Vol. 5, 1960, pp. 17–61, and Erdös, Paul and Alfréd Rényi, "On the Strength and Connectedness of a Random Graph," *Acta Mathematica Scientia Hungary*, Vol. 12, 1961, pp. 261–267.

To see the importance of this work, it will be useful to first consider the work of a man who has little in common with Erdös and Rényi, beyond being his nationality. Frigyes Karinthy, known for his poems and prose, was extremely popular in early 20th-century Hungary. His 1929 short story "Láncszemek" (Chains) contained the following passage.

> To demonstrate that people on Earth today are much closer than ever, a member of the group suggested a test. He offered a bet that we could name any person among earth's one and a half billion inhabitants and through at most five acquaintances, one of which he knew personally, he could link to the chosen one.[95]

Today, this claim is referred to as "six degrees of separation." If it's true, then nearly everyone on the planet is a friend of a friend of a friend of a friend of a friend of a friend of yours. What does this have to do with random graphs? Well, the social network of humans is a graph. Imagine each person on the planet as a vertex with edges connecting pairs who are friends. If friendships form by chance, in a random sort of way, Erdös and Rényi's result shows that people only need to average one friend each for everyone to be connected in the "friendship graph."

But there's a big difference between there being a path connecting apair of people who have never met and there being a *short* path connecting them. Is it really *six* degrees of separation or should we expect the shortest paths between randomly selected people to travel through hundreds of people? It's not known from where Karinthy drew the inspiration to make his bold claim, but the work of Erdös and Rényi backed it up. The mathematicians did not discuss the application of random graphs to modeling a social network, and there is no evidence that they even read Karinthy's short story, but they did show that random points are very likely to be connected by short paths in random graphs.

The very simple formula for the average distance between vertices is

$$d = \frac{\log(N)}{\log(k)},$$

where N is the number of vertices and k is the average degree of the vertices.[96] According to sociologists, the number of people we know by name ranges from 200 to 5,000.[97] While knowing someone by name is a far cry from counting that person as a friend, we can still plug these numbers into the formula above to see the result. Taking $N = 7.7$ billion (the approximate population of the Earth as of this writing) and $k = 200$ (the extreme low end of the sociologists' estimate), we have

$$d = \frac{\log(7,600,000,000)}{\log(200)} \approx 4.29.$$

The fact that values for d are surprisingly small is known as the "small-world" phenomena. When a pair of strangers find themselves in a conversation, say on a plane, it's not unusual

---

95 Taken here from Barabási, Albert-László, *Linked: The New Science of Networks*, Perseus Publishing, Cambridge, MA, 2002, p. 26.

96 Taken here from Barabási, Albert-László, *Linked: The New Science of Networks*, Perseus Publishing, Cambridge, MA, 2002, p. 35 footnote.

97 Barabási, Albert-László, *Linked: The New Science of Networks*, Perseus Publishing, Cambridge, MA, 2002, p. 18.

for them to find that they have a common acquaintance. When this happens, one of them will often remark, "What a small world!" Yet, the mathematics tells us such connections are common. The more links we allow, the greater the chances are of a connection. John Allen Paulos put it like this: "If two strangers sit next to each other on an airplane, more than 99 times out of 100 they will be linked in some way by two or fewer intermediates."[98] Such links are not always easy to discover. While I know my acquaintances and you know yours, we don't know all of our acquaintance's acquaintances, which can prevent us from finding paths that involve two intermediaries.

Although graph theory is an area of mathematics, one needn't be a mathematician to carry out experiments in it. We now turn to a colorful nonmathematician who did exactly that.

Stanley Milgram

American social psychology professor Stanley Milgram (1933-1984) had a reputation for creative, but controversial, experiments. As a Jew, his struggles to understand the holocaust led him, in the early 1960s, to carry out experiments focused on obedience to authority. He found that most people would torture another human, if ordered to do so.[99]

Milgram went on to investigate the much less disturbing small-world phenomena experimentally in 1967. He asked people in Wichita, Kansas and Omaha, Nebraska to try to get letters to target individuals in Massachusetts.[100] Each person was to send his or her letter to someone he or she knew, asking that person to do the same. In many cases, the recipient of a letter declined to forward it to someone else. Only 42 of the 160 letters made it all the way to the intended destinations. However, for these 42 letters, the median number of intermediaries was 5.5.[101] Milgram's result, in this instance, makes the world seem like a smaller friendlier place. Is a randomly selected person really a stranger when such a short path connects him or her to you?

It's likely that, in some cases, shorter paths existed than the ones produced by the people mailing the letters. For example, if Milgram was asked to get a letter to Erdös, he would have to decide whether he should send it to his father, who had immigrated to America from Hungary, or to a colleague in Harvard's mathematics department. One of these paths might be a bit shorter than the other, but how could Milgram know which it was?

Milgram's median value of 5.5 intermediaries rounds up to 6, but he was not the creator of the phrase "six degrees of separation." John Guare's 1990 play "Six Degrees of Separation" marked its first appearance.[102]

---

98 Paulos, John Allen, *Beyond Numeracy: Ruminations of a Numbers Man*, Vintage Books, A Division of Random House, Inc., New York, 1992, p. 38.

99 See Milgram, Stanley, "Behavioral Study of Obedience," *Journal of Abnormal and Social Psychology*, Vol. 67, 1963, pp. 371–378 and Milgram, Stanley, *Obedience to Authority: An Experimental View*, Tavistock Publications, London, 1974.

100 Barabási, Albert-László, *Linked: The New Science of Networks*, Perseus Publishing, Cambridge, MA, 2002, p. 27.

101 Barabási, Albert-László, *Linked: The New Science of Networks*, Perseus Publishing, Cambridge, MA, 2002, p. 29. Watts, Duncan J., *Six Degrees: The Science of a Connected Age*, W. W. Norton & Company, New York, NY, 2003, pp. 132–134 point out some flaws in Milgram's experiment. Taking these into account, the results are not as impressive as they are often portrayed to be.

102 https://en.wikipedia.org/wiki/Six_Degrees_of_Separation_(play)

As a practical application of the small-world phenomenon, you may well be able to greatly improve your life by making use of your social network. In many cases, jobs are found through a friend of a friend. It's also a common way for pairs of people who eventually get married to first meet. If you think about how friends of yours might be able to help other friends of yours, and your friends do the same, you will all benefit!

It's easiest to study random graphs through computer simulations. Accurate data for real-world networks can be hard to obtain. But to test the accuracy of mathematical models, it's necessary to compare them to real-world data. Appropriately enough, one useful dataset originates with Erdös.

The Erdös Collaboration Graph

Someone who has coauthored a paper with Erdös is said to have Erdös number 1. John Conway provides one of the over 500+ examples. If a mathematician has not coauthored a paper with Erdös, but has coauthored a paper with someone who has, that mathematician is said to have Erdös number 2, and so on. Erdös himself has Erdös number 0. A snapshot of the Erdös collaboration graph from July 2004 yielded the following data.[103]

```
Erdös number  0 ---        1 person
Erdös number  1 ---      504 people
Erdös number  2 ---    6,593 people
Erdös number  3 ---   33,605 people
Erdös number  4 ---   83,642 people
Erdös number  5 ---   87,760 people
Erdös number  6 ---   40,014 people
Erdös number  7 ---   11,591 people
Erdös number  8 ---    3,146 people
Erdös number  9 ---      819 people
Erdös number 10 ---      244 people
Erdös number 11 ---       68 people
Erdös number 12 ---       23 people
Erdös number 13 ---        5 people
```

While some mathematicians' Erdös numbers are greater than 6, the overwhelming majority are not. This is impressive, given that the dataset consists of about 1.9 million papers written by about 401,000 distinct authors.[104] The table above doesn't include mathematicians who have never had coauthors. In the collaboration graph, these individuals are represented by isolated nodes (that is, nodes with degree 0) and there are about 84,000 of them. The total number of authors who do not have an Erdös number, which includes the 84,000, is about 133,000.

Matrices can be used to find Erdös numbers. This is not an efficient way to get a single person's Erdös number, for it generates all of them at once. The process begins with the construction of an *n* by *n* matrix, where *n* is the number of mathematicians who

---

103 Taken here from http://www.oakland.edu/enp/trivia/.
104 http://www.oakland.edu/enp/trivia/.

have published papers with coauthors. You may imagine the authors' names running along the top and the left-hand side of the matrix in alphabetical order. Entry i, j of the matrix is a 1 if person i has written a paper with person j, and 0 otherwise. The row corresponding to Erdös has 1s in positions corresponding to mathematicians with Erdös number 1. Now square the matrix. A nonzero entry in position i, j of this new matrix represents a path of length 2 connecting that pair of authors. Entries in Erdös's row that have changed from 0 to some nonzero value represent mathematicians with Erdös number 2. We can continue the process. Raising the original matrix to the nth power reveals whether or not there's a path of length n linking pairs of authors. The number in that position is actually the number of such paths. We must be careful though, it's possible that a path of length k exists for some $k < n$, but no path of length n exists. If we want to be able to tell at a glance whether a mathematician has a path of length n or less to some other mathematician, we can look at the appropriate entry for the sum of the matrices raised to the powers 1 through n.

An interesting feature of matrices used to represent various social networks that exhibit the small-world property is that they can initially consist of nearly all zeros. As was mentioned previously, sociologists estimate the number of people we know by name ranges from 200 to 5,000.[105] Thus, for the matrix representing the worldwide acquaintance graph, which has about 7.7 billion rows and columns, each row would have no more than 5,000 nonzero entries. That is, the 0s constitute well over 99.9999% of the matrix. Yet, as it is raised to higher and higher powers, the nonzero entries quickly begin to dominate. Matrices that overwhelmingly consist of 0s are called **sparse matrices** and the graphs they represent are called **sparse graphs**. It would not be surprising to have the small-world property in graphs that have edges connecting most pairs of vertices. The surprise is that it pops up for so many sparse graphs.

Paul Erdös died in 1996, but a new sort of collaboration graph had already been born.

Six Degrees of Kevin Bacon

In January 1994, Craig Fass, Christian Gardner, Brian Turtle, and Mike Ginelly, students at Albright College in Reading, Pennsylvania, saw an interview actor Kevin Bacon gave in which he said, he had "worked with everybody in Hollywood or someone who's worked with them."[106] They realized that Bacon could indeed be connected to nearly every actor through various films. In graph theoretic terms, the actors were vertices and when two appeared in the same film, those vertices were connected by an edge. Three of the students appeared on *The Jon Stewart Show*, with one proclaiming "We are three men on a mission. Our mission is to prove to the Jon Stewart audience, nay, the world, that Bacon is God."[107] The three went on to have a book published in 1996,[108] as well as a board game based on the concept.

---

105 Barabási, Albert-László, *Linked: The New Science of Networks*, Perseus Publishing, Cambridge, MA, 2002, p. 18.
106 https://www.thestar.com/entertainment/2012/09/13/google_adds_six_degrees_of_kevin_bacon_to_search_engine.html.
107 Barabási, Albert-László, *Linked: The New Science of Networks*, Perseus Publishing, Cambridge, MA, 2002, pp. 58–59.
108 Fass, Craig, Brian Turtle, and Mike Ginelli, *Six Degrees of Kevin Bacon*. Plume, New York, NY, 1996.

The number of films needed to connect an actor to Kevin Bacon is his or her "Bacon number." There are some actors who do not have one. That is, there is no path through films that connects them to Kevin Bacon. However, the vast majority are connected to him in six degrees or less. The Kevin Bacon graph, as it's referred to by mathematicians, provides a great example of the small-world phenomena. And the graph has attracted serious study from mathematicians.

One of the main reasons the Kevin Bacon graph has been heavily used is that the data are reliable and easily accessible through www.imdb.com/. Similarly, the Erdös collaboration graph data can be accessed through https://mathscinet.ams.org. In the early years of such research, there weren't many other good datasets. One more example was provided by the neural network of *C. elegans*, a 1-mm-long worm. With only 302 neurons, this was a much smaller graph.[109]

Bacon is not the most connected actor in film, nor has he appeared in more films than anyone else. Many other actors could lead one to discover the small-world property of the film graph. It just happened to be Bacon who was used to draw the world's attention to it. His reaction was not what I would have expected. He explained,

> A few years ago when I first heard about the six degrees of Kevin Bacon game I was really kind of horrified. I thought it was a joke at my expense. And I was hoping that it would go the way of pet rocks and 8-track cassettes tapes, but it seems to be hanging on. You know, I'm kinda glad actually that it has stuck around, because if you take me out of the equation it's really a sort of beautiful notion, the notion that we're all connected in some kind of way. The notion that when something happens to our friends and neighbors down the block or our friends and neighbors and brothers and sisters on the other side of the planet that it affects us and that we have a responsibility as human beings to take care of this planet that we're all riding on. This is the idea behind sixdegrees.org.[110]

Back in 1624, John Donne expressed this idea in the following poem.

No Man is an Island

No man is an island entire of itself;
every man is a piece of the continent,
a part of the main; if a clod be
washed away by the sea, Europe is the less,
as well as if a promontory were, as well
as if a manor of thy friends or of thine
own were; any man's death diminishes me,
because I am involved in mankind; and
therefore never send to know for whom the
bell tolls; it tolls for thee.

Bacon's nonprofit charity, sixdegrees.org, has raised over $5,000,000, since its launch in January 2007. Bacon gave a much fuller accounting of its origins in a TED Talk.[111]

---

109 Barabási, Albert-László, *Linked: The New Science of Networks*, Perseus Publishing, Cambridge, MA, 2002, p. 50.

110 Taken from https://www.youtube.com/watch?v=vOUdy2J0lfU The quote begins at 5:56. The website for Bacon's charity is https://www.sixdegrees.org/.

111 The Six Degrees, Kevin Bacon, TEDxMidwest, https://www.youtube.com/watch?v=n9u-TITxwoM.

While there don't seem to be as many variants for the Kevin Bacon game as there are for the Tower of Hanoi, some do exist. One, called the "Max Bacon problem" asks, for a given actor, "Does there exist a path to Kevin Bacon that's of length k or more for some large value of k?"[112] For example, the actor could be Mark Blankfield (he's good!) with $k = 100$. The 98 actors between these two men can be anyone we like, but we are not allowed to use a given intermediary more than once. Otherwise, a given edge could be repeatedly traversed to artificial inflate the length of the path. This variant is much harder than the original problem.

Another game that can be considered a variant is played on a completely different graph, the graph whose vertices are Wikipedia pages and whose edges are links that connect them. To play this game, two Wikipedia pages are selected and the player seek paths between them. The shorter path wins.[113]

Erdös-Bacon numbers[114]

There are published mathematicians who have appeared in films and possess both Erdös numbers and Bacon numbers. This led to the concept of an Erdös–Bacon number, which is the sum of the two. As an example, Brian Greene, a Columbia University professor of physics and mathematics has Erdös–Bacon number 5. He coauthored a paper with Shing-Tung Yau, who wrote a paper with Ronald Graham, who wrote many papers with his good friend Paul Erdös. Greene also had a cameo role in *Frequency*, which featured John Di Benedetto, who was in *Sleepers* with Kevin Bacon.[115]

It's a Small World After All

Once the small-world property was recognized, mathematicians began to find it in all sorts of real-world graphs.

The National Security Agency found the telephone metadata graph (with telephone numbers as the nodes and edges representing calls made from one number to another) potentially useful in building on previously garnered intelligence. Once a terrorist has been identified, through whatever means, a closer look at those linked to him by telephone calls could identify other, previously unknown, terrorists. And those nodes could lead to even more, and so on. When the program became publicly known, there was some public outcry, fueled by privacy concerns. Because of the small-world property of such graphs, it wouldn't take many links to get from a terrorist to nearly everyone. In 2014, *The New York Times* reported,

---

112 Lane, Matthew, *Power-Up: Unlocking the Hidden Mathematics in Video Games*, Princeton University Press, Princeton, NJ, 2017, p. 162.
113 Lane, Matthew, *Power-Up: Unlocking the Hidden Mathematics in Video Games*, Princeton University Press, Princeton, NJ, 2017, p. 160.
114 Singh, Simon, https://web.archive.org/web/20050318014622/http://www.simonsingh.com/Erdos-Bacon_Numbers.html. Also see https://en.wikipedia.org/wiki/Erd%C5%91s%E2%80%93Bacon_number.
115 Singh, Simon, https://web.archive.org/web/20050318014622/http://www.simonsingh.com/Erdos-Bacon_Numbers.html.

President Obama has said he wants to alter the program so that only records of callers up to two links from a suspect would be provided to the government, and Congress is working on legislation to authorize such a program.[116]

While two links might not always be enough to identify all of the bad guys, continuing out beyond this point would certainly be low yield territory. The legislation passed by an overwhelming majority.

Does the random graph model put forth by Erdös and Rényi, which has the ubiquitous small-world property, provide the perfect model for real-world graphs? Sadly, no. The situation in the real world was different somehow, for there was another property the real graphs possessed that random graphs do not – clustering.

A cluster is present when there's a vertex of very large degree, compared to the average. Such vertices are often called hubs. They appear in some man-made networks designed with efficiency in mind, such as the one pictured in Figure 20.35. But why should hubs be present in graphs whose growth is not carefully planned, but rather determined by large numbers of people who do not coordinate their actions?

FIGURE 20.35
An artificial graph for airline flights created with efficiency in mind.

Image modified by Josh Gross from https://pixabay.com/vectors/usa-map-united-states-of-america-35713/.

---

116 Savage, Charlie, "Phone Company Pushed Back Against N.S.A.'s Data Collection, Court Papers Show," *The New York Times*, May 14, 2014, available online at https://www.nytimes.com/2014/05/15/us/politics/phone-company-pushed-back-against-nsas-data-collection-court-papers-show.html.

Another feature of many real-world graphs that is not present in the random graph model is the manner in which the degrees of the vertices are distributed. One would expect the distribution to resemble that of the normal curve. That is most vertices should have degree close to the average with about the same number having more or less, but both extremes tapering off quickly as one heads away from the mean.

Instead, the distribution often follows a **power law**. This means that the fraction of vertices in a graph having degree k, denoted $P(k)$, for large k, satisfies the relationship $P(k) \sim k^{-\gamma}$, where γ is a constant, usually between 2 and 3. What this means is that the graph looks a bit like exponential decay, but has a much longer tail. A typical example is given in Figure 20.36.

This graph shows that a few weblogs have a huge number of links and although the graph declines rapidly at first, the rate of decrease slows and the tail is nearly a horizontal line that doesn't appear to be in any hurry to get closer to the x-axis.

A network having a power law distribution for the degrees of its vertices is said to be a **scale-free network**.

While the explanation for the power law distribution was found relatively recently by graph theorists, it's really thousands of years old. In fact, its appearance in the Bible has led to it being called **The Matthew Effect**. The relevant passage is Matthew 25:29.

> For to every one who has will more be given, and he will have abundance; but from him who has not, even what he has will be taken away.[117]

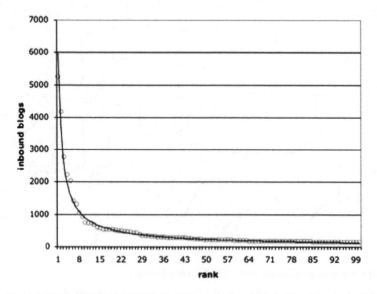

FIGURE 20.36
Distribution of number of links to the most popular weblogs.

Image from https://kottke.org/03/02/weblogs-and-power-laws courtesy of Jason Kottke.

---

117 Taken here from RSV (Revised Standard Version).

Or to put it another way, the rich get richer and the poor get poorer. We don't need a conspiracy for this to be the case. It's simply easier to make money, or gain followers, when you already have a lot of money or followers. Success begets success.

If we wish to model this phenomenon, instead of adding edges to a graph randomly, vertices of higher degree should be proportionally more likely to gain edges, as a graph evolves over time. Modifying the random graph model in this manner leads to both clustering and the power law distribution, while maintaining the small-world property. This result was put forth in 1999 by Albert-László Barabási and Réka Albert, who called the manner in which vertices are favored by new edges "preferential attachment."[118] A measure of how important this paper is considered to be is that, as of this writing, it has earned 35,578 citations on Google Scholar.

Researchers had been aware of the power law distribution and its consequences in other contexts for a long time. One important figure in this history is Vilfredo Pareto (1848–1923), an Italian engineer and economist, after whom the Pareto principle is named. Discovered in 1896 and also known as the 80/20 rule, it states that 80% of the effects arise from 20% of the causes.[119] This is a very general statement and indeed it applies to a very wide class of phenomena. Some examples are

80% of the land is owned by 20% of the people
80% of people live in 20% of the cities
80% of sales come from 20% of the customers
80% of citations are earned by 20% of the research papers
80% of phone calls are received by 20% of the people

This list could be continued for many pages.

A consequence of the power law for languages is that, for a given text, the nth most frequent word will occur about $1/n$ as often as the most frequent word. This is known as Zipf's law and it holds across all natural languages. George Zipf (1902–1950) was a linguist employed by Harvard Univerersity. Although the law bears his name, it was known to others before him.[120] Zipf merely brought it to a wider audience.

The audience included Benoit Mandelbrot, who investigated Zipf's law early in his career and offered a possible explanation for it.[121] Popularizer Michael Stevens of Vsauce fame made an excellent video on Zipf's law and other instances of the Pareto principle, aimed at a general audience.[122] It features Mandelbrot's explanation, as well as others.

The next section serves to illustrate the broad applicability of graph theory by showing how it can help solve a puzzle that seems to have nothing to do with graphs.

---

118 Barabási, Albert-László and Réka Albert, Réka, "Emergence of Scaling in Random Networks," *Science*, Vol. 286, No. 5439, October 15, 1999, pp. 509–512.

119 Pareto, Vilfredo, "*Cours d'économie politique*," two volumes, F. Rouge, Éditeur, Librairie de l'Université, Lausanne, 1896.

120 They include Jean-Baptiste Estoup (1868–1950), a French stenographer, and Felix Auerbach[4] (1856–1933), a German physicist.

121 Mandelbrot, Benoit, "An Informational Theory of the Statistical Structure of Language," in Jackson, Willis, editor, *Communication Theory: Papers Read at a Symposium on "Applications of Communication Theory" held at the Institution of Electrical Engineers, London, September 22nd–26th, 1952*, Academic Press, 1953, pp. 486–502.

122 The Zipf Mystery, https://www.youtube.com/watch?v=fCn8zs912OE, September 15, 2015.

Instant Insanity

The puzzle pictured in color figure 37 **(See color insert)** is called "Instant Insanity." It was created by Franz Owen Armbruster. Parker Brothers began selling it in 1967 and over 12 million have now been purchased.

In this game you have four cubes with colored sides. In the version I own, they're colored blue, red, green, and white. Because each cube has six sides, at least one color must repeat on each cube. The idea of the game is to stack the cubes so that each of the four visible side surfaces of the combined object shows all four colors.

Each cube is depicted in an unfolded form in color figure 38 **(See color insert)**, so that all six sides can be seen at once. We can generate a graph for each cube (Figure 20.37) by letting an edge join two colors if and only if they appear on opposite faces. The weights in these graphs indicate the cube being represented.

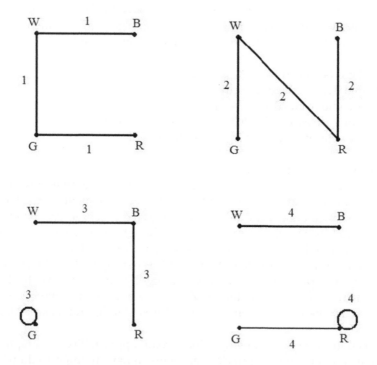

FIGURE 20.37
"Opposite faces" graphs for the four cubes.

Image created by Craig P. Bauer.

Combining these into a single graph gives us Figure 20.38.

We now look for two subgraphs that each completes a circuit of the nodes without having any edges in common. Furthermore, we need to have exactly one edge from each of the original graphs in each of our new graphs. The solution is shown in Figure 20.39.

The solution graphs are labeled front/back and left/right because they provide the orientations of the cubes for the arrangement we're seeking. They do not specify a unique arrangement, but the number of possibilities is now quite small and we may easily solve the puzzle.

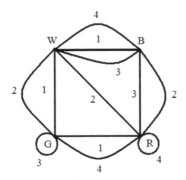

FIGURE 20.38
The combined graph.

Image created by Craig P. Bauer.

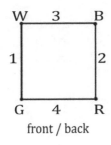 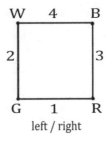

front / back   left / right

FIGURE 20.39
A pair of special subgraphs.

Image created by Craig P. Bauer.

The edges that are left unused from the combined graph (including the loops) represent top/bottom.

An Out-of-this-World Application

Graphs can arise naturally. Examples include "mud cracks, shrinkage cracks of glazed chinaware, cracks in ancient lava beds, rivers, and so on."[123] But there are also man-made

---

123 Examples taken from Gardner, Martin, "Mathematical Games, Thoughts on the task of communication with intelligent organisms on other worlds," *Scientific American*, Vol. 213, No. 2, August 1, 1965, pp. 96–100. This paper also appears in Gardner, Martin, *Martin Gardner's 6th Book of Mathematical Diversions from Scientific American*, The University of Chicago Press, as chapter 25 "Extraterrestrial Communication," pp. 253–262. As Gardner typically did with his collected articles, an addendum is provided at the end of this piece. In this instance it takes the form of a pair of letters, related to the piece, that appeared in the January 1966 issue of *Scientific American*.

graphs such as the internet, road networks, railway systems, and airline connections, as well as graphs made by other creatures such as spider webs and animal trails. By only looking at the graph, ignoring context, we can often determine whether the graph was natural or the work of living creatures. In natural graphs, vertices of order 3 are the most common, while for the artificial graphs order 4 takes the lead. Figures 20.40 through 20.43 illustrate the contrast.

Before satellites and probes began sending back excellent pictures of Mars, astronomers squinted their eyes while peering through telescopes and thought they saw canals. The graph theory, based on the degrees of the vertices, supported the conjecture that they were artificial. Eventually, however, it was recognized to be a sort of optical illusion; the "canal graph" does not actually exist. But for a while it was hotly debated![124]

FIGURE 20.40
A natural graph formed by mud cracking as it dries.

Image created by Famartin. This file is licensed under the Creative Commons Attribution-Share Alike 4.0 International license.

---

124 For more detail on this episode, see Gardner, Martin, "Mathematical Games, Thoughts on the task of communication with intelligent organisms on other worlds," *Scientific American*, Vol. 213, No. 2, August 1, 1965, pp. 96–100.

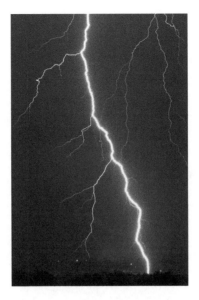

FIGURE 20.41
Lightning.

FIGURE 20.42
Amazon River Basin.

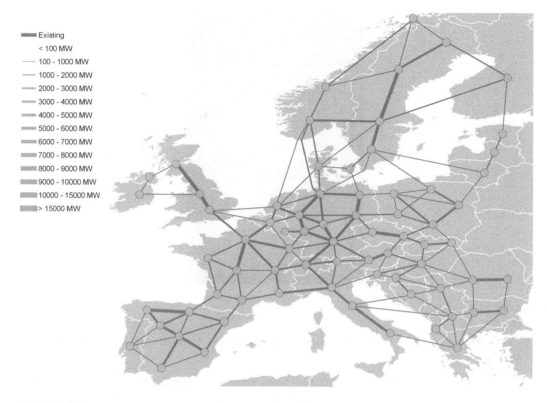

FIGURE 20.43
An integration of renewable energy into Europe's electric grid.

Image from Hewicker, Christian, "Integration of renewable energy in Europe," 2014, available online at
https://blogs.dnvgl.com/energy/integration-of-renewable-energy-in-europe.

Ramsey Theory

Suppose that there are six people and any pair of them must be friends or strangers. Must there be a trio who are all friends or all strangers? Like so many other puzzles in this chapter, this one does not sound like a graph theory problem, yet the tools that graph theory offers are ideal for solving it. To see this, represent the six people as six points and connect pairs of people by gray lines, if they are friends, and black lines, if they are strangers. We don't need to look at all of the possibilities to see that each must have a trio (represented by a triangle with all sides having the same color) of friends or strangers.

The efficient approach to this problem begins with one of the people and considers the five lines that lead to the others (Figure 20.44).

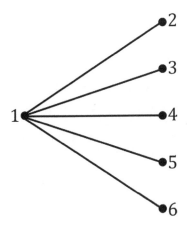

FIGURE 20.44
The friends and strangers of person 1.

Image created by Craig P. Bauer.

Because we have five lines and only two colors, there must be at least three lines that are the same color. This is an application of the Pigeon-hole principle from Chapter 5. Without loss of generality, let's assume that the three edges of the same color connect 1 with 3, 4, and 5 (indicated in gray in Figure 20.45).

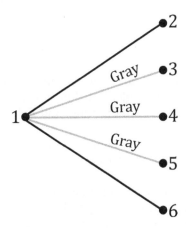

FIGURE 20.45
Three lines must be of the same color.

Image created by Craig P. Bauer.

Whether the other two edges that connect with 1 are black or gray does not matter!.
Now consider the edges connection 3 and 4, 4 and 5, and 3 and 5. If any one of these is gray, we have a gray triangle and, hence a trio of friends. On the other hand, if none of them are gray, then they must all be black and the trio of 3, 4, and 5 are all strangers.

That's all there is to the proof! Any such gathering of six people must contain at least one trio of friends or strangers.

It's easy to show that six people is the minimum needed to guarantee that one of the desired trios exists. All you have to do is exhibit an edge-colored graph for five people in which no such trio exists. You are asked to do this as an exercise.

A slight change makes the problem much harder. Can you prove that to guarantee a group of 4 mutual friends or strangers, a gathering of 18 people is necessary? Or more challenging yet, can you firmly establish the number of people needed to guarantee a group of 5 mutual friends or strangers? If so, you will be the first. The value is not known!

In general, we can ask for the number of people needed such that we are guaranteed to have a group of f mutual friends or a group of s mutual strangers. The answers are called Ramsey numbers. These values can be denoted as $R(f, s)$. A table of all of the known values, along with bounds for some of the unknown values is given below.[125]

$f \backslash s$	1	2	3	4	5	6	7	8	9	10
1	1	1	1	1	1	1	1	1	1	1
2		2	3	4	5	6	7	8	9	10
3			6	9	14	18	23	28	36	40–42
4				18	25	36–41	49–61	59–84	73–115	92–149
5					43–48	58–87	80–143	101–216	133–316	149–442
6						102–165	115–298	134–495	183–780	204–1,171
7							205–540	217–1,031	252–1,713	292–2,826
8								282–1,870	329–3,583	343–6,090
9									565–6,588	581–12,677
10										798–23,556

Because of the symmetry that would be present in the table above, some values were left out. What remains forms a triangle, but don't expect there to be a simple rule to extend it, like for the triangles of Pascal, Stirling, and Bell!

> Calculating precise values for Ramsey numbers, or even close estimates, appears to be among the most fiendishly difficult open combinatorial problems.[126]

Determining Ramsey numbers is an example of a problem in Ramsey theory, but there are many others. Ramsey theory is usually classified as combinatorics, probably because it asks "How many?" questions. I placed it in the present chapter because of the usefulness of graph theory in establishing one of the few simple results in this area. Graph theory and combinatorics often go hand in hand.

Ramsey theory is an area of mathematics that might not exist, if it wasn't for the fact that most great mathematicians do their best work early in their lives, for the man this area is named after died at only 26 years of age. It wasn't suicide or a violent death that claimed British mathematician Frank Ramsey (Figure 20.46), but rather the bad luck of a brilliant

---

125 https://en.wikipedia.org/wiki/Ramsey%27s_theorem#Ramsey_numbers

126 Wilson, Robin, and John J. Watkins, editors, *Combinatorics: Ancient & Modern*, Oxford University Press, Oxford, 2013, p. 357.

FIGURE 20.46
Frank Ramsey (1903–1930).

Front cover of the biography *Frank Ramsey (1903-1930): A Sister's Memoir* **by Margaret Paul, 2012, courtesy of Enfield Publishing & Distribution Co.**

mind being accompanied by a weak body, and complications following surgery. In his short life, Ramsey also made important contributions to philosophy and economics.

Open Problems

Several open problems were encountered in this chapter. Three more follow below.

1. What is the Crossing Number of the complete bipartite graph $K_{9,9}$?[127] A conjecture gives the value as 256, but there's no proof yet. It might be something else.

2. Does every complete graph K_{2n} with $n \geq 2$ have a perfect 1-factorization? Anton Kotzig conjectured the answer to be yes back in 1964, and the problem has been chipped away at in the decades since then, but the question has still not been answered completely.[128]

---

127 Problem found at http://dimacs.rutgers.edu/~hochberg/undopen/graphtheory/graphtheory.html#prob4.
128 https://en.wikipedia.org/wiki/Graph_factorization.

3. Consider the following headline from June 14, 2018.

> Tunnel collapse forces Amtrak to cancel service, delays Union Pacific service in 23 states.[129]

The collapse in the Starlight tunnel only directly affected about 50 feet of track. Why did the problem cascade across 23 states? In a similar manner, a single downed wire initiated a cascade that led to a major blackout in the Western United States and portions of Canada on August 10, 1996. Power was knocked out for about 7.5 million people and the cost was estimated at $2 billon. An important open problem is how to design networks to serve us efficiently while, at the same time, making them robust. Ideally there would be no small set of points or edges that if disrupted, whether by nature or terrorists, could throw major portions of the network into disarray. Some extremes are shown in Figure 20.47.

The network on the left, with the hub, is easy to disrupt. The one on the right is much more robust, but would not work well for applications such as airline flight. These extremes were considered back in 1964 by Paul Baran, working for RAND.[130]

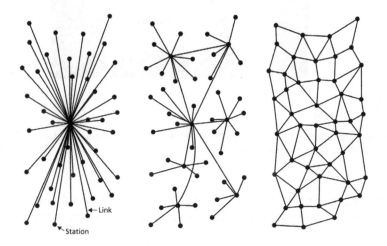

FIGURE 20.47
Centralized, Decentralized and Distributed Networks.

Image from RAND report P-2626, Baran, Paul, "On Distributed Communications: 1. Introduction to Distributed Communications Network," August 1964, 37 pages, https://www.rand.org/pubs/papers/P2626.html, courtesy of RAND Corporation.

---

129 http://kval.com/news/local/tunnel-collapse-forces-amtrak-to-cancel-service-delays-union-pacific-service-in-23-states.
130 Baran, Paul, "On Distributed Communications: 1. Introduction to Distributed Communications Network," August 1964, 37 pages.

Exercise Set 20

I think maybe the most important thing I've learned on this trip is that the simple, romantic life isn't enough for me. I need something complicated to think about.[131]

1. A new bridge was added in Königsberg in 1875. It changed the graph to the one shown in Figure 20.48. Is it possible to cross all eight bridges exactly once in a stroll?

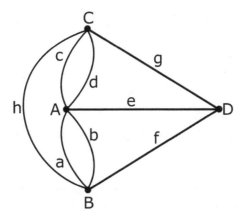

FIGURE 20.48
Bridges of Königsberg, 2.0.

Image created by Josh Gross.

2. Can you navigate the graph in Figure 20.49 Königsberg bridges style?

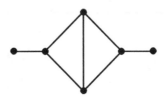

FIGURE 20.49
A navigable network?

Image created by Craig P. Bauer.

---

131 From Knuth, Donald, *Surreal Numbers: How Two Ex-Students Turned on to Pure Mathematics and Found Total Happiness*, 1974, p. 4.

3. Consider Figure 20.50.

FIGURE 20.50
The sign-manual of Muhammad.

Image from Lucas, Édouard, *Récréations Mathématiques I*, Gauthier-Villars et fils, Paris, 1891, p. 36 (public domain).

These crossed crescents, known as the sign-manual of Muhammad, are said to have been drawn in the sand in a single stroke (i.e., an Eulerian path) by the prophet with the tip of his scimitar. Is this possible?

4. Can you find a route around the Star of David (Figure 20.51), Königsberg bridges style?

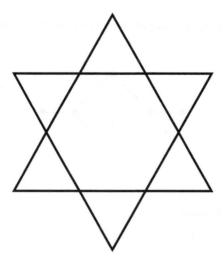

FIGURE 20.51
The Star of David.

Image created by Josh Gross.

5. Does the graph shown in Figure 20.52 have an Eulerian cycle? Either find it or explain why one cannot exist.

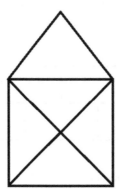

FIGURE 20.52
Is there an Eulerian cycle?

Image created by Craig P. Bauer.

6. Does the graph shown in Figure 20.53 have an Eulerian cycle? Either find it or explain why one cannot exist.

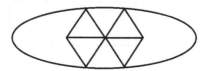

FIGURE 20.53
Is there an Eulerian cycle?

Image created by Craig P. Bauer.

7. Does the graph shown in Figure 20.54 have an Eulerian cycle? Either find it or explain why one cannot exist.

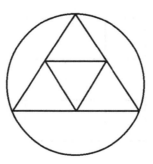

FIGURE 20.54
Is there an Eulerian cycle?

Image created by Craig P. Bauer.

8. For which values of n does K_n have an Eulerian cycle? Justify your result.

9. Draw a graph representing the paths that Pac-Man can follow in the original arcade game. Does this graph have an Eulerian cycle?

10. Is there a curve that passes through each straight line segment of Figure 20.55 exactly once?

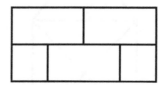

FIGURE 20.55
Can you pass through each line segment exactly once?

Image created by Craig P. Bauer.

To make what is being asked clearer, an unsuccessful attempt is shown in Figure 20.56.

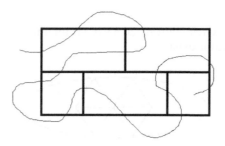

FIGURE 20.56
A failed attempt.

Image created by Craig P. Bauer.

11. Is there a curve that passes through each straight line segment in Figure 20.57 exactly one?

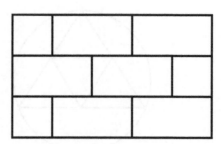

FIGURE 20.57
Can you pass through each line segment exactly once?

Image created by Craig P. Bauer.

12. Can the sum of the valencies of all of the vertices of a graph ever be odd? Give an example or explain why it's impossible.

13. The following is a translation of problem 18 from *Propositiones ad Acuendos Juvenes* (Problems to Sharpen the Young), attributed to Alcuin of York. The work dates back to at least the 9th century.

> **A wolf, a goat and a bunch of cabbages.** A man had to take a wolf, a goat and a bunch of cabbages across a river. The only boat he could find could only take two of them at a time. But he had been ordered to transfer all of these to the other side in good condition. How could this be done?[132]

Although not explicitly stated in the above, it is understood that the wolf and the goat can never be left alone, or the wolf will eat the goat. Similarly, the goat cannot be trusted with the cabbages.

You can solve the using a state graph. This sort of approach was described in Chapter 14. Given that the problem is so old, and reprinted so often, you may already have seen it and be able to solve it without the graph. If this is the case, create it anyway. This is the method of solution you're asked to pursue here.

14. Does Euler's polyhedral formula hold for a cube with a hollow in its center that is also in the shape of a cube?

15. Does Euler's polyhedral formula hold for the solid shown in Figure 20.58?

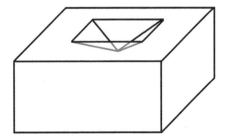

FIGURE 20.58
A solid with an indentation.

Image created by Craig P. Bauer.

The shading is an attempt to show that the solid has an indentation at the top, as if one of the pyramids in Egypt had been turned upside-down and pushed into it.

16. Does Euler's polyhedral formula hold for the solid shown in Figure 20.59?

---

132 This translation is from Hadley, John and David Singmaster, "Problems to Sharpen the Young," *The Mathematical Gazette*, Vol. 76, No. 475, March 1992, pp. 102–126.

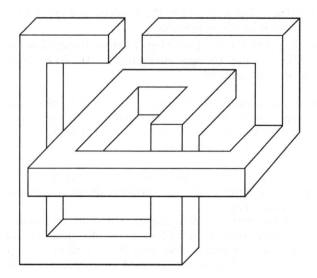

FIGURE 20.59
A solid that doesn't quite meet at the ends.

Image created by Craig P. Bauer.

17. Does Euler's polyhedral formula hold for the solid shown in Figure 20.60?

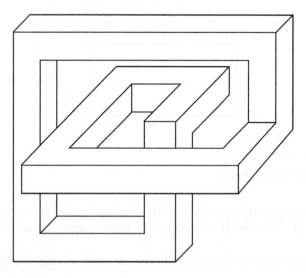

FIGURE 20.60
Closing the gap on Figure 20.59.

Image created by Craig P. Bauer.

18. This chapter showed how a cube can be represented in the plane. Show the corresponding representation for a tetrahedron. Count faces, vertices, and edges, and show that they satisfy Euler's polyhedral formula.

19. Show the planar representation for an octahedron. Count faces, vertices, and edges, and show that they satisfy Euler's polyhedral formula.

20. Can you draw a connected planar graph that has an odd number of faces, an even number of vertices, and an even number of edges? Do so, or prove that it isn't possible.

21. Euler's polyhedral formula can be rewritten as $F + V - E = 2$, and 2 is referred to as the **Euler number** or **Euler characteristic**. For solids with a single hole, the Euler characteristic is 0. That is, in those cases we have $F + V - E = 0$. If there are two holes, we have Euler characteristic -2, which means $F + V - E = -2$. So, Euler's formula can be modified to work on a wider range of solids. The Greek letter χ (chi) is often used to indicate the Euler number of an object like so: $\chi(\text{sphere}) = 2$, $\chi(\text{torus}) = 0$. What is the Euler number for a solid with three holes? Note: today's mathematicians usually refer to "handles" instead of holes and the number of them is called the **genus** of the solid.

22. Make a conjecture concerning the relationship between the genus, g, of a solid and its Euler number, n. Check online to see if your guess is correct.

23. Is $K_{2, 3}$ planar? Draw it as such or prove that it is not planar.

24. Identify which points would be on each side, if the graphs in Figure 20.61 were drawn in the manner typical of bipartite graphs.

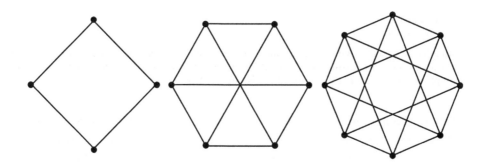

FIGURE 20.61
Some bipartite graphs.

Image created by Josh Gross.

25. Identify which points would be on each side, if the graphs in Figure 20.62 were drawn in the manner typical of bipartite graphs.

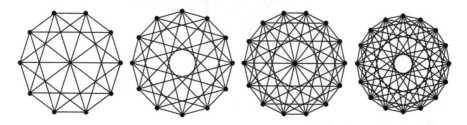

FIGURE 20.62
Some more bipartite graphs.

Image created by Josh Gross.

26. What is the quickest proof that $K_{4,\,3}$ is planar?

27. What is the crossing number of K_5?

28. What is the crossing number of $K_{3,\,3}$?

29. Find a 3-coloring for the Petersen graph shown in Figure 20.63.

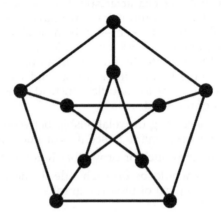

FIGURE 20.63
The Petersen graph.

Adapted by Craig P. Bauer from an image created by Leshabirukov. This file is licensed under the Creative Commons Attribution-Share Alike 3.0 Unported license. Named after Julius Petersen, this graph first appeared in a paper by Alfred Kempe in 1886.

30. What's the minimum number of colors needed for graphs on tori with two holes?

31. What's the minimum number of colors needed for graphs on tori with three holes?

32. The weighted graph in Figure 20.64 has been changed slightly from the one Dijkstra's algorithm was demonstrated on. Apply the algorithm to this new graph, showing the steps as in the example, to determine the shortest path from P to Q.

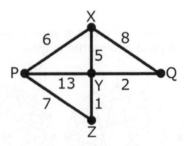

FIGURE 20.64
A weighted graph.

Image created by Josh Gross.

33. There's no shortage of entertaining stories connected with Paul Erdös's personal life. Find one not included in this chapter. Present it in your own words and cite your source.

34. In a world of 7.7 billion people, how many acquaintances would each have to have for average distance between people, along a path of acquaintances, to be 6?

35. How large can a group of people be, where the average number of acquaintances is 500, if we desire the average number of links between people to not exceed 6?

36. How large can a group of people be, where the average number of acquaintances is 500, if we desire the average number of links between people to not exceed 5?

37. How large can a group of people be, where the average number of acquaintances is 500, if we desire the average number of links between people to not exceed 4?

38. Give three examples not seen in this chapter of distributions that follow a power law.

39. Prove that $R(f, 2) = f$ for all f.

40. Exhibit an edge-colored graph for five people in which there is no trio of friends and no trio of strangers.

41. There are many easily stated unsolved problems in graph theory. Find one not included here and explain it in your own words.

References/Further Reading

Television is something the Russians invented to destroy American education.

– Paul Erdös[133]

Adamic, Lada A., Albert-László Barabási, Hawoong Jeong, Bernardo Huberman, Réka Albert, and Ginestra Bianconi, "Power-Law Distribution of the World Wide Web," *Science*, Vol. 287, No. 5461, March 24, 2000, pp. 2115.

Aigner, Martin and Günter M. Ziegler, *Proofs from THE BOOK*, third edition, Springer, Berlin, 2004.

Albert, Réka and Albert-László Barabási, "Statistical Mechanics of Complex Networks," *Review of Modern Physics*, Vol. 74, No. 1, January–March 2002, pp. 47–97.

Alladi, Krishnaswami and Steven Krantz, coordinating editors, "Reflections on Paul Erdös on His Birth Centenary," *Notices of the American Mathematical Society*, Vol. 62, No. 2, February 2015, pp. 121–143.

Alladi, Krishnaswami and Steven Krantz, coordinating editors, "Reflections on Paul Erdös on His Birth Centenary, Part II," *Notices of the American Mathematical Society*, Vol. 62, No. 3, March 2015, pp. 226–247.

Anderson, Philip Warren, "More Is Different," *Science*, Vol. 177, No. 4047, August 1972, pp. 393–396.

Appel, Kenneth and Wolfgang Haken, "Every Planar Map is Four Colorable," *Bulletin of the American Mathematical Society*, Vol. 82, 1976, pp. 711–712.

Appel, Kenneth and Wolfgang Haken, "Every Planar Map is Four Colorable: Part 1, Discharging, Part 2 (With J. Koch) Reducibility," *Illinois Journal of Mathematics*, Vol. 21, 1977a, pp. 429–490 and 491–567.

Appel, Kenneth and Wolfgang Haken, "The Solution of the Four-color-map Problem," *Scientific American*, Vol. 237, No. 4, October 1977b, pp. 108–121.

---

133 The Russians Erdös refers to here are likely Boris Rosing and Vladimir Kosmich Zworykin. See https://en.wikipedia.org/wiki/Vladimir_K._Zworykin for more.

Applegate, David L., Robert E. Bixby, Vasek Chvátal, and William J. Cook, *The Traveling Salesman Problem: A Computational Study*, Princeton University Press, Princeton, NJ, 2007.

Barabási, Albert-László, *Linked: The New Science of Networks*, Perseus Publishing, Cambridge, MA, 2002. This book is written at the popular level.

Barabási, Albert-László and Réka Albert, "Emergence of Scaling in Random Networks," *Science*, Vol. 286, No. 5439, 1999, pp. 509–512.

Barnette, David, *Map Coloring, Polyhedra, and the Four-Color Problem*, Mathematical Association of America, Washington, DC, 1983.

Batagelj, Vladimir and Andrej Mrvar, "Some Analyses of Erdös Collaboration Graph," *Social Networks*, Vol. 22, No. 2, May 2000, pp. 173–186.

Biggs, Norman L, E. Keith Lloyd, and Robin J. Wilson, *Graph Theory, 1736–1936*, Clarendon Press, Oxford, 1976, reissued in paperback in 1998.

Bollobás, Béla, *Random Graphs*, second edition, Academic, New York, 2001.

Brown, Thomas A., "A Note on 'instant Insanity'," *Mathematics Magazine*, Vol. 41, No. 4, September 1968, pp. 167–169. This paper gives the number theory solution.

Carteblanche, F. de (a pseudonym), "The Coloured Cubes Problem," *Eureka*, Vol. 9, 1947, pp. 9–11. This paper gives the graph theory solution to Instant Insanity.

Case, James, "The Continuing Appeal of Small-world Networks," *SIAM News*, Vol. 34, No. 9, 2001, available online at http://72.32.205.185/pdf/news/588.pdf.

Casti, John L., *Reality Rules, Picturing the World in Mathematics, Volume I: The Fundamentals*, Wiley-Interscience, New York, 1992.

Casti, John L., *Reality Rules, Picturing the World in Mathematics, Volume II: The Frontier*, Wiley-Interscience, New York, 1997.

Cayley, Arthur, "On the Partitions of a Close," *The London, Edinburgh, and Dublin Philosophical Magazine and Journal of Science*, Series 4, Vol. 21, No. 142, 1861, pp. 424–428.

Chung, Fan, "About Ron Graham," available online at www.math.ucsd.edu/~fan/ron/.

Colligan, Douglas, "The High Priest of Guerilla Psych," *OMNI*, Vol. 2, No. 6, March 1980, pp. 108–110, 112–115. This article is on Stanley Milgram.

Coxeter, Harold Scott MacDonald, "The Four-color Map Problem, 1840–1890," *Mathematics Teacher*, Vol. 52, No. 4, April 1959, pp. 283–289.

Csicery, George Paul, director, *N Is a Number: a Portrait of Paul Erdös*. This is a documentary about Paul Erdös.

D'alarcao, Hugo and Thomas E. Moore, "Euler's Formula and a Game of Conway's," *Journal of Recreational Mathematics*, Vol. 9, No. 4, 1976–1977, pp. 149–251.

Dantzig, George B., "Reminiscences about the Origins of Linear Programming," *Operations Research Letters*, Vol. 1, No. 2, April 1982, pp. 43–48.

Dantzig, George B., D. Ray Fulkerson, and Selmer Martin Johnson, "Solution of a Large-Scale Traveling-Salesman Problem," *Journal of the Operations Research Society of America*, Vol. 2, No. 4, November 1954, pp. 393–410.

Davis, Gerald F., "The Significance of Board Interlocks for Corporate Governance," *Corporate Governance: An International Review*, Vol. 4, No. 3, July 1996, pp. 154–159.

Davis, Gerald F. and Henrich R. Greve, "Corporate Elite Networks and Governance Changes in the 1980s," *American Journal of Sociology*, Vol. 103, No. 1, July 1997, pp. 1–37.

Dijkstra, Edsger W., "A Note on Two Problems in Connexion with Graphs," *Numerische Mathematik*, Vol. 1, No. 1, December 1959, pp. 269–271.

Dudeney, Henry E., "Perplexities," *The Strand Magazine*, Vol. 46, July 1913, p. 110.

Dudeney, Henry E., "Perplexities," *The Strand Magazine*, Vol. 46, August 1913, p. 221.

Eppstein, David, "Twenty Proofs of Euler's Formula: V-E+F=2," *The Geometry Junkyard*, available online at www.ics.uci.edu/~eppstein/junkyard/euler/.

"The Erdös Number Project," Oakland University, available online at www.oakland.edu/enp/.

Erdös, Paul and Alfréd Rényi, "On Random Graphs I," *Publicationes Mathematicae (Debrecen)*, Vol. 6, 1959, pp. 290–297.

Erdös, Paul and Alfréd Rényi, "On the Evolution of Random Graphs," *Publications of the Mathematical Institute of the Hungarian Academy of Sciences*, Vol. 5, 1960, pp. 17–61.

Erdös, Paul and Alfréd Rényi, "On the Strength and Connectedness of a Random Graph," *Acta Mathematica Scientia Hungary*, Vol. 12, 1961, pp. 261–267.

Euler, Leonhard, *Proof of Some Notable Properties with Which Solids Enclosed by Plane Faces are Endowed*, 1758, translated by Chris Francese and David Richeson, 16 pages, available online at https://pdfs.semanticscholar.org/691b/cc650768d9b05eb5b14c0349a8889c908515.pdf.

Federico, Pasquale J., *Descartes on Polyhedra: A Study of the De Solidorum Elementis*, Vol. 4 of Sources in the history of mathematics and physical sciences, Springer, New York, 1982.

Francese, Christopher and David Richeson, "The Flaw in Euler's Proof of His Polyhedral Formula," *The American Mathematical Monthly*, Vol. 114, April 2007, pp. 286–296.

Franklin, Philip, "The Four Color Problem," *American Journal of Mathematics*, Vol. 44, No. 3, July 1922, pp. 225–236.

Franklin, Philip, "A Six Color Problem," *Journal of Mathematics and Physics*, Vol. 13, No. 1–4, April 1934, pp. 363–369.

Fritsch, Rudolf and Gerda Fritsch, *The Four-Color Theorem: History, Topological Foundations, and Idea of Proof*, translated from the 1994 German original by Julie Peschke, Springer, New York, 1998.

Graham, Ronald L., Jaroslav Nešetřil, and Steve Butler, editors, *The Mathematics of Paul Erdös*, two volumes, second edition, Springer, New York, 2013.

Graves, Robert Perceval, *Life of Sir William Rowan Hamilton*, three volumes, Hodges, Figgis, & Co., Dublin, 1889.

Grossman, Jerrold W., "Paul Erdös: The Master of Collaboration," in Graham, Ronald L., Jaroslav Nešetřil and Steve Butler, editors, *The Mathematics of Paul Erdös*, Vol II, second edition, Springer, New York, 2013, pp. 489–496. This is in a volume already referenced, but I wanted to bring special attention to it.

Grossman, Jerrold W. and Patrick D. F. Ion, "On a Portion of the Well-Known Collaboration Graph," *Congressus Numerantium*, Vol. 108, 1995, pp. 129–131. This is on the Erdös graph.

Guare, John, *Six Degrees of Separation: A Play*, Vintage Books, New York, 1990.

Harary, Frank, "On 'The tantalizer' and 'Instant insanity'," *Historia Mathematica*, Vol. 4, 1977, pp. 205–206.

Hoffman, Paul, *The Man Who Loved Only Numbers: The Story of Paul Erdös and the Search for Mathematical Truth*, Hyperion, New York, 1998.

Hopkins, Brian and Robin J. Wilson, "The Truth about Königsberg," *College Mathematics Journal*, Vol. 35, No. 3, May 2004, pp. 198–207.

Jensen, Tommy R. and Bjarne Toft, *Graph Coloring Problems*, John Wiley & Sons, New York, 1995.

Jin, Emily M., Michelle Girvan, and Mark E. J. Newman, "Structure of Growing Social Networks," *Physical Review E*, Vol. 64, No. 4, October 2001.

Khovanova, Tanya, "A Math Paper by Moscow, U.S.S.R.," *Tanya Khovanova's Math Blog*, available online at https://blog.tanyakhovanova.com/2008/08/a-math-paper-by-moscow-ussr/. This is funny.

Kleinberg, Jon, "The Small-world Phenomena: An Algorithmic Perspective," in *Proceedings of the 32nd Annual ACM Symposium on Theory of Computing*, Association of Computing Machinery, New York, 2000, pp. 163–170.

Kleinberg, Jon M., "Navigation in a Small World," *Nature*, Vol. 406, No. 6798, August 24, 2000, p. 845.

Kleinfeld, Judith S., "The Small World Problem," *Society*, Vol. 39, No. 2, January 2002, pp. 61–66.

Klyve, Dominic and Lee Stemkoski, *The Euler Archive*, available online at http://eulerarchive.maa.org/. Most of Leonhard Euler's papers are available here.

Koch, Richard, *The 80/20 Principle: The Secret of Achieving More with Less*, Doubleday, New York, 1998.

Kochen, Manfred, editor, *The Small World*, Ablex, Norwood, NJ, 1989.

Korte, Charles and Stanley Milgram, "Acquaintance Networks between Racial Groups – Application of the Small World Method," *Journal of Personality and Social Psychology*, Vol. 15, No. 2, 1970, pp. 101–108.

Lakatos, Imre, *Proofs and Refutations: The Logic of Mathematical Discovery*, Cambridge University Press, Cambridge, 1976.

Lawler, Eugene L., Jan Karel Lenstra, Alexander H. G. Rinnooy Kan, and David B. Schmoys, editors, *The Traveling Salesman Problem: A Guided Tour through Combinatorial Optimization*, John Wiley & Sons, New York, 1985.

Lewis, Hazel, "Euler's Formula," *Maths Careers*, 2018, available online at www.mathscareers.org. uk/article/eulers-formula/.

Malkevitch, Joseph, "The First Proof of Euler's Formula," *Mitteilungen aus dem Mathematischen Seminar Universität Giessen*, Vol. 165, 1984, pp. 77–82.

May, Kenneth O., "The Origin of the Four-color Conjecture," *Isis*, Vol. 56, No. 3, Autumn 1965, pp. 346–348.

Mayfield, Kendra, "Kevin Bacon: You've Got Mail," *Wired*, January 15, 2002, available online at www.wired.com/2002/01/kevin-bacon-youve-got-mail/.

McKay, Brendan D., "A Note on the History of the Four-colour Conjecture," *Journal of Graph Theory*, Vol. 72, No. 3, March 2013, pp. 361–363.

Merton, Robert K., "The Matthew Effect in Science," *Science*, Vol. 159, No. 3810, January 5 1968, pp. 56–63.

Milgram, Stanley, "Behavioral Study of Obedience," *Journal of Abnormal and Social Psychology*, Vol. 67, No. 4, 1963, pp. 371–378.

Milgram, Stanley, "The Small World Problem," *Psychology Today*, Vol. 1, No. 1, May 1967, pp. 61–67.

Milgram, Stanley, *Obedience to Authority: An Experimental View*, Harper & Row, New York, 1974.

Milgram, Stanley, *The Individual in a Social World: Essays and Experiments*, second edition, McGraw-Hill, New York, 1992.

Newman, Mark E. J., "Models of the Small World," *Journal of Statistical Physics*, Vol. 101, No. 3–4, November 2000, pp. 819–841.

Newman, Mark E. J., "The Structure of Scientific Collaboration Networks," *Proceedings of the National Academy of Sciences*, Vol. 98, No. 2, January 16, 2001, pp. 404–409.

Newman, Mark E. J., Steven H. Strogatz, and Duncan J. Watts, "Random Graphs with Arbitrary Degree Distributions and Their Applications," *Physical Review E*, Vol. 64, No. 2, August 2001.

Newman, Mark E. J., Duncan J. Watts, and Steven H. Strogatz, "Random Graph Models of Social Networks," *Proceedings of the National Academy of Sciences*, Vol. 99, suppl 1, February 19, 2002, pp. 2566–2572.

O'Connor, John J. and Edmund F. Robertson, "Dénes König," *MacTutor Archive*, January 2014, available online at www-history.mcs.st-andrews.ac.uk/Biographies/Konig_Denes.html. This biography includes an interesting quote from Paul Erdös's introduction to the first issue of *Journal of Graph Theory*, in which he reflects on the rising importance of graph theory.

Padberg, Manfred W. and Giovanni Rinaldi, "Optimization of a 532-city Symmetric Traveling Salesman Problem by Branch and Cut," *Operations Research Letters*, Vol. 6, No. 1, March 1987, pp. 1–7.

Paulos, John Allen, *A Mathematician Reads the Newspaper*, BasicBooks, New York, 1995.

Pegg, Jr., Ed, "Beyond Sudoku: The Icosian Game, Revisited," *The Mathematica Journal*, Vol. 11, No. 3, available online at www.mathematica-journal.com/issue/v11i3/contents/superhamilton/ superhamilton.pdf.

Pool, Ithiel de Sola and Manfred Kochen, "Contacts and Influence," *Social Networks*, Vol. 1, No. 1, 1978, pp. 1–51.

Price, Derek J. de Solla, "Networks of Scientific Papers," *Science*, Vol. 149, No. 3683, July 30 1965, pp. 510–515.

Ramsey, Frank P., "On a Problem in Formal Logic," *Proceedings of the London Math Society*, Series 2, Vol. 30, 1930, pp. 264–286.

Reynolds, Patrick, "The Oracle of Bacon," available online at http://oracleofbacon.org/.

Richeson, David, "The Polyhedral Formula," in Bradley, Robert E. and C. Edward Sandifer, editors, *Leonhard Euler: Life, Work and Legacy*, Vol. 5 of Studies in the history and philosophy of mathematics, Elsevier, Amsterdam, 2007, pp. 421–439.

Richeson, David, *Euler's Gem: The Polyhedron Formula and the Birth of Topology*, Princeton University Press, Princeton, NJ, 2012.

Ringel, Gerhard, *Map Color Theorem*, Springer, New York, 1974.

Robertson, Neil, Daniel Sanders, Paul Seymour, and Robin Thomas, "The Four-colour Theorem," *Journal of Combinatorial Theory*, Series B, Vol. 70, No. 1, May 1997, pp. 2–44.

Sachs, Horst, Michael Stiebitz, and Robin J. Wilson, "An Historical Note: Euler's Königsberg Letters," *Journal of Graph Theory*, Vol. 12, No. 1, Spring 1988, pp. 133–139.

Schechter, Bruce, *My Brain Is Open: The Mathematical Journeys of Paul Erdös*, Simon & Schuster, New York, 1998. This is a biography of Paul Erdös.

Shannon, Claude E., "A Theorem on Coloring the Lines of a Network," *Journal of Mathematics and Physics*, Vol. 28, No. 1–4, April 1949, pp. 148–152.

Soifer, Alexander, *The Mathematical Coloring Book: Mathematics of Coloring and the Colorful Life of Its Creators*, Springer, New York, 2009.

Strogatz, Steven H., "Exploring Complex Networks," *Nature*, Vol. 410, No. 6825, March 8, 2001, pp. 268–276.

Swart, Edward R., "The Philosophical Implications of the Four-color Problem," *American Mathematical Monthly*, Vol. 87, No. 9, November 1980, pp. 697–707.

Temperley, Harold Neville Vazeille, *Graph Theory and Applications*, Horwood, John Wiley & Sons, New York, 1981.

Thomas, Robin, "An Update on the Four-colour Theorem," *Notices of the American Mathematical Society*, Vol. 45, No. 7, August 1998, pp. 848–859.

Travers, Jeffrey and Stanley Milgram, "An Experimental Study of the Small World Problem," *Sociometry*, Vol. 32, No. 4, December 1969, pp. 425–443.

Tutte, William T., "On Hamiltonian Circuits," *Journal of the London Mathematical Society*, Series 1, Vol. 21, No. 2, April, Vol., No., 1946, pp. 98–101.

Tutte, William T., "The Coming of the Matroids," in Lamb, John D. and Donald A. Preece, editors, *Surveys in Combinatorics, London Mathematical Society Lecture Note Series*, Vol 267, Cambridge University Press, Cambridge, 1999, pp. 3–20.

Tutte, William T., *Graph Theory as I Have Known It*, Oxford Lecture Series in Mathematics and its Applications, Book 11, Oxford University Press, Oxford, reprint edition, 2012. The first edition appeared in 1998.

Tymoczko, Thomas, "Computers, Proofs and Mathematicians: A Philosophical Investigation of the Four-color Proof," *Mathematics Magazine*, Vol. 53, No. 3, May 1980, pp. 131–138.

Waldrop, Mitchell M., *Complexity: The Emerging Science at the Edge of Order and Chaos*, Touchstone, New York, 1992.

Watts, Duncan J., *Small Worlds: The Dynamics of Networks between Order and Randomness*, Princeton University Press, Princeton, NJ, 1999.

Watts, Duncan J., "Networks, Dynamics and the Small-world Phenomenon," *American Journal of Sociology*, Vol. 105, No. 2, September 1999, pp. 493–527.

Watts, Duncan J., "A Simple Model of Global Cascades on Random Networks," *Proceedings of the National Academy of Sciences*, Vol. 99, No. 9, April 30 2002, pp. 5766–5771.

Watts, Duncan J., *Six Degrees: The Science of a Connected Age*, W. W. Norton & Company, New York, 2003. This book is written at the popular level.

Watts, Duncan J. and Steven H. Strogatz, "The Collective Dynamics of 'Small-World' Networks," *Nature*, Vol. 393, No. 6684, June 4 1998, pp. 440–442.

Weil, André, "Euler," *American Mathematical Monthly*, Vol. 91, No. 9, November 1984, pp. 537–542.

Wikipedia Contributors, "Petersen Graph," *Wikipedia, The Free Encyclopedia*, available online at https://en.wikipedia.org/w/index.php?title=Petersen_graph&oldid=837570707. See this page for some of the many cool properties of this graph.

Wilson, Robin, *Four Colors Suffice: How the Map Problem Was Solved*, Princeton University Press, Princeton, NJ, 2002.

Wilson, Robin and John J. Watkins, editors, *Combinatorics: Ancient & Modern*, Oxford University Press, Oxford, 2013.

Wilson, Robin J., "An Eulerian Trail through Königsberg," *Journal of Graph Theory*, Vol. 10, No. 3, Autumn 1986, pp. 265–275.

Winn, John A., *Asymptotic Bounds for Classical Ramsey Numbers*, Polygonal Publishing House, Washington, NJ, 1988.

Zipf, George Kingsley, *Human Behavior and the Principle of Least Effort*, Addison-Wesley, Cambridge, MA, 1949.

21

Trees

The idea of family tree is very old, so it's surprising that mathematicians didn't adapt it for mathematical use until the 19th century. Nowadays, they're used very widely, both in the form favored by genealogists and in a generalized form. As the definition below shows, trees are simply a special kind of graph.

Definition – A **Tree** is a connected graph without any cycles as subgraphs.
Recall from Chapter 20 that a graph is connected if there's a path from each vertex to every other vertex. A subgraph is obtained from the original graph by removing vertices and the edges that are incident on those vertices. A cycle is a path from a point to itself.

Example 1
Everyone has two biological parents (disregarding any cloned humans who may exist). Each of the parents has two parents and so on. We can construct a family tree to show these relations. Ideally, it should look something like Figure 21.1.

However, another possibility is shown in Figure 21.2.

Person B would be referred to as **inbred**. He or she would not actually have a family tree; it's just a family graph, because of the cycle. It is not, based on the limited data of Figure 21.1, fair to conclude that Person A is not inbred. Person A could, for example, fail to have the required eight distinct great-grandparents. The family tree of someone who is not inbred is an example of a binary tree. Each generation back must yield twice as many distinct ancestors as the previous generation. Hence, n generations back requires 2^n distinct ancestors from that generation.

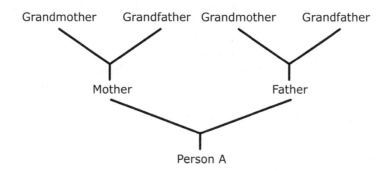

FIGURE 21.1
A family tree.

Image created by Josh Gross.

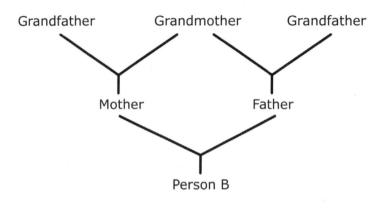

FIGURE 21.2
Another family tree.

Image created by Josh Gross.

Just for fun, let's look back 2,000 years. First we need to estimate how many generations this is. How many years are there between the generations, on average, in your family? I'll use 25, but your answer may be different. With 25 years between generations, a timespan of 2,000 years would amount to 80 generations. So, 2,000 years ago, I must have had $2^{80} \approx 1.2 \times 10^{24}$ distinct ancestors walking around. Given that the population of the Earth is currently about 7.7×10^9, I'm left to conclude that either I'm inbred or the Earth was a lot more crowded 2,000 years ago! Of course, I may have used too small of a value for the average time between generations. Let's say everyone got a late start – 50 years between generations! Then there would only have been 40 generations between myself and my people from the year 20. In this case, my peeps would only have numbered $2^{40} \approx 1.1 \times 10^{12}$, which is still far larger than the Earth's present population. Oh well.

A node of a tree in graph theory can have any number of "parents" while in a family tree two is typical. Also, there cannot be any inbreeding in a mathematical tree. In mathematical terms, this means that there can be no cycles. A few examples of trees are shown below. Note that trees can be drawn in various orientations. They need not branch out in the upward direction (Figure 21.3).

Trees must be connected. A collection of trees is a graph, but it isn't a tree. Such graphs are sometimes referred to as **forests**.

Trees are so handy that, despite my trying to avoid them until the present chapter, one slipped through in Chapter 7, when an enumeration problem that led to the multiplication principle was investigated, and another appeared in Chapter 19 (the Collatz graph). Various other examples are presented in this chapter.

Cayley Counts Trees

Drawing his inspiration from operators in differential calculus and a more concrete problem in chemistry, British mathematician Arthur Cayley (Figure 21.4), in 1857, was the first to put the word "tree" into print as a type of graph The concept, however, had

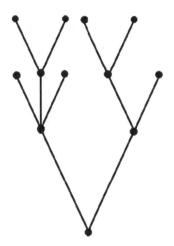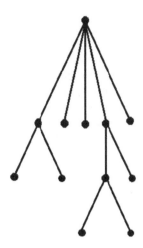

FIGURE 21.3
A pair of trees.

Image created by Craig P. Bauer.

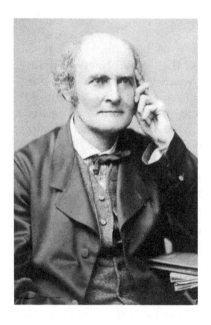

FIGURE 21.4
Arthur Cayley (1821–1895).

Public domain.

already been around for a decade.[1] Instead of using today's graph theory terms such as nodes or vertices, he continued the tree analogy by referring to "knots." Edges were branches. The origin point which all branches can be traced back to was called the **root**. The term root is still used today, as is the term **leaf**, used for the vertices at the extreme end from the root. The leaves always have degree 1. The root is the only other vertex that may have degree 1, although it often does not.

In Cayley's first paper on trees, he showed how to calculate (using generating functions) the number of distinct (unlabeled) rooted trees with n edges.[2] Specific values he included, with his mistakes corrected, are shown in the following table.

Number of edges	Number of rooted trees
1	1
2	2
3	4
4	9
5	20
6	48
7	115
8	286
9	719
10	1,842

In later papers, building on work by Camille Jordan, Cayley counted unrooted trees.[3] To see the difference, consider the two (distinct) rooted trees shown in Figure 21.5.

They are clearly different when viewed in this manner, but if we don't identify the topmost vertex as the root in each graph, then we can redraw either graph to match the other. For example, the graph on the right only needs to have the top edge drawn pointing straight down from the vertex of degree 3 to match the graph on the left. Hence, the number of rooted trees only offers a bound on the number of unrooted trees.

For this new problem, Cayley counted the possibilities not as a function of the number of edges, but rather as a function of the number of vertices. This is not a serious change, as any tree with n vertices must have $n - 1$ edges. Still, because such enumerations are usually given nowadays in terms of the number of vertices, the results are presented in this manner in the following table, along with the number of rooted trees again.

---

1 Biggs, Norman L., E. Keith Lloyd, and Robin J. Wilson, *Graph Theory, 1736–1936*, Clarendon Press, Oxford, 1976, p. 38. See von Staudt, Karl Georg Christian, *Geometrie der Lage*, Nürnberg, 1847 and Kirchoff, Gustav Robert, "Über die Auflösung der Gleichungen, auf welche man bei der Untersuchung der linearen Vertheilung galvanischer Ströme geführt wird," *Annalen der Physik und Chemie*, Vol. 72, 1847, pp. 497–508 for the earlier appearances of the concept.

2 Cayley, Arthur, "On the Theory of the Analytical Forms Called Trees," *The London, Edinburgh, and Dublin, Philosophical Magazine and Journal of Science*, Series 4, Vol. 13, No. 85, 1857, pp. 172–176.

3 See Jordan, Camille, "Sur les assemblages de lignes," *Journal für die reine und angewandte Mathematik*, Vol. 70, 1869, pp. 185–190, Cayley, Arthur, "Über die analytischen Figuren, welche in der Mathematik Bäume genannt werden und ihre Anwendung auf die Theorie chemischer Verbindungen," *Berichte der deutschen chemischen Gesellschaft*, Vol. 8, 1875, pp. 1056–1059, and Cayley, Arthur "On the Analytical Forms Called Trees, with Application to the Theory of Chemical Combinations," *Report of the British Association for the Advancement of Science*, 1875, pp. 257–305.

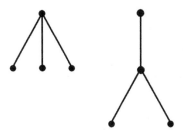

FIGURE 21.5
A pair of rooted trees.

Image created by Craig P. Bauer.

Number of vertices	Number of rooted trees	Number of unrooted trees
1	1	1
2	1	1
3	2	1
4	4	2
5	9	3
6	20	6
7	48	11
8	115	23
9	286	47
10	719	106

The trees shown in Figure 21.6 give the six possibilities with six vertices for the unrooted case.

The fact that a tree with n vertices must have $n-1$ edges can be taken as the definition of a tree, for no non-tree graph has this property. More definitions are possible. Here's a nice theorem that gives five different, but equivalent, ways to define a tree.

Daisy Chain Theorem: Let T be a graph with n vertices. The following are equivalent.[4]

1. T is a tree.
2. T has no cycles and $n-1$ edges.
3. T is connected and has $n-1$ edges.
4. T is connected, but removing any edge disconnects it.
5. There exists exactly one path between any two vertices.
6. T has no cycles, but connecting two vertices with an edge will create a cycle.

---

4 Taken here from McEliece, Robert J., Robert B. Ash, and Carol Ash, *Introduction to Discrete Mathematics*, Random House, New York, NY, 1989, p.107.

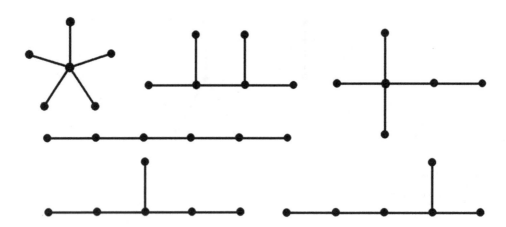

FIGURE 21.6
The six unrooted trees with six vertices.

Image created by Craig P. Bauer.

Cayley's Chemistry

The problem in chemistry that inspired Cayley to make use of trees was that of isomers. These are compounds that have the same chemical formula but possess different physical properties, due to differing arrangements of the atoms in the molecule. An example is given in Figure 21.7.

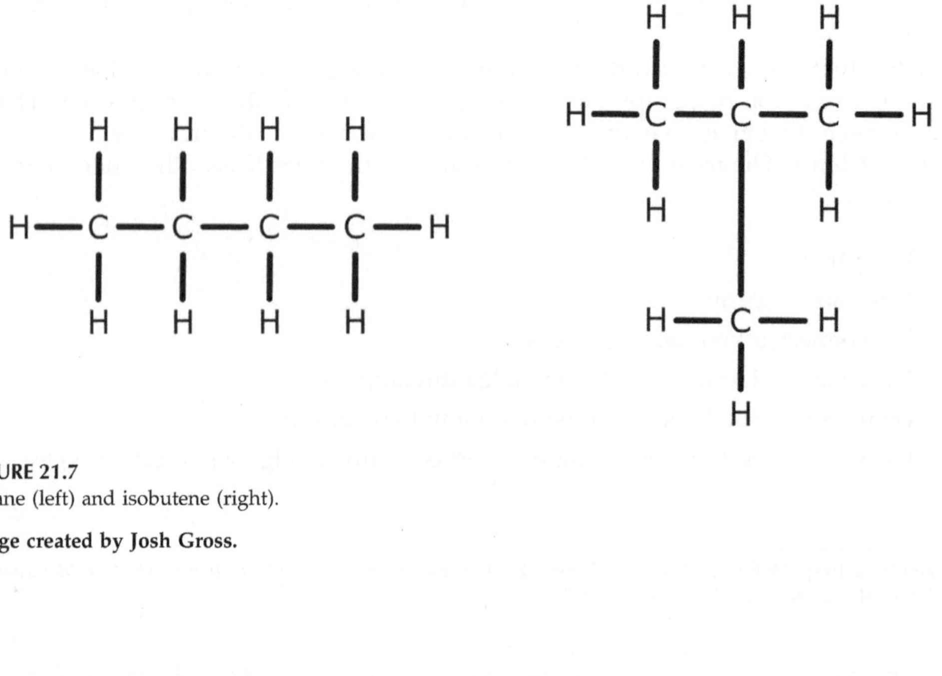

FIGURE 21.7
Butane (left) and isobutene (right).

Image created by Josh Gross.

Both of these illustrations show a molecule with 4 carbon atoms and 10 hydrogen atoms, so they share the chemical formula C_4H_{10}, but their properties differ. The one on the left is called butane (sometimes normal butane or *n*-butane) and the one on the right is called isobutene. At sea level, butane has a boiling point of $30°F/-1°C$, but for isobutene it's $11°F/-12°C$.

The notation used in Figure 21.7 is essentially due to Scottish chemist Alexander Crum Brown, and dates to 1864.[5] The only difference is that Brown placed circles around the letters representing the atoms, making them resemble graphs more strongly. Indeed, this resemblance was noticed by Cayley. In 1874, he wrote a paper bringing the connection to the attention of others.[6] In particular, he calculated the number of isomers of C_nH_{2n+1}, for $n = 1, 2, 3, 4, 5, 6$. In an 1875 paper, he examined C_nH_{2n+2} (the alkanes).[7] They correspond to trees with n vertices, where no vertex has degree greater than 4. The first few values are as follows:

Alkane	Isomers
CH_4	1
C_2H_6	1
C_3H_8	1
C_4H_{10}	2
C_5H_{12}	3
C_6H_{14}	5
C_7H_{16}	9
C_8H_{18}	18

Viewing only the values through C_6H_{14}, it's easy to think that the answer will always be a Fibonacci number, but the next answer is 9, not 8.

It's because of this link with chemistry that the degree of a vertex is sometimes called its valency. Also, Brown's "graphic notation" for representing chemical compounds is what motivated J. J. Sylvester to call the abstract mathematical diagrams **graphs**.[8]

While the papers discussed here are of historical importance, it should be noted that the approach used by Cayley is not the easiest way to count the isomers. In 1935, George Pólya published an easier method.[9]

---

5 Biggs, Norman L, E. Keith Lloyd, and Robin J. Wilson, *Graph Theory, 1736–1936*, Clarendon Press, Oxford, 1976, p. 57.

6 Cayley, Arthur, "On the Mathematical Theory of Isomers," *Philosophical Magazine*, Series 5, Vol. 47, 1874, pp. 444–446.

7 Cayley, Arthur, "On the Analytical Forms Called Trees, with Application to the Theory of Chemical Combinations," *Report of the British Association for the Advancement of Science*, Vol. 45, 1875, pp. 257–305.

8 See Biggs, Norman L., E. Keith Lloyd, and Robin J. Wilson, *Graph Theory, 1736–1936*, Clarendon Press, Oxford, 1976, p. 65. The first use of the word graph in the sense of Chapters 20 and 21 of the present book is in Sylvester, J. J., "Chemistry and Algebra," *Nature*, Vol. 17, 1877–1878, p. 284.

9 Pólya, George, "Un problème combinatoire general sur les groups de permutations et le calcul du nombre des isomers des composes organiques (A general combinatorial problem on groups of permutations and the calculation of the number of isomers of organic compounds)," *Comptes Rendus Hebdomadaires des Séances de l'Académie des Sciences*, Paris, Vol. 201, 1935, pp. 1167–1169. Also see Pólya, George, "Kombinatorische Anzahlbestimmungen für Gruppen, Graphen und chemische Verbindungen," *Acta Mathematica*, Vol. 68, 1937, pp. 145–254 or its English translation Pólya, George, and Ronald C. Read, *Combinatorial Enumeration of Groups, Graphs, and Chemical Compounds*, Springer-Verlag, 1987.

Binary Trees

Cayley also counted binary trees.[10] These are trees in which every vertex, except the leaves, has degree 2. Enumerations of trees can be looked at by the number of vertices, the number of edges, or by two variables: the number of internal vertices and the number of leaves. Another way to view them is by **height**, the maximum length path from the root to a leaf. Counting them this way also gives the number of ways to associate a sum (or product) of n terms (or factors.) That is, it's the number of distinct ways to insert pairs of parentheses. The various possibilities can lead to different results if the operation is nonassociative and noncommutative. This is a sequence that grows very rapidly. The first few terms are as follows.[11]

1, 1, 3, 21, 651, 457653, 210065930571, 44127887745696109598901,
194727047691529644955965931760610302427680340,
379186231026592608286823502802789327737023315030011810784643770115806480
8916492244872560821

A difference equation that generates these values is

$$a_{n+1} = a_n^2 + 2a_n \sum_{i=0}^{n-1} a_i.$$

While there's much more that can be said about enumeration of trees, there are also many practical uses for them, especially for binary trees. Examples follow:

Example 2 – Data compression
Before getting into how binary trees can help compress data, consider the following code that, at first glance, appears to reduce our alphabet to a binary system. Portions of this example are reproduced from my first book, *Secret History: The Story of Cryptology*.[12]

(International) Morse Code

A	._	N	_.
B	_...	O	_ _ _
C	_._.	P	._ _.
D	_..	Q	_ _._
E	.	R	._.
F	.._.	S	...
G	_ _.	T	_
H	....	U	.._
I	..	V	..._
J	._ _ _	W	._ _
K	_._	X	_.._
L	._..	Y	_._ _
M	_ _	Z	_ _..

---

10 Cayley, Arthur, "On the Theory of the Analytical Forms Called Trees. Second Part," *Philosophical Magazine*, Vol. 18, 1859, pp. 374–378.
11 Taken from https://oeis.org/A001699, which also gives a list of references and other relevant information.
12 Bauer, Craig P., *Secret History: The Story of Cryptology*, Chapman and Hall/CRC, Boca Raton, FL, 2013.

There are also combinations of dots and dashes representing the digits 0 through 9, but as these can be spelled out, they aren't strictly necessary. Notice that the most common letters, E and T, have single character representations, while Z, a much rarer letter, is represented by four characters.

Although the representations shown above make Morse code efficient, it isn't actually a binary system, because a pause is needed between letters when transmitting to avoid ambiguity. Without the pause, a string like _..._ _... could be split up to form NEWS or NUDE or several other English words. Nick Berry's DataGenetics blog provided a list of 25,787 ambiguous Morse code words.[13]

Morse code really consists of three symbols, when you consider the pause, which is like the space bar on a keyboard. A longer pause, needed to separate words, can be thought of as hitting the space bar twice. The Huffman code described below is a true binary code.

Huffman codes make use of the same idea as Morse code. Instead of representing each character by eight bits, as is standard for computers, common characters are assigned shorter representations, while rarer characters receive longer representations. The compressed text is then stored along with a key giving the substitutions that were used. This is a simple example of an important area known as **data compression**. High compression rates allow information to be sent more rapidly, as well as take up less space when stored; ZIP files are an example. If not zipped, the download time would be much longer. Huffman coding can also be used to compress images; JPEG images provide an example.

The next level of Huffman coding is to replace strings of characters with bit strings. A common word may be reduced to less space than a single character normally requires, while a rarer word becomes longer after encoding. This method should only be applied to large files. It wouldn't be efficient to encode a short letter in this manner.

Letter	Huffman Code	Letter	Huffman Code	
E	000	M	11010	
T	001	W	11011	
A	0100	F	11100	
O	0101	G	111010	(six characters)
I	0110	Y	111011	
N	0111	P	111100	
S	1000	B	111101	
H	1001	V	111110	
R	1010	K	1111110	(seven characters)
D	10110	J	11111110	(eight characters)
L	10111	X	111111110	(nine characters)
C	11000	Q	1111111110	(ten characters)
U	11001	Z	1111111111	

---

13 http://www.datagenetics.com/blog/march42012/results.html

When using Morse code, it is important to leave spaces between the letters, as was mentioned earlier. For Huffman coding, no spaces are needed. The tree shown in Figure 21.8, along with its explanation, illustrate why.

Huffman coding uses a binary tree with leaves at various levels. Each leaf is assigned a letter, with leaves closer to the root receiving the more common letters. The string of 0s and 1s from the root to the leaf is the binary representation of the given letter. There can be no ambiguity, because the string leading to a given letter cannot be split into strings representing any other letters. The tree structure guarantees this. Starting from the root, a path that leads to a given letter does not encounter any other letters along the way.

Although named after David A. Huffman (Figure 21.9), these codes were originated by Claude Shannon and Robert M. Fano and only improved upon by Huffman, who was one of Fano's students.[14]

The terms encode and decode are used in the context of techniques like this, but data compression is not the same as coding theory. Coding theory lengthens messages in an effort to make garbled bits recoverable. It adds redundancy, while data compression seeks to remove it, as does cryptography. A few simple examples in coding theory were given by the check digit schemes covered in Chapter 7.

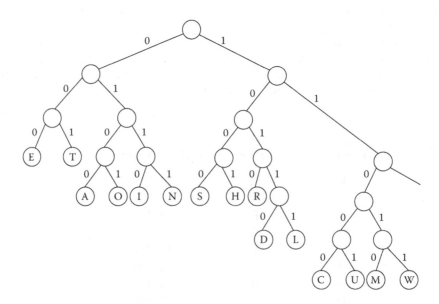

FIGURE 21.8
A Huffman code tree.

Image created by Craig P. Bauer.

14 Soni, Jimmy and Rob Goodman, *A Mind at Play: How Claude Shannon Invented the Information Age*, Simon & Schuster, New York, NY, 2017, pp. 154–156.

FIGURE 21.9
David A. Huffman (1925–1999).

Courtesy of Don Harris, University of California, Santa Cruz.

Example 3 – Trees and probability

Binary trees are very handy for thinking through various problems in probability. A simple example is given here, but there are many more such applications.

Suppose a certain drug test is 97% accurate. Sounds pretty good doesn't it? If someone tests positive for the drug, does that mean that there's a 97% chance that he or she is a user? To answer this question we need one more number, namely the percentage of the population that uses the drug. Let's suppose that the drug isn't very popular. Only 2% of the population uses it. With the information we now have, we can generate a probability tree (Figure 21.10).

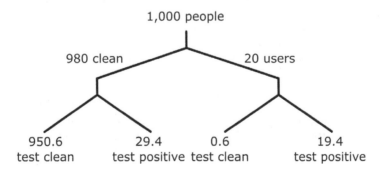

FIGURE 21.10
A tree for determining probabilities.

Image created by Josh Gross.

If all we know is that a person tested positive, he or she could be in the clean group or in the group of users. Out of the 29.4 + 19.4 = 48.8 people who tested positive, only 19.4 are users, so the probability that the accused really uses the drug is 19.4/48.8 ≈ 0.3975. This is a 39.75% chance. The accused is more likely to be clean than not. However, the only number that you'll hear in the media, if a celebrity fails a test, is 97% accurate.

The sort of problem described above is known as a **conditional probability**. It asked what is the probability that someone is a user, *given that* he or she tested positive. In general, such probabilities are written $P(A \mid B)$, where A is the event that we're interested in (being a user) and B is the event we know occurred (testing positive). The calculation that was motivated by the tree above can be expressed algebraically as

$$P(A|B) = \frac{P(B|A)P(A)}{P(B)}$$

This is known as **Bayes' Theorem**, after the British statistician Thomas Bayes (1701-1761).

Drug testing can also be viewed from an economic point of view. It's expensive. Is it worth it? The following passage from *Nickel and Dimed* by Barbara Ehrenreich may help you decide.

> ...according to a 1999 report from the American Civil Liberties Union, "Drug Testing: A Bad Investment." Studies show that preemployment testing does not lower absenteeism, accidents, or turnovers and (at least in the high-tech workplaces studied) actually lowered productivity – presumably due to its negative effect on employee morale. Furthermore, the practice is quite costly. In 1990, the federal government spent $11.7 million to test 29,000 federal employees. Since only 153 tested positive, the cost of detecting a single drug user was $77,000.[15]

Another question that can be posed is does drug testing serve as a deterrent? Again, our conclusions should be based on hard data. Consider the following excerpt from Dr. Michael Colgan's *Optimum Sports Nutrition* before answering!

> In 1984 and 1985, US Olympic Committee drug testing, always announced well beforehand, found less than 1% of athletes positive for steroids. But during the same years with the same athletes, the US Olympic Committee did a number of unannounced tests. For these the athletes were guaranteed that results would not be subject to sanctions, and would not go on the athlete's record. *One in every two* tested positive for steroids. Half of our finest athletes were on the juice! You have to conclude that drug testing is more a deterrent to public criticism of fat cat sports officials than to drug use.[16]

There's some jargon that goes with any test designed to detect something, be it a drug in a person's system, the presence of a disease, a lie, or if a signature is genuine. The term **sensitivity** refers to the probability of a successful detection when it is present. In the

---

15 Ehrenreich, Barbara, *Nickel and Dimed: On (Not) Getting by in America*, Henry Holt and Company, New York, First Owl Books paperback edition, 2002, p. 128 footnote.

16 Colgan, Michael, *Optimum Sports Nutrition*, Advanced Research Press, Ronkonkoma, New York, NY, 1993, p. 399, which, in turn, references Voy, Robert E., *Drugs Sports and Politics*, Human Kinetics, Champaign, IL, 1991. See www.youtube.com/watch?v=BgyqAD5Z6_A for more.

other direction, **specificity** refers to the probability of the test not returning a false-positive, in other words of being specific to what is being sought.

A false-positive is also referred to as a **type I error**. A test that fails to detect what is being sought when it actually is present (a sensitivity failure) is said to have a **type II error**. Here's how these concerns correspond to some social issues.

Issue	Type I Error	Type II Error
Financial support (e.g., welfare, food stamps, unemployment benefits)	A deserving person being denied support	An undeserving person receiving support
Criminal justice	A guilty person going unpunished	An innocent person getting convicted

When it comes to financial support, mathematician John Allen Paulos characterized liberals as being more concerned about avoiding type I errors and conservatives as more concerned with avoiding type II errors. But for the criminal justice issue, he described conservatives as more interested in avoiding type I errors and liberals more interested in avoiding type II errors.[17] Ideally, we would all be concerned about both types of errors, but which is it that gets you more cranked up for each issue? Does how Paulos would label you, based on your answers, match your own perception of your political leaning?

Monty Hall, Again

A tree diagram (Figure 21.11) can be used to help convince people that it's wise to switch in the Monty Hall problem (see Chapter 12), after one of the curtains has been opened.[18] To start, assume that the player initially picked Curtain #1. (This approach can be repeated with any other initial pick and it will lead to the same conclusion – that the player should switch.) The car is equally likely to be behind each of the three curtains, so probabilities of 1/3 are assigned to each. If the car is indeed behind Curtain #1, Monty will open one of the other curtains with equal probability. Hence, the "Car Behind Curtain #1" branch splits into two branches, each having a probability of 1/2. However, if the car is behind Curtain #2, Monty must open Curtain #3. Remember, we're assuming the player picked Curtain #1, so Monty cannot open that curtain. Similarly, if the car is behind Curtain #3, Monty must open Curtain #2. Thus, we have a pair of branches that get assigned probability 1.

---

17 Paulos, John Allen, *Beyond Numeracy: Ruminations of a Numbers Man*, Vintage Books, A Division of Random House, Inc., New York, NY, 1992, p. 58.

18 A diagram like the one in Figure 21.11 appears at https://en.wikipedia.org/wiki/Monty_Hall_problem and in Rosenhouse, Jason, *The Monty Hall Problem: The Remarkable Story of Math's Most Contentious Brain Teaser*, Oxford University Press, Oxford, 2009, p. 53.

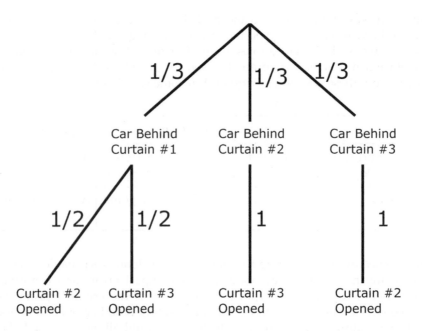

FIGURE 21.11
A Monty Hall problem tree.

Image created by Josh Gross.

If the player sticks with his or her initial pick of Curtain #1, then he or she is at the end of a path that encountered probabilities 1/3 and 1/2. Which of the two such paths it is, depends on which curtain Monty opened. In either case, the probability of being a winner is $(1/3)(1/2) = 1/6$.

However, if the player switches, after Monty has opened either Curtain #1 or Curtain #2, then he or she is at the end of one of the other paths, each of which is a winner with probability $(1/3)(1) = 1/3$. So, the probability of winning is doubled by switching.

We're a Happy Family

A sort of family tree can be made for a mathematician by considering the director of his or her dissertation as a mathematical father. The director of the father's dissertation would be the grandfather. Mathematicians who did their dissertations under the same advisor would be siblings.

You can investigate the graph at http://genealogy.math.ndsu.nodak.edu/. A tiny portion of it is shown in Figure 21.12.

This serves as another example of the small-world phenomena seen in Chapter 20. There's actually a shorter path connecting me to Alan Turing, as I know Alan's nephew, Sir Dermot Turing, an excellent writer and speaker.

At this point, it shouldn't surprise you that virtually all mathematicians are connected by this family tree.

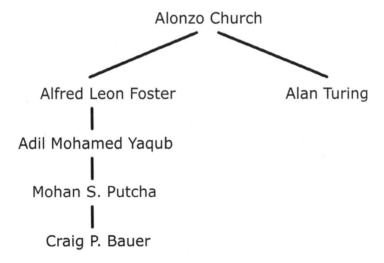

FIGURE 21.12
A piece of the math family tree.

Image created by Josh Gross.

Sierpiński's Carpet (Again)

If we are restricted to drawing trees that can be traced on Sierpiński's carpet, it turns out that we are under no restriction whatsoever. *Every* tree can be sketched in this manner![19]

Open Problem

A graph with n edges is said to be graceful if its vertices can be labeled with some subset of the integers 0 through n (with none used more than once), such that when the edges are labeled using the absolute value of the differences of their vertices, no edge label appears more than once.

The graph shown in Figure 21.13, $K_{3,3}$, has been labeled to demonstrate that it is graceful.[20]

Not all graphs are graceful. For example, the graph shown in Figure 21.14 cannot be labeled in the manner described above.

The **Graceful Tree conjecture** claims that all trees are graceful. It's also referred to as the Ringel–Kotzig conjecture after Gerhard Ringel and Anton Kotzig. It dates back to 1963 and remains open.

---

19 I first saw this in Chamberland, Marc, *Single Digits: in Praise of Small Numbers*, Princeton University Press, Princeton, NJ, 2015, p. 198.

20 Labeling taken from Weisstein, Eric W., "Graceful Graph," From *MathWorld*-A Wolfram Web Resource, available online at http://mathworld.wolfram.com/GracefulGraph.html.

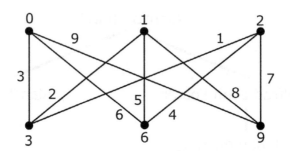

FIGURE 21.13
$K_{3,3}$.

Image created by Josh Gross.

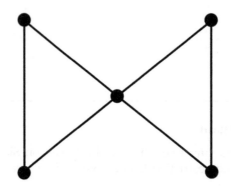

FIGURE 21.14
An ungraceful graph.

Image created by Josh Gross.

Exercise Set 21

Paul Erdös got the Monty Hall problem wrong and stubbornly stuck with his solution.[21]

1. In the film *Good Will Hunting*, the title character, portrayed by Matt Damon, solves the following mathematics problem.

    Draw all the homeomorphically irreducible trees with $n = 10$.

    This is just a fancy way of asking for sketches of all trees with 10 nodes and no vertices of degree 2. Go ahead and do it.

---

21 Rosenhouse, Jason, *The Monty Hall Problem: The Remarkable Story of Math's Most Contentious Brain Teaser,* Oxford University Press, 2009, p. 23.

2. If you need to ease into the problem above with some warm-up exercises, then sketch

a) The 2 homeomorphically irreducible trees with $n = 6$.

b) The 2 homeomorphically irreducible trees with $n = 7$.

c) The 4 homeomorphically irreducible trees with $n = 8$.

d) The 5 homeomorphically irreducible trees with $n = 9$.

Notice that in Exercise 1 I didn't tell you how many trees there are. That's for you to determine!

3. How many homeomorphically irreducible trees are there with

a) $n = 1$?

b) $n = 2$?

c) $n = 3$?

d) $n = 4$?

e) $n = 5$?

Sketch all of the above.

4. Prior to the applications detailed in this chapter, I wrote, "While there's much more that can be said about enumeration of trees, …" Here's an example. Arthur Cayley also found time to count *labeled* trees with n vertices.[22] The 16 possibilities for $n = 4$ are shown in Figure 21.15.

How many possibilities are there for the smaller case $n = 3$? Sketch all of them. Note: In 1889, Cayley gave a very nice formula for the number of possibilities for each value of n, but he didn't offer a proof.[23] The first complete proof didn't appear until 1918, when Heinz Prufer finally provided it.[24]

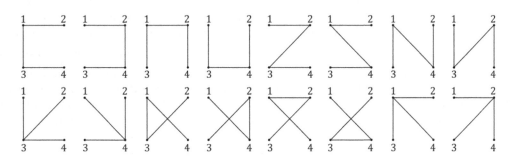

FIGURE 21.15
The 16 labeled trees with 4 vertices.

Image created by Josh Gross.

22 Cayley, Arthur, "A Theorem on Trees," *Quarterly Journal of Pure and Applied Mathematics*, Vol. 23, 1889, pp. 376–378.

23 Cayley, Arthur, "A Theorem on Trees," *Quarterly Journal of Pure and Applied Mathematics*, Vol. 23, 1889, pp. 376–378.

24 Prüfer, Heinz, "Neuer Beweis eines Satzes über Permutationen," *Archiv für Mathematik und Physik*, Vol. 3, No. 27, 1918, pp. 142–144.

5. Represent "THINK BIG" using the Huffman code given in this chapter.

6. Below is a short text created using the Huffman code given in this chapter. What does it say?

   101100101011100111000010101111110001011010110

7. Repeat example 3 with a more popular drug. Assume 50% of the population uses. How does this change the chance of someone failing the test actually being a user? Also, if 50% of people use a particular drug, shouldn't it just be made legal, and the testing done away with?

8. Suppose that the drug test consists of asking about use while the person is hooked up to a polygraph machine (i.e., a lie detector). Such tests can return a result of "inconclusive," which may be handled in different ways when attempting to determine the sensitivity and specificity of the test. For a collection of studies, the values were[25]

   a) 88% sensitive and 86% specific, if inconclusive results are counted as errors;

   b) 97% sensitive and 92% specific, if inconclusive results are ignored.

   Another study obtained

   c) 89% sensitive and 80% specific.

   The percentage of people who lie while taking a polygraph test varies greatly, depending on what group is being questioned and what they are being questioned about. Polygraph tests are periodically administered to some members of the intelligence community with the aim of identifying traitors. Presumably, traitors make up a very small percentage of this group, and hence the percentage of liars is tiny. On the other hand, there may be a high percentage of liars among suspects in criminal investigations. For the sake of this problem, assume that liars account for 1% of those being tested. What is the probability in each case, a, b, and c, that a liar will pass the test? Also, what is the probability in each case, that a truthful person will fail the test?

9. Repeat the exercise above, but now assume that 10% of the people being tested lie.

10. Repeat the exercise above, but now assume that 25% of the people being tested lie.

11. An artificial intelligence (AI) using a deep learning convolutional neural network (CNN) was trained to distinguish between malignant and benign skin lesions. Its skills were tested against a group of 58 dermatologists from around the world. The results of the AI and the average results of the dermatologists were as follows.[26]

	Sensitivity	Specificity
AI program	95.0%	82.5%
Dermatologists	86.6%	75.7%

---

25 Gastwirth, Joseph L., "The Statistical Precision of Medical Screening Procedures: Application to Polygraph and AIDS Antibodies Test Data," *Statistical Science*, Vol. 2, No. 3, August 1987, pp. 213–222, available online at https://projecteuclid.org/download/pdf_1/euclid.ss/1177013215, p. 217 cited here.

26 https://www.theguardian.com/society/2018/may/29/skin-cancer-computer-learns-to-detect-skin-cancer-more-accurately-than-a-doctor and https://academic.oup.com/annonc/advance-article-abstract/doi/10.1093/annonc/mdy166/5004443?redirectedFrom=fulltext.

The AI's greater sensitivity means that fewer of its patients would be sent home thinking they were fine, when they were not. The AI's higher specificity means that fewer of its patients would undergo unnecessary surgeries. Your task is to put numbers to these conclusions. Assume that half of the patients have lesions that require treatment.

a) What's the probability that a person diagnosed by the AI will be recommended for an unnecessary surgery?

b) What's the probability that a person diagnosed by a dermatologist will be recommended for an unnecessary surgery?

c) What's the probability that a person diagnosed by the AI will be cleared when his or her lesion is dangerous?

d) What's the probability that a person diagnosed by a dermatologist will be cleared when his or her lesion is dangerous?

12. Morning sickness is often taken as a sign of pregnancy, yet a study showed that as a test it has a specificity of 85% and a sensitivity of only 39%.[27] If a woman's sexual activity is such that her odds of being pregnant can be estimated (somehow) as 50–50, and she is experiencing morning sickness, what is the probability that she's actually pregnant?

13. Consider the following paragraph concerning a test designed to detect Down syndrome (aka Trisomy 21).

> In recent years, noninvasive prenatal screening tests have been made available. These tests involve blood being taken from the expectant mother as early as 10 weeks of gestation. It relies on the detection of cell-free DNA that circulates between the fetus and the expectant mother. According to the latest research, this blood test can detect up to 98.6% of fetuses with trisomy 21. A "positive" result on the test means that there is a 98.6% chance that the fetus has trisomy 21; a "negative" result on the test means that there is a 99.8% chance that the fetus does not have trisomy 21.[28]

Can the numbers above all be correct?

14. When the HIV blood test (ELISA) became available, many public officials advocated giving everyone the test. At this time, it was believed that 1 out of every 1,000 people in the USA was infected with HIV; $P(\text{HIV}+) = 0.001$. The test had a sensitivity of 95%. That is, 95% of infected people would actually test positive; $P(\text{Test} + \mid \text{HIV}+) = 0.95$. The specificity of the test was 99%. That is, 99% of people who were free of infection would test negative; $P(\text{Test} - \mid \text{HIV}-) = 0.99$.[29]

a) Find the following probabilities:

$$P(\text{HIV}-), \qquad P(\text{Test} - \mid \text{HIV}+), \qquad P(\text{Test} + \mid \text{HIV}-)$$

---

27 Kohlberg, Gavi and Mark Hammer, "Pregnancy: Sensitivity and Specificity," available online at http://getthediagnosis.org/diagnosis/Pregnancy.htm. This website lists the sensitivity and specificity of 17 tests/indicators and provides references to the original sources.

28 https://www.ndss.org/resources/understanding-a-diagnosis-of-down-syndrome/.

29 Data supplied by Frank Serio, Chair, Department of Mathematics, Northwestern State University of Louisiana, Natchitoches, LA.

b) Construct a tree diagram for this situation. Your first branch should be for the person's HIV status, and the second for his or her test result.

c) What is $P(\text{Test} +)$? Hint: Add the results of two branches of your tree.

d) What is $P(\text{HIV+} \mid \text{Test} +)$?

e) What does this last probability mean, in words? Do you think it would have been a good idea to test everyone in the country?

15. This problem came to me from Andy Glen, a professor at the United States Military Academy at West Point. Suppose we wish to save money when testing a large number of samples for HIV. We could split each sample in half and put one half off to the side. Then, the other halves could be mixed together in groups of 10 and tested. If a group tests clean, we've now cleared 10 samples with a single test. If a group tests positive, we then go back and test each of the 10 samples that contributed to it. In a best-case scenario, we get 10 samples tested for the cost of 1 and in a worst-case scenario, we have to do 11 tests to check 10 samples. We could, of course, mix a smaller or larger number of samples together (groups of 5 or 12, etc.). The question is, if 1.1 million Americans currently have HIV, and we wish to test the entire population to identify them, what size groups should the samples be tested in to minimize the cost? Use 330 million for the population of the United States. Don't worry about whether the samples being combined should be of the same blood type or not.

16. As is almost always the case with open problems, mathematicians have chipped away at the Graceful Tree conjecture without finding a complete proof. Search online to find details of partial results and relate them in your own words.

17. Is $K_{2,2}$ graceful?

18. Is K_3 graceful?

19. Is K_4 graceful?

20. Is K_5 graceful?

References/Further Reading

We live for books.
– Umberto Eco

Biggs, Norman L, E. Keith Lloyd, and Robin J. Wilson, *Graph Theory, 1736–1936*, Clarendon Press, Oxford, UK, 1976 (reissued in paperback in 1998).

Cayley, Arthur, "On the Theory of the Analytical Forms Called Trees," *The London, Edinburgh, and Dublin Philosophical Magazine and Journal of Science*, Series 4, Vol. 13, No. 85, 1857, pp. 172–176.

Cayley, Arthur, "On the Theory of the Analytical Forms Called Trees. Second Part," *The London, Edinburgh, and Dublin Philosophical Magazine and Journal of Science*, Series 4, Vol. 18, No. 121, 1859, pp. 374–378.

Cayley, Arthur, "On the Analytical Forms Called Trees, with Application to the Theory of Chemical Combinations," *Report of the British Association for the Advancement of Science*, Vol. 45, 1875, pp. 257–305.

Cayley, Arthur, "Über die analytischen Figuren, welche in der Mathematik Bäume genannt werden und ihre Anwendung auf die Theorie chemischer Verbindungen," *Berichte der deutschen chemischen Gesellschaft*, Vol. 8, 1875, pp. 1056–1059.

Cayley, Arthur, "On the Number of Univalent Radicals C_nH_{2n+1}," *The London, Edinburgh, and Dublin Philosophical Magazine and Journal of Science*, Series 5, Vol. 3, No. 15, 1877, pp. 34–35.

Cayley, Arthur, "On the Analytical Forms Called Trees," *American Journal of Mathematics*, Vol. 4, No. 1, 1881, pp. 266–268.

Cayley, Arthur, "A Theorem on Trees," *Quarterly Journal of Pure and Applied Mathematics*, Vol. 23, 1889, pp. 376–378.

El-Basil, Sherif, *Combinatorial Organic Chemistry: An Educational Approach*, Nova Science Publishers, Inc., Huntington, NY, 2000.

Fujita, Shinsaku, *Symmetry and Combinatorial Enumeration in Chemistry*, Springer, Berlin, 1991.

Gallian, Joseph A., "A Dynamic Survey of Graph Labeling," 22nd edition, *The Electronic Journal of Combinatorics*, December 15, 2019, available online at www.combinatorics.org/ojs/index.php/eljc/article/viewFile/DS6/pdf.

Gastwirth, Joseph L., "The Statistical Precision of Medical Screening Procedures: Application to Polygraph and AIDS Antibodies Test Data," *Statistical Science*, Vol. 2, No. 3, August 1987, pp. 213–222, available online at https://projecteuclid.org/download/pdf_1/euclid.ss/1177013215.

Harary, Frank, "Unsolved Problems in the Enumeration of Graphs," *Publications of the Mathematical Institute of the Hungarian Academy of Sciences*, Vol. 5, 1960, pp. 63–95.

Harary, Frank and Edgar M. Palmer, *Graphical Enumeration*, Academic Press, New York, 1973.

Pólya, George, "Kombinatorische Anzahlbestimmungen für Gruppen, Graphen und chemische Verbindungen," *Acta Mathematica*, Vol. 68, No. 1, December 1937, pp. 145–254. Pólya used generating functions here.

Pólya, George and Ronald C. Read, *Combinatorial Enumeration of Groups, Graphs and Chemical Compounds*, Springer, Berlin, 1987. This is an English translation of the above.

Prüfer, Heinz, "Neuer Beweis eines Satzes über Permutationen," *Archiv für Mathematik und Physik*, Vol. 3, No. 27, 1918, pp. 142–144.

Rains, Eric M. and Neil J. A. Sloane, "On Cayley's Enumeration of Alkanes (Or 4-valent trees)," *Journal of Integer Sequences*, Vol. 2, 1999, Article 99.1.1.

Read, Ronald C., *Some Enumeration Problems in Graph Theory*, PhD thesis, University of London, 1958.

Read, Ronald C., "The Enumeration of Locally Restricted Graphs (I)," *Journal of the London Mathematical Society*, Series 1, Vol. 34, No. 4, October 1959, pp. 417–436.

Read, Ronald C., "The Enumeration of Locally Restricted Graphs (II)," *Journal of the London Mathematical Society*, Series 1, Vol. 35, No. 3, July 1960, pp. 344–351.

Rosa, Alex, "On Certain Valuations of the Vertices of a Graph," *Theory of Graphs, International Symposium*, Rome, July 1966, Gordon and Breach, New York and Dunod, Paris, 1967, pp. 349–355.

Weisstein, Eric W., "Graceful Graph," From *MathWorld*-A Wolfram Web Resource, available online at http://mathworld.wolfram.com/GracefulGraph.html.

22

Relations, Partial Orderings, and Partitions

Recall from Chapter 6 that one way of defining a **function** is as a set of ordered pairs such that no two pairs have the same first value and different second values. We can generalize this concept by defining a **relation** to be a set of ordered pairs without the restriction required to make it a function. Hence, all functions are relations, but not all relations are functions. If a relation is represented by R, one way of indicating that an ordered pair (a, b) belongs to the relation is by writing $a\,R\,b$.

Example 1 – A set of ordered pairs

$$S = \{(1,3),\, (1,7),\, (2,4),\, (3,8),\, (4,8)\}$$

The set S does not give us a function, because $(1, 3)$ and $(1, 7)$ are both in S. However, S is a relation.

Example 2 – A circle

Like functions, relations can be represented algebraically or graphically. The set of points satisfying $x^2 + y^2 = 16$ is expressed graphically in Figure 22.1.

Failing the vertical line test prevents the graph of a circle from being a function, but it is a relation.

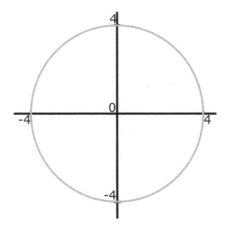

FIGURE 22.1
Graphical representation of a circle.

Image created by Josh Gross.

The concept of a relation is very inclusive, but by placing additional constraints on it we get many interesting special cases. Functions are an example. Another special case is that of partial orderings.

Partial Orderings

A relation R is a **partial order** (sometimes called a **weak partial order**) if it satisfies the following properties:

1. *a R a* (reflexive)
2. if *a R b* and *b R a*, then *a = b* (antisymmetric)
3. if *a R b* and *b R c*, then *a R c* (transitive)

The reflexive property may seem obvious, but it isn't something we can take for granted. For example, if the proposed relation is <, then this property doesn't hold. On the other hand, if the proposed relation is ≤, then we do have a partial order. In fact, the symbol ≤ is often used in place of R when the relation in question is a partial order, even if the relation is not "less than or equal to." Although the usual definition of ≤ gives a partial order, the concept of partial order is much more general.

A relation that fails to be a partial order only because reflexivity doesn't hold is called a **strict partial order**.

Example 3 – Two partial orderings for rational numbers

Georg Cantor noted that "the most different laws" can be used to order the same set. As an example, he mentioned the obvious relation for the rational numbers between 0 and 1, as well as a more exotic possibility: Given two rational numbers, p_1/q_1 and p_2/q_2 in the open interval (0, 1) with p_i and q_i relatively prime integers, we can define an ordering by $p_1/q_1 < p_2/q_2$, if $p_1 + q_1 < p_2 + q_2$. If we happen to have $p_1 + q_1 = p_2 + q_2$, then we take whichever ratio is smaller under the normal definition to also be the smaller for this new definition.[1] Modifying this slightly by using ≤, we have a partial ordering.

The idea of a partial ordering is old. Garrett Birkhoff wrote that they "occur in a fragmentary way in Leibniz' works (circa 1690)."[2] Birkhoff contributed by coining the abbreviation **poset** for a partially ordered set.[3] This was not his only contribution!

Hasse diagrams are sometimes used to depict partial orderings on sets with a small number of elements. They're named after Helmut Hasse (Figure 22.2), who used these diagrams to study the solutions of polynomial equations in his 1926 textbook *Höhere Algebra* (*Higher Algebra*). Hasse taught at Göttingen, a German university considered by some to have been the mathematical capital of the world for a time. This is no longer the case, because

> On April 7, 1933, a new law ... summarily dismissed all those who were Jewish ... The effect on the Mathematical Institute [at Göttingen] was drastic. ... All told, in 1933 eighteen mathematicians left or were driven out from the faculty at the Mathematical Institute in Göttingen.[4]

---

1 Cantor, Georg, *Contributions to the Founding of the Theory of Transfinite Numbers*, Dover, New York, 1955, pp. 110–111.

2 Birkhoff, Garrett, *Lattice Theory*, revised edition, American Mathematical Society, 1948, p. 1, footnote 1.

3 Birkhoff, Garrett, *Lattice Theory*, revised edition, American Mathematical Society, 1948, p. 1, footnote 1.

4 www.gap.dcs.st-and.ac.uk/~history/Mathematicians/Hasse.html.

FIGURE 22.2
Helmut Hasse (1898–1979).

Image created by Konrad Jacobs. This file is licensed under the Creative Commons Attribution-Share Alike 2.0 Germany license.

Hasse was one of the mathematicians found acceptable by the Nazis. This is how he got his position. Follow the link in the footnote to learn more. If you prefer, you may call Hasse diagrams **cover diagrams**. I shall do so here.

Example 4
Let $R = \{(a,a),\ (a,b),\ (a,c),\ (a,d),\ (b,b),\ (b,d),\ (c,c),\ (c,d),\ (d,d)\}$. You may verify the three properties previously stated to see that R is indeed a partial ordering. The cover diagram for R is shown in Figure 22.3.

Note that larger values are placed higher up in the diagram. For this example, we would say that b and c are **incomparable**, as we have neither $(b,c) \in R$ or $(c,b) \in R$. If any two elements are **comparable**, we have a **total ordering**.

Example 5
Another example of a partial ordering arises from the concept of divisibility. We may define $a \le b$ if a divides b. This is also denoted as $a \mid b$. If we restrict our set to the positive divisors of 24, we get the cover diagram shown in Figure 22.4.

It's easy to bring combinatorics into a discussion of partial orderings. Ignoring different labelings, how many possible partial orderings are there for a set of n objects? I'll provide the answer for $n = 3$ and leave the rest to you.

The five possible cover diagrams for a set of three objects are shown in Figure 22.5. From here, it follows that there are 19 possible partial orderings on the set $\{a, b, c\}$. Do you see why? Consider the number of ways each of the cover diagrams can be labeled.

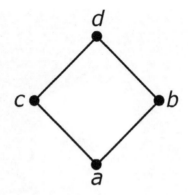

FIGURE 22.3
A cover diagram.

Image created by Josh Gross.

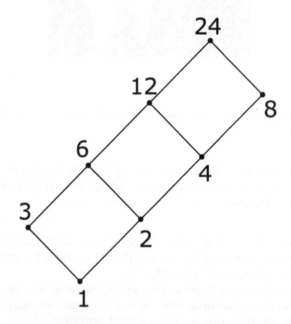

FIGURE 22.4
A cover diagram for a divisibility relation.

Image created by Josh Gross.

Example 6 – The subset poset

Consider the set $S = \{a, b, c\}$. If we look at the power set of S, $P(S) = \{\{a\}, \{b\}, \{c\}, \{a, b\}, \{a, c\}, \{b, c\}, \{a, b, c\}, \oslash\}$, we see that some elements of $P(S)$ are subsets of other elements of $P(S)$. We can use this to define a relation R on $P(S)$ by $x \, R \, y$ if x is a subset of y (proper or not). The cover diagram is shown in Figure 22.6.

FIGURE 22.5
Cover diagrams for a three-element set.

Image created by Josh Gross.

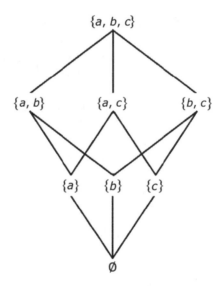

FIGURE 22.6
A cover diagram for the subset relation.

Image created by Josh Gross.

There's nothing special about S having three elements in the example above. The same subset relation defines a partial ordering on any power set.

Connections between different areas of mathematics are the norm. Examples 4 through 6 show how cover diagrams connect partial orders to graph theory.

Equivalence Relations

An **equivalence relation is** another special type of relation. This is a relation on a set S written as ~ and is having the following properties for all a, b, and c in S.

1. $a \sim a$ (reflexive)
2. if $a \sim b$, then $b \sim a$ (symmetric)
3. if $a \sim b$ and $b \sim c$, then $a \sim c$ (transitive)

Equivalence relations are often used to group together objects that are similar in some way.

Example 7 – Equality
A simple but very important equivalence relation is given by =.

Example 8 – Modular arithmetic
In his 1801 book *Disquisitiones Arithmeticae*, Carl Friedrich Gauss introduced an equivalence relation second in importance only to equality.[5] It plays a major role in many areas of mathematics, both pure and applied. He defined it as follows;

$$a \sim b \,(\text{modulo}\, n) \Leftrightarrow n \,\text{divides}\, (a - b)$$

That is, the remainder upon dividing $a - b$ by n must be zero for a to be in the same equivalence class as b. In this particular example, the equivalence classes are often called congruence classes.

Let $n = 4$. Then $0 \sim 4$, since 4 divides the difference of these numbers. Also $0 \sim 8$ and $0 \sim -4$. The equivalence class containing 4 is given by

$$\{\cdots -12, -8, -4, \ 0, \ 4, \ 8, \ 12, \cdots\}$$

There are three other equivalence classes given by

$$\{\cdots -11, -7, -3, \ 1, \ 5, \ 9, \ 13, \cdots\}$$
$$\{\cdots -10, -6, -2, \ 2, \ 6, \ 10, \ 14, \cdots\}$$
$$\{\cdots -9, -5, -1, \ 3, \ 7, \ 11, \ 15, \cdots\}$$

Every integer falls into one of the equivalence classes, so the equivalence relation in this example is said to **partition** the integers. Every equivalence relation is a partition and vice versa.

Modular arithmetic has many applications, some of which you may already be familiar with. For example, if the sum of the digits of a number is divisible by 9, then the number itself is divisible by 9. To verify this, we make use of the following results:

Let $k = ab$, where a and b are integers. If n is also an integer, then $k \bmod n = (a \bmod n)(b \bmod n)$.

From this it follows by induction that if p is a positive integer, then $k^p \bmod n = (k \bmod n)^p$.

Also, if $k = a + b$ (sum instead of a product), then $k \bmod n = (a \bmod n) + (b \bmod n)$.

---

5 Gauss, Carl Friedrich, *Disquisitiones arithmeticae*, 1801. Gauss completed the book in 1798, at the age of 21, but it was a few years before it was published.

Examples:

$$899 \bmod 13 = (29 \bmod 13)(31 \bmod 13) = (3)(5) \bmod 13 = 15 \bmod 13 = 2$$

$$53^{63} \bmod 22 = (53 \bmod 22)^{63} = 9^{63} \bmod 22 = \left(9^3\right)^{21} \bmod 22 = (729)^{21} \bmod 22$$
$$= 3^{21} \bmod 22 = \left(3^7\right)^3 \bmod 22 = 2187^3 \bmod 22 = 9^3 \bmod 22 = 729 \bmod 22 = 3$$

$$135 \bmod 50 = (100 \bmod 50) + (35 \bmod 50) = 35$$

Now, consider a number n with the digits $n_k n_{k-1} \cdots n_3 n_2 n_1 n_0$, where the subscript represents the power of 10 in that position; that is $n = n_k \cdot 10^k + \cdots + n_3 \cdot 10^3 + n_2 \cdot 10^2 + n_1 \cdot 10^1 + n_0$. Then $n \bmod 9 = (n_k + n_{k-1} + \cdots n_3 + n_2 + n_1 + n_0) \bmod 9$. Thus, if the sum of the digits is divisible by 9, so is n.

You're asked to prove some of the properties of modular arithmetic I've made use of here in the exercises at the end of this chapter.

Example 9
Consider the set of words in some dictionary. We may define ~ by Word 1 ~ Word 2 if and only if both words have the same number of letters. Another possibility is to define them to be equivalent if they begin with the same letter.

Equivalence Relations and Partitions

Having digested this book's chapters on combinatorics, you might ask if there's a nice way to count the number of equivalence relations on a set with n elements. There is and it's the nth Bell number (see Chapter 11). This is not a coincidence. Bell numbers give the number of ways to partition a set of items into distinct sets. Equivalence relations also partition objects into distinct sets. As was mentioned previously, any partition defines an equivalence relation and any equivalence relation defines a partition. They are in one-to-one correspondence. Partitions of a different sort are examined later in this chapter.

Nontransitive Relations

Both partial orderings and equivalence relations insist on transitivity, but this is not a property that we can take for granted. There are many examples of relations that violate it. Several are detailed below.

Example 10 – Can we have $A = B$ and $B = C$, but $A < C$
Henri Poincaré took the time to describe a manifestation of the odd scenario stated in the title of this example.[6] In his example, a person cannot distinguish between a weight A of 10 g and a weight B of 11 g. According to Fechner's law, sensation is proportional to the logarithm of the stimulus, so weights that differ by a very small amount will produce essentially the same stimulus and seem identical. Similarly, the weight B of

---

6 Poincaré, Henri, *Science and Hypothesis*, Dover reprint edition, New York, NY, 1952, pp. 22 and 31.

11 g will be indistinguishable from a weight C of 12 g. But, the difference between A and C passes the threshold of distinguishability and a person will say A weighs less than C. Thus, we have $A = B$ and $B = C$, but $A < C$, from the perspective of our senses.

Example 11

Nontransitivity arises in a similar way in this example. Define $a \sim b$ if the distance from a to b is less than 10 miles.

Certainly $a \sim a$ for all points a.

Also, $a \sim b \Rightarrow b \sim a$.

(This may not be true when driving, if one-way streets force different routes depending on the direction!)

However, if $a \sim b$ and $b \sim c$, we don't always have $a \sim c$. Figure 22.7 shows the difficulty.

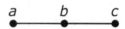

FIGURE 22.7
A few points serve to illustrate a nontransitive relation.

Image created by Josh Gross.

If the distance from a to b is 7 miles and the distance from b to c is 7 miles, then the distance from a to c will be 14 miles, so a is not related to c.

Example 12 – Rock, Paper, Scissors, Spock, Lizard

The rules of this nontransitive game are most easily presented using a directed graph (Figure 22.8). The line between each pair always points to the defeated move.

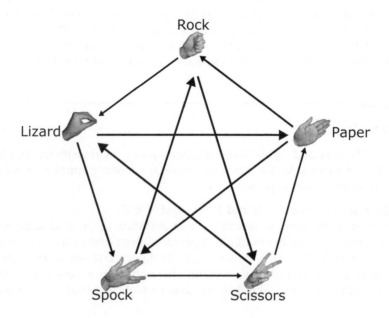

FIGURE 22.8
Rock, Paper, Scissors, Spock, Lizard.

Image created by Josh Gross.

In words, Scissors cuts Paper covers Rock crushes Lizard poisons Spock smashes Scissors decapitate Lizard eats Paper disproves Spock vaporizes Rock crushes Scissors. Stating the rules in this manner gives an Eulerian circuit of the graph above (see Chapter 20).

Example 13 – Weird dice

The following anecdote concerns a pair of the world's richest men, Warren Buffet and Bill Gates.

> Buffett once challenged Gates to a game of dice, using a set of four unusual dice with a combination of numbers from 0 to 12 on the sides. Buffett suggested that each of them choose one of the dice, then discard the other two. They would bet on who would roll the highest number most often. Buffett offered to let Gates pick his die first. This suggestion instantly aroused Gates's curiosity. He asked to examine the dice, after which he demanded that Buffett choose first.
>
> "It wasn't immediately evident that because of the clever selection of numbers for the dice they were nontransitive," Gates said. "The mathematical principle of transitivity, that if A beats B and B beats C, then A beats C, did not apply. Assuming ties were rerolled, each of the four dice could be beaten by one of the others: Die A would beat B an average of 11 out of 17 rolls – almost two-thirds of the time. Die B would beat C with the same frequency. Likewise, C would beat D 11/17 of the time too. And improbable as it sounds, die D would beat A just as often."[7]

A smaller example, involving just three dice, is now detailed. The numbers on the faces of each die are

Die 1: 3, 3, 3, 3, 3, 6
Die 2: 2, 2, 2, 5, 5, 5
Die 3: 1, 4, 4, 4, 4, 4

Table 22.1 shows all of the possible results when Die 1 is rolled against Die 2. The entries show which die wins in each case.

TABLE 22.1

Die 1 vs. Die 2.

		Result of Die 1					
		3	3	3	3	3	6
	2	Die 1	Die 1	Die 1	Die 1	Die 1	Die 1
Result of Die 2	2	Die 1	Die 1	Die 1	Die 1	Die 1	Die 1
	2	Die 1	Die 1	Die 1	Die 1	Die 1	Die 1
	5	Die 2	Die 2	Die 2	Die 2	Die 2	Die 1
	5	Die 2	Die 2	Die 2	Die 2	Die 2	Die 1
	5	Die 2	Die 2	Die 2	Die 2	Die 2	Die 1

---

7 Lowe, Janet, *Bill Gates Speaks: Insight from the World's Greatest Entrepreneur*, John Wiley & Sons, New York, NY, 1998, pp. 148–149.

As you can see, Die 1 wins with probability 21/36 = 7/12.

Next, the results of Die 2 rolled against Die 3 are examined (Table 22.2).

TABLE 22.2

Die 2 vs. Die 3.

		Result of Die 2					
		2	2	2	5	5	5
	1	Die 2	Die 2	Die 2	Die 2	Die 2	Die 2
Result of	4	Die 3	Die 3	Die 3	Die 2	Die 2	Die 2
Die 3	4	Die 3	Die 3	Die 3	Die 2	Die 2	Die 2
	4	Die 3	Die 3	Die 3	Die 2	Die 2	Die 2
	4	Die 3	Die 3	Die 3	Die 2	Die 2	Die 2
	4	Die 3	Die 3	Die 3	Die 2	Die 2	Die 2

The entries show that Die 2 beats Die 3 with probability 21/36 = 7/12.

Given that Die 1 beats Die 2 and Die 2 beats Die 3, we'd expect Die 1 to beat Die 3. Let's look at the results (Table 22.3).

TABLE 22.3

Die 1 vs. Die 3.

		Result of Die 1					
		3	3	3	3	3	6
	1	Die 1	Die 1	Die 1	Die 1	Die 1	Die 1
Result of	4	Die 3	Die 3	Die 3	Die 3	Die 3	Die 1
Die 3	4	Die 3	Die 3	Die 3	Die 3	Die 3	Die 1
	4	Die 3	Die 3	Die 3	Die 3	Die 3	Die 1
	4	Die 3	Die 3	Die 3	Die 3	Die 3	Die 1
	4	Die 3	Die 3	Die 3	Die 3	Die 3	Die 1

It turns out that Die 3 beats Die 1 with probability 25/36, the most decisive victory yet!

So, Die 1 beats Die 2 and Die 2 beats Die 3, but Die 3 beats Die 1. The dice are nontransitive. For this reason, they're sometimes called weird dice.

Example 14 – The Condorcet paradox

Suppose that an election is held with three candidates, a, b, and c. Furthermore, suppose that every voter has a clear ranking of the candidates. A third of voters rank their choices in order or preference as $a > b > c$. Another third rank them as $b > c > a$. And the final third rank them $c > a > b$.

The result is that 2/3 of the voters prefer a to b, 2/3 prefer b to c, and 2/3 prefer c to a!

This paradox dates back to 1785 and is named after a French mathematician, the Marquis de Condorcet (1743–1794). The above is a popular example of the paradox, but there are other ways in which the preferences of the group can be nontransitive, even though the preferences of individuals are transitive.

Charles Dodgson, who wrote *Alice in Wonderland* under his pen name, Lewis Carroll, studied various voting schemes and the cases in which they failed. He independently discovered the Condorcet paradox and designed his own system to avoid it. He published it in 1876.[8] Unfortunately, in 1950, Kenneth Arrow proved that there's no fair way to run an election, if there are more than two candidates and more than two voters. The result is known as Arrow's impossibility theorem.[9]

Another sort of Partition

There's another kind of partition and, like so much else in mathematics, the story of these partitions seems to begin with Gottfried Wilhelm Leibniz. He called them "divulsions" in his 1674 letter to Jacob Bernoulli, but differences in terminology aside, this marks the beginning of what has proven to be a challenging mathematical topic.[10]

A **partition** of an integer is a way of expressing that integer as a sum of positive integers. The order of the terms doesn't matter, but that still leaves a variety of solutions, even for a small number like 4.

$$\begin{aligned} 4 &= 4 \\ &= 3 + 1 \\ &= 2 + 2 \\ &= 2 + 1 + 1 \\ &= 1 + 1 + 1 + 1 \end{aligned}$$

Partitions of integers are not to be confused with the partitions that are another way of looking at equivalence relations. The latter are enumerated by Bell numbers, while those being considered now follow a different pattern. Letting $P(n)$ denote the number of partitions of the integer n, we have the values shown in Table 22.4.

$P(n)$ gets off to a slow start, but eventually shows impressive growth!

Leibniz didn't get as far as 1,000, or even 100, but he did calculate $P(3) = 3$, $P(4) = 5$, $P(5) = 7$, and $P(6) = 11$. This small set of examples led him to suspect that the result is always prime, a conclusion that the table shows is incorrect. Indeed, the very next value, $P(7) = 15$, is a composite.[11]

---

8 Dodgson, Charles, "A method of taking votes on more than two issues," Pamphlet, Clarendon Press, Oxford, headed "not yet published." It was reprinted in Black, Duncan, *The Theory of Committees and Elections*, Cambridge University Press, 1958 and McLean, Iain, and Arnold B. Urken, *Classics of Social Choice*, University of Michigan Press, 1995.

9 Arrow, Kenneth J., "A Difficulty in the Concept of Social Welfare," *Journal of Political Economy*, Vol. 58, No. 4, 1950, pp. 328–346.

10 Wilson, Robin, and John J. Watkins, editors, *Combinatorics: Ancient & Modern*, Oxford University Press, Oxford, 2013, p. 205, which cites Gottfried Wilhelm Leibniz, *Math. Schriften*, Vol. IV, 2, Specimen de divulsionibus aequationum … Letter 3 dated September 2, 1674 (see Mahnke, Dietrich, "Leibniz auf der Suche nach einer allgemeinen Primzahlgleichung," *Bibliotheca Mathematica*, Vol. 13, 1912–1913, pp. 29–61).

11 Wilson, Robin, and John J. Watkins, editors, *Combinatorics: Ancient & Modern*, Oxford University Press, Oxford, 2013, p. 206.

TABLE 22.4

n	P(n)
1	1
2	2
3	3
4	5
5	7
6	11
7	15
8	22
9	30
10	42
⋮	⋮
100	190,569,292
⋮	⋮
1,000	24,061,467,864,032,622,473,692,149,727,991

Leibniz also failed to come up with the right answer for the other kind of partition.

> ... an unpublished manuscript of Leibniz from the late 1600s shows that he had tried to count the number of ways to partition {1, 2, ..., n} into three or four subsets, but with almost no success.[12]

But his greatest error was in not publishing what he had found. His partial results could have saved later mathematicians much time. Still, his results were rediscovered before they saw print, centuries after Leibniz's death.[13] And the later mathematicians managed to outdo him.

Euler Demystified

Partitions of integers are like the other kind of partitions, if all of the objects being considered are identical. That is, if we care about how many objects are in the sets, but not *which objects* are in the sets. Oddly, partitions of integers, despite this simple connection, turned out to be a much more difficult topic. We don't have an answer as nice as Bell numbers! Nevertheless, integer partitions can be used to illuminate other areas of mathematics.

Chapter 17 included the following result obtained by Euler.

---

12 Wilson, Robin, and John J. Watkins, editors, *Combinatorics: Ancient & Modern*, Oxford University Press, Oxford, 2013, p. 27.

13 Knobloch, Eberhard, *Die Mathematischen Studien von G. W. Leibniz zur Kombinatorik*, Textband, im Anschluss an den gleichnamigen Abhandlungsband zum ersten Mal nach den Originalhandschriften herausgegeben, Steiner, 1976.

$$\begin{aligned}
\sigma(n) = {} & \sigma(n-1) + \sigma(n-2) - \sigma(n-5) - \sigma(n-7) \\
& + \sigma(n-12) + \sigma(n-15) - \sigma(n-22) - \sigma(n-26) \\
& + \sigma(n-35) + \sigma(n-40) - \sigma(n-51) - \sigma(n-57) \\
& + \sigma(n-70) + \sigma(n-77) - \sigma(n-92) - \sigma(n-100) + \cdots
\end{aligned}$$

Although the pattern behind the terms in this recursion was revealed, it wasn't made clear what led Euler to it in the first place. As I indicated in Chapter 17, Euler didn't feel a need to create an aura of mystery about his results, so he explained his motivation. The formula was inspired by a result he had previously obtained, a generating function for partitions:

$$\prod_{k=1}^{\infty} \frac{1}{1-x^k}.$$

This led him to examine the infinite product

$$s = (1-x)(1-x^2)(1-x^3)(1-x^4)(1-x^5)(1-x^6)(1-x^7)(1-x^8)\cdots. \tag{I}$$

After much multiplication, he found that the above becomes

$$s = 1 - x - x^2 + x^5 + x^7 - x^{12} - x^{15} + x^{22} + x^{26} - x^{35} - x^{40} + \cdots. \tag{II}$$

The exponents in this expression are the same as the numbers seen in his pentagonal formula for $\sigma(n)$, but there are many steps between this expression and the formula. They involve manipulations of (I) and (II) above.

Euler took the natural logarithm of both sides of (I) to get

$$\ln(s) = \ln\big[(1-x)(1-x^2)(1-x^3)(1-x^4)(1-x^5)(1-x^6)(1-x^7)\cdots\big],$$

which can be rewritten as follows:

$$\ln(s) = \ln(1-x) + \ln(1-x^2) + \ln(1-x^3) + \ln(1-x^4) + \cdots.$$

To get rid of the logarithms, he differentiated both sides of the equation, getting

$$\frac{1}{s}\frac{ds}{dx} = -\frac{1}{1-x} - \frac{2x}{1-x^2} - \frac{3x^2}{1-x^3} - \frac{4x^3}{1-x^4} + \frac{5x^4}{1-x^5} - \cdots.$$

He then multiplied both sides by $-x$ to get

$$-\frac{x}{s}\frac{ds}{dx} = \frac{x}{1-x} + \frac{2x^2}{1-x^2} + \frac{3x^3}{1-x^3} + \frac{4x^4}{1-x^4} + \frac{5x^5}{1-x^5} + \cdots. \tag{III}$$

Turning back to (II), Euler differentiated both sides to get

$$\frac{ds}{dx} = -1 - 2x + 5x^4 + 7x^6 - 12x^{11} - 15x^{14} + 22x^{21} + 26x^{25} - \cdots .$$

Multiplying both sides of the above by $-\frac{x}{s}$ and simplifying gives

$$-\frac{x}{s}\frac{ds}{dx} = \frac{x + 2x^2 - 5x^5 - 7x^7 + 12x^{12} + 15x^{15} - 22x^{22} - 26x^{26} + \cdots}{s}.$$

And replacing s with its expression from (II) gives.

$$-\frac{x}{s}\frac{ds}{dx} = \frac{x + 2x^2 - 5x^5 - 7x^7 + 12x^{12} + 15x^{15} - 22x^{22} - 26x^{26} + \cdots}{1 - x - x^2 + x^5 + x^7 - x^{12} - x^{15} + x^{22} + x^{26} - \cdots}. \tag{IV}$$

Having obtained two expressions for $-\frac{x}{s}\frac{ds}{dx}$, Euler made the substitution $-\frac{x}{s}\frac{ds}{dx} = t$. This allowed the expressions to be referred to more conveniently. It doesn't matter that they are expressions of $-\frac{x}{s}\frac{ds}{dx}$. What is relevant in the work that follows is that they are equal to each other.

Euler expanded each term in (III) and arranged the results so that the same powers of x were aligned in columns. You can double check his expansions by performing long division of polynomials.

For example,

$$\frac{x}{1-x} = x + x^2 + x^3 + x^4 + x^5 + x^6 + x^7 + x^8 + \cdots$$

$$\frac{2x^2}{1-x^2} = 2x^2 + 2x^4 + 2x^6 + 2x^8 + \cdots$$

and so on.

The result is as follows:

$$
\begin{array}{rcccccccccccccccc}
t = x & + & x^2 & + & x^3 & + & x^4 & + & x^5 & + & x^6 & + & x^7 & + & x^8 & +\cdots \\
& + & 2x^2 & & & + & 2x^4 & & & + & 2x^6 & & & + & 2x^8 & +\cdots \\
& & & + & 3x^3 & & & & & + & 3x^6 & & & & & +\cdots \\
& & & & & + & 4x^4 & & & & & & & + & 4x^8 & +\cdots \\
& & & & & & & + & 5x^5 & & & & & & & +\cdots \\
& & & & & & & & & + & 6x^6 & & & & & +\cdots \\
& & & & & & & & & & & + & 7x^7 & & & +\cdots \\
& & & & & & & & & & & & & + & 8x^8 & +\cdots
\end{array}
$$

Looking at the columns, Euler realized that the coefficients are the positive divisors of the exponent, with each divisor appearing exactly once. Because the sum of the positive divisors of n is represented by $\sigma(n)$, t can be expressed very nicely. We have

$$t = \sigma(1)x + \sigma(2)x^2 + \sigma(3)x^3 + \sigma(4)x^4 + \sigma(5)x^5 + \sigma(6)x^6 + \sigma(7)x^7 + \sigma(8)x^8 + \cdots \quad \text{(V)}$$

Turning to (IV), which, after Euler's substitution, is

$$t = \frac{x + 2x^2 - 5x^5 - 7x^7 + 12x^{12} + 15x^{15} - 22x^{22} - 26x^{26} + \cdots}{1 - x - x^2 + x^5 + x^7 - x^{12} - x^{15} + x^{22} + x^{26} - \cdots},$$

we can cross-multiply to get

$$t\left(1 - x - x^2 + x^5 + x^7 - x^{12} - x^{15} + x^{22} + x^{26} - \cdots\right)$$

$$= x + 2x^2 - 5x^5 - 7x^7 + 12x^{12} + 15x^{15} - 22x^{22} - 26x^{26} + \cdots.$$

Moving everything over to the left-hand side, Euler got

$$t\left(1 - x - x^2 + x^5 + x^7 - x^{12} - x^{15} + x^{22} + x^{26} - \cdots\right)$$

$$-x - 2x^2 + 5x^5 + 7x^7 - 12x^{12} - 15x^{15} + 22x^{22} + 26x^{26} - \cdots = 0$$

Euler then replaced t with the expression given in (V), getting

$$\left(\sigma(1)x + \sigma(2)x^2 + \sigma(3)x^3 + \sigma(4)x^4 + \sigma(5)x^5 + \sigma(6)x^6 + \sigma(7)x^7 + \cdots\right)$$

$$\left(1 - x - x^2 + x^5 + x^7 - x^{12} - x^{15} + x^{22} + x^{26} - \cdots\right)$$

$$-x - 2x^2 + 5x^5 + 7x^7 - 12x^{12} - 15x^{15} + 22x^{22} + 26x^{26} - \cdots = 0$$

As he had done before, following the multiplications, Euler arranged the results so that the same powers of x were aligned in columns.

$$\begin{aligned}
0 = {} & \sigma(1)x + \sigma(2)x^2 + \sigma(3)x^3 + \sigma(4)x^4 + \sigma(5)x^5 + \sigma(6)x^6 + \cdots \\
& - x - \sigma(1)x^2 - \sigma(2)x^3 - \sigma(3)x^4 - \sigma(4)x^5 - \sigma(5)x^6 - \cdots \\
& - 2x^2 - \sigma(1)x^3 - \sigma(2)x^4 - \sigma(3)x^5 - \sigma(4)x^6 - \cdots \\
& + 5x^5 + \sigma(1)x^6 + \cdots
\end{aligned}$$

Because the above is equal to 0, the coefficients for the various powers of x must each sum to 0. Looking at the coefficients of x^6, we have

$$\sigma(6) - \sigma(5) - \sigma(4) + \sigma(1) = 0.$$

This can be rewritten as

$$\sigma(6) = \sigma(5) + \sigma(4) - \sigma(1)$$

and again as

$$\sigma(6) = \sigma(6-1) + \sigma(6-2) - \sigma(6-5).$$

Euler found that as the power of x increases, the sequence of numbers subtracted in parentheses is 1, 2, 5, 7, 12, 15, 22, 26, 35, 40, 51, 57, 70, 77, ... Expressed symbolically, he found

$$
\begin{aligned}
\sigma(n) = \ &\sigma(n-1) + \sigma(n-2) - \sigma(n-5) - \sigma(n-7) \\
+\ &\sigma(n-12) + \sigma(n-15) - \sigma(n-22) - \sigma(n-26) \\
+\ &\sigma(n-35) + \sigma(n-40) - \sigma(n-51) - \sigma(n-57) \\
+\ &\sigma(n-70) + \sigma(n-77) - \sigma(n-92) - \sigma(n-100) + \cdots
\end{aligned}
$$

For some columns of like powers such as x^2 and x^5, it's necessary to subtract a term that has nothing to do with σ in order to get the proper coefficient. In every such case the number subtracted is identical to the exponent in question. And these numbers arise where the formula above has a term of the form $\sigma(n-n)$. So, Euler's instructions for using the formula above were to take all terms for which σ has a nonnegative argument and if one arises where the argument is 0, replace that term with n.

The above shows how Euler arrived at his pentagonal formula.[14] At various points, he simply assumed that the patterns he found continued. He happened to be right, but the process by which Euler found the result doesn't satisfy modern demands for rigor. It's my opinion that he was right to race ahead. Other mathematicians, and in some instances Euler himself, would later fill in details and make his work more rigorous. If Euler had slowed down to figure out and fill in missing steps, he would not have been able to write nearly as many papers as he did and the world would have missed out on many gems. As a consequence, mathematics today would probably not be quite as advanced.

Euler was inspired to this work by a result he obtained concerning partitions and the above formula applies to them as well. We have

$$P(n) = P(n-1) + P(n-2) - P(n-5) - P(n-7) + P(n-12) + P(n-15) - \cdots$$

James Joseph Sylvester (Figure 22.9) was one of the first great mathematicians to work in America. Like other early greats in this country, he came from somewhere else (England). In his first years in America, 1841–1843, he had a violent encounter with a student and faced antisemitism. He returned to England, but came to America again in 1876, at which time he was hired by Johns Hopkins University. In 1883 he returned to England for good.

---

14 This presentation was adapted from the English translation of Euler's work that appears in Young, Robert M., *Excursions in Calculus: An Interplay of the Continuous and the Discrete*, The Mathematical Association of America, 1992, pp. 364–367 and Polya, George, *Mathematics and Plausible Reasoning, Volume 1: Induction and Analogy in Mathematics*, Princeton University Press, Princeton, NJ, 1954, pp. 96–98. For Euler's original paper, in French, see Euler, Leonhard, *Opera Omnia*, Series 1, Vol. 2, pp. 241–253.

FIGURE 22.9
J. J. Sylvester (1814–1897)

Public domain.

Sylvester advanced the study of partitions with a paper bearing the colorful title "A Constructive Theory of Partitions, Arranged in Three Acts, an Interact, and an Exodion."[15]

In this paper, Sylvester represented partitions by diagrams. His first example was the partition of 20 given by 5 + 5 + 4 + 3 + 3. It's reproduced below.

```
* * * * *
* * * * *
* * * *
* * *
* * *
```

Each row in the above diagram represents a term in the partition. The rows have 5, 5, 4, 3, and 3 stars, read from top to bottom. Sylvester called such diagrams Ferrers graphs. Norman Macleod Ferrers (1829–1903) was an English mathematician, who came up with the idea of using the graphs, although he never put his idea into print.[16]

---

15 Sylvester, James Joseph, with insertions by Fabian Franklin, "A Constructive Theory of Partitions, Arranged in Three Acts, an Interact and an Exodion," *American Journal of Mathematics*, Vol. 5, No. 1, 1882, pp. 251–330.

16 www.history.mcs.st-and.ac.uk/Biographies/Ferrers.html.

If you mistakenly read the diagram by columns, you found a different partition of 20, namely 5 + 5 + 5 + 3 + 2. This "mistaken" perspective on the diagram is referred to as the **conjugate** and it turns out to be very useful. For example, considering conjugates reveals that the number of partitions of n into exactly k parts is the same as the number of partitions of n in which the largest part is k. The former is the set of diagrams with exactly k rows and the second, represented by the conjugate, is the number of diagrams where at least one column has length k.[17] Imagine how much longer the proof of this fact would be without the use of a diagram!

Sylvester made other important contributions, but the use of Ferrers graphs is the most accessible of them. Instead of detailing his other results, contributions by later mathematicians are now presented.

Hardy and Ramanujan

Researchers chipped away at the problem of integer partitions by finding formulas that gave the number of possibilities when a bound is placed on the number of terms. For example, if we limit ourselves to two terms, we have $p_2(n) = \lfloor \frac{1}{2}(n+1) \rfloor$. Notice that this formula uses the floor function of Chapter 6. If we allow possibilities having 1, 2, or 3 terms, the answer is $p_3(n) = \left\{ \frac{1}{12}(n+3)^2 \right\}$, where $\{\ \}$ indicates that the value should be rounded to the nearest integer.

Finally, G. H. Hardy (Figure 22.10) and Srinivasa Ramanujan (Figure 22.11), working together, found the following asymptotic solution, to the entire problem, in 1918.[18]

$$P(n) \sim \frac{1}{4n\sqrt{3}} e^{\pi\sqrt{2n/3}}$$

Recall that, in this context, $\sim$ means that the ratio of the two sides of the equation converges to 1, as n approaches infinity. The appearance of both e and π in this formula is lovely. The asymptotic formula has a decreasing percentage error as n grows, but it cannot be depended on to give the correct answer. However, what was shown above is only the first term in a series that Hardy and Ramanujan found to estimate $P(n)$. If more terms are taken into consideration, a much greater accuracy is obtained. The pair showed that five terms, with the result rounded to the nearest integer, are sufficient to exactly determine $P(200)$.

In 1942, Paul Erdös, ever the simplifier, showed that the formula of Hardy and Ramanujan could be derived by elementary means.

---

17 This result was first mentioned in Sylvester, James Joseph, "On Mr. Cayley's Impromptu Demonstration of the Rule for Determining at Sight the Degree of any Symmetrical Function of the Roots of an Equation Expressed in Terms of the Coefficients" *The London, Edinburgh, and Dublin Philosophical Magazine and Journal of Science*, Series 4, Vol. 5, No. 31, 1853, pp. 199–202 and attributed to Ferrers.

18 Hardy, Godfrey Harold, and Srinivasa Ramanujan, "Asymptotic Formulae in Combinatory Analysis," *Proceedings of the London Mathematical Society*, Series 2, Vol. 17, No. 1, 1918, pp. 75–115.

FIGURE 22.10
G. H. Hardy (1877-1947).

Public domain.

FIGURE 22.11
Srinivasa Ramanujan (1887–1920).

Public domain.

Simpler Results

If you'd like something simpler, Ramanujan found other formulas. For example, if the number for which you're counting the partitions of is of the form $5n + 4$, then the answer is a multiple of 5. Using modular arithmetic, this relation, and others, can be tersely stated like so.[19]

---

19 The first two were proved in Ramanujan, Srinivasa, "Congruence Properties of Partitions," *Mathematische Zeitschrift*, Vol. 9, No. 1–2, March 1921, pp. 147–153.

$$P(5n + 4) \equiv 0 \,(\text{mod}\,5),$$
$$P(7n + 5) \equiv 0 \,(\text{mod}\,7),$$
$$P(11n + 6) \equiv 0 \,(\text{mod}\,11).$$

Today, 22,474,608,014 such congruences are known.[20] An example is

$$P(28995244292486005245947069n + 28995221336976431135321047) \equiv 0 \bmod 29.$$

However, the three congruences that Ramanujan found represent the complete list where the multiplier of n and the modulus are the same prime number.

Still in the category of simpler results, there's a nice connection with a topic from Chapter 13, although we have to change the problem a bit to get it. When looking at partitions of 4, the possibilities included 3 + 1, but not 1 + 3. If we choose to count different orderings of the terms as distinct (as long as the terms that are moved are not identical), then we have what is known as a **composition**. While there are 5 partitions of 4, the list of compositions is

$$
\begin{aligned}
4 &= 4 \\
&= 3 + 1 \\
&= 1 + 3 \\
&= 2 + 2 \\
&= 2 + 1 + 1 \\
&= 1 + 2 + 1 \\
&= 1 + 1 + 2 \\
&= 1 + 1 + 1 + 1
\end{aligned}
$$

We get 8 possibilities altogether. Of these, there are only two that don't use the number 1. Back in 1876, Arthur Cayley proved that for any positive integer n, the number of compositions without 1s is the $n - 1$st Fibonacci number.[21] That is the number of compositions of 4 without any 1s is F_3 and, in general, the number of compositions of n without 1s is F_{n-1}.

Open Problems

$P(n)$ is sometimes a prime number. For example, $P(2) = 2$, $P(3) = 3$, $P(4) = 5$, $P(5) = 7$, and $P(6) = 11$. But does this happen infinitely often? That is, are there infinitely many values of n such that $P(n)$ is prime? Nobody knows![22]

---

20 Johansson, Fredrik, "Efficient Implementation of the Hardy–Ramanujan–Rademacher Formula," *LMS Journal of Computation and Mathematics*, Vol. 15, 2012, pp. 341–359.

21 Wilson, Robin, and John J. Watkins, editors, *Combinatorics: Ancient & Modern*, Oxford University Press, Oxford, 2013, p. 223.

22 Wilson, Robin, and John J. Watkins, editors, *Combinatorics: Ancient & Modern*, Oxford University Press, Oxford, 2013, pp. 205–206.

And the biggest problem of all, the nature of our universe, discrete or continuous, remains open. The following passage suggests that posets may help answer the question.

> But there are now hints that discreteness plays an even more fundamental role. One of the goals of physics at present is the construction of a theory which could reconcile the two pillars of 20th-century physics: general relativity and quantum mechanics. In describing string theory, loop quantum gravity, and a variety of other approaches (including non-commutative geometry and causal set theory), Smolin argued that all of them involve discreteness at a fundamental level (roughly, the Planck scale, which is much too small and fleeting to be directly observed). Causal set theory is based on discrete partially ordered sets and has already attracted the attention of combinatorialists. Indeed, developments such as the holographic principle suggest that the basic currency of the universe may not be space and time, but information measured in bits. Maybe the 'theory of everything' will be combinatorial![23]

Exercise Set 22

> What are numbers, really? I've been studying them for 50-odd years and I don't think I know!
>
> John H. Conway[24]

1. How many relations can be defined on a set of 2 objects?
2. How many relations can be defined on a set of 3 objects?
3. How many relations can be defined on a set of 4 objects?
4. How many relations can be defined on a set of n objects?
5. Is the set $M = \{(a, b), (b, c), (c, d)\}$ a partial order?
6. Is the set $S = \{(1, 1), (2, 1), (2, 2)\}$ a partial order?
7. Is the set $T = \{(a, a), (b, b), (c, c), (a, b), (b, c), (c, a)\}$ a partial order?
8. Suppose shapes are cut out of a piece of paper. For two shapes, a and b, we can define $a \, R \, b$ if shape a can be positioned to completely cover shape b (without folding or any other tricks). Does R define a relation? Does R define a partial ordering?
9. Sketch the cover diagram for the poset for the divisors of 36.
10. Sketch the cover diagram for the poset for the divisors of 70.
11. Sketch the cover diagram for the subset poset for $S = \{a, b, c, d\}$.

23 Wilson, Robin, and John J. Watkins, editors, *Combinatorics: Ancient & Modern*, Oxford University Press, Oxford, 2013, pp. 363–364.
24 Roberts, Siobhan, *Genius at Play – The Curious Mind of John Horton Conway*, Bloomsbury USA, 2015, p. 173.

12. Use the cover diagram shown in Figure 22.12 to write out the poset it represents as a set of ordered pairs.

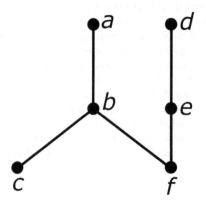

FIGURE 22.12
What poset does this represent?

Image created by Josh Gross.

13. Use the cover diagram shown in Figure 22.13 to write out the poset it represents as a set of ordered pairs.

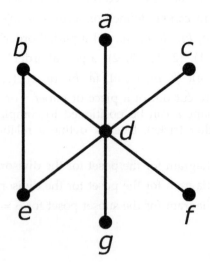

FIGURE 22.13
What poset does this represent?

Image created by Josh Gross.

14. Does the diagram shown in Figure 22.14 represent a poset?

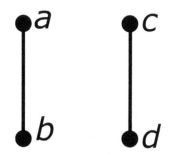

FIGURE 22.14
Does this represent a poset?

Image created by Josh Gross.

15. Does the diagram shown in Figure 22.15 represent a poset?

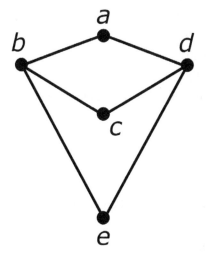

FIGURE 22.15
Does this represent a poset?

Image created by Josh Gross.

16. Is < an equivalence relation? Prove or disprove.

17. Is ≤ an equivalence relation? Prove or disprove.

18. Define words to be equivalent if they begin *or* end with the same letter? Is this an equivalence relation?

19. Let $M = \{(a, b) \in \mathbb{Z} \times \mathbb{Z} \mid b \neq 0\}$. Define a relation ~ on M by:

$$(a, b) \sim (c, d) \text{ if } ad = bc.$$

Show that ~ is an equivalence relation.

20. Let $M = \{a, b, c, d\}$.

    a. How many functions are there from M to M?

    b. How many of these functions are onto?

    c. How many of these functions are 1:1.

    d. How many of these functions are 1:1 and onto?

    e. How many relations are possible on M?

    f. How many equivalence relations are possible on M?

21. Can you find relations where exactly one of the properties fails and prevents it from being an equivalence relation? Try this for reflexivity, symmetry, and transitivity. Do the second two properties imply the first?

22. Let $k = ab$, where a and b are integers. Prove that if n is also an integer, then $k \bmod n = (a \bmod n)(b \bmod n)$.

23. Use the result of Exercise 22 and induction to prove that if p is a positive integer, then $k^p \bmod n = (k \bmod n)^p$.

24. Prove that if $k = a + b$, then $k \bmod n = (a \bmod n) + (b \bmod n)$.

25. Calculate $266^{32} \bmod 23$.

26. Calculate $34^{78} \bmod 14$.

27. Calculate $794^{21} \bmod 81$.

28. Determine whether the dice described below constitute a transitive or nontransitive set.

Die 1: 3, 3, 3, 3, 3, 3

Die 2: 2, 2, 2, 2, 6, 6

Die 3: 1, 1, 1, 5, 5, 5

Die 4: 0, 0, 4, 4, 4, 4

29. Determine whether the dice described below constitute a transitive or nontransitive set.

Die 1: 4, 4, 4, 4, 4, 9

Die 2: 3, 3, 3, 3, 8, 8

Die 3: 2, 2, 2, 7, 7, 7

Die 4: 1, 1, 6, 6, 6, 6

Die 5: 0, 5, 5, 5, 5, 5

30. Does the game Rock–Paper–Scissors violate transitivity, or did I have to go to the larger game, Rock, Paper, Scissors, Spock, Lizard, to get this weird result?

31. Can you make up a game like Rock–Paper–Scissors with exactly four hand gestures that violates transitivity and also has equal chances of being won with any move, if one's opponent selects moves randomly?

32. Can you make up a game like Rock–Paper–Scissors with exactly six hand gestures that violates transitivity and also has equal chances of being won with any move, if one's opponent selects moves randomly?

33. For which values of n can you make up a game like Rock–Paper–Scissors with exactly n hand gestures that violates transitivity and also has equal chances of being won with any move, if one's opponent selects moves randomly?

34. Borromean Rings are shown in color figure 39 **(See color insert)**. Explain whether you think they violate transitivity in some way or not.[25]

35. List all of the integer partitions of 5.

36. List all of the integer partitions of 6.

37. List all of the integer partitions of 7.

38. Draw the Ferrers graphs for the partitions you found in Exercise 35.

39. Draw the Ferrers graphs for the partitions you found in Exercise 36

40. Use

$$P(n) \sim \frac{1}{4n\sqrt{3}} \, e^{\pi\sqrt{2n/3}}$$

to estimate the number of partitions of 100. Compare this with the true value given in this chapter. What is the absolute error? What is the relative error?

41. Use

$$P(n) \sim \frac{1}{4n\sqrt{3}} \, e^{\pi\sqrt{2n/3}}$$

to estimate the number of partitions of 1,000. Compare this with the true value given in this chapter. What is the absolute error? What is the relative error?

42. Use the formula below to calculate $P(5)$ in terms of previous values.

$$P(n) = \frac{1}{n} \sum_{k=0}^{n-1} \sigma(n-1)P(k)$$

43. Use the formula below to calculate $P(6)$ in terms of previous values.

$$P(n) = \frac{1}{n} \sum_{k=0}^{n-1} \sigma(n-1)P(k)$$

44. What can you say about a partition, if it's known to be self-conjugate?

---

25 For more information see Wikipedia contributors, "Borromean Rings," *Wikipedia, The Free Encyclopedia*, https://en.wikipedia.org/w/index.php?title=Borromean_rings&oldid=847673021 or Chamberland, Marc, *Single Digits: in Praise of Small Numbers*, Princeton University Press, Princeton, NJ, 2015, pp. 74–76.

45. Find a neat formula connected with partitions in some way that does not appear in this chapter. Explain it in your own words.

46. How many compositions are there for 10?

47. How many compositions are there for 11?

48. How many compositions are there for 12?

49. Can you define mathematics? Give it your best shot! This question was posed in Exercise set 0. Has your answer changed? If so, how?

References/Further Reading

In 1999, A&E aired a special episode of their Biography series, namely "Biography of the Millennium: 100 People - 1000 Years." It counted down the 100 most influential people of the previous 1,000 years. Three out of the top 10 were professors of mathematics, but #1 was Johannes Gutenberg.

Alder, Henry L., "Partition Identities – From Euler to the Present," *The American Mathematical Monthly*, Vol. 76, No. 7, August–September 1969, pp. 733–746.

Andrews, George E., "Partition Identities," *Advances in Mathematics*, Vol. 9, No. 1, August 1972, pp. 10–51.

Andrews, George E., *The Theory of Partitions, Encyclopedia of Mathematics and its Applications*, Vol. 2, Addison-Wesley, 1976, reissued by Cambridge University Press, Cambridge, 1984, paperback edition 1998.

Andrews, George E., "Euler's Pentagonal Number Theorem," *Mathematics Magazine*, Vol. 56, No. 5, November 1983, pp. 279–284.

Angel, Levi and Matt Davis, "A Direct Construction of Nontransitive Dice Sets," *Journal of Combinatorial Designs*, Vol. 25, No. 11, November 2017, pp. 523–529.

Berndt, Bruce C., *Ramanujan's Notebooks, Parts I–V*, Springer, Berlin, 1985, 1989, 1991, 1994, 1998.

Birkhoff, Garrett, *Lattice Theory*, revised edition, American Mathematical Society, New York, 1948.

Chamberland, Marc and Eugene A. Herman, "Rock-Paper-Scissors Meets Borromean Rings," *The Mathematical Intelligencer*, Vol. 37, No. 2, June 2015, pp. 20–25.

Condorcet, Jean-Antoine-Nicolas de Caritat, *Essai sur l'application de l'analyse à la probabilité des décisions rendues à la pluralité des voix*, Imprimerie Royale, Paris, 1785, available online at http://gallica.bnf.fr/ark:/12148/bpt6k417181/f4.image.

Dickson, Leonard Eugene, *History of the Theory of Numbers*, Vol. 2, Carnegie Institution, Washington, D. C., 1920, reprinted by Chelsea, 1952. See Chapter III for partitions.

Gardner, Martin, "Mathematical Games: The Paradox of the Nontransitive Dice and the Elusive Principle of Indifference," *Scientific American*, Vol. 223, December 1970, pp. 110–114.

Grime, James, "Non-transitive Dice," http://singingbanana.com/dice/article.htm.

Gupta, Hansraj, "Partitions – A Survey," *Journal of Research of the National Bureau of Standards – B. Mathematical Sciences*, Vol. 74B, No. 1, January–March 1970, pp. 1–29.

Gupta, Hansraj, C. E. Gwyther, and Jeffrey Charles Percy Miller, "Tables of Partitions," *Royal Society Mathematical Tables*, Vol. 4, Cambridge University Press, Cambridge, 1958.

Hardy, Godfrey Harold, *Ramanujan*, Cambridge University Press, Cambridge, 1940, reprinted by Chelsea, 1999.

Kang, Yibin, Qiuhui Pan, Xueting Wang, and Mingfeng He, "A Golden Point Rule in Rock–Paper–Scissors–Lizard–Spock Game," *Physica A: Statistical Mechanics and its Applications*, Vol. 392, No. 11, June 1, 2013, pp. 2652–2659.

Kass, "Sam, Rock Paper Scissors Spock Lizard," www.samkass.com/theories/RPSSL.html.

Kimberling, Clark, "The Origin of Ferrers Graphs," *The Mathematical Gazette*, Vol. 83, No. 497, July 1999, pp. 194–198.

Klein, Douglas J., "Prolegomenon on Partial Orderings in Chemistry," *MATCH Communications in Mathematical and in Computer Chemistry*, Vol. 42, October 2000, pp. 7–21.

Knuth, Donald E., "Subspaces, Subsets and Partitions," *Journal of Combinatorial Theory*, Series A, Vol. 10, No. 2, March 1971, pp. 178–180.

Peltomäki, Matti and Mikko Alava, "Three- and Four-state Rock-paper-scissors Games with Diffusion," *Physical Review E*, Vol. 78, No. 3, September 2008.

Rival, Ivan, editor, *Ordered Sets*, Proceedings of the NATO Advanced Study Institute held at Banff, Canada, August 28 to September 12, 1981, D. Reidel Publishing Company, Dordrecht, Holland, 1982. The emphasis of this volume is on applications of posets.

Savage, Jr., Richard P., "The Paradox of Nontransitive Dice," *The American Mathematical Monthly*, Vol. 101, No. 5, May 1994, pp. 429–436.

Schaefer, Alex and Jay Schweig, "Balanced Nontransitive Dice," *The College Mathematics Journal*, Vol. 48, No. 1, January 2017, pp. 10–16.

Sylvester, James Joseph, with insertions by Fabian Franklin, "A Constructive Theory of Partitions, Arranged in Three Acts, an Interact and an Exodion," *American Journal of Mathematics*, Vol. 5, No. 1, 1882, pp. 251–330.

Tenney, Richard L. and Caxton C. Foster, "Non-Transitive Dominance," *Mathematics Magazine*, Vol. 49, No. 3, May 1976, pp. 115–120.

Conclusion (for the whole book, not just this chapter!)

What does the future bring for discrete mathematics? I suspect the same as it has in store for mathematics in general – more and more connections between topics and greater simplicity. The connections presented in this book are only a tiny sample of what has already been found. They were emphasized in the hope that it makes the material easier to grasp. And with luck, future connections will make math simpler yet. Andrew Wiles, who proved Fermat's last theorem, explained this in a bonus feature on the DVD of the musical *Fermat's Last Tango*.

> Mathematics does sometimes give the impression of being spread over such a large area that even one mathematician can't understand another, but if you think back to 18th-century mathematics, most modern mathematicians would understand it all, and in a much more unified way than the 18th-century mathematicians. I think this dispersion that one senses is really just because we don't understand it well enough yet and over the next 200 years all our current methods and proofs will be simplified and people will see it as a whole and it will be much easier. I mean, nowadays most high school students will study calculus. That would have been unthinkable in the 17th century, but now it's routine and that will happen to current mathematics in 300 years' time.[26]

We've seen examples of long difficult proofs that were replaced by shorter simpler proofs. Perhaps the proof of Fermat's last theorem, with the benefit of a couple more centuries of mathematics, will indeed one day fit in a margin.

---

26 Rosenblum, Joshua, composer, and Joanne Sydney, Lessner, author, *Fermat's Last Tango*, Recorded at The York Theatre Company, New York, NY, December 29, 2000, Clay Mathematics Institute, 2001.

Index

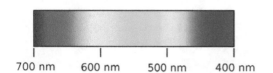

FIGURE 1
The Visible Light Spectrum.

FIGURE 2
"No Restrictions Allowed."

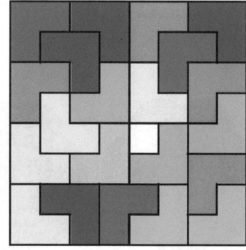

FIGURE 3
A Tromino Tiling.

Image created by Josh Gross.

FIGURE 4
Another Tromino Tiling.

Image created by Josh Gross.

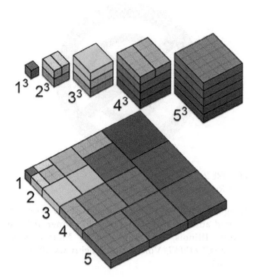

FIGURE 6
"Adelaide"

FIGURE 5
A Proof Without Words.

Image created by Cmglee. This file is licensed under the Creative Commons Attribution-Share Alike 3.0 Unported license.

Courtesy of A. W. F. Edwards from from Edwards, A. W. F., *Cogwheels of the Mind: The Story of Venn Diagrams*, Foreword by Ian Stewart, The Johns Hopkins Press, Baltimore and London, 2004, p. 90 (Figure 7.4).

FIGURE 7
Symmetric seven-set diagrams.

Courtesy of Frank Ruskey.

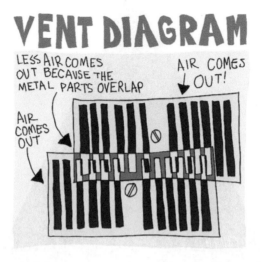

FIGURE 8
A Mike riley "I Taste Sound" Cartoon.

Courtesy of Mike Riley, http://www.itastesound.com/.

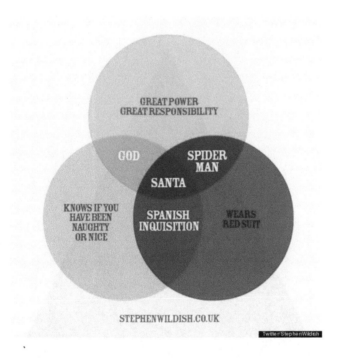

FIGURE 9
A Stephen Wildish Venn Diagram.

Courtesy of Steve Wildish, http://stephenwildish.co.uk/.

FIGURE 10
Pascal's Triangle mod 2 using the color assignment 0 = black and 1= blue.

Image created by Garrett Ruths

FIGURE 11
Pascal's Triangle mod 3 using the color assignment 0 = black, 1= blue, and 2 = red.

Image created by Garrett Ruths.

FIGURE 12
Pascal's Triangle mod 4 using the color assignment 0 = black, 1= blue, 2 = red, and 3 = cyan.

Image created by Garrett Ruths.

FIGURE 13
Pascal's Triangle mod 5 using the color assignment 0 = black, 1= blue, 2 = red, 3 = cyan, and 4 = yellow.

Image created by Garrett Ruths.

FIGURE 14

Pascal's Triangle mod 6 using the color assignment 0 = black, 1= blue, 2 = red, 3 = cyan, 4 = yellow, and 5 = white.

Image created by Garrett Ruths.

FIGURE 15

Pascal's Triangle mod 7 using the color assignment 0 = black, 1= blue, 2 = red, 3 = cyan, 4 = yellow, 5 = white, and 6 = green.

Image created by Garrett Ruths.

FIGURE 16
Stirling Numbers of the second kind mod 2 using the color assignment 0 = black and 1 = blue.

Image created by Garrett Ruths.

FIGURE 17
Stirling Numbers of the second kind mod 3 using the color assignment 0 = black, 1= blue, and 2 = red.

Image created by Garrett Ruths.

FIGURE 18
Stirling Numbers of the second kind mod 4 using the color assignment 0 = black, 1= blue, 2 = red, and 3 = cyan.

Image created by Garrett Ruths.

FIGURE 19
Stirling Numbers of the second kind mod 5 using the color assignment 0 = black, 1= blue, 2 = red, 3 = cyan, and 4 = yellow.

Image created by Garrett Ruths.

FIGURE 20
Stirling Numbers of the second kind mod 6 using the color assignment 0 = black, 1= blue, 2 = red, 3 = cyan, 4 = yellow, and 5 = white.

Image created by Garrett Ruths.

FIGURE 21
Stirling Numbers of the second kind mod 7 using the color assignment 0 = black, 1= blue, 2 = red, 3 = cyan, 4 = yellow, 5 = white, and 6 = green.

Image created by Garrett Ruths.

FIGURE 22
$k = 1$ Triangle mod 2.

Image created by Craig P. Bauer and Joe Molinar.

FIGURE 23
$k = 1$ Triangle mod 3.

Image created by Craig P. Bauer and Joe Molinar.

FIGURE 24
$k = 1$ Triangle mod 4.

Image created by Craig P. Bauer and Joe Molinar.

FIGURE 25
$k = 1$ Triangle mod 5.

Image created by Craig P. Bauer and Joe Molinar.

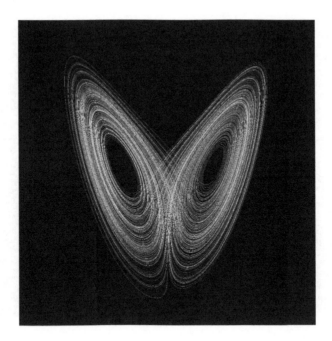

FIGURE 26
The Lorenz attractor.

The initial conditions that Lorenz used to generate his images are given in Lorenz, Edward N., *The Essence of Chaos*, University of Washington Press, Seattle, Washington, 1993, p. 188 as $b = 8/3$, $\sigma = 10$, and $r = 28$. He used the fourth-order Runge-Kutta scheme to approximate future values.

FIGURE 27
Dragon curves fitted together to make a tiling.

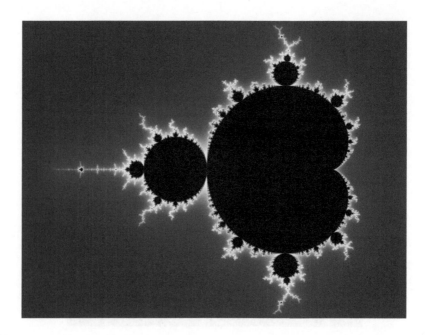

FIGURE 28
The Mandelbrot set.

Image created by Wolfgang Beyer with the program *Ultra Fractal 3*. This file is licensed under the Creative Commons Attribution-Share Alike 3.0 Unported license.

 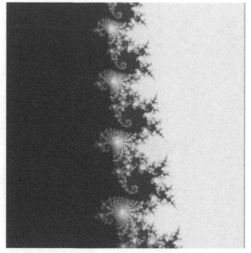

FIGURES 29 AND 30
Close-ups of portions of the Mandelbrot Set.

Copyright (c) 2019, Dr Andrew Burbanks, University of Portsmouth, UK.

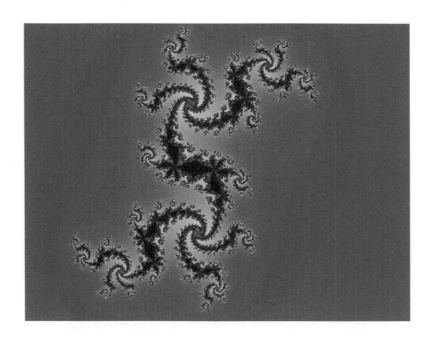

FIGURE 31
The Julia set for c = 0.300283 + 0.48857i.

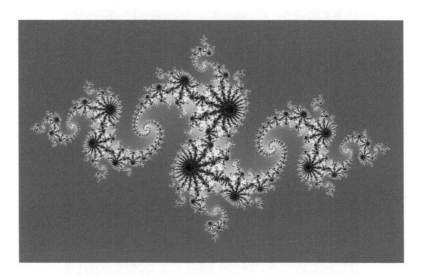

FIGURE 32
The Julia set for c = -0.8 + 0.156i.

FIGURE 33
A fractal arising from Newton's method.

Image created by Cody Spath.

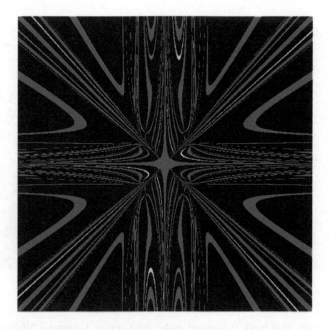

FIGURE 34
Beauty arising from another root approximation algorithm.

Image created by Craig P. Bauer.

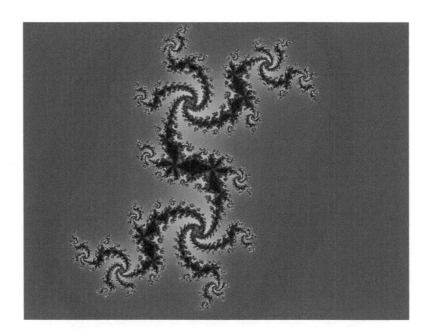

FIGURE 31
The Julia set for c = 0.300283 + 0.48857i.

Courtesy of Neal Ziring, http://users.erols.com/ziring/mandel.html.

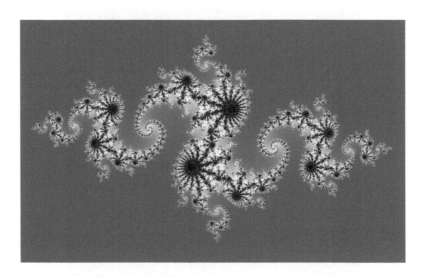

FIGURE 32
The Julia set for c = -0.8 + 0.156i.

FIGURE 33
A fractal arising from Newton's method.

Image created by Cody Spath.

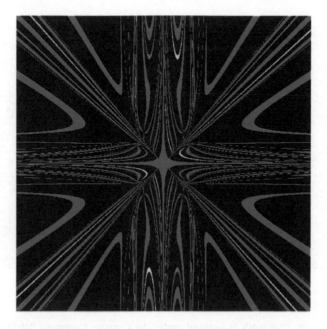

FIGURE 34
Beauty arising from another root approximation algorithm.

Image created by Craig P. Bauer.

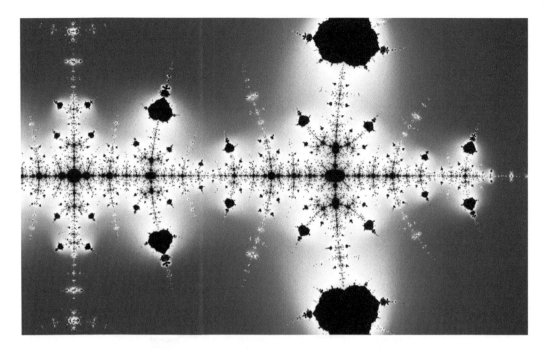

FIGURE 35
A view of the $3n + 1$ fractal.

Public domain.

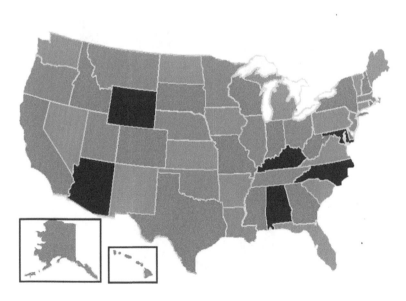

FIGURE 36
A four-colored map of the United States.

FIGURE 37
Instant Insanity.

Image created by Craig P. Bauer.

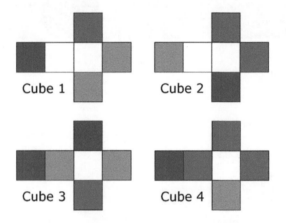

FIGURE 38
Unfolded Instant Insanity Cubes.

Image created by Josh Gross.

FIGURE 39
Borromean Rings.

Public domain.